Communications in Computer and Information Science 1287

More information about this series at http://www.springer.com/series/7899

Marcin Hernes · Krystian Wojtkiewicz ·
Edward Szczerbicki (Eds.)

Advances in Computational Collective Intelligence

12th International Conference, ICCCI 2020
Da Nang, Vietnam, November 30 – December 3, 2020
Proceedings

Editors
Marcin Hernes
Wroclaw University of Economics and Business
Wrocław, Poland

Krystian Wojtkiewicz
Wrocław University of Science and Technology
Wrocław, Poland

Edward Szczerbicki
University of Newcastle
Newcastle, Australia

ISSN 1865-0929 ISSN 1865-0937 (electronic)
Communications in Computer and Information Science
ISBN 978-3-030-63118-5 ISBN 978-3-030-63119-2 (eBook)
https://doi.org/10.1007/978-3-030-63119-2

This Springer imprint is published by the registered company Springer Nature Switzerland AG
The registered company address is: Gewerbestrasse 11, 6330 Cham, Switzerland

Preface

This volume contains the proceedings of the 12th International Conference on Computational Collective Intelligence (ICCCI 2020), which was at first planned to be held in Danang, Vietnam. However, due to the COVID-19 pandemic, the conference date was first postponed to November 30 – December 3, 2020, and then moved to a virtual space.

The conference was co-organized jointly by the University of Danang, Vietnam - Korea University of Information and Communication Technology, Vietnam, Wrocław University of Science and Technology, Poland, International University - VNU-HCM, Vietnam, and the Wrocław University of Economics and Business, Poland, in cooperation with the IEEE SMC Technical Committee on Computational Collective Intelligence, European Research Center for Information Systems (ERCIS), and Nguyen Tat Thanh University, Vietnam.

Following the successes of the First ICCCI (2009) held in Wrocław, Poland, the Second ICCCI (2010) in Kaohsiung, Taiwan, the Third ICCCI (2011) in Gdynia, Poland, the 4th ICCCI (2012) in Ho Chi Minh City, Vietnam, the 5th ICCCI (2013) in Craiova, Romania, the 6th ICCCI (2014) in Seoul, South Korea, the 7th ICCCI (2015) in Madrid, Spain, the 8th ICCCI (2016) in Halkidiki, Greece, the 9th ICCCI (2017) in Nicosia, Cyprus, the 10th ICCCI (2018) in Bristol, UK, and the 11th ICCCI (2019) in Hendaye, France, this conference continued to provide an internationally respected forum for scientific research in the computer-based methods of collective intelligence and their applications.

Computational collective intelligence (CCI) is most often understood as a subfield of artificial intelligence (AI), dealing with soft computing methods that facilitate group decisions or processing knowledge among autonomous units acting in distributed environments. Methodological, theoretical, and practical aspects of CCI are considered as the form of intelligence that emerges from the collaboration and competition of many individuals (artificial and/or natural). The application of multiple computational intelligence technologies such as fuzzy systems, evolutionary computation, neural systems, consensus theory, etc., can support human and other collective intelligence, and create new forms of CCI in natural and/or artificial systems. Three subfields of the application of computational intelligence technologies to support various forms of collective intelligence are of special interest but are not exclusive: the Semantic Web (as an advanced tool for increasing collective intelligence), social network analysis (as the field targeted at the emergence of new forms of CCI), and multi-agent systems (as a computational and modeling paradigm especially tailored to capture the nature of CCI emergence in populations of autonomous individuals).

The ICCCI 2020 conference featured a number of keynote talks and oral presentations, closely aligned to the theme of the conference. The conference attracted a substantial number of researchers and practitioners from all over the world, who submitted their papers for the main track and four special sessions.

The main track, covering the methodology and applications of CCI, included: knowledge engineering and Semantic Web, social networks and recommender systems, collective decision-making, applications of collective intelligence, data mining methods and applications, machine learning methods, computer vision techniques, biosensors and biometric techniques, natural language processing, as well as innovations in intelligent systems. The special sessions, covering some specific topics of particular interest, included: applications of collective intelligence, deep learning and applications for Industry 4.0, experience enhanced intelligence to IoT, intelligent management information systems, intelligent modeling and simulation approaches for games and real world systems, low resource languages processing, computational collective intelligence and natural language processing, computational intelligence for multimedia understanding, intelligent processing of multimedia in Web systems.

We received more than 310 submissions from 47 countries all over the world. Each paper was reviewed by two to four members of the International Program Committee (PC) of either the main track or one of the special sessions. Finally, we selected 70 best papers for oral presentation and publication in one volume of the *Lecture Notes in Artificial Intelligence* series and 68 papers for oral presentation and publication in one volume of the *Communications in Computer and Information Science* series.

We would like to express our thanks to the keynote speakers: Richard Chbeir from Université de Pau et des Pays de l'Adour (UPPA), France, Thanh Thuy Nguyen from VNU University of Engineering and Technology, Vietnam, Klaus Solberg Söilen from Halmstad University, Sweden, and Takako Hashimoto from Chiba University of Commerce, Japan, for their world-class plenary speeches.

Many people contributed toward the success of the conference. First, we would like to recognize the work of the PC co-chairs and special sessions organizers for taking good care of the organization of the reviewing process, an essential stage in ensuring the high quality of the accepted papers. The workshop and special session chairs deserve a special mention for the evaluation of the proposals and the organization and coordination of the work of seven special sessions. In addition, we would like to thank the PC members, of the main track and of the special sessions, for performing their reviewing work with diligence. We thank the Local Organizing Committee chairs, publicity chair, Web chair, and technical support chair for their fantastic work before and during the conference. Finally, we cordially thank all the authors, presenters, and delegates for their valuable contribution to this successful event. The conference would not have been possible without their support.

Our special thanks are also due to Springer for publishing the proceedings and sponsoring awards, and to all the other sponsors for their kind support.

It is our pleasure to announce that the ICCCI conference series continues to have a close cooperation with the Springer journal *Transactions on Computational Collective Intelligence*, and the IEEE SMC Technical Committee on *Transactions on Computational Collective Intelligence*.

Finally, we hope that ICCCI 2020 contributed significantly to the academic excellence of the field and will lead to the even greater success of ICCCI events in the future.

December 2020

Marcin Hernes
Krystian Wojtkiewicz
Edward Szczerbicki

Organization

Organizing Committee

Honorary Chairs

Pierre Lévy	University of Ottawa, Canada
Cezary Madryas	Wrocław University of Science and Technology, Poland

General Chairs

Ngoc Thanh Nguyen	Wrocław University of Science and Technology, Poland
Bao-Hung Hoang	Center of Information Technology of Thua Thien Hue Province, Vietnam
Cong-Phap Huynh	University of Danang, Vietnam - Korea University of Information and Communication Technology, Vietnam

Program Chairs

Costin Bădică	University of Craiova, Romania
Dosam Hwang	Yeungnam University, South Korea
Edward Szczerbicki	The University of Newcastle, Australia
The-Son Tran	University of Danang, Vietnam - Korea University of Information and Communication Technology, Vietnam
Gottfried Vossen	University of Münster, Germany

Steering Committee

Ngoc Thanh Nguyen	Wrocław University of Science and Technology, Poland
Shyi-Ming Chen	National Taiwan University of Science and Technology, Taiwan
Dosam Hwang	Yeungnam University, South Korea
Lakhmi C. Jain	University of South Australia, Australia
Piotr Jędrzejowicz	Gdynia Maritime University, Poland
Geun-Sik Jo	Inha University, South Korea
Janusz Kacprzyk	Polish Academy of Sciences, Poland
Ryszard Kowalczyk	Swinburne University of Technology, Australia
Toyoaki Nishida	Kyoto University, Japan
Manuel Núñez	Universidad Complutense de Madrid, Spain
Klaus Solberg Söilen	Halmstad University, Sweden
Khoa Tien Tran	International University - VNU-HCM, Vietnam

Special Session Chairs

Bogdan Trawiński	Wrocław University of Science and Technology, Poland
Marcin Hernes	Wrocław University of Economics and Business, Poland
Sinh Van Nguyen	International University - VNU HCM, Vietnam
Thanh-Binh Nguyen	University of Danang, Vietnam - Korea University of Information and Communication Technology, Vietnam

Organizing Chairs

Quang-Vu Nguyen	University of Danang, Vietnam - Korea University of Information and Communication Technology, Vietnam
Krystian Wojtkiewicz	Wrocław University of Science and Technology, Poland

Publicity Chairs

My-Hanh Le-Thi	University of Danang, Vietnam - Korea University of Information and Communication Technology, Vietnam
Marek Krótkiewicz	Wrocław University of Science and Technology, Poland

Webmaster

Marek Kopel	Wrocław University of Science and Technology, Poland

Local Organizing Committee

Hai Nguyen	University of Danang, Vietnam - Korea University of Information and Communication Technology, Vietnam
Van Tan Nguyen	University of Danang, Vietnam - Korea University of Information and Communication Technology, Vietnam
My-Hanh Le-Thi	University of Danang, Vietnam - Korea University of Information and Communication Technology, Vietnam
Marcin Jodłowiec	Wrocław University of Science and Technology, Poland

Bernadetta Maleszka	Wrocław University of Science and Technology, Poland
Marcin Maleszka	Wrocław University of Science and Technology, Poland
Artur Rot	Wrocław University of Economics and Business, Poland
Anna Chojnacka-Komorowska	Wrocław University of Economics and Business, Poland

Keynote Speakers

Richard Chbeir	Université de Pau et des Pays de l'Adour (UPPA), France
Thanh Thuy Nguyen	VNU University of Engineering and Technology, Vietnam
Klaus Solberg Söilen	Halmstad University, Sweden
Takako Hashimoto	Chiba University of Commerce, Japan

Special Session Organizers

ACI 2020 – Special Session on Applications of Collective Intelligence

Quang-Vu Nguyen	University of Danang, Vietnam - Korea University of Information and Communication Technology, Vietnam
Van Du Nguyen	Nong Lam University, Vietnam
Van Cuong Tran	Quang Binh University, Vietnam

CCINLP 2020 – Special Session on Computational Collective Intelligence and Natural Language Processing

Ismaïl Biskri	University of Québec à Trois-Rivières, Canada
Thang Le Dinh	University of Québec à Trois-Rivières, Canada

DDISS 2020 – Special Session on Data Driven IoT for Smart Society

P. W. C. Prasad	Charles Sturt University, Australia
S. M. N. Arosha Senanayake	University of Brunei, Brunei

DLAI 2020 – Special Session on Deep Learning and Applications for Industry 4.0

Anh Duc Le	Center for Open Data in the Humanities, Japan
Tho Quan Thanh	Ho Chi Minh City University of Technology, Vietnam
Tien Minh Nguyen	Hung Yen University of Technology and Education, Vietnam
Anh Viet Nguyen	Le Quy Don Technical University, Vietnam

EEIIOT 2020 – Experience Enhanced Intelligence to IoT

Edward Szczerbicki	The University of Newcastle, Australia
Haoxi Zhang	Chengdu University of Information Technology, China

IMIS 2020 – Special Session on Intelligent Management Information Systems

Marcin Hernes	Wrocław University of Economics and Business, Poland
Artur Rot	Wrocław University of Economics and Business, Poland

IMSAGRWS 2020 – Special Session on Intelligent Modeling and Simulation Approaches for Games and Real World Systems

Doina Logofătu	Frankfurt University of Applied Sciences, Germany
Costin Bădică	University of Craiova, Romania
Florin Leon	Gheorghe Asachi Technical University of Iaşi, Romania

IWCIM 2020 – International Workshop on Computational Intelligence for Multimedia Understanding

Behçet Uğur Töreyin	Istanbul Technical University, Turkey
Maria Trocan	Institut Supérieur d'électronique de Paris, France
Davide Moroni	Institute of Information Science and Technologies, Italy

LRLP 2020 – Special Session on Low Resource Languages Processing

Ualsher Tukeyev	Al-Farabi Kazakh National University, Kazakhstan
Madina Mansurova	Al-Farabi Kazakh National University, Kazakhstan

WEBSYS 2020 – Intelligent Processing of Multimedia in Web Systems

Kazimierz Choroś	Wrocław University of Science and Technology, Poland
Maria Trocan	Institut Supérieur d'électronique de Paris, France

Program Committee

Muhammad Abulaish	South Asian University, India
Sharat Akhoury	University of Cape Town, South Africa
Stuart Allen	Cardiff University, UK
Ana Almeida	GECAD-ISEP-IPP, Portugal
Bashar Al-Shboul	University of Jordan, Jordan
Adel Alti	University of Setif, Algeria
Taha Arbaoui	University of Technology of Troyes, France
Mehmet Emin Aydin	University of the West of England, UK
Thierry Badard	Laval University, Canada
Amelia Badica	University of Craiova, Romania

Costin Badica	University of Craiova, Romania
Hassan Badir	Ecole nationale des sciences appliquees de Tanger, Morocco
Dariusz Barbucha	Gdynia Maritime University, Poland
Paulo Batista	Universidade de Evora, Portugal
Khalid Benali	University of Lorraine, France
Morad Benyoucef	University of Ottawa, Canada
Leon Bobrowski	Bialystok University of Technology, Poland
Abdelhamid Bouchachia	Bournemouth University, UK
Peter Brida	University of Zilina, Slovakia
Krisztian Buza	Budapest University of Technology and Economics, Hungary
Aleksander Byrski	AGH University of Science and Technology, Poland
David Camacho	Universidad Autonoma de Madrid, Spain
Alberto Cano	Virginia Commonwealth University, USA
Frantisek Capkovic	Institute of Informatics, Slovak Academy of Sciences, Slovakia
Richard Chbeir	Université de Pau et des Pays de l'Adour (UPPA), France
Shyi-Ming Chen	National Taiwan University of Science and Technology, Taiwan
Raja Chiky	Institut Supérieur d'electronique de Paris (ISEP), France
Amine Chohra	Paris-East University (UPEC), France
Kazimierz Choros	Wrocław University of Science and Technology, Poland
Jose Alfredo Ferreira Costa	Universidade Federal do Rio Grande do Norte, Brazil
Rafal Cupek	Silesian University of Technology, Poland
Ireneusz Czarnowski	Gdynia Maritime University, Poland
Paul Davidsson	Malmo University, Sweden
Camelia Delcea	Bucharest University of Economic Studies, Romania
Tien V. Do	Budapest University of Technology and Economics, Hungary
Habiba Drias	University of Science and Technology Houari Boumedienne, Algeria
Abdellatif El Afia	ENSIAS, Mohammed V University in Rabat, Morocco
Nadia Essoussi	University of Tunis, Tunisia
Rim Faiz	University of Carthage, Tunisia
Marcin Fojcik	Western Norway University of Applied Sciences, Norway
Anna Formica	IASI-CNR, Italy
Naoki Fukuta	Shizuoka University, Japan
Mohamed Gaber	Birmingham City University, UK
Faiez Gargouri	University of Sfax, Tunisia
Mauro Gaspari	University of Bologna, Italy

K. M. George	Oklahoma State University, USA
Janusz Getta	University of Wollongong, Australia
Daniela Gifu	Alexandru Ioan Cuza University, Romania
Daniela Godoy	ISISTAN Research Institute, Argentina
Antonio Gonzalez-Pardo	Universidad Autonoma de Madrid, Spain
Manuel Grana	University of the Basque Country, Spain
Foteini Grivokostopoulou	University of Patras, Greece
William Grosky	University of Michigan, USA
Kenji Hatano	Doshisha University, Japan
Marcin Hernes	Wroclaw University of Economics, Poland
Huu Hanh Hoang	Hue University, Vietnam
Bonghee Hong	Pusan National University, South Korea
Tzung-Pei Hong	National University of Kaohsiung, Taiwan
Frédéric Hubert	Laval University, Canada
Maciej Huk	Wrocław University of Science and Technology, Poland
Dosam Hwang	Yeungnam University, South Korea
Lazaros Iliadis	Democritus University of Thrace, Greece
Agnieszka Indyka-Piasecka	Wrocław University of Science and Technology, Poland
Dan Istrate	Université de Technologie de Compiègne, France
Mirjana Ivanovic	University of Novi Sad, Serbia
Jaroslaw Jankowski	West Pomeranian University of Technology, Poland
Joanna Jedrzejowicz	University of Gdańsk, Poland
Piotr Jedrzejowicz	Gdynia Maritime University, Poland
Gordan Jezic	University of Zagreb, Croatia
Geun Sik Jo	Inha University, South Korea
Kang-Hyun Jo	University of Ulsan, South Korea
Christophe Jouis	Université Sorbonne Nouvelle Paris 3, France
Przemysław Juszczuk	University of Economics in Katowice, Poland
Petros Kefalas	CITY College, International Faculty of the University of Sheffield, Greece
Marek Kisiel-Dorohinicki	AGH University of Science and Technology, Poland
Attila Kiss	Eötvös Loránd University, Hungary
Marek Kopel	Wrocław University of Science and Technology, Poland
Leszek Koszalka	Wrocław University of Science and Technology, Poland
Ivan Koychev	Sofia University, "St. Kliment Ohridski," Bulgaria
Jan Kozak	University of Economics in Katowice, Poland
Adrianna Kozierkiewicz	Wrocław University of Science and Technology, Poland
Ondrej Krejcar	University of Hradec Kralove, Czech Republic
Dariusz Krol	Wrocław University of Science and Technology, Poland

Marek Krotkiewicz	Wrocław University of Science and Technology, Poland
Jan Kubicek	VSB -Technical University of Ostrava, Czech Republic
Elzbieta Kukla	Wrocław University of Science and Technology, Poland
Marek Kulbacki	Polish-Japanese Academy of Information Technology, Poland
Piotr Kulczycki	Polish Academy of Science, Systems Research Institute, Poland
Kazuhiro Kuwabara	Ritsumeikan University, Japan
Halina Kwasnicka	Wrocław University of Science and Technology, Poland
Hoai An Le Thi	University of Lorraine, France
Sylvain Lefebvre	Toyota ITC, France
Philippe Lemoisson	French Agricultural Research Centre for International Development (CIRAD), France
Florin Leon	Gheorghe Asachi Technical University of Iasi, Romania
Doina Logofatu	Frankfurt University of Applied Sciences, Germany
Edwin Lughofer	Johannes Kepler University Linz, Austria
Juraj Machaj	University of Zilina, Slovakia
Bernadetta Maleszka	Wrocław University of Science and Technology, Poland
Marcin Maleszka	Wrocław University of Science and Technology, Poland
Adam Meissner	Poznań University of Technology, Poland
Héctor Menéndez	University College London, UK
Mercedes Merayo	Universidad Complutense de Madrid, Spain
Jacek Mercik	WSB University in Wroclaw, Poland
Radosław Michalski	Wrocław University of Science and Technology, Poland
Peter Mikulecky	University of Hradec Kralove, Czech Republic
Miroslava Mikusova	University of Zilina, Slovakia
Javier Montero	Universidad Complutense de Madrid, Spain
Manuel Munier	Université de Pau et des Pays de l'Adour (UPPA), France
Grzegorz J. Nalepa	AGH University of Science and Technology, Poland
Laurent Nana	University of Brest, France
Anand Nayyar	Duy Tan University, Vietnam
Filippo Neri	University of Napoli Federico II, Italy
Linh Anh Nguyen	University of Warsaw, Poland
Loan T. T. Nguyen	International University - VNU-HCM, Vietnam
Sinh Van Nguyen	International University - VNU-HCM, Vietnam
Adam Niewiadomski	Lodz University of Technology, Poland
Adel Noureddine	Université de Pau et des Pays de l'Adour (UPPA), France

Agnieszka Nowak-Brzezinska	University of Silesia, Poland
Alberto Núnez	Universidad Complutense de Madrid, Spain
Manuel Núnez	Universidad Complutense de Madrid, Spain
Tarkko Oksala	Aalto University, Finland
Mieczyslaw Owoc	Wrocław University of Economics, Poland
Marcin Paprzycki	Systems Research Institute, Polish Academy of Sciences, Poland
Isidoros Perikos	University of Patras, Greece
Marcin Pietranik	Wrocław University of Science and Technology, Poland
Elias Pimenidis	University of the West of England, UK
Nikolaos Polatidis	University of Brighton, UK
Hiram Ponce Espinosa	Universidad Panamericana, Mexico
Piotr Porwik	University of Silesia, Poland
Radu-Emil Precup	Politehnica University Timisoara, Romania
Ales Prochazka	University of Chemistry and Technology, Czech Republic
Paulo Quaresma	Universidade de Evora, Portugal
Mohammad Rashedur Rahman	North South University, Bangladesh
Ewa Ratajczak-Ropel	Gdynia Maritime University, Poland
Tomasz M. Rutkowski	University of Tokyo, Japan
Virgilijus Sakalauskas	Vilnius University, Lithuania
Khouloud Salameh	Université de Pau et des Pays de l'Adour (UPPA), France
Imad Saleh	Université Paris 8, France
Ali Selamat	Universiti Teknologi Malaysia, Malaysia
Andrzej Sieminski	Wrocław University of Science and Technology, Poland
Paweł Sitek	Kielce University of Technology, Poland
Vladimir Sobeslav	University of Hradec Kralove, Czech Republic
Klaus Söilen	Halmstad University, Sweden
Stanimir Stoyanov	Plovdiv University, Bulgaria
Libuse Svobodova	University of Hradec Kralove, Czech Republic
Martin Tabakov	Wroclaw University of Science and Technology, Poland
Muhammad Atif Tahir	National University of Computer and Emerging Sciences, Pakistan
Yasufumi Takama	Tokyo Metropolitan University, Japan
Trong Hieu Tran	VNU-University of Engineering and Technology, Vietnam
Diana Trandabat	Alexandru Ioan Cuza University, Romania
Bogdan Trawinski	Wrocław University of Science and Technology, Poland
Jan Treur	Vrije Universiteit Amsterdam, The Netherlands

Chrisa Tsinaraki	European Commission - Joint Research Center (EC - JRC), Europe
Ualsher Tukeyev	Al-Farabi Kazakh National University, Kazakhstan
Olgierd Unold	Wrocław University of Science and Technology, Poland
Natalie Van Der Wal	Vrije Universiteit Amsterdam, The Netherlands
Bay Vo	Ho Chi Minh City University of Technology, Vietnam
Thi Luu Phuong Vo	International University - VNU-HCM, Vietnam
Lipo Wang	Nanyang Technological University, Singapore
Roger M. Whitaker	Cardiff University, UK
Adam Wojciechowski	Lodz University of Technology, Poland
Krystian Wojtkiewicz	Wrocław University of Science and Technology, Poland
Farouk Yalaoui	University of Technology of Troyes, France
Slawomir Zadrozny	Systems Research Institute, Polish Academy of Sciences, Poland
Drago Zagar	University of Osijek, Croatia
Danuta Zakrzewska	Lodz University of Technology, Poland
Constantin-Bala Zamfirescu	Lucian Blaga University of Sibiu, Romania
Katerina Zdravkova	Ss. Cyril and Methodius University, Macedonia
Aleksander Zgrzywa	Wrocław University of Science and Technology, Poland
Haoxi Zhang	Chengdu University of Information Technology, China
Jianlei Zhang	Nankai University, China
Adam Ziebinski	Silesian University of Technology, Poland

Special Session Program Committees

ACI 2020 – Special Session on Applications of Collective Intelligence

Quang-Vu Nguyen	University of Danang, Vietnam - Korea University of Information and Communication Technology, Vietnam
Van-Du Nguyen	Nong Lam University, Vietnam
Van-Cuong Tran	Quang Binh University, Vietnam
Adrianna Kozierkiewicz	Wrocław University of Science and Technology, Poland
Marcin Pietranik	Wrocław University of Science and Technology, Poland
Chando Lee	Daejeon University, South Korea
Cong-Phap Huynh	University of Danang, Vietnam - Korea University of Information and Communication Technology, Vietnam
Thanh-Binh Nguyen	University of Danang, Vietnam - Korea University of Information and Communication Technology, Vietnam

Tan-Khoi Nguyen	University of Danang - University of Science and Technology, Vietnam
Tuong-Tri Nguyen	Hue University, Vietnam
Minh-Nhut Pham-Nguyen	University of Danang, Vietnam - Korea University of Information and Communication Technology, Vietnam
Xuan-Hau Pham	Quang Binh University, Vietnam

CCINLP 2020 – Special Session on Computational Collective Intelligence and Natural Language Processing

Ismaïl Biskri	Université du Québec à Trois-Rivières, Canada
Mounir Zrigui	Université de Monastir, Tunisia
Anca Pascu	Université de Bretagne Occidentale, France
Éric Poirier	Université du Québec à Trois-Rivières, Canada
Adel Jebali	Concordia University, Canada
Khaled Shaalan	The British University in Dubai, UAE
Vladislav Kubon	Charles University, Czech Republic
Louis Rompré	Cascades Papier Kingsey Falls, Canada
Thang Le Dinh	Université du Québec à Trois-Rivières, Canada
Usef Faghihi	Université du Québec à Trois-Rivières, Canada
Nguyen Cuong Pham	University of Science, Vietnam
Thuong Cang Phan	University of Cantho, Vietnam

DDISS 2020 – Special Session on Data Driven IoT for Smart Society

Minoru Sasaki	Gifu University, Japan
Michael Yu Wang	The University of Hong Kong, Hong Kong
William C. Rose	University of Delaware, USA
Le Hoang Son	Vietnam National University, Vietnam
Darwin Gouwanda	Monash University Malaysia, Malaysia
Owais A. Malik	Universiti Brunei Darussalam, Brunei
Ashutosh Kumar Singh	National Institute of Technology, India
Lau Siong Hoe	Multimedia University, Malaysia
Amr Elchouemi	Forbes School of Business and Technology, USA
Abeer Alsadoon	Charles Sturt University, Australia
Sabih Rehman	Charles Sturt University, Australia
Nectar Costadopoulos	Charles Sturt University, Australia
K. S. Senthilkumar	St. George's University, Grenada
Yuexian Zou	Peking University, China

DLAI 2020 – Special Session on Deep Learning and Applications for Industry 4.0

Anh Le Duc	Center for Open Data in the Humanities, Japan
Minh-Tien Nguyen	Hung Yen University of Technology and Education, Vietnam

EEIOT 2020 – Experience Enhanced Intelligence to IoT

IMIS 2020 – Special Session on Intelligent Management Information Systems

Eunika Mercier-Laurent	Jean Moulin University Lyon 3, France
Małgorzata Pankowska	University of Economics in Katowice, Poland
Mieczysław Owoc	Wrocław University of Economics and Business, Poland
Bogdan Franczyk	University of Leipzig, Germany
Kazimierz Perechuda	Wrocław University of Economics and Business, Poland
Jan Stępniewski	Université Paris 13, France
Helena Dudycz	Wrocłw University of Economics and Business, Poland
Jerzy Korczak	International University of Logistics and Transport in Wroclaw, Poland
Andrzej Bytniewski	Wrocław University of Economics and Business, Poland
Marcin Fojcik	Western Norway University of Applied Sciences, Norway
Monika Eisenbardt	University of Economics in Katowice, Poland
Dorota Jelonek	Czestochowa University of Technology, Poland
Paweł Weichbroth	WSB University in Gdańsk, Poland
Jadwiga Sobieska-Karpinska	Witelon State University of Applied Sciences in Legnica, Poland
Marek Krótkiewicz	Wrocław University of Science and Technology, Poland
Paweł Siarka	Wrocław University of Economics and Business, Poland
Łukasz Łysik	Wrocław University of Economics and Business, Poland
Adrianna Kozierkiewicz	Wrocław University of Science and Technology, Poland
Karol Łopaciński	Wrocław University of Economics and Business, Poland
Marcin Maleszka	Wrocław University of Science and Technology, Poland
Ingolf Römer	University of Leipzig, Germany
Martin Schieck	University of Leipzig, Germany
Anna Chojnacka-Komorowska	Wrocław University of Economics and Business
Krystian Wojtkiewicz	Wrocław University of Science and Technology, Poland
Jacek Winiarski	University of Gdańsk, Poland
Wiesława Gryncewicz	Wrocław University of Economics and Business, Poland
Tomasz Turek	Czestochowa University of Technology, Poland
Marcin Jodłowiec	Wrocław University of Science and Technology, Poland

Anna Sołtysik-Piorunkiewicz	University of Economics in Katowice, Poland
Paula Bajdor	Czestochowa University of Technology, Poland
Dorota Jelonek	Czestochowa University of Technology, Poland
Ilona Pawełoszek	Czestochowa University of Technology, Poland
Ewa Walaszczyk	Wrocław University of Economics and Business, Poland
Krzysztof Hauke	Wrocław University of Economics and Business, Poland
Piotr Tutak	Wrocław University of Economics and Business, Poland
Andrzej Kozina	Cracow University of Economics, Poland

IMSAGRWS 2020 – Special Session on Intelligent Modeling and Simulation Approaches for Games and Real World Systems

Alabbas Alhaj Ali	Frankfurt University of Applied Sciences, Germany
Costin Bădică	University of Craiova, Romania
Petru Caşcaval	Gheorghe Asachi Technical University of Iaşi, Romania
Gia Thuan Lam	Vietnamese-German University, Vietnam
Florin Leon	Gheorghe Asachi Technical University of Iaşi, Romania
Doina Logofătu	Frankfurt University of Applied Sciences, Germany
Fitore Muharemi	Frankfurt University of Applied Sciences, Germany
Julian Szymański	Gdańsk University of Technology, Poland
Pawel Sitek	Kielce University of Technology, Poland
Daniel Stamate	Goldsmiths, University of London, UK

IWCIM 2020 – International Workshop on Computational Intelligence for Multimedia Understanding

Enis Cetin	Bilkent University, Turkey, and UIC, USA
Michal Haindl	Institute of Information Theory and Automation of the CAS, Czech Republic
Andras L. Majdik	Hungarian Academy of Sciences, Hungary
Cristina Ribeiro	University of Porto, Portugal
Emanuele Salerno	National Research Council of Italy (CNR), Italy
Ales Prochazka	University of Chemistry and Technology, Czech Republic
Anna Tonazzini	National Research Council of Italy (CNR), Italy
Gabriele Pieri	National Research Council of Italy (CNR), Italy
Gerasimos Potamianos	University of Thessaly, Greece
Gorkem Saygili	Ankara University, Turkey
Josiane Zerubia	Inria, France
Maria Antonietta Pascali	National Research Council of Italy (CNR), Italy
Marie-Colette Vanlieshout	CWI Amsterdam, The Netherlands

Marco Reggiannini	National Research Council of Italy (CNR), Italy
Nahum Kiryati	Tel Aviv University, Israel
Rozenn Dahyot	Trinity College Dublin, Ireland
Sara Colantonio	National Research Council of Italy (CNR), Italy
Shohreh Ahvar	Institut Supérieur d'Électronique de Paris (ISEP), France
Tamás Szirányi	Institute for Computer Science and Control, Hungary

LRLP 2020 – Special Session on Low Resource Languages Processing

Miguel A. Alonso	Universidade da Coruña, Spain
Pablo Gamallo	University of Santiago de Compostela, Spain
Nella Israilova	Kyrgyz State Technical University, Kyrgyzstan
Marek Kubis	Adam Mickiewicz University, Poland
Belinda Maia	University of Porto, Portugal
Madina Mansurova	Al-Farabi Kazakh National University, Kazakhstan
Gayrat Matlatipov	Urgench State University, Uzbekistan
Marek Miłosz	Lublin University of Technology, Poland
Diana Rakhimova	Al-Farabi Kazakh National University, Kazakhstan
Altynbek Sharipbay	L. N. Gumilyov Eurasian National University, Kazakhstan
Ualsher Tukeyev	Al-Farabi Kazakh National University, Kazakhstan

WEBSYS 2020 – Intelligent Processing of Multimedia in Web Systems

Shohreh Ahvar	ISEP, Paris, France
Frédéric Amiel	ISEP, Paris, France
František Čapkovič	Slovak Academy of Sciences, Slovakia
Kazimierz Choroś	Wrocław University of Science and Technology, Poland
Patricia Conde-Cespedes	ISEP, Paris, France
Marek Kopel	Wrocław University of Science and Technology, Poland
Mikołaj Leszczuk	AGH University of Science and Technology, Poland
Bożena Kostek	Gdańsk University of Technology, Poland
Alin Moldoveanu	Politehnica University of Bucharest, Romania
Tarkko Oksala	Helsinki University of Technology, Finland
Andrzej Siemiński	Wrocław University of Science and Technology, Poland
Maria Trocan	ISEP, Paris, France
Aleksander Zgrzywa	Wrocław University of Science and Technology, Poland

Contents

Data Mining and Machine Learning

Deep Learning and Applications for Industry 4.0

Recommender Systems

Computer Vision Techniques

Decision Support and Control Systems

Intelligent Management Information Systems

Innovations in Intelligent Systems

Intelligent Modeling and Simulation Approaches for Games and Real World Systems

Experience Enhanced Intelligence to IoT

Data Driven IoT for Smart Society

Applications of Collective Intelligence

Natural Language Processing

Low Resource Languages Processing

Computational Collective Intelligence and Natural Language Processing

Data Mining and Machine Learning

Rule Induction of Automotive Historic Styles Using Decision Tree Classifier

Hung-Hsiang Wang[1] and Chih-Ping Chen[2](✉)

[1] Department of Industrial Design, National Taipei University of Technology, Taipei, Taiwan
[2] Doctoral Program in Design, College of Design, National Taipei University of Technology, Taipei, Taiwan
roychen092@hotmail.com

Abstract. For industrial designers, how to classify vehicle styles was a challenge, and at a large depends on the designer's internal knowledge. Although data mining and machine learning technologies had become mature and affordable today, the applications using machine learning in car styling were few. This investigation focuses on using the decision tree method to discuss the relationship between automotive styles and the design features of 35 cars produced by the automotive manufacturer, Dodge between 1942 and 2017. The study summarized 8 design features from previous literature: the length, fender design, number of headlamps, rear form, the position of the quarter glass, engine hood scoop, rocket tail, and side decoration design while the styles are chosen were streamlined style, popular style, and modern style. The decision tree algorithm (C5.0) was employed to obtain the optimal rule of decision tree to compare the historic design styles from ten sets of decision tree rules. The result showed that there was a clear relationship between the key design features and the historic style of vehicles. The average accuracy of the ten sets of decision trees is 90.6%. The highest accuracy of the optimal model is 97%. However, the variation between the predicted accuracies of decision tree models calculated is high, ranging from 80% to 97%. Based on the decision tree and statistics method, the design features include the length, fender design, rocket tail design, rear form, and position of the quarter glass was more important than the others. This method had the potential to identify automotive historic styles based on key features.

Keywords: Decision tree analysis · Automotive-style · C4.5/C5.0 · Data mining

1 Introduction

The process of design thinking for industrial designers is often a black box. The designer engages in design work through their intuition, sensibility, analogy, or metaphor methods. Although machine learning and artificial intelligence have developed rapidly. But few industrial designers use rational and scientific methods to classify and predict car styles. And style is generally considered a popular phenomenon and cannot be summarized into rule-base understandable by designers. However, we believe that technology and science progress can help the designer's work and find out the rules that can be understood behind

M. Hernes et al. (Eds.): ICCCI 2020, CCIS 1287, pp. 3–14, 2020.
https://doi.org/10.1007/978-3-030-63119-2_1

the style. Therefore, the motivation of this research is to find out the correlation between style and features and discover the induction of rules in the design field.

The goal of the investigation is to introduce a process with preliminary data mining techniques for car stylists or designers. To obtain hidden decision rules from a brand's car style data set in a period. And two contributions using this rule-based method as follow. One is to identify the new car style of the specific brand, and the other is to compare with the car styles of other competitors.

Thus, the car designer can induct the rules from the decision tree to reflect their design decisions on car styling, while building design knowledge that belongs to the designer.

2 Artificial Intelligence and Style

2.1 Research Relevant to the Application of AI in Design

In 1992, two academics coincidentally suggested the application of artificial intelligence in design. Because of the rapid development of computer technology, it is possible to build computer programs to model the human mind and test psychological theories of human performance, to understand human reasoning, and to exploit the knowledge used by human beings in solving a particular problem [1]. Another design researcher mentioned that in the article "Design ability". Although AI techniques may be meant to supplant human thinking, research in AI can also be a means of trying to understand human thinking [2].

The new generation of design research is often closely linked with other courses, having frequent interaction with other disciplines like a net. Meredith Davis once mentioned that it is difficult to define content boundaries even for research within the design discipline. She believed that design research can have even more opportunities and mentioned the concept of a digital design research database [3]. Digital storage of design data and information can be realized by introducing a computer system into the design process [4]. Another example is the case using an automotive database for analysis [5]. For the "data-product-data" design model, in terms of the value of data, the product category stems from the fact that the value of the data can be used for the design of the new product [6]. Furthermore, the study and analysis of the data can become tools for design research.

2.2 The Definition and Classification of Design

Style is related to history. In an article published in 1960, Ernst Gombrich mentioned the history of taste and fashion is the history of preference. People make various choices amongst the given alternatives. Simon Herbert also mentioned in his 1975 article that the order of the search also determines a type of style. Human designers have a set of specific procedures to initiate design units, design restrictions, or goals [1]. For example, the design technique used in the garlic press [7] designed by Alessi designers in 1996 was used in the product design of Guido Venturini.

Literature related to automotive-style classification and characteristics suggests that there is a rudimentary form of the expert system for style classification developed using external characteristics of the cars [8]. In addition, Chris Dowlen also conducted a series of research related to automotive design history using the automotive design database. An example is the development of the style and layout of cars [9]. Triz was used to investigate the functionality assessment and measurement in automotive history [10]. The cluster analysis was employed to uncover the creative and innovative automotive-styles in history and classified them before extracting the characteristics of the automotive design to assess the innovativeness of automotive design history [11].

2.3 Potential of Using Data Mining and Decision Tree to Classify Styles

Artificial intelligence covers various topics including data analysis, machine learning, and data mining. although deep learning has already been able to identify and classify the style of artworks and reconstruct artworks of specified features [12]. However, deep learning fails to point out the principle behind style classification. The data mining system can be used to identify, amend, and extract information [4]. The decision tree is an algorithm in machine learning and is a commonly used data mining technique, often used for classification and prediction. It is easy to implement and explain and suits the intuitive thinking style of human beings.

The common decision tree algorithm includes ID3, C4.5, C5.0, CART, random forest [13] and so on. ID3 is a simple decision tree algorithm [14] while C4.5 improved the

Table 1. Comparison of four kinds of decision trees

	ID3	C4.5/C5.0	CART	Random Forest
Time	1986	1993/1994	1984	1995
Inventor	Ross Quinlan	Ross Quinlan	Leo Breiman	Tin Kam Ho
Type of attributes	Only the categorical variable	Both categorical and numerical variable	Both categorical and numerical variable	Both categorical and numerical variable
Classification criteria	Entropy & Information Gain	Gain Ratio	Gini or twoing criteria	Ensemble Method
Advantage(s)	Very simple decision tree algorithm	Improve the shortcoming of ID3 and handle numerical attributes	Produce regression tree, can handle outlier	Formed by multiple trees and is the most accurate classification method, suitable for big data
Shortcoming(s)	Not suitable for any trimming and cannot handle numeric attributes and missing values	Trees become larger and more complex, and have overfitting problems	Decision tree may be unstable	Training can be slow

shortcomings of ID3 and is a subsequent edition of ID3 [15]. Some sources (e.g. Weka) name this algorithm J48. C5.0 is the commercial version of C4.5 and is faster [16]. CART is a classification and regression tree [17] while the random forest is the extension of CART, and is made from multiple trees [18]. The relevant comparison is summarized (see Table 1).

In summary, ID3 was developed in the olden days and cannot handle numerical attributes and data with missing values and hence is not considered for this investigation. Although C4.5 is a newer edition of ID3, it is also not the latest version. C5.0 is the latest version. Despite having the ability to handle categorical and numerical variables, CART also belongs to the earlier period when compared with other decision tree algorithms. The random forest algorithm is the newer decision tree algorithm, but it is made of multiple trees and hence faces the problem of slower training speed. As a result, the C5.0 decision tree was chosen as the optimal decision tree for this investigation. To solve the classification of automobile styles and discuss the rules between design styles and design features.

3 Method

3.1 Choice of Features and Style

Figure 1 depicts the meanings of features using the alphabet as shown above. Feature selection for design style is not easy, but this investigation tries to define it in the following ways. The first is the previous literature review, including the 2004 design journal paper [8], and the 2013 research paper of Chris Dowlen [11]. Secondly, a comparison of two different styles of representative vehicle observation. Hence, we set the external features

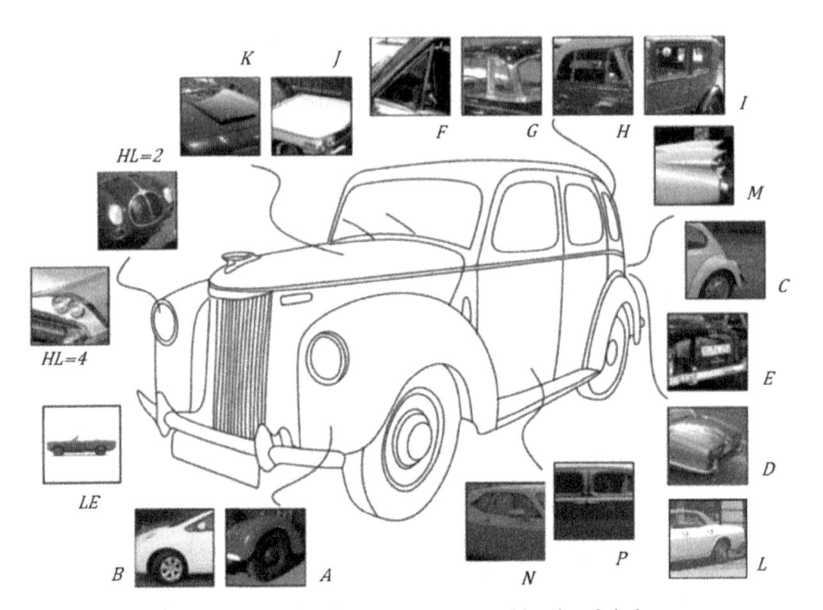

Fig. 1. The design features represented by the alphabets

such as length (LE), fender (FE), number of headlamps (HL), rear form (RF), the position of quarter glass (QG), rocket tail (RT), engine hood scoop (EH) and side decoration (SD) as classification feature.

Based on the literature, some features changed with design style and were sufficient to differentiate them. The chosen samples were produced between 1942 and 2017. This 75 years period witnessed the change of three periods of design styles namely streamlined style (SS), popular style (PS), and modern style (MS). The style for cars produced between 2000 and 2017 was also classified as a modern style. Thus the three styles of automotive-style classification were used as the target features.

3.2 Choice of Case Study

The American automotive industry is the most representative [8], the brand was chosen from there. By considering the representativeness of the brand, the number of cases, and fame, the American vehicle brand Dodge was chosen as the case study with sufficient samples to cover the three periods of interest. Table 2 represented the design features of the vehicles.

Regarding the choice of the test samples of the investigation, the database that referenced the automotive handbook "The Complete Book of Collectible Cars" [19] and relevant literature online was the main sample of interest.

Table 2. The list of thirty-five Dodge vehicles produced between 1942 and 2017

Name	LE	FE	HL	RF	QG	EH	RT	SD	Name	LE	FE	HL	RF	QG	EH	RT	SD
Custom	203.3	A	2	C	F	J	L	N	Challenger II	191.3	B	4	D	H	K	L	P
Custom II	204.6	A	2	C	F	J	L	N	Magnum	215.8	B	4	D	G	J	L	N
Wayfarer	196	A	2	C	F	J	L	N	Mirada	209.5	B	4	D	G	J	L	N
Wayfarer II	196.3	A	2	C	F	J	L	N	600 ES	180.7	B	4	D	H	J	L	N
Coronet	189.6	B	2	D	F	J	M	N	Shelby	174.8	B	4	D	G	K	L	N
Royal 500	196	B	2	D	F	J	M	N	Dayona	175	B	2	D	G	K	L	N
Custom Royal	212.1	B	2	D	F	J	M	N	Lancer	180.4	B	4	D	G	J	L	N
Custom Royal II	212.1	B	2	E	F	J	M	N	Spirit	181.2	B	2	D	G	J	L	N
D-500	212.2	B	2	E	F	J	M	N	Stealth	179.1	B	2	D	G	J	L	N
Custom Royal III	213.8	B	4	E	F	J	M	N	Dayona II	179.8	B	2	E	G	J	L	N
Polara	202	B	4	E	F	J	L	N	Viper RT/10	175.1	B	2	D	H	J	L	N
Dart GT	196.4	B	2	D	F	J	L	N	Viper GTS	176.7	B	2	D	G	J	L	N
Charger	208	B	0	D	F	J	L	P	Neon I	171.8	B	2	D	G	J	L	N
Coronet II	209.2	B	4	E	F	J	L	P	Neon II	174.4	B	2	D	H	J	L	N
Charger II	208	B	0	D	F	J	L	P	Dart I	178.9	B	2	D	G	J	L	N
Coronet	209.2	B	4	E	I	K	L	P	Viper	175.7	B	2	D	G	J	L	N
Charger II	220	B	0	D	F	K	L	P	Dart II	184.2	B	2	D	H	J	L	N
Challenger	191.3	B	4	D	H	K	L	P									

3.3 Classification Methods and Tools

The data analysis tool was the decision tree C5.0 package, a classification technology from data mining software that uses R 3.6.1. A randomly selected 25 samples (71.4%) were used as the training samples while the remaining 10 samples (28.6%) were used as the test samples. It is hoped that the two stages including the analysis process of the training and testing samples can obtain the most optimal mining model.

In order to prevent deviation amongst the accuracies of the decision trees, then models were generated (n = 10) and the average accuracy ($\overline{X}$) was calculated. The formula is as shown below.

$$\overline{x} = \frac{\sum_{i=1}^{n} Xi}{n} \tag{1}$$

The data set S for the decision tree has c different classifications. The entropy of the data set is S and P_i is the probability of classification i appearing in data set S where i = 1,2,3 ~ c. Attribute A in information gain of data set S is defined as Gain(S,A), which refers to the gain from splitting of data set S by attribute A and is used to measure the classification ability of design attributes. The formula is as shown below.

$$\text{Entropy}(S) = \sum_{i=1}^{c} -P_i \log_2 P_i \tag{2}$$

$$\text{Gain}(S,A) = \text{Entropy}(S) - \text{Entropy}(S,A) \tag{3}$$

Split information value is $\text{SplitInfo}_A(S)$ where $|S_i|/|S|$ is the weightage of i subset. The gain ratio is GainRatio(A) where A is the design characteristic attribute and the formula is as shown below.

$$\text{SplitInfo}_A(S) = \sum_{i=1}^{c} \frac{|S_i|}{|S|} \times \log_2\left(\frac{|S_i|}{|S|}\right) \tag{4}$$

$$\text{GainRatio}(A) = \frac{\text{Gain}(S, A)}{\text{SplitInfo}_A(S)} \tag{5}$$

The definition of accuracy is the overall ability of the model to perform the correct classification for the test sample. The matrix of the report includes the quantity of true positive (TP), true negative (TN), false positive (FP) and false-negative (FN). The formula is as shown below.

$$\text{Accuracy} = \frac{TP + TN}{TP + TN + FP + FN} \tag{6}$$

4 Results

4.1 Decision Tree Classification Model Diagram

Based on the level distribution of the node in the decision tree diagram of the most optimal decision tree (Model A), one can know that the order of information gain for the decision tree (Fig. 2).

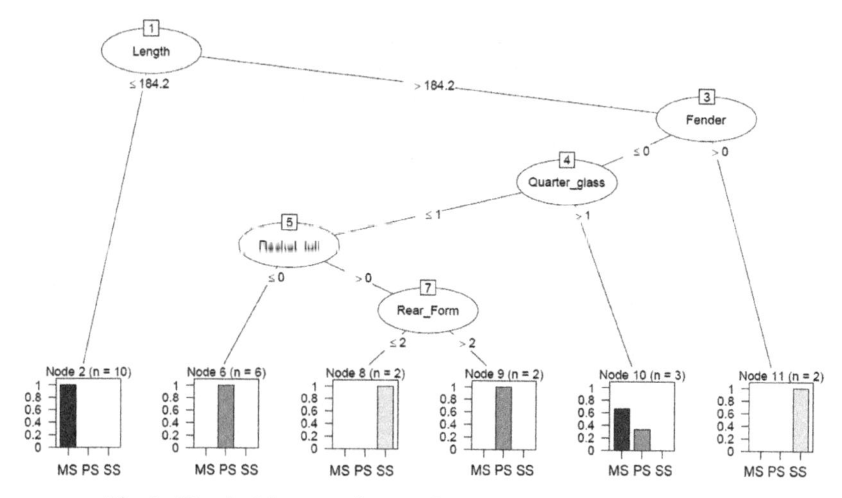

Fig. 2. The decision tree diagram for automotive style classification

This order of information gain also represents the extent to which the feature affects the automotive design style classification. This means that LE is the main factor determining the different automotive design style, followed by FE. The third would be QG, fourth was RT, fifth was RF.

(A) **Rules**

There are six rules of the decision tree (Model A) as shown below Table 3.

Table 3. Rules of the most optimal decision tree (Model A)

Rule 1	LE less than 184.2in is MS
Rule 2	LE greater than 184.2in and has FE is SS
Rule 3	LE greater than 184.2in, without FE and has QG positioned at the back or both at the front and the back can be MS (possibility of 2/3) or PS (possibility of 1/3)
Rule 4	LE greater than 184.2in, without FE, has QG positioned at the front or without QG and without RT is PS
Rule 5	LE greater than 184.2in, without FE, has QG positioned at the front or without QG, has RT and has a fin-shaped RE is PS
Rule 6	LE greater than 184.2in, without FE, has QG positioned at the front or without QG, without RT and has a rectangular or curved RE is SS

(B) **Accuracy and Confusion Matrix of Decision Tree (Model A)**

The sample size includes the training mode and verification mode in Model A is 35 with an accuracy of 97.00%. The sample size of the training mode is 25 with an accuracy of 96%, while the verification mode is 10 with an accuracy of 100%. The confusion matrix of the decision tree based on Eq. 6 is shown in Table 4. There is a classification error which is the Custom Royal produced in 1955. It is classified as PS in the design history but is classified as SS by the decision tree.

Table 4. The confusion matrix of decision tree

	Train & Verification prediction		
Style	Modern	Popular	Streamlined
Modern	16	0	0
Popular	0	12	1
Streamlined	0	0	6

4.2 The Average Accuracy of Ten Decision Trees

The statistics on the average accuracy of the ten decision trees showed in Table 5. The models were created by the computer automatically to prevent deviation amongst the accuracies of the decision trees. Every decision tree had different design features and shapes. The design features have consisted of the aforementioned several of eight features.

Table 5. The statistics of average accuracies tested by ten decision trees

Item	1	2	3	4	5	6	7	8	9	10	Mean
Train(%)	88	96	96	100	96	88	96	100	96	88	94.4
Test(%)	60	100	80	90	80	90	80	90	60	80	81
Accuracy(%)	80	97	91	97	91	89	91	97	86	86	90.6
Classification Features	QG FE RT	LE FE QG RT RF	LE FE RT RF	QG LE FE RT RF	QG LE FE RT	LE FE RT	LE FE RT RF	FE RT RF LE QG	LE FE RT	LE FE	
Example	Model B	Model A						Model C			

Next, we calculate the average accuracy of the data based on Eq. 1. Then comparing the style determined by the computer with the style in the vehicle design history to obtain the accuracy of prediction. The accuracy of the training mode ranged from 88% to 100%, while the accuracy of the test mode ranged from 60% to 100%. The statistics showed that the training samples had an average accuracy of 94.4%, while the test samples had an average accuracy of 81%. The 35 car samples had an average accuracy of 90.6%.

4.3 Correlation Between Design Features and Accuracy

The design features adopted by 10 different decision trees in Table 5. It can be found that the three high-accuracy (97%, item 2, 4, and 8) decision trees adopt up to 5 design features and all have the feature of the vehicle length. The decision tree with the lowest accuracy (80%, item 1) adopts only three design features and does not include the vehicle length feature. We use regression to analyze the correlation between accuracy and the number of design features. The result that P-value is 0.0012. Thus, the accuracy of the decision tree is highly correlated statistically with the number of design features by the system created.

4.4 The Accumulated Number of Statistical Design Features

Based on the statistics of each of the design features, FE has the highest accumulated number followed by LE, and RT with the number of nine. RF and QG have the same accumulated number of five. Three of the design features HL, EH, and SD are not considered as classification features from the computer-generated decision trees. The result is shown in Table 6.

Table 6. The accumulated number of statistical design features

Feature	LE	FE	HL	SD
Accumulated number	9	10	0	0
Feature	RF	QG	EH	RT
Accumulated number	5	5	0	9

4.5 Entropy, Information Gain and Gain Ratio

For the most optimal decision tree (Model A), the target carriable is the style and hence there are three styles, SS, PS, and MS. There is a design feature known as "vehicle length" and is either greater than 184.2 inches or less than 184.2 inches. This produced 25 observations under the training mode and 4 of which belong to SS, 9 belong to PS and 12 belong to MS. As a result, Eq. 2~5 can be employed to calculate the entropy, information gain, information split, and gain ratio with automotive length as the target variable.

(1) Entropy before the split, Entropy (Dodge 25Cars) = 1.4594
(2) Weighted average entropy after the split, Entropy (Dodge 25Cars, Length) = 0.7963
(3) Information Gain, Gain (Dodge 25Cars, Length) = 0.6631
(4) Split Information, $SplitInfo_{Length}$ (Dodge 25Cars) = 0.9707
(5) Gain Ratio, Gain Ratio (Length) = 0.6831

The following is the comparison of Model A, B, and C (see Table 5) whose first feature attributes are LE, QG, and FE respectively. The information gain for LE is the largest, suggesting the messiness of information within the attribute is smaller, and the information for classification is better. LE is also the attribute with the largest gain ratio, making it suitable as the first attribute to split. The result is shown in Table 7.

Table 7. The comparison of entropy, information gain, information split and gain ratio of the attributes for the three models

	Attribute	Entropy(S)	Entropy(S, A)	Gain(S,A)	$SplitInfo_A(S)$	Gain Ratio(A)
Model A	Length	1.46	0.8	0.66	0.97	0.68
Model B	Quarter glass	1.52	0.97	0.54	0.99	0.55
Model C	Fender	1.57	1.49	0.079	0.63	0.12

5 Discussion

5.1 Case Study of Automotive-Style Classification

(A) **Dodge Charger-new sample not in the database**

The Charger produced by Dodge in 2005, vehicle length is 200.1 inches and is classified as MS or PS by the decision tree. We use the rule3 "LE greater than 184.2in, without FE and has QG positioned at the back or both at the front and the back can be MS (possibility of 2/3) or PS (possibility of 1/3)". The features of the car with long length and engine hood scoop look like a modern car with a retro style.

(B) **Dodge Custom Royal**

The Custom Royal was produced by Dodge in 1955 and was classified as SS by the decision tree while it should be under PS (1955–1975) according to the design history. Although we concluded that the system made an error in classification, Custom Royal is lacking in the fender which is present in SS. However, it also does not have the luxury style and rocket tail feature of PS. As a result, it is a product of the transition period. It is thus reasonable to classify it as the start of PS or the end of SS.

(C) **Chrysler PT Cruiser–new sample of other competitors**

PT Cruiser was produced by Chrysler in 2002. Since Dodge and Chrysler share similar background and origin, PT Cruiser is 168in long with two headlamps, fender, quarter glass positioned at the back, a curved rear form, no engine hood scoop, no rocket tail, and no side decoration. It was classified as MS by the decision tree. Although the judgment of the period was correct, the style classification was wrong. In simple terms, due to the energy crisis in late 70s, the MS cars are largely compact or mid-sized. As a result, even though the retro car itself has many retro design features, the length was influenced by the MS. It is no longer like those cars before the energy crisis, The PS cars are more grand, luxurious and have longer designs.

5.2 Summary

1. The classification accuracy of the decision tree is extremely high, and an average accuracy can be obtained through gen consecutive decision tree calculations. The average sample in training mode has an accuracy of 94.4% while the average test sample has an accuracy of 81%. The variation in accuracy automatically executed by the computer is large and although we believed that the computer can use the existing feature conditions to make an accurate classification, the reality was that not all decision trees can accomplish this. This suggested that the decision method for every decision tree is different. After testing the 35 original data, we found one dispute which was the classification of Custom Royal, produced in 1955, as SS.
2. The main feature affecting the judgment of automotive-style is automotive length, followed by fender, rocket tail, the position of the quarter glass and rear form. But the number of headlamps, the presence of engine hood scoop and side decoration are not important features.
3. Evolution of style: The evolution of style is gradual, it is impossible to suddenly jump from SS design to MS design. There must be several years in between for the styles to evolve. The ultimate goal is to use certain design features to understand the rules of design style. However, a decision tree was still unable to classify accurately the style of cars produced during the transition period.
4. The meaning in design science: For the design science field, the use of decision tree methods has a symbolic meaning. hen designers design cars or products, they are all sensuous or even working in the black box. There is no way to know what the designer is thinking. But with the C5.0 decision tree method, 10, 50, or 100 decision trees can be created. Each decision tree represents a way of thinking by design features to design style. It is possible to stimulate the designer's thinking.

6 Conclusion

The rules, which determine the design style, found in this investigation is not consistent with the knowledge, intuition, and experience of a practicing designer. The first rule for the most optimal decision tree (Model A) was the automotive length. However, the designers would not consider length as the top priority in design. As the current modern family cars are drastically different from the grand, luxurious and long cars half a century ago, the designers would focus on the retro style details such as fender and the curved rear form. There is also no rule in the decision tree that places weight on these features that the designers emphasize. The Data mining approach to style classification could help us discover the key features and rules that human designers ignore or misunderstand. This shows the deviation between the logical view of the decision tree and the emotional view of the designers and is a potential topic for future studies.

References

1. Chan, C.-S.: Exploring individual style in design. Environ. Plann. B: Plann. Des. **19**(5), 503–523 (1992)
2. Cross, N.: Design ability. Nord. Artitekturforskning **1992**, 4 (1992)
3. Davis, M.: Why do we need doctoral study in design? Int. J. Des. **2**(3), 71–79 (2008)
4. Haffey, M.K.D., Duffy, A.H.B.: Knowledge discovery and data mining within a design environment. In: Cugini, U., Wozny, M. (eds.) From Knowledge Intensive CAD to Knowledge Intensive Engineering. ITIFIP, vol. 79, pp. 59–74. Springer, Boston, MA (2002). https://doi.org/10.1007/978-0-387-35494-1_5
5. Dowlen, C.: Design Paradigms in Car History (PhD Thesis). London South Bank University (2017)
6. Yu, C., Zhu, L.: Product design pattern based on big data-driven scenario. Adv. Mech. Eng. **8**(7), 1–9 (2016)
7. Zuffi, S.: The Dream Factory Alessi Since 1921, Könemann (1998)
8. Wang, H.-H., Chen, C.-P.: STYRULE-a classification system of car style. J. Des. **9**(2 1004/06), 107–121 (2004)
9. Dowlen, C., Shackleton, J.: Design history of the car: an empirical overview of the development of layout and form. In: International Conference on Engineering Design (ICED 2003), Stockholm, August 19–21 (2003)
10. Dowlen, C.: Measuring history: does historical car performance follow the triz performances curve? In: International Conference on Engineering Design (ICED 2011), 15–18 August 2011. Technical University of Denmark (2011)
11. Dowlen, C.: Automobile design history – what can we learn from the behavior at the edges? Int. J. Des. Creat. Innov. **1**(3), 177–192 (2013)
12. Lecoutre, A., Negrevergne, B., Yger, F.: Recognizing art style automatically in painting with deep learning. Proc. Mach. Learn. Res. **77**, 327–342 (2017)
13. Rokach, L., Maimon, O.: Data mining with Decision Trees: Theory and Applications 2nd Edition. World Scientific (2005)
14. Quinlan, J.R.: C4.5: Programs for Machine Learning. Morgan Kaufmann, Los Altos (1993)
15. Quinlan, J.R.: Bagging, boosting, and C4.5. In: Proceedings of the Thirteenth National Conference on Artificial Intelligence, pp. 725–730 (1996)
16. Quinlan, J.R.: Induction of decision trees. Mach. Learn. **1**, 81–106 (1986)
17. Breiman, L., Friedman, J., Olshen, R., Stone, C.: Classification and Regression Trees. Wadsworth Int. Group (1984)
18. Ho, T.K.: The random subspace method for constructing decision forests. IEEE Trans. Pattern Anal. Mach. Intell. **20**(8), 832–844 (1998)
19. Langworth, R.M. et al.: The Complete Book of Collectible Cars, Publications International, Ltd (2001)

Deep Learning for Multilingual POS Tagging

Alymzhan Toleu(✉), Gulmira Tolegen, and Rustam Mussabayev

Institute of Information and Computational Technologies, Almaty, Kazakhstan
alymzhan.toleu@gmail.com, gulmira.tolegen.cs@gmail.com, rmusab@gmail.com

Abstract. Various neural networks for sequence labeling tasks have been studied extensively in recent years. The main research focus on neural networks for the task are range from the feed-forward neural network to the long short term memory (LSTM) network with CRF layer. This paper summarizes the existing neural architectures and develop the most representative four neural networks for part-of-speech tagging and apply them on several typologically different languages. Experimental results show that the LSTM type of networks outperforms the feed-forward network in most cases and the character-level networks can learn the lexical features from characters within words, which makes the model achieve better results than no-character ones.

Keywords: LSTM · Neural networks · Pos tagging

1 Introduction

Neural networks have taken the fields of natural language processing (NLP) [7,26,27], computer vision [4,9] and signal processing [1,14,18] by storm. In particular, the general neural network (NN), recurrent neural network (RNN), LSTM network have been applied to the different NLP tasks such as part-of-speech tagging (POS tagging), named entity recognition (NER)[24,25], semantic role labeling (SRL) and language modeling, etc. These neural models [13,17,20] have been producing impressive results on many other tasks over its advantage of learning features through vector space automatically.

Widely used traditional linear statistical models (e.g.., conditional random fields, HMM) compares with the neural networks, there are two changes: i) the neural networks extend the traditional models from a linear to non-linear architecture and ii) replace the discrete feature representation with the distributed feature representation with continuous space. Wang and Manning (2013)[28] conducted an empirical study on the effect of non-linearity and the results showed that the non-linear models were effective in distributed continuous input space for NER and syntactic chunking. Compare with discrete feature representation, distributed representation is dense, compact and more feasible for more complex computation. Unlike traditional linear models, neural networks capable of representing a more complicated decision surface thanks to the non-linearity.

M. Hernes et al. (Eds.): ICCCI 2020, CCIS 1287, pp. 15–24, 2020.
https://doi.org/10.1007/978-3-030-63119-2_2

For NLP, the traditional approaches rely on a rich set of hand-crafted features from input sentences and feature design is an entirely empirical, manual, and time-consuming process. Instead, the neural network represents the characters/words/sentences to distributed embedding and takes the representation as input to learn several neural layers of feature in order to avoid the task-specific feature engineering as much as possible.

In this paper, we present an empirical study which compares the performances of different neural network architectures for multilingual POS tagging. Firstly, we summarize the different neural architectures for sequence labeling according to its functionality and characteristics. Then, we take four neural architectures for comparison, their architecture complexity ranges from simple to complex: i) multilayer perceptron (MLP) with a conditional random field (CRF) layer, it denotes for DNN; ii) LSTM network; iii) bi-direction LSTM (BiLSTM); iv) character-level BiLSTM. The first three models are word-based taggers that represent word into embedding and extract the different level of features through its layers. The last model takes characters as atomic units to derive the embeddings for the words. This way can make the model learn the lexical features within words. We tested the models on four typologically different languages: English, Russian, Finnish and Kazakh. Experimental results show that the LSTM network performs better than DNN, and the use of characters leads to significant improvements in performance, especially for out-of-vocabulary words tagging.

The paper is organized as follows: Section 2 summarizes the neural architectures for sequence labeling tasks. Section 3 describes the models applied in this work. Section 4 reports the experimental results. Section 5 concludes the paper.

2 Related Work

In this section, we briefly summarize the different neural network architectures in NLP. Table 1 lists the neural network architectures.

2.1 Deep Neural Network

Feed-forward neural network was first proposed by (Bengio et al., 2003) [2] for probabilistic language model, and reintroduced later by (Collobert et al., 2011)[5] for multiple NLP tasks. The architecture can be summarized as follows:

- the first layer use to mapping the word/character to embedding;
- the second layer extracts features from a window of words/characters;
- the following layers are standard NN layers for feature extraction.

This model is a fully-connected neural network. For sequence labeling tasks, there are strong dependencies between labels. For example, in the NER task, it is impossible for some tags to follow other tags (e.g.., the tag "I-PER" cannot follow the tag "B-LOC"). In order to model the tag dependencies, Collobert et al.(2011)[5] introduced a transition score A_{ij} for jumping form tag $i \in T$ to tag

Table 1. Neural architectures mostly used for sequence labeling.

Models	Description
Deep neural network (DNN)	A typical multilayer feed-forward neural network with a number of hidden units whose weights are fully connected.
Max margin tensor neural network (MMTNN)	MMTNN can explicitly model the interactions between tags and context words/characters by exploiting tag embeddings and tensor-based transformations.
Convolution neural network (CNN)	A convolution layer of NN can be considered as a generalization of a window approach that extracts local features around each window of the given sequence. The convolution layer is able to handle the sequence of variable length.
Recurrent neural network (RNN)	RNN is a time series model and the output is fed back as input.
Long short term memory (LSTM)	An LSTM network has multiplicative gates that allow the memory cells to store and access information over long periods of time.
Bidirectional LSTM (BiLSTM)	A bidirectional recurrent neural network with memory block.
Bidirectional LSTM-CNN	BiLSTM combine with CNN

$j \in T$, where T is tag set. For a input sentence $x = x_1, ..., x_N$ with a tag sequence $y = y_1, ..., y_N$, the DNN computes a sentence-level score summing the transition and neural network scores:

$$score(x, y, \theta) = \sum_{i=1}^{N} (A_{t_{i-1},t_i} + f(t_i|i)) \tag{1}$$

where $f(t_i|i)$ denotes the output score by the DNN with parameters θ for tag t_i of the i-th word/character.

Collobert et al. (2008) [5] applied DNN for many sequence labeling tasks such as POS tagging, Chunking and NER. The results show that DNN outperforms traditional approaches in many experiments for different tasks. Using pre-trained word embedding could improve model performance significantly.

2.2 Max-Margin Tensor Neural Networks

Max-margin tensor neural network (NMTNN) is a tensor layer trained with max-margin criteria as the objective function which was proposed by (Pei et al., 2014) [23] for sequence labeling tasks, and this model capable of capturing the complicated interactions between tags and context words/characters. The architecture of MMTNN can be summarized as follows:

- the first is a lookup table layer for word/character and tag embedding;
- the second is tensor transformation layer that used to extract more high-level interaction between tag and word/character embedding features;
- the following layers are standard NN layers.

MMTNN feeds the tag embedding combined with words as input that enables the model can capture interactions between tags and context words/characters. It models the tag vs. tag and tag vs. context together within the neural network, not model the tag dependencies individually. Compared with MNTNN, one of the limitations of DNN is that it uses a transition score for modeling the interaction between the tags and the transition score is modeled separately from neural network parameters. Moreover, MMTNN contains a tensor layer in order to extract high dimensional feature interactions. The sentence-level score of DNN (Eq. 1) can be rewritten as follows for NMTNN:

$$score(x, y, \theta) = \sum_{i=1}^{N}(f(t_i|i, t_{i-1})) \tag{2}$$

where $f(t_i|i, t_{i-1})$ is the network score for tag t_i at the i-th word/character with previous tag t_{i-1}.

2.3 Convolutional Neural Network

DNN and MMTNN use a window of fixed-sized words/characters as input to predict the corresponding labels. DNN works well on simple tasks like POS tagging, but it fails on more complex tasks like SRL[6]. The main advantage of Convolutional Neural Network (CNN) compared to DNN is that it uses special convolution layer and pooling operation to extract important local features. CNN is also computationally efficient due to its parameter sharing technique. The architecture of CNN is summarized as follows:

- the first layer is a lookup-table layer;
- the second layer is a convolution layer that extracts local features around each word and combines these features into a global feature that can be fed to the following layers;
- the next layer is max-pooling layer;
- the following layers are standard NN layers.

Collobert et al. (2011) [6] used CNN for SRL task, and the comparison results showed that CNN gave around 20% improvements over DNN for SRL.

2.4 Recurrent Neural Network

As described, DNN, MMTNN, and CNN, the data flow of them does not form the cycles. If we relax this condition and allow the cyclic connections, then it becomes a recurrent neural network (RNN)[8]. A simple RNN contains a single self-connected hidden layer and a "memory" cell that captures information about what is calculated before. RNN have shown great success in time series tasks such as language modeling (Mikolov et al., 2012) [19]. In practice, RNNs are limited to looking back only a few steps. In other words, the RNN may"forget" the early inputs (the vanishing gradient problem). Therefore, the commonly used type of RNNs is LSTM which is better at capturing long-term dependencies.

LSTM architecture consists of a set of recurrently connected sub nets, known as memory blocks. Each block contains one or more self-connected memory cells and three multiplicative units: input, output, forget gates that provide write, read, and reset operations for the cells. LSTM has been applied in wide range of applications, and has shown great success in many NLP tasks such as NER [3], POS tagging [11] and SRL [29]. Huang et al.(2015)[12] proposed a variety of bi-directional LSTM for English POS tagging including LSTM, BiLSTM, CRF, LSTM-CRF and BiLSTM-CRF models, and the reported accuracy results for those models are respectively 94.63%, 96.04%, 94.23%, 95.62%, and 96.11% . Ma et al.(2016)[16] introduced a novel neural network of combined biLSTM, CNN and CRF. Ling et al.(2015)[15] proposed a new character to word (C2W) model which has two parts: 1) construct word embedding by composing characters using BiLSTM; 2)and fed this word embeddings to another BiLSTM for tagging. The authors tested C2W on the two tasks: language model and POS tagging. The results showed that the C2W outperformed other models for language model (Word-based model) and the C2W's improvements were especially pronounced in Turkish that is a agglutinative language similar to Kazakh. For POS tagging, C2W can achieve comparable or better results than state-of-the-art systems.

3 Part-of-Speech Taggers

Following the neural architectures described above, in this work, we implemented and tested four architectures as POS taggers. The architectures applied in this work can be summarized as follows:

- Deep Neural Network (DNN): A general fully-connected neural network[5] that takes a window of words as input and learns several layers of feature extraction that process the inputs. To model the POS tag dependencies, we use a CRF layer which captures the tag transition scores[1].
- Long Short Term Memory (LSTM): a recurrent neural network that capable of learning long-term dependencies[10]. LSTM architecture consists of a set of recurrently connected sub-nets that can be viewed as memory blocks. Each block contains one or more self-connected memory cells and three multiplicative unit: the input, output and forget gates that provide continuous analogues of write, read and reset operation for the cells. The computation for an LSTM forward pass is given below:

[1] For all those LSTM type of models, we did not use the CRF layer, since LSTM can captures sentence-level information.

$$
\begin{aligned}
z^t &= g(\mathbf{W_z}x^t + \mathbf{R_z}y^{t-1} + b_z) && \textit{block input} \\
i^t &= g(\mathbf{W_i}x^t + \mathbf{R_i}y^{t-1} + \mathbf{P_i} \odot c^{t-1} + b_i) && \textit{input gate} \\
f^t &= g(\mathbf{W_f}x^t + \mathbf{R_f}y^{t-1} + \mathbf{P_f} \odot c^{t-1} + b_f) && \textit{forget gate} \\
c^t &= i^t \odot z^t + f^t \odot c^{t-1} && \textit{cell state} \\
o^t &= g(\mathbf{W_o}x^t + \mathbf{R_o}y^{t-1} + \mathbf{P_o} \odot c^{t-1} + b_o) && \textit{output gate} \\
y^t &= o^t \odot h(c^t) && \textit{block output}
\end{aligned}
$$

where x^t is input vector at time t, $\mathbf{W_k}$ are input weight, $\mathbf{P_k}$ are peephole weights, k can be any of $[z, i, f, o]$. The activation function of gates is the logistic sigmoid function and hyperbolic tangent is used for block input and output.
- bi-directional LSTM (biLSTM): Bi-directional LSTM[12] that processes the input sentence in both directions for tagging the current input word.
- character-level bi-directional LSTM (CharBiLSTM)[15]: it contains two layers of Bi-directional LSTMs i) the BiLSTM uses the characters to generate the word embedding and ii) the other BiLSTM uses generated word embedding to do POS tagging.

4 Experiments

We report the experimental results of applying the various neural architectures for four typologically different languages: two agglutinative languages - Kazakh and Finnish; an analytic language - English; and a fusional - Russian.

For each language, we train four different neural networks: DNN, LSTM, BiLSTM, and CharBiLSTM. For evaluation, we report accuracy for all words and out-of-vocabulary (OOV) words.

4.1 Data Set

Table 2 lists the evaluation data-sets of Universal Dependency treebank [21] for four languages. For each language, the data-set is divided into the train, development and test sets. The sentence number of each data-sets and the OOV rates are reported.

4.2 Model Setup

As listed in Table 2, the size of training data used in this work is not very large compared with the popular Wall Street Journal corpus [22] (the training data is about 38,219 sentences), thus, we intentionally made the LSTM-based models to be small enough in order to reduce the training time.

Table 2. Statistics of the corpora.

	English	Finnish	Russian	Kazakh
Train	10,734	12,217	4029	10,151
Test	1,647	648	499	1,268
Dev	1,647	716	502	1,268
OOV(%)	7.67	24.50	26.01	12.43

For the different neural architectures, there are several hyper-parameters that should be tuned since the different values impact the model's results. For the four neural networks, we set the learning rate to 0.02 and decrease the learning rate according to the training epoch. For other hyper-parameters, such as number of the hidden units, window size, etc. We used the same values for all languages' DNN models (window size - 3, word dimension - 50, hidden units - 200). For LSTM-based models, we set word dimension to 50 and set the number of the hidden units to 100. All four taggers are implemented in Java language and to compare the training and decoding times, we run all experiments on the same test machine, which featured with Xeon(R) CPU E5620, 2.40GHz, 8 cores with 8GB of memory.

Table 3. Tagging results on the UD data sets. `Test` is the accuracy of all tokens; `OOV` is the accuracy of OOV words; `time` is the running time per iteration.

Models		English	Finnish	Kazakh	Russian
DNN	Test	86.12	84.57	86.37	84.57
	OOV	**66.11**	62.69	**70.91**	67.79
	time(s)	70.26	81.15	67.79	44.39
LSTM	Test	88.39	85.70	90.00	83.16
	OOV	48.06	53.72	54.28	40.78
	time(s)	103.42	88.35	76.00	52.89
BiLSTM	Test	88.92	84.80	90.03	83.47
	OOV	43.95	51.21	55.19	44.53
	time(s)	171.59	142.2	129.94	78.58
CharBiLSTM	Test	**90.08**	**87.82**	**92.23**	**90.77**
	OOV	62.99	**63.21**	70.49	**82.18**
	Time(s)	256.67	180.2	218.34	131.70

4.3 Results

Table 3 lists the accuracy results and decoding speed of the various POS taggers for four languages. Unless stated otherwise we refer to the general (all tokens) accuracy when comparing model performances. It can be seen that the DNN model performs consistently worse than the LSTM for three languages (English, Finish, and Kazakh), and LSTM gives $\approx$2%, $\approx$1%, $\approx$4% improvements over DNN for those languages (in the same order). For Russian, DNN performs better than LSTM since it has a small training data compared with other languages. LSTM requires large training data to achieve good generalization due to its large number of parameters. BiLSTM slightly outperforms LSTM as it applies bi-directional LSTMs for modeling the sequence of words.

It can be seen from Table 3 that the model CharBiLSTM outperforms all other models and achieves over 90% accuracy for English, Kazakh and Russian. Compared with LSTM and BiLSTM, CharBiLSTM uses characters as atomic units, and this feature gives significant improvements about 2%∽3% over others. One possible explanation of this is that CharBiLSTM contains double layers of LSTM: the one for learning the features from characters that more comprehensively captures the lexical features, and the other for extracting the higher-level context embedding. It should be noted that these models were trained without using any external features in all experiments.

For OOV results, CharBiLSTM gives better results in most cases, and it is interesting that DNN also gives better OOV accuracy for English and Russian. On average, the model CharBiLSTM gives $\approx$ 20% improvements over BiLSTM for the OOV accuracy. Models' decoding time depends on many factors, we here focus on the architecture. It can be seen that DNN is the fastest model and the CharBiLSTM is the slowest one. It reflects the fact that the model's decoding speed decreases as its architecture become more complex.

5 Conclusion

In this paper, we summarized various neural architectures for the task of sequence labeling and implemented four (DNN, LSTM, BiLSTM, and CharBiLSTM) of them for the comparisons in the case of multilingual POS tagging. The four typologically different languages: Kazakh, Russian, Finnish and English are used as the evaluation data-sets. Experimental results showed that the character-level BiLSTM outperformed other models in most cases on all token accuracy, especially on OOV term. CharBiLSTM benefits from being sensitive to lexical aspects within words as it takes characters as atomic units to derive the embeddings for the word. In terms of decoding speed, the fastest model is DNN which also achieves better OOV results than LSTM and BiLSTM. The slowest model is CharBiLSTM that has a more complex structure than others.

Acknowledgments. This research has been conducted within the framework of the grant num. BR05236839 "Development of information technologies and systems for stimulation of personality's sustainable development as one of the bases of development of digital Kazakhstan".

References

1. Baba Ali, B., Wójcik, W., Orken, M., Turdalyuly, M., Mekebayev, N.: Speech recognizer-based non-uniform spectral compression for robust MFCC feature extraction. Przegl. Elektrotechniczny **94**, 90–93 (2018)
2. Bengio, Y., Ducharme, R., Vincent, P., Janvin, C.: A neural probabilistic language model. J. Mach. Learn. Res. **3**, 1137–1155 (2003)
3. Chiu, J.P., Nichols, E.: Named entity recognition with bidirectional LSTM-CNNs. Trans. Assoc. Comput. Linguist. **4**, 357–370 (2016)
4. Cohen, T., Geiger, M., Köhler, J., Welling, M.: Spherical CNNS. ArXiv abs/1801.10130 (2018)
5. Collobert, R., Weston, J.: A unified architecture for natural language processing: deep neural networks with multitask learning. In: Proceedings of the 25th International Conference on Machine Learning (ICML 2008), pp. 160–167. ACM, New York, NY, USA (2008)
6. Collobert, R., Weston, J., Bottou, L., Karlen, M., Kavukcuoglu, K., Kuksa, P.: Natural language processing (almost) from scratch. J. Mach. Learn. Res. **12**, 2493–2537 (2011)
7. Duong, L., Cohn, T., Verspoor, K., Bird, S., Cook, P.: What can we get from 1000 tokens? a case study of multilingual POS tagging for resource-poor languages. In: Proceedings of the 2014 Conference on Empirical Methods in Natural Language Processing (EMNLP), pp. 886–897. Association for Computational Linguistics, Doha, Qatar (Oct 2014). https://doi.org/10.3115/v1/D14-1096, https://www.aclweb.org/anthology/D14-1096
8. Elman, J.L.: Finding structure in time. Cogn. Sci. **14**(2), 179–211 (1990)
9. Elsayed, G.F., et al.: Adversarial examples that fool both computer vision and time-limited humans. In: Proceedings of the 32Nd International Conference on Neural Information Processing Systems (NIPS 2018), pp. 3914–3924. Curran Associates Inc., USA (2018)
10. Hochreiter, S., Schmidhuber, J.: Long short-term memory. Neural Comput. **9**(8), 1735–1780 (1997)
11. Horsmann, T., Zesch, T.: Do LSTMs really work so well for PoS tagging? – a replication study. In: Proceedings of the 2017 Conference on Empirical Methods in Natural Language Processing. Association for Computational Linguistics, Copenhagen, Denmark (Sep 2017)
12. Huang, Z., Xu, W., Yu, K.: Bidirectional LSTM-CRF models for sequence tagging (2015), cite arxiv:1508.01991
13. Kalimoldayev, M., Mamyrbayev, O., Kydyrbekova, A., Mekebayev, N.: Voice verification and identification using I-vector representation. Int. J. Math.Phys. **10**(1), 66–74 (2019)
14. Kalimoldayev, M.N., Alimhan, K., Mamyrbayev, O.J.: Methods for applying VAD in Kazakh speech recognition systems. Int. J. Speech Technol. **17**(2), 199–204 (2014). https://doi.org/10.1007/s10772-013-9220-6
15. Ling, W., et al.: Finding function in form: compositional character models for open vocabulary word representation. In: Proceedings of the 2015 Conference on Empirical Methods in Natural Language Processing, pp. 1520–1530. Association for Computational Linguistics, Lisbon, Portugal (Sep 2015)
16. Ma, X., Hovy, E.: End-to-end sequence labeling via bi-directional LSTM-CNNs-CRF. In: Proceedings of the 54th Annual Meeting of the Association for Computational Linguistics (Vol. 1: Long Papers), pp. 1064–1074. Association for Computational Linguistics, Berlin, Germany (Aug 2016)

17. Mamyrbayev, O., Turdalyuly, M., Mekebayev, N., Alimhan, K., Kydyrbekova, A., Turdalykyzy, T.: Automatic recognition of Kazakh speech using deep neural networks. In: Nguyen, N.T., Gaol, F.L., Hong, T.-P., Trawiński, B. (eds.) ACIIDS 2019. LNCS (LNAI), vol. 11432, pp. 465–474. Springer, Cham (2019). https://doi.org/10.1007/978-3-030-14802-7_40
18. Mamyrbayev, O., et al.: Continuous speech recognition of Kazakh language. ITM Web of Conferences 24, 01012 (2019)
19. Mikolov, T.: Statistical Language Models Based on Neural Networks. Ph.D. Thesis, Brno University of Technology (2012)
20. Mikolov, T., Sutskever, I., Chen, K., Corrado, G., Dean, J.: Distributed representations of words and phrases and their compositionality. In: Proceedings of the 26th International Conference on Neural Information Processing Systems (NIPS 2013), Vol. 2, pp. 3111–3119, Curran Associates Inc., USA (2013)
21. Nivre, J., et al.: Universal dependencies v1: a multilingual treebank collection. In: Proceedings of the Tenth International Conference on Language Resources and Evaluation (LREC 2016). European Language Resources Association (ELRA), Portorož, Slovenia (May 2016)
22. Paul, D.B., Baker, J.M.: The design for the wall street journal-based CSR corpus. In: Speech and Natural Language: Proceedings of a Workshop Held at Harriman, New York, February 23–26, 1992 (1992)
23. Pei, W., Ge, T., Chang, B.: Max-margin tensor neural network for Chinese word segmentation. In: Proceedings of the 52nd Annual Meeting of the Association for Computational Linguistics, (Vol. 1: Long Papers), pp. 293–303. Association for Computational Linguistics, Baltimore, Maryland (Jun 2014)
24. Tolegen, G., Toleu, A., Mamyrbayev, O., Mussabayev, R.: Neural named entity recognition for Kazakh. In: Proceedings of the 20th International Conference on Computational Linguistics and Intelligent Text Processing. CICLing, Springer Lecture Notes in Computer Science (2019)
25. Tolegen, G., Toleu, A., Zheng, X.: Named entity recognition for Kazakh using conditional random fields. In: Proceedings of the 4-th International Conference on Computer Processing of Turkic Languages TurkLang 2016, pp. 118–127. Izvestija KGTU im.I.Razzakova (2016)
26. Toleu, A., Tolegen, G., Makazhanov, A.: Character-aware neural morphological disambiguation. In: Proceedings of the 55th Annual Meeting of the Association for Computational Linguistics, (Vol. 2: Short Papers), pp. 666–671. Association for Computational Linguistics, Vancouver, Canada (Jul 2017). 10.18653/v1/P17-2105
27. Toleu, A., Tolegen, G., Makazhanov, A.: Character-based deep learning models for token and sentence segmentation. In: Proceedings of the 5th International Conference on Turkic Languages Processing (TurkLang 2017). Kazan, Tatarstan, Russian Federation (October 2017)
28. Wang, M., Manning, C.D.: Effect of non-linear deep architecture in sequence labeling. In: IJCNLP (2013)
29. Zhou, J., Xu, W.: End-to-end learning of semantic role labeling using recurrent neural networks. In: Proceedings of the 53rd Annual Meeting of the Association for Computational Linguistics and the 7th International Joint Conference on Natural Language Processing (Vol. 1: Long Papers), pp. 1127–1137. Association for Computational Linguistics, Beijing, China (Jul 2015)

Study of Machine Learning Techniques on Accident Data

Zakaria Shams Siam, Rubyat Tasnuva Hasan, Soumik Sarker Anik, Ankit Dev, Sumaia Islam Alita, Mustafizur Rahaman, and Rashedur M. Rahman(✉)

Department of Electrical and Computer Engineering, North South University, Plot-15, Block-B, Bashundhara Residential Area, Dhaka, Bangladesh
{zakaria.siam,rubyat.tasnuva,soumik.anik,ankit.dev,islam.sumaia, mustafizur.rahaman171,rashedur.rahman}@northsouth.edu

Abstract. Road crash is one of the major burning issues for Bangladesh. There are several factors that are responsible for occurring road crashes. If we can understand the causes and predict the severity level of a particular type of accident upfront, we can take necessary steps in the proper time to lessen the damages. In this study, we have built some predictive models of different homogeneous road crash groups of Bangladesh using machine learning methods that can predict that particular road crash severity level based on the environmental factors and road conditions. We have applied Agglomerative Hierarchical Clustering to find different clusters of road crashes and then applied Random Forest technique to extract the significant predictors of each cluster and then applied C5.0 to build predictive models of each cluster. Finally we have discussed the patterns of fatal and non-fatal accidents of Bangladesh through rule generation technique.

Keywords: Machine learning · Road accident · Agglomerative hierarchical clustering · C5.0 · Random forest · PART

1 Introduction

Traffic accident occurs when a vehicle crashes with another vehicle, pedestrian, animal, debris, tree or any other substances in the road. Traffic collisions can result severe property damage, moderate or serious injury and even loss of life in some acute cases. Traffic accident is considered as a serious manmade disaster that causes severe harm to our earth [1]. Bangladesh (BD) is a developing country and that is why the car ownership rate in our country is really low. According to the World Bank, only 3 people in about 1000 had a personal vehicle in BD in 2010 [2]. But due to rapid urbanization, this number is increasing and the number of traffic accidents in Bangladesh is rising at an alarming rate.

According to WHO's reports, almost 1.25 million people die worldwide because of road accidents averaging 3287 deaths per day; almost 20-50 million people get injured. Traffic accident is listed as the 9th leading cause of death worldwide, and it costs almost 518 billion USD globally [3]. Bangladesh is now undergoing a severe Road Traffic

M. Hernes et al. (Eds.): ICCCI 2020, CCIS 1287, pp. 25–37, 2020.
https://doi.org/10.1007/978-3-030-63119-2_3

Accident (RTA) problem. 43% increase in the number of accidents has been observed between 1982 and 2000 and around the same period, the number of fatalities has been increased by 400%. At least 4284 people died and 9112 people were injured in 2017 alone, which was a 25.56% increase in death compared to 2016 [4]. RTAs are costing our country nearly 2% of GDP [5].

Previously in Bangladesh, to tackle this problem, most of the researchers conducted their studies on the descriptive statistics and temporal analysis on road crashes based on a few risk factors or specific road users or certain type of accidents. Besides, some studies have applied advanced statistical approaches such as Poisson Regression, Ordered Probit Model etc. [6, 7]. Instead of using only traditional statistical approaches, different data mining techniques can be used to extract useful information and pattern from a large set of raw data which is a much more fruitful approach to gain information about hidden patterns than any other traditional statistical approaches [8]. A study conducted by Liling Li et al. in 2017, has applied both statistical analysis and different data mining techniques to find out factors that are directly related to the fatal accidents [9]. Different studies have applied different machine learning classification algorithms to predict accident severity on different road crash datasets of different countries [10–12]. To increase the accuracy of the models, different studies of developed countries have applied clustering algorithms before applying classification or other models to address data heterogeneity issue [13]. A recent study in Bangladesh conducted by Md Asif Raihan et al. applied different data mining techniques to find hazardous clusters, extract predictors and finally applied classification model, classification and regression trees (CART) on 2006-2010 accident data of Bangladesh [8]. However, the RTA related obtained dataset is often underreported and incomplete in the case of developing countries [14] and the underreporting rate in the case of traffic accident in Bangladesh could go even beyond 50% [15]. According to the Setu Bhaban-HQ of Bangladesh Bridge Authority (JMBA), the number of accidents in 2004, 2005 and 2006 were 85, 108 and 80 respectively; whereas according to the MAAP 5.0 software (which stores and processes all the accident data) of Accident Research Institute, BUET [16], the number of accidents in 2004, 2005 and 2006 were 6, 18 and 25 [17]. This statistics clearly shows the problem of data underreporting in Bangladesh. Md Asif Rayhan [8] has indicated this problem of underreporting in Bangladesh in his 2017 study. However, no technique was applied to tackle this underreporting problem of the dataset and thus the dataset remained bias for being classified. Generally the non-fatal accidents (grievous, simple injury and slight accidents) are more underreported than fatal accidents in our country. It has been kind of a socially established fact for the traffic system of Bangladesh which indicates a serious problem. According to the estimation of WHO, the yearly actual fatalities could be 20,000 whereas according to the police reported statistics, it is approximately 3000 each year [8]. So, taking consideration of this underreporting problem, we have applied up-sampling technique on the data to reduce biasness before applying classification model. Moreover, we have conducted this study on the latest data (2015) from ARI and we have used C5.0 classification model which is very much newer model than CART and most other decision tree algorithms. Not only that, we have also conducted Projective Adaptive Resonance Theory (PART) analysis on C5.0 model. In summary, in this study, we have used Agglomerative Hierarchical Clustering (AHC) to extract the more similar type of accidents (clusters) from each crash group (pedestrian, single vehicle and multi-vehicle crash), and then applied Random Forest (RF) to find the most influential attributes of each cluster and finally applied C5.0 (a machine learning based classification model) on each cluster and then generated rules through PART analysis to find the causalities and patterns of different fatal and non-fatal accidents in our country.

The structure of this paper is as follows: Sect. 2 describes the collected dataset. Section 3 deals with the methodology. Section 4 discusses about experiments, result analysis and discussion. Lastly, Sect. 5 focuses on more possible future works and conclusion.

2 Dataset

We were provided with a sample dataset from Accident Research Institute (ARI), BUET [16], that contained the traffic accidents occurred in the year of 2015 in Bangladesh. This dataset is the latest published dataset of Bangladesh so far from ARI. Table 1 represents the description of the dataset.

Table 1. Dataset description

Number of Rows	1585 (1523 after removing the missing rows)
Number of Variables	21
Number of Missing Values	62
Variable Names	Thana, District, No. of Vehicles, No. of Driver, Passenger and Pedestrian Casualties (3 different variables), Junction Type (Cross/Tee etc.), Traffic Control(Uncontrolled/Controlled by road divider etc.), Collision Type (Head on/Rear end etc.), Movement (One way/Two way), Divider(Yes/No), Weather Condition (Rain/Fog etc.), Light Condition (Daylight/Dawn/Dusk etc.), Road Geometry(Straight/Crest etc.),Surface Condition(Wet/Dry etc.), Surface Type(Paved/Brick Road etc.), Surface Quality(Good/Rough etc.), Road Class(National/Regional/City road etc.), Road Feature(Bridge/Culvert/Normal road etc.), Location Type(Urban/Rural) and Accident Severity(Fatal, Grievous, Simple Injury and Motor/Property Damage Only).
Actual Class Labels in the Dataset	Fatal (F) Grievous (G) Slight(S) Motor/Property Damage Only (M)
Modified Class Labels and their Quantity	Non-Fatal (NF = G + S + M) (305) Fatal (F)(1218)

3 The Methodology

Figure 1 illustrates the overall research design of our study. The dataset is preprocessed first. Data cleaning is done by pruning the 62 rows that contained missing values. We have discussed the steps of this whole design in the next sections.

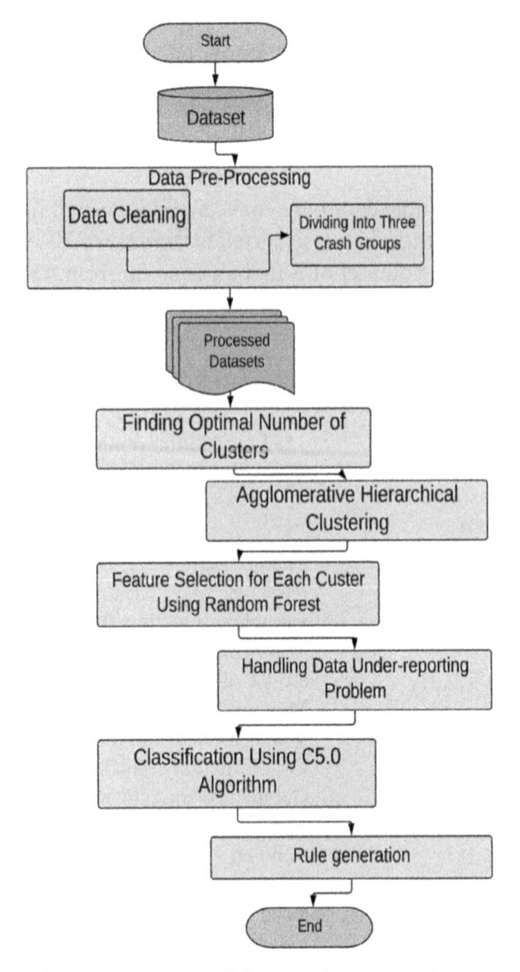

Fig. 1. Flowchart of the overall research design

3.1 Clustering to Subgroup Similar Types of Accidents

Our first objective is to find the most similar types of road accidents in our collected dataset since we are interested in finding the underlying relationship of accident severity with different groups of accidents. For that, we have started first by finding three different crash groups- 1. Pedestrian Crash Group, 2. Single Vehicle Crash Group and

3. Multi-Vehicle Crash Group. Based on the attribute 'No. of Pedestrian Casualties', we have differentiated between pedestrian crash group and single/multi-vehicle crash group. Again, we have differentiated between single vehicle crash group and multi-vehicle crash group based on the variable 'No. of Vehicles involved'. After that, we have discarded the variable 'No. of Vehicles involved' from our further study. Then, having 3 different crash groups, we have extracted the most similar clusters of road accidents from each of the 3 different crash groups.

Clustering method is a very good example of unsupervised machine learning technique that is useful for extracting a set of homogeneous clusters from a dataset [18]. In any clustering algorithm, three following steps are executed to find the end result:

a. At first, the dissimilarity matrix has to be calculated. Basically, it can mathematically measure the distance between the data points in the dataset, so that the closest points of the dataset can later be grouped together [19].
b. Then, a specific type of clustering algorithm has to be chosen. Our study has used the Agglomerative Hierarchical Clustering (AHC) algorithm.
c. At last, the validation of the cluster analysis has to be checked or assessed.

AHC algorithm considers every tuple of the dataset as a single individual cluster and then it finds the most similar points and then joins them recursively. Thus, clusters are starting to form. At the end, this algorithm produces some number of clusters. We have chosen AHC because hierarchical clustering is more exploratory in nature than K-means clustering or Partition Around Medoids (PAM) clustering, and also, in general, AHC is better in finding small sized clusters than Divisive Clustering Algorithm. So, for our data, AHC will best fit our objective. We have used complete linkage approach.

Since our data contains mixed type attributes (the class label is nominal and all other variables are numeric), we have used Gower Distance metric to calculate our dissimilarity matrix to handle this mixed type attributes. The average of partial distances/differences across the records are calculated, and thus Gower distance is measured [20]. The formula for calculating Gower distance is:

$$\mathrm{d(i,j)} = \frac{1}{\mathrm{p}} \Sigma_{\mathrm{i=1}}^{\mathrm{p}} \mathrm{d}_{\mathrm{ij}}^{(\mathrm{f})} \tag{1}$$

where $d_{ij}^{(f)}$ are the partial dissimilarities that are dependent upon the type of the variable that is evaluated [20]. After calculating the dissimilarity matrix, we have applied AHC algorithm on each crash group to further divide them into more similar subsets. One drawback of clustering approach is that almost every clustering algorithm can return clusters even if there is no natural cluster structure in the dataset. Therefore, we must validate the quality of the result of our cluster analysis afterwards.

After extracting the optimal number of clusters for each crash group, we have selected the most important variables for each cluster. These important variables are known as predictors and can best explain their corresponding cluster. We have utilized Random Forest technique to select the most important variables from each cluster.

Random forest is an example of ensemble learning technique for building prediction models. Random forests construct many decision trees at the time of training the dataset, and they generate the class which is the mode of the classes that are given as the output by the individual trees [21]. Random forests are very good at ranking the importance of predictors/variables in a prediction problem [21]. This is the reason why we have chosen Random Forest to rank the importance of every variables in each cluster. After that we have collected the best variables in each cluster based on their Mean Decrease Accuracy (MDA) value. We have also evaluated how good Random Forest did at predicting the predictors by calculating the Out of Bag error (OOB error) for every cluster. If the error is less, then we can say that the fit is decent and Random Forest has done a good job at finding important predictors for each cluster.

3.2 Classification/Predictive Models for Each Cluster

After oversampling the datasets to handle the underreporting problem, we have applied Decision Tree (C5.0) on each individual oversampled cluster to build their corresponding predictive models. We have only considered the predictors that are given highest rank from Random Forest algorithm into the classification models. We have generated rules of each predictive model from each cluster of crashes through PART analysis. From there, we can detect various underlying relationship between the environmental and the road geometrical factors with their corresponding accident severity level in the form of rules. These rules actually form the decision system, where we can predict accident severity upfront by knowing the predictors.

We have used Microsoft Excel 2019 to perform the necessary preprocessing parts of the data. We have used R programming language for clustering, extracting important predictors and building classification models. We have used the Synthetic Minority Oversampling Technique (SMOTE) function in Waikato Environment for Knowledge Analysis (WEKA) to up-sample the data for classification part.

4 Experiments, Result Analysis and Discussion

4.1 Results of Cluster Analysis

The question that arises first in cluster analysis is that how many clusters will be optimal. We have applied Silhouette method to extract the optimal number of clusters for each of the 3 crash groups. We have produced the silhouette plot that exhibits the closeness of every point in a cluster compared to the points in the adjacent clusters [19]. We have taken 2 minimum clusters and 7 maximum clusters for our calculation of determining the optimal number of clusters for each crash group. After dividing the dataset, pedestrian crash group had 781 tuples, single vehicle crash group had 140 tuples and multi-vehicle crash group had 602 tuples. We have plotted the Average Silhouette Width vs Number of Clusters for each of the three crash groups in Fig. 2.

We should choose the number of clusters that corresponds to the highest silhouette width in each plot. The silhouette coefficient ranges in the interval [-1,1]; where -1 indicates not so good consistency within clusters and 1 indicates good consistency within clusters. From the figure, it is apparent that the number of optimal clusters for pedestrian crash group is 2, since it produces maximum average silhouette width > 0.325; for single

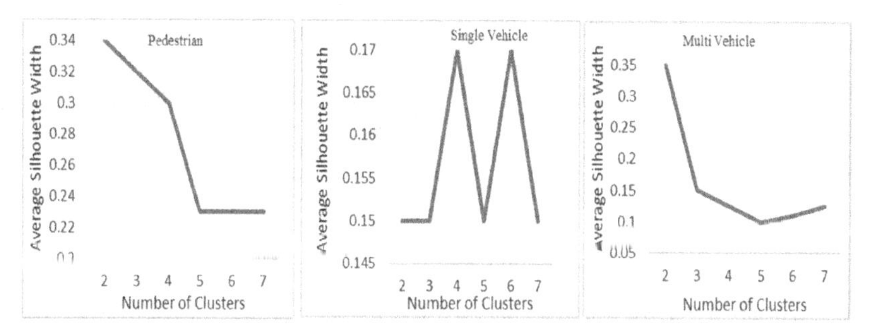

Fig. 2. Average Silhouette Width plots for each crash group

vehicle crash group, it is 4, since it produces maximum average silhouette width, but due to the low sample size of our collected dataset, this particular crash group contains only 140 observations. So, obviously it would not be a good idea to make 4 clusters within only 140 samples. So, we have taken the optimal number of clusters for single vehicle crash group as 2. The number of optimal clusters for multi vehicle crash group is 2, since it produces maximum average silhouette width = 0.35.

Now, we have applied AHC in R programming language to find out the corresponding optimal number of clusters for each of the 3 crash groups. We have used the daisy() function with metric = c("gower") from the cluster library/package to calculate the dissimilarity matrix with gower distance to handle our mixed attribute types. We have used hclust() function of cluster package to perform AHC with complete linkage method and cutree() function to retrieve the corresponding optimal number of clusters for each crash group.

After extracting the clusters from each crash group, we have found that pedestrian cluster 1 had 322 cases (266 fatal cases and 56 non-fatal cases). Pedestrian cluster 2 had 459 records (413 fatal and only 46 non-fatal). Single Vehicle (SV) cluster 1 had 107 records (81 fatal and 26 non-fatal) and cluster 2 got only 33 records (19 fatal and 14 non-fatal). Due to this low sample size, we have discarded SV cluster 2 from our further analysis. Multi Vehicle (MV) cluster 1 got 447 tuples (364 were fatal and 83 were non-fatal) and cluster 2 got 155 tuples (75 fatal and 80 non-fatal accidents). So, the number of non-fatal accidents in all the clusters except MV cluster 2 is very less against the number of fatal accidents, which clearly indicates the data underreporting problem. We have tackled this problem later in the eve of classification analysis part. Table 2 shows the internal clustering validation statistics for all the clusters.

The average distance within clusters (cohesion) should be as small as possible and the average distance between clusters (separation) should be as large as possible to have a good clustering structure. From Table 2, we can see that for each of the 3 crash groups, the average distance between clusters is comparatively larger than average distance within clusters. So, the clustering structures for each crash group is quite good.

4.2 Selecting Influential Attributes by Random Forest Analysis

In the next step, we have applied Random Forest (RF) on 5 different clusters from 3 different crash groups. We have used randomForest() function from randomForest library.

Table 2. Result for cluster validation

	Pedestrian (2 clusters)	Single vehicle (2 clusters)	Multi vehicle (2 clusters)
within clusters sum of squares	11.02	4.2	11.19
average distance within clusters	0.15	0.23	0.18
average distance between clusters	0.23	0.27	0.28
wb ratio	0.65	0.85	0.65
Dunn Index	1.34	1.16	1.24
avg. Silhouette Width	0.34	0.15	0.35

We have used importance() and varImpPlot() functions to extract the most important predictors for each cluster.

We have discarded the number of pedestrian, driver and passenger casualties from our further analysis in RF because, these variables are obviously correlated with crash severity. We have also discarded thana and district variables because we do not need these for our objective. We have only considered the variables that discuss about the environmental and the road conditions. After that, we have evaluated the most important variables from each cluster by their mean decrease accuracy (MDA) score. In general, the higher the MDA score of a variable is, the more importance the variable possesses to build a predictive model. This is because the MDA score of a particular variable tells us how much removing that particular variable decreases the accuracy of the model. The overall results of RF are shown in Table 3.

Now, we need to validate the performance of RF in extracting the most influential attributes through an evaluation metric. We have used Out of Bag (OOB) error for this validation part. In the code, we have taken the number of trees = 500. From the last column of Table 3, we see that, all the clusters got OOB errors of less than 28% except multi-vehicle crash group cluster 02. This suggests a decent fit. Since multi-vehicle cluster 2 got 42.58% of OOB error, so we have discarded this particular cluster for our further analysis. Now, we are ready for the next step- to build predictive models for each of the 4 clusters.

4.3 Classification and Rule Generation

Now when we have four clusters and high impact variables for each cluster, we have applied C5.0 algorithm in the clusters. There are various types of decision tree algorithm but we have chosen C5.0 because it is one of the newest as well as one of the best algorithms. The decision trees by C5.0 is easy to build and easily interpretable compared to other advanced and sophisticated machine learning models. In order to classify the clusters, we first oversampled the clustered dataset by 15 times using SMOTe function

Table 3. Summary table for random forest with MDA & OOB error

Homogenous Subset	Cluster No.	Most Influential Attributes				OOB (%)
Pedestrian	Cluster 1	Road Class	Divider	Traffic Control		17.39
	MDA	5 4	4.61	4.09		
	Cluster 2	Movement	Traffic Control	Divider	Location Type	9.75
	MDA	15.49	14.89	13.54	12.48	
Single Vehicle	Cluster 1	Junction Type	Traffic Control	Road Feature		19.51
	MDA	18.55	9.96	6.51		
	Cluster 2	Not Done due to low Sample size				
Multi Vehicle	Cluster 1	Weather	Movement	Road Feature	Collision Type	18.56
	MDA	9.66	9.37	8.75	8.14	
	Cluster 2	Movement	Road Geometry	Road Class	Divider	42.58
	MDA	8.34	6.68	6.65	6.28	

in WEKA due to underreporting problem which is mentioned earlier. After being up sampled by 15 times, the datasets of each cluster are now more conforming to the statistics mentioned in the introduction part [15].

Now using R tool, we have applied C5.0 algorithm. We have used MASS, Caret, caTools, recipes, C50, ROCR, e1071 and tidyverse packages. We have used 10-fold cross validation technique for classification. To validate our classification, we have calculated some performance metrics like accuracy, precision, recall and F1 measure from the confusion matrix of our model. We have used the following equation to calculate the 95% confidence interval.

$$\frac{2 \times \mathrm{N} \times \mathrm{acc} + \left(\mathrm{Z}_{\frac{\alpha}{2}}\right)^2 \pm \mathrm{Z}_{\frac{\alpha}{2}}\sqrt{\left(\mathrm{Z}_{\frac{\alpha}{2}}\right)^2 + 4 \times \mathrm{N} \times \mathrm{acc} - 4 \times \mathrm{N} \times \mathrm{acc}^2}}{2\left(\mathrm{N} + \left(\mathrm{Z}_{\frac{\alpha}{2}}\right)^2\right)} \tag{2}$$

Where, N = number of test records, acc = accuracy of the given model and at 95% confidence level, $Z_{\frac{\alpha}{2}} = 1.96$. Thus, we have calculated confidence interval at 95% confidence. Table 4 represents the performance metrics and statistics for C5.0 model for each cluster of each crash group. From Table 4, it can be calculated that the overall accuracy of the model for pedestrian crash group is around 77.08%; for single vehicle crash group, it is 81.43% and for multi vehicle crash group, it is 80%.

Table 4. Summary table for various performance metrics & statistics for C5.0 algorithm

Homogeneous Subset	Accuracy	Precision	Recall	F1 Measure	95% CI
Ped_Cluster 1	84.55%	84.54%	97.62%	90.61%	76.41%–90.73%
Ped_cluster 2	69.60%	65.66%	100%	79.25%	59.57%–77.55%
SV_Cluster 1	81.43%	81.4%	100%	89.76%	70.34%–89.72%
MV_Cluster 1	80%	79.4%	100%	88.57%	72.96%–85.9%

4.4 Rule Generation Using PART

Next, we have generated prediction rules from the C5.0 predictive models of each cluster using PART algorithm. The main goal of these rules is to show the model in more interpretable form. We have found 7 rules from pedestrian cluster 1 and 4 rules from pedestrian cluster 2. Table 5 represents the PART generated rules for pedestrian cluster 1.

Table 5. Rule table for pedestrian cluster 1

Rule no	Rules	No. of Instances	Class
1	Divider > 1 AND RoadClass <= 4	46	F
2	TrafficControl <= 3 AND RoadClass <= 4	79	
3	RoadClass <= 1	81	
4	TrafficControl > 5 AND Divider > 1	12	
5	RoadClass <= 2	36	NF
6	TrafficControl > 3 AND Divider <= 2 AND RoadClass > 1	582	
7	RoadClass > 4	981	

From these PART generated rules, rule 1 says that if Divider is not Present and Road Class is national/regional/feeder/rural then chances of fatal accidents are higher and it is true for 46 instances in the dataset, which is actually true in respect with the traditional belief because in non-urban road with no divider, accidents tend to be more severe. Rule 2 says that if Traffic control is uncontrolled/divider controlled/crosswalk and Road Class is national/regional/feeder/rural then rate of fatal accident is higher and is true for 79 instances. Rule 3 is very straightforward and says that if Road class is national then rate of fatal accident is higher and is true for 81 instances. So, it is apparent that most traffic accidents happen on national road and this is also supported by a study conducted by Labib et al. [22]. The next rule says that if traffic control is controlled by traffic light/both traffic police & light/signal and divider is absent then accidents are fatal and is true for 12 instances. Among all, the last rule defers from the traditional belief. Generally these traditional beliefs are set according to the data of developed countries and their data is very highly monitored and rate of underreporting is very low. But a developing country like Bangladesh has more than 50% underreported data. So it is expected that some prediction will be different from traditional beliefs. Rule 5 indicates that if road class is national/rural then non-fatal accidents are higher which is not similar to the traditional beliefs. Next rule says if traffic is controlled by traffic police/ traffic light/both traffic police & light/signal and divider is present and Road class is not national the number of non-fatal accidents are higher. Last rule says that if Road Class is city, road accidents are usually non-fatal.

In pedestrian cluster 2, among the 4 rules 2 are for Non-fatal and 2 are for fatal. If traffic control is controlled by traffic light/both traffic police & light/signal the fatal accident rate is high which is very much unlikely according to the traditional beliefs. But it is true for only 19 instances which we can consider as outliers. If movement is both sided then fatal accident rate is high and is true for 73 instances. Again, if location type is urban then non-fatal rate is higher and is true for 93 instances. Moreover, if movement is one way then non-fatal accident is high. These rules are conforming to the traditional rules.

PART algorithm has generated 4 rules for single vehicle cluster 1. If the junction type is crossing of four roads then fatal accident rate is higher and is true for only 8 instances, so we can take this rule as the rule of outliers. Furthermore, if it is bridge/culvert/narrow road then chances of occurring fatal accidents are higher and is true for 145 instances. This is also supported by a study conducted by Sadeek et al. [6]. Again, if road feature is normal road then chances of non-fatal accident is higher and is true for 272 instances. Moreover, if junction type is actually not crossing of any number of road then chances of non-fatal accidents are high and is true for 342 instances.

For multi-vehicle cluster 1, PART has generated only three rules. If movement is both sided and weather is not dry (rain/storm/foggy) then number of fatal accidents are high and it is true for 22 instances. This study complies with the findings from a study conducted by Md. Asif Rayhan et al. [8]. Also, if movement is one sided then chances of non-fatal accidents are high and it is true for 249 instances. And in dry weather, the chances of non-fatal accidents are higher and this is true for 1715 instances. All these three rules are very much similar to the traditional beliefs.

5 Conclusion and Possible Future Work

In this study, we were interested in finding a pattern for the fatal and non-fatal road crashes in Bangladesh and also their corresponding causality. Our study provided this patterns and causalities as the end result through the PART analysis on C5.0 models for each crash clusters. These results can help us understand the underlying factors and issues hidden in our road traffic system and hence can help the decision authority to make better decisions to reduce road accidents in Bangladesh. However, in our analysis, two classes have been trained individually with respect to the collected dataset and we applied the model based on them individually. In the future work, we hope to merge the two classes as a single class by using Multiclass-Multilabel Classification to achieve more accurate result [23]. Besides, as discussed earlier, underreported dataset is a big problem while conducting this type of research. In this work, we applied up sampling technique to address this issue. In the future, we hope to apply advanced methods to address this problem to make this research even more accurate by reducing the biasness [24]. The future study will be more explorative and more new knowledge may be identified which is our main concern as our future work.

References

1. Abirami, K.: "Academia". https://academia.edu/11542944/MAN_MADE_DISASTER
2. "The World Bank". https://data.worldbank.org/indicator/IS.VEH.NVEH.P
3. "Association for Safe International Road Travel". https://www.asirt.org/safe-travel/road-safety-facts/
4. "JBCPS". https://www.banglajol.info
5. Hoque, M., Anowar, S., Raihan, M.: Towards sustainable road safety in Bangladesh. In: International Conference on Sustainable Transport for Developing Countries: Concerns, Issues and Options, Dhaka (2008)
6. Sadeek, S., Anik, Md.: Effect of Road Infrastructures on Casualty Occurrence in Bangladesh. (2018)
7. Anowar, S., Yasmin, S., Tay, R.: Factors influencing the severity of intersection crashes in bangladesh. Asian Transp. Stud. **3**, 1–12 (2014)
8. Raihan, M., Hossain, M., Hasan, T.: Data mining in road crash analysis: the context of developing countries. Int. J. Inj. Control Saf. Promot. **25**(1), 1–12 (2017)
9. Li, L., Shrestha, S., Hu, G.: Analysis of road traffic fatal accidents using data mining techniques, pp. 363–370. (2017). https://doi.org/10.1109/sera.2017.7965753
10. Al-Radaideh, Q.A., Daoud, E.J.: Data mining methods for traffic accident severity prediction (2018)
11. Bahiru, T.K., Singh, D.K., Tessfaw, E.A.: Comparative study on data mining classification algorithms for predicting road traffic accident severity. In: 2018 Second International Conference on Inventive Communication and Computational Technologies (ICICCT 2018), Coimbatore, pp. 1655–1660 (2018)
12. Ramya, S., Reshma, SK., Manogna, V., Saroja, Y., Gandhi, G.: Accident severity prediction using data mining methods. Int. J. Sci. Res. Comput. Sci. Eng. Inf. Technol. 528–536 (2019). https://doi.org/10.32628/cseit195293
13. Sohn, S.Y., Lee, S.H.: Data fusion, ensemble and clustering to improve the classification accuracy for the severity of road traffic accidents in Korea. Saf. Sci. **41**, 1–14 (2003). https://doi.org/10.1016/S0925-7535(01)00032-7
14. Thomas, A., Jacobs, G., Saxton, B., Gururaj, G., Rahman, F.: The Involvement and Impact of Road Crashes on the Poor. TRL Limited, UK (2004)
15. Alam, M., Karim, D., Hoque, M., Islam, Q., Alam, M.: Initiatives regarding road accident database in Bangladesh. In: International Conference on Road Safety in Developing Countries, Dhaka (2006)
16. Accident Research Institute(ARI). http://ari.buet.ac.bd/
17. A. R. Institute: Road Safety Training Course For partitioners
18. Clustering Validation Statistics: 4 Vital Things Everyone Should Know - Unsupervised Machine Learning, STHDA. http://www.sthda.com/english/wiki/wiki.php?id_contents=7952
19. Reusova, A.: Hierarchical Clustering on Categorical Data in R, Medium, 1 April 2018. https://towardsdatascience.com/hierarchical-clustering-on-categorical-data-in-r-a27e578f2995
20. Filaire, T.: Clustering on mixed type data, Medium, 17 July 2018. https://towardsdatascience.com/clustering-on-mixed-type-data-8bbd0a2569c3

21. Bhalla, D.: A Complete Guide To Random Forest in R, Listen DATA, https://www.listendata.com/2014/11/random-forest-with-r.html#Random-Forest-R-Code
22. Labib, M.F., Rifat, A.S., Hossain M.M., Das, A.K., Nawrine, F.: Road accident analysis and prediction of accident severity by using machine learning in Bangladesh. In: 2019 7th International Conference on Smart Computing & Communications (ICSCC), Sarawak, Malaysia (2019)
23. Shahriar, M.M., Iqubal, M.S., Mitra, S., Das, A.K.: A deep learning approach to predict malnutrition status of 0–59 month's older children in Bangladesh. In: IEEE International Conference on Industry 4.0, Artificial Intelligence, and Communications Technology (IAICT 2019), BALI, Indonesia, pp. 145–149 (2019)
24. Sechidis, K., Sperrin, M., Petherick, E.S., Luján, M., Brown, G.: Dealing with under-reported variables: An information theoretic solution. Int. J. Approximate Reasoning **85**, 159–177 (2017). https://doi.org/10.1016/j.ijar.2017.04.002

Soil Analysis and Unconfined Compression Test Study Using Data Mining Techniques

Abdullah Md. Sarwar, Sayeed Md. Shaiban, Suparna Biswas, Arshi Siddiqui Promiti, Tarek Ibne Faysal, Lubaba Bazlul, Md. Sazzad Hossain, and Rashedur M. Rahman(✉)

Department of Electrical and Computer Engineering, North South University, Plot-15, Block-B, Bashundhara Residential Area, Dhaka, Bangladesh
{abdullah.sarwar,sayeed.shaiban,suparna.biswas,siddiqui.promiti, tarek.faysal,lubaba.bazlul,sazzad.hossain09, rashedur.rahman}@northsouth.edu

Abstract. In this study, Random Forest Regressor, Linear Regression, Generalized Regression Neural Network (GRNN) and Fully connected Neural Network (FCNN) models are leveraged for predicting unconfined compression coefficient with respect to standard penetration test (N-value), depth and soil type. The study is focused on a particular correlation of undrained shear strength of clay (Cu) with the standard penetration strength. The data used is from 14 no. ward in Mymensingh and Rangamati districts which are situated in Bangladesh. By using this data, the study tries to solidify the correlation of SPT (N-value) with Cu. It also tries to check the goodness of the relationship by comparing it with unconfined compression strength values gained from the unconfined compression test calculated from the field by experts.

Keywords: Standard penetration test (SPT) · Unconfined compression test · Random Forest Regressor · Linear Regression · General Regression Neural Network (GRNN) · Fully Connected Neural Network (FCNN)

1 Introduction

In foundation design, standard penetration test's (SPT) blow counts (N-value) are used to approximate shear strength characteristics of soils. In soil mechanics, shear strength is the amount of stress a soil can sustain. Here, we focus on a particular correlation between standard penetration test and undrained shear strength of clay. Hara et al. [5] recommended the following relationship between the undrained shear strength of clay (Cu) and N_{60}:

$$\frac{C_u}{P_a} = 0.29 \times N_{60}^{0.72} \tag{1}$$

where P_a = atmospheric pressure (≈100 kN/m2; ≈2000 lb/in2). $N_{60}^{0.72}$ is the blow count of SPT and Cu is the undrained shear strength of clay. In this paper, we leverage this

M. Hernes et al. (Eds.): ICCCI 2020, CCIS 1287, pp. 38–48, 2020.
https://doi.org/10.1007/978-3-030-63119-2_4

equation, shown in the methodology section and use it to predict unconfined compression strength (q_u) from various parameters like soil type, N-value and depth.

Investigating soil with in-situ field testing is a process of placing an instrument at an exact point in a borehole to quantify the in-situ qualities of the soil. The standard penetration test (SPT), cone penetration test (CPT), piezocone (CPTu), flat dilatometer (DMT), pressure meter (PMT), and vane shear test (VST) are all common in-situ field tests. SPT test starts with the force of a slide hammer into the ground to measure the quantity of hits required to create a separation of 300 mm. The poundings are then recorded and last two records of the poundings are known to be the N-value (blow-count). This value requires an adjustment of energy efficiency for a particular drill rig. An energy corrected N-value (adjusted to 60%) which is advised by ASTM D-4633 (Standard Test Method for Energy Measurement for Dynamic Penetrometers) is represented as N_{60}.

The goal of this study is to find out the calculated unconfined compression strength of soil by training machine learning models with our datasets and analyze how the models trained with calculated values would stack up against the values gained from unconfined compression tests from the field. Thus, simplifying the current geotechnical approaches. To this cause, we investigate using four machine learning methods on the collected dataset of Bangladesh.

2 Related Work

Most researchers have pointed out their concerns regarding the energy correlation to the value of SPT. During this test, the energy carried to the rods is proportional to the hypothetical free fall potential energy, which can vary from 30% to 90% [5, 7]. Schmertmann and Palacios [11] have shown that the SPT blow check is relative to the vitality passed on. In [2, 7, 9] it was stated that there should be a correction of energy level of the SPT-N value, at a level of 60% (CFEM 2006). This SPT-N value is then named as N_{60}. This SPT-N value is estimated to be a normal energy proportion of 60% (ERR = 60%) in the United States/Canada as per ASTM D1586-11 (2014). In this investigation vitality proportion of 60% (ERR = 60%) is accepted. Balachandran et al. [1] worked on finding statistical correlations between undrained shear strength (C_u) and SPT-N value. They conducted their study in a site in Toronto, Ontario, Canada. Based on the conducted tests, the soil was categorized as cohesive glacial till, with various surfaces like silty clay till and clayey silt till. The authors first gathered the sets of field vane shear strength (C_u) and SPT-N value at similar depths in the same boreholes. The field estimated SPT-N values were then corrected by the CFEM (2006). Because of the changeability in gear and working conditions, direct utilization of SPT-N esteems for geo-technical configuration was not recommended.

Few works have been done to find various kinds of correlation to unconfined compressive strength to various geotechnical properties. In [13], Yılmaz and Sendır in Turkey worked on a correlation of Schmidt rebound number (N) with q_u and Young's modulus (Et) of the gypsum by empirical association. They found a possibility of approximating q_u and Et, from their Schmidt hammer rebound number using their proposed empirical relationships of q_u and Et. Another study by Grima and Babuška [4] from Netherlands predicted unconfined compressive strength of rock samples using fuzzy models. They used an information driven approach to the modeling (q_u) and then numerical conclusions were compared with a conventional statistical model. It turned out that fuzzy model did not work better than Neural Network based models. But the authors reported that neural network conclusions were more complicated to analyze [4]. Bui et al. [3] proposed an integrated machine learning approach in order to predict the compression coefficient of soil; the approach integrated particle swarm optimization (PSO) technique and Multi-Layer Perceptron Neural Network to propose the PSO-MLP Neural Nets method. 12 factors or soil parameters were used as input to predict the soil compression coefficient, which was an efficient measurement of the soil strength, and 154 soil samples were used to train the model. Comparison of the PSO-MLP method was made with other standard methods, and result analysis and accuracy obtained showed significant improvement compared to the other methods. Machine learning and data mining techniques appear to be competent methods for the simulation, prediction and determination of geo-technical parameters, and several approaches have been undertaken in order to determine and predict soil type, soil strength, etc.

There is not much work related to the aspect of N_{60}-C_u relationship with respect to data mining. But there are significant amount of research works that have been carried out in order to investigate various relationships on soil properties, soil types and on standard penetration test using various data mining techniques. These topics indirectly in various ways affect our work here. One such work was done by Sarkar et al. [10] where they built a neural network called GRNN in order to predict the Standard penetration test (SPT) and soil types. It focused on the soils of a city located in the country of Bangladesh called Khulna. The soils comprised mainly of silt and clay. They accumulated 2326 N-values from SPT occurring in 42 clusters containing 143 boreholes spread over an area of 37 km^2. At first the borehole locations were trained with the soil types and then the soil types were used to predict SPT values.

In another paper, Puri, Prasad and Jain [8] investigated and determined the relationships between several soil properties using various machine learning models. The investigation established correlations between several soil parameters, including SPT value with soil type, compression index (Cc) using liquid limit (LL) and void ratio (e), cohesion (c) and many others. The investigation used models of Linear Regression Analysis, Support Vector Machine (SVM), Random Forest (RF), Artificial Neural Network (ANN), and M5 Model trees, and established the correlations.

3 Dataset

The dataset we used came from two geographical locations situated in the country of Bangladesh. They are 14 no. ward in the division of Mymensingh which is situated 120 km north of Dhaka, the capital of Bangladesh and Rangamati in the division of Chittagong, the port city of Bangladesh.

3.1 Mymensingh

This data set was collected to prepare a contingency plan for dealing with the socio-economic and physical vulnerability of the area. Various kinds of data related to geological, building and socio-economic aspect were collected. For the geological data collection part, borehole tests and micro tremor tests were conducted. From various test results, we only chose the data that was relevant to our correlation $N_{60} - C_u$ which is the Borehole log. We found a total of 12 Borehole points in it. Figure 1 shows the behavior of N-values gained from SPT with increasing depth in the Mymensingh dataset.

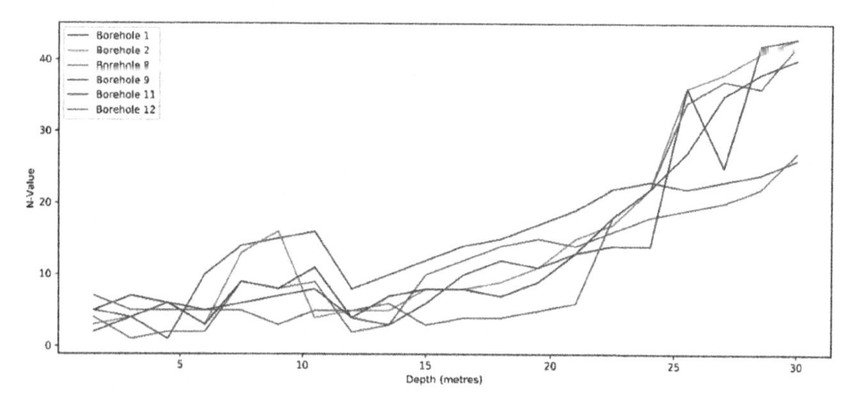

Fig. 1. Trend of N-values in different depths of each borehole in Mymensingh.

3.2 Rangamati

A Geo-technical investigation was performed for the construction of a 500.00 m long P.C. Girder Bridge Work Over Kaptai Lake at NaniarChar, Rangamati. The dataset obtained consisted of data obtained through the exploratory boring drilling, with subsoil stratification, and standard penetration test (SPT).

When we collected data, we found the attributes depth, SPT and soil type common in the two types of data sets from the two places mentioned above. But the dataset from Mymensingh also contained values of unconfined compression tests calculated by experts on field that we could not find for the other data set of Rangamati.

4 Methodology and Results

The collected dataset was divided into three segments. First segment contains only the Mymensingh dataset. Second one contains both Mymensingh and Rangamati data. Third segment contains the real values from Unconfined Compression Tests. This data is used for testing purpose only.

We decided to follow a specific format while collecting the data. The data is accumulated to a table from the available bore logs that had the columns "Depth", "N-value", and the dominating type of material in the soil in that depth for the certain borehole area. For the "Material type", the numerical values in Table 1 stand as follows: fine SAND: 1. silty CLAY: 2, clayey SILT with FINE SAND: 3, FINESAND with SILT: 4, clayey SILT: 5, silty FINE SAND: 6, sandy SILT: 7, Rubbish: 8, silty SAND: 9, coarse SAND: 10.

Table 1. Sample of accumulated dataset

Borehole	Depth	Material type	N-Value	Cu	qu
1	1.5	8	5	92.39669	184.7934
1	3	2	4	78.68305	157.3661
1	4.5	2	1	29	58
1	6	3	10	152.1942	304.3883
1	7.5	2	14	193.9145	387.8289
1	9	3	15	203.7904	407.5807
1	10.5	3	16	213.4835	426.9670
1	12	2	8	129.6053	259.2106
1	13.5	2	10	152.1942	304.3883
1	15	6	12	173.5436	347.0871
1	16.5	6	14	193.9145	387.8289
2	1.5	2	3	63.96248	127.9250
2	3	2	4	78.68305	157.3661
2	4.5	2	6	105.358	210.7157
2	6	3	3	63.9625	127.9250

Afterwards, using the N-values we gained the value of "Cu" which is the Undrained Shear strength of soil from Eq. (1). And then, we used the relation shown in Eq. (2) to find "q_u" or Unconfined Compression strength.

$$q_u = 2 \times Cu \tag{2}$$

For Mymensingh data alone, we were able to collect 241 rows. And for Rangamati, we got 77 rows. And hence, we got a total of 318 data. We worked on two aspects with these values. Firstly, predicting the calculated value of q_u to check the validity of the correlation. Secondly, the error of prediction that is found if we train a machine learning model with calculated values and test it on the values found on the field using Unconfined Compression Test.

4.1 Models

In this study, we have tested out four kinds of machine learning models. As our problem is a regression-based problem, we have tried out Linear Regression, Random Forest Regressor, Generalized Regression Neural Networks (GRNN) and Fully Connected Neural Network (FCNN) to help our cause.

Linear Regression: Regression is a system of modeling a target "y" value based on independent "x" values. Hence it is able to measure a cause and effect relationship. A linear regression is a type of regression evaluation that maps a linear relationship between the independent(x) and dependent(y) variable(s).

Random Forest Regressor: A Random Forest Regressor is known to be an ensemble technique and a bagging algorithm that is able to do both Regression and Classification tasks using decision trees. For our model, we have kept the maximum depth of the tree to be 10, the randomness of the bootstrapping of the samples to be 5 and mean absolute error as the criterion to be used to measure the quality of the split.

Generalized Regression Neural Networks (GRNN): Inspired by the paper of Sarkar et al. [10], we have implemented a simple neural network called Generalized Regression Neural Networks (GRNN). Generalized Regression Neural Networks (GRNN) is a special case of Radial Basis Networks (RBN) which is a single-pass neural network using a Gaussian activation function in the hidden layer [12]. GRNN consists of input, hidden, summation, and division layers. For our GRNN model we have used standard deviation of 1.

Fully Connected Neural Network (FCNN): Neural Networks are modeled loosely on the neurons in the human brain, designed to identify trends. Fully Connected Neural Networks (FCNN) are several stacked Neural Network layers that are fully connected. In FCNN, the function of an output neuron depends on each input dimension. Our FCNN consists of input layer having 3 inputs as features (depth, soil type, N-value), 2 hidden layers having 3 and 8 neurons or nodes respectively and 1 output layer that predicts the q_u values. Each of the layers consists of linear functions followed by the non-linearity, Rectified Linear Unit (ReLU). For loss function Mean squared Error is used with the optimizer Adam [6]. Adam is used with a decay of $1e^{-3}/200$. This is decided on a trial and error basis.

4.2 Accuracy Metric

The accuracies of the models are judged using two performance metrics for regression. These are R-squared value and Mean Squared Error.

R Squared Value: R-squared (R^2) value measures statistically the closeness of fitness of the data towards the regression line. It is the ratio of explained variation to the total variation. R-squared value in the case of regression provides us with an estimate of how well the model predicts values compared with the actual data. It usually ranges from 0 to 1, the closer it is to 1, and the more accurate the model is able to make predictions. R-squared value normally checks how much better the regression model is to a horizontal line generated to the mean of the data. But if the model is worse than the mean of the data, R-squared value can give negative value which indicates that the model predictions compared with the expected outcomes do not fit the model.

Mean Squared Error: The mean squared error tells us how close our regression model is to a set of values. It measures the distance between the regression line and the given values. The lower the value of "MSE", the more accurate the model. As the squared error is averaged, the increase in data results in less error.

$$MSE = \frac{1}{N}\left[\Sigma\left(\widehat{Y} - Y\right)^2\right] \tag{3}$$

Where, N is the total amount of data, $\widehat{Y}$ is the predicted values and Y is the actual value.

4.3 Results

We have seen that our models perform relatively well on the calculated q_u values when trained on both single Mymensingh and combined Mymensingh-Rangamati dataset. The scores are shown in Table 2. Here, FCNN seems to have the highest MSE score in the Mymensingh dataset, followed by GRNN, Random Forest Regressor and Linear Regression. FCNN also had the best MSE score on the calculated q_u values for the combined dataset. The R-squared value seems to be quite similar for all the models in both the dataset. But in general, the Neural Network based regression models seem to perform better than conventional machine learning models.

Table 2. Scores on Test split from Mymensingh and Mymensingh-Rangamati combined dataset.

Scores on Mymensingh dataset		
Models	R^2 Score	MSE Score
Linear Regression	0.9894	107.8833
Random Forest Regressor	0.9995	18.1698
GRNN	0.9747	0.0017
FCNN	0.9977	0.0002
Scores on Mymensingh-Rangamati dataset		
Models	R^2 Score	MSE Score
Linear Regression	0.9910	147.1977
Random Forest Regressor	0.9998	3.4732
GRNN	0.9900	0.0013
FCNN	0.9920	0.0007

Table 3. Predicted q_u values on models (Linear Regression and Random Forest Regressor) trained and tested with the combined dataset of Mymensingh-Rangamati.

Actual values	Predicted values	
	Linear Regression	Random Forest Regressor
484.9047083	495.8978704	484.9047083
311.1572382	299.328183	311.3275766
311.1572382	306.6188681	310.9496172
250.6914918	222.2082204	250.4249058

(*continued*)

Table 3. *(continued)*

Actual values	Predicted values	
	Linear Regression	Random Forest Regressor
29	61.39368044	30.25490062
484.9047083	493.0159781	484.9047083
484.9047083	495.8978704	484.9047083
193.9144548	189.2897434	193.9043792
129.605308	124.0536387	129.605308
193.9144548	189.2897434	193.9043792
484.9047083	493.9766089	484.9047083
213.4835349	205.2713271	213.6315701
92.39668736	91.21313442	92.39668736

Table 4. Predicted q_u values on models (Generalized Regression Neural Network, Fully Connected Neural Network) trained and tested with the combined dataset of Mymensingh-Rangamati.

Actual values	Predicted values	
	Generalized Regression nEural Networks	Fully Connected Neural Network
969.8094166	969.8519642	969.9978
622.3144764	606.2135705	621.0875
622.3144764	596.846037	615.06476
501.3829837	457.5690968	518.4589
58	140.0193478	106.062416
969.8094166	972.0230653	939.6435
969.8094166	969.8519642	969.9978
387.8289097	391.7779154	374.0966
259.210616	205.1737078	245.72409
387.8289097	391.7779154	374.0966
969.8094166	970.913559	946.6136
426.9670699	381.9677024	407.0509

Table 3 and Table 4 shows some of the calculated values when predicted using the mentioned models. All the models seem to present relatively good predicted values compared to the actual values.

Table 5. Predicted q_u values on models trained on combined datasets of Mymensingh-Rangamati and tested on the data gained from the unconfined compression tests.

Actual values	Predicted values		
	Linear Regression	Random Forest Regressor	Generalized Regression Neural Network
37	66.09655475	47.59679041	41.89010
146	95.20753684	92.39668736	60.39685
53	78.51208035	63.96248105	45.72606
94	78.51208035	63.96248105	45.72606
69	79.64684639	63.96248105	45.91526
154	93.02209111	78.68305099	56.84788
100	119.7725806	117.7249695	78.71311
88	97.48050602	92.39668736	60.49208

Table 5 shows the predictions on the real values achieved from the unconfined compression test. The models do not perform well on the unconfined compression test values that we have gained from the Mymensingh dataset. But this is to be expected, as the values gained from the test in the field considers a lot of variables that we do not take into account here. We only trained our models on calculated values from the correlation N_{60}-C_u. Hence adding variations to the dataset, which we currently lack with respect to the unconfined compression test values, could significantly improve the scores to the performance of our models.

Figure 2 shows a visual representation of the model's performance using the values from the Table 4 and Table 5. Here, the x-axis represents the number of instances we have chosen to compare each model, and the y-axis represents the N-value. The values are pretty similar for diagram (a), (c), (e), (g) in Fig. 2, which are the results against the calculated values from the correlation. And diagram (b), (d), (f), (h) in Fig. 2, which are the results against real values gained from the field, shows though there's a significant difference in the actual and predicted values. The predictions do seem to recognize the trend of the real values using models trained on calculated values of the correlation. We refrain from showing scores achieved against the unconfined compression test values, as the number of values we were able to gather were insignificant for testing. From all the experiments, Random Forest Regressor seem to be the most suitable model in this case, as it performs the best among the real values and sufficiently good among the calculated values.

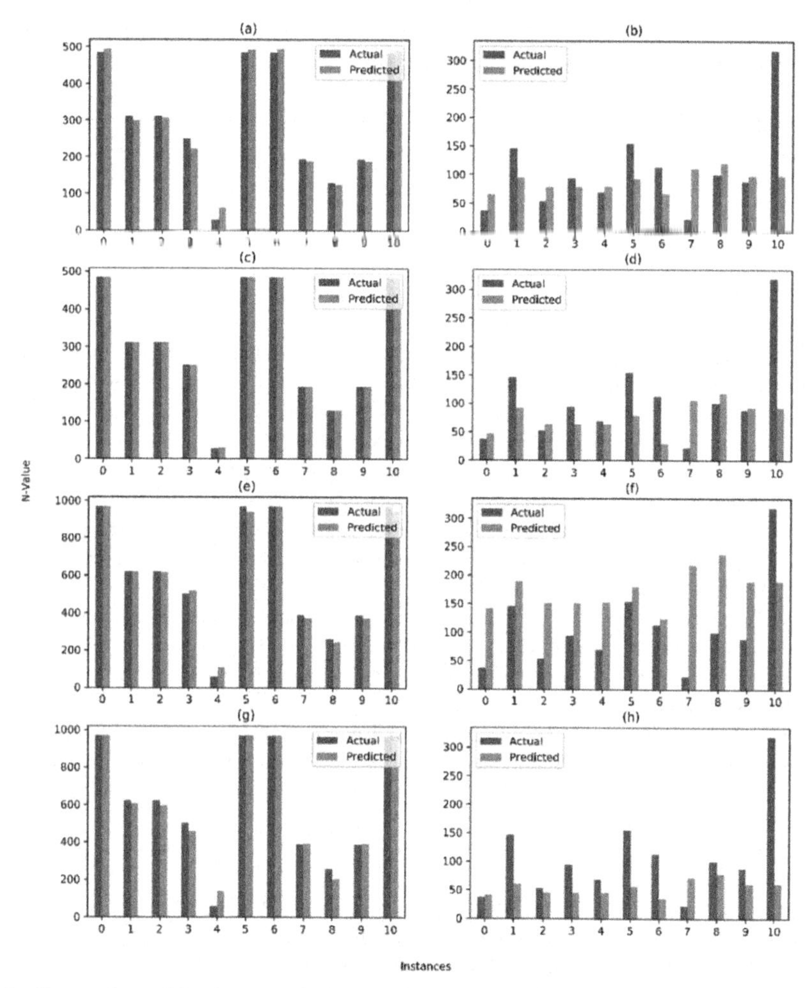

Fig. 2. Comparison of few instances between actual and predicted values of the combined dataset of Mymensingh-Rangamati and real values from unconfined compression test using Linear Regression for (a) and (b), Random Forest Regressor for (c) and (d), FCNN for (e) and (f), GRNN for (g) and (h)

5 Conclusion

In conclusion, we came to the realization that the amount of data was a big issue for us. As a result, the models perform poorly on the real unconfined compression test values. But we have impressive scores on calculated q_u values. These results further solidify the correlation between N_{60}-C_u. Hence, we can say that the relationship of unconfined compressive strength is solid and can be used for further studies. We have used four models, tested on three kinds of data and trained on two kinds of data. For future work, we would like to check the variation of q_u on different regions in the same country. We also hope to collect more data and test it on actual field values gained from unconfined compression tests. This should introduce more variation to our data and help build a more robust model that can handle outliers easily.

Acknowledgment. The authors would like to thank the Department of Urban and Regional Planning from Bangladesh University of Engineering and Technology for providing us with the related datasets from Mymensingh Ward 14.

References

1. Balachandran, K., Liu, J., Cao, L., Peaker, S.: Statistical correlations between undrained shear strength (CU) and both SPT-N value and net limit pressure (PL) for cohesive glacial tills
2. Bolton Seed, H., Tokimatsu, K., Harder, L., Chung, R.M.: Influence of SPT procedures in soil liquefaction resistance evaluations. J. Geotech. Eng. **111**(12), 1425–1445 (1985)
3. Bui, D.T., Nhu, V.H., Hoang, N.D.: Prediction of soil compression coefficient for urban housing project using novel integration machine learning approach of swarm intelligence and multi-layer perceptron neural network. Adv. Eng. Inform. **38**, 593–604 (2018)
4. Grima, M.A., Babuška, R.: Fuzzy model for the prediction of unconfined compressive strength of rock samples. Int. J. Rock Mech. Min. Sci. **36**(3), 339–349 (1999)
5. Hara, A., Ohta, T., Niwa, M., Tanaka, S., Banno, T.: Shear modulus and shear strength of cohesive soils. Soils Found. **14**(3), 1–12 (1974)
6. Kingma, D.P., Ba, J.: Adam: A method for stochastic optimization (2014). arXiv preprintarXiv:1412.6980
7. Kovacs, W.D., Salomone, L.A.: Closure to "SPT hammer energy measurement" by William D. Kovacs and Lawrence A. Salomone (April, 1982). J. Geotech. Eng. **110**(4), 562–564 (1984)
8. Puri, N., Prasad, H.D., Jain, A.: Prediction of geotechnical parameters using machine learning techniques. Procedia Comput. Sci. **125**, 509–517 (2018)
9. Robertson, P., Campanella, R., Wightman, A.: SPT-CPT correlations. J. Geotech. Eng. **109**(11), 1449–1459 (1983)
10. Sarkar, G., Siddiqua, S., Banik, R., Rokonuzzaman, M.: Prediction of soil type and standard penetration test (SPT) value in Khulna city, Bangladesh using general regression neural network. Q. J. Eng. Geol. Hydrogeol. **48**(3), 2014-2108 (2015). Geological Society of London
11. Schmertmann, J.H., Palacios, A.: Energy dynamics of SPT. J. Geotechnical Eng. Div. **105**(8), 909–926 (1979)
12. Specht, D.F., et al.: A general regression neural network. IEEE Trans. Neural Netw. **2**(6), 568–576 (1991)
13. Yılmaz, I., Sendır, H.: Correlation of schmidt hardness with unconfined compressive strength and young's modulus in gypsum from Sivas (Turkey). Eng. Geol. **66**(3–4), 211–219 (2002)

Self-sorting of Solid Waste Using Machine Learning

Tyson Chan[1], Jacky H. Cai[1], Francis Chen[1], and Ka C. Chan[1,2](✉)

[1] The University of New South Wales, Sydney, Australia
[2] University of Southern Queensland, Springfield, Australia
kc.chan@usq.edu.au

Abstract. In waste recycling, the source separation model, decentralises the sorting responsibility to the consumer when they dispose, resulting in lower cross contamination, significantly increased recycling yield, and superior recovery material quality. This recycling model is problematic however, as it is prone to human error and community-level participation is difficult to incentivise with the greater inconvenience being placed on consumers. This paper aims to conceptualise a solution by proposing a unique mechatronic system in the form of a self-sorting smart bin. It is hypothesised that in order to overcome the high variability innate to disposed waste, a robust supervised machine learning classification model supported by IoT integration needs to be utilised. A dataset comprising of 680 samples of plastic, metal and glass recyclables was manually collected from a custom-built identification chamber equipped with a suite of sensors. The dataset was then split and used to train a modular neural network comprising of three concurrent individual classifiers for images (CNN), sounds (MLP) and time series (KNN-DTW). The output class probabilities were then integrated by one combined classifier (MLP), resulting in a prediction time of 0.67 s per sample, a prediction accuracy of 100%, and an average confidence of 99.75% averaged over 10 runs of an 18% validation split.

Keywords: Waste automation · Recycling · Neural network

1 Introduction

1.1 Waste Recycling

Global trends in waste generation are forecasted to drastically increase over the coming century [1]. In 2008, waste production within Australia alone was over 46 million tonnes per year, equating to 5.8 kg of unwanted rubbish per Australian, per day [2]. Despite our nation's recycling efforts, a staggering 49% (2011) is either dumped in the environment or directed to landfill [2]. This is largely attributed to imperfect separation techniques and cross-contamination between recycling streams.

M. Hernes et al. (Eds.): ICCCI 2020, CCIS 1287, pp. 49–60, 2020.
https://doi.org/10.1007/978-3-030-63119-2_5

Currently, the dominant model within Australia is single stream recycling, where all recyclables are collected in commingled bins before being separated at centralised material recovery facilities (MRFs) [3]. Where urban sprawl is prominent, waste is typically generated at a lower density and over a larger surface area, hence the storage and collection of mixed recyclables is logistically viable. However, with rapid urbanisation and the rise of smart cities, waste generation will begin to concentrate around densely-populated metropolitan centres [4]. In this case, the model is likely to shift toward source separation recycling; where waste is sorted into different streams at the point of disposal [5]. By decentralising the separation of recyclables, it is possible to increase recycling rates and improve the overall quality of recovered materials. Despite this, current source separation practices only marginally outperform single stream due to manual sorting efforts inconveniencing users and being more prone to human error.

This project aims to optimise the source separation recycling model by automating the waste sorting process. High levels of variation in disposed waste makes material identification a very complex classification problem. As such, the proposed smart bin intends to leverage the global infrastructure of the Internet of Things (IoT) and the robust capabilities of Machine Learning (ML). Users will be able to dispose recyclables into the smart bin, which will automatically scan the item with a suite of sensors. The resulting data will then be relayed to a server, where a statistical classifier will fuse the information sources, and determine the most suitable recycling stream; plastic, metal, glass, or landfill. Finally, the server will return the predicted label back to the smart bin for further sorting and storage into separate receptacles.

1.2 Literature Review of Self-sorting Bins

A niche application of technology in waste that has not made its way to market yet is in the form of a self-sorting bin. Three examples of self-sorting smart bins projects are the GreenCan [6], the Automated Waste Sorting Receptacle (AWSR) [7], and the Commingled Waste Stream Separation System [8].

The GreenCan from the University of Michigan is a self-sorting bin that can identify recyclables from non-recyclables [6]. The prototype uses the analysis of impact acoustics to classify the material. The object is dropped on a wood plank and the resulting impact sound is recorded and analysed. Once the material is classified it rotates the plank about an axle in the centre to drop the object into either the recyclables or non-recyclables section. The system is impressive as it accurately classifies sample items and also completes the whole process in a very short time. The GreenCan has limitations in that it can only determine if an object is recyclable or non-recyclable, this limitation is caused due to the use of only one sensor for object classification. However, the single method employed to classify the object was and shows itself to be promising.

The Automated Waste Sorting Receptacle (AWSR), designed at Georgia Institute of Technology designed, is a full-scale self-sorting recycle bin that utilises computer vision, artificial neural networks and machine learning to classify the sample objects [7]. The prototype can determine if an object is a plastic bottle or an aluminium can. The design and execution of the project is impressive, although, there are a few areas features that could be improved. As the identification chamber is on slides from side to side on a linear actuator, it is quite slow in its sorting process. Also it can only take one item at a time to sort which can, again, slow down the whole process. The most glaring limitation on the AWSR is its ability to only sort into two material categories; plastic and metal. Being able to sort into only two streams of recyclables is unrealistic for a real-world application. This limitation on the amount of materials it can sort is due to the limited number of sensors used for identification. Like the GreenCan, only having one sensor as the input for the machine learning algorithm reduces its capabilities in identifying a diverse range of materials. The two material classification methods used in the mentioned student projects demonstrate the powerful capabilities of each approach but also exposes areas that could be improved upon such as using a combination of different material identifications sensors.

In 1992, Tarbell et al. first hypothesised a statistical approach to the waste classification problem, proposing the Commingled Waste Stream Separation System (CWSSS) [8]. This paper proposes an "intelligent" automation system for characterising materials by feeding radiation, acoustic and electromagnetic sensor data into a dynamic machine learning classification model. The paper states that training is done on series of glass, plastic and metal containers of various shapes, sizes, deformation patters and contamination levels. Each object is then passed through an array of sensors and their respective responses are recorded. Our machine learning approach was largely inspired by the CWSSS. The system serves to demonstrate a conceptual decision making and quality control system for product classification and sorting. A major aspect of this lies in the fusion of information from multiple independent data sources. It can be implied from this paper that spectroscopic, acoustic and inductive data streams can provide relevant features for characterising the materials of items on a conveyor belt for the purposes of identification.

Durrant-Whyte from the Australian Centre for Field Robotics defines data fusion as the process of combining information from a number of different sources to provide a robust and complete description of an environment or process of information [9]. Data fusion is significant in any application where large amounts of data must be combined, fused and distilled to obtain information of appropriate quality and integrity on which decision can be formulated and executed autonomously. Indeed, such models are probabilistic by nature and could provide a powerful and consistent means of describing uncertainty, leading naturally into ideas of information fusion and decision making. This paper intends to approach the waste material identification problem by means of statistical classification. However, the very nature of how this will be achieved is through data fusion. Each individual sensor does not provide definite data streams for describing the material a disposed item. Rather the fusion of all of the data streams will be leveraged in order to provide the statistical classification model with as large of a knowledge base as possible from which it will learn the relevant correlations and make decisions.

2 Self-sorting Bin Design

The nature of this project is one of product design, as such a well-defined engineering design methodology was followed. This engineering design recipe closely aligns with the methodology outlined in [10], which resembles the waterfall model by sequentially stepping through requirements, product concept, solution concept, embodiment, and detailed design. The primary purpose of utilising the engineering design process was to methodically lay out a plan of execution in order to manifest a simple idea into a functional prototype.

This first step is to evaluate design options using the morphological matrix, a useful tool that details design options for a specific component or subsystem in an easy to read table. This allows for complex concepts and ideas to be brought together by compiling various options for each component. As such the matrix is very powerful for generating a range of "low-fidelity" prototype deigns that are a combination of such options, leading to the decisions on the mechanical and electrical design of the self-sorting bin.

2.1 Mechanical Design

The self-sorting smart bin design consists of three subsystems: input, identification, and sorting. The CAD drawings of the subsystems and the assembled system are shown together in Fig. 1.

The **input subsystem** serves the purpose of detecting when objects are input into the bin (top left in Fig. 1). Logically, this will be the point at which the sensor suite in the identification is turned on. This is achieved with an ultrasonic sensor located near the input chute from which items would be able to slide down and drop through to the identification chamber. At the bottom of the slide, there is a servo motor with a 400mm length arm attachment. This is used primarily as the mechanism which times the impact of the disposed item in the identification chamber during a three second sampling window from the sensors (for the microphone and time series recordings).

The **identification chamber** is the subsystem which contains the sensor suite for collecting data for characterising the material of disposed items (middle left in Fig. 1). A major design aspect was to leverage off-the-shelf sensors as a means of reducing the cost, and making the software implementation more challenging. This subsystem was built first and prioritised as it is a bottleneck for dataset collection and the machine learning software implementation.

The **sorting subsystem** serves the purpose of sorting recyclable waste into one of four receptacles arranged in a 2×2 grid (bottom left Fig. 1). It leverages a flap system which is powered by two servo motors. The identification chambers base plate is attached to an axle which will allow it to swivel depending on the label of the item. When sorting occurs, a servo motor will rotate the sorting planks according to the specified label. If the item is metal or landfill, the item falls through to the left side of the sorting system, while the glass and plastic case fall through to the right.

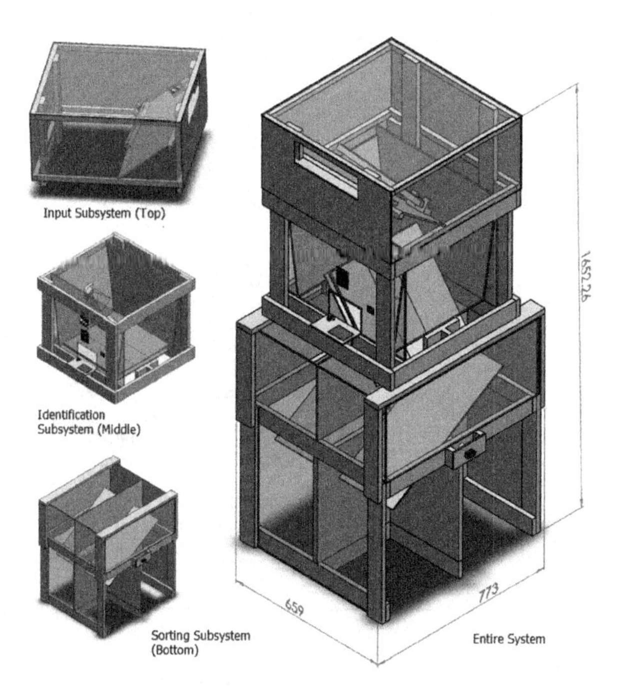

Fig. 1. CAD of three subsystems (left) and entire system (right).

2.2 Electrical Design

The electrical design of the prototype is largely concentric around the selection of hardware facilitating embedded processing for the self-sorting smart bin. It also details the sensors selected for the identification chamber. A high-level electrical diagram at the end serves to details how each of the electrical components are wired together is show in Fig. 2.

A key requirement for the software classification of disposed recyclables is the system's ability to process raw sensor data and send and receive information over a network. A Raspberry Pi 3 (Pi) was selected for this purpose, acting as the "intelligent brain" for the smart bin. In the context of the smart bin, the Pi facilitates the manipulation and storage of data, predominantly for relaying sensor data to a server for classification, and receiving back the predicted label for sorting.

A dedicated microcomputer is incorporated in the electrical design as a means of interfacing with sensors and actuators. The microcontroller selected was Arduino Uno (Uno), largely due to its extensive support for GPIO with support for analogue, digital, and pulse width modulation (PWM) capabilities. The Arduino acts as the slave in the master-slave configuration, responsible for performing low-level interactions with sensors and actuators.

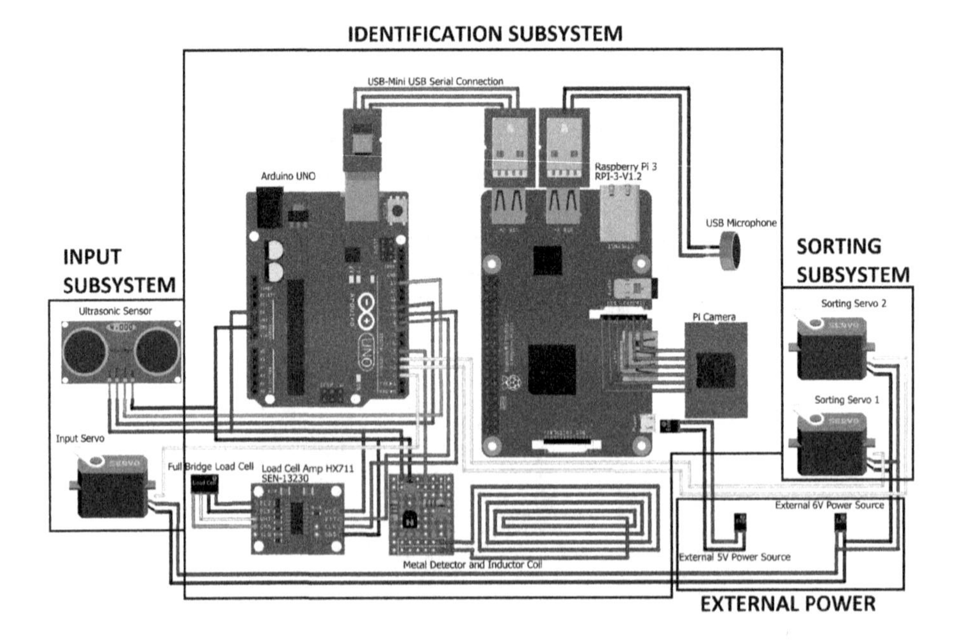

Fig. 2. Electrical wiring diagram.

2.3 Sensors

Sensors include the camera, microphone, metal detector and load cell.

The **Raspberry Pi Camera V2** was selected for capturing image data. This is a digital camera which can be connected directly into the Pi, capable of capturing RGB images. The Camera V2 includes a Sony IMX219 8-megapixel sensor facilitating high definition photographs. Interfacing with the camera is achieved via the PiCamera Python library.

The **stereo microphone** connected to a USB audio adapter was used for recording audio data. This simple microphone setup ensured that no custom electrical wiring or hardware development was required to have it work. Interfacing with the microphone (recording) is achieved via the PyAudio Python library.

The **metal detector** implemented in the identification chamber is the only sensor which was custom built. Its purpose is to detect the presence of nearby metals and send the electronic signal (inductive frequency) from the Uno slave to the Pi master. The metal detector operates on a tank circuit and Colpitts oscillator connected to the Arduino. The way in which this works is by creating an oscillating current, that results in a changing magnetic field. During initialization, the current frequency is stored to a variable, and subtracted from each frequency reading as a means of zeroing. When an electromagnetically inductive object enters the proximity of the metal detector, the frequency will change, resulting in a frequency reading which is sampled 10 Hz programmatically from the Pi.

The **load cell** is responsible for measuring the mass and impact force of items dropped into the identification chamber. A 5kg straight bar resistive load cell was selected due to its relative ease being integrated beneath the base plate of the identification chamber. To read data from the load cell, a load cell amplifier was used. The way in which this works is by converting mechanical load into resistance. The load cell has resistors attached to it, which increase in resistance when the bar is bent, changing the circuit voltage. By comparing the input voltage with the output voltage, it is possible to infer the mass. The amplifier in this case is needed because the change in voltage is in the millivolt range.

3 Software Architecture

The incorporation of four unique sensors in the identification chamber allows the system to characterise disposed items based on their visual appearance, acoustic resonance, electromagnetic inductance, and weight. Individual classifiers were created for each of the sensor data streams as a means of extracting information in the form of probabilistic predictions. The classifier outputs were then combined into a neural network, forming an overall modular neural network architecture. This section describes the modular neural network model and the software architecture.

3.1 Classification Models

The machine learning problem is one of multilabel statistical classification of recyclable items based on their material. Information about the disposed item is sourced from four independent data streams recording the system when the recyclable item impacting the identification chamber. The task is to predict the most suitable label for the disposed item in a statistical fashion, with the possible outputs being plastic, metal, glass. There are three types of sensor data streams: (1) JPG for raw RGB pixel image data, (2) WAV for lossless recorded audio data, (3) CSV for mass and inductive frequency time series data.

The way in which this was approached was by a means of supervised machine learning. This implies that the dataset from which the algorithm learns contains both features and labels for training. The dataset used was manually collected by dropping 34 unique recyclables items into the identification chamber, 20 times each Labels were added to the samples manually by hand. Each sample comprises of three files (JPG, WAV, CSV). This totalled at 680 samples which was then split into a training set (506 samples), testing set (54 samples), and validation set (120 samples). There is no overlap of samples in the training, testing, and validation sets. The training/testing set represents the knowledge base which the classifiers will be trained on, while the validation set is how the performance of the model will be gauged. As initial experiments, the dataset was generated in a laboratory environment. Raw data was used in the experiments without any data cleaning performed.

In implementation, an individual classifier for each of the three data streams is built. Each classifier takes in one of the three data files as input and output a prediction probability for each of the class labels. The purpose of the classifier is simply to reduce the dimensionality of the input vector to just 3-dimensions, whereby all of the extracted information is embedded into 3 probability variables.

The **image classifier** is responsible for classifying JPG images. It takes in a raw RGB pixel image, passes it through a retrained CNN, and outputs a prediction probability for the three class labels. This implementation is based on the output of Google's Inception Model "retrain.py" script [11] and Raval's TensorFlow implementation [12].

The **sound classifier** is responsible for processing the audio data stream. It takes in a WAV file, performs manual feature extraction, passes the resulting feature vector into a MLP NN, and outputs a prediction probability for the three class labels. MLPs are one of the simplest feedforward ANNs comprising only of deeply connected hidden layers. The features are extracted using the Librosa Python library while the MLP is built using TensorFlow. The exact implementation is based on Saeed's approach to Urban Sound Classification [13], while the deep NN codebase is derived from Kinsley's deep learning with NN tutorial series [14].

The **time series classifier** targets CSV data and outputs the predicted probabilities for each of the class labels. The implementation used is K-Nearest Neighbours (KNN) with Dynamic Time Warping (DTW). The series classifier is the only classifier in the entire neural network architecture which does not require a training or learning phase. Instead, when the classifier receives a CSV file, it will apply DTW to calculate the difference measure between the input time series and all other time series in the training set. The KNN algorithm is then used to find the 10 nearest neighbours (most similar time series in the training set), and approximate the predicted probability based on the labels of those neighbours. This implementation is derived from Regan's "KnnDtw" class available on GitHub [15]. It should also be noted that the series classifier actually processes the frequency and mass data streams by running two instantiations of the class.

3.2 Combined Classifier

With the individual classifiers up and running, it is now possible to sample a disposed object with the sensor suite, process the collected data files with each of the three classifiers, and obtain a total of 12 predicted class probabilities. This comprises of the three class probabilities from the image classifier, sound classifier, frequency series classifier, and mass series classifier. These output probabilities can be seen as the information which the classifiers have extracted via dimensionality reduction. They essentially represent how confident the classifier is in the sampled item belonging to each of the output classes.

The final goal for the machine learning backend is to integrate each of the classifier outputs into a final overall prediction. The way this is approached, is by designing and implementing the overall neural network architecture in a modular and concurrent fashion. With this modular definition in place, each of the classifiers previously discussed can now be seen as a "module" within the overall neural network. This is visualised in the fully integrated modular neural network architecture (Fig. 3).

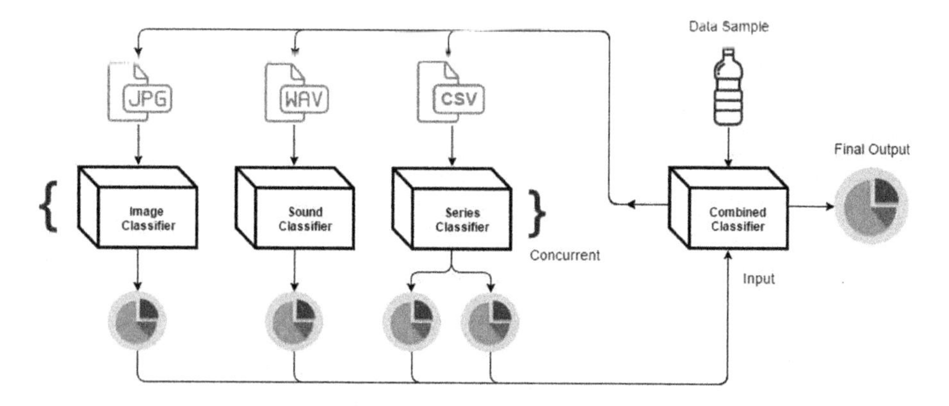

Fig. 3. Modular neural network architecture.

The combined classifier takes in as input a feature vector comprising of the 12 output probabilities from the individual classifiers, processes it through a feedforward MLP, and outputs the final prediction probabilities. The MLP is comprised of a single hidden layer containing 50 nodes, a number which was arbitrarily assigned. The underlying principle behind why this works is because each of the classifier outputs are treated as a weighted vote in probabilistically determining the final combined class prediction.

The modular neural network introduces a fourth and final classifier; the combined classifier. Essentially, this acts as the "overseer" module which manages and directs the individual classifiers or "worker" modules. In the modular neural network architecture, each classifier is abstracted into its own python script for concurrent execution. To facilitate the modularity and concurrency of the network, inter-process communication is established between the overseer and each of its workers via TCP/IP connection. When the combined classifier is first executed, it will expect three classifiers to connect via the socket. When a sample is to be predicted, the data begins at the combined classifier. As the overseer, the combined classifier will delegate the classification of each of the sample's data files by sending out requests to the worker modules. The complete integration diagram of the hardware and software components is shown in Fig. 4.

4 Classifiers Performance

To obtain the performance statistics, the validation set was run 10 times and averaged (accounting for the random 90% train 10% split). The confidence refers

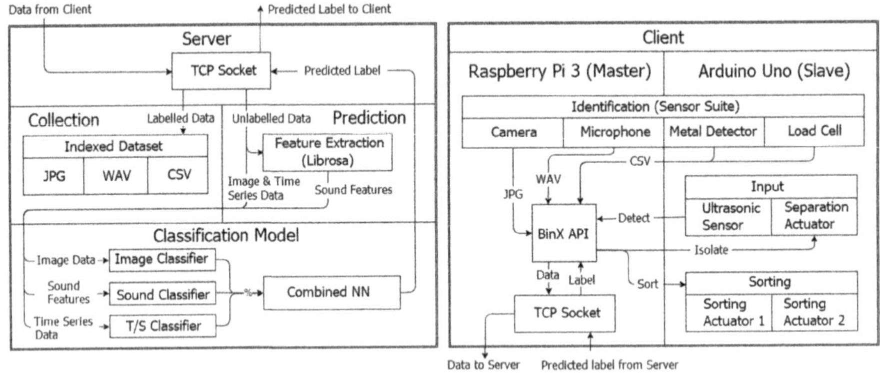

Fig. 4. Detailed integration diagram.

to the confidence the classifier is in the correct class (Table 1). The prediction accuracy is the number of samples where the maximum confidence matches the actual label (Table 2). The voting accuracy has a 99% confidence threshold filter applied, where if the confidence is above 99%, the vote will be the most confident label, but if the confidence is below 99%, the vote is "landfill" Table 3.

Table 1. Confidence in correct class.

Classifier	Plastic	Metal	Glass	Average
Image	98.19%	87.54%	86.12%	90.62%
Sound	65.58%	92.10%	100.00%	85.89%
Frequency	64.25%	97.00%	53.50%	71.58%
Mass	98.25%	100.00%	100.00%	99.08%
Combined	99.59%	99.91%	99.73%	99.75%

When examining the overall performance of the combined classifier implemented, we see that the modular neural network performs extremely well across all cases. Over 10 validation runs, the classifier achieved an average confidence in the correct label of 99.75%, capable of making a prediction every 0.641 s. This confidence does not appear to vary too much between classes, with the lowest average confidence being 99.59% for plastics and the highest being 99.91% for metals. Such performance suggests that the machine learning model is not only accurate, but reasonably acceptable in terms of classification speed as well, validating the suitability for commercial deployment.

From the experimental results, the image classifier incorrectly predicts 3% of metal as glass; and 10% of glass as plastic. The sound classifier incorrectly predicts 34% of plastic, of which 17.65% predicted as metal, 82.35% predicted

Table 2. Prediction accuracy.

Classifier	Plastic	Metal	Glass	Average
Image	100.00%	97.00%	90.00%	95.67%
Sound	66.00%	92.00%	100.00%	86.00%
Frequency	95.00%	97.50%	70.00%	87.50%
Mass	100.00%	100.00%	100.00%	100.00%
Combined	100.00%	100.00%	100.00%	100.00%

as glass; and incorrectly predicts 8% of metal; of which 80.65% predicted as plastic, 19.35% predicted as glass. The frequency classifier incorrectly predicts 5% of plastic as glass; 2.5% of metal as glass; and 30% of glass as plastic.

Table 3. Voting accuracy.

Threshold	Image	Sound	Frequency	Mass	Combined
Above 99% Confidence	38.33%	72.50%	32.50%	96.67%	95.00%
Below 99% Confidence	61.67%	27.50%	67.50%	3.33%	5.00%

With the 99% confidence threshold filter applied for landfill votes, the combined classifier was able to achieve a voting accuracy of 95%. Of the 5% where the vote was changed to landfill, 100% of the cases were marked as false negatives. In other words, the classifier's most confident class was always the correct class, attributing to the 100% prediction accuracy. Had the threshold been set to 98% instead of 99%, voting accuracy in the validation set would have been 100%. This serves to support the validity of the machine learning approach implemented, given the staggeringly high accuracy of the machine learning model.

In terms of processing speed, the average classification times for image, sound, time series, and the combined neural network are 0.382, 0.630, 0.603, and 0.641 s, respectively.

5 Conclusion

Recycling is often associated with the popular single stream recycling model. This centralised approach collects all recyclables in one bin for mass processing. However, due to high levels of cross contamination between streams, only three-quarters of waste processed this way is actually recycled. The alternative is source separation, which collects waste in separate bins at the point of disposal. This decentralised model is proven to have lower levels of cross contamination, resulting in increased recycling yield and higher quality recovered materials.

This paper proposes a unique mechatronic system in the form of an autonomous recycling system. Such a system intends to improve the efficiency of source separation recycling in tomorrow's smart cities, with the goal of reducing the amount of waste directed to landfills. This paper describes the development of the software architecture for a self-sorting smart bin, capable of identifying and sorting disposed waste based on material. Due to high variability innate to waste, a robust statistical classification model has been built by a means of supervised machine learning and fusing information from four independent sensors; a camera, microphone, metal detector and loadcell. The proposed bin also intends to leverage the global infrastructure of the internet of things with the classification model being built and trained using Google's TensorFlow machine intelligence framework. With excellent results achieved, this work has contributed to the underlying vision of a sustainable future, where humanity's waste management is automated and efficient.

References

1. Hoornweg, D., Bhada-Tata, P.: What A Waste - A Global Review of Solid Waste Management. The World Bank, Washington (2012)
2. Smith, K., O'Farrell, K., Brindley, F.: Department of Sustainability, Environment, Water, Population and Communities - Waste and Recycling in Australia 2011. Hyder Consulting Pty Ltd, North Sydney (2011)
3. Peacock, D.: ReLoop: What is Single Stream Recycling. http://greenblue.org/reloop-what-is-single-stream-recycling/
4. Khanna, P.: Urbanisation, Technology, and the Growth of Smart Cities. Singapore Management University. https://cmp.smu.edu.sg/ami/article/20161116/urbanisation-technology-and-growth-smart-cities
5. Peacock, D.: ReLoop: What is Source Separated Recycling. http://greenblue.org/reloop-what-is-source-separated-recycling/
6. GreenCan: MHacks. https://mhacks.devpost.com/submissions/17562-greencan
7. Bradley, H.: Automated Waste Sorting Receptacle (Full Demo). https://www.youtube.com/watch?v=v95Ifjz9sSg
8. Tarbell, K.A., Tcheng, D.K., Lewis, M.R., Newell, T.A.: Applying Machine Learning to the Sorting of Recyclable Containers. University of Illinois Urbana-Champaign, Urbana (1992)
9. Durrant-Whyte, H.: Multi Sensor Data Fusion. Australian Centre for Field Robotics, Sydney (2001)
10. Haik, Y., Shahin, T.M.: Engineering Design Process, 2nd edn. Cengage Learning, Stamford (2011)
11. Tensorflow - retrain.py. https://github.com/tensorflow/tensorflow/blob/master/tensorflow/examples/image_retraining/retrain.py
12. Raval, S.: Build a TensorFlow Image Classifier in 5 Min. https://www.youtube.com/watch?v=QfNvhPx5Px8
13. Saeed, A.: Urban Sound Classification, Part 1 - Feature extraction from sound and classification using Neural Networks. https://aqibsaeed.github.io/2016-09-03-urban-sound-classification-part-1/
14. Kinsley, H.: Deep Learning with TensorFlow - How the Network will run. https://pythonprogramming.net/tensorflow-neural-network-session-machine-learning-tutorial/?completed=/tensorflow-deep-neural-network-machine-learning-tutorial/
15. Regan, M.: K Nearest Neighbours & Dynamic Time Warping. https://github.com/markdregan/K-Nearest-Neighbors-with-Dynamic-Time-Warping

Clustering Algorithms in Mining Fans Operating Mode Identification Problem

Bartosz Jachnik[1(✉)], Paweł Stefaniak[1], Natalia Duda[1], and Paweł Śliwiński[2]

[1] KGHM Cuprum Research and Development Centre, gen. W. Sikorskiego 2-8, 53-659 Wrocław, Poland
bjachnik@cuprum.wroc.pl

[2] KGHM Polska Miedź S.A., M. Skłodowskiej-Curie 48, 59-301 Lubin, Poland

Abstract. Most of the machinery and equipment of the mine infrastructure is controlled by an industrial automation system. In practice, SCADA (Supervisory Control And Data Acquisition) systems very often acquire many operational parameters that have no further analytical use. The variability of recorded signals very often depends on the machine load, as well as organizational and technical aspects. Therefore, SCADA systems can be a practically free source of information used to determine KPI (e.g. performance, energy and diagnostic) for a single object as well as given mining process. For example, the ability to reliably identify different operational modes of the mining industrial fans based on data from SCADA gives a wide range of potential applications. Accurate information on this subject could be used - apart from basic monitoring and reporting needs (e.g. actual work vs. schedule comparison) – also in more complex problems, like power consumption predictions. Given the variety of industrial fans used in the mining industry and the different operational data collected, this is yet not a trivial task in a general case. The main aim of this article is to provide reliable algorithms solving fans operational mode identification issue, which will be possible to apply in a wide range of potential applications.

Keywords: Data clustering algorithms · Energy efficiency · Underground mining · Industrial fans

1 Introduction

Industrial fan monitoring systems can differ significantly from one case to another. Often they are a vast source of operational data, which sometimes is not used to its full extent. In other cases, a set of monitored parameters is very limited, providing just some basic information. Often, these systems are not uniform even within one company. This leads to many difficulties in data integration and effective management of mine ventilation, often due to a lack of access to reliable information and inefficient reporting systems.

In the article methods for automatic identification of operational modes of industrial fans have been presented. These methods are based on data from the SCADA (Supervisory Control And Data Acquisition) system from a multi-site mining enterprise, with

M. Hernes et al. (Eds.): ICCCI 2020, CCIS 1287, pp. 61–73, 2020.
https://doi.org/10.1007/978-3-030-63119-2_6

different data standards applied [2–5]. Due to differences in the data availability, three methods have been proposed for the operating modes classification issue. Presented solutions are based on two common clustering algorithms: k-means method and DBSCAN. These algorithms are widely used in various fields of science, business and industry [6, 16, 18–20]. In many cases, successful implementation of these methods necessitates significant changes in these clustering algorithms, and therefore numerous variations of these algorithms were developed [8–15, 17], which can improve algorithms performance in specific cases. In this article though, implementation of algorithms in their classic form was enough to obtain satisfactory results.

Implementation of algorithms presented in this article would improve the overall ventilation system monitoring, and especially would allow controlling actual operational regime of the fan stations in relation to the schedule. This solution can also be used in energy consumption prediction, fan stations work optimization and their efficiency assessment. An optimal efficiency rate can be established, which could be further used in technical diagnostics problem.

This paper is organized as follows: firstly the problem has been described and fans operational modes have been defined. In the next section, exemplary fan station functioning in the mining industry have been described. Following methodology, which is the main part of this article has been split into three subsections: source data characteristics and preprocessing, algorithms description and applications to real-life data/algorithms comparison. In the fifth section, application of the algorithms to the real-life data has been presented and algorithms results. In the final section, some of the most important conclusions have been underlined.

2 Problem Description

In the mining industry, a variety of different industrial fans are used. These differences are mostly noticeable in design solutions, work configuration, technical condition, control systems etc. In many cases, these devices do not provide a wide range of operational data, and the set of information provided by sensors very often differs from one case to another. In the case of bigger mining companies fans are operating as a part of bigger stations, which are often equipped in different sets of sensors. Therefore multiple methods based on different input data have to be implemented to solve this problem in the general case.

Underground mine operations are not always uniformly distributed in time, and therefore industry fan operation has to be adjusted accordingly to ensure adequate ventilation on site. Therefore work intensity of the fans, understood primarily as airflow speed in the ventilation duct, is limited when possible (i.e. weekends, holidays) to reduce power consumption. Additionally, there are safety issues influencing fans operating mode, which are often established by internal/external regulations. For example, in the case of Polish mines, airflow speed in the shaft has to be limited to 12 m/s during the mining cage (lift) movement. Very often, there is a need for unscheduled lift movement due to unplanned events (accidents, equipment breakdowns, inspections etc.), which requires manual adjustments of the fan operating modes. Such events often are not reported, and can have a significant impact on the ventilation systems performance KPIs.

Reliable methods to control power consumption of the ventilation systems have special significance, due to their share importance in total mine energy usage, which in case of some coal mines can reach up to 26% [21]. Experience from Polish copper mines confirms this conclusion. In the case of KGHM Polish Copper, three processes have been identified as most energy intensive: air conditioning, ventilation and conveyors based transport. It is worth mentioning, that ventilation in this case was more energy intensive, than air-conditioning, water drainage, belt-conveyors based and vertical transport systems combined. Additionally, data suggests dynamic increase of ventilation power consumption, along with the time of performing mining activities, caused by the increase of the area necessary for ventilation and the distance of mining faces from ventilation shafts. Fans power consumption is expected to vary in time quite significantly, due to changes in configuration and cross-sections of excavations which influence airflow resistance, different temperature and humidity of air during the seasons, technical condition of the fan and its potential failures, external loses due to duct leaks, etc.

In practical applications four main operating modes of industrial fans can be distinguished:

- Off mode (OM),
- Normal operation mode (NM),
- Limited operation mode (LM),
- Transition mode (TM).

Off mode can be characterized as the one with airflow speed measured in front of the fan of 0 m/s, or small positive number, to take into account some natural air movements or sensor measurement errors. The normal operation mode can be defined as fans airflow occurring during a normal working day at the mine. It can differ from one case to another, and is often defined by the equipment manufacturer in the fan technical description. The limited mode has an airflow speed level much lower than the normal mode, which is typical for the weekend days in the mines or the cage moving conditions. When the Polish case is considered, a limited mode can be defined as a mode with airflow speed measured in front of the fan being less than 12 m/s, but greater than 0 m/s. To transition mode were assigned all the observations accompanying the change of the mode from normal to limited, or in the opposite direction.

3 Description of the Industrial Fan Station

In the case of most KGHM underground mines, usually there are a couple of main vent stations functioning per mine, consisting of 3, 4 or 6 industrial fans each. Very often fans functioning as a part of the same station are the same type, but since in methods presented in this article each fan has been considered separately, that assumption is not necessary for presented algorithms to work correctly.

Fans can be defined by a wide set of technical and operational data, but to the main one describing their productivity and power consumption belongs:

- Nominal effectivity (expressed in m^3/min of airflow),
- Speed (number of turns per second),
- Engine power (kW),
- Voltage (kV).

These parameters can be used to determine the target work efficiency of the fans used, especially in normal working mode. The most basic KPI used for fan efficiency monitoring is defined as a ratio of the total airflow to power consumption.

Industrial vent stations considered in this case are used to drain hot air from the mine, which is then replaced due to pressure difference occurring between underground and the surface. These shafts are also often used for people, machine and other goods transport.

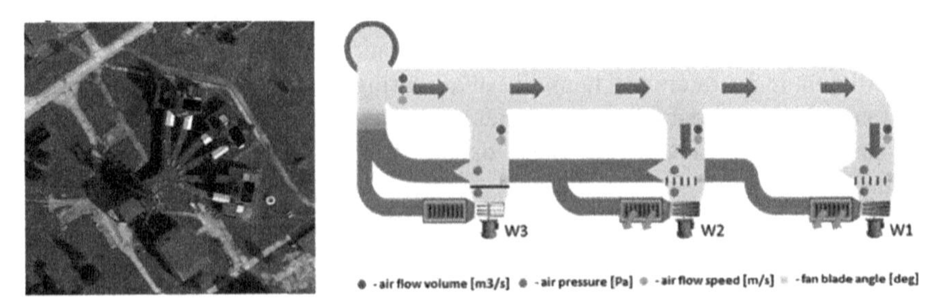

Fig. 1. External view and a scheme of one of the KGHM's main vent stations

Drained air from the shaft is flowing through the main duct of the fan station where it is split into separate ducts of each fan (Fig. 1). In some cases, there is a set of sensors located just outside of the ventilation shaft, allowing easy insight into the operation of the whole station. Otherwise, the data has to be aggregated using sensors placed next to each fan. In order to control the airflow through each fan one of the two methods can be used, depending on the fan model:

- fan blades angle manipulation,
- partial closure of the duct in front of the fan.

In this second case, a set of two air sensors are often used, placed on each side of the locking system. This solution allows the calculation of the pressure difference between these two points (Fig. 2).

The scheme presented in Fig. 1 (right panel) shows location of some basic sensors, which readings will be used in algorithms described in the next section. Some of the stations do not have a full set of sensors presented in the figure above, others have sensors system much more extensive, including measurements of various atmospheric parameters like outside temperature and atmospheric pressure, or detailed engine data (oil pressure and temperature, vibration data, bearing temperature etc.). The only data used in described algorithms which is acquired outside of the airflow ducts is the power consumption data, which in the case considered was stored in other IT system dedicated to energy billing in the mine.

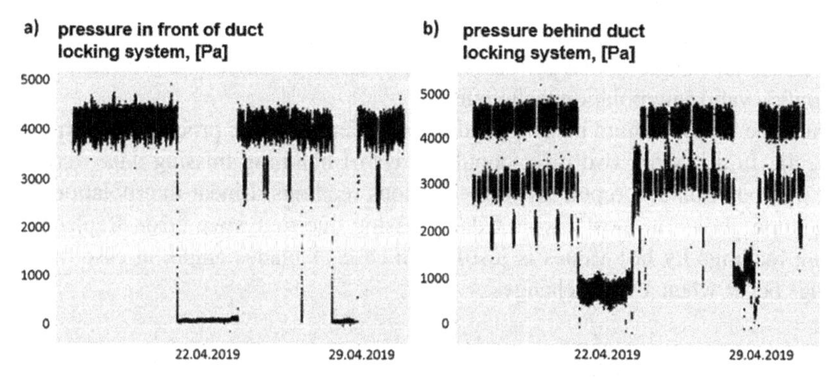

Fig. 2. Raw data example: pressure in front of (a) and behind (b) the fan station locking system.

4 Methodology

4.1 Source Data Characteristics and Preprocessing

The data from monitoring systems of vent stations is stored in the data warehouse of the company. In case of some stations the list of available variables is very long, reaching several dozen entries. In other cases, this list contains just a couple of variables. In this article three work identification algorithms are presented, using – apart from date and time – four parameters in total (see Table 1).

Table 1. List of the parameters monitored in vent stations used in the proposed methods

No	Parameter	Usage Method 1	Method 2	Method 3	Description	Units	Sampling
1	DATE	✔	✔	✔	Date	yyyy-mm-dd	-
2	TIME	✔	✔	✔	Time	hh-mm-ss	1 s
3	PRESSURE	✔	✖	✖	Air pressure data	Pa	1 s
4	AIRFLOW	✔	✔	✖	Airflow data	m3/s	1 s
5	BLADESANG	✖	✔	✖	Fan blades angle	deg	changeable (when value change by a certain Δ)
6	POWERUS	✖	✖	✔	Power consumption	kWh	15 min

Unfortunately, the only parameter available for all the vent stations that were considered in this case, was power consumption, used by the third method. The problem arose due to the low sampling frequency of this variable is described later in this work. Blades angle data is updated with a change of its value.

As a part of data preprocessing two independent blocks can be distinguished: data cleansing and interpolation. Methods of data cleaning are used to remove potential errors and detect missing data. While singular missing values are usually easy to identify error, detecting and replacing missing data in the general case cannot be based on the one considered variable alone. For example, there were cases, in which airflow data contained Null values for longer periods of time. This situation could indicate two possible scenarios: either the ventilator was shut down, or due to some reason (e.g. sensor damage, connection error) the readings were not correctly saved in the database. To identify which one of the situation occurred firstly the values of other parameters should be checked. If there were no significant changes in the pressure levels, power usage and blades angle in that period of time, it can be assumed that ventilator was on, and apply a valid interpolation technique.

Once the reason behind missing reading was identified, the process of interpolation can begin. In this work two most popular ways of handling missing data were used: linear interpolation and replacement by previous readings. Linear interpolation would be a natural choice in most cases of data missing due to system error. Replacing the missing readings by last values is justified in case of blades angle, in case of which readings occur when a value changes.

4.2 Algorithms Description

Due to a variety of measured values of the fan sensors, approach to the operating mode identification problem will vary from one application to another. The main goal of this research is to provide reliable algorithms solving this problem, which can be applied in the widest possible range of industrial fans currently available. In the paper three algorithms have been described, which are using various sets of initial parameters:

1. Airflow measurements (expressed as m^3 per second) and stagnation (total) pressure (Pa).
2. Airflow measurements and fan blades angle (deg).
3. Energy consumption (kWh).

The output of these algorithms should be very similar (comparison on real data described in next chapter), various input data ensure a wide range of possible applications. Described algorithms are using two popular data clustering methods: density-based spatial clustering of applications with noise (DBSCAN) and k-means. These algorithms are implemented in most data-analysis packages, and have been chosen due to their simplicity and performance. DBSCAN was chosen due to the fact that in case method 1 and 3, in some cases a non-linearly separable clusters have to be dealt with. After proper data preparation, in mode identification based on airflow and blade angle *k-means* method turned out to be sufficient. In this particular case, all the developed algorithms were programmed in R, but due to the specificity of this language, in this article they were presented in the form of pseudocode.

Since the levels of statistics can vary depending on the type of the day (working/weekend/holiday) it is suggested to apply these algorithms to specific working days/shifts.

```
Method No. 1 (data[,(airflow, pressure)], AF_NM, AF_LM)
   result[data.airflow = 0, ] <- OM
   data.remove[data.airflow = 0, ]
   eps <- 2.5% * (max(data[, airflow]) - min(data[, airflow]))
   data.add[cluster] <- DBSCAN(data, eps)[cluster, ]
   calculate average values of airflow (AF avg) within clusters
   if data.count_unique(cluster) < 3 then
      for each cluster
         if cluster.AF_avg <=AF_LM
            set operational mode of cluster as LM
         elseif cluster.AF_avg >= AF_NM
            set operational mode of cluster as NM
         else
            set operational mode of cluster as TM
         endif
      next cluster
   else
      Choose three most numerous clusters
      Set operational mode:
         -   class with lowest AF_avg (AF_avg_min) as LM
         -   class with highest AF_avg (AF_avg_max) as NM
         -   other as TM
      for each unasigned cluster
         if cluster.AF_avg <= AF_avg_min
            set operational mode of cluster as LM
         elseif cluster.AF_avg >= AF_avg_max
            set operational mode of cluster as NM
         else
            set operational mode of cluster as TM
         endif
      next cluster
   endif
end
```

Arguments of the described algorithms – apart from the airflow and pressure measurements – also average values of airflow for the normal and limited operation mode. These values are used in cases, where the DBSCAN algorithm could distinguish only one or two clusters, and therefore it is difficult to establish - without given reference point – to which operational modes these clusters correspond to. These values can be easily determined using the described above Method No. 1 algorithm on big enough data set, containing observations from all four considered operational modes. In fact, method I as a whole, can be seen as a simple modification of the DBSCAN algorithm so that it will return a predefined number of clusters, for which purpose the comparison of mean values in each cluster is considered.

The other issue that is worth mentioning is the determination optimal value of an epsilon for DBSCAN algorithm. In the described methods very simplistic approach to this problem has been presented, with epsilon value set as 2.5% of total airflow data spread. This value was based on historical data analysis and may vary, depending on the application. There are of course more sophisticated, automatic methods of determining the optimal value of epsilon, which could potentially improve the performance of the methods. Some of these methods were described in [1, 7] (Fig. 3).

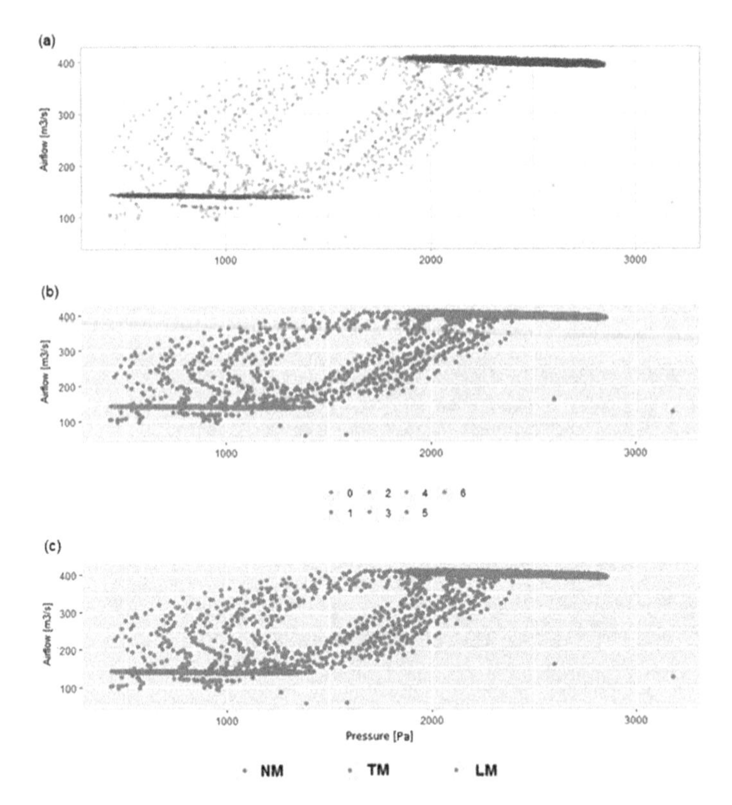

Fig. 3. Method No. 1 applied to exemplary data. (a) raw data, (b) the result of the DBSCAN classification algorithm, (c) result of method No. 1.

```
Method No. 2 (data[,(airflow, angle)], period, AF_NM, AF_LM)
   result[airflow = 0, ] <- OM
      data.remove[airflow = 0, ]
      aggregate data by period using the average function
      perform k-means clustering with 3 centers on data[,angle]
      calculate average value of airflow (AF_avg) within clusters
      for each cluster
         if cluster.AF_avg <= AF_LM
            set operational mode of cluster as LM
         elseif cluster.AF_avg >= AF_NM
            set operational mode of cluster as NM
         else
            set operational mode of cluster as TM
         end if
      next cluster
end
```

In this method, based on the blade angles of the industrial fan, a simpler *k-means* clustering algorithm has been used. Similarly to the method No. 1, reference points to the average values of airflow for normal and limited operating modes have to be provided, in order to correctly identify clusters. The period parameter is a measure of data aggregation, which in this case was set to 1 min, but it also may vary depending on the installed sensors and sampling rate (Fig. 4).

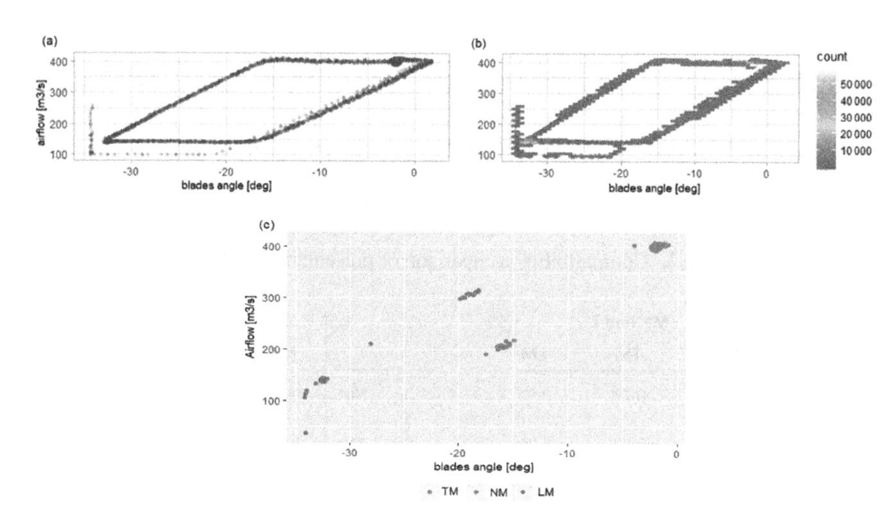

Fig. 4. Result of method No. 2 application for real data. (a) raw data, (b) density plot, (c) result of k-means algorithm on aggregated data.

```
Method No. 3 (data[,power consumption], PC_NM, PC_LM)
  result[power consumption = 0, ] <- OM
  data.remove[power consumption = 0, ]
  eps <- 2.5% * (max(data[, power consumption]) - min(data[, power consumption]))
  perform DBSCAN algorithm on data[,power consumption] with eps
  calculate average value of power consumption (PC_avg) within clusters
  threshold = 5% * data.count()
  identify clusters whose number is larger than threshold (clusters_sig)
  for each cluster in clusters_sig
    if cluster.PC_avg <= PC_LM
      set operational mode of cluster as LM
    elseif cluster.PC_avg >= PC_NM
      set operational mode of cluster as NM
    end if
  next cluster
  set operation mode of all remaining clusters as TM
end
```

The third presented method is – in similarity to method No. 1 - based on the DBSCAN algorithm. The necessary epsilon value for the algorithm has been calculated in a similar fashion to Method No. 1. Since this method is based only on one variable, it may be least precise out of the three presented methods. Yet in some cases, the power consumption data are one of the few variables available, and therefore this algorithm may become useful.

5 Applications to Real-Life Data and Algorithms Comparison

A simple variant of confusion matrices for comparing results of the three methods has been presented in Table 2.

Table 2. Compatibility comparison of presented algorithms

Method 1 \ Method 2	NM	TM	LM
NM	49,8%	0,2%	0,0%
TM	0,0%	0,5%	0,2%
LM	0,0%	0,3%	49,0%

Method 1 \ Method 3	NM	TM	LM
NM	41,9%	5,4%	2,8%
TM	0,3%	0,0%	0,4%
LM	3,2%	2,7%	43,4%

In those matrices, values on the diagonal of the matrix represent the percentage of observations to which both algorithms assigned the same operational mode. Thus, methods No. 1 and No. 2 gave the same results for over 99% of the observations.

Therefore, for the purpose of preparing periodic reports, these two methods can be considered pretty much identical. The second matrix compares the results obtained by using first and third methods. Here the compatibility of the methods is over 85%. It can be seen, that the third algorithm has particularly significant problems recognizing the transitional mode. This is due to the power consumption data aggregation to 15 min periods in considered case, while the transitional mode usually lasted only a few minutes. Therefore, 15-minutes periods even if contained the transitional mode observations, were mostly composed out of the observations from normal or limited modes. Please notice, that from those matrices the off mode has been excluded. Since its identification is very similar between the described algorithms (identical in methods 1 and 2) its inclusion would increase the compatibility level and would make the algorithm comparison less clear.

Given the work mode of each ventilator it is possible to use those to define the operation mode of the whole station as an intersection using *if-then-else* statement. For example, in case of ventilation station consisting of 4 ventilators, a normal mode can be defined as at least 2 vents working in a normal mode. Else if at least one ventilator is working in a limited mode, the whole station is working in limited mode etc. (Fig. 5).

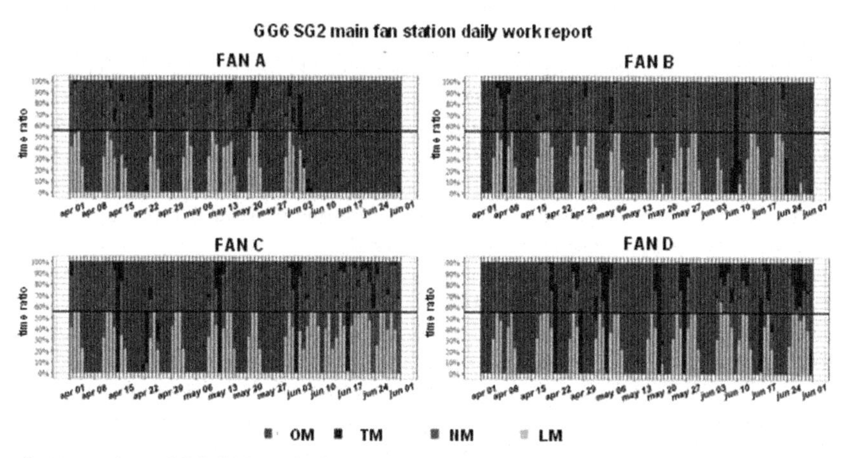

Fig. 5. Exemplary GG6 SG2 main fan station work report. The black vertical line represents limited mode participation resulting from the schedule.

6 Conclusions

In the article three different methods have been presented for the industrial fan stations operation mode recognition problem. Methods are based on different input data, enabling the application to a wide possible range of used sensors. Variables used in developed algorithms are total pressure, airflow, fan blades angle and power consumption. Two of the presented methods ensure a high level of convergence (> 99%). The third method, based on the power consumption level will provide slightly lower accuracy, due to bigger data aggregates. Basing on the identified working mode of each fan, it is possible to determine the working mode of the whole industrial fan station. Developed methods can be implemented as an automatically refreshed daily procedure over the historical data of the industrial fan stations functioning in the mines. Used classification algorithms (*k-means*, *DBSCAN*) are easily accessible and implemented in most programming languages

References

1. Rahmah, N., Sitanggang, I.S.: Determination of optimal epsilon (Eps) value on DBSCAN algorithm to clustering data on Peatland hotspots in sumatra. In: IOP Conference Series: Earth and Environmental Science, vol. 31, no. 1. IOP Publishing (2016)
2. Stefaniak, P., Kruczek, P., Śliwiński, P., Gomolla, N., Wyłomańska, A., Zimroz, R.: Bulk material volume evaluation and tracking in belt conveyor network based on data from SCADA. In: Widzyk-Capehart, E., Hekmat, A., Singhal, R. (eds.) Proceedings of the 27th International Symposium on Mine Planning and Equipment Selection-MPES 2018, pp. 335–344. Springer, Cham (2019)
3. Stefaniak, P., Wodecki, J., Zimroz, R.: Maintenance management of mining belt conveyor system based on data fusion and advanced analytics. In: Timofiejczuk, A., Łazarz, B., Chaari, F., Burdzik, R. (eds.) International Congress on Technical Diagnostic, pp. 465–476. Springer, Cham (2016)
4. Sawicki, M., Zimroz, R., Wyłomańsk, A., Obuchowski, J., Stefaniak, P., Żak, G.: An automatic procedure for multidimensional temperature signal analysis of a SCADA system with application to belt conveyor components. Proc. Earth Planet. Sci. **15**, 781–790 (2015)
5. Wodecki, J., Stefaniak, P., Polak, M., Zimroz, R.: Unsupervised anomaly detection for conveyor temperature SCADA data. In: Timofiejczuk, A., Chaari, F., Zimroz, R., Bartelmus, W., Haddar, M. (eds.) Advances in Condition Monitoring of Machinery in Non-Stationary Operations, vol. 9, pp. 361–369. Springer, Cham (2018)
6. Benabdellah, A.C., Benghabrit, A., Bouhaddou, I.: A survey of clustering algorithms for an industrial context. Proc. Comput. Sci. **148**, 291–302 (2019)
7. Soni, N.: Aged (automatic generation of eps for dbscan). Int. J. Comput. Sci. Inf. Secur. **14**(5), 536 (2016)
8. Yu, X., Zhou, D., Zhou, Y.: A new clustering algorithm based on distance and density. In: Proceedings of ICSSSM'05. 2005 International Conference on Services Systems and Services Management, 2005, vol. 2. IEEE (2005)
9. Birant, D., Kut, A.: ST-DBSCAN: An algorithm for clustering spatial–temporal data. Data Knowl. Eng. **60**(1), 208–221 (2007)
10. Viswanath, P., Pinkesh, R.: l-DBSCAN: a fast hybrid density based clustering method. In: 18th International Conference on Pattern Recognition (ICPR'06), vol. 1. IEEE (2006)
11. Xiaoyun, C., et al.: GMDBSCAN: multi-density DBSCAN cluster based on grid. In: 2008 IEEE International Conference on e-Business Engineering. IEEE (2008)
12. Ram, A., et al.: An enhanced density based spatial clustering of applications with noise. In: 2009 IEEE International Advance Computing Conference. IEEE (2009)
13. Borah, B., Bhattacharyya, D.K.: An improved sampling-based DBSCAN for large spatial databases. In: Proceedings of the International Conference on Intelligent Sensing and Information Processing, 2004. IEEE (2004)
14. Uncu, O., et al.: GRIDBSCAN: grid density-based spatial clustering of applications with noise. In: 2006 IEEE International Conference on Systems, Man and Cybernetics, vol. 4. IEEE (2006)

15. Liu, P., Zhou, D., Wu, N.: VDBSCAN: varied density based spatial clustering of applications with noise. In: 2007 International Conference on Service Systems and Service Management. IEEE (2007)
16. Khan, K., et al.: DBSCAN: past, present and future. In: The Fifth International Conference on the Applications of Digital Information and Web Technologies (ICADIWT 2014). IEEE (2014)
17. Wang, T., et al.: NS-DBSCAN: a density-based clustering algorithm in network space. ISPRS Int. J. Geo-Inf. **8**(5), 218 (2019)
18. Zhang, M.: Use density-based spatial clustering of applications with noise (DBSCAN) algorithm to identify galaxy cluster members. In: IOP Conference Series: Earth and Environmental Science, vol. 252, no. 4. IOP Publishing (2019)
19. Jing, W., Zhao, C., Jiang, C.: An improvement method of DBSCAN algorithm on cloud computing. Proc. Comput. Sci. **147**, 596–604 (2019)
20. Ester, M., et al.: A density-based algorithm for discovering clusters in large spatial databases with noise. In: KDD, vol. 96, no. 34 (1996)
21. Efremenko, V., Belyaevsky, R., Skrebneva, E.: The increase of power efficiency of underground coal mining by the forecasting of electric power consumption. In: E3S Web of Conferences. EDP Sciences, vol. 21 (2017)

K-Means Clustering for Features Arrangement in Metagenomic Data Visualization

Hai Thanh Nguyen[1(✉)], Toan Bao Tran[2,3], Huong Hoang Luong[4], Trung Phuoc Le[4], Nghi C. Tran[5], and Quoc-Dinh Truong[1]

[1] College of Information and Communication Technology, Can Tho University, Can Tho 900000, Vietnam
{nthai,tqdinh}@cit.ctu.edu.vn

[2] Center of Software Engineering, Duy Tan University, Da Nang 550000, Vietnam
tranbaotoan@dtu.edu.vn

[3] Institute of Research and Development, Duy Tan University, Da Nang 550000, Vietnam
huonghoangluong@gmail.com, trunglp2@fe.edu.vn

[4] Department of Information Technology, FPT University, Can Tho 900000, Vietnam

[5] National Central University, Taoyuan, Taiwan
tcnghivn@gmail.com

Abstract. Personalized medicine is one of the most concern of the scientists to propose successful treatments for diseases. This approach considers patients' genetic make-up and attention to their preferences, beliefs, attitudes, knowledge and social context. Deep learning techniques hold important roles and obtain achievements in bioinformatics tasks. Metagenomic data analysis is very important to develop and evaluate methods and tools applying to Personalized medicine. Metagenomic data is usually characterized by high-dimensional spaces where humans meet difficulties to interpret data. Visualizing metagenomic data is crucial to provide insights in data which can help researchers to explore patterns in data. Moreover, these visualizations can be fetched into deep learning such as Convolutional Neural Networks to do prediction tasks. In this study, we propose a visualization method for metagenomic data where features are arranged in the visualization based on K-means clustering algorithms. We show by experiments on metagenomic datasets of three diseases (Colorectal Cancer, Obesity and Type 2 Diabetes) that the proposed approach not only provides a robust method for visualization where we can observe clusters in the images but also enables us to improve the performance in disease prediction with deep learning algorithms.

Keywords: Personalized medicine · Bioinformatics · Visualization · Convolutional neural networks · K-means · Clustering · Disease prediction

M. Hernes et al. (Eds.): ICCCI 2020, CCIS 1287, pp. 74–86, 2020.
https://doi.org/10.1007/978-3-030-63119-2_7

1 Introduction

Personalized medicine or precision medicine is a medical procedure that separates patients who have the same diagnosis into different groups with interventions, treatments, or other medical decisions, specifically designed for the individual patient, based on their medical symptoms [1]. Understanding and treating disease with personalized medicine may be examined as an addition to traditional approaches. With more precise tools, the doctors can select a therapy or treatment protocol that relies on the patient's medical record that may not only reduce the disadvantageous side effects and make sure a better result but also help to prevent the other diseases and reduce trial-and-error prescriptions. In recent years, personalized medicine is much more accurate and more personalized due to the tremendous progress in genetics. The success of the Human Genome Project in [2] laid an important role in understanding the mechanisms of genetic disease at the molecular level across the genome.

Metagenomics (also referred to as environmental and community genomics) is the genomics analysis of microorganisms based on the analysis of DNA sequences. The uncultured microorganisms represent the huge majority of organisms in most habitats on this planet proving by the analysis of 16rRNA sequences, it is the beginning for the development of metagenomics and led to the discovery of vast new lineages of microbial life [3]. Additionally, metagenomic sequencing is being used to characterize the microbial communities, several diseases have also been shown to be associated with a disturbed microbiome. The human body has 1.3 times more microbes than body cells which have been studied in many different environments by the Human Microbiome Project (HMP) [4]. These microbial communities have been assayed using high-throughput DNA sequences. The HMP indicated that dissimilar to individual microorganisms, many metabolic processes have been present among all body habitats with various frequencies and revealed the advantages of metagenomics in diagnostics and evidence-based medicine. As a consequence, metagenomics is a powerful tool to handle many of the pressing issues in the field of Personalized Medicine or relatives [5].

Recently, diagnostic metagenomics has been used widely to detect an unknown pathogen on clinical samples or outbreaks of disease [6]. In virus discovery, revealing the existence of numerous "orphan viruses" that have not been associated with any disease and thereby establishing the existence of a normal human virus microbiome or "virome" [7]. Furthermore, The first genome sequence of *SARS-CoV-2* was conducted with metagenomic RNA sequencing [8].

In this study, we present a visualization method using K-means clustering for arranging features in images. Then, these images are fetched into a deep learning algorithm such as Convolutional Neural Networks for classification tasks. Our contributions in this study include:

- We present a features clustering approach with Classic k-means algorithm and compare other features arrangement methods. Mini-Batch K-Means algo-

rithm is also carried to compare. The numbers of clusters in K-means are also explored to analyze the performance of the visualization.
- The efficient of the proposed visualization methods is evaluated on three types of diseases including Colorectal Cancer, Obesity and Type 2 diabetes. The performance on the datasets with the considered diseases obtains better results in prediction tasks comparing to the-state-of-the-art such as MetAML and Fill-up with phylogenetic ordering.
- The proposed method not only provides better visualizations which exhibit clearly clusters in data but also produces promising results in prediction tasks with a shallow Convolutional Neural Network.

2 Related Work

The metagenomic analysis is an interesting subject for the scientific community, it characterizes bacterial community composition from a particular environment, and avoiding the use of cell cultures [9]. To resolve the problems in metagenomics studies, in recent years, the contributions of machine learning, deep learning techniques have attempted to propose and develop increasingly. Several potential machine learning model e.g., Linear Regression, Random Forests (RFs), Support Vector Machines (SVMs), K-nearest neighbors, Gradient Boosting, k-Nearest Neighbors or deep learning architectures with Convolutional Neural Networks (CNN) demonstrated the potential of developing the microbial biomarker for the disease or the host phenotype prediction [9,10,14]. Generally, it has obtained numerous potential results in classification, clustering, dimensionality reduction, or data augmentation tasks on the biological databases.

Several machine learning fundamentals were proposed in [11] and the authors presented the solutions for Operational Taxonomic Unit (OTU) clustering, binning, taxonomic profiling and assignment, comparative metagenomics, and gene prediction. The authors in [11] analyzed the DNA sequence data by grouping it into OTUs for exploring the genetic diversity of a microbial community. Sequences can be clustered according to a computed distance between two sequences and a certain threshold which are considered to be sufficiently close. OTU clustering has the double advantage of being a light-weight approach in terms of input data (amplicon sequencing) and consequent computation, as well as of producing high-quality groups precisely because amplicon sequences are by definition taxa-specific and different between species [11]. Despite those advantages, the OTU-based methods also have certain restrictions. The analysis and biologically meaningful can be problematic. Furthermore, due to sequencing errors, the wrong choice of the threshold can result in artificial inflation of OTUs [12]. Moreover, several studies indicated the different clustering methods may have non-equivalent clusters on the same dataset.

3 Features Clustering in Synthetic Metagenomic Images

In this section, we introduce K-means clustering to arrange features in metagenomic visualization by fill-up. A description on datasets for the experiments are also introduced.

3.1 Metagenomic Datasets Description for the Analysis

Three species abundance datasets are taken into account in the experiments to evaluate our approach. The set of datasets consists of samples of three diseases such as Obesity (OBE), Colorectal Cancer (COL), and Type 2 Diabetes samples from Western women (WT2) [10,20]. The features in the considered datasets are bacterial species, the samples are either patients or healthy individuals. Each sample includes species abundances that indicate relative proportion of bacterial species existing in human gut. Abundance of each feature is represented as a real number and the total abundance of all species in a sample sums to 1. The classification task in this work is to predict whether a sample gets the disease or not (binary classification).

Table 1 shows details of three considered datasets with the number of samples, features and other information. In order to arrange all features into a 2D matrix for visualizing onto an image, we calculate the ceiling of Square Root of the number of features. For example, we have 503 features in COL dataset, so the square root of 503 is 22.43. The ceiling of 22.43 is 23; hence, we need a matrix of 23×23 to contain 503 features. If we would like to illustrate a feature by a pixel, and ensure that all features are visible in the image, we need a minimum size of 23×23 pixels for samples in COL dataset (as shown in Table 1).

Table 1. Information on three considered datasets.

Diseases	**Colorectal Cancer**	**Obesity**	**Type 2 diabetes**
Datasets name	**COL**	**OBE**	**WT2**
#Features	503	465	381
#Samples	121	253	96
Ratio of patients	0.40	0.65	0.55
Ratio of Controls (healthy)	0.60	0.35	0.45
Minimum size of images	23×23	22×22	20×20

3.2 K-Means Clustering

Data visualization is a robust method to interpret data. In our experiments, we consider clustering methods to arrange features onto images. Clustering is unsupervised learning method which is the process of grouping similar entities together. We expect that placing features which have similar characteristics close together can help to observe patterns in data easier and also improve the performance of classification tasks on generated visualizations.

K-means is a common unsupervised learning algorithm and used widely in data science. In principle, the goal of k-means is to divide the input data into the k clusters in such a way the data in the same cluster have the same properties. K-means clustering minimizes within-cluster variances (squared Euclidean distances). The Euclidean distances are computed by the following formula:

$$\partial_{ji} = \sqrt{\sum_{s=1}^{m}(x_{is} - x_{js})^2}$$

With:

- $a_i = (x_{i1}, x_{i2}, ...x_{im})$, i = 1..n - the ith object to be clustered.
- $c_j = (x_{j1}, x_{j2}, ...x_{jm})$, j = 1..k - central element group j.

We also consider the *Mini-Batch K-means* algorithm which has the same goal as K-means but its main idea is to use a small batch of samples. Reducing the computational cost and the stored memory are the advantages of Mini-Batch K-means since it does not use all samples on the dataset. The clusters are updated at every mini-batch and repeated until convergence. In [17], the authors indicated that the number of clusters as know as k has an important impact on the difference between the partition obtained by k-means and mini-batch k-means. It is not possible for applying the mini-batch k-means algorithm to the large dataset with a big k, it will result in equivalent partitions to the ones from k-means. Based on that, with k-means and Mini-Batch K-means, we may acquire some important information on our datasets. Moreover, "Cluster-Then-Predict" approach directs towards an improved overall prediction accuracy with an increased collected sample data size leading to better clustering and improved classification of each cluster [18].

3.3 Fill-Up with K-Means Clustering Algorithms

The synthetic metagenomic images are generated by applying the "Fill-up" [16] method. The purpose of this method is to arrange features into a squared matrix of a given size which is minimum size to fit all features. More specifically, this matrix contains arranged abundance or presence values in a right-to-left order by row top-to-bottom. The order to arrange species can be either random or following a specific order, for instance phylogenetic-sorting [16]. But the distribution of features is relatively dissimilar, by grouping the equivalent features with clustering algorithms into the discriminative clusters with almost non-overlapping boundaries, we may have some promising results. Suppose we set k clusters, this approach is performed as shown in Algorithm 1. When the features clustering process is done, we receive L array (For example, $L = (5, 2, 4, 2, 1, ...)$ that means the first feature is assigned to cluster 5, the second feature belongs to cluster 2 while the third feature is in cluster 4 and so on) to rearrange order of features

Algorithm 1. Algorithm for features clustering based on K-Means algorithms

Input:

- k: number of clusters
- D: a data matrix where each row is a sample and each column represents a feature
- *Algo*: The algorithm for clustering (Classic K-means or Mini Batch K-means)
- *batch_size*: if we use Mini Batch K-means

Output:

- L: an array containing a list of cluster labels of all features

Begin

Step 1: Transpose D: $D_t = t(D)$. Because we want to group features into clusters, we transpose D so that each feature at this time is considered as a "data point" for clustering.

Step 2: Run a clustering algorithm on the transposed data (D_t) to indicate which clusters the features belong to.
. **if** *Algo*=='Mini batch K-means' **then**
. Using the K-means algorithm with k clusters.
. **else**
. Using the Mini batch K-means algorithm with k clusters and a batch size of *batch_size*.
. **end if**
. The labels of clusters contained features are saved to L. L includes information on clusters which each feature belongs to. For example, the 1st feature is assigned to the 3rd cluster, the 2nd feature is labeled to cluster of 1st, and so on.
Return L

End

based on the order of labeled clusters so that features in the same group (cluster) are arranged side by side. In order to visualize features onto images, we use 10 colors (in gray scale or using rainbow colormap) with Species Bin (SPB) in [16].

Figure 1 exhibits the comparison between clustered features with Mini batch K-means and non-clustered features (phylogenetic ordering). The first image on the top-left is the global image which is generated by the average values of each features of all samples visualized by Fill-up and clustering method. Similarly, the top-right image represents the global map of Fill-up with phylogenetic ordering. The bottom-left image is from a specific patient from WT2 dataset with clustered features generated by fill-up using Mini batch K-means for features arrangement. The last image exhibits the same patients but with phylogenetic ordering fill-up. The global map is an images which visualizes features in a dataset with each pixel shows an average value of a feature in training dataset. In our study, we only use samples of training set to cluster features and build coordinates for all features. These coordinates conducted from training set are carried out to generate all images for samples of both training set and test sets. Comparison between two

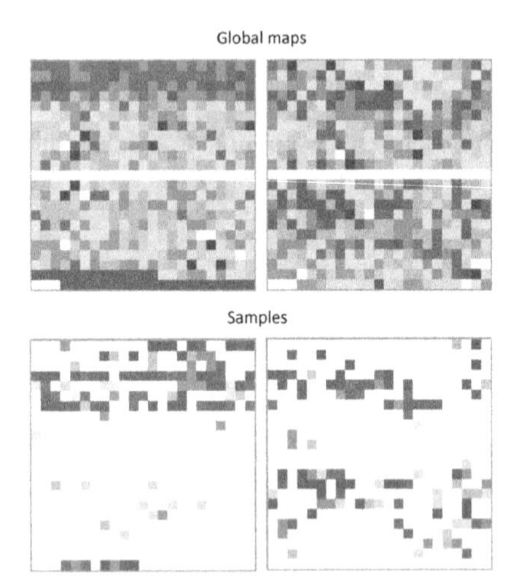

Fig. 1. Visualization comparison between features arrangement based on Mini batch K-means clustering and features ordered phylogenetically for COL dataset. Left: Top-bottom: global and sample images with clustered features. Right: Top-bottom: global and sample images with phylogenetic ordering.

Fig. 2. Visualization Comparison between gray scale and rainbow colormap, and comparison among the considered numbers of clusters for K-means clustering algorithms (k = 3, 5, 10 and 20) for Colorectal Cancer disease dataset.

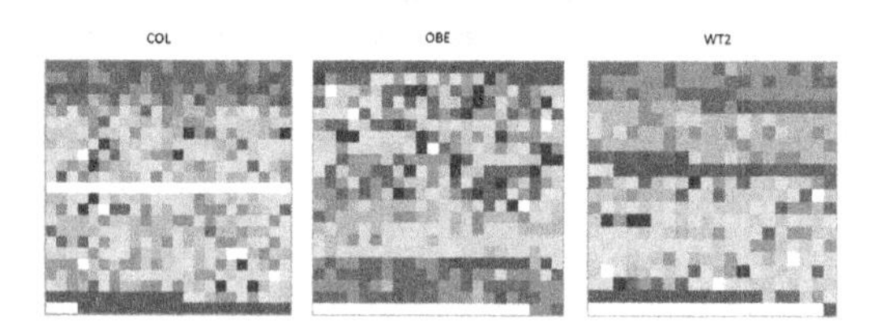

Fig. 3. Visualization of global images of three datasets (Colorectal Cancer, Obesity, Type 2 Diabetes) with k = 10 based on Mini-batch K-means.

global images, the difference of features distribution is easy to observe, with clustering, the similar features have a propensity to close to each other. In other words, the features are represented newly efficiently. Otherwise, the features are distributed discretely in the other, and get a bit confused to recognize which one is similar to the others and it also consists of some overlapped points. We see the same thing on sample images (Fig. 2).

For comparison among various k clusters, we choose the number of clusters k = 3, 5, 10, and 20 for both k-means and mini-batch k-means (batch size = 32 for mini-batch k-means) to visualize features. The different features distributions of each k are illustrated using rainbow images and gray images. With k = 3 or 5, the similar of features distribution are visualized clearly, and k = 10 or 20, it is also not much different on both K-means and Mini-batch K-means. Obviously, with Mini-batch K-means, we can observe the clustered features are arranged almost exactly and better than K-means due to the number of features with the same properties are grouped together accurately. Three global maps examples of 3 considered disease datasets are also presented in Fig. 3.

These synthetic images then are fetched into a convolutional neural network to do prediction tasks with the learning model configurations and results described in Sect. 4.

4 Experimental Results

We present settings and configurations for the experiments and describe metrics for the comparison. Then, the results of experiments using K-means clustering in Fill-up visualization on three considered datasets are analyzed in this section. A performance comparison between the proposed visualization and state-of-the-art is exhibited.

4.1 Settings for Learning Models and Visualization Methods

In order to evaluate the efficiency of the proposed visualization method for supporting disease prediction, we run a shallow deep learning architecture (exhibited in Fig. 4), a convolutional neural network (CNN) with one convolutional layer including 64 filters of 3×3 followed by a max pooling of 2×2 (stride of 2) and

```
      OPERATION            DATA DIMENSIONS   WEIGHTS(N)   WEIGHTS(%)

          Input   #####     22   22    3
     InputLayer     |   -------------------         0     0.0%
                  #####     22   22    3
         Conv2D    \|/  -------------------      1792    18.8%
           relu   #####     22   22   64
   MaxPooling2D   Y max -------------------         0     0.0%
           relu   #####     11   11   64
        Flatten   ||||| -------------------         0     0.0%
                  #####        7744
          Dense   XXXXX -------------------      7745    81.2%
        sigmoid   #####           1
```

Fig. 4. An illustration of components of the proposed CNN for classifying color metagenomic synthetic images of 22 × 22 from OBE dataset.

fully connected layer to classify synthetic images. The shallow CNN is implemented with Adam optimization function using a default learning rate of 0.001 and a batch size of 16. We deploy the technique of Early Stopping to reduce the overfitting issue. If the loss is not improved after 5 consecutive epochs, the learning will be stopped. Otherwise, the learning can run to 500 epochs. We use loss function of binary cross entropy for the binary classification task. These parameters were suggested in [16] for metagenomic synthetic images. We generate images using two colormaps which mentioned above on both K-means and Mini-batch K-means with k = 3, 5, 10, and 20; batch size = 32 with Mini-batch K-means. The results of WT2, COL, and OBE are visualized in Fig. 5.

4.2 Metrics for Comparison

The efficiency of the proposed methods are assessed on three diseases including Colorectal Cancer, Obesity and Type 2 Diabetes. The performance of is evaluated by computing the average accuracy on 10-folds stratified-cross validation.

4.3 Different Parameters with Visualizations Based on K-Mean Clustering Algorithms

The result indicated the number of clusters plays an important role and affects the performance on each dataset. For more specific, the accuracy difference among k = 3, 5, 10 and k = 20 is illustrated in Fig. 5 due to the number of clusters comparison mentioned above. In general, there are no significant difference between the performance of K-means and Mini-batch K-means. We note that rainbow colormap outperforms gray scale on 2 out 3 considered diseases with the highest performance on the COL dataset visualized by rainbow colormap. OBE dataset also reaches the highest performance on rainbow images. Otherwise, gray images are advantages of WT2 dataset, the best accuracy is up to 76.8%. The numbers of clusters for K-means clustering algorithms reach high performances depending on the datasets. The low number of k (3 and 5) is efficient for COL dataset while WT2 dataset requires lager k (10 and 20).

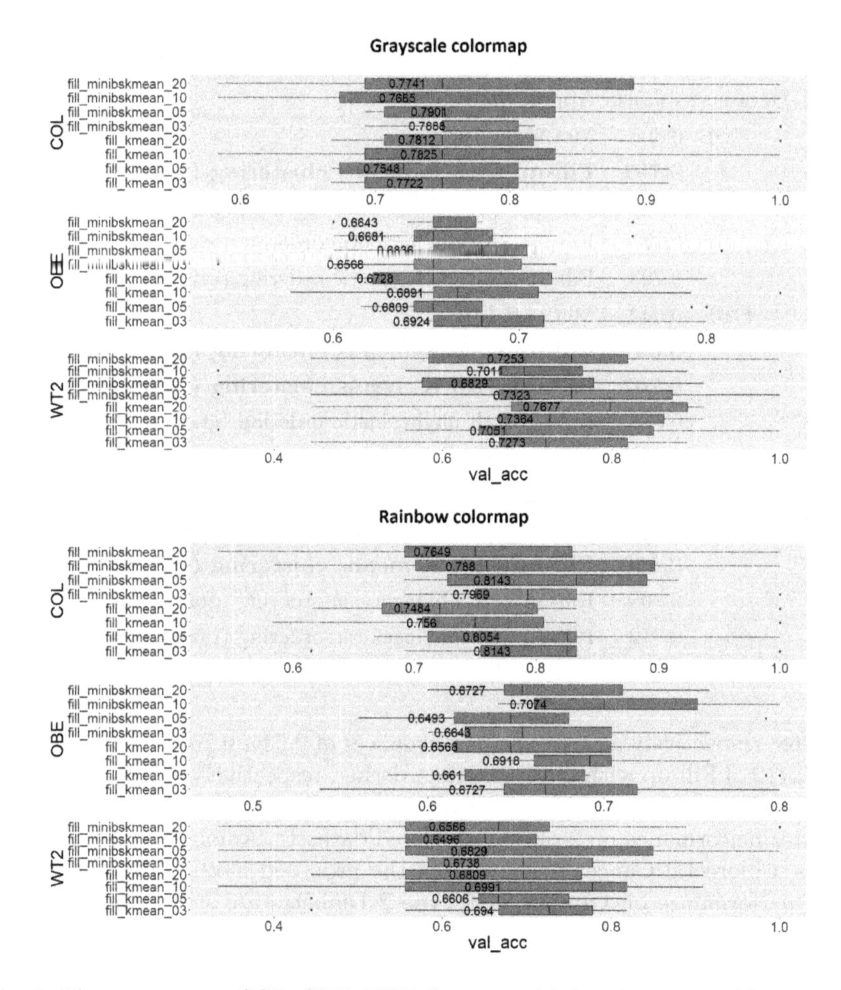

Fig. 5. The accuracy on COL, OBE, WT2 datasets with k = 3, 5, 10, and k = 20 using gray and rainbow colormap images. The numbers in the boxplots illustrate average accuracy. Dots in the chart reflect outliers in the results (Color figure online)

4.4 Compared to the State-of-the-Art

We compare our method to some state-of-the-art including MetAML [10] and Fill-up with phylogenetic ordering [16]. MetAML is metagenomic computation framework using classic machine learning such as Random Forests while the work in [16] run the disease prediction by the Convolutional Neural Network on Fill-up presents features onto images with phylogenetic ordering. Table 2 shows that our method outperforms the state-of-the-art. The proposed method obtains the highest accuracy of 0.815 for disease prediction on COL dataset while MetAML and Fill-up using phylogenetic ordering obtained 0.805 and 0.793, respectively. The results on OBE and WT2 dataset illustrate average performances of 0.707

Table 2. Comparison with the-state-of-the-art

Datasets	val_acc	Approaches
COL	0.805	MetAML [10]
	0.791	**Fill-up with K-means clustering (gray)**
	0.815	**Fill-up with K-means clustering (rainbow)**
	0.775	Fill-up with phylogenetic ordering (gray) [16]
	0.793	Fill-up with phylogenetic ordering (rainbow) [16]
OBE	0.644	MetAML [10]
	0.697	**Fill-up with K-means clustering (gray)**
	0.707	**Fill-up with K-means clustering (rainbow)**
	0.680	Fill-up with phylogenetic ordering (gray) [16]
	0.668	Fill-up with phylogenetic ordering (rainbow) [16]
WT2	0.703	MetAML [10]
	0.768	**Fill-up with K-means clustering (gray)**
	0.699	**Fill-up with K-means clustering (rainbow)**
	0.705	Fill-up with phylogenetic ordering (gray) [16]
	0.712	Fill-up with phylogenetic ordering (rainbow) [16]

and 0.768, respectively against the performances of 0.644, 0.703 of MetAML and 0.680, 0.712 of Fill-up with phylogenetic ordering, respectively. Furthermore, the rainbow colormap seems to be more efficient than gray. The rainbow achieves the better performance on 2 out 3 considered datasets. Among three considered diseases, Colorectal Cancer benefits from the proposed method. Although prediction performance on Obesity and Type 2 Diabetes are still poor, but there can be potential solutions to improve the performance.

5 Conclusion

We introduced a method which clusters features by K-means algorithms and then visualizes them into images. The proposed method helps to group features which have similar characteristics to be close each other. The images generated by such arrangement enables Convolutional Neural Network to improve the performance comparing to the arrangement using phylogenetic order. The performance depends on the number of clusters of k-means clustering algorithm. If we set this parameter too high, the clusters in the image will not be clear. As observed, we noted that $k = 5$ and $k = 10$ clusters with k-means give the best. Further research can investigate, evaluate, and experiment to indicate good numbers of k.

The features are visualized with gray and rainbow colormaps. As shown from the results, the images using rainbow colormap exhibits better performances than images with gray scale. Shallow architectures of Convolutional Neural Networks are leveraged to classify synthetic images, more researches should investigate deeper architectures to improve the performance. Obesity and Type 2 Diabetes are still challenging prediction tasks. In the future, sophisticated approaches should be investigated to improve prediction performance on these diseases.

References

1. Moscow, J.A., et al.: The evidence framework for precision cancer medicine. Nat. Rev. Clin. Oncol. **15**(3), 183–192 (2017)
2. Chial, H.: DNA sequencing technologies key to the Human Genome Project. Nat. Educ. **1**(1), 219 (2008)
3. Handelsman, J.: Metagenomics: application of genomics to uncultured microorganisms. Microbiol. Mol. Biol. **69**(1), 195–195 (2005)
4. Turnbaugh, P., Ley, R., Hamady, M., et al.: The human microbiome project. Nature **449**, 804–810 (2007). https://doi.org/10.1038/nature06244
5. Chen, H., et al.: An assessment of the functional enzymes and corresponding genes in chicken manure and wheat straw composted with addition of clay via metagenomic analysis. Ind. Crops Prod. **153**, 2020 (2020). https://doi.org/10.1016/j.indcrop.2020.112573
6. Nakamura, S., et al.: Direct metagenomic detection of viral pathogens in nasal and fecal specimens using an unbiased high-throughput sequencing approach. PLoS ONE **4**(1), e4219 (2009)
7. Li, L., Delwart, E.: From orphan virus to pathogen: the path to the clinical lab. Curr. Opin. Virol. **1**(4), 282–288 (2011)
8. Udugama, B., et al.: Diagnosing COVID-19: the disease and tools for detection. ACS Nano **14**(4), 3822–3835 (2020)
9. Shah, S.H.J., Malik, A.H., Zhang, B., Bao, Y., Qazi, J.: Metagenomic analysis of relative abundance and diversity of bacterial microbiota in Bemisia tabaci infesting cotton crop in Pakistan, May 2020 (2020). https://doi.org/10.1016/j.meegid.2020.104381
10. Pasolli, E., et al.: Machine learning meta-analysis of large metagenomic datasets: tools and biological insights. PLoS Comput. Biol. **12**(7), e1004977 (2016). https://doi.org/10.1371/journal.pcbi.1004977
11. Soueidan, H., Nikolski, M.: Machine learning for metagenomics: methods and tools. Metagenomics **1**(1) (2017)
12. Patwardhan, A., Ray. S., Roy, A.: Molecular markers in phylogenetic studies-a review. J. Phylogenetics Evol. Biol. **02**(02) (2014)
13. Reiman, D., Metwally, A., Sun, J., Dai, Y.: PopPhy-CNN: a phylogenetic tree embedded architecture for convolutional neural networks to predict host phenotype from metagenomic data. IEEE J. Biomed. Health Inform. (2020). https://doi.org/10.1109/JBHI.2020.2993761
14. Zhou, F., et al.: Bayesian biclustering for microbial metagenomic sequencing data via multinomial matrix factorization. arXiv:2005.08361 (2020)
15. Asnicar, F., et al.: Precise phylogenetic analysis of microbial isolates and genomes from metagenomes using PhyloPhlAn 3.0. Nat. Commun. **11**, 2500 (2020). https://doi.org/10.1038/s41467-020-16366-7
16. Nguyen, T.H., et al.: Disease prediction using synthetic image representations of metagenomic data and convolutional neural networks. In: IEEE-RIVF, pp 231–236. IEEE Xplore (2019). ISBN 978-1-5386-9313-1

17. Alonso, J.B.: K-means vs mini batch k-means: a comparison (2013)
18. Soni, R., James Mathai, K.: An innovative 'cluster-then-predict' approach for improved sentiment prediction. In: Choudhary, R.K., Mandal, J.K., Auluck, N., Nagarajaram, H.A. (eds.) Advanced Computing and Communication Technologies. AISC, vol. 452, pp. 131–140. Springer, Singapore (2016). https://doi.org/10.1007/978-981-10-1023-1_13
19. Liang, Q. et al.: DeepMicrobes: taxonomic classification for metagenomics with deep learning. NAR Genomics Bioinform. **2**(1) (2020)
20. Reiman, D., Dai, Y.: Using Conditional Generative Adversarial Networks to Boost the Performance of Machine Learning in Microbiome Datasets. bioXiv:2020.05.18.102814 (2020). https://doi.org/10.1101/2020.05.18.102814

Small Samples of Multidimensional Feature Vectors

Leon Bobrowski[1,2](✉)

[1] Faculty of Computer Science, Białystok University of Technology, Białystok, Poland
l.bobrowski@pb.edu.pl
[2] Institute of Biocybernetics and Biomedical Engineering, PAS, Warsaw, Poland

Abstract. A small sample of multidimensional feature vectors appears when the number of features is much greater than the number of objects (feature vectors).

For example, such circumstances appear typically in genetic data sets. In such cases, feature clustering can become a useful tool in classification or prognosis tasks. Feature clustering can be performed through the minimization of the convex and piecewise linear (*CPL*) criterion functions.

Keywords: Data mining · Linear dependence · Feature clustering · Convex and piecewise linear functions

1 Introduction

Data sets composed of small numbers of high dimensional feature vectors are often met in practice [1]. Typical examples in this area are given by genetic data sets when each feature vector represents a huge number of genes. Learning data sets composed of small numbers of high dimensional feature vectors are usually linearly separable [2]. The linear separability of learning sets can be destroyed by tye existence of linear relationships between feature vectors [3].

Computational tools for expoloring the linear separability of learning sets can be based on the minimization of the perceptron criterion function with the regularization [4]. The perceptron criterion function belongs to a large family of the *convex and piecewise linear* (*CPL*) criterion functions. The collinearity criterion function also belongs to the *CPL* family. The minimization of the collinearity criterion function gives possibility to explore collinearity relations in data sets and to modeling multiple interactions between selected features [5].

Optimal parameters constituting the minimum value of the *CPL* criterion functions can also be used in feature clustering. Properties of the *CPL* criterion functions are theoretically analyzed in this context in the presented paper.

2 Linearly Independent Feature Vectors

Let us consider m objects (patients) O_j ($j = 1,\ldots, m$) represented as n-dimensional feature vectors $\mathbf{x}_j = [x_{j,1},\ldots,x_{j,n}]^T$, or as points $\mathbf{x}_j$ in the n-dimensional feature space $\boldsymbol{F}[n]$ ($\mathbf{x}_j$

M. Hernes et al. (Eds.): ICCCI 2020, CCIS 1287, pp. 87–98, 2020.
https://doi.org/10.1007/978-3-030-63119-2_8

$\in \boldsymbol{F}[n]$). The components $x_{j,i}$ of the feature vector $\mathbf{x}_j$ represent individual features X_i ($i = 1,\ldots, n$) as the numerical results of specific measurements on a given object O_j ($x_{j,i} \in \{0,1\}$ or $x_{j,i} \in R^1$).

Data set $\boldsymbol{C}$ is composed of m feature vectors $\mathbf{x}_j$:

$$C = \{\mathbf{x}_j : j = 1, \ldots, m\}, \tag{1}$$

The feature vector $\mathbf{x}_{j'}$ from set $\boldsymbol{C}$ (1) *depends linearly* on other vectors $\mathbf{x}_j$ ($j \neq j'$) from this set if the following relation occurs [4]:

$$\mathbf{x}_{j'} = \Sigma_{j \neq j'} \alpha_j \mathbf{x}_j, \text{ where } \alpha_j \in R^1 \tag{2}$$

The vector $\mathbf{x}_{j'}$ is a *linear combination* of other vectors $\mathbf{x}_j$ ($j \neq j'$) if the relation (2) occurs.

Definition 1: Feature vectors $\mathbf{x}_j$ from the data set $\boldsymbol{C}$ (1) are linearly independent in the feature space $\boldsymbol{F}[n]$ ($\mathbf{x}_j \in \boldsymbol{F}[n]$) if none of the vectors $\mathbf{x}_{j'}$ can be represented as a linear combination (2) of other vectors $\mathbf{x}_j$ ($j \neq j'$) in this set.

Remark 1: The maximal number m of feature vectors $\mathbf{x}_j$ (4) that can by linearly independent in the n-dimensional feature space $\boldsymbol{F}[n]$ is equal to n.

Remark 2: If data set $\boldsymbol{C}$ (1) is composed of a small number m of n-dimensional feature vectors $\mathbf{x}_{jj}$ ($m \ll n$) then these vectors are usually linearly independent [4].

Remark 3: Feature vectors $\mathbf{x}_j$ from the set $\boldsymbol{C}$ (1) that are linearly independent in the n-dimensional feature space $\boldsymbol{F}[n]$ are also linearly independent in a larger feature space $\boldsymbol{F}[n']$ ($\boldsymbol{F}[n] \subset \boldsymbol{F}[n']$, where $n < n'$).

Feature vectors $\mathbf{x}_j$ from the data set $\boldsymbol{C}$ (1) are often divided into K *learning sets* $\boldsymbol{C}_k$ [2]:

$$C_k = \{x_j(k) : j \in J_k\} \tag{3}$$

where J_k is the set of indices j of the feature vectors $\mathbf{x}_j$ assigned to the k-th category (*class*) ω_k ($k = 1,\ldots, K$) of j - th the object O_j.

The k-th learning set C_k (3) is composed of m_k examples of labeled feature vectors $\mathbf{x}_j(k)$ assigned to the k-th category ω_k.

Definition 2: The learning sets C_k (3) are *separable* in the feature space $\boldsymbol{F}[n]$, if they are disjoined in this space $C_k \cap C_{k'} = \emptyset$, if $k = k'$. This means that the feature vectors $\mathbf{x}_j(k)$ and $\mathbf{x}_j'(k')$ belonging to different learning sets C_k and $C_{k'}$ cannot be equal:

$$(k \neq k') \Rightarrow (\forall j \in J_k)\, and \, (\forall j' \in J_{k'}) \quad \mathbf{x}_j(k) \neq \mathbf{x}_{j'}(k') \tag{4}$$

The separation of the sets C_k (3) by hyperplanes $\boldsymbol{H}(\mathbf{w}_k,\theta_k)$ in the feature space $\boldsymbol{F}[n]$ is also considered:

$$H(\mathbf{w}_k, \theta_k) = \{\mathbf{x} : \mathbf{w}_k^T\mathbf{x} = \theta_k\}. \tag{5}$$

where $\mathbf{w}_k = [w_{k,1},\ldots,w_{k,n}]^T \in \boldsymbol{R}^n$ is the k-th weight vector, $\theta_k \in R^1$ is the threshold, and $\mathbf{w}_k^T\mathbf{x}$ is the inner product.

Definition 3: The learning sets C_k (3) are *linearly separable* in the n-dimensional feature space $F[n]$ if each of these sets can be completely separated from the sum of other sets $C_{k'}$ $(k' \neq k)$ by some hyperplane $H(\mathbf{w}_k, \theta_k)$ (5):

$$(\forall k \in \{1,\ldots,K\})(\exists \mathbf{w}_k, \theta_k)\left(\forall \mathbf{x}_j(k) \in C_k\right) \ \mathbf{w}_k^T\mathbf{x}_j(k) > \theta_k$$
$$and \ \left(\forall \mathbf{x}_{j'} \in C_{k'}, k' \neq k\right) \ \mathbf{w}_k^T\mathbf{x}_{j'}, \left(k'\right) < \theta_k \qquad (6)$$

According to the inequalities (6), all labelled vectors $\mathbf{x}_j(k)$ from the learning set C_k (3) are situated on the positive side $(\mathbf{w}_k^T\mathbf{x}_j(k) > \theta_k)$ of the separating hyperplane $H(\mathbf{w}_k, \theta_k)$ (5) and all vectors $\mathbf{x}_{j'}(k')$ from the remaining sets $C_{k'}$ are situated on the negative side $(\mathbf{w}_k^T\mathbf{x}_{j'}(k') < \theta_k)$ of this hyperplane. The separating hyperplane $H(\mathbf{w}_k, \theta_k)$ (5) can be efficiently found through the minimization of the perceptron criterion function $\Phi_k(\mathbf{w}, \theta)$ defined on elements $\mathbf{x}_j(k)$ of the k-th learning set C_k (1) and elements $\mathbf{x}_{j'}(k')$ of the remaining sets $C_{k'}$ $(k' \neq k)$ [2].

3 Perceptron Criterion Function $\Phi_{P,K}(W, \theta)$

The linear separability (6) of the learning sets C_k (3) can be verified by minimizing the perceptron criterion functions $\Phi_{p,k}(\mathbf{w}, \theta)$ [5]. The perceptron criterion function $\Phi_{p,k}(\mathbf{w}, \theta)$ is defined as the weighted sum of the *positive* penalty functions $\varphi_{k,j}^+(\mathbf{w},\theta)$ and of the *negative* penalty functions $\varphi_{k,j}^-(\mathbf{w}, \theta)$. The positive penalty functions $\varphi_{k,j}^+(\mathbf{w}, \theta)$ are linked to m_k elements $\mathbf{x}_j(k)$ of the k-th learning set C_k (3):

$$(\forall \mathbf{x}_j(k) \in C_k)$$
$$\varphi_{k,j}^+(\mathbf{w}, \theta) = \begin{cases} 1 + \theta - \mathbf{x}_j(k)^T\mathbf{w} & if \ \mathbf{x}_j(k)^T\mathbf{w} \leq \theta + 1 \\ 0 & if \ \mathbf{x}_j(k)^T\mathbf{w} > \theta + 1 \end{cases} \qquad (7)$$

The negative penalty functions $\varphi_{k,j}^-(\mathbf{w},\theta)$ are defined similarly as:

$$(\forall k' \neq k) \ (\forall \mathbf{x}_j(k') \in C_{k'})$$
$$\varphi_{k,j}^-(\mathbf{w}, \theta) = \begin{cases} 1 - \theta + \mathbf{x}_j(k')^T\mathbf{w} & if \ \mathbf{x}_j(k')^T\mathbf{w} \geq \theta - 1 \\ 0 & if \ \mathbf{x}_j(k')^T\mathbf{w} < \theta - 1 \end{cases} \qquad (8)$$

Each of n unit vectors $\mathbf{e}_i$ allows to define the i-th *cost function* $\varphi_i^0(\mathbf{w})$ in the n-dimensional parameter space R^n, where $\mathbf{w} = [w_1,\ldots,w_n]$:

$$(\forall i \in \{1,\ldots,n\})$$
$$\varphi_i^0(\mathbf{w}) = \gamma_i \mid \mathbf{e}_i^T\mathbf{w} \mid = \gamma_i \mid w_i \mid = \begin{cases} -\gamma_i w_i & if \ w_i \leq 0 \\ \gamma_i w_i & if \ w_i > 0 \end{cases} \qquad (9)$$

The *CPL* cost functions $\varphi_i^0(\mathbf{w})$ (9) are equal to the costs γ_i $(\gamma_i > 0)$ of particular features X_i multiply by the absolute values $\mid w_i \mid$ of the components w_i of the weight vector $\mathbf{w} = [w_1,\ldots, w_n]^T$ [4].

The perceptron criterion function $\Phi_{p,k}(\mathbf{w}, \theta)$ is defined in the below manner [5]:

$$\Phi_{p,k}(\mathbf{w}, \theta) = \sum_{j \in J_k^+} \beta_j^+ \varphi_j^+(\mathbf{w}, \theta) + \sum_{j' \in J_k^-} \beta_{j'}^- \varphi_{j'}^-(\mathbf{w}, \theta) \tag{10}$$

where J_k^+ is the set of indices j of the feature vectors $\mathbf{x}_j(k)$ from the learning set C_k (3) and, J_k^- is the set of indices j' of the vectors $\mathbf{x}_{j'}(k')$ from the remaining sets $C_{k'}$ $(k' \neq k)$.

The positive parameters β_j of the function $\Phi_{p,k}(\mathbf{w}, \theta)$ (10) represent *prices* of individual feature vectors $\mathbf{x}_j$ with standard values $\beta_j^+ = 1/(2\, m_k)$ and $\beta_{j'}^- = 1/(2\, (m - m_k))$ [5].

Minimization of the convex and piecewise linear (*CPL*) criterion function $\Phi_{p,k}(\mathbf{w}, \theta)$ (10) allows to determine optimal parameters $(\mathbf{w}_k^*, \theta_k^*)$:

$$\left(\forall (\mathbf{w}, \theta) \in R^{n+1}\right) \quad \Phi_{p,k}(\mathbf{w}, \theta) \geqslant \Phi_{p,k}\left(\mathbf{w}_k^*, \theta_k^*\right)^- \tag{11}$$

Theorem 1: The minimal value $\Phi_{p,k}(\mathbf{w}_k^*, \theta_k^*)$ (11) of the k-th perceptron criterion function $\Phi_{p,k}(\mathbf{w}, \theta)$ (11) is equal to zero ($\Phi_{p,k}(\mathbf{w}_k^*, \theta_k^*) = 0$) if and only if the k-th learning set $\boldsymbol{C}_k$ (3) can be completely separated from the sum of the other sets $\boldsymbol{C}_{k'}$ by a certain hyperplane $\boldsymbol{H}(\mathbf{w}_k, \theta_k)$ (5) in the n-dimensional feature space $\boldsymbol{F}[n]$ [5]:

$$\begin{gathered}(\exists \mathbf{w}_k, \theta_k)(\forall \mathbf{x}_j(k) \in \mathbf{C}_k) \;\; \mathbf{w}_k^T \mathbf{x}_j(k) > \theta_k, \; and \\ (\forall k' \neq k)(\forall \mathbf{x}_{j'}(k') \in \boldsymbol{C}_{k'}) \;\; \mathbf{w}_k^T \mathbf{x}_{j'}(k') < \theta_k\end{gathered} \tag{12}$$

The regularized criterion function $\Psi_{p,k}(\mathbf{w}, \theta)$ is defined as the sum of the perceptron criterion function $\Phi_{p,k}(\mathbf{w}, \theta)$ (10) and the cost functions $\varphi_i^0(\mathbf{w})$ (9):

$$\Psi_{p,k}(\mathbf{w}, \theta) = \Phi_{p,k}(\mathbf{w}, \theta) + \lambda \sum_{i \in \{1,\ldots,n\}} \varphi_i^0(\mathbf{w}) = \Phi_{p,k}(\mathbf{w}, \theta) + \lambda \sum_{i \in \{1,\ldots,n\}} \gamma_i |w_i| \tag{13}$$

where $\lambda \geq 0$ is the *cost level*. The standard values of the cost parameters γ_i are equal to one $((\forall i \in \{1,\ldots,n\})\; \gamma_i = 1)$.

We can calculate the conditional minimum value $\Psi_{p,k}(\mathbf{w}_{k,\lambda}^*, \theta_k^*)$ of the *CPL* criterion function $\Psi_{p,k}(\mathbf{w}, \theta)$ (13) with the threshold value θ_k^* specified by the condition (11):

$$\left(\exists \mathbf{w}_{k,\lambda}^*\right)\left(\forall \mathbf{w} \in R^n\right) \Psi_{p,k}(\mathbf{w}, \theta) \geqslant \Psi_{p,k}\left(\mathbf{w}_{k,\lambda}^*, \theta_k^*\right) = \Psi_{p,k}^* > 0 \tag{14}$$

The minimal value $\Psi_{p,k}^*$ (14) of the regularized criterion function $\Psi_{p,k}(\mathbf{w}, \theta)$ (13) is used, among others, in the *relaxed linear separability* (*RLS*) method of feature subsets selection [6].

The *basis exchange algorithm* can be used for minimization of the convex and piecewise linear (*CPL*) criterion functions $\Phi_{p,k}(\mathbf{w}, \theta)$ (8) and $\Psi_{p,k}(\mathbf{w}, \theta)$ (11) [7]. Basis exchange algorithm is based on the Gauss - Jordan transformation and, for this reason, is similar to the *Simplex* algorithm of linear programming [8]. High efficiency and precision of the basis exchange algorithm allow minimizing *CPL* criterion functions even in the case of large, multidimensional sets $\boldsymbol{C}_k$ (3).

The high efficiency of the *CPL* procedures has allowed, among others, the application of the *RLS* method to select such optimal subsets of genes that are characterized by high discriminative power [6]. The *RLS* method was applied to the *Breast Cancer* data set which contains descriptions of $m_1 = 46$ cancer and $m_2 = 51$ non-cancer patients. Each patient in this set was characterized by $n = 24481$ genes. The *RLS* method allowed to select the optimal subset of $n_1 = 12$ genes and such a linear combination of these genes, which correctly distinguishes cancer from non-cancer patients in this set [4]. This example demonstrates the ability to use data mining techniques based on the *CPL* criterion functions also in the case when the number n of features X_i ($i = 1, \ldots, n$) is many times greater than the number m of objects O_j ($j = 1, \ldots, m$).

4 Collinearity Criterion Functions $\Phi_C(\mathbf{W})$

The collinearity criterion functions $\Phi_c(\mathbf{w})$ is defined as the weighted sum of the penalty functions $\varphi_j(\mathbf{w})$ linked to particular feature vectors $\mathbf{x}_j = [x_{j,1}, \ldots, x_{j,n}]^T$ [9]:

$$(\forall \mathbf{x}_j \in \boldsymbol{C}\,(1))$$

$$\varphi_j(\mathbf{w}) = |\,1 - \mathbf{x}_j^T\mathbf{w}\,| = \begin{cases} 1 - \mathbf{x}_j^T\mathbf{w} & \text{if } \mathbf{x}_j^T\mathbf{w} \le 1 \\ \mathbf{x}_j^T\mathbf{w} - 1 & \text{if } \mathbf{x}_j^T\mathbf{w} > 1 \end{cases} \tag{15}$$

The collinearity criterion function $\Phi_c(\mathbf{w})$ is defined as the weighted sum of the penalty functions $\varphi_j(\mathbf{w})$ (15) determined by the feature vectors $\mathbf{x}_j$ from the data set $\boldsymbol{C}$ (1):

$$\Phi_c(\mathbf{w}) = \sum_{j \in J} \beta_i \varphi_i(\mathbf{w}) \tag{16}$$

where $J = \{j: \mathbf{x}_j \in \boldsymbol{C}\,(1)\}$ and the positive parameters β_j ($\beta_j > 0$) in the function $\Phi_c(\mathbf{w})$ can be treated as the *prices* of particular feature vectors $\mathbf{x}_j$. The standard choice of the parameters β_j values is one (($\forall j \in \{1, \ldots, m\}$) $\beta_j = 1.0$).

The basis exchange algorithm allows finding the minimal value $\Phi_c(\mathbf{w}_k^*)$ of the collinearity criterion function $\Phi_c(\mathbf{w})$ (16) efficiently [7]:

$$(\exists \mathbf{w}_k^*)(\forall \mathbf{w})\Phi_c(\mathbf{w}) \geqslant \Phi_c(\mathbf{w}_k^*) = \Phi_c^* \geqslant 0 \tag{17}$$

Definition 3: The data subset $\boldsymbol{C}_k$ ($\boldsymbol{C}_k \subset \boldsymbol{C}\,(1)$) is *collinear* when all feature vectors $\mathbf{x}_j$ from this subset can be located on some hyperplane $H(\mathbf{w}, \theta)$ (5) with $\theta \neq 0$.

Theorem 2: The minimal value $\Phi_c(\mathbf{w}_k^*)$ (17) of the collinearity criterion function $\Phi_c(\mathbf{w})$ (16) defined on m elements $\mathbf{x}_j$ of data subset $\boldsymbol{C}_k$ ($\boldsymbol{C}_k \subset \boldsymbol{C}\,(1)$) is equal to zero ($\Phi_k(\mathbf{w}_k^*) = 0$) when the subset $\boldsymbol{C}_k$ is collinear [9].

Minimizing the criterion function $\Phi_c(\mathbf{w})$ (16) in the case of a large number m of feature vectors $\mathbf{x}_j$ ($m >> n$) allows extracting many collinear subsets $\boldsymbol{C}_k$ from a given data set $\boldsymbol{C}$ (1).

5 Dual Planes and Vertices in the Parameter Space

Let us use *m augmented* feature vectors $\mathbf{y}_j$ ($\mathbf{y}_j \in \boldsymbol{F}[n+1]$) in examing properties of the *k*-th perceptron criterion function $\Phi_{p,k}(\mathbf{w}, \theta)$ (10) [2]:

$$\begin{aligned} (\forall \mathbf{x}_j \in C_{\mathbf{k}}) \qquad \mathbf{y}_j &= \left[\mathbf{x}_j^T, 1\right]^T, \textit{ and} \\ (\forall k' \neq k)(\forall \mathbf{x}_j \in C_{k'})\ \mathbf{y}_j &= -\left[\mathbf{x}_j^T, 1\right]^T \end{aligned} \tag{18}$$

and the *augmented* weight vector $\mathbf{v}$:

$$\mathbf{v} = \left[\mathbf{w}^T, -\theta\right]^T = [w_1, \ldots, W_n, -\theta] \tag{19}$$

The linear separability inequalities (6) can now be given in the below form:

$$(\exists \mathbf{v}_k)\left(\forall j \in \{1, \ldots, m\}\right)\ \mathbf{v}_k^T \mathbf{y}_j > 0 \tag{20}$$

The linear separability inequalities (20) can be replaced by the below set of *m* inequalities with the *margin* equal to one [5]:

$$(\exists \mathbf{v}_k)\left(\forall j \in \{1, \ldots, m\}\right)\ \mathbf{v}_k^T \mathbf{y}_j \geq 1 \tag{21}$$

Let us remark that the sharp inequalities (20) with the symbol "$>$" (*greater*) have been replaced by the soft inequalities (21) with the symbol "$\geq$" (*greater or equal*).

The inequalities (21) can be used in the *perceptron penalty functions* $\varphi_j(\mathbf{v})$ (7) and (8):

$$(\forall j \in \{1,\ldots, m\}) \quad \varphi_j(\mathbf{v}) = \begin{cases} 1 - \mathbf{y}_j^T \mathbf{v} & \textit{if} \ \ \mathbf{y}_j^T \mathbf{v} < 1 \\ 0 & \textit{if} \ \ \mathbf{y}_j^T \mathbf{v} \geq 1 \end{cases} \tag{22}$$

The convex and piecewise-linear (*CPL*) penalty functions $\varphi_j(\mathbf{v})$ (22) are used for reinforcing the linear separability inequalities (6) through minimizing the perceptron criterion function $\Phi_{p,k}(\mathbf{v}) = \Phi_{p,k}(\mathbf{w}, \theta)$ (10) [4]:

$$\Phi_{p,k}(\mathbf{v}) = \sum_{j \in \{1,\ldots,m\}} \beta_j \varphi_j(\mathbf{v}) \tag{23}$$

and (11):

$$(\exists \mathbf{v}_{\mathbf{k}}^*)\left(\forall \mathbf{v} \in R^{n+1}\right) \Phi_{p,k}(\mathbf{v}) \geqslant \Phi_{p,k}\left(\mathbf{v}_{\mathbf{k}}^*\right) = \Phi_{p,k}^* \geqslant 0 \tag{24}$$

Augmented feature vectors $\mathbf{y}_j$ (18) allow to define below hyperplanes h_j^1 in the parameter space $\boldsymbol{R}^{n+1}$ ($\mathbf{v} \in \boldsymbol{R}^{n+1}$):

$$\left(\forall j \in \{1, \ldots, m\}\right)\ h_j^1 = \left\{\mathbf{v} : \mathbf{y}_j^T \mathbf{v} = 1\right\} \tag{25}$$

Each of $n + 1$ unit vectors $\mathbf{e}_i$ defines the hyperplane h_i^0 in the parameter space $\boldsymbol{R}^{n+1}$:

$$(\forall i = 1, \ldots, n+1) \quad h_i^0 = \left\{\mathbf{v} : \mathbf{e}_i^T \mathbf{v} = 0\right\} = \{\mathbf{v} : v_i = 0\} \tag{26}$$

Definition 4: The vertex $\mathbf{v}_k$ in the parameter space $\boldsymbol{R}^{n+1}$ ($\mathbf{v}_k \in \boldsymbol{R}^{n+1}$) is located at the intersection of m_k hyperplanes h_j^1 (25) ($j \in J_k$) and $n + 1$ - m_k hyperplanes h_i^0 (26):

$$(\forall j \in J_k) \mathbf{v}_k^T \mathbf{y}_j = 1 \tag{27}$$

and

$$(\forall i \in I_k) \quad \mathbf{v}_k^T \mathbf{e}_i = 0 \tag{28}$$

where J_k is the k-th subset of indexes j of l_k feature vectors $\mathbf{y}_j$ (18) and I_k is the k-th subset of indexes i of $n + 1 - l_k$ unit vectors $\mathbf{e}_i$.

Let us assume at the beginning that the linear Eqs. (27) and (28) can be represented by using the following *vertexical* matrix $\boldsymbol{B}_k$ [4]:

$$\boldsymbol{B}_k \mathbf{v}_k = \mathbf{1}_k \tag{29}$$

where $\boldsymbol{B}_k = [\mathbf{y}_{j(1)},\ldots, \mathbf{y}_{j(lk)}, \mathbf{e}_{lk+1},\ldots, \mathbf{e}_n]^T$, and $\mathbf{1}_k = [1,\ldots, 1, 0,\ldots, 0]$.

Definition 5: The *rank* r_k of the vertex $\mathbf{v}_k$ in the parameter space $\boldsymbol{R}^{n+1}$ ($\mathbf{v}_k \in \boldsymbol{R}^{n+1}$) is equal to the number l_k of feature vectors $\mathbf{y}_j$ (18) in the nonsingular matrix $\boldsymbol{B}_k$ (29).

The symbol $\mathbf{y}_{j(i)}$ means such augmented vector $\mathbf{y}_j$ (18) (j =1,…, m) in the matrix $\boldsymbol{B}_k$ (29) which replaced the i-th unit vector $\mathbf{e}_i$ (i =1,…, l_{rk}) and constitutes the i-th row of this matrix. The vector $\mathbf{1}_k$ has the first l_k components equal to one and the last $n + 1 - l_k$ components equal to zero.

If the vertexical matrix $\boldsymbol{B}_k$ (29) containing l_k feature vectors $\mathbf{y}_j$ (18) is not singular, then it forms the basis of the augmented feature space $\boldsymbol{F}[n + 1]$) ($\mathbf{y}_j \in \boldsymbol{F}[n + 1]$), and the vertex $\mathbf{v}_k = [v_{k,1},\ldots, v_{k.n}]^T$ can be computed as the sum of the first l_k columns of the inverse matrix $(\boldsymbol{B}_k^T)^{-1}$:

$$\mathbf{v}_k = \left(\mathbf{B}_k^T\right)^{-1} \mathbf{1}_k = \mathbf{r}_1 + \ldots + \mathbf{r}_{lk} \tag{30}$$

where $\mathbf{r}_i$ is the i-th column (i = 1,…, l_k) of the inverse matrix $(\boldsymbol{B}_k^T)^{-1}$:

$$\left(B_k^T\right)^{-1} = \left[\mathbf{r}_1, \ldots, \mathbf{r}_{lk}, \mathbf{r}_{lk+1}, \ldots, \mathbf{r}_{n+1}\right] \tag{31}$$

Remark 4: The vertexical matrix $\boldsymbol{B}_k = [\mathbf{y}_{j(1)},\ldots, \mathbf{y}_{j(lk)}, \mathbf{e}_{lk+1},\ldots, \mathbf{e}_n]^T$ (29) can be designed from m augmented feature vector $\mathbf{y}_j$ (18) and n unit vectors $\mathbf{e}_i$ by using the basis exchange algorithm [7].

Remark 5: Each component $v_{k.i}$ of the vector $\mathbf{v}_k = [v_{k,1},\ldots, v_{k,n+1}]^T$ (30) linked to the unit vector $\mathbf{e}_i$ in the basis $\boldsymbol{B}_k$ (29) is equal to zero:

$$(\forall i \in \{l_k + 1, \ldots, n + 1\}) \quad v_{k \cdot i} = 0 \tag{32}$$

The above property is related to the presence of unit vectors $\mathbf{e}_i$ ($i \in I_k$) in the Eq. (28) that defines the vertex $\mathbf{v}_k$. The last $n + 1 - l_k$ columns of the matrix $\boldsymbol{B}_k$ (29) are formed by unit vectors $\mathbf{e}_i$. The condition $v_{k.i} = 0$ (32) results from the equation $\mathbf{v}_k^T\mathbf{e}_i = 0$ (28).

Theorem 3: If m ($m \leq n$) feature vectors $\mathbf{x}_j$ ($j = 1,\ldots, m$) are linearly independent, then the learning sets C_k (3) are linearly separable (6) in the n-dimensional feature space $\boldsymbol{F}[n]$ regardless of the assignment of these vectors to individual sets C_k.

Proof: Linear independence of m feature vectors $\mathbf{x}_j$ also means linear independence of the augmented vectors $\mathbf{y}_j$ (18). Increasing the dimension n of feature vectors $\mathbf{x}_j$ maintains their linear independence. The matrix $\boldsymbol{B}_k$ (29) containing m linearly independent vectors $\mathbf{y}_j$ (18) is non-singular and the vertex $\mathbf{v}_k$ can be computed (30) on the basis of the inverse matrix $(\boldsymbol{B}_k^T)^{-1}$. The inequalities (21) are met in the vertex $\mathbf{v}_k$ (30). □

6 Vertexical Feature Subspaces $\boldsymbol{F}_k[r_k]$

The vertexical matrix (*basis*) $\boldsymbol{B}_k = [\mathbf{y}_{j(1)},\ldots, \mathbf{y}_{j(lk)}, \mathbf{e}_{lk+1},\ldots, \mathbf{e}_n]^T$ (29) has been connected to the vertex $\mathbf{v}_k$ (30) of the rank l_k through the Eqs. (27) and (28). In accordance with the Eqs. (32), such component $v_{k.i}$ of the vector $\mathbf{v}_k = [v_{k,1},\ldots, v_{k,n+1}]^T$ (30) which are linked to the unit vector $\mathbf{e}_i$ in the basis $\boldsymbol{B}_k$ (29) are equal to zero (32).

The augmented weight vector $\mathbf{v}_k$ was defined by means of the Eq. (19):

$$\mathbf{v}_k = \left[\mathbf{w}_k^T, -\theta_k\right]^T = \left[w_{k,1}, \ldots, w_{k,n}, -\theta_k\right] \tag{33}$$

Definition 6: The k-th *vertexical subspace* $\boldsymbol{F}_k[r_k]$ ($\boldsymbol{F}_k[r_k] \subset \boldsymbol{F}[n]$) is based on such r_k features X_i constituting the feature subset $F_k = \{X_{i(1)}, \ldots, X_{i(rk)}\}$ which are linked to the non-zero components $w_{k,i}$ ($w_{k,i} \neq 0$) of the optimal weight vector $\mathbf{w}_k^* = [w_{k,1}^*,\ldots,w_{k,n}^*]^T$ which constitutes the minimal value $\Phi_{p,k}(\mathbf{w}_k^*, \theta_k^*)$ (11) of the perceptrom criterion function $\Phi_{p,k}(\mathbf{w}, \theta)$ (10):

$$(\forall i \in \{1, \ldots, n\})\, \mathit{if} \quad w_{k,1}^* \neq 0, \ \mathit{then} \ X_i \in F_k \tag{34}$$

Theorem 4: If m ($m \leq n$) feature vectors $\mathbf{x}_j$ ($j = 1,\ldots, m$) (1) are linearly independent in the feature space $\boldsymbol{F}[n]$ ($\mathbf{x}_j \in \boldsymbol{F}[n]$), then the maximal dimension r_k of the k-th vertexical subspace $\boldsymbol{F}_k[r_k]$ can be equal to m ($r_k \leq m$).

The proof of this theorem can be based on the vectoral Gauss – Jordan transformation used in the basis exchange algorithm sketched below [7].

7 Basis Exchange Algorithm

The *fundamental theorem of linear programming* allows to shown that the minimum $\Phi_{p,k}(\mathbf{v}_k^*)$ (24) of the *CPL* perceptron criterion function $\Phi_{p,k}(\mathbf{v})$ (23) can always be located in one of the vertices $\mathbf{v}_k$ (30) [8]. A similar property has also the collinearity criterion function $\Phi_c(\mathbf{w})$ (16), another function of the *CPL* type [9].

The basis exchange algorithm allows for efficient minimization of the *CPL* criterion functions [7]. The new properties of the basis exchange algorithm in the case of a small number m of multidimensional feature vectors $\mathbf{x}_j$ are examined on the example of the collinearity criterion function $\Phi_c(\mathbf{w})$ (16).

The collinearity criterion function $\Phi_c(\mathbf{w})$ (16) is defined on m feature vectors $\mathbf{x}_j = [x_{j,1},\ldots,x_{j,n}]^T$ (1) of dimension n. Particular attention is paid to the case when the number m of feature vectors $\mathbf{x}_j$ is much smaller than the dimension n $(m < < n)$.

In accordance with the basis exchange algorithm, the optimal vertex $\mathbf{w}_k^*$ constituting the minimal value $\Phi_c(\mathbf{w}_k^*)$ (17) of the criterion function $\Phi_c(\mathbf{w})$ (16) is found by building a sequence of m nonsingular matrices (*bases*) $\boldsymbol{B}_l$ with the rank equal to l:

$$B_0, B_1, \ldots, B_{m-1}, \quad B_m \tag{35}$$

The zero matrix $\boldsymbol{B}_0$ in this sequence is equal to the unit matrix $\boldsymbol{I}$ $(\boldsymbol{B}_0 = \boldsymbol{I})$ of the dimension n x n. The first base matrix $\boldsymbol{B}_1$ $(l = 1)$ is composed of the feature vector $\mathbf{x}_{j(1)}$ $(j(1) \in \{1,\ldots, n\}))$ and $m - 1$ unit vectors $\mathbf{e}_i$ $(i = 2,\ldots, m)$. The last matrix $\boldsymbol{B}_m = [\mathbf{x}_{j(1)},\ldots, \mathbf{x}_{j(m)}, \mathbf{e}_{m+1},\ldots, \mathbf{e}_n]^T$ in the sequence (35) can consist of all m feature vectors $\mathbf{x}_j$ $(j = 1,\ldots, m)$ from the data set $\boldsymbol{C}$ (1).

The Gauss-Jordan transformation allows to generate the following sequence of inverted matrices $\boldsymbol{B}_l^{-1}$ based on non-singular matrices $\boldsymbol{B}_l$ (35):

$$\boldsymbol{B}_0^{-1}, \boldsymbol{B}_1^{-1}, \ldots, \boldsymbol{B}_{m-1}^{-1}, \boldsymbol{B}_m^{-1} \tag{36}$$

The l-th inverted matrix $\boldsymbol{B}_l^{-1}$ is represented in the following manner (31):

$$(\forall l \in \{1, \ldots, m\}) \quad \boldsymbol{B}_l^{-1} = [\mathbf{r}_1(l), \ldots, \mathbf{r}_n(l)] \tag{37}$$

where the symbol $\mathbf{r}_i(l)$ stands for the i-th column of the l-th inverse matrix $\boldsymbol{B}_l^{-1}$.

The first matrix $\boldsymbol{B}_0^{-1}$ in the sequence (36) is equal to the unit matrix $\boldsymbol{I}$ $(\boldsymbol{B}_0^{-1} = \boldsymbol{I})$. The last matrix in the sequence (36) can be equal to the inverse matrix $\boldsymbol{B}_m^{-1}$.

The sequence (36) of the inverse matrices $\boldsymbol{B}_l^{-1}$ (37) is obtained in the multistage process of the matrices $\boldsymbol{B}_l$ (35) transformations. During the l-th stage the matrix basis $\boldsymbol{B}_k$ (4) is transformed into the basis $\boldsymbol{B}_{k+1}$:

$$(\forall l \in \{1, \ldots, m - 1\}) \quad \boldsymbol{B}_l \rightarrow \boldsymbol{B}_{l+1} \tag{38}$$

The basis $\boldsymbol{B}_{l+1}$ is obtained as a result of the replacing the $(l + 1)$-th unit vector $\mathbf{e}_{l+1}$ in the matrix $\boldsymbol{B}_l = [\mathbf{x}_{j(1)},\ldots, \mathbf{x}_{j(l)}, \mathbf{e}_{l+1},\ldots, \mathbf{e}_n]^T$ by the $j(l + 1)$-th feature vector $\mathbf{x}_{j(l+1)}$. In accordance with the Gauss-Jordan vector transformation the replacement of the unit vector $\mathbf{e}_{l+1}$ by the feature vector $\mathbf{x}_{j(l+1)}$ causes the following modifications of the columns $\mathbf{r}_i(l)$ of the inverse matrix $\boldsymbol{B}_l^{-1}$ [5]:

$$\begin{aligned} \mathbf{r}_{l+1}(l+1) &= (1 / \mathbf{r}_{l+1}(l)^T\mathbf{x}_{j(l+1)})\, \mathbf{r}_{l+1}(l) \\ \textit{and } (\forall i \neq l+1)\;\; \mathbf{r}_i(l+1) &= \mathbf{r}_i(l) - (\mathbf{r}_i(l)^T\mathbf{x}_{j(l+1)})\, \mathbf{r}_{l+1}(l+1) = \\ &= \mathbf{r}_i(l) - (\mathbf{r}_i(l)^T\mathbf{x}_{j(l+1)} / \mathbf{r}_{l+1}(l)^T\mathbf{x}_{j(l+1)})\, \mathbf{r}_{l+1}(l) \end{aligned} \tag{39}$$

where $j(l + 1)$ is the index of the feature vector $\mathbf{x}_{j(l+1)}$ inserted into the basis $\boldsymbol{B}_l$.

Remark 1: The Gauss-Jordan vector transformation (39) linked to the replacement of the unit vector $\mathbf{e}_{l+1}$ in the basis $\boldsymbol{B}_l = [\mathbf{x}_{j(1)},\ldots, \mathbf{x}_{j(l)}, \mathbf{e}_{l+1},\ldots, \mathbf{e}_n]^T$ by the feature vector $\mathbf{x}_{j(l+1)}$ cannot be performed when the below *collinearity condition* is met:

$$\mathbf{r}_{l+1}(l)^T\mathbf{x}_{j(l+1)} = 0 \tag{40}$$

The collinearity condition (40) causes division by zero in the Eq. (39).

Let the symbol $\mathbf{r}_{l+1}[l]$ denote the column $\mathbf{r}_{l+1}(l) = [\mathrm{r}_{l+1,1}(l),\ldots, \mathrm{r}_{l+1,n}(l)]^T$ of the inverse matrix $\boldsymbol{B}_l^{-1}$ (6) after the reducing of the last $n - l$ components $\mathrm{r}_{l+1,i}(l)$:

$$\mathbf{r}_l[l] = \left[\, r_{l,1}(l), \ldots, r_{l,l}(l)\right]^T \tag{41}$$

Similarly, the symbol $\mathbf{x}_j[l] = [\mathrm{x}_{j,1},\ldots,\mathrm{x}_{j,l}]^T$ means the reduced vector obtained from the feature vector $\mathbf{x}_j = [\mathrm{x}_{j,1},\ldots,\mathrm{x}_{j,n}]^T$ after reducing of the last $n - l$ components $\mathrm{x}_{j,i}$:

$$(\forall j \in \{1, \ldots, m)\mathbf{x}_j[l] = \left[\mathrm{x}_{j,1}, \ldots, \mathrm{x}_{j,l}, l\right]^T \tag{42}$$

Lemma 1: The collinearity condition (40) *appears* during the l-th step when the reduced vector $\mathbf{x}_{j(l+1)}[l]$ is a linear combination of the basis *reduced* vectors $\mathbf{x}_{j(i)}[l]$ with $i \leq l$:

$$\mathbf{x}_{j/l+1}[l] = \alpha_1\mathbf{x}_{j(1)}[l] + \ldots + \alpha_l\mathbf{x}_{j(l)}[l] \tag{43}$$

where $(\forall i \in \{1,\ldots, l\})\; \alpha_l \in R^1$.

The proof of this lemma can be based on the collinearity condition (41).

8 Feature Clustering

The basis exchange algorithm allows for the effective minimization of the *CPL* criterion functions. The optimal vertex $\mathbf{w}_k^*$ of the rank r_k constituting the minimal value $\Phi_{p,k}(\mathbf{w}_k^*, \theta_k^*)$ (11) of the perceptron criterion function $\Phi_{p,k}(\mathbf{w}, \theta)$ (10) or the minimal value $\Phi_c(\mathbf{w}_k^*)$ (17) of the collinearity criterion function $\Phi_c(\mathbf{w})$ (16) can be found in this way. Each optimal vertex $\mathbf{w}_k^*$ determines the k-th vertexical subspace $\boldsymbol{F}_k[r_k]$ ($\boldsymbol{F}_k[r_k] \subset \boldsymbol{F}[n]$). The subspace $\boldsymbol{F}_k[r_k]$ is composed of such r_k ($r_k \leq m$) features X_i from the feature subset (*cluster*) $F_k = \{X_{i(1)}, \ldots, X_{i(rk)}\}$that are linked to the non-zero components $\mathrm{w}_{k,i}^*$ of the optimal weight vector $\mathbf{w}_k^* = [\mathrm{w}_{k,1}^*,\ldots,\mathrm{w}_{k,n}^*]^T$.

Feature clusters $F_k = \{X_{i(1)}, \ldots, X_{i(rk)}\}$ are indicated by non-zero components $\mathrm{w}_{k,i}^*$ ($\mathrm{w}_{k,i}^* \neq 0$) of the optimal weight vectors $\mathbf{w}_k^* = [\mathrm{w}_{k,1}^*,\ldots,\mathrm{w}_{k,n}^*]^T$. The optimal vector $\mathbf{w}_k^*$

constitutes the minimal value $\Phi_{p,k}(\mathbf{v}_k^*)$ (24) of the *CPL* criterion function $\Phi_{p,k}(\mathbf{v})$ (23). The composition of the feature cluster F_k depends both on the structure of the data set $\boldsymbol{C}$ (1) and parameters of the criterion function $\Phi_{p,k}(\mathbf{v})$ (23).

Many feature clusters $F_{k(l)}$ ($F_{k(l)} \subset F$) can be extracted in accordance with the following procedure from a given feature set $F = \{X_1, ..., X_n\}$ composed of a large number n of features X_i ($m << n$):

Feature clustering procedure (44)

1. $l = 1$
2. the minimum value $\Phi_{p,k(l)}(\mathbf{w}_{k(l)}^*, \theta_{k(l)}^*)$ (11) (or $\Phi_c(\mathbf{w}_{k(l)}^*)$ (17)) is found and the feature cluster $F_{k(l)}$ ($F_{k(l)} \subset F$) is extracted
3. if $\Phi_{p,k(l)}(\mathbf{w}_{k(l)}^*, \theta_{k(l)}^*) = 0$, then the feature set F is reduced by the feature cluster $F_{k(l)}$ ($F \rightarrow F\ /F_{k(l)}$), otherwise the procedure is stopped.
4. the step 2. is repeated with the l counter increased by one ($l \rightarrow l + 1$)

Feature clusters $F_{k(l)}$ ($F_{k(l)} \subset F$) enable defining (*Definition* 6) vertexical subspaces $\boldsymbol{F}_{k(l)}[r_{k(l)}]$ ($\boldsymbol{F}_{k(l)}\ [r_{k(l)}] \subset \boldsymbol{F}[n]$) where $l = 1,\ldots, L$. Local classifiers or prognostic models may be built in each of the feature subspaces $\boldsymbol{F}_{k(l)}[r_{k(l)}]$. Global classifiers or prognostic models with a good quality can be obtained in the whole feature space $\boldsymbol{F}[n]$ by averaging local models.

9 Concluding Remarks

This paper presents a new concept for designing classifiers or prognostic models from data in the case when the data set $\boldsymbol{C}$ (1) consists a small number m high - dimensional feature vectors $\mathbf{x}_j$ ($m << n$).

It has been shown that the learning sets C_k (3) composed of small numbers m_k of linearly independent feature vectors $\mathbf{x}_j$ are linearly separable (6) regardless of the assignment of these vectors to individual learning sets C_k (*Theorem* 3). Linear separability (6) of the learning sets C_k (3) can be destroyed by the existence of collinearity relations between selected features X_i [4].

Linear separability (6) of the learning sets C_k (3) can be explored by the minimization of the convex and piecewise linear (*CPL*) criterion functions $\Phi_{p,k}(\mathbf{w}, \theta)$ (10) or $\Psi_{p,k}(\mathbf{w}, \theta)$ (13). Minimizing the collinear criterion function $\Phi_c(\mathbf{w})$ (14) allows extracting linear relations between some features X_i or feature vectors $\mathbf{x}_j$ [9].

The basis exchange algorithm can be used in feature clustering. The presented concept of feature clustering gives possibility for the decomposition of whole feature space $\boldsymbol{F}[n]$ into vertexical feature subspaces $\boldsymbol{F}_{k(l)}[r_{k(l)}]$. Such decomposition may allow for the replacement of feature vectors $\mathbf{x}_j$ agregation by the averaging of selected features X_i. The last possibility can be particularly important in the cases of data sets $\boldsymbol{C}$ (1) composed from a small number m of high - dimensional feature vectors $\mathbf{x}_j$ ($m << n$).

Acknowledgments. The presented study was supported by the grant S/WI/2/2020 from Bialystok University of Technology and funded from the resources for research by Polish Ministry of Science and Higher Education.

References

1. Hand, D., Smyth, P., Mannila, H.: Principles of Data Mining. MIT Press, Cambridge (2001)
2. Duda, O.R., Hart, P.E., Stork, D.G.: Pattern Classification. John Wiley, New York (2001)
3. Bishop, C.M.: Pattern Recognition and Machine Learning. Springer, Heidelberg (2006)
4. Bobrowski L.: Data Exploration and Linear Separability, pp. 1–172. Lambert Academic Publishing (2019)
5. Bobrowski, L.: Data mining based on convex and piecewise linear (CPL) criterion functions (in Polish). Bialystok University of Technology Press (2005)
6. Bobrowski, L., Łukaszuk, T.: Relaxed linear separability (RLS) approach to feature (Gene) subset selection. In: Xia, X. (ed.) Selected Works in Bioinformatics, pp. 103–118. INTECH (2011)
7. Bobrowski, L.: Design of piecewise linear classifiers from formal neurons by some basis exchange technique. Pattern Recognit. **24**(9), 863–870 (1991)
8. Simonnard, M.: Linear Programming, Prentice Hall, Englewood Cliffs (1966)
9. Bobrowski, L.: Discovering main vertexical planes in a multivariate data space by using CPL functions. In: Perner, P. (ed.) ICDM 2014. LNCS (LNAI), vol. 8557, pp. 200–213. Springer, Cham (2014). https://doi.org/10.1007/978-3-319-08976-8_15

Using Fourier Series to Improve the Discrete Grey Model (1, 1)

Van-Thanh Phan[1,2](✉), Zbigniew Malara[3](✉), and Ngoc Thang Nguyen[4](✉)

[1] Institute of Research and Development, Duy Tan University, Da Nang 550000, Vietnam
thanhkem2710@gmail.com
[2] Quang Binh University, 312 Ly Thuong Kiet Street, Dong Hoi, Quang Binh, Vietnam
[3] Wroclaw University of Science and Technology, 27 Wybrzeże Wyspiańskiego Street, 50-370 Wrocław, Poland
zbigniew.malara@pwr.edu.pl
[4] Faculty of Economics, Tay Nguyen University, 567 Le Duan Street, Buon Ma Thuot City, DakLak, Vietnam
nnthang@ttn.edu.vn

Abstract. Discrete grey model (1, 1) (abbreviates as DGM (1, 1)), is a version of grey forecasting model. Since appeared, its has been attracted by many scientists in dealing with the problem related to uncertainly information and small sample data. In recent years, this model has been improved the accuracy in forecast by scientifics. However, the existing DGM (1, 1) model cannot be used in some special scenarios such as the significant fluctuation or noise in data. Solving this issue, this paper propose a novel grey forecasting model named as Fourier Discrete Grey Model (1, 1) (abbreviated as F-DGM (1, 1)). This model was built by combined the Fourier series and DGM (1, 1) model. Through the example in Xie and Liu's paper (Xie and Liu [28]) and practical application, these simulation outcomes demonstrated that the F-DGM (1, 1) model provided remarkable prediction performance compared with the other grey forecasting models. Future direction, the authors will use different equations or different methodologies to improve the DGM (1, 1) model. The other direction is applied the proposed model for dealing with the highly fluctuation data in different industries.

Keywords: Discrete grey model · Fluctuation time series · Fourier series · Accuracy

1 Introduction

Grey forecasting model is an effective method for working on system forecasting with poor or uncertain information. In the early 1980s, Prof. Deng based on control theory to propose the grey model GM (1, 1) [1, 2]. This model is a time series prediction model. Based on the grey generating approach, it reduce the fluctuations of raw data series by transforming the data become linearly. Because the GM (1, 1) need small sample data to built the behavior function, has simple calculate process and higher prediction accuracy [3–5]. That reason why many scientists applied the GM (1, 1) model in order to solve

M. Hernes et al. (Eds.): ICCCI 2020, CCIS 1287, pp. 99–109, 2020.
https://doi.org/10.1007/978-3-030-63119-2_9

problems related to uncertain information in many disciplines such as tourism industry [6, 7], logistics and transportation industry [8–10], financial and economic [11–13], IC industry [14–17], energy industry [18–20].

In order to increase the accuracy of the forecasting model, many scholars tries to improve the precision accuracy by proposed new procedures or new models. Based on the background values, Lin et al. [21] and Wang et al. [22] proposed the new calculation to estimate the coefficiency parameter of forecasting model. In order to modify the internal parameter estimation, Hsu [17] and Wang et al. [23] proposed a new approach in calculation the development coefficient and grey input coefficient of grey forecasting model. Some scholars concentrate on the residuals modification had established the new GM (1, 1) model [15, 24]. The other direction is considered by combined the different methodologies to make the hybrid models based on GM (1, 1). GM (1,1) combined with econometric model in [25], the Markov chain combine with GM (1, 1) become the grey Markov model in [26, 27], or using the fuzzy number applied in grey model to built the grey fuzzy model [21], and so on. Despite these model has been improved. However, the accuracy of the GM (1, 1) is not suitable in some scenarios.

In among directions mentioned, Xie and Liu [28] had used grey different equation to built the new ones named as Discrete Grey Model (1, 1) (abbreviated as DGM (1, 1)) to improved the accuracy of GM (1, 1) model. Recently, this model was developed and widely applied in different industries. To solve the growth of prediction of discrete grey model, Qi et al. [29] used quadratic time-varying parameters to established a new grey discrete parameters prediction model. The simulation result demonstrated that the new model achieved the greatly improvement the prediction precision. To forecast the China's per capita gross domestic product (GDP) during the period time 2001 to 2009, Liu et al. [30] give the concrete calculation formula to improve prediction accuracy by optimization the ending-point of discrete grey (1, 1) model, the results show that the accuracy of the proposed model is superior to the traditional discrete GM (1, 1) and ending-point fixed discrete grey model (EDFGM (1,1)). Furthermore, research results indicated that the one step and two step prediction accuracy has more dominant advantage. Fei et al. [31] improve the form discrete GM (1, 1) model to forecast the real estate prices. Research results showed that proposed model can be considered as the precise form of grey forecasting model.

Although DGM (1, 1) model have been improved and have provided us with promising results, However, in some situations like the data are highly fluctuating, or nonlinear, the DGM (1, 1) is not always satisfactory in some scenarios. That reason why this study put forward to solve these issues. Based on the characteristics of Fourier series and the mathematical algorithm of DGM (1, 1), this study to proposed a novel model aim to increase the performance of forecast. The overall process of proposed model has implemented through two parts. The first part is based on the mathematical algorithm of DGM (1, 1) to calculated the predicted and error value, then using Fourier series to modify the error values. To express the advantages of the proposed model, this study uses the numerical example in Xie and Liu's paper [28] and the closed price in Taiwan as an example. Simulation results indicated that the proposed model provided remarkable prediction performance compared with the other grey forecasting models in the highly fluctuation sequence.

The rest of the article is organized as follows. A Sect. 2 briefly introduces the original DGM (1, 1) and proposes a novel grey forecasting model named F-DGM (1, 1). Section 3 illustrates the efficiency performances of F-DGM (1, 1) compare with the others forecasting model in several examples. And the conclusions are made in the final section.

2 Methodology

2.1 A Brief Introduction to the Discrete Grey Model

According to Xie and Liu [28], The Discrete Grey Model (1, 1) was detail introduced as follows.

Step 1: Assume that we have the *non-negative* original sequence data $X^{(0)}$

$$X^{(0)} = \left\{x^{(0)}(1), x^{(0)}(2), \ldots, x^{(0)}(n)\right\},\ n \geq 4 \tag{1}$$

Step 2: We used *one time accumulated generating operation* (*1-AGO*) to construct $X^{(1)}$ aim to make increasingly sequence $X^{(0)}$ by equation belows:

$$x^{(1)}(k) = \sum_{i=1}^{k} x^{(0)}(i),\ k = 1, 2, \ldots, n \tag{2}$$

$$X^{(1)} = \left\{x^{(1)}(1), x^{(1)}(2), .., x^{(1)}(n)\right\} \tag{3}$$

Step 3: The Discrete grey model has the forms below:

$$x^{(1)}(k+1) = \beta_1 x^{(1)}(k) + \beta_2 \tag{4}$$

Where β_1 and β_2 are model parameters.

Step 4: In oder to calculate the parameters β_1 and β_2 we use ordinary least squares method (OLS) as follows function.

$$\begin{bmatrix} \beta_1 \\ \beta_2 \end{bmatrix} = (B^T B)^{-1} B^T Y_n \tag{5}$$

Where B and Y_n are defined as follows:

$$B = \begin{bmatrix} x^{(1)}(1) & 1 \\ x^{(1)}(2) & 1 \\ \cdots & \cdots \\ x^{(1)}(n-1) & 1 \end{bmatrix} \text{ and } Y_n = \begin{pmatrix} x^{(0)}(2) \\ x^{(0)}(3) \\ \cdots \\ x^{(0)}(n) \end{pmatrix}$$

Particular solution of Discrete grey model is

$$\hat{x}^{(1)}(k+1) = \beta_1^k x^{(0)}(1) + \frac{1-\beta_1^k}{1-\beta_1} \times \beta_2;\ k = 1, 2, \ldots, n-1, \tag{6}$$

Where initial condition: $x^{(1)}(1) = x^{(0)}(1)$

Step 5: The predicted datum of $x^{(0)}(k)$ can be estimated by apply the *inverse accumulated generating operation* (*I-AGO*) $\hat{x}^{(1)}(k)$ follow equation belows:

$$\hat{x}^{(0)}(k+1) = \hat{x}^{(1)}(k+1) - \hat{x}^{(1)}(k),\quad k = 1, 2, 3, \ldots \tag{7}$$

2.2 The Error Filter of DGM (1, 1) by Fourier Series

Based on the characteristics of Fourier series, its transform the residuals error into frequency spectra and then select the low-frequency terms. Moreover, in the researches of Wang and Phan [24] and Huang et al. [6], the Fourier technique was used to filter out high-frequency terms, removed the noise in the sequence data aim for better performance. That reason why this study uses the Fourier series to solve this problem in the DGM (1, 1) model aim to improve the prediction accuracy. The process are given belows:

Step 1: Based on the actual data series of x and the predicted series of $\hat{x}$, we calculated the residual series of ε by function belows:

$$\varepsilon(k) = x(k) - \hat{x}(k), \ k = 2, 3, \ldots n \tag{8}$$

$$\varepsilon = \{\varepsilon(k)\}, \ k = 2, 3, \ldots n$$

According to the definition of the Fourier series [6], the residual sequence of DGM (1, 1) can be approximately expressed as

$$\hat{\varepsilon}(k) = \frac{1}{2}a_{(0)} + \sum_{i=1}^{Z}\left[a_i \cos\left(\frac{2\pi i}{n-1}(k)\right) + b_i \sin\left(\frac{2\pi i}{n-1}(k)\right)\right], \ k = 1, 2, 3, .., n \tag{9}$$

Where $z = (\frac{n-1}{2}) - 1, \ (z \in Z^*)$

z is called the minimum deployment frequency of Fourier series.

Step 2: Therefore, the residual series is rewritten as

$$\varepsilon = P \times C \tag{10}$$

Where

$$P = \begin{bmatrix} \frac{1}{2} & \cos\left(\frac{2\pi\times 1}{n-1}\times 2\right) & \sin\left(\frac{2\pi\times 1}{n-1}\times 2\right) & \ldots & \cos\left(\frac{2\pi\times z}{n-1}\times 2\right) & \sin\left(\frac{2\pi\times z}{n-1}\times 2\right) \\ \frac{1}{2} & \cos\left(\frac{2\pi\times 1}{n-1}\times 3\right) & \sin\left(\frac{2\pi\times 1}{n-1}\times 3\right) & \ldots & \cos\left(\frac{2\pi\times z}{n-1}\times 3\right) & \sin\left(\frac{2\pi\times z}{n-1}\times 3\right) \\ \ldots & & \ldots & \ldots & & \ldots \\ \frac{1}{2} & \cos\left(\frac{2\pi\times 1}{n-1}\times n\right) & \sin\left(\frac{2\pi\times 1}{n-1}\times n\right) & \ldots. & \cos\left(\frac{2\pi\times z}{n-1}\times n\right) & \sin\left(\frac{2\pi\times z}{n-1}\times n\right) \end{bmatrix}$$

And $C = [a_0, a_1, b_1, a_2, b_2, \ldots, a_Z, b_Z]^T$

Step 3: To solve the parameter a_0, a_1, b_1, a_2, b_2… a_Z, b_Z, we used the ordinary least squares method (OLS) by the equation belows:

$$C = \left(P^T P\right)^{-1} P^T [\varepsilon]^T \tag{11}$$

Once the parameters are calculated, the modified residual series is calculate by following equation

$$\hat{\varepsilon}(k) = \frac{1}{2}a_0 + \sum_{i=1}^{Z}\left[a_i \cos\left(\frac{2\pi i}{n-1}(k)\right) + b_i \sin\left(\frac{2\pi i}{n-1}(k)\right)\right]$$

Step 4: we got the final predicted value $\hat{v}$ from the predicted series $\hat{x}$ and $\hat{\varepsilon}$, as follow equation:

$$\hat{v} = \begin{cases} \hat{v}_1 = \hat{x}_1 \\ \hat{v}_k = \hat{x}_k + \hat{\varepsilon}_k \; (k = 2, 3, \ldots, n) \end{cases} \tag{12}$$

2.3 Prediction Capability Analysis

In order to evaluate the performance and reliability of forecasting model, this study used Means Absolute Percentage Error (MAPE) and prediction accuracy indexes [33]. Their calculation is introduced as follows:

$$MAPE = \frac{1}{n}\sum_{k=2}^{n}\left|\frac{x^{(0)}(k) - \hat{x}^{(0)}(k)}{x^{(0)}(k)}\right| \times 100\% \tag{13}$$

Where $x^{(0)}(k)$ is actual value and $\hat{x}^{(0)}(k)$ is the forecasted values in time period k, and n is the total number of predictions.

The other index is ρ (called as the prediction accuracy or precision rate). And This index is calculated by equation belows:

$$\rho = 100 - MAPE \tag{14}$$

Based on the value of MAPE, Wang and Phan [32] classified into four grades which are The MAPE value of forecasting model greater than 10% indicated that the forecast models is an inaccurate, from 5% to 10% is a reasonable forecast, from 1% to 5% is a good forecast, and less than 1% is an excellent forecast.

3 Verification of the F-DGM (1, 1)

In this section, the numerical examples and the real practical are given to illustrated the performance of the proposed model with the other grey forecasting models, which are the GM (1, 1), DGM (1, 1) and OSDGM (1, 1) [in Xie and Liu's paper]. The first example is the numerical example was proposed in the Xie and Liu's paper [28], and the second example is the forecast the close price of Taiwan Stock Exchange Capitalization Weight Stock Index.

The procedure of the GM (1, 1) model was comprehensively illustrated in Wang and Phan [24, 32] and DGM (1, 1) model, OSDGM (1, 1) model illustrated in Xie and Liu [28]. In term of the proposed prediction model, the overall procedure was above mentioned (see Sects. 2.1 and 2.2). Specifically as follows.

3.1 Numerical Example in Xie and Liu's Paper

The same sample from Xie and Liu's paper [28] is applied here to compare the precision. In Xie and Liu study, they used the raw data sequence jumps randomly $X^{(0)} = (21.1, 26.6, 36.1, 52.3, 80.1, 126.8)$ to demonstrate the higher performance of proposed model named as DGM (1, 1) and OSDGM (1, 1) compared with GM (1, 1). In this study, we used the same sequence above to compare the forecasting performance of the F-DGM (1, 1) with the GM (1, 1), and the OSDGM (1 1) model in Xie and Liu's paper. Based on the mathematical algorithm of each model to calculated the results, the forecasting results are shown in Table 1 and Fig. 1.

Table 1. Forecasted results from the grey models

Original value		GM (1, 1)		DGM (1, 1)		OSDGM (1, 1)		**F-DGM (1, 1)**	
	$x^{(0)}(k)$	$\hat{x}^{(0)}(k)$	MAPE (%)	$\hat{x}^{(0)}(k)$	MAPE (%)	$\hat{x}^{(0)}(k)$	MAPE (%)	$\hat{x}^{(0)}(k)$	**MAPE (%)**
k = 1	21.1	21.1	0	21.1	0	21.1	0	21.1	**0**
k = 2	26.6	22.45	15.61	22.90	13.93	23.48	11.73	26.6	**0**
k = 3	36.1	33.91	6.06	34.79	3.64	35.67	1.19	36.1	**0**
k = 4	52.3	51.24	2.03	52.85	1.06	54.20	3.63	52.3	**0**
k = 5	80.1	77.41	3.36	80.30	0.25	82.35	2.81	80.1	**0**
k = 6	126.8	116.96	7.76	122.00	3.78	125.12	1.32	126.8	**0**
MAPE (%)			6.96		3.77		3.44		**0.00**
Precision accuracy (%)			93.04		96.23		96.56		**100**

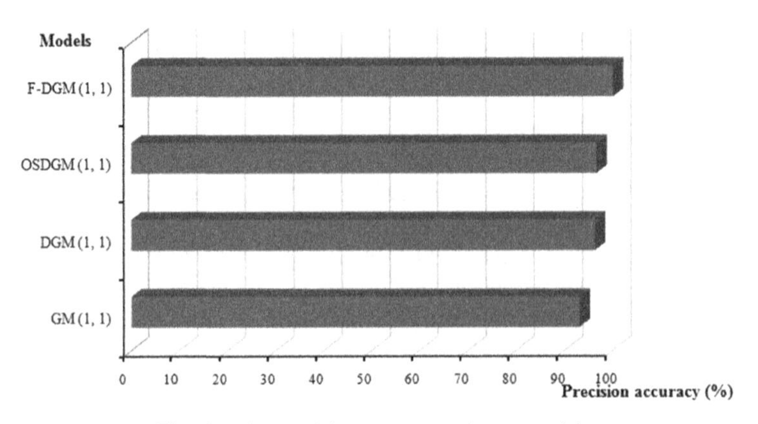

Fig. 1. The precision accuracy of grey models.

Table 1 indicated that the DGM (1, 1) has a higher accuracy than the original GM (1, 1) with MAPE decreased from 6.96% to 3.77%. Furthermore, The OSDGM (1, 1) model proposed by Xie and Liu also get the better forecasted performance than DGM. The MAPE reduced from 3.77% to 3.44%. In this study, by adopting the F-DGM (1, 1), the MAPE decreased from 3.44% to 0%. This result indicated that the perturbation of F-DGM (1, 1) is smaller than the other forecasting models.

3.2 The Closed Price of Taiwan Stock Exchange Capitalization Weight Stock Index Forecasting

To verify the effectiveness of F-DGM (1, 1) model in the real case, this paper uses the historical data for the closed price of TSEC weighted index from Aug 25, 2015 to Sep 11, 2015. This sequence data is obtained from the website of the Index Book [34]. There are totally 14 observations in the sequence data and illustrated in Fig. 2. Figure 2 reveals that close price of TSEC weighted index is significant fluctuation over the period. In this case, study divided the sequence data become two period times to show the superiority of proposed model. The first sequence data sample from Aug 25, 2015 to Sep 08, 2015 (11 data points) for interpolation data testing. And the second data sample during period time from Sep 09, 2015 to Sep 11, 2015 is the out - of sample data for extrapolation data testing.

In this study, we used Microsoft Excel in order to calculate the parameters of forecasting models. This is useful software for matrix calculation such as in order to matrix multiplication of two relevant arrays, we used the function MMULT (array 1, array 2). To solves the inverse matrix we used the Minverse function. Because of the computation

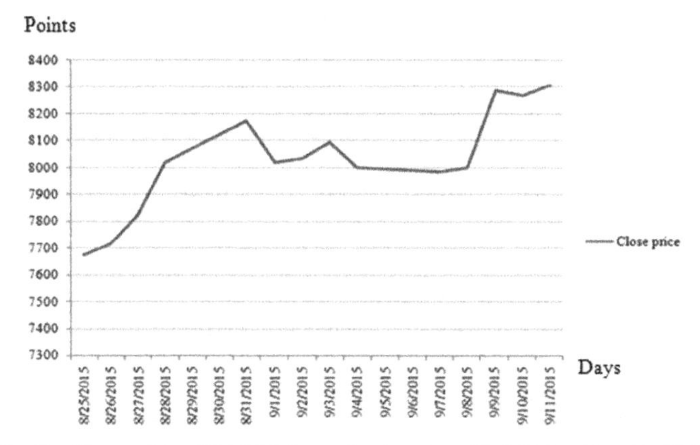

Fig. 2. Close price of TAIEX from the day Aug 25, 2015 to Sep 11, 2015.

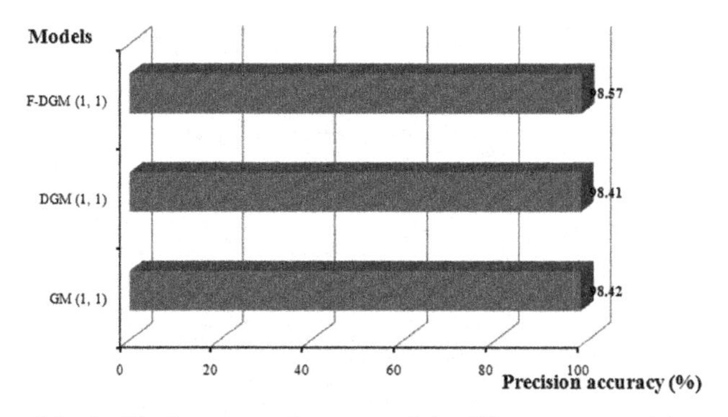

Fig. 3. The forecast performance of the different grey models.

is too long so, the calculation process are omitted here. The forecasted results and the error value of these models for in-sample, out-of sample forecasted are shown in Table 2.

Table 2. In samples and out-of-samples comparison among three grey forecasting models.

Date	Actual value	Forecasted value					
		GM (1, 1)	Error (%)	DGM (1, 1)	Error (%)	F-DGM (1, 1)	Error (%)
In sample							
Aug 25,2015	7,675.64	7675.64	0	7675.64	0	7675.64	0
Aug 26,2015	7,715.59	7885.27	2.20	7896.16	2.34	7725.64	0.13
Aug 27,2015	7,824.55	7905.48	1.03	7916.25	1.17	7814.50	0.13
Aug 28,2015	8,019.18	7925.74	1.17	7936.39	1.03	8029.23	0.13
Aug 31,2015	8,174.92	7946.05	2.80	7956.58	2.67	8164.87	0.12
Sep 01, 2015	8,017.56	7966.42	0.64	7976.82	0.51	8027.61	0.13
Sep 02, 2015	8,035.29	7986.83	0.60	7997.12	0.48	8025.24	0.13
Sep 03, 2015	8,095.95	8007.30	1.09	8017.46	0.97	8106.00	0.12
Sep 04, 2015	8,000.60	8027.83	0.34	8037.86	0.47	7990.55	0.13
Sep 07, 2015	7,986.56	8048.40	0.77	8058.31	0.90	7996.61	0.13
Sep 08, 2015	8,001.50	8069.03	0.84	8078.81	0.97	7991.45	0.13
Average MAPE (%)			1.15		1.15		0.11
Out-of-sample							
Sep 09, 2015	8286.92	8099.85	2.26	8099.37	2.26	7928.85	4.32
Sep 10, 2015	8268.68	8120.55	1.79	8119.97	1.80	8018.22	3.03
Sep 11, 2015	8305.82	8141.30	1.98	8140.63	1.99	8233.48	0.87

(*continued*)

Table 2. *(continued)*

Date	Actual value	Forecasted value					
		GM (1, 1)	Error (%)	DGM (1, 1)	Error (%)	F-DGM (1, 1)	Error (%)
Average MAPE (%)			2.01		2.02		2.74
Average MAPE of In and out-of-samples			**1.58**		**1.59**		**1.43**
Precision accuracy (%)			98.42		98.41		98.57
Performance			Good		Good		Good

Table 2 indicated that the simulation outcomes of proposed model is more accuracy with the average MAPE indexes for in and out-of-sample forecast is 1.43%, much smaller than among three forecasting models. More clearly visualization was shown in Fig. 3.

4 Conclusion

This paper proposed an effectiveness grey forecasting model for dealing with the data have a significant fluctuations based on the advantages of Fourier series. The proposed model is done in two stages. First, we based on the mathematical algorithm of DGM (1, 1) model to estimate the prediction and then modify the error value by Fourier series. The simulation results from Table 1 and Table 2 indicated that the performance of F-DGM (1, 1) model can forecast better than existing DGM (1, 1) model. In the future, the author can used different equations or different methodologies to reduce the error of DGM (1, 1) model. Or applied the proposed model to assess some special cases.

References

1. Deng, J.L.: Grey Prediction and Decision. Huazhong University of Science and Technology, Wuhan (1986)
2. Deng, J.L.: Solution of grey differential equation for GM (1, 1| τ, r) in matrix train. J. Grey Syst. **13**, 105–110 (2002)
3. Deng, J.L.: Control problems of grey systems. Syst. Control Lett. **1**, 288–294 (1982)
4. Lin, Y., Liu, S.: A historical introduction to grey systems theory. In: 2004 IEEE International Conference on Systems, Man and Cybernetics, vol. 3, pp. 2403–2408 (2004)
5. Liu, S., Jeffrey, F., Yang, Y.: A brief introduction to grey systems theory. Grey Syst. Theory Appl. **2**, 89–104 (2012)
6. Huang, Y.L., Lee, Y.H.: Accurately forecasting model for the stochastic volatility data in tourism demand. Mod. Econ. **2**, 823–829 (2011)
7. Chu, F.L.: Forecasting tourism demand in Asian-Pacific countries. Ann. Tour. Res. **25**, 597–615 (1998)
8. Jiang, F., Lei, K.: Grey prediction of port cargo throughput based on GM (1, 1, a) model. Logist. Technol. **9**, 68–70 (2009)

9. Guo, Z.J., Song, X.Q., Ye, J.: A Verhulst model on time series error corrected for port throughput forecasting. J. Eastern Asia Soc. Transp. Stud. **6**, 881–891 (2005)
10. Lu, I.J., Lewis, C., Lin, S.J.: The forecast of motor vehicle, energy demand and CO_2 emission from Taiwan's road transportation sector. Energy Policy **37**, 2952–2961 (2009)
11. Kayacan, E., Ulutas, B., Kaynak, O.: Grey system theory-based models in time series prediction. Expert Syst. Appl. **37**, 1784–1789 (2010)
12. Askari, M., Askari, H.: Time series grey system prediction-based models: gold price forecasting. Trends Appl. Sci. Res. **6**, 1287–1292 (2011)
13. Wang, Y.F.: Predicting stock price using fuzzy grey prediction system. Expert Syst. Appl. **22**, 33–39 (2002)
14. Tsai, L.C., Yu, Y.S.: Forecast of the output value of Taiwan's IC industry using the grey forecasting model. Int. J. Comput. Appl. Technol. **19**, 23–27 (2004)
15. Hsu, L.C.: Applying the grey prediction model to the global integrated circuit industry. Technol. Forecast. Soc. Chang. **70**, 563–574 (2003)
16. Hsu, L.C.: A genetic algorithm based nonlinear grey Bernoulli model for output forecasting in integrated circuit industry. Expert Syst. Appl. **37**, 4318–4323 (2010)
17. Hsu, L.C.: Using improved grey forecasting models to forecast the output of opto-electronics industry. Expert Syst. Appl. **38**, 13879–13885 (2011)
18. Li, D.C., Chang, C.J., Chen, C.C., Chen, W.C.: Forecasting short-term electricity consumption using the adaptive grey-based approach-an Asian case. Omega **40**, 767–773 (2012)
19. Hsu, C.C., Chen, C.Y.: Application of improved grey prediction model for power demand forecasting. Energy Convers. Manag. **44**, 2241–2249 (2003)
20. Kang, J., Zhao, H.: Application of improved grey model in long-term load forecasting of power engineering. Syst. Eng. Procedia **3**, 85–91 (2012)
21. Lin, Y.H., Chiu, C.C., Lee, P.C., Lin, Y.J.: Applying fuzzy grey modification model on inflow forecasting. Eng. Appl. Artif. Intell. **25**, 734–743 (2012)
22. Wang, Z.X., Dang, Y.G., Liu, S.F.: The optimization of background value in GM (1, 1) model. J. Grey Syst. **10**, 69–74 (2007)
23. Wang, C.H., Hsu, L.C.: Using genetic algorithms grey theory to forecast high technology industrial output. Appl. Math. Comput. **195**, 256–263 (2008)
24. Wang, C.N.; Phan, V.T.: Enhancing the accurate of grey prediction for GDP growth rate in Vietnam. In: 2014 International Symposium on Computer, Consumer and Control (IS3C), pp. 1137–1139 (2014)
25. Liu, S.F., Lin, Y.: Grey Information Theory and Practical Applications. Springer, London (2006)
26. Dong, S., Chi, K., Zhang, Q.Y., Zhang, X.D.: The application of a grey Markov model to forecasting annual maximum water levels at hydrological stations. J. Ocean Univ. China **11**, 13–17 (2012)
27. Hsu, Y.T., Liu, M.C., Yeh, J., Hung, H.F.: Forecasting the turning time of stock market based on Markov-Fourier grey model. Expert Syst. Appl. **36**, 8597–8603 (2009)
28. Xie, N.M., Liu, S.F.: Discrete grey forecasting model and its optimization. Appl. Math. Model. **33**, 1173–1186 (2009)
29. Qi, Y.J., Wu, Z.P., Li, Y., Yu, J.: Grey discrete parameters model and its application. In: 2014 IEEE/ACIS 13th International Conference on Computer and Information Science (ICIS), pp. 399–404 (2014)
30. Liu, J.F., Liu, S.F., Fang, Z.G.: Predict China's per capita GDP based on ending point optimized discrete grey (1, 1) model. In: 2013 IEEE International Conference on Grey Systems and Intelligent Services, pp. 113–117 (2013)
31. Fei, Z.H., Jiang, Z.Y., He, Z.F.: Discrete GM (1, 1) model and its application for forecasting of real estate prices. In: 2011 International Conference on Management Science and Industrial Engineering (MSIE), pp. 1195–1197 (2011)

32. Wang, C.N., Phan, V.T.: An improvement the accuracy of grey forecasting model for cargo throughput in international commercial ports of Kaohsiung. Int. J. Bus. Econ. Res. **3**, 1–5 (2014)
33. Makridakis, S.: Accuracy measures: theoretical and practical concerns. Int. J. Forecast. **9**, 527–529 (1993)
34. Website of index book. http://www.indexbook.net/

Studying on the Accuracy Improvement of GM (1, 1) Model

Van Đat Nguyen[1], Van-Thanh Phan[2,3](✉), Ngoc Thang Nguyen[1](✉), Doan Nhan Dao[4](✉), and Le Thanh Ha[1]

[1] Faculty of Economics, Tay Nguyen University, 567 Le Duan Street, Buon Ma Thuot City, Dak Lak, Vietnam
{nvdat,nnthang,ltha}@ttn.edu.vn
[2] Institute of Research and Development, Duy Tan University, Da Nang 550000, Vietnam
thanhkem2710@gmail.com
[3] Quang Binh University, 312 Ly Thuong Kiet Street, Dong Hoi, Quang Binh, Vietnam
[4] National Academy of Public Administration, 77 Nguyen Chi Thanh Street, Dong Da Ward, Ha Noi City, Vietnam
doannhandao@gmail.com

Abstract. In order to expand the application of GM (1, 1) in the condition of fluctuation data and incomplete information, this paper proposed the new systematic optimization based on three steps as follows. First step, we used parameters c_1 to transform any sequence data into the non-negative sequence data. The second, we used moving average operation method on the new sequence data to smooth the sequence data aim to satisfy the quasi-exponential condition and quasi-smooth condition. The final, we adopt Fourier series to modify residual error of model a grey sequence. To demonstrate the superiority of the proposed model, the numerical example in the research of Wang and Hsu and the raw data sequence are used. The simulation outcomes show that the proposed approach provides a better forecast results than several different kinds of grey forecasting models with the lowest average of MAPE for in and out-of-samples in two cases. For future direction, the author will applied different methodologies to improve the performance of GM (1, 1) or use proposed model to analyse the issues with high fluctuation data.

Keywords: GM (1, 1) · Fluctuation data · Negative numbers · Moving average · Residual error modification · Fourier series

1 Introduction

Grey system theory was proposed by Prof. Deng in early 1908s. It is a new methodology working on model uncertainty and information insufficiency system. This theory include the conditional analysis, controlling, forecasting and decision-making based on six axioms are principle of information differences, principle of non-uniqueness, principle of minimal information, principle of recognition base, principle of new information priority and principle of absolute grayness [1, 2]. The grey generating, grey relational analysis, grey forecasting, grey decision making, and grey control are five main groups

M. Hernes et al. (Eds.): ICCCI 2020, CCIS 1287, pp. 110–121, 2020.
https://doi.org/10.1007/978-3-030-63119-2_10

of grey system theory. Among them, the grey forecasting one of the most important part and widely used to deal with problem in various industries. The general function of grey forecasting model is GM (n, m). Because of the easily calculation and required the limited data. therefore, Grey model has become than effective method to study uncertainty problems and incomplete information. Over the last two decades, the GM (1, 1) model has been aroused the interest of the scientists and scholars in the world. In early 1990s, some universities have started teaching grey system theory in education program such as Australia, China, Japan, Taiwan, as well as in USA. Every year, A conference on grey system theory and applications is held by Chinese Grey System Theory Association CGSA. Some journal also was established to published the research results related to grey system theory. Recently, more than 750000 articles related the grey system is accept and published in the world [2].

In the research of Lin and Liu [2], the authors indicated the advantages of grey forecasting model compared with three methods (probability and statistics, fuzzy theory, and rough sets) was shown under two dimensions, the first is the higher prediction accuracy and the second is dealing with the uncertain systems. Furthermore, Lin and Yang [9] also figured out that Grey system theory can be deal with the problems of a uncertainty information and a small amount of data while both of probability and statistics and fuzzy mathematics method can be unsolvable. That reason why grey forecasting model has widely and successfully applied in various fields such as tourism industry [3, 4], logistics industry [5, 6], financial and economic [7, 8], IC industry [9–11], energy industry [12, 13] and so forth.

Although those GM (1, 1) models have been improvement the performance in forecasting but the GM (1, 1) still cannot be used for some actual situations. Some reasons that the grey forecasting model has restriction like the raw time series data must be satisfy the non-negative, to perform the grey forecasting model, the sequence data must be satisfy the quasi-smooth and quasi-exponential condition. The causes internal errors in constructing the grey forecasting model by using a the first time AGO and mean generating operation (MGO) as well as inverse accumulated generating operation (IAGO). And finally, the fluctuating data set, or stochastic volatility data is also the main caused the grey forecasting model is inappropriate. Therefore, many scholars try to put forward proposed the new procedures or new models to reduce the forecast error of grey model. In 2011, Zeng et al. [14] adopted moving average operation method to smooth the fluctuation data. In the establish process of GM (1, 1) model, Tan [23] adopted the integral method to get the background values and expanded the equal interval into the unequal interval by changed the value of $z(1)$ into $z^{(1)}(k+1) = \frac{1}{2m}\left[(m+1)x^{(1)}(k) + (m-1)x^{(1)}(k+1)\right]$ with $k = 2, 3, \ldots n$ and $m = \left(\sum_{i=2}^{n} \frac{x^{(1)}(i)}{x^{(1)}(i-1)}\right)^{\frac{1}{n-1}}$ n: the number of element in $x^{(0)}$. Wang et al. [15] also provided the different way to calculate the value of $z^{(1)}$. Wu et al. [16] used the fractional accumulating operation replace to the traditional accumulating operation (one time accumulating operation). Shen and Qin [17] also used fractional accumulating operation replace to the traditional accumulating operation. All above studied concentrate studying on improve the accuracy of the GM (1, 1) model by using different methodologies to calculate the background values.

Other scholars concentrate on combined the advantages of different forecasting models with GM (1, 1) to improve the accuracy like fuzzy in Wei et al. [18]. Some researchers focus on reduced the residual error by modified the residual error. Li et al. [19] proposed the Grey-Markov model aim to reduce random fluctuation data to forecast the customer of Chinese international airlines. Huang and Lee [3] used Fourier series to modified the residual error of grey forecasting model (GM (1, 1)) to solve the stochastic volatility data in tourism industry. Wang and Phan [20] proposed an effective model named as FMGM (1, 1) by combine the traditional GM (1, 1) with Fourier series aim to reduce the residual error of GM (1, 1). Zhou et al. [21] integrate the traditional grey model GM (1, 1) and the trigonometric residual modification technique to forecast the electricity demand in China. Chen [22] used the nonlinear grey Bernoulli model (NGBM) to predict the annual unemployment rate and foreign exchange rates.

Based on the above background settings and the restriction existing of grey forecasting model, it is necessary to put forward a new systematic approach in order to improve the prediction performance. This study developed from the *mathematical algorithm* of traditional GM (1, 1) and combined with *quintessential ideas* of previous studied [3, 14, 16, 21, 23, 24] to establish the effectiveness systematic approach for improving the prediction performance. The proposed systematic approach illustrated its effectiveness under three remarkable improvements. The first remarkable is dealing with the negative value aim to expand the application of grey forecasting model suitable for many actual systems, The second remarkable is smooth the sequence data in which satisfy the quasi-exponential condition and quasi-smooth condition to get the good estimate performance and the final is focused on reduce the residual error.

The remain parts of this paper are structured as follows: Section 2 covers the algorithm mathematical of GM (1, 1). Section 3 provides and explanations the procedure of proposed systematic approach. Section 4 illustrates some numerical simulations to verify the effectiveness of proposed systematic approach compared with the others grey forecasting models. Finally, section five concludes the whole research.

2 Traditional Grey Forecasting Model and Systematic Optimization Approach

2.1 Traditional Grey Forecasting Model "GM (1, 1)"

GM (1, 1) model is one of the most frequently used grey forecasting models. Its a time series forecasting model. The sequence data need to construct the GM (1, 1) model must be more than or equal four. The constructing process of GM (1, 1) model is described below:

Step 1: Assume that: the *non-negative* original sequence data $X^{(0)}$

$$X^{(0)} = \left\{x^{(0)}(1), x^{(0)}(2), \ldots, x^{(0)}(n)\right\}, \ n \geq 4 \tag{1}$$

Step 2: Construct $X^{(1)}$ by *one time accumulated generating operation (1-AGO)*, which is

$$X^{(1)} = \left\{x^{(1)}(1), x^{(1)}(2), \ldots, x^{(1)}(n)\right\} \tag{2}$$

$$\text{where } x^{(1)}(k) = \sum_{i=1}^{k} x^{(0)}(i),\ k = 1, 2, \ldots, n \tag{3}$$

Step 3: Generate a background value z by Mean generating operation *(MGO).*

$$z^{(1)}(k) = 0.5x^{(1)}(k+1) + 0.5x^{(1)}(k),\ k = 2, 3.. \tag{4}$$

Step 4: The result of *1-AGO* is monotonic increase sequence which is similar to the solution curve of first order linear differential equation. Therefore, the solution curve of following differential equation represents the approximation of *1-AGO* data:

$$\frac{dx^{(1)}}{dt} + ax^{(1)} = b \tag{5}$$

Where the parameter a and b are model parameters. $\hat{x}^{(0)}(1) = x^{(0)}(1)$ is the corresponding initial condition.

Step 5: Calculating the parameters a and b can be by functions "ordinary least square method (OLS)".

$$\begin{bmatrix} a \\ b \end{bmatrix} = (B^T B)^{-1} B^T Y_n \tag{6}$$

Where B and Y_n are defined as follows:

$$B = \begin{pmatrix} -z^{(1)}(2) & 1 \\ -z^{(1)}(3) & 1 \\ \ldots & \ldots \\ -z^{(1)}(n) & 1 \end{pmatrix} \text{ and } Y_n = \begin{pmatrix} x^{(0)}(2) \\ x^{(0)}(3) \\ \ldots \\ x^{(0)}(n) \end{pmatrix}$$

We solve the Eq. (5) together with initial condition, and the particular solution is

$$\hat{x}^{(1)}(k+1) = \left(x^{(0)}(1) - \frac{b}{a}\right)e^{-a(k)} + \frac{b}{a},\ k = 2, 3, \ldots \tag{7}$$

Step 6: Applying *inverse accumulated generating operation* (*I-AGO*) to $\hat{x}^{(1)}(k)$, the predicted datum of $x^{(0)}(k)$ can be estimated as:

$$\hat{x}^{(0)}(k+1) = \hat{x}^{(1)}(k+1) - \hat{x}^{(1)}(k),\quad k = 1, 2, 3, \ldots \tag{8}$$

Or

$$\hat{x}^{(0)}(k+1) = (1 - e^{-a})\left(x^{(0)}(1) - \frac{b}{a}\right)e^{-ak},\quad k = 1, 2, 3, \ldots \tag{9}$$

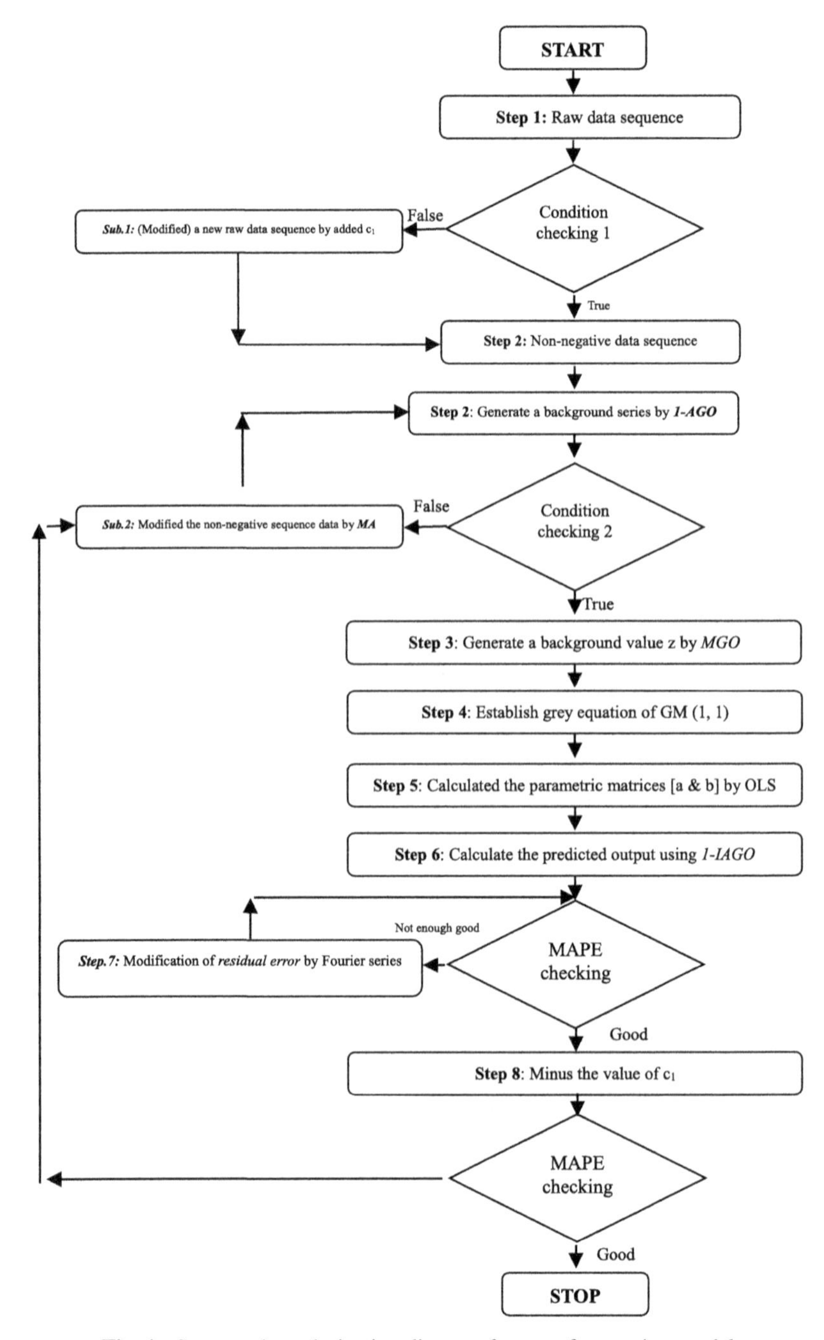

Fig. 1. Systematic optimization diagram for grey forecasting model

2.2 Design of Systematic Optimization for Grey Forecasting Model

Overall systematic approach are explained as follow (Fig. 1):

Step 1: We have any raw data sequence $X^{(0)} = \{x^{(0)}(1), x^{(0)}(2), \ldots, x^{(0)}(n)\}$ with $n \geq 4$.

First condition checking: **IF**: any value data in raw sequence data is greater than or equal to 0 (satify condition 1 all data value in raw sequence data is non-negative value) **THEN** go to step 2. **ELSE**: go to sub step 1.

Sub step 1: Make the data sequence become the non-negative data sequence by added two constant number c_1, c_1 is defined as:

$$c_1 = \begin{bmatrix} 0 & If: \forall k \in (1, .., n): x^{(0)}(k) \geq 0 \\ \max\limits_{k=1,..,n} \{|x^{(0)}(k)|\} & If: \exists k \in (1, .., n): x^{(0)}(k) < 0 \end{bmatrix}, \; k = 1, 2, .., n, \; n \geq 4$$

and the new sequence data is $X^{(0)'} = \{x^{(0)}(1) + c_1, x^{(0)}(2) + c_1, \ldots, x^{(0)}(n) + c_1\} x^{(0)'}(k) = x^{(0)}(k) + c_1$

Step 2: We use *the first-order accumulating generating operation (1-AGO)*;

$$X^{(1)} = \left\{x^{(1)'}(1), x^{(1)}(2)', \ldots, x^{(1)'}(n)\right\}, \text{ where } x^1(k) = \sum_{i=1}^{k} x'^{(0)}(i), k = 1, 2, \ldots, n$$

The result of *1-AGO* is monotonic increase sequence which is similar to the solution curve of first order linear differential equation. Therefore, the solution curve of differential equation as $\frac{dx^{(1)}}{dt} + ax^{(1)} = b$.

Second conditions checking: **IF**: the new series satisfy the (*conditions checking 2)* "which are quasi-smooth and quasi-exponential law checking" as follows:

$$\begin{cases} 0 \leq \xi(k) = \frac{x^{(0)'}(k)}{x^{(1)}(k-1)} \leq 0.5 \quad \text{Quasi - smooth checking} \\ 1 \leq \xi^1(k) = \frac{x^{(1)}(k)}{x^{(1)}(k-1)} \leq 1.5 \; \text{Quasi - exponential checking} \end{cases}, \; k = 4, \ldots, n$$

THEN Go to step 3. **ELSE**, go to sub. 2

To perform the sequence data suitable the quasi-smooth checking and quasi-exponential checking, we used the moving average operation by the formulae follows:

The first data $x'(1) = \frac{3x^{(0)'}(1)+x^{(0)'}(2)}{4}$, the final data: $x'(n) = \frac{x^{(0)'}(n-1)+3x^{(0)'}(n)}{4}$, and the intermediate data: $x'(k) = \frac{x^{(0)'}(k-1)+2x^{(0)'}(k)+x^{(0)'}(k+1)}{4}$.

Step 3: Generate a background value z by *MGO* as follows Eq. (4).

Step 4: Establish grey equation of GM (1, 1).

Step 5: Calculating the model parameters *a* and *b* in Eq. (6) based on the vector Y and matrix B. and get the $\hat{x}^{(1)}(k)$

Step 6: Applying inverse accumulated generating operation (*I-AGO*) to get the predicted value of $\hat{x}^{(0)\prime}(k)$,

MAPE or (ρ) checking: **IF** the predicted value is satisfy in the (Table 1) "which are greater than good condition checking" **THEN** Go to step 8. **ELSE**, go to Step 7.

Step 7: In this step, the author used the Fourier series to modify the residual error [all algorithm of process was illustrated in Thang and Phan].

Step 8: After calculation all above steps and get the predicted value, we need to minus the same amount (c_1) (if the first step added), we can get the predicted value. $\hat{x}^{(0)}(k) = \hat{x}^{(0)\prime}(k) - c_1$.

2.3 Prediction Capability Analysis

To test the quality of these models in forecasting, this research using mean absolute percentage error index (*MAPE*) [25]. The *MAPE* is the mean of the absolute percentage differences between the forecasts and the actual value. In 1979, Makridakis et al. [25] mentioned that the *MAPE* measures are helpful in making comparison between different forecasting models. Moreover, the *MAPE* index is the most commonly used accuracy measures in forecasting studies [13, 20]. Therefore, the *MAPE* was selected to evaluate the performance of the forecasting method and it's defined as follow:

$$MAPE = \left(\frac{1}{n-1} \sum_{k=2}^{n} \left| \frac{x^{(0)}(k) - \hat{x}^{(0)}(k)}{x^{(0)}(k)} \right| \right) \times 100\% \tag{10}$$

Where $x^{(0)}(k)$ and $\hat{x}^{(0)}(k)$ are actual and forecasting values in time period k, respectively, and n is the total number of predictions.

When the value of *MAPE* is close to 0, its means that the forecasting model is high accuracy, and has given good performance and *vice versa*. Based on the values of *MAPE*, the level of forecasting accuracy can be classified into four grades, as shown in Table 1 [26]:

Table 1. Criteria of *MAPE* [26]

MAPE (%)	Forecasting power
<10	Excellent
10–20	Good
20–50	Reasonable
>50	Incorrect

3 Results and Discussion

*In this section, the example with negative values in the sequence data and the fluctuation data of numerical examples came from the previous research papers (the is high technology industrial output forecasting in Taiwan by Wang and Hsu [24] are used to show the accuracy of proposed approach with the traditional model GM (1, 1) in [3], optimized model GM (1, 1) in [*14, 23*], FMGM (1,1) in [*20*]. All the computational required for the present studies have been carried out on Microsoft Excel and (MAPE)* index was used *to evaluate the forecast capability of these models:*

3.1 Numerical Example with Negative Value in the Data Sequence

To demonstrates the accuracy of the proposed approach in the negative value environment, this article uses the example was designed by themselves. There are in total 12 observations available as illustrated in Table 2. As can be seen the Table 2 and Fig. 2, have negative value in the raw data sequence (k_1 and k_3), that reason why the trend of data sequence is a wild fluctuation, and not following the rule of exponential growth.

Table 2. The example data

k	1	2	3	4	5	6	7	8	9	10	11	12
Data	−2	1	−4	4	3	5	11	17	15	20	30	33

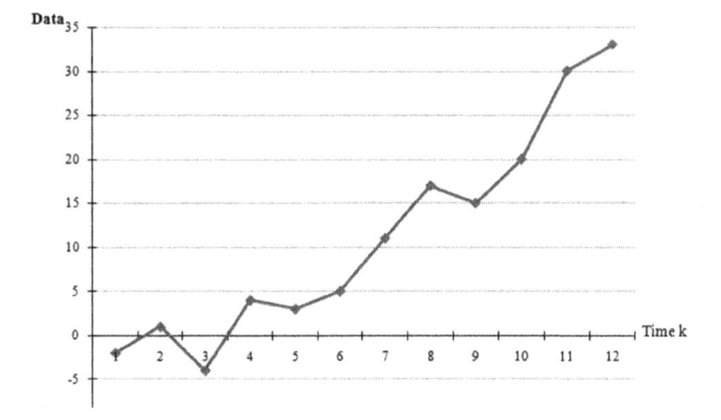

Fig. 2. The curve of actual data sequence

In this situation, the traditional model GM (1, 1) in [3], optimized model GM (1, 1) in [14, 23], FMGM (1, 1) in [20] cannot deal with the negative value. But when applied the first steps of proposed approach, they can solve the negative value. In this case, this study sets the samples from k = 1 to 11 (11 data points) for in-sample estimation and k = 12 for out-of-sample for interpolation data testing and extrapolation data testing, respectively. Following the algorithm of all forecasting models, the modeling process is given below:

Step 1: Raw sequence data:

$X^{(0)} = \{-2, 1, -4, 4, 3, 5, 11, 17, 15, 20, 30\}$, we can see that the raw sequence data include the negative value (not satisfaction of condition 1) next to the sub step 1.

Convert the raw sequence data become the nonnegative value by added the $c_1 = \max\{\text{abs}\,(x(k))\}$: in this situation $c_1 = 30$.

The new sequence data is:

$X_{s1}^{(0)} = \{28, 31, 26, 34, 33, 35, 41, 47, 45, 50, 60\}$ is suitable with condition 1.

And then checking the condition 2: After checking the sequence data $X^{(0)'}$ in the condition 2, we see that, from time k = 4 onwards. The $X^{(0)}$ is satisfy of condition 2.

Then next step, we can see that the new sequence data is satisfy with condition 1 and 2, So all grey forecasting models were mentioned above can solve: For the rest steps computation following the same algorithm mathematical of grey model, all of four models have been carried out on Microsoft Excel to calculate the parameters based on all algorithms mathematical of each model. Because of the computation is too long so, the calculation process are omitted here. The *MAPE* of these models for both in-sample and out-of sample of these models are shown in Figs. 3 and 4.

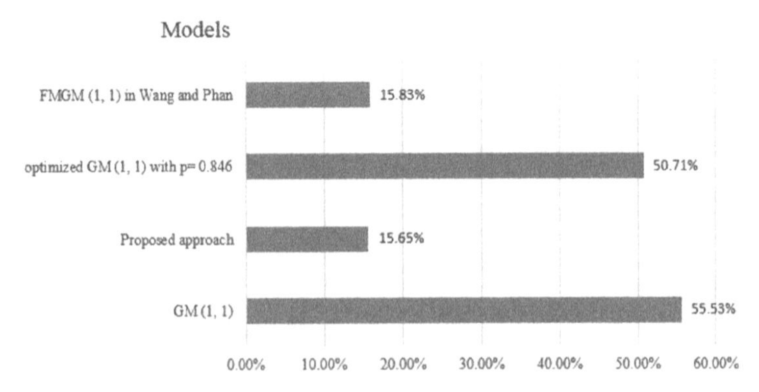

Fig. 3. The MAPE of the different grey models forecasting in samples

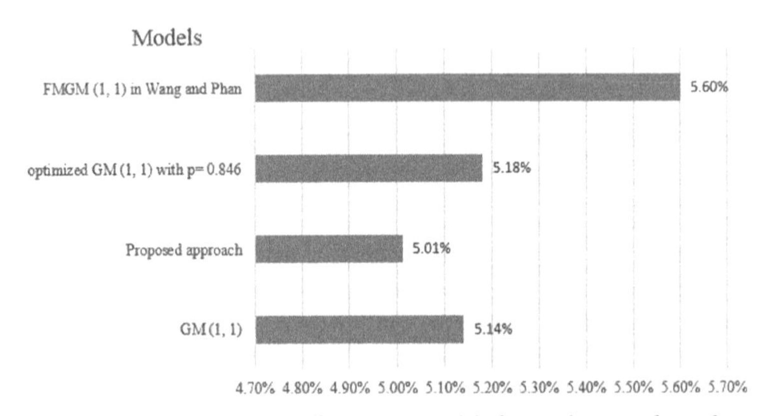

Fig. 4. The MAPE of the different grey models forecasting out of sample

As can be seen from the Figs. 3 and 4, among four grey forecasting models comparison, just only proposed approach provides an excellent result with the MAPE 15.65% (less than 20%). The others models can get a better performance with the lowest of average of MAPE for in and out-of-samples. More specifically, The proposed approach is the first order ranking in this situation with the MAPE is 15.65% in sample and 5.01% for out-of sample. Follow by are the FMGM (1, 1) in Wang and Phan with MAPE in sample is 15.83% and 5.60% for out-of sample data. The position of three and four are optimal GM (1, 1) with p = 0.846 and the traditional GM (1, 1), respectively.

3.2 High Technology Industrial Output Forecasting in Taiwan (Wang and Hsu, 2008)

The second example was used to test the performance of the proposed approach over the others grey forecasting models are coming from the reference in (Wang and Hsu [24]). The historical data from 1990 to 2005 was represented in Fig. 5. As can be seen from the Fig. 5, the historical data also are significant fluctuation over the period time. To show the effectiveness of the proposed approach over the others grey forecasting models, this study setting the data from 1990 to 2001 are as the training data (in-sample), and the value from 2002 to 2005 (out-of-sample) are predicted. In order to find out the parameters of r forecasting models above which are traditional GM (1, 1), optimized GM (1, 1), FMGM (1,1), and the proposed model. All parameters estimation results, forecasted values and *MAPE* of these models for both in-sample and out-of sample of these models are presented in Fig. 5.

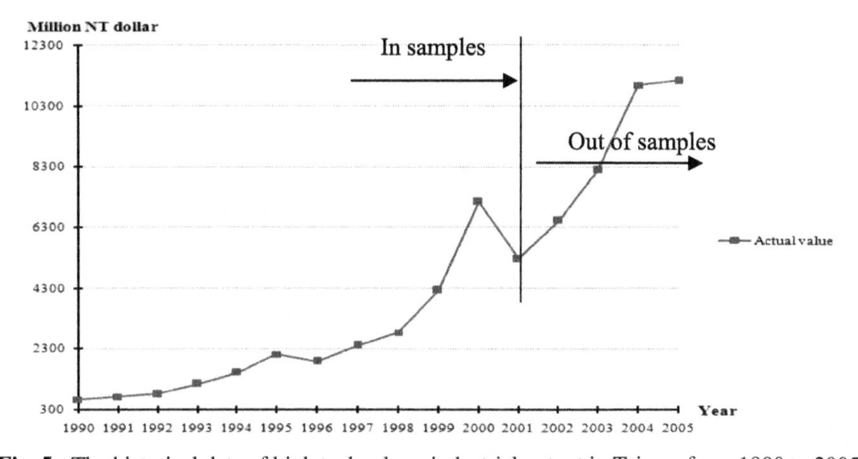

Fig. 5. The historical data of high technology industrial output in Taiwan from 1990 to 2005

As can be seen from Table 3, both two forecasting models which are FMGM (1,1) and proposed approach provided an excellent result with the *MAPE* nearly 0% while the other forecasting model illustrate the good condition in estimation in samples data training. In term of out-of sample data. The proposed approach provides considerably better precise forecast than others grey forecasting models with the lowest average of MAPE. More specifically, the first order ranking in prediction accuracy is the proposed approach with MAPE in and out-of sample are 0% and 19.67%, respectively. The second ranking is FMGM (1, 1) with MAPE in and out-of sample are 0% and 23.56%, Follow by are Optimized GM (1,1) with $p = 0.47$ and traditional GM (1, 1).

All forecasted value of different grey forecasting model in forecast the high technology industrial output are show in Fig. 6:

Table 3. The average MAPE of different grey forecasting models

Models	Average of MAPE in sample	Average MAPE out- of sample
GM (1,1)	13.74	24.15
FMGM (1,1)	0.00	23.56
Optimized GM (1,1) p = 0.47	13.059	21.17
Proposed approach	**0.00**	**19.67**

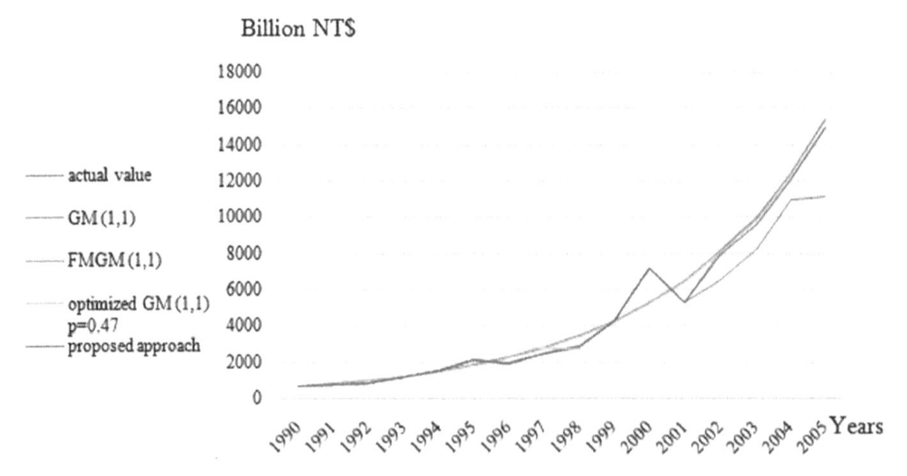

Fig. 6. The actual and forecasted value of the different grey models

4 Conclusions

The new systematic approach is proposed to improve the performance of GM (1, 1). Based on combine the mathematical algorithm of the grey forecasting model with the inheritance quintessential ideas of some previous highlight studies. In order to illustrate effectiveness of proposed method, two numerical example are used. All results show that the proposed approach not only overcome the limitations of GM (1, 1) but also increase the prediction accuracy in the highly fluctuation data. For future direction, the proposed approach will be applied in various areas to dealing with similar problems. An the other direction is use different equations or different methodologies to improve the performance of GM (1, 1) model.

References

1. Deng, J.-L.: Introduction to grey system theory. J. Grey Syst. **1**(1), 1–24 (1989)
2. Liu, S., Lin, Y.: Grey Information Theory and Practical Applications. Springer, London (2006). https://doi.org/10.1007/1-84628-342-6
3. Huang, Y.-L., Lee, Y.-H.: Accurately forecasting model for the stochastic volatility data in tourism demand. Mod. Econ. **2**(05), 823–829 (2011)

4. Chu, F.-L.: Forecasting tourism demand in Asian-Pacific countries. Ann. Tour. Res. **25**(3), 597–615 (1998)
5. Wang, C.-N., Phan, V.-T.: An improvement the accuracy of grey forecasting model for cargo throughput in international commercial ports of Kaohsiung. Int. J. Bus. Econ. Res. 1–5 (2014)
6. Jiang, F., Lei, K.: Grey prediction of port cargo throughput based on GM (1, 1, a) model. Logist. Technol. **9**, 68–70 (2009)
7. Askari, M., Askari, H.: Time series grey system prediction-based models: gold price forecasting. Trends Appl. Sci. Res. **6**(11), 1287 (2011)
8. Kayacan, E., Ulutas, B., Kaynak, O.: Grey system theory-based models in time series prediction. Expert Syst. Appl. **37**(2), 1784–1789 (2010)
9. Lin, C.-T., Yang, Shih-Yu.: Forecast of the output value of Taiwan's opto-electronics industry using the grey forecasting model. Technol. Forecast. Soc. Chang. **70**(2), 177–186 (2003)
10. Hsu, L.-C.: A genetic algorithm based nonlinear grey Bernoulli model for output forecasting in integrated circuit industry. Expert Syst. Appl. **37**(6), 4318–4323 (2010)
11. Hsu, L.-C.: Using improved grey forecasting models to forecast the output of opto-electronics industry. Expert Syst. Appl. **38**(11), 13879–13885 (2011)
12. Hsu, C.-C., Chen, C.-Y.: Applications of improved grey prediction model for power demand forecasting. Energy Convers. Manag. **44**(14), 2241–2249 (2003)
13. Hung, K.-C., Chien, C.-Y., Kuo-Jung, W., Hsu, F.-Y.: Optimal alpha level setting in GM (1, 1) model based on genetic algorithm. J. Grey Syst. **12**(1), 23–31 (2009)
14. Bin, Z., Zeng, W.-J.: GM (1, 1) model of moving average operation to the original sequence and its application. Adv. Mater. Res. **225**, 381–384 (2011)
15. Wang, Z.-X., Dang, Y.-G., Liu, S.-F., Zhou, J.: The optimization of background value in GM (1, 1) model. J. Grey Syst. **10**(2), 69–74 (2007)
16. Wu, L., Liu, S., Yao, L., Yan, S., Liu, D.L.: Grey system model with the fractional order accumulation. Commun. Nonlinear Sci. Numer. Simul. **18**(7), 1775–1785 (2013)
17. Shen, Y., Qin, P.: Optimization of grey model with the fractional order accumulation. J. Grey Syst. **17**(3), 127–132 (2014)
18. Wei, L., Fei, M., Huosheng, H.: Modeling and stability analysis of grey–fuzzy predictive control. Neurocomputing **72**(1), 197–202 (2008)
19. Li, G.-D., Yamaguchi, D., Nagai, M.: A GM (1, 1)–Markov chain combined model with an application to predict the number of Chinese international airlines. Technol. Forecast. Soc. Chang. **74**(8), 1465–1481 (2007)
20. Wang, C.-N., Phan, V.-T.: Enhancing the accurate of grey prediction for GDP growth rate in Vietnam. In: 2014 International Symposium on Computer, Consumer and Control (IS3C), pp. 1137–1139. IEEE (2014)
21. Zhou, P., Ang, B.W., Poh, K.L.: A trigonometric grey prediction approach to forecasting electricity demand. Energy **31**(14), 2839–2847 (2006)
22. Chen, C.-I.: Application of the novel nonlinear grey Bernoulli model for forecasting unemployment rate. Chaos, Solitons Fractals **37**(1), 278–287 (2008)
23. Tan, G.-J.: The structure method and application of background value in grey system GM (1, 1) model (I). Syst. Eng.-Theory Pract. **20**(4), 98–103 (2000)
24. Wang, C.-H., Hsu, L.-C.: Using genetic algorithms grey theory to forecast high technology industrial output. Appl. Math. Comput. **195**(1), 256–263 (2008)
25. Makridakis, S., et al.: The accuracy of extrapolation (time series) methods: results of a forecasting competition. J. Forecast. **1**(2), 111–153 (1982)
26. Lewis, C.D.: Industrial and Business Forecasting Methods: A Practical Guide to Exponential Smoothing and Curve Fitting. Butterworth-Heinemann, Oxford (1982)

Deep Learning and Applications for Industry 4.0

An Evaluation of Image-Based Malware Classification Using Machine Learning

Tran The Son[1(✉)], Chando Lee[1], Hoa Le-Minh[2], Nauman Aslam[2], Moshin Raza[2], and Nguyen Quoc Long[3]

[1] Vietnam – Korea University of Information and Communication Technology, Da Nang, Vietnam
ttson@vku.udn.vn, bright-way@naver.com

[2] Faculty of Engineering and Environment, Northumbria University, Newcastle-upon-Tyne, UK
{hoa.le-minh,nauman.aslam}@northumbria.ac.uk, mohsinraza119@gmail.com

[3] IT Faculty, FTP University, Da Nang, Vietnam
longnq9@fe.edu.vn

Abstract. This paper investigates the image-based malware classification using machine learning techniques. It is a recent approach for malware classification in which malware binaries are converted into images (i.e. malware images) prior to feeding machine learning models, i.e. *k*-nearest neighbour (*k*-NN), Naïve Bayes (NB), Support Vector Machine (SVM) or Convolution Neural Networks (CNN). This approach relies on image texture to classify a malware instead of signatures or behaviours of malware collected via malware analysis, thus it does not encounter a problem if the signatures of a new malware variant has not been collected or the behaviours of a new malware variant has not been updated.

This paper evaluates classification performance of various machine learning classifiers (i.e. *k*-NN, NB, SVM, CNN) fed by malware images in various dimensions (i.e., $128 \times 128, 64 \times 64, 32 \times 32, 16 \times 16$). The experiment results achieved on three different datasets including Malimg, Malheur and BIG2015 show that *k*-NN outperforms others on three datasets with high accuracy (i.e. 97.9%, 94.41% and 95.63% respectively). On the contrary, NB showed its weakness on image-based malware classification. Experiment results also indicate that the accuracy of the *k*-NN reaches the highest value at the input image size of 32×32 and tends to reduce if too many feature information provided by large input images, i.e. 64×64, 128×128.

Keywords: Deep Learning · CNN · *k*- NN · Naïve Bayes · SVM · Image-Based malware classification

1 Introduction

Nowadays, anti-virus software applications (such as Symantec, Kaspersky, Kingsoft, etc. [1]) are considered as a crucial protection and should be installed for any computers and mobile devices in order to avoid malware infection and attack. Malware is observed to

M. Hernes et al. (Eds.): ICCCI 2020, CCIS 1287, pp. 125–138, 2020.
https://doi.org/10.1007/978-3-030-63119-2_11

be generated rapidly and severely than those generated last decade. It is not only capable to destroy the infected computers and networks, but also steals users' information (e.g. bank account information) or control the infected computer for conducting other harmful actions (distributed denial of service – DDoS attack, advanced persistent threat – APT attack) [2]. According to Cybersecurity Ventures, the global cybercrime damages are estimated to reach $6 trillion annually by 2021 [3] in which DDoS and malware attacks are dominant.

To detect and remove a malware from an infected computer, malware classification is an important step to determine the name, type (or family), signatures and behaviors of the malware before taking the necessary actions (e.g. to remove, quarantine and so on). There are two approaches commonly used for malware classification are signature-based and behavior-based [4] approaches based on the signatures and behaviors which are collected via static and dynamic analysis. The signature-based classification is known as very fast and precise, but it can be easily bypassed by malware variants generated by applying obfuscation techniques (e.g. encryption, packing, polymorphism, and metamorphism) [1, 4]. This problem can be solved by applying the behavior-based classification since malware behaviors of variants are almost the same. However, collecting malware behaviors is considered as time and space complexity because it must be collected during activation of the malware [1].

In recent year, a new approach for malware classification based on the image processing [5–9] is introduced. It allows a classifier to detect and classify the existence of a malware by studying of the texture of the malware image, which is easily converted from the collected binary malware [5]. This approach does not rely on the signatures or behaviors of malware collected via static or dynamic analysis as the traditional malware classifiers do, thus overcoming some of weakness of traditional signature-based and behavior-based approaches. In particular, the image-based malware approach is able to recognize a new variant of a malware based on the similarity of texture of the malware image while it does not need to activate the malware as the behavior-based approach does.

Machine learning has emerged as one of the promising solutions for image processing [6], hence applying machine learning for image-based malware classification has garnered much interest in research community [6–9]. A malware instance which is usually in portable executable (PE) format (an EXE file) is constructed by a sequence of bits. To convert it into a grayscale image, a group of eight sequence bits (one byte) is treated as a 8-bit grayscale pixel value and stored in a 2D matrix (corresponding to the height and the width of an image). Therefore, each grayscale pixel is represented by a value ranging from 0 to 255.

The rest of the paper is structured as follows. Section 2 covers the literature review. Section 3 introduces an image-based malware classification model applying four commonly used machine learning techniques, i.e. k-NN; CNN; NB and SVM, working with various dimensions of the input malware images. Experiment setups and results are presented in Sect. 4. Finally, Sect. 5 concludes the paper.

2 Related Work

Nataraj *et at.* [5] are known as one of the first authors who proposed a model for image-based malware classification using machine learning technique, i.e. *k*-NN. In this model, the GIST descriptor [8] was used for extracting the features of the input images applied for training *k*-NN. Their experiment was conducted on Malimg dataset containing 9339 malware instances collected from 25 different malware families. The accuracy was achieved at high value of almost 98%. However, utilising the GIST descriptor is known as very complicated and time consuming [9].

In another work, S. Yajamanam *et al.* selected only 60 highest ranking features from 320 features provided by GIST for training. As the size of the feature vector is reduced, the accuracy of this model just achieved 92% which is not high compared to others in the literature.

Instead of using GIST, the authors in [10] adopted and used the descriptor published in [11] to extract the needed features for pre-training the deep learning image recognition model on the ImageNet dataset consisting of 1.2 million images in 1,000 classes to improve the classification performance. However, the accuracy achieved by this model was only 92% approx.

Without the use of image descriptors, Quan Le *et al.* [12] applied input images for training the deep learning model. Raw input images were converted into 1D fixed-size vectors before feeding into a CNN model. The accuracy of their classification model achieved was relatively high (95–98%) on various datasets. However, converting images into 1D fixed size vector might cause a loss of image information. Similarly, the author in [13] also applied raw malware images for training the CNN for malware classification. However, their employed dataset has been balanced before training to improve the accuracy.

Above-mentioned related studies reveal that grayscale pixel values of malware images can be treated as features and applied for training the employed machine learning model, instead of using features extracted by an image descriptor (e.g. GIST, SIFT, SUFT, KAZE [14]). It helps reduce complexity and time consuming of an image-based malware classification model using machine learning. However, the dimension of the input images should be small enough such that a classifier does not spend too much time for learning but still provides sufficient information for classification.

Inspired by aforementioned challenge, this paper investigates the impact of the dimension of the input malware images and the impact of the learning techniques, i.e. *k*-NN, NB, SVM, and CNN, applied on the performance of image-based malware classification. The experiment is conducted on various datasets, i.e. Malimg, Malheur [15] and BIG2015 [9].

3 System Model

This section introduces a model applied for image-based malware classification using four learning techniques, i.e. *k*-NN, NB, SVM and CNN. As shown in the Fig. 1, malware binaries (i.e. malware executable files) are *converted* into grayscale images prior to feeding the image-based malware classifier at which they are classified into a specific malware such as Trojan, Backdoor, Worm and so on.

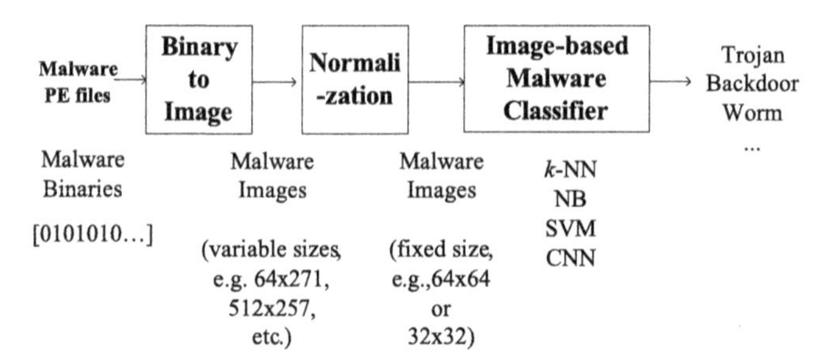

Fig. 1. The system model

One of the simple methods for converting a binary file to a grayscale image is to treat the sequence bytes (8-bit group) of the malware binary as the pixel values of a grayscale image (encoded by 8 bits). As the width of an image is usually fixed at 32, 64, 128 pixels [5, 13, 16], the height of the malware image generated by the above-mentioned method is variable depending on the size the malware binary.

As a result, different malware binaries generate different malware images which have different shapes as illustrated in Fig. 2(a), (b), and (c) for three malware families of Yuner.A, Dialplatform.B, and Swizzor.gen!E respectively.

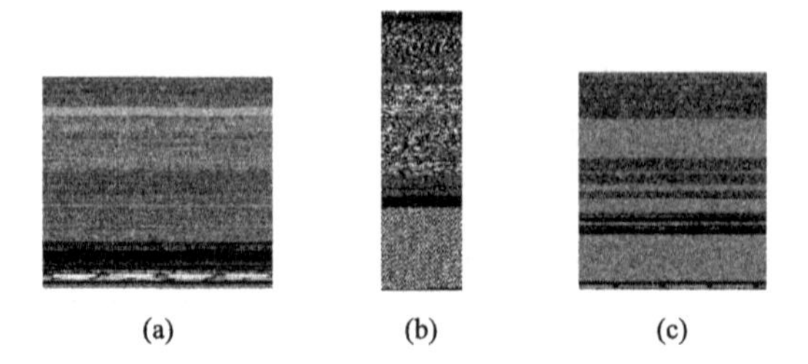

Fig. 2. Malware image of (a) Yuner.A; (b) Dialplatform.B; (c) Swizzor.gen!E

However, variants in a malware family are observed as very similar to the original one. As illustrated in Fig. 3(a), (b), and (c), there are three variants of the Instantaccess malware, which are randomly picked-up among 431 variants store in the Malimg dataset [5]. Their textures are observed as very much similar. That allows the classifier to recognize the similarity among variants of a malware family.

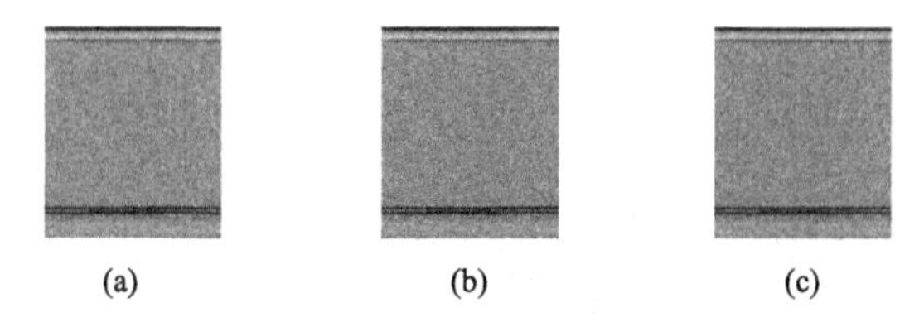

Fig. 3. Three variants of the Instantaccess family

All input images are re-shaped into a fixed dimension, i.e. 32 × 32 (or 64 × 64) as depicted in Fig. 4. This step also known as a *normalization* of input images needed for a classifier to compare the similarity among images when doing a classification. Once the image's dimension is determined, it is fixed for all input images. Also, the employed classifiers (i.e. k-NN, NB, SVM and CNN) have a pre-defined structure with a given number of inputs and outputs. Therefore, the dimension of the input images should be fixed to match the number of inputs of the classifier. It is noted that resizing the malware image might lead to the loss of information (i.e. machine instruction, opcode, etc.), however, the texture of the image is still mainly retained as depicted in Fig. 4. Therefore, it does not have much impact on the result of the classification.

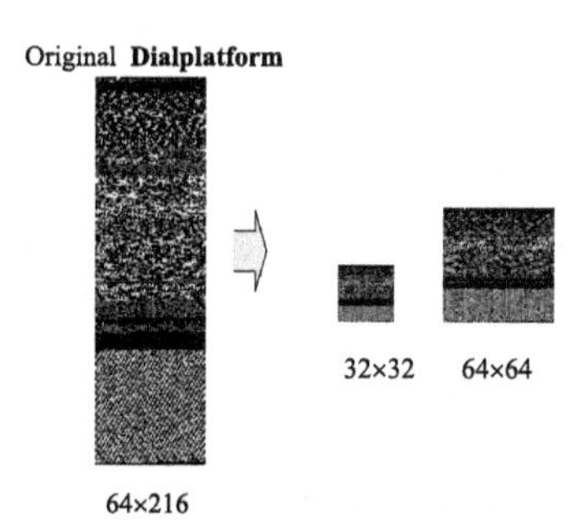

Fig. 4. The images of Dialplatform malware after normalization

As previously discussed, grayscale malware images can be fed into the machine learning models for training. In other words, grayscale pixel values can be treated as the features of the input images instead of using features extracted by image descriptors.

In this paper, four commonly used machine learning models have been applied, i.e. k-NN, NB, SVM and CNN in order to compare the performance of classification among learning techniques.

The k-NN is known as one of the simplest and most intuitive techniques among machine learning techniques [17]. It is based on the distance between the new sample (need to be classified) and k nearest neighbours (in the training set of observed data) for classification. The Euclidean distance is commonly used to determine the distance among samples for k-NN. Several related studies applied k-NN for their image-based malware classification such as [5, 10] and obtained good results.

The SVM is another simple algorithm in the field of machine learning [1]. It is first designed for binary classification to which it searches for a hyperplane to separate two classes with the maximum margin. But the later versions of SVM can handle multi-class classification. One of the advantages of using the SVM is capable to solve high-dimensional dataset without overfitting. SVM is known to be more precise than k-NN [1], but it requires more time for training than k-NN.

Unlike k-NN and SVM, the NB is based on the conditional probabilities calculated via joint probabilities of observed samples to classify a new sample. NB Classifier is known as fast and easy to implement but it is not popular for image classification because its discriminative power is known as not strong compared to others [1]. This paper still employs NB for a comparison.

CNN is one of the most commonly used deep learning architecture, especially for computer vision [6] and natural language processing [18]. A CNN has three different layers [19]: convolutional, pooling and fully connected used for training and finding weights and biases that minimizes certain loss function in order to map the inputs to the outputs as expected. The number of convolutional, pooling and fully connected layers can be more than one depending on the expected outcomes of designers.

4 Experiments and Results

4.1 Dataset

As discussed, this paper conducted the experiment on 3 different datasets, i.e. Malimg [5], Malheur [15] and BIG2015 [9]. They are listed in details in Table 1, Table 2 and Table 3 respectively.

Table 1. Malimg dataset

No.	Family name	Variants
1	Allaple.L	1591
2	Allaple.A	2949
3	Yuner.A	800
4	Lolyda.AA 1	213
5	Lolyda.AA 2	184
6	Lolyda.AA 3	123
7	C2Lop.P	146
8	C2Lop.gen!G	200
9	Instantaccess	431
10	Swizzor.gen!I	132
11	Swizzor.gen!E	128
12	VB.AT	408
13	Fakerean	381
14	Alueron.gen!J	198
15	Malex.gen!J	136
16	Lolyda.AT	159
17	Adialer.C	125
18	Wintrim.BX	97

(*continued*)

Table 1. (*continued*)

No.	Family name	Variants
19	Dialplatform.B	177
20	Dontovo.A	162
21	Obfuscator.AD	142
22	Agent.FYI	116
23	Autorun.K	106
24	Rbot!gen	158
25	Skintrim.N	80
	Total	**9,339**

Table 2. Malheur dataset

No.	Family name	Variants
1	Adultbrowser	262
2	Allaple	300
3	Bancos	48
4	Casino	140
5	Dorfdo	65
6	Ejik	168
7	Flystudio	33
8	Ldpinch	43
9	Looper	209
10	Magiccasino	174
11	Podnuha	300
12	Posion	26
13	Porndialer	98
14	Rbot	101
15	Rotator	300
16	Sality	85
17	Spygames	139
18	Swizzor	78
19	Vapsup	45
20	Vikingdll	158

(*continued*)

Table 2. (*continued*)

No.	Family name	Variants
21	Vikingdz	68
22	Virut	202
23	Woikoiner	50
24	Zhelatin	41
	Total	**3133**

Table 3. BIG2015 dataset

	Family name	Variants		Family name	Variants
1	Ramnit	1541	6	Tracur	751
2	Lollipop	2478	7	Kelihos_ver1	398
3	Kelihos_ver3	2942	8	Obfuscator.ACY	1228
4	Vundo	475	9	Gatak	1031
5	Simda	42		**Total**	**10886**

Among above-mentioned datasets, malware instances stored in the Malimg have been already converted to images (including 9,339 images), thus this work just adopted it for experiment.

The Malheur dataset stores its 3133 malware instances in PE format (EXE format). Hence, in order to conduct the experiment, this work converts them into images following the rule given in [5] and [13].

Unlike above-mentioned datasets, BIG2015 provides its malware instances (i.e. 10886) in hex dump format. In this experiment, they are converted into binaries before converting into images for getting clear malware images.

4.2 Experimental Setups

The *k*-NN, NB, SVM and CNN employed in this experiment are constructed upon the libraries of Tensorflow [20], Sci-kit learn [21] and Keras [22]. Those libraries are widely used for research in the field of machine learning [23, 24]. The experiment runs on a Linux machine (Ubuntu 18.04) Intel Xeon 3.3 GHz, 16 GB RAM.

For training and testing the employed models, the dataset is divided into the training set and the testing set which are 70% and 30% of the dataset respectively. The experiment runs 10 times with different random seeds for each run to avoid overfitting phenomenon. The average accuracy and corresponding standard deviation are calculated based on the accuracies obtained via 10 runs.

For the CNN, over-fitting can be caused by redundant information, hence the drop-out technique is applied for which it drops some random neurons from the network at each iteration of training with the given probability p (i.e. 0.25 and 0.5 after pooling and flattening respectively). The experiment on CNN runs in 10 epochs.

To evaluate the experiment results achieved in this work, following metrics are utilised [25, 26].

Accuracy: It is the total number of samples that are correctly predicted over the total number of samples. It is given by

$$Accuracy = \frac{\text{Number of correct predictions}}{\text{Total number of prediction made}} = \frac{TP + TN}{TP + FP + TN + FN} \tag{1}$$

where TP is true positive, TN is true negative, FP is false positive, and FN is false negative [25].

Loss (or Log-loss)*:* It takes into account the uncertainty of the prediction based on how much it varies from the actual label. It is defined as

$$Loss = \frac{-1}{n} \sum_{i=1}^{n} \sum_{j=1}^{m} y_{ij} * \log(p_{ij}) \tag{2}$$

where n is the total number of samples, m is the total number of classes.

Also, the *training time* and *testing time* are used for evaluation of the effectiveness of the employed models.

4.3 Experiment Results

The Impacts Of Classifiers And Datasets

For this investigation, the input image size is fixed to 32×32 pixels which is considered as the smallest image size used in the area of image processing [16]. It is because the texture of a given image still recorded to be remained almost 70% at this size.

From Table 4, it can be seen that the experiment results obtained by four different classifiers, i.e. k-NN, NB, SVM, CNN on the Malimg dataset look very good (97.9%, 95.51%, 97.4%, 96.2% of accuracy respectively) and almost similar. Whereas they are shown as different among classifiers in Table 5 and Table 6 on Malheur and BIG2015 datasets in which there exist a bad classification, i.e. 61.8% of accuracy. It can be explained that images of a given malware family provided by Malimg dataset are very much similar in texture and dimension (see Fig. 5), therefore it is easy for a classifier to do classification. While variants of some malware families in the Malheur dataset are observed as very much different as illustrated Fig. 6. It is a big challenge for a classifier to learn and classify a new instance of malware.

Hence, this investigation supposes that though malware classification based on image processing is a new approach and to be claimed that it could overcome some of weakness of traditional signature-based and behavior-based approaches [5, 12, 27], it much depends on the similarity of images converted from malware binaries, which might be

Table 4. Achieved results on MALIMG Dataset

	Accuracy (%)		Loss	Time (s)	
	Average	Std(*)		Training	Testing
k-NN	97.90507	0.350423	0.020949	0.336979	15.11978
NB	95.51035	0.298935	0.044897	0.060728	0.467068
SVM	97.40186	0.296367	0.025981	4.901894	4.39065
CNN(**)	96.2955	0.451375	0.115675	88.49583	1.056793

(*) Std: Standard deviation; (**) 10 epochs for each run

Table 5. Achieved results on MALHEUR Dataset

	Accuracy (%)		Loss	Time (s)	
	Average	Std		Training	Testing
k-NN	94.41489	0.698005	0.055851	0.064312	1.807809
NB	76.12412	0.770271	0.238759	0.308044	0.715957
SVM	91.31915	0.480964	0.086809	2.027232	1.123164
CNN	86.10638	2.477696	0.545019	30.57552	0.371535

Table 6. Achieved results on BIG2015 Dataset

	Accuracy (%)		Loss	Time (s)	
	Average	Std		Training	Testing
k-NN	95.63631	0.295743	0.043637	0.509225	11.40247
NB	61.80313	0.61594	0.381969	0.068486	0.213256
SVM	84.97393	0.51496	0.150261	24.26418	11.92428
CNN	94.44342	0.551091	0.216621	103.6633	1.228915

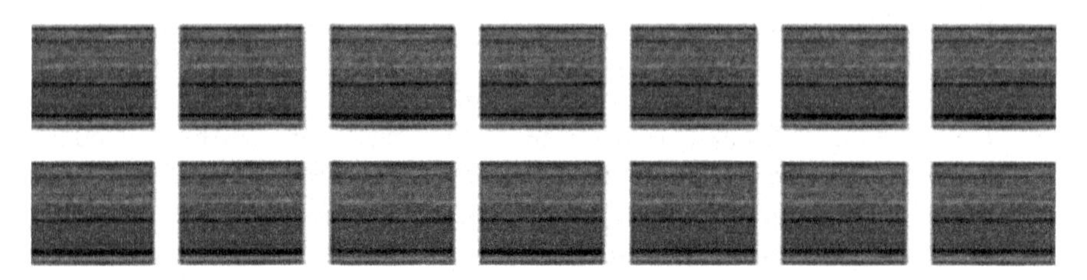

Fig. 5. 14/125 malware images of the Adialer.C in Malimg

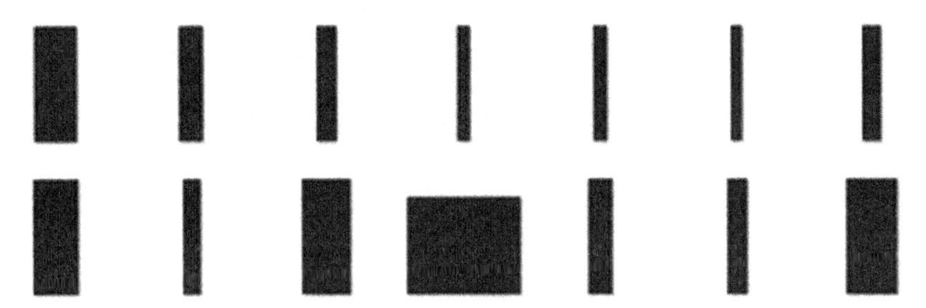

Fig. 6. 14/33 malware images of the FlyStudio family in Malheur

different among variants in the same family. It could impact on the accuracy of a classifier using machine learning.

The type of classifier (or learning technique) is another factor that affects the accuracy of image-based malware classification. As we can see in Table 4, Table 5, Table 6, k-NN provides the best performance compared to others, whereas NB is observed as the worst. This indicates why NB is not very popular for image processing including image-based malware classification as discussed in Sect. 3.

Among four employed classifiers, CNN is known as the most complicated one but in this experiment its accuracy is reported as similar to SVM on three datasets and learning time is observed as much longer than that of others (see Table 4, Table 5, Table 6). It is because grayscale malware images are very simple (see also Fig. 2, Fig. 3), thus, using a complicated network for learning and classifying might not be effective.

Therefore, it is said that k-NN outperforms others for image-based malware classification using 32×32 grayscale input images.

The Impact Of The Dimension Of The Input Images

To investigate the impact of the dimension of the input images, the experiment is conducted with different fixed sizes of the input images, i.e. 128×128, 64×64, 32×32, 16×16 pixels, using k-NN model on the Malimg dataset.

As shown in Fig. 7, the accuracy of k-NN reaches the highest value at 97.9% with the input image's size of 32×32 and tends to decrease significantly when the size of the input image increases, i.e. 83.33% and 52.71% of accuracy with the input images' size of 64×64, 128×128 pixels respectively. It is due to the fact that k-NN relies on the Euclidean distance to decide the nearest neighbour(s), too many pixels make it get confused rather than helping to improve the accuracy.

In contrast, the accuracy of k-NN is observed as slightly changes when the dimension of the input images reduces from 32×32 to 16×16, i.e. 97.9% to 97.2% respectively. It is because the smaller image provides a lack of important features needed for classification. However, training time (dotted-line) and testing time of k-NN with 16×16 input images is recorded as very quick as depicted in Fig. 8.

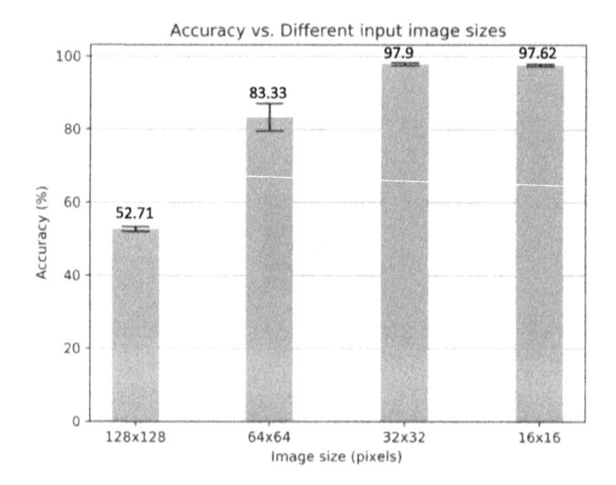

Fig. 7. The accuracy of k-NN with different sizes of input images

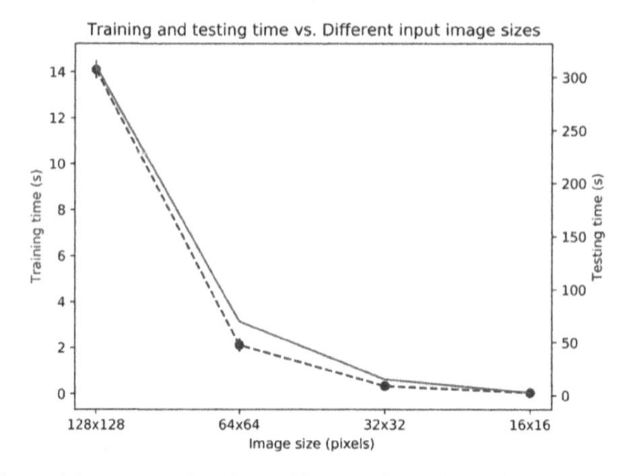

Fig. 8. Training and testing time of k-NN with different input images' sizes

5 Conclusion

This paper investigated the image-based malware classification using various kinds of machine learning techniques, i.e. k-NN, NB, SVM and CNN, which is a new approach for malware classification. The experiment has shown that accuracy of the classification depended on the kind of classifier employed, and the similarity of malware images in a family provided by the dataset. Among the classifiers under test, k-NN achieved the highest accuracy on three datasets (i.e. 97.9%, 94.41% and 95.63% respectively) while NB showed its weakness on image-based malware classification, especially on the datasets in which the similarity of malware images in a family is low. The experiment results also indicated that the accuracy of the k-NN tends to reduce if too many feature information provided by enlarging the input images (such as 64×64 or 128×128 pixels) makes the classifier get confused.

References

1. Ye, Y., Li, T., Adjeroh, D., Iyengar, S.S.: A Survey on malware detection using data mining techniques. ACM Comput. Surv. (CSUR) **50**(3), 1–40 (2017). Article No. 41
2. Kaspersky Security Bulletin 2019, Kaspersky (2019). https://securelist.com/kaspersky-security-bulletin-threat-predictions-for-2019/88878/
3. Cybersecurity Ventures (2018). https://cybersecurityventures.com/-cybercrime-damages-6-trillion-by-2021/
4. Souri, A., Hosseini, R.: A state-of-the-art survey of malware detection approaches using data mining techniques. Hum.-Centric Comput. Inf. Sci. **8**(1), 1–22 (2018). https://doi.org/10.1186/s13673-018-0125-x
5. Nataraj, L., Karthikeyan, S., Jacob, G., Manjunath, B.: Malware images: visualization and automatic classification. In: Proceedings of the 8th International Symposium on Visualization for Cyber Security, Pittsburgh, Pennsylvania, USA (2011)
6. Farabet, C., Couprie, C., Najman, L., LeCun, Y.: Learning hierarchical features for scene labeling. IEEE Trans. Pattern Anal. Mach. Intell. **35**(8), 1915–1929 (2013)
7. Han, K.S., Lim, J.H., Kang, B., Im, E.G.: Malware analysis using visualized images and entropy graphs. Int. J. Inf. Secur. **14**(1), 1–14 (2014). https://doi.org/10.1007/s10207-014-0242-0
8. Douze, M. et al.: Evaluation of GIST descriptors for web-scale image search. In: Proceedings of the ACM International Conference on Image and Video Retrieval, Article No. 19, Greece (2009)
9. Ahmadi, M., Ulyanov, D., Semenov, S., Trofimov, M., Giacinto, G.: Novel feature extraction, selection and fusion for effective malware family classification. In: Proceedings of the 6th ACM Conference on Data and Application Security and Privacy, Louisiana, USA (2016)
10. Bhodia, N., Prajapati, P., Troia, F.D., Stamp, M.: Transfer learning for image-based malware classification. In: Proceedings of the 5th International Conference on Information Systems Security and Privacy, pp. 719–726 (2015)
11. Alex, T.: Malware-detection-using-Machine-Learning. https://github.com/tuff96/Malware-detection-using-Machine-Learning
12. Le, Q., Boydell, O., Mac Namee, B., Scanlon, M.: Deep learning at the shallow end: Malware classification for non-domain experts. Digit. Invest. **26**(1), 5118–5126 (2018)
13. Cui, Z., et al.: Detection of malicious code variants based on deep learning. IEEE Trans. Ind. Inform. **14**(7), 3187–3196 (2018)
14. Tareen, S.A.K., Saleem, Z.: A comparative analysis of SIFT, SURF, KAZE, AKAZE, ORB, and BRISK. In: International Conference on Computing, Mathematics and Engineering Technologies (iCoMET 2018), Sukkur, Pakistan (2018)
15. Rieck, K., Trinius, P., Willems, C., Holz, T.: Automatic analysis of malware behavior using machine learning. J. Comput. Secur. (JCS) **19**(4), 639–668 (2011)
16. Torralba, A.: How many pixels make an image? Vis. Neurosci. **26**(1), 123–131 (2009)
17. Orava, J.: k-nearst neighbour kernel density estimation, the choice of optimal k. Tatra Mountains Math. Publ. **50**(1), 39–50 (2011)
18. Kalchbrenner, N., Grefenstette, E., Blunsom, P.: A convolutional neural network for modelling sentences. In: Proceedings of the 52nd Annual Meeting of the Association for Computational Linguistics, Maryland, USA, pp. 655–665 (2014)
19. Albelwi, S., Mahmood, A.: A framework for designing the architectures of deep convolutional neural networks. Entropy **19**(6), 242 (2017)
20. Google Brain Team: TensorFlow. https://www.tensorflow.org/. Accessed 18 Nov 2019
21. Pedregosa, F., et al.: Machine learning in python. J. Mach. Learn. Res. **12**, 2825–2830 (2011)
22. Keras: Keras Documentation (2015). https://keras.io/

23. Abadi, M. et al.: TensorFlow: a system for large-scale machine learning. In: Proceedings of the 12th USENIX conf. on Operating Systems Design and Implementation, Savannah, GA, USA, pp. 265–283 (2016)
24. Van den Bossche, J., et al.: Scikit-learn. https://scikit-learn.org/stable/. Accessed 18 Nov 2019
25. Powers, D.M.W.: Evaluation: from precision, recall and f-measure to ROC, informedness, markedness & correlation. J. Mach. Learn. Technol. **2**(1), 37–63 (2011)
26. Stamp, M.: Data analysis. In: Introduction to Machine Learning with Applications in Information Security. CRC Press, Taylor & Francis Group (2018). ISBN-13: 978-1-138-62678-2
27. Yajamanam, S., Selvin, V., Troia, F.D., Stamp, M.: Deep learning versus gist descriptors for image-based malware classification. In: Proceedings of the 4th International Conference on Info. Systems Security and Privacy (ICISSP 2018), pp. 553–561 (2018)

Automatic Container Code Recognition Using MultiDeep Pipeline

Duy Nguyen, Duc Nguyen, Thong Nguyen, Khoi Ngo, Hung Cao, Thinh Vuong, and Tho Quan(✉)

Faculty of Computer Science and Engineering, Ho Chi Minh City University of Technology, Ho Chi Minh City, Vietnam
qttho@hcmut.edu.vn

Abstract. Identification of license plates on intermodal containers (or containers) while entering and departing from the yard provides a wide range of practical benefits, such as organizing automatic opening of the rising arm barrier at the entrance and exit to and from the site. In addition, automatic container code recognition can also assist in thwarting the entrance of unauthorized vehicles into the territory. With the recent development of AI, this process is preferably automatic. However, the poor quality of images obtained from surveillance cameras might have detrimental effects on AI models. To deal with this problem, we present a pipeline dubbed as MultiDeep system, which combines several state-of-the-art deep learning models for character recognition and computer vision processes to solve problems of real camera data. We have also compared our results with other pipeline models on real data and accomplished fairly positive results. In this paper, without further references, we will only consider intermodal containers when referring to them as containers.

Keywords: Container code · Optical Character Recognition · MultiDeep pipeline

1 Introduction

The term *Industry 4.0* originates from major changes in industrial history, which is considered to be the fourth largest technological revolution. Digitization introduced by Industry 4.0 creates a vision in which an integrated process of cross-technology, technologies and systems would unite everything - production, services, logistics, personnel. And resource planning. Hence comes the concept of the digital factory, which is essentially based on four principles: technical support, automated decisions, information transparency and complete network connectivity. Thus, Industry 4.0 also requires digital identification of workpieces, tools, containers, machines and equipment. In this paper, we focus on container code identification.

M. Hernes et al. (Eds.): ICCCI 2020, CCIS 1287, pp. 139–153, 2020.
https://doi.org/10.1007/978-3-030-63119-2_12

As the manufacturing and shipping industry is growing, and also introducing some new challenges to container management, which is currently handled manually in many sites, without standards and without the ability to integrate into sophisticated computing systems. Containers sometimes are not placed properly and therefore get lost, adversely having detrimental impact on the ability to transport materials and goods. In addition to the significant lack of efficiency across the entire container management system, there is a lack of transparency in the transportation of containers within the system. As a result, the cost of transportation is boosted or management is not effective.

Currently, many logistics enterprises have applied *Artificial Intelligence* techniques in managing containers in ports and warehouses. Especially, since the optimization of management and transportation at the port is still a problem, an abundance of corporations are interested in. Much attention in this field is drawn to the problem of *container code recognition*. In practice, each container will include a unique *identification code* of ISO standards. ISO6346[1] is the international container standard used to denote the identity of that container, which is divided into 2 parts, as illustrated in Fig. 1. The first is the container identifier information including the owner code, the container list (identifier), the serial number and the check digit. The rest is the shipping size code for that container.

Businesses have set up multiple cameras at ports, on control axes to monitor the status of containers based on the above code, and a large number of machine learning algorithms have been developed for the *Optical Character Recognition (OCR)* problem, which researchers have an intention of applying them for container code recognition problem. However, when putting into use in reality, due to the harsh condition of the vast terrain, the arrangement of surveillance cameras to acquire comprehensive coverage of vehicle details faces difficulties and is intractable. Therefore, the quality of the resulting camera image is poorly affected by practical factors called *hindering*, which leads to the problems encountered identification processes, as illustrated in Fig. 2.

There are a majority of machine learning models, especially the recently developed *deep learning* ones, which have been introduced for *Optical Character Recognition (OCR)* problem. However, the practical difficulties mentioned above hinder those academic models from achieving high accuracy when applied to real data. To tackle this, we propose a pipeline architecture, named *MultiDeep*.

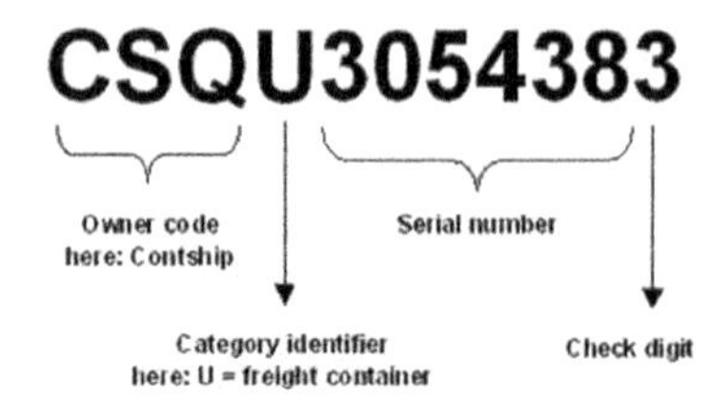

Fig. 1. Container code of the ISO6346 standard

[1] https://en.wikipedia.org/wiki/ISO_6346.

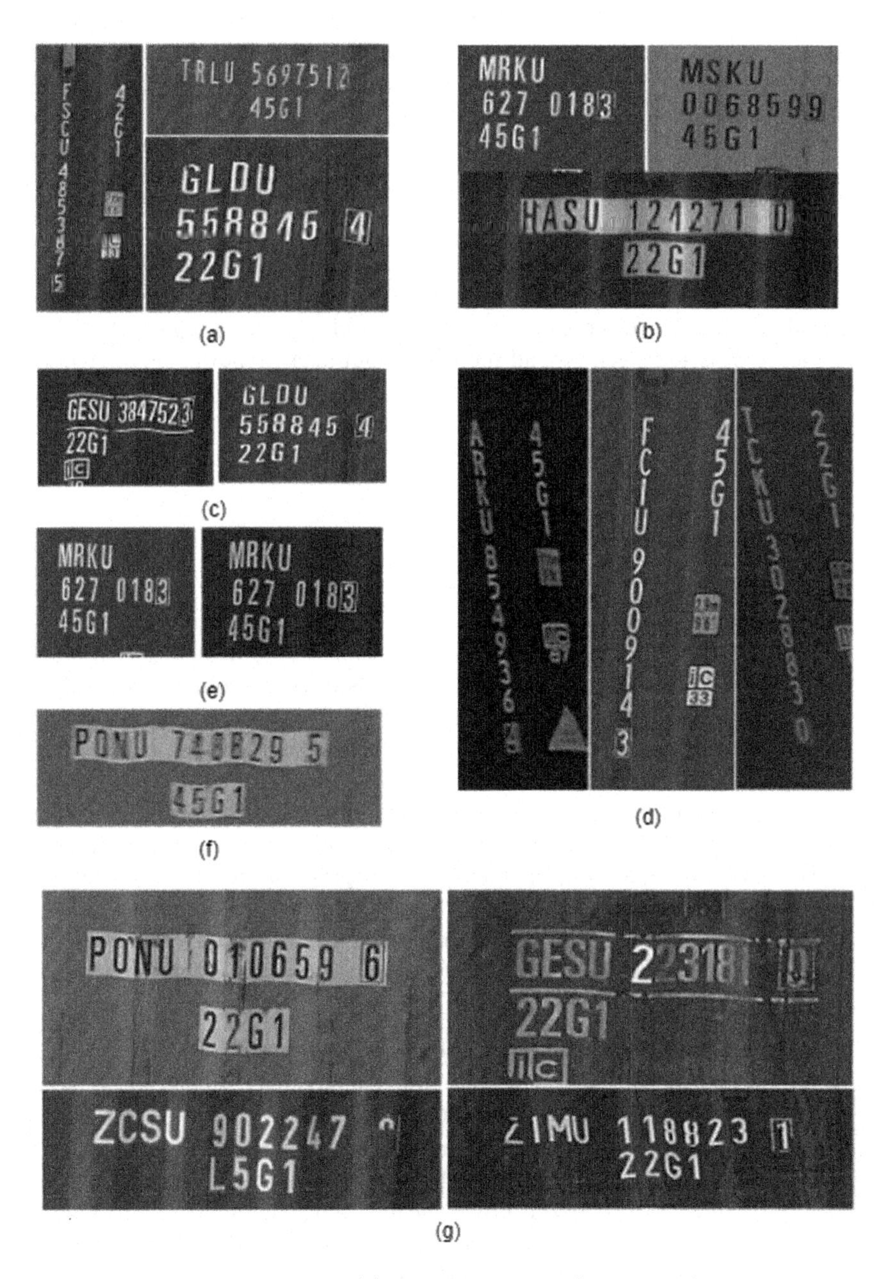

Fig. 2. Hindering factor examples (a) Different formats (horizontal/vertical letters, 2/3 lines) (b) Color of background (dark/bright), color of letters (white/black) and letters in a box (c) Inclined letters due to the rough surface (d) Different camera angles (e) Different brightness (day/night time, artificial lighting) (f) Blurry, noisy images (g) Letters' quality (rusted, deformed, discolored) (Color figure online)

This architecture takes advantage of various deep learning models and computer vision techniques to address those specific hindering factors. We have made a great deal of comparison relating to the performance of our approach with other baseline models and our method was proven to obtain encouraging initial results.

2 Related Work

Nowadays, there have been a wide range of studies related to character recognition, such as recognition in text, handwriting and especially, text boxes in reality, in which bounding boxes can be strongly affected by angles, contrast and last but not least, elasticity. Almost all models accomplish its power by aggregating the deepness, which means that by combining a bunch of layers, in particular convolutional filters, with interplayed max-pooling ones.

2.1 Scene Text Detection

2.1.1 Character Region Awareness for Text Detection

One of the most promising approaches is affinity measurement, along with region detection for every character. Particularly, the most prevalent model with this idea is [1]. As a deep learning model, CRAFT is devised from VGG-16 [2], cooperated with batch normalization. Utilizing the advantages of deep-learning and affinity assessment method allows CRAFT to deal with text regions which are escalated due to inclining or curvature, since it has realized relations among characters. Notwithstanding, as for container number images, CRAFT might have difficulties recognizing those relationships. Not only are number plates displayed vertically, in other words, each row comprises at most 1 or 2 numbers but also speeds of trucks when entering harbors might have important impacts on sensitivity of cameras. Additionally, this model is still far from approaching plates in which digits are damaged due to decaying or paint removal.

2.1.2 Efficient and Accuracy Scene Text Detector

Zhou et al. provides a time-saving method but still attains high efficiency. In fact, by putting only convolutional layers (FCN) into use, not only does EAST [3] remove intermediate steps, sometimes sorely unnecessary, for instance candidate proposal, text formation and text partition but also aggregate its time to learn a bunch of features, and propose boxes with a high likelihood of consisting of letters. The only weakness of EAST lies in text proposal phase, since it is only able to obtain rectangles or more generally, quadrangles. This might provide us with practical difficulties, yet some cameras have the capability of taking convex images with regards to wind direction and letter decaying processes in the past.

2.2 Optical Character Recognition

2.2.1 Spatial Attention Residue Network

This method has proved its benefits in assisting recognition models in avoiding noises made by image qualities. In spite of focusing on improving image input, STAR-Net [4] still includes a recognition component in its architecture in order to prove its effectiveness. Beside its strength of increasing image qualities, STAR Net also has the ability to learn dependencies of characters as in CRAFT [1]. Despite that, it differentiates from CRAFT in the point that instead of utilizing affinity, STAR-Net seeks to learn it implicitly through LSTM cells [5]. Nonetheless, though promising, STAR-Net faces similar obstacles of CRAFT. It shows poor results when being tested with vertical container number plates. In addition, images with a wide range of contrast might cause STAR-Net troubles, especially images taken at late night.

2.2.2 Attentional Scene Text Recognizer with Flexible Rectification

ASTER's [6] core idea is at the whole analogous to STAR-Net [4]. Its main goal is to adjust directions of images after traversing through the network. Despite the same goal, ASTER's architecture is separate from STAR-Net. It applies Thin-Plate-Spline transformation and then recurrent architecture to achieve eventual output. One recognizable advantage is that modules are all differentiable, which leads to the fact that in order to train rectification network, there is no need to carry out human labelling but take advantage of back-propagated results of recognition component. However, the other side of the coin is that the problem of disqualified images remains unsolved, for it is still only concerned with tackling optical situations.

2.3 End-to-End Text Extraction from Image

2.3.1 Tesseract

Tesseract [7] is an optical character recognition engine for various operating systems. It includes many steps for preprocessing image and postprocessing result after feeding image into OCR engine. From its recent version, Tesseract adds LSTM as its primary OCR engine. Due to large dataset trained, it can score a very good result on clear document-like images. But in our experiment, Tesseract can not detect container code area because of the image quality and the ratio between text area and full image. Therefore, for a fair comparison, we only use it in text recognition step.

2.3.2 Feature-Based Local Intensity Gradient (LIG) and Adaptive Multi-threshold Methods

The combination of LIG and adaptive multi-threshold [8] gives out very fast results with minimum hardware requirements because it only relies on basic image processing. And as a result, the performance is only somewhat acceptable on clear images. Therefore, applying this method in real-life problems will cause issues.

2.3.3 Segmentation-Based and HMM-Based Method

This is a method that does not depend on machine learning but rather on probability and image processing [9]. Same as LIG, the model works fast and efficiently on datasets with horizontal clear images. However, the model accuracy is greatly reduced on blurry or tilted images. The problem of the model lies in the segmentation step on each code character, where it uses a high pass filter and scan each line of pixel from top to bottom. Thus, this model is ineffective in real-life problems.

2.3.4 Text-Line Region Location Algorithm Combined with Isolate Characters and Character Recognition

This method [10] combines image processing and machine learning. It scans each pixel from top to bottom to find lines, then uses image processing to isolate each character, and finally feeds each character into an SVM to get the result. However, the paper shows that the model only trained on one type of container image: horizontal letters on the same background. When applied with our real problem, the accuracy will be reduced on images with inclined letters, multicolor background, or vertical letters.

2.3.5 Spatial Transformer Networks and Connected Component Region Proposals

This is a newly introduced model with a high performance provided by the author [11]. The dataset used in the training are images of one or two sides of a container, with a lot of noises. However, when looking at the datasets, their images are of high resolution, and only daylight images are used to train the model. Thus, the model will definitely less effective on our dataset with blurry and nighttime images.

2.4 Summary

To conclude, almost all of the prominent models maintain their own advantages and disadvantages. Nevertheless, they are yet far from tackling a wide variety of situations which people can encounter in real world, especially in industry, in which most of the time images include inclining directions and optical obstacles. As a consequence, we propose the *MultiDeep* pipeline to address this.

3 The MultiDeep Pipeline

As stated in Related Works section, there have been a wide variety of studies related to text and digit recognition. Almost all of these models have been developed based on techniques of RNN ones. As a consequence, they will achieve efficiency when texts are demonstrated in a specified order (from left to right and line by line). Notwithstanding, in practice, container code can be demonstrated on 2, 3 lines or even a vertical one with multiple directions and optical conditions.

Consequently, applying a single model for both training and detection causes a wide range of difficulties. Moreover, because of the variety of container images, concentrating on only one model makes it hard to learn all of the features. As a result, we propose MultiDeep model which combines various Machine Learning models, computer vision techniques and some heuristics and algorithms with a view to accomplishing the most promising outcome. The main idea is that MultiDeep divides the container code recognition through two phases

- *Phase 1:* detect regions which comprise texts.
- *Phase 2:* determine texts in those regions

In both phases, we utilize a diversity of heuristics to tackle practical obstacles. The first obstacle is that beside container codes, on container surfaces are there other kinds of text regions. The second one is that there might be errors in misclassifying texts which are caused by local recognition. Those heuristics are included in post-processing components in Phase 1 and Phase 2. In addition, with a view to attaining high accuracy in recognition, we utilize some computer vision approaches to preprocess images right before Phase 1 and Phase 2. As a result, we can handle problems such as container code images can be both horizontal and vertical, colorful, peculiar optical conditions, oversized and vacuous boxes. Details of the MultiDeep pipeline has been represented in Fig. 3, includes 6 processes. Regarding to our discussed 2 phases, Process 1 and Process 2 are conducted for environment preparation and processing. Phase 1 includes Process 3 and Process 4 and Phase 2 is involved in Process 5 and Process 6.

3.1 Process 1: Model Initiation

Two deep learning models, ASTER and CRAFT, are loaded for next processes.

3.2 Process 2: Preprocessing

The original image will be preprocessed in order to improve accuracy in the future steps. The main task includes the following ones.

1. *Increasing brightness:* In this task, we modify the value of all pixels by a scale constant α and additive one β (Eq. 1). By applying appropriate values to all of the image pixel values, the image obviously becomes brighter. The values of α and β are picked in trial-and-error method. In our dataset, $\alpha = 1.5$ and $beta = -10$ yields the best results.

$$P'(i,j) = \alpha P(i,j) + \beta(\alpha, \beta > 0) \tag{1}$$

2. *Scaling:* It has a connection with the resizing process of a digital image, comprised of Raster ImageScaling and Vector Image Scaling. Raster graphic is a bitmap image made of individual pixels. Pixels are mapped to a new grid, which may possess different dimensions from the original matrix. In

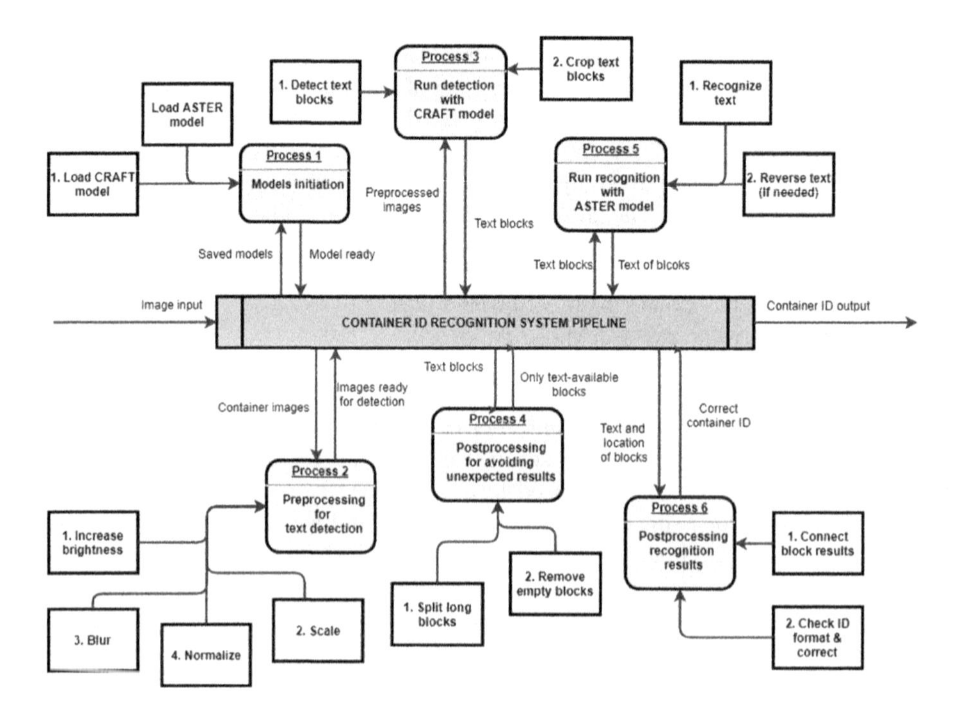

Fig. 3. MultiDeep pipeline

particular, we utilize both nearest neighbor (Eq. 2) and linear method (Eq. 3), and the results obtained are nearly analogous.

$$P'(i,j) = P(i',j') \text{ where } i',j' = \arg\min_{x,y} distance((i,j),(x,y)) \tag{2}$$

$$P'(i,j) = \sum_{k,t} w(k,t,i,j)P(k,t) \tag{3}$$

$w(k,t,i,j)$: the weighted coefficient calculated for (k,t) based upon the distances of pixels to (i,j)

3. *Blurring:* this process involves in boosting the smoothness of the image. The low pass filter, in form of a kernel, allows low frequency elements to enter and thwarts high frequency ones from dominating the whole image, being depicted in 3×3 as

$$\text{filter} = \frac{1}{9}\begin{pmatrix} 1 & 1 & 1 \\ 1 & 1 & 1 \\ 1 & 1 & 1 \end{pmatrix} \tag{4}$$

4. *Normalization:* the final stage is mostly linked to the histogram stretching algorithm, which is a procedure to boost the image's contrast measures. The

adjustment for each pixel can be described as

$$P(i) = \frac{cdf(i) - cdfmin}{(width \times height) - cdfmin}(L - 1) \tag{5}$$

cdf: cumulative distribution function of pixel values specified on each image
L: the range of pixel value.

The result of this process is illustrated in Fig. 4.

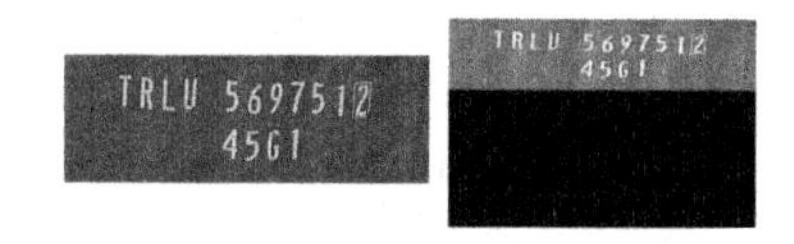

Fig. 4. Original picture and the preprocessed one.

3.3 Process 3: Text Area Recognition

In this process, we employ the CRAFT model to detect text areas on the image. Notice that the result of process 2 has significant effects on the increase of CRAFT accuracy, as illustrated at Fig. 5.

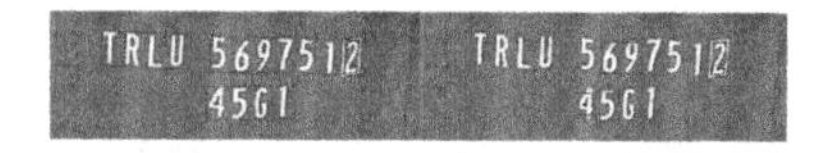

Fig. 5. Text detection with preprocessing (left) and without preprocessing (right)

3.4 Process 4: Text Area Recognition Postprocessing

We perform postprocessing for the resulting text areas from Process 3. It includes two steps as follows.

1. *Long block splitting*: Splitting is an image processing technique used to segment an image. The image is successively split into quadrants based on a homogeneity criterion. Here we split long blocks whose length is greater than 7 characters, in accordance with the ISO6346 discussed above. Cutting text region on scaled image which is the input image for CRAFT model in previous process has been scaled to a certain width. The image with long block will have more than 7 characters. Measuring a specific width, we can define a threshold. If block's width is greater than threshold, we consider that is long block.
2. *Empty block removing*: In our model, we need to detect some texts which have blanks. For example, "PONU 010659 6" is a text which has a blank area. Unfortunately, ASTER model is not trained for images whose contents include blank areas. As a consequence, we remove that blank area in this postprocessing step.

The result of this process is illustrated in Fig. 6.

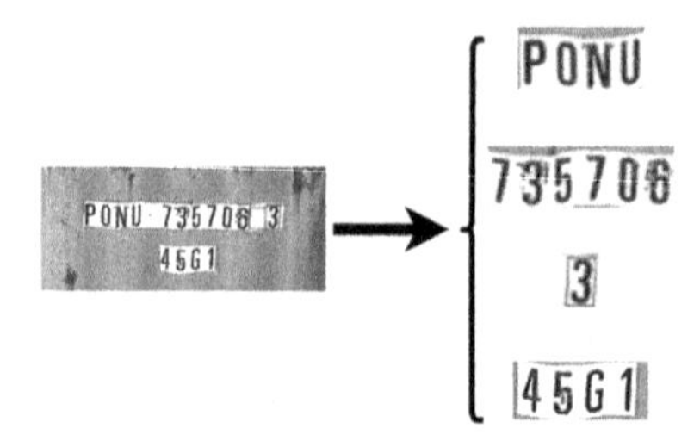

Fig. 6. Illustration of text area recognition postprocessing step

3.5 Process 5: Character Recognition

At this stage, we undertake character recognition using the ASTER model. However, once executing the ASTER model, result we got may be the opposite of our expectation, e.g. the result may be "321" whereas expected as "123". Thus, we also carry on the *text inversion job* (once needed), as illustrated in Fig. 7.

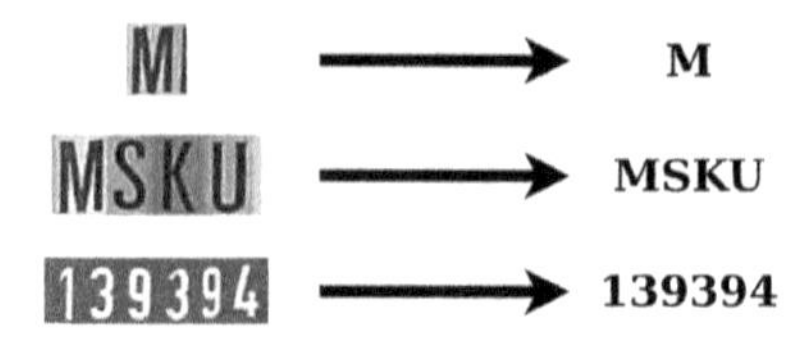

Fig. 7. Illustration the character recognition step, when the code is broken down into sub-parts and recognized accordingly

3.6 Process 6: Character Recognition Postprocessing

Last but not least, postprocessing is carried out with the input which is the results from Process 5. It comprises two steps:

1. Block result connection: In this step, we connect those relevant segmented image blocks to get an intact and clear foreign target image.
2. ID format correction: In this step, we utilize heuristics and experiments to correct mistakes caused by the recognition process. Since the format of the container is well-defined as discussed, we use heuristics to correct the ID when detecting a mismatching.

The result of the final step is illustrated in Fig. 8.

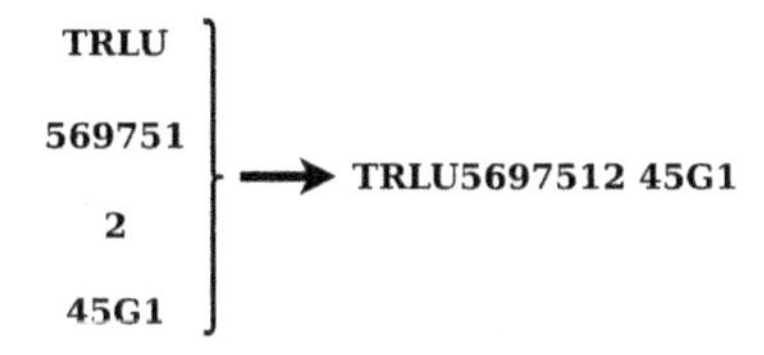

Fig. 8. Illustration of the postprocessing step

3.7 Summary

In the summary, an image is passed into our pipeline will go through all above step and be returned as in Fig. 9:

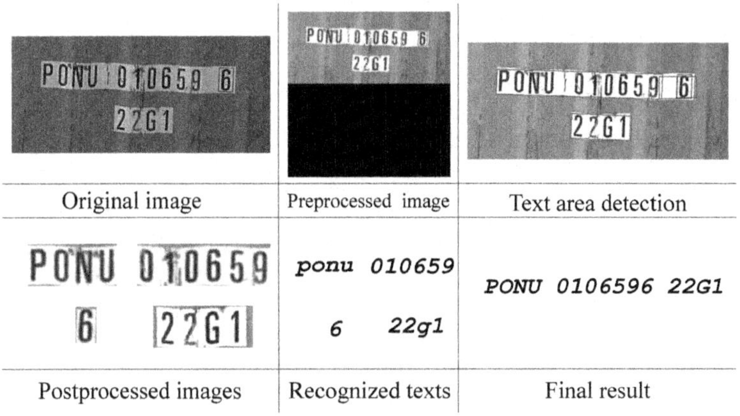

PONU 010659 6 22G1	PONU 010659 6 22G1	PONU 010659 6 22G1
Original image	Preprocessed image	Text area detection
PONU 010659 6 22G1	ponu 010659 6 22g1	PONU 0106596 22G1
Postprocessed images	Recognized texts	Final result

Fig. 9. An overview of returned results after each step.

4 Experiment

In this experiment, we put into use a dataset with a total of 700 images capturing real-life container images, provided by the CyberLogitec company. The whole dataset consists of 7 divided ones (100 images per set) as described below:

- Dataset 1: Images that are clear, easy to recognize, with black letters and bright background.
- Dataset 2: Images that are clear, easy to recognize, with white letters and dark background.
- Dataset 3: Images with inclined letters (Fig. 2c).

- Dataset 4: Images captured with diverse camera angles (Fig. 2d).
- Dataset 5: Images that have different brightness (Fig. 2e).
- Dataset 6: Blurry, unclear images due to the far distance of camera when taking picture (Fig. 2f).
- Dataset 7: Images with deformed or discolored letters (Fig. 2g).

For comparison being fair and most suitable, we divide the whole process into 2 steps: detection and recognition. In detection step, we use F1-score computed using precision and recall to compare between methods, and in recognition step, it is Character Error Rate (CER).

The experiment results show that the deep learning models suffer from a great deal of difficulty when dealing with real images captured from surveillance cameras. We use CER (character error rate) to evaluate the accuracy of the models.

In our experiments, even though the most powerful models and approaches are deployed, including ones popularly applied in industry, the results are still poor without proper pre-and post-processing stages. The Tesseract-OCR model obtain imperfect results because the training of this model does not match the test images. Meanwhile, with EAST-Tesseract, the result of text detection is clear, but the model often misses a number of characters. Despite that, unfortunately, the text recognition phase is not effective. For the EAST-StarNet model, the result of text detection is close to EAST-Tesseract model, but the StarNet model is better than Tesseract. When we combined CRAFT with other character recognition models, the image needs scaling to the appropriate size. Notwithstanding, due to its low resolution, the image became pixelated and the experimental results suffer poor performance. When we enhanced them with post-processing process, the detection of the letters shows positive results, however, the recognition does not meet our expectation.

Our MultiDeep pipleline, once fully deployed, enjoys the most accurate results, which are sufficient to be applied in the real application. The result accuracy is not high, but it is up to our expectation. It is also noted that, among all literature, MultiDeep is the sole model that can handle the container code recognition in an end-to-end manner.

We can see that, our failed cases can interfere human to read. In 6th dataset, failed cases have some character's color mixed with background's, and in 7th dataset, they are very blur and skew, so that pretrained CRAFT model can not detect them. Due to the above errors, Aster model and our postprocessing methods gives wrong outputs. These result is showed in Fig. 10 (Tables 1, 2 and 3).

Table 1. Text detection evaluation with F-measure.

	Dataset 1	Dataset 2	Dataset 3	Dataset 4	Dataset 5	Dataset 6	Dataset 7
East	0.9205	0.7713	0.9499	0.5055	0.7458	0.7963	0.9373
Pre + East	0.8971	0.5961	0.9283	0.3492	0.6087	0.8107	0.9408
Craft	0.4021	0.7079	0.3715	0.8548	0.6079	0.4019	0.3474
Pre + Craft	0.9879	0.9794	0.9819	0.9817	0.9447	0.7808	0.9683

Table 2. Text regconition evaluation using character error rate.

	Dataset 1	Dataset 2	Dataset 3	Dataset 4	Dataset 5	Dataset 6	Dataset 7
Tesseract	99.6%	100%	99.9332%	99.7898%	99.5506%	99.6916%	99.8667%
StarNet	13.2%	17.8667%	17.4465%	18.5004%	16.7041%	22.2051%	12.4%
Aster	5.6667%	11%	24.4652%	27.3301%	12.6592%	26.2143%	12.7333%

Table 3. End-to-end evaluation.

	Dataset 1	Dataset 2	Dataset 3	Dataset 4	Dataset 5	Dataset 6	Dataset 7
Multi-deep	65%	61%	35%	65%	40%	15%	21%

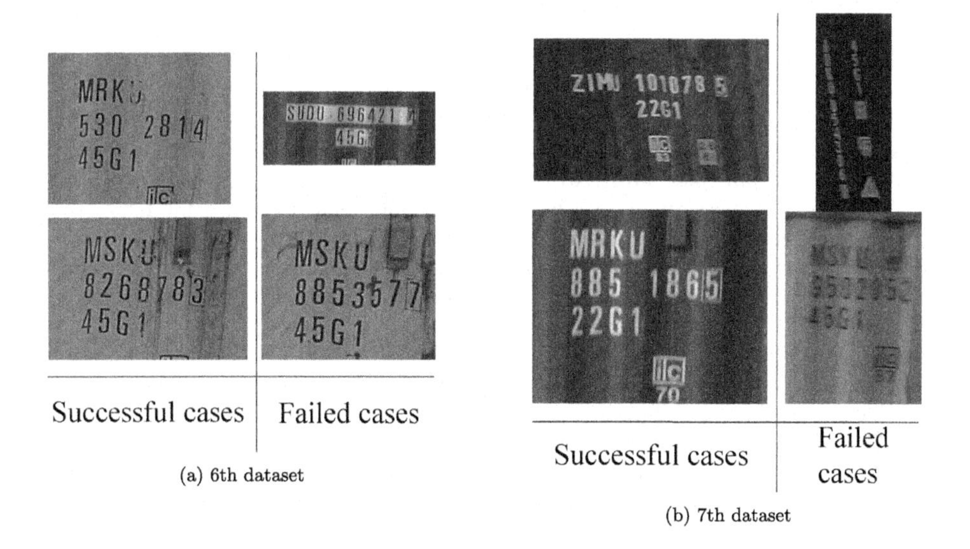

Fig. 10. The results from 6th and 7th dataset

5 Conclusion

Container code recognition is a realistic problem appealing to huge attention due to its practicality. One of emerging trends to address this puzzle is employing deep learning models trained for text recognition. Nonetheless, these models are sensitive to instability of real images' quality captured at the sites. By presenting MultiDeep pipeline, we are capable of recognizing the container identification and obtaining its right format. We have implemented MultiDeep with real dataset of an industry company and obtained promising initial results. With the current results, we need to optimize the model to seek better recognition so as to better adapt to practical problems.

In our experiments, we can take into account real-life images which are not always helpful for the machine learning models to yield high results. There are multiple obstacles when performing character recognition in real life. Besides, there are approaches which can help to boost the accuracy, some of which are listed below.

- Providing re-trained deep models of CRAFT and ASTER with larger dataset.
- Trying other Computer Vision algorithms in preprocessing and postprocessing steps.
- Improving the heuristics algorithms in postprocessing after recognition.

Those are also the directions we intend to pursue in future researches.

Acknowledgement. We are grateful to CyberLogitec company for funding and providing real dataset for this research.

References

1. Baek, Y., Lee, B., Han, D., Yun, S., Lee, H.: Character region awareness for text detection. In: Proceedings of the IEEE Conference on Computer Vision and Pattern Recognition, pp. 9365–9374 (2019)
2. Simonyan, K., Zisserman, A.: Very deep convolutional networks for large-scale image recognition. arXiv preprint arXiv:1409.1556 (2014)
3. Zhou, X., et al.: East: an efficient and accurate scene text detector. In: Proceedings of the IEEE Conference on Computer Vision and Pattern Recognition, pp. 5551–5560 (2017)
4. Liu, W., Chen, C., Wong, K.Y.K., Su, Z., Han, J.,: Star-net: a spatial attention residue network for scene text recognition. In: BMVC, vol. 2, p. 7 (2016)
5. Hochreiter, S., Schmidhuber, J.: Long short-term memory. Neural Comput. **9**(8), 1735–1780 (1997)
6. Shi, B., Yang, M., Wang, X., Lyu, P., Yao, C., Bai, X.: Aster: an attentional scene text recognizer with flexible rectification. IEEE Trans. Pattern Anal. Mach. Intell. **41**(9), 2035–2048 (2018)
7. Smith, R.: An overview of the tesseract OCR engine. In: Ninth International Conference on Document Analysis and Recognition (ICDAR 2007), vol. 2, pp. 629–633. IEEE (2007)
8. Pan, W., Wang, Y., Yang, H.: Robust container code recognition system. In: Fifth World Congress on Intelligent Control and Automation (IEEE Cat. No. 04EX788), vol. 5, pp. 4061–4065. IEEE (2004)
9. Wu, W., Liu, Z., Chen, M., Liu, Z., Wu, X., He, X.: A new framework for container code recognition by using segmentation-based and hmm-based approaches. Int. J. Pattern Recognit. Artif. Intell. **29**(01), 1550004 (2015)

10. Wu, W., Liu, Z., Chen, M., Yang, X., He, X.: An automated vision system for container-code recognition. Exp. Syst. Appl. **39**(3), 2842–2855 (2012)
11. Verma, A., Sharma, M., Hebbalaguppe, R., Hassan, E., Vig, L.: Automatic container code recognition via spatial transformer networks and connected component region proposals. In: 2016 15th IEEE International Conference on Machine Learning and Applications (ICMLA), pp. 728–733. IEEE (2016)

An Efficient Solution for People Tracking and Profiling from Video Streams Using Low-Power Compute

Marius Eduard Cojocea[1,2](✉) and Traian Rebedea[1,2]

[1] University Politehnica of Bucharest, 313 Splaiul Independentei, Bucharest, Romania
iedi.cojocea@gmail.com, traian.rebedea@cs.pub.ro
[2] Open Gov SRL, 95 Blvd. Alexandru Ioan Cuza, Bucharest, Romania

Abstract. Balancing between performance and speed is vital for real-time applications. Given some of the latest edge devices, such as Raspberry Pi 4, Intel Neural Compute Stick 2, or Nvidia Jetson series, edge processing can become a valid choice for deploying computer vision algorithms in real-time scenarios. Object detection and tracking are two common problems that can be solved using such algorithms, which can be deployed with reasonable performance and speed on edge devices. In this paper, we show that the YOLO architecture can be successfully used for object detection and DeepSORT for object tracking on edge devices. The objects of interest in our scenario are persons, thus indicating face detection and tracking as another problem that is solved in the scope of the paper. Using Raspberry Pi 4 and Intel Neural Compute Stick 2, object detection and tracking models can be run on edge devices, though at around half the performance and more than 10 times slower than on a GPU server.

Keywords: People detection · People tracking · People counting · Face detection · Face tracking · Low-power compute

1 Introduction

In the last years, especially since 2012, when AlexNet [1] paved the way for Convolutional Neural Networks (CNN) based models, Computer Vision has seen an exponential increase in popularity and performance. The number of Computer Vision models that solve real world problems greatly increased thanks to CNNs, which also offered significantly increased performance, usually at the cost of larger computational costs.

Convolutional Neural Networks represent a great tool in Computer Vision due to their feature extraction capability and their ability to accept raw images as input. The automatic feature extraction is very important in Computer Vision, since it makes it possible to learn representations of visual objects without having to describe them symbolically or mathematically (e.g. a person is hard to describe mathematically, but a CNN can learn such a description and use it to detect persons).

In this paper we describe a real-time system that employs edge processing to detect, track, count and classify (e.g. age, gender) persons in videos taken by cameras. Tracking

M. Hernes et al. (Eds.): ICCCI 2020, CCIS 1287, pp. 154–165, 2020.
https://doi.org/10.1007/978-3-030-63119-2_13

the faces can also be useful, since it could indicate the persons' interest regarding the objects at which they are looking at. This solution can be deployed in commercial spaces, subway stations, concert halls, and more. For example, a shopping mall could use such a system for profiling its customers and adapt its stores and merchandise accordingly, while also having an estimated count of the number of unique people passing by.

To achieve this, the first step is detecting objects of interest. Thus, a model is required that can reliably detect and classify objects as persons. There are several deep learning models for this task and we will discuss the top candidates in the next section. A tracking algorithm is then used for identifying unique objects in a real-time video. The actual tracking is of less importance, but using such an algorithm is the most reliable way to count the number of unique objects in a video in a certain interval of time. Since each object is attached a unique id number, the total number of unique objects will be equal to the largest id number.

Face detection and tracking are more difficult problems, especially in videos with many persons, mainly due to the small size of the faces. The existing models, used on such videos, have high false positive rate and false negative rate. Regarding the face tracking problem, the solution is to use the person's tracking id resulted from the object tracking algorithm, assigning this id to the face identified at any frame inside the person's bounding box.

These solutions require significant computational power, which would not be a problem when using a computer with one or more dedicated high end GPUs, but it is an issue when using edge devices. There are many advantages when using edge devices such as no need for remote transferring the video streams, cheaper acquisition prices, ease of install to the small size and a low power consumption.

Every few months, a new and more powerful device is created, with more computational power or better optimization. Thus, despite the algorithms being tens of times slower on our proposed edge system and running at half performance, these numbers will be significantly improved in the near future, just by sheer hardware performance.

2 Related Work

2.1 Object Detection and Object Tracking

Object detection requires to detect multiple objects from an image, classify them into their some classes of interest, and differentiate between objects of the same class.

Prior to 2012, heuristic models and fully connected networks were used, but they achieved poor performance in real-world scenarios. This was in part due to smaller computational power as well. Despite the idea of CNNs being around for more than a decade, these models began to be popular among Machine Learning researchers and developers since 2012, when AlexNet [1] achieved significantly greater performance than any previous models. Since then, many models have been created, achieving ever increasing performance and speed.

RCNN [2], Fast RCNN [3], Faster RCNN [4] and Mask RCNN [5] represent a series of object detection models that use regions in order to localize and classify objects in images. Despite incremental improvements, they are still lacking in both performance and speed. You Only Look Once (YOLO) [6] is a model using a different approach than region based models. It divides the image into grid cells and for each cell it generates bounding boxes and class probabilities for them. The bounding boxes with the class probability above a certain threshold are selected and used in order to locate the object. This is a very fast model, being able to run at real-time speed, but having slightly poorer performance that the state of the art at that moment.

YOLOv2 [7] and YOLOv3 [8] represent successive improvements of YOLO, increasing both the speed and the performance of the first version. YOLO-LITE [9] is a shallower version of YOLO. It is intended to be used on devices with reduced computational power, such as edge devices. Despite having lower performance, it can actually be used at reasonable frame rates on edge devices.

On the other hand, object tracking represents a problem where given some object detections, it is required to uniquely tag and track objects alongside multiple frames. A simple method for tracking multiple objects is using the Euclidean distance between the centers of bounding boxes at successive frames. Thus, an object in a frame is considered the same with an object in a previous frame if the Euclidean distance between their bounding boxes centers is small. This method is fast, but has low performance as it uses no visual features to check for the same object in different frames. It is also vulnerable to errors in detection, occlusions and objects leaving the scene.

An improvement to this method can be made using a Kalman filter [10], which is able to predict the future position of objects based on their past trajectory. Simple Online and Realtime Tracking (SORT) [11] uses this approach, being a fast model, able to run in real-time. Also, this is an online model, which means that it does not delay the tracking, in order to have more information, using only the frames up until the current frame.

DeepSORT [12] is an improved version of SORT, focused on tracking persons. The DeepSORT model does not rely solely on the objects' position and movement, but introduces a metric which compares two persons regarding their visual features. This means that two persons in successive frames need to have a small Euclidean distance between their bounding boxes' centers, match the trajectory and be similar visually. The association metric uses a CNN architecture that generates a feature vector of dimensionality 128 based on the pixels inside the bounding box. Two objects are considered the same if the cosine distance between the two feature vectors projected onto the unit hypersphere is smaller than a threshold. Thus, the model has greatly improved performance compared to previous methods, being able to correctly tag objects even when they are occluded or leave the scene.

2.2 Face Detection

Face detection is a particular case of the object detection problem. Face detection models work very well with images where few medium or large faces are present, but struggle with images with high number of small faces. Two classical models used for face and other human features detection, are Histogram of Oriented Gradients [13], which is a feature descriptor using gradient orientation in localized regions of images in order to detect objects, and Haar Cascades [14], which is a classifier using a cascade of features in order to identify objects. These models lack severely in performance when used on images with many persons and small faces.

A modern approach is Multi-Task Cascaded Convolutional Neural Network (MTCNN) [15], which uses a cascaded architecture with three stages of deep CNNs to detect faces and landmark locations, where face detection and face alignment are done jointly. This model, albeit slower, it has significantly improved performance over the previous models.

Qi et al. [16] present a face detection method capable of detecting faces of various levels of granularity. This is achieved by using a cascade of three Deep Convolutional Neural Networks based on separable residual convolution, where non maximum suppression is applied on the bounding box regressors.

Usually, the face detection models have increased performance when applied to the detected persons' bounding boxes, due to the fact that no faces will be detected outside of a person's bounding box, reducing the false positive rate drastically. Doing this is equivalent to simplifying the problem by making the face detector only look where faces are possible to exist, provided the person detector is reliable.

2.3 Face Tracking

Face tracking is a particular case of object tracking and there is extensive work done to solve this problem. There are approaches using Markov Decision Process [17], others that rely on data association [18], and some that even generalize the problem by trying to recognize faces that have been seen before in the lack of a continuous representation [19]. But these methods require significant processing power, thus are not suitable for edge processing. More, in many real-time scenarios the increase in performance over a simple person tracker will not be large enough to account for the increased processing and deployment costs.

In the proposed solution, face tracking combines the person tracker together with the face detector. Instead of creating a separate model for face tracking, it is more efficient to rely on the person tracker and then to detect faces in the bounding boxes of persons. If a face is detected, it will share the id with the person. Since some persons will not be facing the camera for the whole time, face detection will not be possible at every frame, especially when a person is facing the opposite direction as the camera. But this should not pose a problem to the algorithm's performance regarding data association, since any erroneous association will be caused by the people tracker.

2.4 Object Detection and Tracking on Edge Devices

In recent years, the number of Computer Vision applications based on edge devices has increased significantly, mainly due to increased computational power. Thus, object detection and tracking models, which would normally require high end GPUs, can be run on edge devices, albeit at significantly lower speeds. This issue can be approached by using a lightweight version of the models, which will offer a decent tradeoff between performance and speed.

Many low compute solutions rely on more powerful devices than those in the Raspberry series, such as devices in the NVIDIA Jetson series. Tijtgat et al. [20] present a solution for object detection using Jetson TX2, attached to a low-altitude Unmanned Aerial Vehicle (UAV). YOLOv2 and TinyYOLO are used for inference, reporting more than 80% mAP (mean average precision) for YOLOv2 at around 9 frames per second and around 70% mAP for TinyYOLO at 11 frames per second, using 416 × 416 pixels images as input. The results are evaluated on UAV123 dataset [21], which consists mostly of videos containing few scattered objects of interest.

Jaramillo-Avila & Anderson [22] propose a solution for object detection using the same hardware as the previous solution, using YOLOv3 for inference. The model is evaluated on Microsoft COCO database, using foveated image sampling. They report around 35% recall and around 25% precision at 4–5 frames per second, using 416 × 416 images as input.

Cojocea, Hornea & Rebedea [23] make a comparison between various edge devices and a high-end GPU, regarding object detection and tracking. Using YOLOv3 and TinyYOLOv3 for inference and 600 × 600 images as input, they report 29.8% mAP for TinyYOLOv3 and 53.2% mAP for YOLOv3. The frame rate varies significantly, depending on the device. Using TinyYOLOv3, they report 0.33 frames per second using Raspberry Pi 3, 2.9 fps using Raspberry Pi 3 and Intel NCS accelerator, 37.8 fps using NVIDIA Jetson TX2 and 343 fps using a high end GPU.

The above mentioned solutions and many more indicate that edge devices and software solutions are both converging to a point where object detection and tracking in real-time is possible using edge processing.

3 Proposed Method Focusing on Low-Power Compute

3.1 Models

Due to the edge processing choice, running very complex neural models on the existing hardware is not desirable, as they would achieve a slow speed. Lighter versions of such models need to be run in order to have a reasonable speed and allow real-time and low-compute edge processing.

Thus, shallow architectures like the tiny versions of YOLO or YOLO-LITE are natural choices for object detection. They give a good tradeoff between performance and speed and can be easily tuned to achieve better performance at the cost of speed and vice versa. The shallow versions of YOLO have between 9–24 layers and experiments have shown that batch normalization is not necessary due to the small size of these networks. Being small, they tend not to suffer from covariate shift. Also, reducing the size of input images can offer a significant boost in terms of frames per second at the cost of mean average precision (mAP). This was considered acceptable for the purpose of our experiments and YOLOv3-tiny was used for the proposed system.

For object tracking both SORT and DeepSORT have been experimented with, but the latter was preferred due to its increased performance at a low compute cost. The network used for generating the feature vector for each detection was pretrained on a dataset containing over 1.1 million images of 1261 persons and is able to make new predictions very fast.

Regarding face detection, MTCNN was necessary, due to its much better performance than the other models, despite its slower speed. The resulting detections are used to track faces in real-time, even for complex scenarios with large crowds, giving each face the id of the tracked person.

3.2 Hardware

With regards to hardware, edge processing was selected for our real-time people tracking solution. The main goal of our paper is to check the performance of running the aforementioned neural models for people tracking on present day edge devices or the edge devices in the near future. Despite the existence of other more powerful edge devices, such as Nvidia Jetson series, Raspberry Pi 4 [24] is much cheaper and has been released fairly recently, in June 2019. It has a 1.5 GHz quad-core CPU and 4 GB of RAM, which are a significant improvement over the previous version.

The Intel Neural Compute Stick 2 (Intel NCS2) [25] has been released in November 2018 and can be used conjointly with Raspberry Pi 4 in order to significantly increase the speed of the implemented models. Intel NCS2 is a device that can be connected to any device via USB and can be used for running neural models, but only for inference and not training. The boost in speed is remarkable, being able to run neural inference up to 30 times faster than the Raspberry Pi 4 alone, and 8 times faster than the previous version of Intel Neural Compute Stick.

Thus our hardware solution for low-power edge processing people tracking uses Raspberry Pi 4 (4 GB RAM) and Intel NCS2. Results for running the same models on a server with dedicated GPUs will be provided for comparison in the next section. The server has the following configuration: Intel Xeon CPU E5-2637 v3 @ 3.50 GHz, 2 × GeForce GTX 1080 Ti GPUs, 64 GB of RAM.

4 Results

Before going over the results, it is important to mention the datasets used for the experiments. The first dataset is provided by the Multi Object Tracking Challenge 2017 [26], containing 14 videos from fixed and mobile cameras, with multiple persons on streets and shopping malls. The videos have lengths varying from 15 s to 1 min, and a frame rate of 30. The second dataset is composed of 20 videos, with lengths of several minutes to tens of minutes, taken from a real-world scenario in a shopping mall in Bucharest, Romania, at the cinema entrance and bar. This dataset contains videos with all kinds of people density. The videos are fed into the model one frame at a time, after each frame has been resized to 600 × 600 pixels.

4.1 Detections

In Fig. 1 and Fig. 2 are presented two captions of the output of the pipeline formed by the object detection and object tracking models. In white bounding boxes are highlighted the detected persons and the numbers in green in the center of each box represent the person's tracking id, which is used for counting the total number of unique persons in the video stream. In the first image, the most relevant persons are detected and even a couple of more distant persons. There is a person in the forefront that is subject to heavy occlusion (by the person with the id number 25) which is not detected, but this person would be hard to detect even by humans. In the second image though, this person is less occluded and is clearly visible to the human eye, but it is still not detected by the object detection model. Also, in the right part of the image, the model detects a person with the id number 31, but it represents a false positive. Besides these type of errors, few people of small size in the background are detected.

Fig. 1. An image representing the result of detection and tracking on a fixed camera outdoor video stream. The white rectangles represent the bounding boxes of detected people and the numbers in green represent that person's tracking id. (Color figure online)

Fig. 2. An image from the same video stream as in Fig. 1, sampled a few frames later. (Color figure online)

In Fig. 3 and Fig. 4 are presented two images from the same video stream, where the MTCNN model for face detection has been used without relying on the object detector. Although it detects some of the faces in the image, as in Fig. 3, it is prone to many false positives such as in Fig. 4. By using the face detection model only on the cropped images of people detected, it enables itself to detect new faces and it reduces the number of totally erroneous detections. Despite this, false positives are still occurring. In Fig. 5, the blue rectangles represent the bounding boxes of the detected persons and the red rectangles represent the faces detected. In the case of the person closest to the camera, it detects the person's elbow as a face, and the other detection while being in the general region of the person's face, it has some significant offsets both in position and size. Fortunately, the other two face detections are accurate, but there are still relevant faces that remain undetected.

4.2 Object Detection

In Table 1 we can see the results for people detection, where performance is measured using mean average precision (mAP). Looking at speed (fps), the Raspberry Pi 4 device

Fig. 3. Image representing the face detection model used on its own. The green rectangles represent the faces detected by the model. (Color figure online)

Fig. 4. Image representing the face detection model used on its own. The green rectangles represent the faces detected by the model. (Color figure online)

Fig. 5. Image representing the face detection model running only for the persons detected by the object detection model. Blue rectangles represent people detected, red rectangles represent the faces detected inside the blue rectangle. (Color figure online)

on its own is not powerful enough to enable an object detector to run at frame rates near real-time. Yet, the results are promising, since this is not a dedicated device, but a general purpose one, meaning that the next version might have enough computational power for real-time speed. But significantly greater frame rate can be achieved by using the Intel NCS2, going up to ten times faster. Furthermore, multiple Intel NCS2 devices could be used in parallel for better speed. We also added, in the first two rows, some results reported in [23] for comparison. We can see that Raspberry Pi 4 runs the same model, with the same mAP, 2.63 times faster than Raspberry Pi 3, indicating an important computational power for the Raspberry Pi series.

Table 1. People detection performance and speed on edge devices

Device	Algorithm	FPS	mAP
Raspberry Pi 3 [23]	Tiny YOLOv3	0.33	29.8
Raspberry Pi 3 + Intel NCS [23]	Tiny YOLOv3	2.90	29.8
Raspberry Pi 4	Tiny YOLOv3	0.87	29.8
	YOLOv3	0.07	53.2
Raspberry Pi 4 + Intel NCS2	Tiny YOLOv3	7.86	29.8
	YOLOv3	0.32	53.2
Computer with dedicated GPUs	Tiny YOLOv3	343.00	29.8
	YOLOv3	47.44	53.2

At the same time, Raspberry Pi 4 with Intel NCS2 accelerator achieves a similar speedup in comparison with Raspberry Pi 3 with Intel NCS (first version). This indicates a similar leap forward regarding computational power in the Intel NCS series.

4.3 Object Tracking

In Table 2, we present the results for the people tracking methods, where we measure the multiple object tracking accuracy (MOTA) and multiple object tracking precision (MOTP) [27]. While DeepSORT has half the frame rate of SORT, it achieves significantly better performance.

Table 2. People tracking performance and speed on edge devices

Device	Algorithm	FPS	MOTA	MOTP
Raspberry Pi 4	SORT	2.89	29.8	65.4
	DeepSORT	1.23	57.4	77.5
Computer with dedicated GPUs	SORT	67.45	29.8	65.4
	DeepSORT	35.48	57.4	77.5

4.4 Object Counting Results

The people counting method depends strictly on the people tracking algorithm's performance, but is not sensible to errors such as identity switches, temporary losing a track or track fragmentation. In Table 3 we see great differences in counting estimation based on the video type. There are four types of videos: outdoor scene with fixed camera, indoor scene with fixed camera, outdoor scene with moving camera, and indoor scene with moving camera. As it was expected, for the videos taken by a moving camera the error was greater due to the continuous changing point of view and the generally small height of the camera. Also, indoor scenes tend to have lower error due to the smaller change in lighting conditions and generally smaller distances. These results show that the proposed method for fixed indoor cameras offers good performance for people counting in commercial areas like shopping malls.

Table 3. People counting error rate

Video type	Error rate
Fixed camera indoor	5.06%
Fixed camera outdoor	9.54%
Mobile camera indoor	12.08%
Mobile camera outdoor	33.53%

4.5 Face Detection

Taken into consideration the results in Table 4, it is clear that the face detection algorithms are still lacking. This is due to the nature of the videos, where there are many people at the same time, with different orientations, and faces represented with few pixels. But it is worth noting that using a deep learning method yields better results.

Table 4. Face detection results

Device	Algorithm	FPS	mAP
Raspberry Pi 4	Haar Cascades	1.27	24.9
	MTCNN	0.12	43.5
Computer with dedicated GPUs	Haar Cascades	21.57	24.9
	MTCNN	10.78	43.5

5 Conclusions

The results show that running the proposed system using Raspberry Pi 4 has a small framerate, even when using Intel NCS2. Despite this, edge processing for real-time scenarios is promising, taking into account that there are much more powerful edge devices, albeit more expensive and power consuming. Furthermore, there are possible optimizations to the used models and to the hardware configuration. Multiple Intel NCS2 devices could be used in parallel to provide a boost in frame rate.

It is likely that the next version of Raspberry Pi could handle neural models at almost real-time frame rates, taking into consideration the hardware leaps this device has made in the last few years. A recent comparison between various versions of Raspberry Pi devices [28] emphasizes the great improvements in computational power, reporting ~4× performance boost in Raspberry Pi 4 in comparison with Raspberry Pi 3, which align with the results obtained in our study for people detection.

The main conclusion is that using edge devices for computer vision is plausible today for small scale projects, but for large scale projects it is still lacking, mainly in terms of computing speed. The models used are tens of times slower and run at half performance on such devices compared to high end GPU systems. However, this speed gap will be significantly reduced by the edge devices in the near future.

Acknowledgements. This research was funded by the MARKSENSE project "Real-time Analysis Platform For Persons Flows Based on Artificial Intelligence Algorithms and Intelligent Information Processing for Business and Government Environment", contract no. 124/13.10.2017, MySMIS 2014 code 119261.

References

1. Krizhevsky, A., Sutskever, I., Hinton, G.E.: Imagenet classification with deep convolutional neural networks. In: Advances in Neural Information Processing Systems, pp. 1097–1105 (2012)
2. Girshick, R., Donahue, J., Darrell, T., Malik, J.: Rich feature hierarchies for accurate object detection and semantic segmentation. In: Proceedings of the IEEE Conference on Computer Vision and Pattern Recognition, pp. 580–587 (2014)
3. Girshick, R.: Fast R-CNN. In: Proceedings of the IEEE International Conference on Computer Vision, pp. 1440–1448 (2015)
4. Ren, S., He, K., Girshick, R., Sun, J.: Faster R-CNN: towards real-time object detection with region proposal networks. In: Advances in Neural Information Processing Systems, pp. 91–99 (2015)
5. He, K., Gkioxari, G., Dollár, P., Girshick, R.: Mask R-CNN. In: Proceedings of the IEEE International Conference on Computer Vision, pp. 2961–2969 (2017)
6. Redmon, J., Divvala, S., Girshick, R., Farhadi, A.: You only look once: unified, real-time object detection. In: Proceedings of the IEEE Conference on Computer Vision and Pattern Recognition, pp. 779–788 (2016)
7. Redmon, J., Farhadi, A.: YOLO9000: better, faster, stronger. In: Proceedings of the IEEE Conference on Computer Vision and Pattern Recognition, pp. 7263–7271 (2017)
8. Redmon, J., Farhadi, A.: Yolov3: an incremental improvement. arXiv preprint arXiv:1804.02767 (2018)
9. Huang, R., Pedoeem, J., Chen, C.: YOLO-LITE: a real-time object detection algorithm optimized for non-GPU computers. In: 2018 IEEE International Conference on Big Data (Big Data), pp. 2503–2510 (2018)
10. Welch, G., Bishop, G.: An introduction to the Kalman filter (1995)

11. Bewley, A., Ge, Z., Ott, L., Ramos, F., Upcroft, B.: Simple online and realtime tracking. In: 2016 IEEE International Conference on Image Processing (ICIP), pp. 3464–3468 (2016)
12. Wojke, N., Bewley, A., Paulus, D.: Simple online and realtime tracking with a deep association metric. In: 2017 IEEE International Conference on Image Processing (ICIP), pp. 3645–3649 (2017)
13. Déniz, O., Bueno, G., Salido, J., De la Torre, F.: Face recognition using histograms of oriented gradients. Pattern Recognit. Lett. **32**(12), 1598–1603 (2011)
14. Viola, P., Jones, M.: Rapid object detection using a boosted cascade of simple features. In: Proceedings of the 2001 IEEE Computer Society Conference on Computer Vision and Pattern Recognition. CVPR 2001, vol. 1, pp. I (2001)
15. Zhang, K., Zhang, Z., Li, Z., Qiao, Y.: Joint face detection and alignment using multitask cascaded convolutional networks. IEEE Signal Process. Lett. **23**(10), 1499–1503 (2016)
16. Qi, R., Jia, R.S., Mao, Q.C., Sun, H.M., Zuo, L.Q.: Face detection method based on cascaded convolutional networks. IEEE Access **7**, 110740–110748 (2019)
17. Xiang, Y., Alahi, A., Savarese, S.: Learning to track: online multi-object tracking by decision making. In: Proceedings of the IEEE International Conference on Computer Vision, pp. 4705–4713 (2015)
18. Yu, H., et al.: Groupwise tracking of crowded similar-appearance targets from low-continuity image sequences. In: Proceedings of the IEEE Conference on Computer Vision and Pattern Recognition, pp. 952–960 (2016)
19. Lin, C.C., Hung, Y.: A prior-less method for multi-face tracking in unconstrained videos. In: Proceedings of the IEEE Conference on Computer Vision and Pattern Recognition, pp. 538–547 (2018)
20. Tijtgat, N., Van Ranst, W., Goedeme, T., Volckaert, B., De Turck, F.: Embedded real-time object detection for a UAV warning system. In: Proceedings of the IEEE International Conference on Computer Vision Workshops, pp. 2110–2118 (2017)
21. Mueller, M., Smith, N., Ghanem, B.: A benchmark and simulator for UAV tracking. In: Leibe, B., Matas, J., Sebe, N., Welling, M. (eds.) ECCV 2016. LNCS, vol. 9905, pp. 445–461. Springer, Cham (2016). https://doi.org/10.1007/978-3-319-46448-0_27
22. Jaramillo-Avila, U., Anderson, S.R.: Foveated image processing for faster object detection and recognition in embedded systems using deep convolutional neural networks. In: Martinez-Hernandez, U., et al. (eds.) Living Machines 2019. LNCS (LNAI), vol. 11556, pp. 193–204. Springer, Cham (2019). https://doi.org/10.1007/978-3-030-24741-6_17
23. Cojocea, E., Hornea, S., Rebedea, T.: Balancing between centralized vs. edge processing in IoT platforms with applicability in advanced people flow analysis. In: 2019 18th RoEduNet Conference: Networking in Education and Research (RoEduNet), pp. 1–6. IEEE (2019)
24. Raspberry Pi 4. https://www.raspberrypi.org/products/raspberry-pi-4-model-b/. Accessed 31 Jan 2020
25. Intel Neural Compute Stick 2. https://software.intel.com/en-us/neural-compute-stick. Accessed 31 Jan 2020
26. Multiple Object Tracking Benchmark MOT17. https://motchallenge.net/data/MOT17/. Accessed 31 Jan 2020
27. Bernardin, K., Stiefelhagen, R.: Evaluating multiple object tracking performance: the CLEAR MOT metrics. EURASIP J. Image Video Process. **2008**, 1–10 (2008)
28. https://magpi.raspberrypi.org/articles/raspberry-pi-4-vs-raspberry-pi-3b-plus. Accessed 30 Apr 2020

Simple Pose Network with Skip-Connections for Single Human Pose Estimation

Van-Thanh Hoang[1,2](✉) and Kang-Hyun Jo[2]

[1] Department of Engineering-Technology, Quang Binh University, Dong Hoi, Vietnam
thanhhv@qbu.edu.vn

[2] School of Electrical Engineering, University of Ulsan, Ulsan, Korea
acejo@ulsan.ac.kr

Abstract. Recently, following the success of deep convolutional neural networks, human pose estimation problem has been largely improved. This paper introduces an improved version of the Simple Pose network for single human pose estimation. It adds the skip-connections between the same-resolution layers of the backbone and up-sampling stream to fuse low-level and high-level features. To make the depth of features from low-level and high-level are same, this paper uses 1 × 1 convolutional layer. The experiments show that this naive technique makes the new networks better over 1% mAP scores with just a small increment in model size.

1 Introduction

The single human pose estimation problem is to detect the position of key-points of a person inside an image or frame of video/webcam. This task can be applied in many applications like video surveillance system, human action recognition/prediction, human-system interaction, and 3D pose estimation [8,9].

Recently, following the development of the convolution neural networks (CNN) [14], the performance of recognizing the pose of single person inside an input image or frame has been largely enhanced. Usually, the single person pose estimation method need to use a person detector [16,22] to get the position of people inside. Then, they estimated the pose for each person. For example, the Stacked Hourglass Network [18] generates the score-maps for every key-point by using a stack of eight Hourglass modules then output the pose of person based on the max-activation across the score-maps. Similarly, the Simple Pose [26] also uses a detector to have bounding boxes then estimates the score-maps of every key-point by using transposed convolution layers following backbone network (Residual [7]) and does not have skip-connections.

This paper improves the Simple Pose Network by adding the skip-connections between the same-resolution layers of the backbone and up-sampling stream to

Most of this work was done when Van-Thanh Hoang studied at University of Ulsan.

M. Hernes et al. (Eds.): ICCCI 2020, CCIS 1287, pp. 166–174, 2020.
https://doi.org/10.1007/978-3-030-63119-2_14

fuse low-level and high-level features. To make the depth of features from low-level and high-level are same, this paper uses a 1×1 convolutional layer. The experiments show that this naive technique makes the new networks have over 1% mAP scores better with just a small increment in model size for both ResNet-18 and ResNet-50 backbones.

2 Related Work

Classical approaches [5,17,21,24,28] tackled human pose estimation problem as graphical model or a tree-structured problem and located keypoint locations based on hand-crafted features.

Recently, thanks to the development of the deep convolution neural network (CNN) [14], state-of-the-art methods [1,10,18] sharply improve the performance of single- and multi- human pose estimation. Because of the powerful of methods based on CNN. This paper mainly focuses on them. There are 2 kinds of approaches for this topic. They are bottom-up and top-down approach. The bottom-up methods firstly located the position of all key-points and then ensemble them into full poses of all people. The top-down methods predicted the location of joints of a human with the given bounding box from a human detector. This paper just discusses about the top-down methods.

Top-Down Approach. Methods in this approach divide up the process of generating pose of all persons inside an image or frame into a two-step system. Firstly, they localize the position of all persons in the input image. And then, based on the bounding boxes, they solve the single person pose estimation problem in the cropped patches. Papandreou *et al.* [19] generated both score-maps and the offset-maps and then obtained the final predicted location of key-points based on both of them. Mask-RCNN [6] generates the proposals for people first, and then predict the pose based on the cropped feature map of the corresponding region. Cascaded pyramid network (CPN) [2] is the leading method on COCO 2017 keypoint challenge [15]. It also involves skip-layer features concatenation and an online hard keypoint mining step. Hourglass [18] is the dominant approach on the MPII benchmark as it is the basis for all leading methods [3,4,27]. It features in a multi-stage architecture with repeated bottom-up, top-down processing and skip-layer features concatenation. Both these works [2,18] use nearest neighbor interpolation method following some convolution layers to upsample the feature map resolution. In contrary, the Simple Pose Network [26] uses the transposed convolutional layers as the combination of the upsampling and convolutional parameters to do upsample in a much simpler way. However, it did not use skip layer connections.

This paper improves the Simple Pose Network by adding the skip-connections between the same-resolution layers of backbone and up-sampling stream to fuse low-level and high-level features. To make the depth of features from low-level and high-level are same, this paper uses a 1×1 convolutional layer. The experiments show that this naive technique can help the new networks have over 1% mAP scores better with just a small increment in model size.

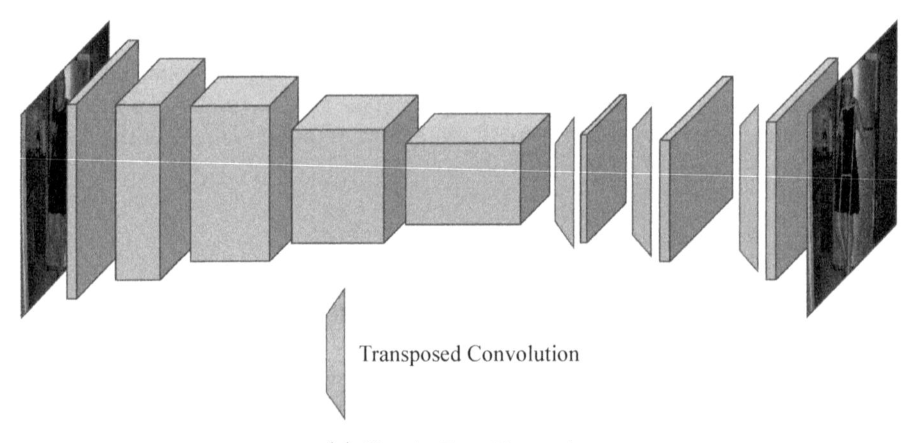

(a) Simple Pose Network.

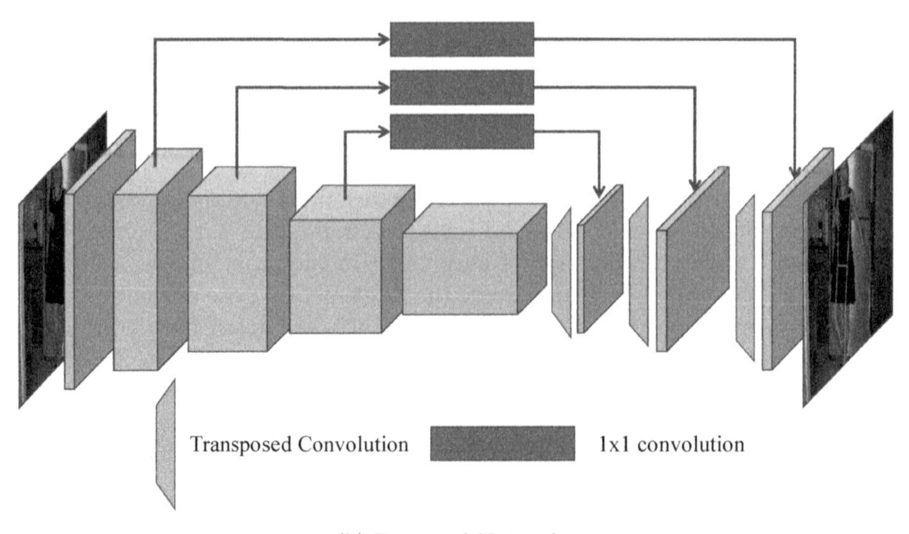

(b) Proposed Network.

Fig. 1. Architecture of the Simple Pose Network and the proposed network.

3 Our Approach

Simple Pose Network. The architecture of Simple Pose Network is shown in Fig. 1a. As can be seen, its architecture is quite simple. It uses ResNet [7], the most common backbone network for image feature extraction, as its backbone. ResNet is also used in [2,19] for pose estimation. Simple Pose simply adds a few transposed convolutional layers over the last convolution stage in the ResNet, called C_5. This structure is already adopted in Mask R-CNN [6], a state-of-the-art method for not only human pose estimation but also object segmentation. It

Table 1. Precision scores of Simple Pose and proposed network on Coco dataset. The input size is 256 × 192. The proposed ($K = 3$) means kernel size of skip-connections is 3 × 3, otherwise the kernel size is 1 × 1.

Network	Backbone	#Param	AP	AP_{50}	AP_{75}	AP_M	AP_L
Simple Pose	ResNet-18	15.3M	66.8	89.5	74.8	63.9	70.9
Proposed		15.5M	68.0	90.4	75.0	64.9	72.3
Proposed ($K = 3$)		16.4M	67.6	89.5	75.0	64.6	72.0
Simple Pose	ResNet-50	34.0M	70.4	88.6	78.3	67.1	77.2
Proposed		34.5M	71.9	91.5	79.2	68.6	76.6

is arguably the simplest way to generate score-maps from deep and low-resolution features.

By default, three transposed convolutional layers with batch normalization [11] and ReLU activation [13] are used. Each layer has 256 filters with 4 × 4 kernel. The stride is 2. A 1 × 1 convolutional layer is added at last to generate predicted score-maps $\{H_1, \ldots, H_K\}$ for all K key points ($k = 17$ for Coco dataset).

Same as in [18,25], the ground truth score-map H_k for joint k is generated by applying a 2D gaussian centered on the k_{th} joint's ground truth location. The L2 Loss is used to compute the loss between the predicted score-maps $\hat{H}_k$ and ground-truth score-maps H_k. The L2 Loss can be computed as follow:

$$\text{Loss}_{L2} = \frac{1}{K} \sum_{k=1}^{K} \|H_k - \hat{H}_k\|_2, \tag{1}$$

It is clear that Simple Pose differs from [18,25] in how high-resolution feature maps are generated. Both works [18,25] use upsampling to increase the feature map resolution and put convolutional parameters in other blocks. In contrary, Simple Pose combines the upsampling and convolutional parameters into transposed convolutional layers in a much simpler way, without using skip layer connections.

A commonality of the three methods is that three upsampling steps and also three levels of non-linearity (from the deepest feature) are used to obtain high-resolution feature maps and score-maps.

Proposed Network. The proposed network can be considered as an improved version of the Simple Pose Network. To do that, it adds the skip-connections between the same-resolution layers of the backbone and up-sampling stream to fuse low-level and high-level features. To make the depth of features from low-level and high-level are same, it adopts a 1 × 1 convolutional layer before combination. The experiments show that this naive technique can help networks have over 1% mAP scores better with just a small increment in model size.

Table 2. Recall scores of Simple Pose and proposed FPPose on Coco dataset. The input size is 256 × 192. The FPPose ($K = 3$) means kernel size of skip-connections is 3 × 3, otherwise the kernel size is 1 × 1.

Network	Backbone	#Param	AR	AR_{50}	AR_{75}	AR_M	AR_L
Simple Pose	ResNet-18	15.3M	70.1	90.6	77.0	66.9	74.9
Proposed		15.5M	71.1	91.0	77.2	67.6	76.3
Proposed (K = 3)		16.4M	70.7	90.6	77.1	67.3	75.7
Simple Pose	ResNet-50	34.0M	76.3	92.9	83.4	72.1	82.4
Proposed		34.5M	74.9	92.4	81.3	71.4	80.3

4 Experiments

4.1 Dataset

This paper uses the Coco dataset [15] for experiments. This dataset requires localization of person keypoints in challenging, uncontrolled conditions. It consists of 105,698 training and around 80,000 validating human instances. The training set contains over 1 million total labeled keypoints.

4.2 Implementation Details

For the ResNet backbone, a pre-trained model that already trained on ImageNet [23] dataset is used. This paper adopts two configurations of ResNets to show the efficiency of the improved technique. They are ResNet-18 (ResNet with 18 layers) and ResNet-50 (ResNet with 50 layers). The remained parameters of lateral convolutions layers, transposed convolution layers, and the last 1 × 1 convolution layer are initialized by Normal distribution for weight and 0 for bias.

All the input images are resized to 256 × 192 pixels. The network is trained using Pytorch [20] and for optimization, the Adam optimizer [12] with a learning rate of 1e−3 is used. The models are trained on a personal computer with a Core i7-8700K 3.70 GHz CPU, 32-GB RAM, and an NVIDIA Titan RTX GPU device for 140 epochs. The learning rate is dropped once by a factor of 10 at epochs of 90 and 120.

4.3 Experiment Results

Quantitative Results. Table 1 and Table 2 shows the AP and AR scores of Simple Pose and proposed networks with 2 kind of backbone: ResNet-18 and ResNet-50 on Coco dataset.

As can be seen, after adding the skip-connections between the same-resolution layers of backbone and up-sampling stream to fuse low-level and high-level features and adopting 1 × 1 convolutional layer before combination to make

Fig. 2. Qualitative results of the proposed FPPose on validating samples from Coco dataset.

the depth of features from low-level and high-level are same, the scores of the new network are better over 1% in comparing to the original. While the model size just increases by 1%.

This paper also do an ablation study to check which kernel size should be used for lateral convolution. As illustrated, the kernel size $K = 1$ is better than $K = 3$ while it has smaller model size. The reason is maybe the $K = 3$ convolution also captures the local spatial information which may harm the high-level features.

Qualitative Results. Some visual examples of multi-person poses predicted by the proposed model are shown in Fig. 2. As you can see, the proposed model can generate good poses for all people inside the input image.

5 Conclusion

This paper modifies the Simple Pose Network to have a better network with just small modifications. It adds the skip-connections between the same-resolution layers of the backbone and up-sampling stream to fuse low-level and high-level features and adopts a 1×1 convolutional layer before combination to make the depth of features from low-level and high-level are same. The experiments show that this naive technique can help the new networks have over 1% mAP score better with just a small increment in model size.

In the future, it is necessary to improve the performance of the model. Additionally, this model should be further optimized to be faster.

Acknowledgments. This work was supported by the National Research Foundation of Korea (NRF) grant funded by the government (MSIT) (No. 2020R1A2C2008972).

References

1. Bulat, A., Tzimiropoulos, G.: Human pose estimation via convolutional part heatmap regression. In: Proceedings of the European Conference on Computer Vision, pp. 717–732 (2016)
2. Chen, Y., Wang, Z., Peng, Y., Zhang, Z., Yu, G., Sun, J.: Cascaded pyramid network for multi-person pose estimation. In: Proceedings of the IEEE Conference on Computer Vision and Pattern Recognition (2018)
3. Chen, Y., Shen, C., Wei, X.S., Liu, L., Yang, J.: Adversarial posenet: a structure-aware convolutional network for human pose estimation. In: Proceedings of the IEEE International Conference on Computer Vision, pp. 1212–1221 (2017)
4. Chu, X., Yang, W., Ouyang, W., Ma, C., Yuille, A.L., Wang, X.: Multi-context attention for human pose estimation. In: Proceedings of the IEEE Conference on Computer Vision and Pattern Recognition, pp. 1831–1840 (2017)
5. Gkioxari, G., Arbelaez, P., Bourdev, L., Malik, J.: Articulated pose estimation using discriminative armlet classifiers. In: Proceedings of the IEEE Conference on Computer Vision and Pattern Recognition, pp. 3342–3349 (2013)
6. He, K., Gkioxari, G., Dollár, P., Girshick, R.: Mask R-CNN. In: Proceedings of the IEEE International Conference on Computer Vision, pp. 2961–2969 (2017)

7. He, K., Zhang, X., Ren, S., Sun, J.: Deep residual learning for image recognition. In: Proceedings of the IEEE Conference on Computer Vision and Pattern Recognition, pp. 770–778 (2016)
8. Hoang, V.T., Hoang, V.D., Jo, K.H.: An improved method for 3D shape estimation using cascade of neural networks. In: Proceedings of the IEEE International Conference on Industrial Informatics, pp. 285–289 (2017)
9. Hoang, V.T., Jo, K.H.: 3D human pose estimation using cascade of multiple neural networks. IEEE Transactions on Industrial Informatics **15**(4), 2064–2072 (2019). https://doi.org/10.1109/TII.2018.2864824
10. Insafutdinov, E., Pishchulin, L., Andres, B., Andriluka, M., Schiele, B.: Deepercut: a deeper, stronger, and faster multi-person pose estimation model. In: Proceedings of the European Conference on Computer Vision, pp. 34–50 (2016)
11. Ioffe, S., Szegedy, C.: Batch normalization: accelerating deep network training by reducing internal covariate shift. In: Proceedings of the International Conference on Machine Learning, pp. 448–456 (2015)
12. Kingma, D.P., Ba, J.: Adam: a method for stochastic optimization. In: Proceedings of the International Conference on Learning Representations (2015)
13. Krizhevsky, A., Sutskever, I., Hinton, G.E.: Imagenet classification with deep convolutional neural networks. In: Proceedings of Advances in Neural Information Processing Systems, pp. 1097–1105 (2012)
14. LeCun, Y., Bottou, L., Bengio, Y., Haffner, P.: Gradient-based learning applied to document recognition. Proc. IEEE **86**(11), 2278–2324 (1998). https://doi.org/10.1109/5.726791
15. Lin, T.Y., et al.: Microsoft coco: common objects in context. In: Proceedings of the European Conference on Computer Vision, pp. 740–755 (2014)
16. Liu, W., et al.: Ssd: single shot multibox detector. In: Proceedings of the European Conference on Computer Vision, pp. 21–37 (2016)
17. Luo, R.C., Chen, S.Y.: Human pose estimation in 3-D space using adaptive control law with point-cloud-based limb regression approach. IEEE Trans. Ind. Inform. **12**(1), 51–58 (2016)
18. Newell, A., Yang, K., Deng, J.: Stacked hourglass networks for human pose estimation. In: Proceedings of the European Conference on Computer Vision, pp. 483–499 (2016)
19. Papandreou, G., et al.: Towards accurate multi-person pose estimation in the wild. In: Proceedings of the IEEE Conference on Computer Vision and Pattern Recognition, pp. 4903–4911 (2017)
20. Paszke, A., et al.: Pytorch: an imperative style, high-performance deep learning library. In: Proceedings of the Neural Information Processing Systems (2019)
21. Pishchulin, L., Andriluka, M., Gehler, P., Schiele, B.: Poselet conditioned pictorial structures. In: Proceedings of the IEEE Conference on Computer Vision and Pattern Recognition, pp. 588–595 (2013)
22. Ren, S., He, K., Girshick, R., Sun, J.: Faster R-CNN: towards real-time object detection with region proposal networks. IEEE Trans. Pattern Anal. Mach. Intell. **39**(6), 1137–1149 (2017)
23. Russakovsky, O., et al.: Imagenet large scale visual recognition challenge. Int. J. Comput. Vis. **115**(3), 211–252 (2015)
24. Sapp, B., Taskar, B.: Modec: multimodal decomposable models for human pose estimation. In: Proceedings of the IEEE Conference on Computer Vision and Pattern Recognition, pp. 3674–3681 (2013)

25. Tompson, J.J., Jain, A., LeCun, Y., Bregler, C.: Joint training of a convolutional network and a graphical model for human pose estimation. In: Proceedings of the Advances in Neural Information Processing Systems, pp. 1799–1807 (2014)
26. Xiao, B., Wu, H., Wei, Y.: Simple baselines for human pose estimation and tracking. In: Proceedings of the European Conference on Computer Vision (2018)
27. Yang, W., Li, S., Ouyang, W., Li, H., Wang, X.: Learning feature pyramids for human pose estimation. In: Proceedings of the IEEE International Conference on Computer Vision, pp. 1281–1290 (2017)
28. Yu, J., Hong, C., Rui, Y., Tao, D.: Multitask autoencoder model for recovering human poses. IEEE Trans. Ind. Electron. **65**(6), 5060–5068 (2018)

Simple Fine-Tuning Attention Modules for Human Pose Estimation

Tien-Dat Tran, Xuan-Thuy Vo, Moahamammad-Ashraf Russo, and Kang-Hyun Jo(✉)

School of Electrical Engineering, University of Ulsan, Ulsan 44610, South Korea
{tdat,xthuy,russo}@islab.ulsan.ac.kr,
acejo@ulsan.ac.kr

Abstract. The convolution neural networks (CNNs) have achieved the best performance not only for human pose estimation but also for other computer vision tasks (e.g., object detection, semantic segmentation, image classification). Then this paper focuses on a useful attention module (AM) for feed-forward CNNs. Firstly, feed the feature map after a block in the backbone network into the attention module, split into two separate dimensions, channel and spatial. After that, the AM combines these two feature maps by multiplication and gives it to the next block in the backbone. The network can capture the information in the long-range dependencies (channel) and the spatial data, which can gain better performance in accuracy. Therefore, our experimental results will illustrate how different between when using the attention module and the existing methods. As a result, the predicted joint heatmap maintains the accuracy and spatially better with the simple baseline. Besides, the proposed architecture gains 1.0 points in AP higher than the baseline. Moreover, the proposed network trained on COCO 2017 benchmarks, which is an accessible dataset nowadays.

Keywords: Deep learning · Attention module · Human pose estimation

1 Introduction

In today's modern world, 2D human pose estimation plays a crucial yet challenging task in computer vision, which can serve for many applications such as human pose estimation [4,5,23,26,29], activity recognition [10,15], human re-identification [17,32] or 3D pose estimation [3]. The main purpose of the human pose is localizing body parts for human body joints. Spatial and channel information plays an important role in making more accuracy in keypoint regression. Hence, this paper will focus on how to make the network learn more about the attention information.

Nowadays, significant progress on human pose has been archived by deep convolution neural networks [12,21,23,24]. However, these networks still have many problems to discuss. Firstly, how to improve the accuracy in many types

M. Hernes et al. (Eds.): ICCCI 2020, CCIS 1287, pp. 175–185, 2020.
https://doi.org/10.1007/978-3-030-63119-2_15

of networks (e.g., real-time network, accuracy network). Secondly, the speed of the network when changing or modify also needs to consider. Last but not least, the new network needs to keep higher accuracy while keeping speed faster as much as it can. So this paper presents a novel of network and the effectiveness of the attention module for speed and accuracy. The proposed experiment shows a comparison between used and not used the attention module. The experiment also compares with the Simple Baseline [31] do not used the attention module and utilized the transpose convolution [6] for upsampling. Our experiment will focus on how efficient for each case and their cost for the network.

To make good use of underlying the attention method, our network increase 1.0 point in AP for accuracy and reduces around 8.2% of parameters compared with the simple baseline when used ResNet-50 [8] as a backbone network. Hence this paper illustrates a new module for the network, which will quickly adapt in many systems for many challenges such as object detection, image classification, and human pose estimation. The proposed method estimates joints for human pose estimation based on the recovery of feature maps after using the up-sampling network.

2 Related Work

Human Pose Estimation. The leading part of human pose estimation lies in joint detection and their relationship with spatial space, which illustrates in Fig. 1. Deeppose [27], Simple baseline utilizes joints prediction thought an end to end network with higher parameter. Later, Newell with Stacked hourglass network [22] decreases the number of the setting while still keep high accuracy. All of the methods used Gaussian distribution to represent local joints. Then used a convolution neural network to estimate human pose estimation. To decrease the employment cost, they need to reduce the number of parameters, and applying suitable up-sampling methods will lower the network's parameter. So, the proposed method used interpolation as up-sampling module.

On the other hand, to enhance the speed of the network, interpolation shows a lot of benefits than the transpose convolution. However, in some complex and higher cost architecture, the transpose convolution gives better accuracy. In comparison, our up-sampling module provides an adequate view for designing the network, with a small number in parameter and high speed or higher parameter and lower speed. Then this paper shows how up-sampling will work in each method and each result.

Upsampling. This module plays an essential role in human pose estimation, especially with not only recovering the loss information from down-sampling (e.g., convolution, pooling, max-pooling, etc.) [11]. It also created a new size that makes suitable for another stage, such as the heat map regression or human pose estimation. However, there are many up-sampling methods used for this task. Then this paper introduces what way was used for the network in our experiment, followed by the result of it in the Experiment section.

Transpose Convolution [6] The transpose convolution is a popular way to up-sampling while the network still keeps the learnable parameters. However, this will make the speed of the system slower because of the added settings. In some cases, this parameter is necessary, but some are just meanness, therefore using another for up-sampling is better. Moreover, to utilize the transpose convolution, the network needs to understand more information than what is necessary while this maybe not required for the low-cost system such as real-time network

Bi-linear Interpolation [20] Bi-linear Interpolation uses a weighted average of the four nearest cell centers. The closer of an output cell center to the input cell center, the higher the influence of its value is on the output cell value. This problem means that the output value could be different than the four nearest input cells, but is always within the same range of values as the four input. Because the index was changed but remained the critical value, Bi-linear was used for up-sampling in the network while it did not consume many parameters.

Attention Mechanism. Human visualization plays an important role in computer vision, and there are several attempts related to attention processing to improve the performance of CNNs. Wang et al. [28] proposed a non-local network to capture long-range dependencies. Inspired by SENet [9] and Inception [25], then SKNet [18] combined the channel attention module from SENet and the multi-branch convolution from the inception. Moreover, the spatial attention module come from STN [14], which proposed by Google, this module aggregates the context information of the feature maps.

In this paper, the proposed method was inspired by CBAM network [30] to make the effective between both channel and spatial module by using element-wise multiplication. After that, the feature map takes an addition to the last feature map to combine the original information and new information from the AT module.

3 Methodology

3.1 Network Architecture

Backbone Network. In the backbone network, there are ResNet-152, ResNet-101, and ResNet-50 [8], which can see in Fig. 1 for full architecture. Each ResNet has four blocks, which included convolution layers and shortcut connections. The input RGB image reduces the size to 256×192 (ResNet-50, ResNet-101) and 384×288 (ResNet-152), the feature maps will move through each backbone block, and the resolution $W \times H$ decreased two times for each block. Finally, after going through the backbone, it makes the size of the feature map reduced to $\frac{W}{16} \times \frac{H}{16}$ with 2048 channels at the end of the backbone. Moreover, the size of the channels also acquires double for each block. It from 256 after the first block come to 2048 in the last layer. The backbone network has a mission to extract the information and feature maps from the input image and feed it to the Up-sampling system for the training.

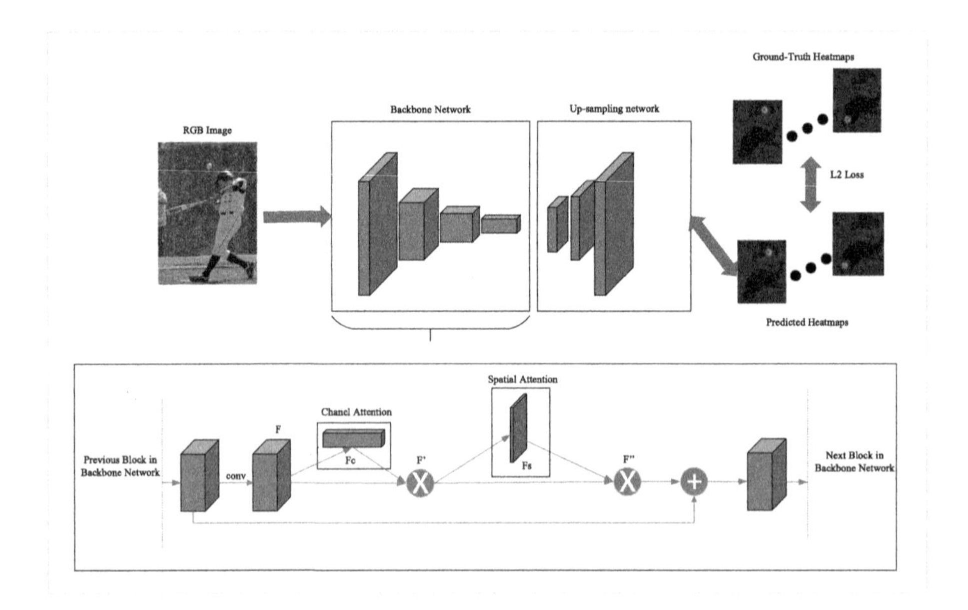

Fig. 1. Illustrating the architecture of the proposed network for human pose estimation. The proposed method divided the system into two sub-networks, Backbone, and Up-sampling network. Backbone extracts the feature map while Up-sampling recovers the feature map for regression. Moreover, this figure show the overview of the attention module which included channel and spatial module in the bottom

Up-Sampling Network. After extracting the information by utilized the backbone network, the upsampling network takes the feature map from the last layer of the backbone network and upsampling to recover the information. Then the feature map will be training with the ground-truth heat maps, which introduces in Fig. 4. The default size of heat maps is 64×48 for 256×192 images and 96×72 for 384×288. These heat maps need to consider the size of the image to fit the size of feature maps in the training process. To measure the predict keypoint, the network will use these heatmaps and ground truth heatmap for regression. For the Up-sampling network, This paper utilizes the upsampling module, which includes one layer of bilinear [20] and one layer of convolution (in Fig. 2). Then

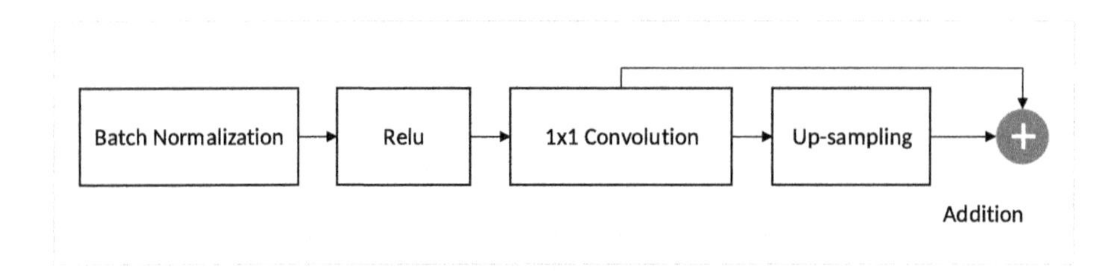

Fig. 2. Architecture of one block in Up-sampling Network

it takes an addition to this two-layer. Batch normalization and ReLU [13] are also inside the up-sampling block.

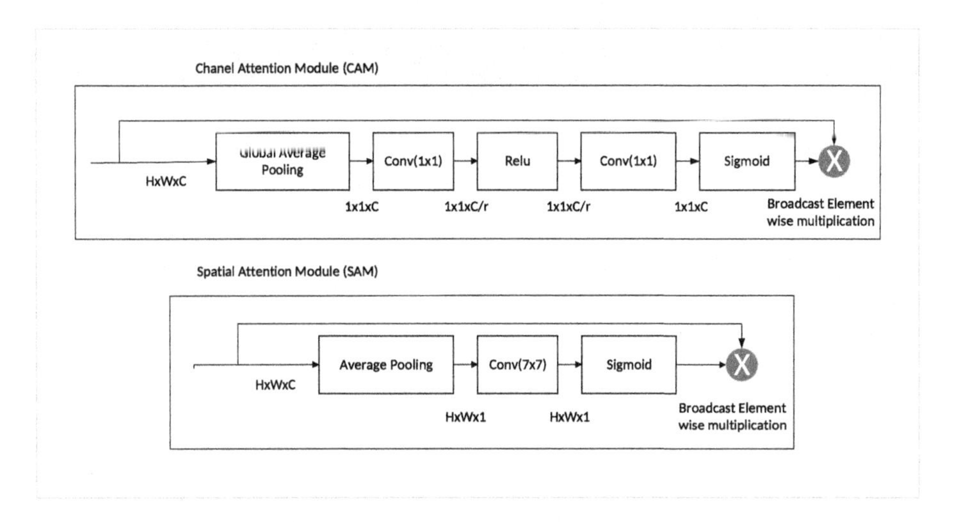

Fig. 3. Architecture of Channel Attention Module (CAM) and Spatial attention Module (SAM)

Attention Module. The attention module includes two main modules, which show in Fig. 3. Firstly, after block one in the backbone network, the feature map was feed to the channel attention module (CAM). In CAM, the feature map take global average pooling to squeeze the feature map from $H \times W \times C$ into $1 \times 1 \times C$. Next, it goes through the convolution layer, which makes the feature map into $1 \times 1 \times \frac{C}{r}$, which r is the reduction ratio and r is set to 16. After that, the CAM used ReLU to activate the weight. The final step in CAM is used 1×1 convolution layer again to regain the channel into $1 \times 1 \times C$ and utilized the sigmoid to normalize the feature map. After that, element-wise multiplication was used for combined the information for CAM.

When the feature map completely go through the CAM, it will be feed into the spatial attention module (SAM). In SAM, the feature map take average pooling for the channel which from $H \times W \times C$ to $H \times W \times 1$. After pooling, a 7×7 convolution layer was used to extract the feature map for spatial information, and the final step in SAM is similar to CAM, which can see in Fig. 3. Finally, the proposed method used element-wise addition for the original feature map and the feature map after AT to combined and give a new feature map for the next block in the backbone network.

3.2 Loss Function

This paper utilizes heat maps to represent the body joint locations for the loss function. As the ground-truth location in Fig. 4 by $a = \{a_k\}_{k=1}^{K}$, where $x_k =$

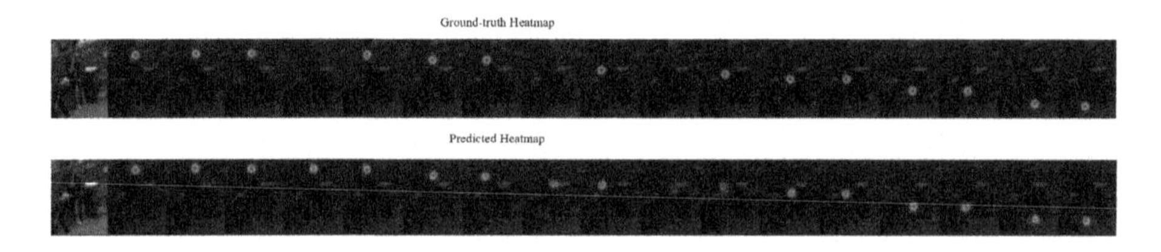

Fig. 4. Heat map for ground-truth (top) and it's prediction (bottom) for human body joints

(x_k, y_k) represent the spatial coordinate of the kth body joint in the image. Then the ground-truth score heat map H_k is generated from a Gaussian distribution with mean a_k and variance $\sum$ as follows,

$$H_k(p) \sim N(a_k, \sum) \tag{1}$$

where $\mathbf{p} \in \mathbb{R}^2$ denotes the coordinate, and $\sum$ is empirically set as an identity matrix $\mathbf{I}$. Final layer of neural network predicts K heat maps, *i.e.*, $\hat{S} = \left\{\hat{S}_k\right\}_{k=1}^{K}$ for K body joints. A loss function is defined by the mean square error, calculated as:

$$L = \frac{1}{NK} \sum_{n=1}^{N} \sum_{k=1}^{K} \left\| S_k - \hat{S}_k \right\|^2 \tag{2}$$

where N is the number of samples in training. the network generated the predict heat maps from the ground-truth heat maps using information from last layer of backbone network.

4 Experiments

4.1 Experiment Setup

Dataset. The proposed method used Microsoft COCO 2017 dataset [19] during the experiments. This dataset has around 200K images and 250K person samples, which have 17 keypoints label for one person. The research was set dataset included three folder train2017/val2017/test-dev2017, respectively, with training, validation, and testing images. Moreover, the annotations of validation and training are public and followed by the original one. Additionally, MPII human pose benchark [1] also utilized. This dataset included 25K scene images and 40K annotated persons (29K for training and 11K for testing), which consist of 16 labeled human body joints.

Evaluation Metrics. This paper used Object Keypoint Similarity (OKS) for COCO [19] with $OKS = \frac{\sum_i exp(-{d_i}^2/2s^2k_i^2)\delta(v_i>0)}{\sum_i \delta(v_i>0)}$. Here d_i is the Euclidean distance between the predicted keypoint and the groundtruth while v_i is the visibility flag of the target, s is the object scale and k_i is a keypoint for each joints. Then calculate the standart average precision and recall score. In Table 1, AP and AR is the average from OKS = 0.5 to OKS = 0.95, while AP^M for medium object and AP^L for large object.

Implementation Details. In model training, The proposed method used data augmentation such as flip, rotation with 40 degrees at the default, and scale, which put the factor for 0.3. The setting set the batch size is 4 and used the shuffle for the training images. The total of the epoch is 100, while the based learning-rate at 0.001 and multiple by 0.1 (learning rate factor) at the 80-th and 90-th epoch in our experiment. The momentum is 0.9, and the Adam optimizer [16] was used.

All experiments are implemented with Pytorch framework and testing in two datasets. The input resolution of images resized to 256 × 192 and 384 × 288. The model was trained on two NVIDIA Titan GPUs with CUDA 10.2 and CuDNN 7.3.

To show clearly about performance of each modules in AT, The proposed method compares each situation when used the module or not, which show in Table 1. The Average Precision (AP) shows that CAM and SAM gain 0.5 AP than not used while the number of parameter only increase 7% in ResNet50 (Table 2).

4.2 Experiment Result

Table 1. Result on using CAM and SAM module. Tick mean used the module in network and blank for not used.

Backbone	CAM	SAM	#Param	AP	AR
ResNet-50			28.2M	70.9	76.2
ResNet-50	✓		30.9M	71.1	76.3
ResNet-50	✓	✓	31.2M	71.4	76.3
ResNet-101	✓	✓	51.3M	72.3	77.1

COCO Datasets Result. The AP in the proposed method is higher than the Simple baseline in all cases, which 1.0 AP, 0.9 AP, and 0.5 in ResNet-50, ResNet-101, and ResNet-152, respectively. The number of parameters also decreases than the baseline because of the difference in the Up-sampling network. While the benchmark used transpose convolution cost a lot parameter, and the proposed method used bi-linear interpolation with no cost for settings. Hence, our approach adds a new module (AT) but change the up-sampling module, so the number of parameters was different. In this experiment the number of parameter was smaller 8.2%, 5.2% and 4.0% in case or ResNet-50, ResNet-101 and ResNet-152, respectively. Besides, the average recall (AR) gain better result in case of

Fig. 5. Example of human pose estimation in video cases: The image was taking each after 20 frames, and the video has the length in 17 s

Fig. 6. Qualitative result for human pose estimation in COCO2017 test-dev set

Table 2. Comparison On COCO TEST-DEV Dataset. AT is mean using attention module for network.

Method	Backbone	Input size	#Params	AP	AP^{50}	AP^{75}	AP^{M}	AP^{L}	AR
CMU-Pose [2]	–	–	–	61.8	84.9	67.5	57.1	68.2	66.5
Mask-RCNN [7]	ResNet-50-FPN	–	–	63.1	87.3	68.7	57.8	71.4	–
SimpleBaseline [31]	ResNet-50	256 × 192	34.0M	70.4	88.6	78.3	67.1	77.2	76.3
SimpleBaseline	ResNet-101	256 × 192	53.0M	71.4	89.3	79.3	68.1	78.1	77.1
SimpleBaseline	ResNet-152	384 × 288	68.6M	73.7	91.9	81.1	70.3	80.0	79.0
Our	ResNet-50	256 × 192	28.2M	70.9	91.5	78.2	68.0	75.4	76.2
Our+AT	ResNet-50	256 × 192	31.2M	71.4	91.6	78.6	68.2	75.7	76.3
Our	ResNet-101	256 × 192	47.2M	71.8	91.7	79.2	68.1	76.6	76.9
Our+AT	ResNet-101	256 × 192	50.2M	72.3	92.0	79.4	68.3	77.1	77.1
Our	ResNet-152	384 × 288	63.4M	73.5	92.6	81.5	70.5	79.5	78.6
Our+AT	ResNet-152	384 × 288	65.8M	74.2	92.7	81.7	70.6	79.8	79.2

ResNet-152 with 0.2 points higher. In conclusion, our method only trains in 100 epochs while the simple baseline train in 140 times in total, so the proposed approach is better than the baseline. Nowadays, many states of the art method is trained with over 200 epoch for COCO dataset.

However, as similar to many architectures nowadays, human pose estimation still has many challenges, which need to be solved. The first problem is that images were containing invisible joints which hard to train and predict. Secondly, images containing low-resolution persons need carefully to extract the feature for human body joints. Next, are images having crowd scenes, which also hard to determine all coordinates of joints for all people. Finally, images containing partial parts lack much information for human pose estimation (Figs. 5 and 6).

5 Conclusion

This paper illustrates the affection of attention module for the CNNs, which show utilized the attention module gain better result while not changing much the number of parameters. Additionally, the attention module highlighted the important feature maps instead of the other part. Hence, the network can improve performance, especially for many tasks in computer vision. The future work is to find out some applications or environments to apply our research, such as in the surveillance system. Another job is dual with the challenges of human pose estimation, which makes the network's accuracy decrease.

Acknowledgement. This work was supported by the National Research Foundation of Korea (NRF) grant funded by the Korea government. (MSIT)(2020R1A2C2008972)

References

1. Andriluka, M., Pishchulin, L., Gehler, P., Schiele, B.: 2D human pose estimation: new benchmark and state of the art analysis. In: 2014 IEEE Conference on Computer Vision and Pattern Recognition, pp. 3686–3693 (2014)

2. Cao, Z., Simon, T., Wei, S.E., Sheikh, Y.: Realtime multi-person 2D pose estimation using part affinity fields (2016)
3. Chen, C., Ramanan, D.: 3D human pose estimation = 2D pose estimation + matching. In: 2017 IEEE Conference on Computer Vision and Pattern Recognition (CVPR), pp. 5759–5767, July 2017. https://doi.org/10.1109/CVPR.2017.610
4. Chou, C.J., Chien, J.T., Chen, H.T.: Self adversarial training for human pose estimation (2017)
5. Chu, X., Yang, W., Ouyang, W., Ma, C., Yuille, A.L., Wang, X.: Multi-context attention for human pose estimation (2017)
6. Dumoulin, V., Visin, F.: A guide to convolution arithmetic for deep learning (2016)
7. He, K., Gkioxari, G., Dollár, P., Girshick, R.: Mask R-CNN (2017)
8. He, K., Zhang, X., Ren, S., Sun, J.: Deep residual learning for image recognition (2015)
9. Hu, J., Shen, L., Albanie, S., Sun, G., Wu, E.: Squeeze-and-excitation networks (2017)
10. Hussain, Z., Sheng, M., Zhang, W.E.: Different approaches for human activity recognition: a survey (2019)
11. Indolia, S., Goswami, A., Mishra, S., Asopa, P.: Conceptual understanding of convolutional neural network- a deep learning approach. Proc. Comput. Sci. **132**, 679–688 (2018). https://doi.org/10.1016/j.procs.2018.05.069
12. Insafutdinov, E., Pishchulin, L., Andres, B., Andriluka, M., Schiele, B.: Deepercut: a deeper, stronger, and faster multi-person pose estimation model (2016)
13. Ioffe, S., Szegedy, C.: Batch normalization: accelerating deep network training by reducing internal covariate shift (2015)
14. Jaderberg, M., Simonyan, K., Zisserman, A., Kavukcuoglu, K.: Spatial transformer networks (2015)
15. Kim, E., Helal, S., Cook, D.: Human activity recognition and pattern discovery. IEEE Pervasive Comput. **9**(1), 48–53 (2010). https://doi.org/10.1109/MPRV.2010.7
16. Kingma, D., Ba, J.: Adam: a method for stochastic optimization. In: International Conference on Learning Representations, December 2014
17. Li, W., Zhao, R., Wang, X.: Human reidentification with transferred metric learning. In: Asian Conference on Computer Vision (ACCV), pp. 31–44, November 2012
18. Li, X., Wang, W., Hu, X., Yang, J.: Selective kernel networks (2019)
19. Lin, T., et al.: Microsoft COCO: common objects in context. CoRR abs/1405.0312 (2014). http://arxiv.org/abs/1405.0312
20. Mastyło, M.: Bilinear interpolation theorems and applications. J. Funct. Anal. **265**, 185–207 (2013). https://doi.org/10.1016/j.jfa.2013.05.001
21. Moon, G., Chang, J.Y., Lee, K.M.: Posefix: model-agnostic general human pose refinement network (2018)
22. Newell, A., Yang, K., Deng, J.: Stacked hourglass networks for human pose estimation. CoRR abs/1603.06937 (2016). http://arxiv.org/abs/1603.06937
23. Ning, G., Zhang, Z., He, Z.: Knowledge-guided deep fractal neural networks for human pose estimation (2017)
24. Sun, K., Xiao, B., Liu, D., Wang, J.: Deep high-resolution representation learning for human pose estimation (2019)
25. Szegedy, C., Ioffe, S., Vanhoucke, V., Alemi, A.: Inception-v4, inception-resnet and the impact of residual connections on learning (2016)
26. Tang, Z., Peng, X., Geng, S., Wu, L., Zhang, S., Metaxas, D.: Quantized densely connected u-nets for efficient landmark localization (2018)

27. Toshev, A., Szegedy, C.: Deeppose: Human pose estimation via deep neural networks. CoRR abs/1312.4659 (2013). http://arxiv.org/abs/1312.4659
28. Wang, X., Girshick, R.B., Gupta, A., He, K.: Non-local neural networks. CoRR abs/1711.07971 (2017). http://arxiv.org/abs/1711.07971
29. Wei, S.E., Ramakrishna, V., Kanade, T., Sheikh, Y.: Convolutional pose machines (2016)
30. Woo, S., Park, J., Lee, J.Y., Kweon, I.S.: Cbam: convolutional block attention module (2018)
31. Xiao, B., Wu, H., Wei, Y.: Simple baselines for human pose estimation and tracking. CoRR abs/1804.06208 (2018). http://arxiv.org/abs/1804.06208
32. Yang, X., Wang, M., Tao, D.: Person re-identification with metric learning using privileged information. CoRR abs/1904.05005 (2019). http://arxiv.org/abs/1904.05005

Human Eye Detector with Light-Weight and Efficient Convolutional Neural Network

Duy-Linh Nguyen(✉), Muhamad Dwisnanto Putro(✉), and Kang-Hyun Jo(✉)

School of Electrical Engineering, University of Ulsan, Ulsan, Korea
{ndlinh301,dwisnantoputro}@mail.ulsan.ac.kr, acejo@ulsan.ac.kr

Abstract. The human eye detection plays an important role in computer vision. Along with face detection, it is widely applied in practical security, surveillance, and warning systems such as eye tracking system, eye disease detection, gaze detection, eye blink, and drowsiness detection system. There have been many studies to detect eyes from applying traditional methods to using modern methods based on machine learning and deep learning. This network is deployed with two main blocks, namely the feature extraction block and the detection block. The feature extraction block starts with the use of the convolution layers, C.ReLU (Concatenated Rectified Linear Unit) module, and max pooling layers alternately, followed by the last six inception modules and four convolution layers. The detection block is constructed by two sibling convolution layers using for classification and regression. The experiment was trained and tested on CEW (Closed Eyes In The Wild), BioID Face and GI4E (Gaze Interaction for Everybody) dataset with the results achieved 96.48%, 99.58%, and 75.52% of AP (Average Precision), respectively. The speed was tested in real-time by 37.65 fps (frames per second) on Intel Core I7-4770 CPU @ 3.40 GHz.

Keywords: Deep learning · Convolutional Neural Network (CNN) · Eye detection · Drowsiness detection system

1 Introduction

In computer vision, the human body is an object that has been focused on research in areas such as human pose, human interaction, human detection, and human segmentation. With small parts like the human face and other parts of the face, it poses many challenges, especially to human eye detection. The eye is a vital sensory organ of humans. The human eye position and state detection provide important information for many applications including psychological analysis, development of biomedical devices, medical diagnostics and driver assistance devices [9]. However, in practical applications when detecting eye position, there are many obstacles depending on lighting conditions and camera distance. They can make the eyes deformed and difficult to handle. Moreover, real-time systems

M. Hernes et al. (Eds.): ICCCI 2020, CCIS 1287, pp. 186–196, 2020.
https://doi.org/10.1007/978-3-030-63119-2_16

require fast and efficient processing speed on mobile or low processors devices. Inspired by driver assistance devices, specifically the drowsiness detection system, this research has proposed a convolutional neural network to detect the position of the human eye. It is a lightweight and highly efficient neural network deployed in real-time systems. In addition, it can be combined with other networks to perform tasks involving the human eye.

The main contributions of this paper are as follows:

1 - Proposed a light-weight and efficient convolutional neural network architecture for human eye detection.
2 - Built the dataset for human eye detection based on CEW, BioID, and GI4E dataset following PASCAL VOC benchmark format.

The organization of this paper as follows: Sect. 2 presents the previous methodologies relative to human eye detection. Section 3 discusses more detail about the proposed network architecture. Section 4 describes and analyzes the results. Finally, Sect. 5 concludes the paper and presents future work.

2 Related Work

In this section, the paper will present several methodologies implement on human eye detection. These methodologies can be divided into the traditional and machine learning methodologies.

2.1 Traditional Methodologies

Traditional methods are built primarily on the geometric structure of the human eye. These methods aim to extract the characteristics of the eyes and pupils based on geometric algorithms. In [16], the authors used the isophote curve to generate the position of the eyes and pupils. A set of random regressors were used in [8] to localize the position of the eye pupils. The method of detecting the pupil of the eye is proposed by using image gradients and squared dot products in [15]. In addition, many studies have focused on the use of pattern matching to localize the eye and pupils of the human eye. Several methods include using the elliptical equation [13] to match the eye area, using the filters described in the Inner Product Detector [1] to locate the eye region. These traditional methods can achieve entirely accurate results and are simple to implement. Their practical application can be limited by external conditions such as lighting conditions, image quality, and deployment equipment.

2.2 Machine Learning Methodologies

With the strong development of machine learning especially in the field of computer vision, the application of these modern methods to object detection has been widely applied. Detecting the eyes and pupils is no exception. These methods can be divided into traditional machine learning and deep learning. The traditional machine learning method is mainly based on extracting the feature map of the object using classical machine learning algorithms. The algorithms mentioned here are the Haar Wavelet and Support Vector Machine (SVM) [2], Histogram of Oriented Gradients (HOG) features combine with SVM [12], self-similarity information and shape analysis [7]. The support of the OpenCV open source library has made these algorithms easy to deploy on many different platforms. Today, many applications for detecting faces, eyes, and eye-related parts are developed based on deep learning, especially the common application of convolution neural networks. These modern machine learning methods exploit the convenience of convolution neural networks to quickly extract feature maps, which apply to eye detection, eye segmentation, and eye classification. Recent studies have focused on the central location of the eye area or the pupil detection. In [3], simple CNN was built to detect human eye centers, the combined use of convolution neural networks applied in [6] to identify the position of the pupils. Another method that is also widely applied in the localization of faces and related parts is the facial landmark detector. The face is trained with 68 feature points and the eye area described by 12 feature points [17–19] or using 3D facial landmark detection [5,20]. From these feature points, it is possible to generate regions that need to be predicted by using transform operations [4]. Deep learning methods demonstrate superior ability in accuracy and flexibility compared to traditional methods. However, in order to increase the accuracy, convolutional neural networks must be built deeper with many different network layers. This reduces the processing speed of the entire network and requires high hardware configuration for training and application deployment.

3 Proposed Network

Using a different approach than the ones mentioned above, this paper proposes a convolutional neural network that allows to locate eye area on the image based on predefined bounding boxes. This network exploits the function of the basic layers and modules in CNN such as convolution and max pooling layers, C.ReLU and Inception modules to extract the feature map. Then, it uses two sibling convolution layers to classify and regress eye bounding box coordinates. The proposed network is described in detail, as shown in Fig. 1.

Most deep convolutional neural networks for object detection are often limited by the special computation time on the CPU. In other words, convolution operation on CPU has high time-consuming when the input image size, kernel size, and output are large [21]. To extract the feature map quickly and efficiently, the blocks in this architecture use the sequence of layers as follows:

3.1 Shrinking Block

The Shrinking block is designed to quickly shrink the input image space by selecting the appropriate kernel size. As described in Fig. 1, Conv1, Pool1, Conv2, and Pool2 use stride of 1, 2, 1 and 2, respectively. Here, this block mainly use a

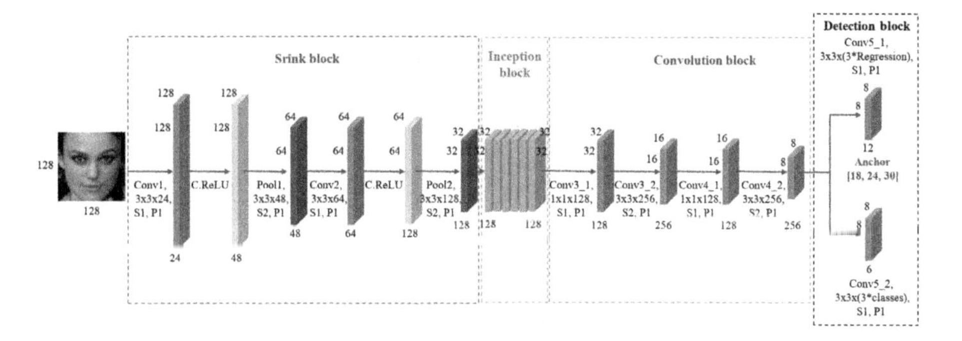

Fig. 1. The proposed convolutional neural network for human eye detection. It is built based on four blocks that are Shrinking, Inception, Convolution, and Detection block.

3×3 kernel size and input image size is 128×128. In addition, C. ReLu [11] is also used to increase the necessary speed while ensuring accuracy. C.ReLU module is described in detail in Fig. 2. With this flexible use, the network can train on various input sizes in different datasets. After going through this block, the image size is shrunk down from the original image size to 32×32. That means the input image size is reduced by four times while preserving the important information of the original image.

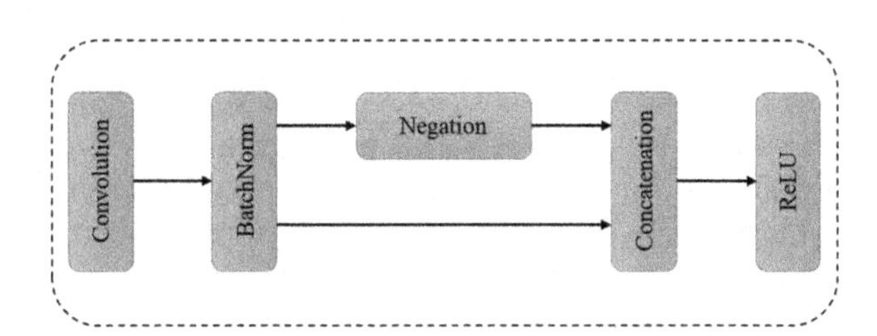

Fig. 2. C.ReLU module.

3.2 Inception Block

Inception block is a combination of six Inception modules [14]. Each Inception module consists of multiple convolution branches with many different kernel sizes. Specifically, this module is designed with four branches, using convolution operations with kernel size 1×1 or 3×3 and the number of kernels is 24 or 32. In addition, it uses a max pooling operation and concatenation plays to combine the results of branches. As a multi-scale block according to the width of the network, these branches can enrich receptive fields. Figure 3 shown detail about Inception module. The feature map with size is $32 \times 32 \times 128$ will be maintained and enriched when processed by this block.

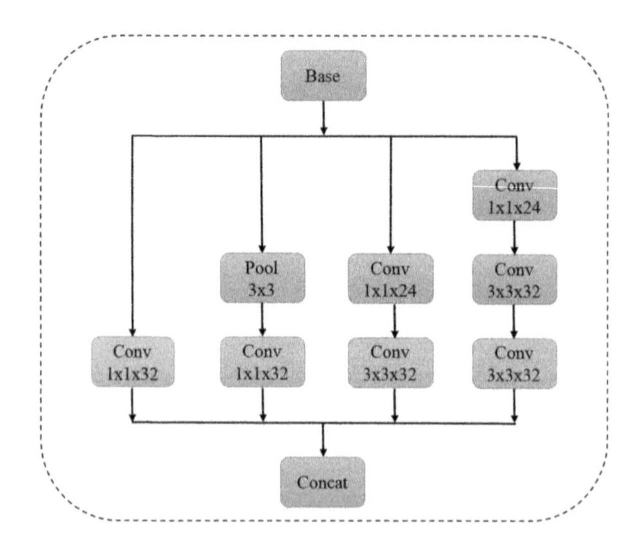

Fig. 3. Inception module.

3.3 Convolution Block

The Convolution block is the final stage of the feature map extraction process. In this block mainly using common convolution operations with kernel size of 1×1 and 3×3 to shrink the size and increase the dimension of the feature map. As described in Fig. 1, Conv3_1 and Conv4_1 use 1×1 kernel size with 128 kernels, Conv3_2 and Conv4_2 use 3×3 kernel size with 256 kernels. The output of this block is the feature map size of $8 \times 8 \times 256$, which means the size of the feature map is further reduced by four times and the number of kernels is doubled from 128 to 256. This block is the bridge between the feature map extraction process and the detection process.

3.4 Detection Block

The end of the network in this paper is the detection block. This block uses two siblings convolution operations with kernel size is 3×3 for classification and bounding box regression. These layers apply on feature map with size 8×8 which is output feature map from previous block. The detector uses square anchors of various sizes to predict the position of the corresponding eye in the original image. In this case, it uses three anchors with sizes 18, 24, 30 for small eye sizes, medium eye sizes and large eye sizes, respectively. Finally, the detector generates four-dimensional vector (x, y, w, h) as location offset and a two-dimensional vector (eye or not eye) as label classification.

3.5 Loss Function

The loss function used in this paper is same as RPN in Faster R-CNN [10], a 2-classes softmax-loss for classification task and the smooth $L1$ loss for regression.

The loss function for an image is defined as follows:

$$L(\{p_i\},\{t_i\}) = \frac{1}{N_{cls}}\sum_i L_{cls}(p_i,p_i^*) + \lambda\frac{1}{N_{reg}}\sum_i p_i^* L_{reg}(t_i,t_i^*), \quad (1)$$

Where i is the index of an anchor in a mini-batch and p_i is the predicted probability of anchor i being an object, p_i^* is ground-truth label and $p_i^* = 1$ if the anchor is positive and is 0 if the anchor is negative. t_i is the center coordinates and dimension of the prediction and t_i^* is the ground truth coordinates of bounding box. $L_{cls}(p_i,p_i^*)$ is the classification loss using the softmax-loss shown as in Eq. (2), $L_{reg}(t_i,t_i^*) = R(t_i - t_i^*)$ with R is the Smooth loss $L1$ defined as in Eq. (3). The two terms are normalized by N_{cls} and N_{reg} and weighted by a balancing parameter λ. N_{cls} is normalized by the mini-batch size, N_{reg} is normalized by the number of anchor locations and λ is assigned by 10.

$$L_{cls}(p_i,p_i^*) = -\sum_{i\in Pos} x_i^p log(p_i) - \sum_{i\in Neg} log(p_i^0), \quad (2)$$

Where $x_i^p = \{0,1\}$ is indicator for matching the $i-th$ default box to ground-truth of category p, p_i^0 is the probability for non-object classification.

$$R(x) = \begin{cases} 0.5x^2 & if\,|x| < 1 \\ |x| - 0.5 & otherwise \end{cases} \quad (3)$$

4 Experimental Result

4.1 Dataset Preparation

This network is trained on the CEW, BioID Face and GI4E dataset. CEW dataset contains 2,423 subjects, among which 1,192 subjects with both eyes closed are collected directly from Internet, and 1,231 subjects with eyes open are selected from the Labeled Face in the Wild (LFW) database. The image size is cropped and resized from coarse faces images to the size 100×100 (pixels) and then extract eye patches of 24×24 centered at the localized eye position. BioID Face dataset consists of 1,521 gray level images with a resolution of 384×286 pixel. Each one shows the frontal view of a face of one out of 23 different test persons. The eye position label is assigned manually and generated coordinate of ground-truth bounding box based on this position with size is 36×36. GI4E is dataset of images for iris center and eye corner detection. The database consists of a set of 1,339 images acquired with a standard webcam, corresponding to 103 different subjects and 12 images each. The images in this dataset have a resolution of 800×600 pixels in PNG format. The images are associated to a ground-truth text file. It contains manually annotated 2D iris and corner points (in pixels). The coordinate of ground-truth of the bounding box also generated from the position of iris by size is 46×46. Each original data set is divided into two basic data sets with 80% for training and 20% for testing phase.

4.2 Experimental Setup

The experiments in this paper are trained on GeForce GTX 1080Ti GPU and tested on Intel Core I7-4770 CPU @ 3.40 GHz, 8 GB of RAM. In order to train the eye detection network, many configurations have been used to improve eye detection quality. The Stochastic Gradient Descent optimization method used, the batch size of 16, the weight decay is 5.10^{-4}, the momentum is 0.9, the learning rate is 10^{-3}. The threshold of IoU (Intersection over Union) is 0.5 to produce the best bounding box. The number of predefined bounding box selected depends on the input image. Therefore, different datasets will have different definitions.

4.3 Experimental Result

The comprehensive network is trained and tested on the mentioned above image datasets. This network architecture is also tested on a real-time system using a camera connecting to the PC (personal computer) run on CPU. For the image datasets, eye detection network achieved results on CEW, BioID Face and GI4E dataset with 96.48%, 99.58% and 75.52% of AP, respectively. For real-time testing, it achieved 37.65 fps on Intel Core I7-4770 CPU @ 3.40 GHz. The testing result of the network on CEW, BioID Face and GI4E dataset shown as in Table 1 and Fig. 4. The results showed that this network performed the best eye detection on the BioID Face dataset with 99.58% of AP and was still limited with GI4E dataset with 75.52% of AP. This means that the performance of the network depends on the difficulty of the dataset. The performance result is also compared with FaceBoxes network [21] on CEW dataset and the result shown as in Table 2. This result proves that the proposed network outperforms the FaceBoxes network in terms of both eye location detection and speed with fewer parameters.

Table 1. The testing result of the proposed network on CEW, BioID Face and GI4E test dataset.

Dataset	Average precision (%)
CEW	96.48
BioID Face	99.58
Gi4E	75.52

During the training, the proposed network is trained with a set of predefined bounding boxes to select the most suitable ones. In addition, this network architecture is also trained with variable input image sizes to demonstrate the flexibility of the network when working with different data sets. Table 3 shows the testing results with various sizes of bounding boxes changes on the CEW test dataset. However, when the number of Inception modules changed from six to nine or more, the result did not change much.

Fig. 4. The qualitative result on CEW dataset (first row), on BioID Face dataset (second row), and on GI4E dataset (third row). The proposed network can detect eye location at many different pose shown as in CEW dataset and when wearing the glasses shown as in BioID Face and Gi4E dataset.

Table 2. The testing result of the proposed network and FaceBoxes network on CEW test dataset.

Network architecture	AP (%)	FPS	Number of parameters
Proposed	96.48	37.43	937,266
Faceboxes	96.31	34.75	1,007,330

Table 3. The testing results with various sizes of bounding boxes changes on the CEW test dataset.

Number of anchor	AP (%)	FPS
3 ([18, 24, 30])	96.48	37.43
4 ([16, 24, 32, 40])	95.77	35.55
5 ([8, 16, 24, 32, 40])	94.81	45.69

From the above experiments, can see that the proposed network results in the best eye position detection when designing the appropriate size and number of anchor boxes (here selected three anchor boxes). However, sometimes the proposed network detects wrong areas such as the background image or several areas of the face that are similar in structure to the eye. It also may not detect the eye area due to the light, color, and contrast conditions. As the confusion results are shown in Fig. 5.

Fig. 5. The confusion results on CEW dataset (first row), on BioID Face dataset (second row), and on GI4E dataset (third row). (Color figure online)

5 Conclusion and Future Work

The proposed network is designed based on the basic layers of the convolutional neural network combined with C.ReLU and Inception module to extract the feature map. Then, it applies two sibling convolution layers to detect eye position on the facial image. This network is implemented training and testing on three datasets with results achieved 96.48%, 99.58%, and 75.52% of AP on CEW, BioID Face, and GI4E dataset, respectively. It is also deployed real-time testing on camera running on the CPU with speed by 37.65 fps. The optimal number of parameters and computation makes it can be deployed on portable devices. In the future, the eye detection network will be developed with new techniques to achieve a higher quality of eye RoI (Region of interest). In addition, the eye detector is continued to optimize and increase applicability in real-time systems.

Acknowledgement. This work was supported by the National Research Foundation of Korea(NRF) grant funded by the Korea government. (MSIT) (2020R1A2C2008972)

References

1. Araujo, G., Ribeiro, F., da Silva, E., Goldenstein, S.: Fast eye localization without a face model using inner product detectors. In: 2014 IEEE International Conference on Image Processing, ICIP 2014, pp. 1366–1370 (2015). https://doi.org/10.1109/ICIP.2014.7025273
2. Chen, S., Liu, C.: Eye detection using discriminatory HAAR features and a new efficient SVM. Image Vision Comput. **33**(C), 68–77 (2015). https://doi.org/10.1016/j.imavis.2014.10.007
3. Chinsatit, W., Saitoh, T.: CNN-based pupil center detection for wearable gaze estimation system. Appl. Comput. Intell. Soft Comput. **2017**, 1–10 (2017). https://doi.org/10.1155/2017/8718956
4. Cortacero, K., Fischer, T., Demiris, Y.: RT-BENE: a dataset and baselines for real-time blink estimation in natural environments. In: 2019 IEEE/CVF International Conference on Computer Vision Workshop (ICCVW), pp. 1159–1168 (2019)
5. Deng, J., et al.: The Menpo benchmark for multi-pose 2D and 3D facial landmark localisation and tracking. Int. J. Comput. Vis. **127**(6–7), 599–624 (2019). https://doi.org/10.1007/s11263-018-1134-y
6. Fuhl, W., Santini, T., Kasneci, G., Kasneci, E.: PupilNet: convolutional neural networks for robust pupil detection. CoRR abs/1601.04902 (2016). http://arxiv.org/abs/1601.04902
7. Leo, M., Cazzato, D., Marco, T., Distante, C.: Unsupervised approach for the accurate localization of the pupils in near-frontal facial images. J. Electr. Imaging **22**, 033033 (2013). https://doi.org/10.1117/1.JEI.22.3.033033
8. Markuš, N., Frljak, M., Pandžić, I., Ahlberg, J., Forchheimer, R.: Eye pupil localization with an ensemble randomized trees. Pattern Recognit. **47**, 578–587 (2014). https://doi.org/10.1016/j.patcog.2013.08.008
9. Mosa, A.H., Ali, M., Kyamakya, K.: A computerized method to diagnose strabismus based on a novel method for pupil segmentation. In: Proceedings of the International Symposium on Theoretical Electrical Engineering (ISTET 2013) (2013)
10. Ren, S., He, K., Girshick, R.B., Sun, J.: Faster R-CNN: towards real-time object detection with region proposal networks. CoRR abs/1506.01497 (2015). http://arxiv.org/abs/1506.01497
11. Shang, W., Sohn, K., Almeida, D., Lee, H.: Understanding and improving convolutional neural networks via concatenated rectified linear units. CoRR abs/1603.05201 (2016). http://arxiv.org/abs/1603.05201
12. Sharma, R., Savakis, A.: Lean histogram of oriented gradients features for effective eye detection. J. Electr. Imaging **24**, 063007 (2015). https://doi.org/10.1117/1.JEI.24.6.063007
13. Swirski, L., Bulling, A., Dodgson, N.: Robust real-time pupil tracking in highly off-axis images. In: Eye Tracking Research and Applications Symposium (ETRA), pp. 173–176 (2012). https://doi.org/10.1145/2168556.2168585
14. Szegedy, C., et al.: Going deeper with convolutions. CoRR abs/1409.4842 (2014). http://arxiv.org/abs/1409.4842
15. Timm, F., Barth, E.: Accurate eye centre localisation by means of gradients. In: VISAPP (2011)
16. Valenti, R., Gevers, T.: Accurate eye center location through invariant isocentric patterns. IEEE Trans. Pattern Anal. Mach. Intell. **34**, 1785–98 (2012). https://doi.org/10.1109/TPAMI.2011.251

17. Wu, Y., Ji, Q.: Constrained joint cascade regression framework for simultaneous facial action unit recognition and facial landmark detection. In: Proceedings of the IEEE Conference on Computer Vision and Pattern Recognition, pp. 3400–3408 (2016)
18. Wu, Y., Ji, Q.: Facial landmark detection: a literature survey. CoRR abs/1805.05563 (2018). http://arxiv.org/abs/1805.05563
19. Xiao, S., Feng, J., Xing, J., Lai, H., Yan, S., Kassim, A.: Robust facial landmark detection via recurrent attentive-refinement networks. In: Leibe, B., Matas, J., Sebe, N., Welling, M. (eds.) ECCV 2016. LNCS, vol. 9905, pp. 57–72. Springer, Cham (2016). https://doi.org/10.1007/978-3-319-46448-0_4
20. Zadeh, A., Chong Lim, Y., Baltrusaitis, T., Morency, L.P.: Convolutional experts constrained local model for 3D facial landmark detection. In: Proceedings of the IEEE International Conference on Computer Vision Workshops, pp. 2519–2528 (2017)
21. Zhang, S., Zhu, X., Lei, Z., Shi, H., Wang, X., Li, S.Z.: Faceboxes: a CPU real-time face detector with high accuracy. CoRR abs/1708.05234 (2017). http://arxiv.org/abs/1708.05234

Recommender Systems

Robust Content-Based Recommendation Distribution System with Gaussian Mixture Model

Dat Nguyen Van[1(✉)], Van Toan Pham[1], and Ta Minh Thanh[1,2]

[1] Research and Development Department, Sun Asterisk, Ha Noi, Viet Nam
{nguyen.van.dat,pham.van.toan}@sun-asterisk.com
[2] Le Quy Don Technical University, Ha Noi, Viet Nam
thanhtm@mta.edu.vn

Abstract. Recommendation systems play an very important role in boosting purchasing consumption for many manufacturers by helping consumers find the most appropriate items. Furthermore, there is quite a range of recommendation algorithms that can be efficient; however, a content-based algorithm is always the most popular, powerful, and productive method taken at the begin time of any project. In the negative aspect, somehow content-based algorithm results accuracy is still a concern that correlates to probabilistic similarity. In addition, the similarity calculation method is another crucial that affect the accuracy of content-based recommendation in probabilistic problems. Face with these problems, we propose a new content-based recommendation based on the Gaussian mixture model to improve the accuracy with more sensitive results for probabilistic recommendation problems. Our proposed method experimented in a liquor dataset including six main flavor taste, liquor main taste tags, and some other criteria. The method clusters n liquor records relied on n vectors of six dimensions into k group ($k < n$) before applying a formula to sort the results. Compared our proposed algorithm with two other popular models on the above dataset, the accuracy of the experimental results not only outweighs the comparison to those of two other models but also attain a very speedy response time in real-life applications.

Keywords: Recommendation · Content-based · Gaussian-mixture-model (GMM) · Distribution-recommendation

1 Introduction

Due to the proliferation of internet, it has brought tremendous chance for people's lives. On the other hand, the myriad and abundance of information on the web has determined a rapidly increasing diffculty in nding what we actually need in a way that can fit the best our requirements [2,7,11]. Recommendation

Supported by Sun Asterisk Inc.

M. Hernes et al. (Eds.): ICCCI 2020, CCIS 1287, pp. 199–211, 2020.
https://doi.org/10.1007/978-3-030-63119-2_17

systems can be effective way to solve such problems without requiring users provide explicit requirements [31,33]. Instead, the system can analysis the content data of item properties, which actively recommend information on users that can satisfy their needs and interests [15,17]. The general content-based architecture is shown in Fig. 1.

Content-based filtering algorithm is widely used because of its simplicity and effectiveness at the begin time of any recommendation systems. According to Pasquale *et al.* [14], there are many benefits reaped from content-based recommendation (CB) systems compared to the other Collaborative Filtering (CF) one such as user independence, transparency, cold-start problems, and so on. Beside, there are still some shortcoming existing as limited content for analyzing, over-specialization or lack of rating data of new users and adequate accuracy for some specific problems. Hangyu *et al.* [29] used GMM for CF recommendation algorithm to solve the sparse users rating data. Chen *et al.* [4] proposed a hybrid model, which combines GMM with item-based CF recommendation algorithm and predicted the ratings on items from users to improve the recommendation accuracy. Rui Chen *et al.* [3] took GMM with enhanced matrix factorization to reduce the negative effect of sparse and high dimension data. In the context of music recommender systems, Yoshii *et al.* [30] proposed a hybrid recommender system that combines collaborative filtering using user ratings and content-based features modeled via GMM over MFCCs by utilizing a Bayesian network. However, CF or hybrid systems require behaviour history of users that the reason for the need of CB. Furthermore, CB based on distribution of item features have not been solved yet. A telling of example is using CB for automatically find similar items based on distribution and distance of its features in Fig. 2. These kind of probabilistic problems in recommendation systems is quite different which cannot be solved by usual common methods. Furthermore, the description of content data of items features is sometimes unreliable, inadequate that detrimentally affect to the accuracy of CB systems [11]. Due to two problems mentioned above, we propose a new approach for solving these problems by using GMM [25] to cluster all items into different groups before applying a gaussian filter function (GFF) as a calculation similarity method for sorting results. To demonstrate our effective model, we experiment and compare to two other popular methods, Bag of Word [1] with GFF (BOW + GFF), and GMM with euclidean distance (ED) [13] (GMM + ED). Our propose model not only outperforms the accuracy of the two others, but also get better in prediction time response.

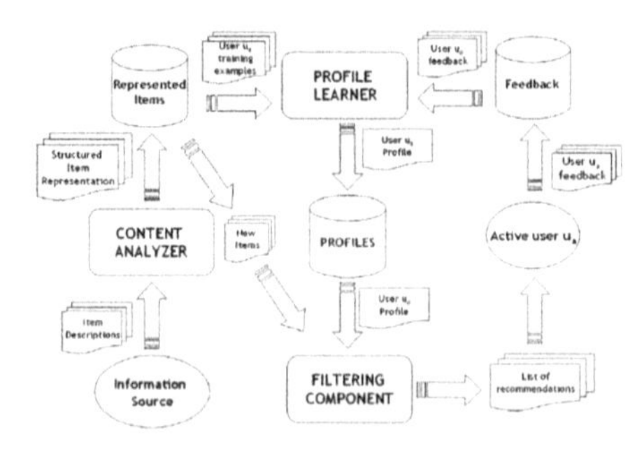

Fig. 1. High level architecture of a content-based recommendation system.

The paper is organized as follows. Related work is introduced in Sect. 2 while dataset in Sect. 3. In Sect. 4, the architecture and details of proposed model is given. Experiments and evaluations are shown in Sect. 5. The conclusion are discussed in Sect. 6.

2 Related Work

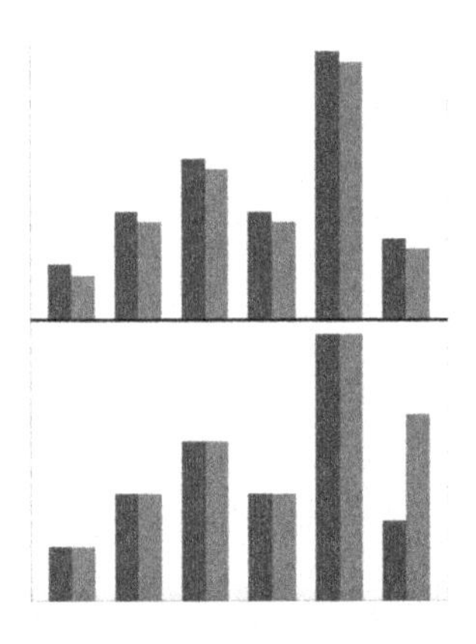

Fig. 2. An example between using distribution and distance calculation for distribution recommendation.

We introduce some preliminary knowledge that needs to be used. The following is the detailed information of them.

2.1 Content-Based Recommendation

`Content-Based Recommendation Systems` is one of the most common method in building recommendation systems. The algorithm is born from the idea of using the content descriptions of each item for recommending purposes. It can be divided into two approaches: Analysing the description of item properties only, and building user profile for individuals based on feature's content of items and personal rating data [14,33].

2.2 Popular Similarities

In the `Content-based` algorithm, the similarity calculation method directly affects the accuracy of results. Some similarity calculation methods have been widely used which are listed below:

`euclidean distance`: One of the most popular methods to measure the similarity between two vectors by calculating the sum of square distance of each element respectively in those vectors. Read [13] for more information.

`Cosin`: The main idea is to measure two vectors by calculating the cosine of angle between the two vectors [21].

`Pearson`: The pearson correlation coefficient reflects the degree of linear correlation between two vectors [26],

`Jaccard`: The Jaccard Similarity is often used to compare similarity and different between two finite sample set [19],

2.3 Gaussian Mixture Model (GMM)

Gaussian Mixture Model is a function that is comprised of several Gaussians. GMM can t any types of distribution, which is usually used to solve the case where the data in the same set contains multiple different distributions [5,24], each identified by $k \in \{1..K\}$ where K is the number of clusters of our dataset.

GMM is defined as:

$$p(x) = \sum_{i=1}^{k} \alpha_i.N(x|\mu_i, \Sigma_i), \tag{1}$$

where $N(x|\mu_i, \Sigma_i)$ is the ith component of the hybrid model, which is a probability density function of the n dimensional random vector x obeying Gaussian distribution. It can be defined as below:

$$N(x) = \frac{1}{(2\pi)^{\frac{n}{2}} \mid \Sigma \mid^{\frac{1}{2}}} \epsilon^{-\frac{1}{2}(x-\mu)^T \sum^{-1}(x-\mu)} \tag{2}$$

and

$$\sum_{i=1}^{k} \alpha_i = 1 \tag{3}$$

We assume that a sample set $D = \{x_1, x_2, x_3, ..., x_m\}$ is given that obey gaussian distribution mixture distribution. We use the random variable $z_j \in \{1, 2, ..., k\}$ to represent the mixed component of the generated sample x_j, whose value is unknown. It can be seen that the prior probability $P(z_j = i)$ of z_j corresponds to $\alpha_i (i = 1, 2, 3, ..., k)$. According to Bayes' theorem [12], we can get the posterior probability of z_j as follows:

$$\begin{aligned} p(z_j - i|x_j) &= \frac{P(z_j = i).p(x_j|z_j = i)}{p(x_j)} \\ &= \frac{\alpha_i.N(x_j|\mu_i, \Sigma_i)}{\sum_{l=1}^{k} \alpha_l.N(x_j|\mu_l, \Sigma_l)} \end{aligned} \tag{4}$$

In the above formula, $p(z_j = i|x_j)$ represents the posterior probability of sample x_j generated by the ith Gaussian mixture. Assuming $\gamma_{ij} = \{1, 2, 3, ..., k\}$ represents $p(z_j = i|x_j)$. When the model parameters $\{(\alpha_i, \mu_i, \Sigma_i)|1 \leq i \leq k\}$ in the Eq. (4) are known, the GMM clusters divide the sample set D into k clusters $C = \{C_1, C_2, ..., C_k\}$ [24]. The cluster label λ_j of each sample x_j can be determined according to equation below:

$$\lambda_j = \arg max_{i \in 1,2,3,...,k} \gamma_{ji}$$

We get the cluster label λ_j to which x_j belongs and divide x_j into cluster C_{λ_j}. The model parameters $\{(\alpha_i, \mu_i, \Sigma_i)|1 \leq i \leq k\}$ is solved by applying EM algorithm [16].

3 Dataset

Our proposed model is implemented on a dataset about liquor, more specifically, about sake which is one of the most prevalent kind of liquor in Japan. The dataset was collected from Sakenowa[1] being one of the most well-known and reputed website selling the sake[2]. The dataset totally contains 1072 records characterized by 19 properties such as liquor name, liquor brand, year of manufacture, liquor images, liquor flavour tags, liquor six axis flavour taste ($f_1, f_2, ..., f_6$ stands for fruity, mellow, rich, mild, dry and light (Fig. 3). Noticeably, liquor six axis flavour taste and liquor flavour taste would play more important role than the others. The range value of six flavour taste ($f_1 - f_6$) axis is in [0, 1], meanwhile the dominant parts belong to [0.2, 0.6]. The text fields in the dataset all is written behind Japanese form. However, this is a real challenging dataset due to lack of many fields that lead to sparse in data, especially in six main fields $f_1, ..., f_6$. Therefore, our task of recommendation become more difficult and be negatively affect the recommendation results. More specifically, a disappearance or null value of 6-axis fields is greater than 30%, a nearly 2% of null value flavour tags. Further more, many tags value is unreliable, untrust and incorrect that need to be clean and pre-processing (Table 1).

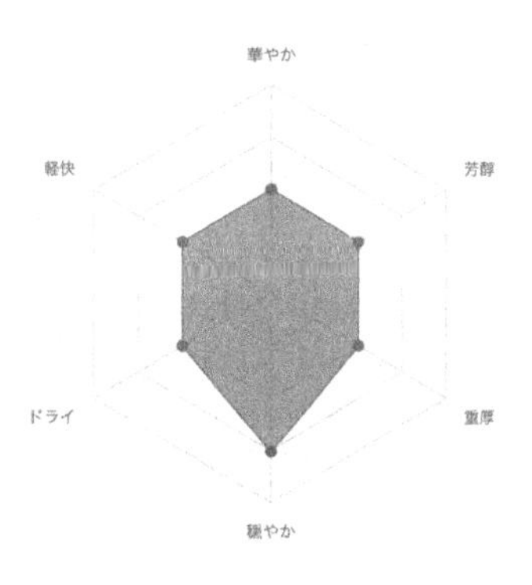

Fig. 3. A visualization about 6-axis flavour taste

4 Proposed Model

We introduce and explain our proposed model more in detail. As it was mentioned in previous part, we have to return the most similar products based on 19 metadata fields. In particular, 6-axis flavour taste and flavour tags are the main factors mostly affecting to the results both in the sensibility and accuracy side. Therefore, we just select 6-axis flavour taste and flavour tags for better results. The more similar in 6-axis flavour taste, the better results will be.

Table 1. Dataset blank fields statistic

$f_{1..6}$	Flavour tags	Product name(en)
Float	String	String
30.4%	1.77%	13.4%

More detail, we initially use Gaussian mixture model to cluster all items into $K = \{1, 2, ..., k\}$ group, then sorting results in each group with each item. Whenever finding top similar items of a item, we just jump up to the group the item belongs to and sort the group's items to return top m similar items. To sort the results, it is also possible to use some popular similarity calculation such as `cosine` or `euclidean distance`, but for better accuracy, we use a equation that calculate the distribution weight between two vectors obeying Gaussian

[1] https://sakenowa.com.
[2] https://en.wikipedia.org/wiki/Sake.

distribution (normal distribution). The results illustrate that the more similar in 6-axis amongst items the bigger weights will be germinated. The flow of our proposed model is shown in Fig. 4.

4.1 Data Pre-processing

It is a fact that text mining is very important in every text-related problems, and CB is not an exception. Previously mentioned, we only choose flavour tags and 6-axis flavour taste as the features for compute the similar between items. The flavour tags are the set of text document written behind Japanese form which require to be cleaned. We convert 6-axis into float and need to do some pre-process techniques for such flavour tags text fields like tokenization, stemmings, stop word removal, find and replace synonyms, lemmatization, and so on [6, 22,23] before utilizing it. Moreover, the flavour tags field has been splitted into different semantic words, so we disregard the tokenization step and move forward with the other steps.

4.2 Clustering

As we recognize that the final recommendation items depend too much on 6-axis flavour taste and flavour tags. In the common and traditional way, there is a way to build a vector representing for all properties of each item, then utilizing a similarity calculation method like cosine or euclidean to sort and return top m results. However, in some case, the flavour tags are not enough adequate and precise that adversely affect to the final recommendation. Moreover, there is always an unseen problem of using cosine or euclidean that a compensate between each element of 6-axis flavour taste ($f_1 - f_6$) leads to unequal among those elements ($f_1 - f_6$) of results. Therefore, we decide to group all items based on it's distribution 6-axis flavour taste into different clusters to ensure items which have the same distribution will be in the same cluster that is the foundation for sorting afterwards (Fig. 5).

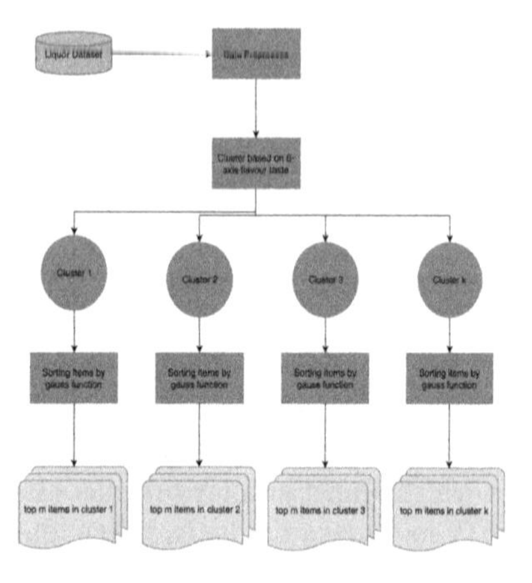

Fig. 4. The model Activities Diagram

4.3 Gaussian Function for Sorting

As we have $K = \{1, 2, ..., k\}$ clusters, we assume a query item is the center of the cluster we want to find. Our destination is figure out top m items that have the same distribution as much as possible, so Gaussian filter function (GFF) is the better choice than cosine or euclidean. The Gaussian function equation is defined as follows:

$$G_{kl}(f_{il}, f_{jl}) = \exp -\frac{(f_{il} - f_{jl})^2}{2\sigma_{kl}^2} \tag{5}$$

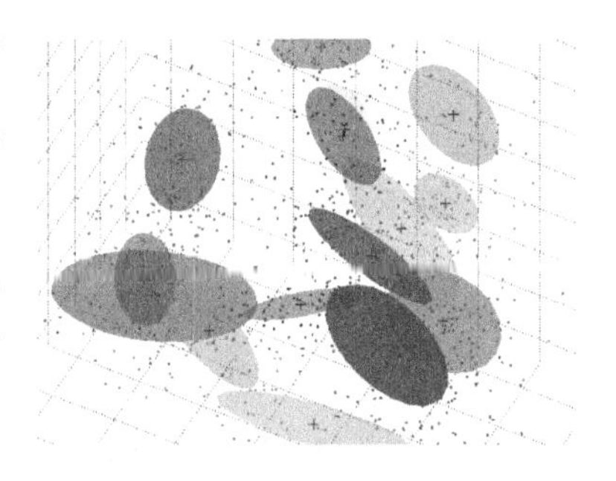

Fig. 5. GMM visualization

where $G_k(f_{il}, f_{jl})$ is considered as a weight between each pair of element l^{th} in 6-axis flavour taste of two different items (i, j) in cluster $k, l = \{1, 2, ..., 6\}$, and σ_{kl} is the standard deviation of the lth element in 6-axis flavour taste in group k. Equation for σ_{kl} is defined as below:

$$\sigma_{kl} = \sqrt{\frac{\sum_{i=1}^{n_k} (f_{ilk} - \mu)^2}{n_k - 1}} \tag{6}$$

where n_k is the number of items belong to cluster k, f_{ilk} is the value of lth of f in six flavour taste of ith item and μ is the mean value of all f_l, in group k. We calculate $G(x, y)$ 6 times for 6 field $f_1 - f_6$ for each pair items over all items of a group to sort in descending to find top best results.

4.4 Levenshtein Distance for Comparison

The flavour tags also play a quite significant role in the final results. We treat tags as vital as each element in 6 flavour taste. To compare and measure the similarity between two tags of string type, we use a good of levenshtein distance to solve it [8,32]. The equation of the levenshtein distance is defined below:

$$lev_{a,b}(i,j) = \begin{cases} max(i,j), \text{if min(i,j)=0} \\ min = \begin{cases} lev_{a,b}(i-1,j)+1 \\ lev_{a,b}(i,j-1)+1, \text{otherwise} \\ lev_{a,b}(i-1,j-1)+1_{(a_i \neq b_j)} \end{cases} \end{cases} \tag{7}$$

4.5 Final Sorting Formula

Combine weight calculating function for 6-axis flavour taste and tags comparison with levenshtein distance (LD), we establish a equation for sorting to get final results as below:

$$S(i,j) = \sum_{k=1}^{K}\sum_{l=1}^{6} G_{kl}(i,j) + lev_{tags}(i,j) \tag{8}$$

where G_{kl} is the weight function corresponding l^{th} in 6-axis flavour taste between $item_i$ and $item_j$ in cluster k (k = {1, ..., K} K groups), $lev_{tags}(i,j)$ is the levenshtein function to compare tags similarity of those two items. We determine that the bigger $S(i,j)$, the better similar between those two items, so we sort by descending order all items of a cluster and return top m items having bigger S(i, j) value.

4.6 Proposed Model Pseudo Code

For clearly, we give the proposed model its algorithm execution process to help readers more easily visualize and imagine our entire process. Let see pseudo code below:

Algorithm 1: Framework model proposal

Input: number of clusters k
Output: Top m other similar items of each item
Data: Dataset L
1. Data pre-processing for text fields
2. Build a matrix for 6-dimension vectors representing for six flavour taste ($f_1 - f_6$)
3. Taking the matrix as an input of GMM to train and save corresponding cluster of each item into dataset
4. **for** `item in dataset` **do**
   - Get cluster number of item
   - Find all items that have the same clusters
   - Applying equation $S(i,j)$ "**(8)**" for each pairs items
   - Return top m similar items by sorting decreasingly

end

5 Experiments

To prove the validity of our proposed model, we compare our proposed model to two other popular algorithms widely used in CB systems such as BOW+GFF and GMM+ED. We also illustrate the impact of GMM cluster into the accuracy and the efficiency of Gaussian filter equation in sorting results rather than those of cosine or euclidean distance.

5.1 Evaluation Method

The evaluation method of recommendation systems commonly used is Root Mean Square Error (MSE) that is the average of the square errors [27]. It is defined as:

$$MSE = \frac{1}{N}\sum_{1}^{n}(r_i - \widehat{r}_i)^2, \tag{9}$$

where r_i is the predicted representing vector item, and $\widehat{r}_i$ is the original representing vector item.

We also use the recommendation results in Sakenowa as the standard measure to compare with our three algorithms because the sake website has so much reputation, popularity, being well-known for commercial purpose in Japan for many years. Beside, the recommendation results of the Sakenowa is also very impressive.

5.2 Experimental Analysis

Some experiments will be conducted to verify the impact of GMM, GFF on probabilistic recommendation problem. Our main proposed model was implemented through such steps as data statistic, data cleaning, data missing value filling, clustering all items into different clusters and eventually using Gaussian filter + levenshtein distance to sort the results. To verify the effective impact of GMM and GFF on better prediction, we divided our experiments into three parts. Firstly, we use Bag-of-Word (BOW) [1] algorithm on some properties like flavour tags before applying GFF for sorting results. In the second way, we apply GMM + ED to clarify the influence of GMM. Finally, we implemented our main proposed model to prove the impact of GMM+GFF then give some comparison. All experiments will be unraveled in detail below:

Experiment 1: BOW+GFF. The reason for this experiment is to verify the impact on result accuracy of GMM compared to BOW algorithm. Therefore, in the experiment, we will implement BOW algorithm comprised with GFF used for sorting on our liquor dataset. Firstly, we do some data preprocessing for text data like stemming, replace synonyms, filling missing data, etc [22]. As it was mentioned above, all important text fields were written behind Japanese form, so we use some tools offered for Japanese preprocessing like Ginza [9], Janome [10], JapaneseStemmer [18] was inspired by Porter Stemming Algorithm [28], etc. Before using GFF for sorting, we use BOW on these preprocessed properties to find the vector matrix representing for the item. The next step, we feed the vector matrix into K-nearest neighbors (k-NN) algorithm using unsupervised k-NN Scikit-Learn [20] to find top similar items based on these vectors. In these top items, we apply equation S (8) to get the best similar items.

Experiment 2: GMM+ED. To demonstrate the impact of GMM, firstly, we still apply some preprocessing steps for text fields as the experiment above. After that, we build a matrix of 6 dimensions representing for 6 flavour taste, then feed it into GMM for training, save all cluster result for each item. Next step, we convert a collection of text flavour tags into a matrix of token counts using CountVectorizer of Scikit-Learn [20] and concatenate along same axis with the matrix of 6 dimensions for sorting. Finally, to return the best similar items of a given item, we just jump up to the cluster containing it and apply ED for sorting the results and get top best similar items of the given item.

Experiment 3: GMM+GFF. Our two above experiment to prove the important role of GMM and GFF in our proposed model. In this experiment, in the first place, we also do preprocess for text fields as same steps in two previous experiments. After that, we build a matrix of 6 dimensions representing for 6 flavour tastes and feed the matrix into GMM for training purpose, then save cluster results for each items. To find top best similar items of a given item, we jump up into the cluster the query item lied in, consider the query item as center then apply (8) equation pair in pair with all items in the cluster, then sorting discerningly to return the top best similar items.

5.3 Experimental Results and Comparison

In this section, we compare our proposed algorithm with the results from the Sakenowa website and two other popular CB algorithms. The recommendation results from Sakenowa for each item are returned from an api[3]; therein, $f_{1 \ldots 6}$ in the api are the value for each flavour taste, respectively. We conclude that our result accuracy outweights the Sakenowa and these two algorithm counter parts. Let see some charts below:

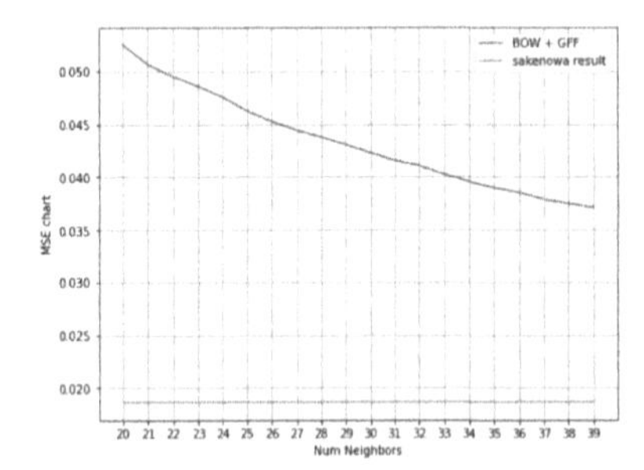

Fig. 6. MSE applied BOW+GFF

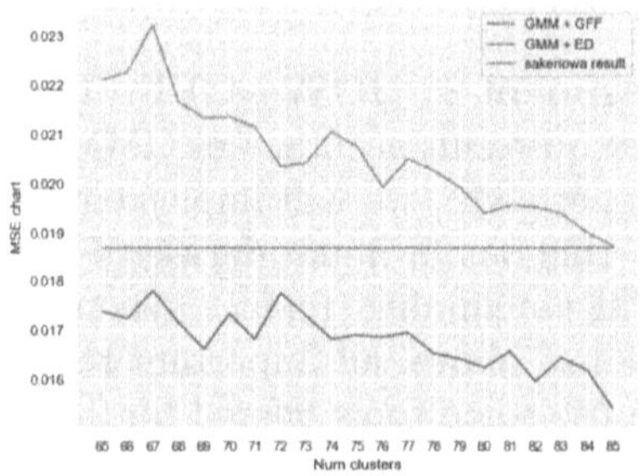

Fig. 7. MSE applied GMM+GFF and GMM+ED

All three experiments return top ten(10) best similar items for each item in dataset. In Fig. 6, list values of MSE are shown through an array of number of neighbors ranging from [25–39] in k-NN algorithm. Despite the tendency of decrease, but it is insignificant and the time response is extremely slow due to bigger number of neighbors.

[3] https://sakenowa.com/api/v1/brands/flavor?f=0&fv=f_1,f_2,f_3,f_4,f_5,f_6.

In Fig. 7 the gap of MSE between GMM+ED and GMM+GFF is shown. It is very clearly seen that GMM+GFF generate better results than the other that verify the effect of GMM in sorting results. Both these two experiments show the effect of the number of clusters of GMM ranging from [65–85]. In Fig. 8, we compare our prediction results of all items in dataset to the recommendation results from the Sakenowa and construct a list of similarity proportion affected by the number of clusters.

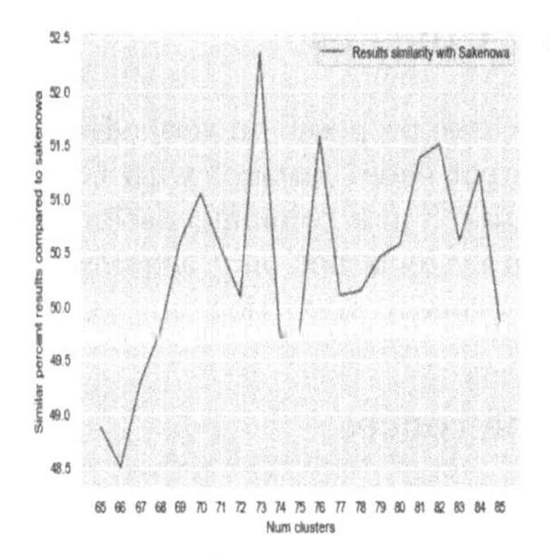

Fig. 8. Similar percent statistic compare to Sakenowa

Table 2. MSE affected by number of clusters

N clusters	GMM+ED	GMM+GFF	Sakenowa results
65	0.02211	**0.01739**	0.01868
70	0.02135	**0.01734**	0.01868
75	0.02074	**0.01691**	0.01868
80	0.01939	**0.01625**	0.01868
85	0.01873	**0.01541**	0.01868

In Table 2 and Table 4, we build a table of statistic of MSE generated from GMM+ED, BOW+GFF, GMM+GFF and recommendation results from Sakenowa. It is matter of fact that our GMM+GFF algorithm outperforms all the others method that demonstrate the effective of our algorithm. Further more, our time response in Table 3 also beat these two others, GMM+ED and BOW+GFF.

Table 3. Time response per query

BOW+GFF	GMM+ED	GMM+GFF
0.1856 s	0.0174 s	**0.0156 s**

Table 4. MSE affected by number of neighbors

Num neighbors	BOW+GFF	Sakenowa results
20	0.05254	0.01868
25	0.04624	0.01868
30	0.04228	0.01868
35	0.03895	0.01868
39	0.03709	0.01868

6 Conclusion

We have proposed an very effective algorithm for recommendation system using content-based features with GMM. We have applied our proposed method for solving liquor recommendations. Further, our probabilistic-based recommendation systems not only acquire a remarkable prediction accuracy, but also has very speedy prediction time response for real-time application.

References

1. Bhattacharya, S., Lundia, A.: Movie recommendation system using bag of words and Scikit-learn (2019)
2. Bollen, D., et al.: Understanding choice overload in recommender systems, pp. 63–70 (2010). https://doi.org/10.1145/1864708.1864724
3. Chen, R., et al.: A hybrid recommender system for Gaussian mixture model and enhanced social matrix factorization technology based on multiple interests. Math. Prob. Eng. 1–22 (2018). https://doi.org/10.1155/2018/9109647
4. Fan-sheng, K.: Hybrid Gaussian pLSA model and item based collaborative filtering recommendation. Comput. Eng. Appl. (2010)
5. Görür, D., Rasmussen, C.: Dirichlet process Gaussian mixture models: choice of the base distribution. J. Comput. Sci. Technol. **25**, 653–664 (2010). https://doi.org/10.1007/s11390-010-9355-8
6. Gurusamy, V., Kannan, S.: Preprocessing techniques for text mining (2014)
7. Guy, I., Carmel, D.: Social recommender systems, pp. 283–284 (2011). https://doi.org/10.1145/1963192.1963312
8. Haldar, R., Mukhopadhyay, D.: Levenshtein distance technique in dictionary lookup methods: an improved approach. In: Computing Research Repository - CORR (2011)
9. Mai, O., Hiroshi, M., Masayuki, A.: Simultaneous learning of ambiguity resolution and dependency labeling for short-unit part of speech usage. In: 25 (2019). http://www.anlp.jp/proceedings/annual_meeting/2019/pdf_dir/F2-3.pdf
10. Janomepy$_{py}$ Janome (2019). https://github.com/mocobeta/janome
11. Khusro, S., Ali, Z., Ullah, I.: Recommender systems: issues, challenges, and research opportunities, pp. 1179–1189, ISBN 978- 981-10-0556-5 (2016). https://doi.org/10.1007/978-981-10-0557-2_112
12. Lee, D.-S., Hull, J., Erol, B.: A Bayesian framework for Gaussian mixture background modeling, vol. 3, pp. III-973 (2003). https://doi.org/10.1109/ICIP.2003.1247409
13. Liberti, L., et al.: Euclidean distance geometry and applications. SIAM Rev. **56**, 3–69 (2012). https://doi.org/10.1137/120875909
14. Lops, P., de Gemmis, M., Semeraro, G.: Content-based recommender systems: state of the art and trends. In: Ricci, F., Rokach, L., Shapira, B., Kantor, P.B. (eds.) Recommender Systems Handbook, pp. 73–105. Springer, Boston, MA (2011). https://doi.org/10.1007/978-0-387-85820-3_3
15. Linyuan, L., et al.: Recommender systems. Phys. Rep. **519**(1), 1–49 (2012). https://doi.org/10.1016/j.physrep.2012. ISSN 0370–1573
16. Lu, Y., Bai, X., Wang, F.: Music recommendation system design based on Gaussian mixture model. In: ICM 2015 (2015)

17. Melville, P., Sindhwani, V.: Recommender systems, pp. 829–838 (2011). https://doi.org/10.1007/978-0-387-30164-8_705
18. MrBrickPanda: Japanese Stemmer (2019). https://github.com/MrBrickPanda/Japanese-stemmer
19. Niwattanakul, S., et al.: Using of Jaccard coefficient for keywords similarity (2013)
20. Pedregosa, F., et al.: Scikit-learn: machine learning in Python. J. Mach. Learn. Res. **12**(Oct), 2825–2830 (2011)
21. Philip, S., Shola, P., Abari, O.: Application of content based approach in research paper recommendation system for a digital library. Int. J. Adv. Comput. Sci. Appl. **5** (2014). https://doi.org/10.14569/IJACSA.2014.051006
22. Rahutomo, R., et al.: Preprocessing methods and tools in modelling Japanese for text classification, p. 8843796 (2019). https://doi.org/10.1109/ICIMTech.2019
23. Rajman, M., Besancçon, R.: Text mining: natural language techniques and text mining applications. In: Proceedings of the 7th IFIP Working Conference on Database Semantics (DS-7) (1997). https://doi.org/10.1007/978-0-387-35300-5_3
24. Rasmussen, C.: The Infinite Gaussian mixture model, vol. 12, pp. 554–560 (2000)
25. Reynolds, D.: Gaussian mixture models. In: Encyclopedia of Biometrics (2008). https://doi.org/10.1007/978-0-387-73003-5_196
26. Sedgwick, P.: Pearson's correlation coefficient. BMJ **345**, e4483–e4483 (2012). https://doi.org/10.1136/bmj.e4483
27. Shani, G., Gunawardana, A.: Evaluating recommendation systems, vol. 12, pp. 257–297 (2011). https://doi.org/10.1007/978-0-387-85820-3_8
28. Jones, K.S., Willett, P. (eds.): Readings in Information Retrieval. Morgan Kaufmann Publishers Inc., San Francisco (1997). ISBN 1558604545
29. Yan, H., Tang, Y.: Collaborative filtering based on Gaussian mixture model and improved Jaccard similarity. IEEE Access **7**, 118690–118701 (2019). https://doi.org/10.1109/ACCESS.2019.2936630
30. Yoshii, K., et al.: Hybrid collaborative and content-based music recommendation using probabilistic model with latent user preferences, pp. 296–301 (2006)
31. Zhu, B., Bobadilla, J., Ortega, F.: Reliability quality measures for recommender systems. Inf. Sci. **442**, 145–157 (2018)
32. Ziolko, B., et al.: Modified weighted Levenshtein distance in automatic speech recognition (2010)
33. Zisopoulos, H., et al.: Content-based recommendation systems (2008)

Incremental SVD-Based Collaborative Filtering Enhanced with Diversity for Personalized Recommendation

Minh Quang Pham[1], Thi Thanh Sang Nguyen[1](✉), Pham Minh Thu Do[1], and Adrianna Kozierkiewicz[2]

[1] School of Computer Science and Engineering, International University, Vietnam National University, HCMC, Ho Chi Minh City, Vietnam
quangphamm1902@gmail.com, {nttsang,dpmthu}@hcmiu.edu.vn
[2] Faculty of Computer Science and Management, Wroclaw University of Science and Technology, Wroclaw, Poland
adrianna.kozierkiewicz@pwr.edu.pl

Abstract. Along with the rapid rise of the internet, an e-commerce website brings enormous benefits for both customers and vendors. However, many choices are given at the same time makes customers have difficulty in choosing the most suitable products. A rising star solution for this is the recommender system which helps to narrow down the amount of suitable and relevant products for each customer. Matrix factorization is one of the most popular techniques used in recommender systems because of its effectiveness and simplicity. In this paper, we introduce a matrix factorization-based recommender system using Singular Value Decomposition (SVD) with some improvements in collaborative filtering and incremental learning. The SVD-based collaborative filtering methods can help generate personalized recommendations by combining user profiles. Moreover, the recommendation lists generated by the system are enhanced with diversity, which might attract more customer interests. Amazon's Electronic data set is used to evaluate our proposed framework of the SVD-based recommender system. The experimental results show that our framework is promising.

Keywords: Recommender system · Collaborative filtering · Matrix factorization · Singular value decomposition · Explicit query aspect diversification

1 Introduction

An effective recommender system is not only the one that could give high accuracy rating-prediction for a user-item pair, but also give the least biased, most related, and reliable recommendations for each user.

In this paper, we propose a recommender system using collaborative filtering methods based on Singular Value Decomposition (SVD) [1]. To improve the limitation of SVD algorithm on producing online recommendations for new users, we build a more

M. Hernes et al. (Eds.): ICCCI 2020, CCIS 1287, pp. 212–223, 2020.
https://doi.org/10.1007/978-3-030-63119-2_18

flexible version of SVD, called incremental SVD. This extended version helps the model to be able to fit new data without re-doing all the computation all over data again. Moreover, the popularity bias problem of the recommendation list is solved using a method called Explicit Query Aspect Diversification (xQuAD) [2]. For recommendation model evaluation, we use two metric types, one for measuring the prediction accuracy and one for measuring recommendation ranking accuracy. A number of experiments have been carried out to evaluate our proposal, compared with other collaborative filtering methods using K-Nearest Neighbors (KNN) [3] (Item-based and User-based) and Alternating Least Squares (ALS) [4].

The following sections will present related work (Sect. 2), methodology (Sect. 3), and experimental results (Sect. 4). Conclusions and future work are presented in Sect. 5.

2 Collaborative Filtering

Recommendation can be classified into two ways [5]: content-based filtering predicts items the user might like based on the item description, and collaborative filtering uses the knowledge about preferences of users for some items to make recommendations. In this paper, we will focus on collaborative filtering methods.

Collaborative filtering techniques which are most used [6–8] look at user previous ratings or transactions to make recommendations. Moreover, it uses the inter-dependent relationships between users and products for generating recommendations. Two popular methods of collaborative-filtering are neighborhood-based models and latent-factor models. While the neighborhood-based models only focus on users or products at a time, the latent-factor model tries to predict the ratings based on both product and user characteristics, called "latent factors". These latent factors are not always defined in the dataset but are inferred from the rating patterns.

2.1 Neighborhood Model [3]

The neighborhood model assumes users or items that are closed to each other would have similar ratings. The neighborhood models making predictions based on "k" nearest neighbors are called Item-based K-Nearest Neighbor (if neighbors are items) and User-based K-Nearest Neighbor (if neighbors are users) [9–11].

Item-Based K-Nearest Neighbor. Item-based K-Nearest Neighbor (Item-based KNN) predicts the rating of a user *u* on item *i* by finding items similar to *i*. The similarity score between two items can be computed using Cosine Similarity as:

$$similarity(i,j) = \frac{r_i.r_j}{\|r_i\|\|r_j\|} \tag{1}$$

where, r_i and r_j are the rating vectors of the two items *i* and *j*, respectively, made by users.

Then, to predict rating on an item, the model will select *K* nearest neighbors for that item. Such prediction can be inferred from neighbors using the weighted mean:

$$\widehat{r_{ui}} = \frac{\sum_{j \in N_u^k(i)} similarity(i,j).r_{uj}}{\sum_{j \in N_u^k(i)} similarity(i,j)} \tag{2}$$

where $N_u^k(i)$ is the k nearest neighbors of item i that are rated by user u, r_{uj} is the rating of user u on item j, and $\widehat{r_{ui}}$ is the predicted rating of user u on item i.

User-Based K-Nearest Neighbor. Similarly, the prediction can be made using a user-based model as:

$$\widehat{r_{ui}} = \frac{\sum_{v \in N_i^k(u)} similarity(u, v).r_{vi}}{\sum_{v \in_i^k(u)} similarity(u, v)} \tag{3}$$

where, $similarity(u, v)$ is the similarity between user u and v, and r_{vi} is the rating of user v on item i.

It is worth notice that in a real e-commerce system, the number of customers and items increases rapidly, there are many problems that need to be addressed, such as, cold-star [12], sparsity and scalability [13, 14].

2.2 Matrix-Factorization Model

Matrix Factorization approaches [15] are latent factor models that base on previous user-item interactions to characterize users and items at the same time.

For matrix factorization models, each item i could be represented by a vector $q_i \in \mathbb{R}^f$ and each user u could be represented by a vector $p_u \in \mathbb{R}^f$ where f is the dimension of the joint latent factor space [15]. The vector q_i values measure how "close" the item is to those factors. Similarly, the vector p_u measures the level of interest of the user on items close to those corresponding factors. The approximated rating of user u on item i could then be inferred as:

$$\widehat{r_{ui}} = q_i^T p_u \tag{4}$$

Singular Value Decomposition (SVD). SVD is the most popular matrix factorization methods that can be applied in recommender systems to *solve sparsity and scalability problem* [16–18], used as dimensionality reduction technique and a powerful mechanism [19]. SVD [1] factorizes a matrix R into three matrices M, Σ, U, where q_i makes up the columns of U^T, p_u makes up the rows of M, and the diagonal matrix that acts as a scaler on M or U^T.

For recommendation, because of the sparsity of the rating matrix, such matrix R is impossible to construct. So instead of directly finding M and U, we find all the vectors q_i, p_u such that Eq. (4) is satisfied for all users u and items i. To find all the vectors q_i, p_u, we try to minimize the squared error on the set of known ratings:

$$min_{q^*,b^*} \sum_{(u,i \in \kappa)} \left(r_{ui} - q_i^T p_u\right)^2 + \lambda\left(\|p_u\|^2 + \|q_i\|^2\right) \tag{5}$$

where, λ is the regularization parameter, κ is the set of (u, i) pairs for which r_{ui} is from the training set. This optimization is not convex, in other words, it may be impossible to find the pair of p_u and q_i that minimize the sum. Instead, we can use Stochastic Gradient Descent (SGD) or Alternating Least Squares (ALS) [4].

The SGD algorithm iterates over the ratings in the training set, for each case, it computes the error as:

$$e_{ui} = \left(r_{ui} - q_i^T . p_u\right) \tag{6}$$

Then, vectors p_u and q_i can be updated simultaneously as:

$$q_i = q_i + \alpha(e_{ui}.p_u - \lambda q_i); \tag{7}$$

$$p_u = p_u + \alpha(e_{ui}.q_i - \lambda p_u) \tag{8}$$

where, α is learning rate.

An alternative algorithm for SGD is to use Alternating Least Squares, which, in some cases, is more effective. ALS techniques tend to fix vectors q_i and treat them as constants, optimize p_u, then fix p_u and optimize q_i, and repeat until convergence [4]. One advantage of this approach is that the computation could be done in parallel because the system computes each p_u, q_i independently of each other.

We can refine our matrix factorization model by adding biases for both users and items. The predicted rating then is computed as:

$$\widehat{r_{ui}} = \mu + b_u + b_i + q_i^T . p_u \tag{9}$$

where b_u is the user bias and b_i is the item bias, and μ is the average rating. The observed rating is then computed by adding the biases and average rating, the system now tries to minimize the new loss function:

$$\min_{p*,q*,b*} \sum\nolimits_{(u,i\in\kappa)} \left(r_{ui} - \mu - b_u - b_i - q_i^T . p_u\right)^2 + \lambda\left(\|q_i\|^2 + \|p_u\|^2 + b_u^2 + b_i^2\right) \tag{10}$$

Incremental Singular Vector Decomposition [20]. The workflow of a recommender system is divided into two different stages: the offline stage or training stage and the online stage. In the online stage, the system gives real-time recommendations to users. A real-time recommendation generated by the SVD model is high-quality and reliable, but the training stage of SVD is computationally expensive and time-consuming. For a $m \times n$ user-item matrix, the time complexity of SVD is almost $\Theta(m)^3$ [20]. The fold-in technique is used to project new users to the space of existed user-item matrix. It assumes that the projection of the additional user (or rating) can also provide a reliable and good approximation of the trained model. The Incremental SVD allows the model to incrementally learn the new ratings and makes more reliable and accurate recommendation for new users without the expensive computation.

2.3 Long-Tail Phenomenon

There exists a phenomenon in recommender systems called long-tail phenomenon [2] which is the distance between physical and online stores. All of the items in the storage can be divided into three different parts: the short head, the long tail, and the distant tail. The short head part is extremely popular items which are often offered to customers by physical stores. The tail part consists of two different areas. The long tail is the most distribution of items which are potential items that should be included by the recommender system, even though most recommender systems often ignore them. The distant tail includes items with few ratings as known as cold-star items which may have a bad effect on a collaborative filtering recommender system. As the goal of the recommender system is to build personalized recommendations for each user, the regular problem caused by the biased popularity is that the recommendation lists are very identical amongst users since they include a lot of popular items. A recommender system that only produces very similar recommendations for every user should not be considered as a reliable system.

Explicit Query Aspect Diversification. A solution to the above described problem is the Explicit Query Aspect Diversification (xQuAD) method, which has been used in the context of information retrieval, such as web search engines. The goal of this algorithm [2] is to use the existing ranked recommendation list L to produce a new recommend list S ($|S| < |L|$) that manages to control the popularity bias while keeping the acceptable accuracy. The new list is built iteratively using the formula:

$$P(v|u) + \lambda P(v, S'|u) \tag{11}$$

where $P(v|u)$ is the probability item v is interesting user u and $P\left(v, S'|u\right)$ denotes the probability of user u being interested in item v which is not currently in the recommend list S. The parameter λ is used to adjust the popularity bias. There are two different versions of xQuAD, one is the *Binary xQuAD* and another is *Smooth xQuAD* [2]. When selecting an item to be added to list S, the ranking score for each item in the recommendation list is computed as:

$$score = (1 - \lambda)P(v|u) + \lambda \sum\nolimits_{c \in \left\{\Gamma, \Gamma'\right\}} P(c|u)P(v|c) \prod\nolimits_{i \in S} (1 - P(i|c, S) \tag{12}$$

Briefly, for an item $v' \in c$, if the recommend list S does not cover (included in) area c (c represents either long-tail (Γ) or short-head (Γ') area), then the estimated user rating or preference will include the additional positive term. Thus, this increases the chance of an item to be selected, balancing the accuracy and the popularity bias.

3 Methodology

In this paper, the process of building the recommender system consists of two primary parts. Firstly, we will train and optimize the recommendation model for predicting the user ratings (Fig. 1). Then, we improve the relevance and reliability of the top-K recommendation generated by the recommender system in the first phase and use it in the final application to recommend products for users (Fig. 2).

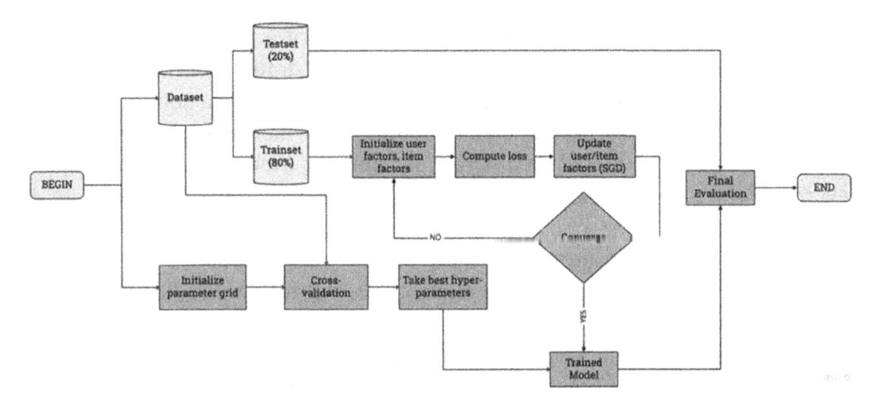

Fig. 1. The off-line stage of training and tuning the recommendation model.

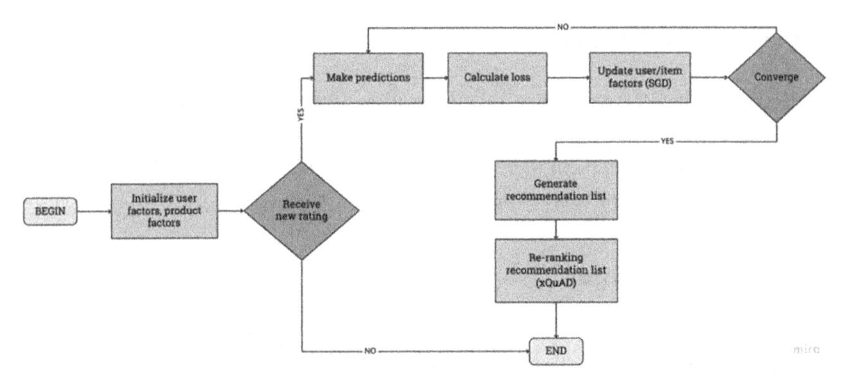

Fig. 2. The online stage of producing real-time recommendations.

To optimize the model for prediction accuracy, experiments are conducted with same training dataset and testing dataset on four different models which are Item-based, User-based K-Nearest Neighbors, SVD, and ALS. Random Search [21] is used to find the optimal hyperparameters. Two evaluation metrics Root Mean Squared Error (RMSE) and Mean Absolute Error (MAE) are used to evaluate prediction accuracy. The mathematical definition of RMSE and MAE are in Eq. (13) and (14).

$$MAE = \frac{1}{n}\sum\nolimits_{u,i} \left|\hat{r}_{ui} - r_{ui}\right| \tag{13}$$

$$RMSE = \sqrt{\frac{1}{n}\sum\nolimits_{u,i} \left(\hat{r}_{ui} - r_{ui}\right)^2} \tag{14}$$

where, $\hat{r}_{ui}$ and r_{ui} are respectively the predicted and actual rating of user u on item i and n is the number of ratings.

By experiments, the SVD model, which has the best performance is used for producing personalized recommendations. As mentioned before, an incremental version of SVD will also be proposed. The approach to implementing Incremental SVD is slightly different from the method in [20]. To predict the rating of a pair of user and product, either the user or the product is new (i.e., it isn't in the training set), the model first initializes its bias to 0. Then, its factor is randomly initialized from a normal distribution using the model's mean and standard deviation rating, which are set to 0 and 0.1 by default, respectively. The model then iteratively runs the Stochastic Gradient Descent for several epochs to update the new user or product biases and factors. After the partial fitting part, for each user, the model predicts the rating of that user and all products. The top-N highest rated products are used to build the recommendation list.

Then, to produce less biased recommendations, we follow the flexible approach to controlling the balance of the recommended items distribution in different areas, introduced in [2]. The framework is built using the xQuAD algorithm introduced in Sect. 2.3. The re-ranked recommendation list, whose size is smaller than the original list, is the final recommendation list.

4 Experimental Results

This section describes our experimental verification of the proposed framework for SVD-based recommender system. We first present the overview of the data set, the evaluation metrics for different aspects of the framework. Finally, we display the result of the experiment and propose our discussion on that.

4.1 Experimental Platform

In this section, four models are evaluated using Amazon's Electronic dataset [22]. The dataset has 7,824,482 ratings of 4,201,696 distinct users and 476,002 distinct products. After taking a deep look into the dataset, we observe that most users have rated less than 10 times (the 99th percentile of number of ratings per user is 12 ratings). Since we need to evaluate the top-N recommendation, we need an appropriate number of ratings for each user, therefore, users with less than 50 ratings are removed. The size of the dataset is reduced to 125,871 ratings of 1,540 users and 48,190 items. Without the pre-processing, the dataset can adversely affect the quality of the model. The training set size is 80% of the dataset and the rest are for the testing set.

4.2 Evaluation Metrics and Results

For this experiment, we evaluate our framework on two different types of metrics. The prediction accuracy metric will measure how good the model is in making prediction. Besides, an additional experiment is done to demonstrate the quality of the incremental approach of the SVD-based model. Later, using the ranking accuracy metric, we measure the effectiveness of the xQuAD algorithm on enhancing the top-N recommendation quality as well as its affection on the ranking accuracy of the system.

Prediction Accuracy. As mentioned in Sect. 3, the two metrics used for evaluating prediction accuracy are Root Mean Squared Error and Mean Absolute Error. Using the cross validation, the results in Fig. 3 is the average values of 5-fold cross validation. The high prediction accuracy of the SVD model stands out with the corresponding RMSE and MAE of approximate 0.998 and 0.735, respectively. The results of the two neighborhood models seem identical since their predictions are produced similarly. The ALS model performs worst amongst all.

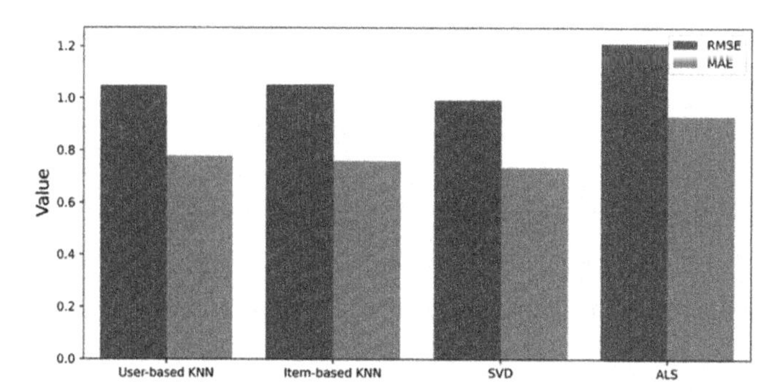

Fig. 3. The results of root mean square error and mean average error.

Besides, to observe the performance of the incremental SVD approach, we divide the training set into two subsets with the ratio of x and $1 - x$, respectively. Since we do not want the model to learn on too small data nor incrementally learn too big data, we choose to do experiments for x from 0.2 to 0.5. First, we train the model on the larger subset. Later, follow the incremental approach, we fold-in the other sub-set (called the fold-in set) into the model. Finally, we test the model on the testing set and observe how the prediction accuracy varies compared to the original model (trained on the full training set).

Figure 4 plots the prediction quality result with the incremental model on different fold-in data ratio. It shows that even with a high ratio of fold-in set, the model is still be able to obtain good prediction. With the ratio of fold-in set and train-set of 0.5 (meaning that 50% of the original training data is used for fitting and the rest is used for folded-in), the RMSE accuracy is very close to the original model (0.990 compared to 0.998, only 0.8% drop). This suggests that the incremental technique offers a good quality model even when the fold-in set size is large.

Recommendation Ranking Accuracy. To evaluate the recommendation ranking accuracy, we use two different types of ranking metrics. One is the decision support metrics, in which we compute Precision@M, Recall@M, and F1@M Another one measures how much the trade-off is between balancing popularity bias and ranking accuracy, in which we compute the recommendation popularity, its coverage of long-tail items, and Normalized Discounted Cumulative Gain [23].

Precision@M, Recall@M, F1@M. In this paper, a product is considered relevant to a user if it is rated higher than 3.0 and is recommended when the system predicts its rating to be high than 3.0. By that, Precision@M is computed as the proportion of the first M recommended items that are relevant and Recall@M is the proportion of relevant items

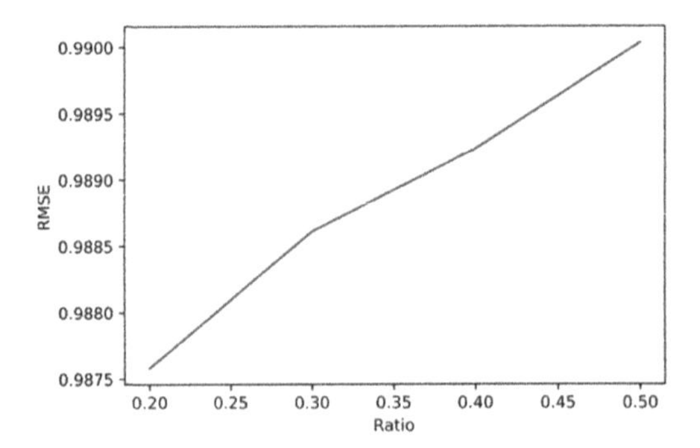

Fig. 4. Prediction accuracy over different fold-in set ratio.

that are recommended in the first M recommended items. The F1@M is then computed as:

$$F1@M = 2 * \left(\frac{Precision@M * Recall@M}{Precision@M + Recall@M} \right) \tag{15}$$

The result illustrates in Fig. 5 is measured by using each model to predict the rating of all users on the items they rated in the test set. The result is the average of Precision@M, Recall@M, and F1@M for M from 5 to 20.

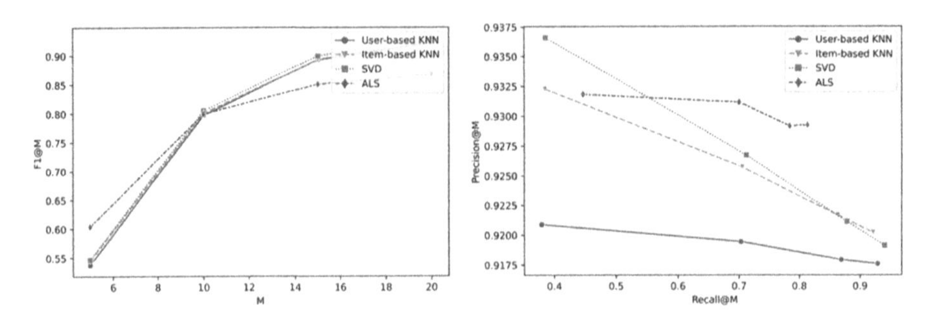

Fig. 5. The results of Precision@M, Recall@M, F1@M

The accuracy for all models seems to be very high. The trade-off between the precision and recall is high for Item-based KNN and SVD models, and low for User-based KNN and ALS models. Amongst all, the most well perform model is the SVD model, however, the trade-off is high when M increases. As expected, the Recall@M increases when M increases since we are increasing the numerator of the Recall@M while keeping the dominator fixed. While the Recall@M tends to increase, the Precision@M decreases, once again, we can see the trade-off.

To solve the popularity bias problem, we apply metrics in [2]: Average Recommendation Popularity (ARP), Average Percentage of Long-tail Items (APLT), and Average Coverage of Long-tail Items (ACLT), to adjust popular items in the recommendation list and make it more personalized. Besides, Normalized Discounted Cumulative Gain (NDCG) [23], a popular ranking metric, is computed for a user's recommendation list.

To build the top-N recommendation, the recommender system uses the trained model to predict each rating in the testing set. In this experiment, the top-50 highest rated products for each user are considered as their recommendation. From the top-50 recommendation list, we apply the xQuAD algorithm to get the final ranked top-10 recommendation list.

Figure 6 is measured using two different versions of xQuAD algorithm on re-ranking the recommendation list. The *reg* in the x-axis label is the regularization parameter of the ranking scoring function, it adjusts how strongly the biased popularity is controlled. As the regularization gets bigger, more long-tail items are included in the recommendation list (measured by APLT and ACLT), and the average popularity of the recommendation decreases (measured by ARP). However, as the popularity bias is getting controlled more strongly, the ranking accuracy of the recommendation seems to be lower (Fig. 7).

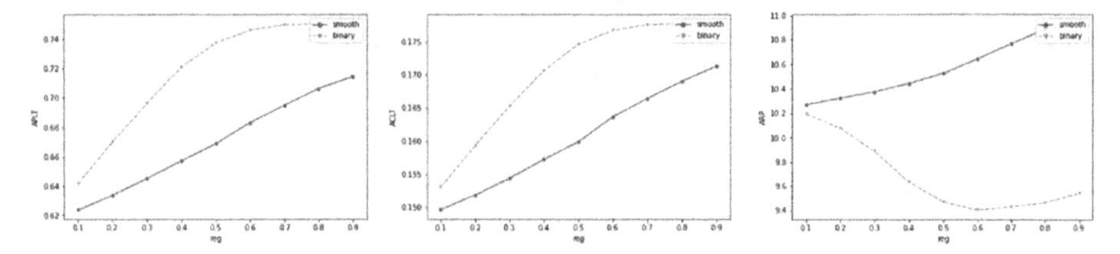

Fig. 6. Results of APLT, ACLT, ARP.

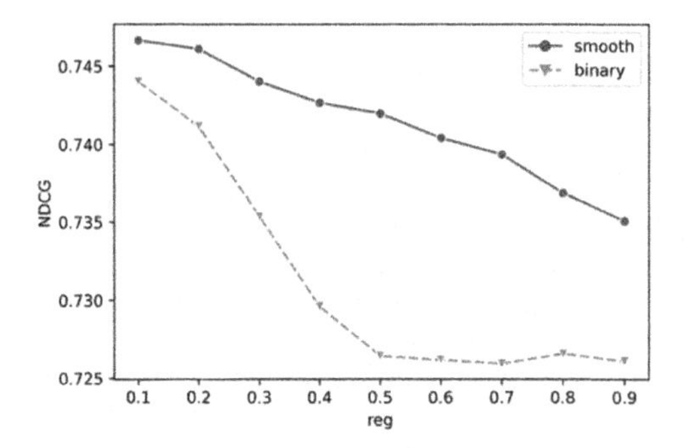

Fig. 7. Popularity bias and ranking accuracy trade-off.

Choosing the appropriate regularization parameter is crucial when using the xQuAD algorithm for recommender system. For the Amazon's Electronic dataset used in this paper, the optimal value of λ we choose is 0.4. With λ of 0.4, we prevent the algorithm from affecting adversely to the accuracy of the recommender system while keeping the power of the algorithm on controlling the popularity bias.

5 Conclusions and Future Work

In this paper, our proposed method can overcome some limitations of usual recommender methods. First, we experimented on many algorithms and choose the one that fits best, which was the SVD-based recommendation model. After that, to deal with the expensive computation of the SVD algorithm in the online stage, we upgrade the SVD by applying the Incremental SVD method. Then, to produce high-quality recommendations, we balance the popularity bias of personalized recommendations while keeping an acceptable ranking accuracy (measured by NDCG) using the xQuAD algorithm. Finally, we apply different metrics to evaluate the effectiveness of the recommendation model on different aspects. As shown in the experiments, the proposed recommendation model can gain accuracy above 90% and include more popular items in the recommendation lists. This makes the recommender system achieve more effective performance.

In future, we plan to build an automatic pipeline for combining all of the methods and algorithms proposed in this paper. Such pipeline can be used with flexibility to fasten the process of build a recommender system. Besides, we also plan to use implicit data along with the existing explicit data to build a more effective model. Also, we will try to optimize the implementation of the Incremental SVD for a faster, less expensive fold-in technique.

Acknowledgment. This research is funded by Vietnam National Foundation for Science and Technology Development (NAFOSTED) under grant number: 06/2018/TN.

References

1. Alter, O., Brown, P.O., Botstein, D.: Singular value decomposition for genome-wide expression data processing and modeling. Proc. Natl. Acad. Sci. U.S.A. **97**, 10101–10106 (2000)
2. Abdollahpouri, H., Burke, R., Mobasher, B.: Managing popularity bias in recommender system with personalized re-ranking (2019)
3. Koren, Y.: Factorization meets the neighborhood: a multifaceted collaborative filtering model. In: 14th ACM SIGKDD International Conference on Knowledge Discovery and Data Mining, USA (2008)
4. Zhou, Y., Wilkinson, D., Schreiber, R., Pan, R.: Large-scale parallel collaborative filtering for the netflix prize. In: Fleischer, R., Xu, J. (eds.) AAIM 2008. LNCS, vol. 5034, pp. 337–348. Springer, Heidelberg (2008). https://doi.org/10.1007/978-3-540-68880-8_32
5. Lops, P., de Gemmis, M., Semeraro, G.: Content-based recommender systems: state of the art and trends. In: Ricci, F., Rokach, L., Shapira, B., Kantor, P.B. (eds.) Recommender Systems Handbook, pp. 73–105. Springer, Boston, MA (2011). https://doi.org/10.1007/978-0-387-85820-3_3
6. Naumov, M., et al.: Deep learning recommendation model for personalization and recommendation systems. arXiv preprint arXiv:1906.00091 (2019)
7. Xiong, R., et al.: Deep hybrid collaborative filtering for web service recommendation. Expert Syst. Appl. **110**, 191–205 (2018)
8. Jiang, L., et al.: A trust-based collaborative filtering algorithm for E-commerce recommendation system. J. Ambient Intell. Human. Comput. **10**(8), 3023–3034 (2018). https://doi.org/10.1007/s12652-018-0928-7

9. Herlocker, J.L., Konstan, J.A., Borchers, A., Riedl, J.: An algorithmic framework for performing collaborative filtering. In: SIGIR Forum, vol. 51, pp. 227–234 (2017)
10. Sarwar, B., Karypis, G., Konstan, J., Riedl, J.: Item-based collaborative filtering recommendation algorithms. In: Proceedings of the 10th International Conference on World Wide Web, pp. 285–295. ACM, New York (2001)
11. Jannach, D., Lerche, L., Kamehkhosh, I., Jugovac, M.: What recommenders recommend: an analysis of recommendation biases and possible countermeasures. User Model. User-Adap. Interact. **25**(5), 427–491 (2015). https://doi.org/10.1007/s11257-015-9165-3
12. Cleger-Tamayo, S., Fernández-Luna, J.M., Huete, J.F.: Top-N news recommendations in digital newspapers. Knowl. Based Syst. **27**, 180–189 (2012)
13. Claypool, M., et al.: Combing content-based and collaborative filters in an online newspaper (1999)
14. Sarwar, B., et al.: Analysis of recommendation algorithms for e-commerce. In: Proceedings of the 2nd ACM Conference on Electronic Commerce (2000)
15. Bell, R.M., Koren, Y., Volinsky, C.: Matrix factorization techniques for recommender system (2009)
16. Barathy, R., Chitra, P.: Applying matrix factorization in collaborative filtering recommender systems. In: 2020 6th International Conference on Advanced Computing and Communication Systems (ICACCS). IEEE (2020)
17. Nilashi, M., Ibrahim, O., Bagherifard, K.: A recommender system based on collaborative filtering using ontology and dimensionality reduction techniques. Expert Syst. Appl. **92**, 507–520 (2018)
18. Teodorescu, O.M., Popescu, P.S., Mihaescu, M.C.: Taking e-assessment quizzes - a case study with an SVD based recommender system. In: Yin, H., Camacho, D., Novais, P., Tallón-Ballesteros, A.J. (eds.) IDEAL 2018. LNCS, vol. 11314, pp. 829–837. Springer, Cham (2018). https://doi.org/10.1007/978-3-030-03493-1_86
19. Zarzour, H., et al: A new collaborative filtering recommendation algorithm based on dimensionality reduction and clustering techniques. In: 2018 9th International Conference on Information and Communication Systems (ICICS). IEEE (2018)
20. Sarwar, B., Karypis, G., Konstan, J., Reidl, J.: Incremental singular value decomposition algorithms for highly scalable recommender system. In: 5th International Conference on Computer and Information Science (2002)
21. Bergstra, J., Bengio, Y.: Random search for hyper-parameter optimization. J. Mach. Learn. Res. **13**, 281–305 (2012)
22. He, R., McAuley, J.: Ups and downs: modeling the visual evolution of fashion trends with one-class collaborative filtering. In: Proceedings of the 25th International Conference on World Wide Web (WWW 2016), pp. 507–517. International World Wide Web Conferences Steering Committee, Republic and Canton of Geneva, CHE (2016)
23. Wang, Y., Wang, L., Li, Y., He, D., Liu, T., Chen, W.: A theoretical analysis of NDCG type ranking measures. J. Mach. Learn. Res. **30** (2013)

Collaborative Filtering Recommendation Based on Statistical Implicative Analysis

Hiep Xuan Huynh[1], Nghia Quoc Phan[2(✉)], Nghia Duong-Trung[3], and Ha Thu Thi Nguyen[4]

[1] Can Tho University, Can Tho 94000, Vietnam
hxhiep@ctu.edu.vn
[2] Tra Vinh University, Tra Vinh 87000, Vietnam
nghiatvnt@tvu.edu.vn
[3] FPT University, Can Tho 94000, Vietnam
duong-trung@ismll.de
[4] Electric Power University, Hanoi 10000, Vietnam
hantt@epu.edu.vn

Abstract. In recent studies on recommender models, association rules have been applied in many studies to improve the effectiveness of recommender models. However, these studies also reveal some drawbacks, such as the models take a considerable amount of time to generate association rules for large datasets; generation algorithms can ignore rules with the significant implication that affect the quality of recommender models. This study proposes collaborative filtering recommender models (CF models) based on association rules following an asymmetric approach of the statistical implicative analysis method to enhance the precision of recommender models. Through experiments on standard datasets and quality comparison with other CF models, we conclude that the proposed models based on the asymmetric relationship achieve better accuracy on the experimental datasets.

Keywords: Implication intensity · Statistical implicative analysis · Association rules

1 Introduction

The statistical implicative analysis is data analysis method [8,9,21] formed in the process of teaching math in high school to evaluate the learning behavior of students. This data analysis method has been affecting many research areas like education, psychology, computer science, and so on. It has become a methodology that allows the evaluation of the implication relationship formed through techniques for acquiring knowledge of the human (natural processes and artificial processes). Especially in data mining, extracting knowledge from a large set of association rules serves to make the decision process, this method is used to detect asymmetric rules $a \to b$. The CF models [14,15,24] have been successfully applied in many e-commerce service systems such as Amazon [16], Netflix

M. Hernes et al. (Eds.): ICCCI 2020, CCIS 1287, pp. 224–235, 2020.
https://doi.org/10.1007/978-3-030-63119-2_19

[7]. The CF is an effective solution to solve the information explosion problem for online systems with a rapidly increasing number of users [1]. These models recommend items to users based on the assumption that similar users will have the same preferences on items or users have identical choices on related items. Therefore, the CF models depend entirely on rating data of users for items. For example, in the movie recommender models for viewers, the models find a group of viewers with the same interests as users who need recommendations based on rating data. Then, the models recommend the movies which were highly appreciated by this group for the viewer who needs recommendations. However, the CF models encounter limitations: new users/new items (cold-Start), sparsity, scalability. This paper proposes CF models based on association rules according to the asymmetric approach of the statistical implicative analysis method to improve the precision of recommender models. These models find items to recommend users based on a set of association rules built from measures that define the implicative relationships among users through rating data.

The content of the article is presented in 7 sections. Section 1 gives an overview of the research content. Section 2 introduces the CF models based on the association rules. Section 3 presents statistical implicative analysis method. Section 4 presents the CF model based on Implication Intensity (II) measures. Section 5 presents the CF model based on Statistical Implicative Similarity (SIS) measures. Section 6 presents the results of evaluating the accuracy of models on two datasets and compares the results with other models. And the summary of the achieved results is presented in Sect. 7.

2 The CF Model Based on Association Rules

The CF models based on association rules (AR model) recommends items for users from the association rule set generated by the user's binary rating matrix [5,10,17,23]. To generate the association rule set, the rating matrix is seen as a database with each user being a transaction containing items rated by 1 in the set of items I of the rating matrix. Hence, transaction k is defined: $T_k = \{i_j \in I | r_{jk} = 1\}$, where r_{jk} is the rating value of user k (u_k) for the item j (i_j). From there, the transactional database used to generate association rules is defined: $D = \{T_1, T_2, \ldots, T_U\}$ where U is the number of users n the rating dataset. The association rules are generated from the transaction database D in the form $\{X \rightarrow Y | X, Y \subseteq I,\ X \cap Y = \emptyset\}$ [2]. The model uses minimum support (s), minimum confidence (c) and the maximum length of the rule (l) to filter association rules (Support($X \rightarrow Y$) $> s$; Confidence($X \rightarrow Y$) $> c$); $|X \cup Y| \leq l$). From the association rule set, the CF-AR model is defined [10]:

- RU is the association rule set for the CF-AR model;
- T_a is an item set rated by user u_a;
- $RU_a = \{X \rightarrow Y | X \subseteq T_a, X \rightarrow Y \in RU\}$ finds all rules in the RU association rule set with the left side containing items rated by user u_a;

- $I_a = \{i_k \in I | i_k \subseteq Y, X \rightarrow Y \in RU_a, \text{highest}(\text{confidence}(X \rightarrow Y))\}$ choose the items from the right of the association rule set RU_a with confidence($X \rightarrow Y$) getting the highest value (u_a not rated yet for I_a) and recommend I_a to user u_a.

3 Statistical Implication Analysis Method

The data mining method based on statistical implication analysis [8] is a data analysis method that studies implicative relationships of items or users in a research dataset. This method allows the detection of asymmetric rules $a \rightarrow b$ as "if a then that almost b" or "consider to what extent that b will meet implication of a". The goal is to detect trends in an item set or a user set by using statistical implication measures. Let us consider a population E of n objects or individuals described by a finite set K of binary variables (attributes, criteria, scores). The authors denote that $A \subset E$ is the subset of individuals for which a is true; $B \subset E$ is the subset of individuals for which b is true; $\bar{A}$ is the complement of A; $\bar{B}$ the complement of B; $n_A = \text{card}(A)$ is the number of elements of set A; $n_B = \text{card}(B)$ is the number of elements of set B. The rejection or acceptance $a \rightarrow b$ is consider the number of counter-examples ($n_{A\bar{B}} = \text{card}(A \cap \bar{B})$). We randomly draw two subsets X and Y of, respectively, n_A and n_B elements. The distribution of card($X \cap \bar{Y}$) follows the Poisson distribution with the parameter $\lambda = \frac{n_A n_B}{n}$. As a result, the probability of $\text{card}(A \cap \bar{B}) \leq \text{card}(X \cap \bar{Y})$ is defined in Eq. (1).

$$\Pr[\text{card}(X \cap \bar{Y}) \leq \text{card}(A \cap \bar{B})] = \sum_{s=0}^{\text{card}(A \cap \bar{B})} \frac{\lambda^s}{s!} e^{-\lambda} \tag{1}$$

Let us consider, in the case of $n_{\bar{B}} \neq 0$, the standardized random variable $Q(a, \bar{b})$ as in Eq. (2).

$$Q(a, \bar{b}) = \frac{\text{card}(X \cap \bar{Y}) - \frac{n_A(n-n_B)}{n}}{\sqrt{\frac{n_A(n-n_B)}{n}}} \tag{2}$$

We denote by $q(a, \bar{b})$ the observed value of $Q(a, \bar{b})$ in the experimental realization. It is defined in Eq. (3).

$$q(a, \bar{b}) = \frac{n_{A\bar{B}} - \frac{n_A(n-n_B)}{n}}{\sqrt{\frac{n_A(n-n_B)}{n}}} \tag{3}$$

The implication intensity of the rule $a \rightarrow b$ is defined by Eq. (4). This measure uses to determine the unlikelihood of the counter-examples $n_{A\bar{B}}$ in the set E.

$$\varphi(a, b) = \begin{cases} 1 - \Pr(Q(a, \bar{b}) \leq q(a, \bar{b})) = \frac{1}{2\pi} \int_{q(A,\bar{B})}^{\infty} e^{-\frac{t^2}{2}} dt & \text{if } n_B \neq n \\ 0 & \text{otherwise} \end{cases} \tag{4}$$

4 The CF Model Based on II Measures (CFIIR Model)

4.1 Association Rules Based on II Measures

Definition of Association Rules Based on II Measures. Association rules based on II measures are asymmetric association rules having form $a \rightarrow b$ generated from a user rating matrix with a given threshold value of II measures [8]. The association rules based on II measures are defined as follows: Let $U = \{u_1, u_2, \ldots, u_n\}$ is a user set; $I = \{i_1, i_2, \ldots, i_m\}$ is an item set; $R = \{r_{i,j}\}$ is a rating matrix, with each row represents one user, each column describes one item. $r_{i,j}$ is the rating value of user u_i for item i_j; t_i is an item set rated by u_i, t_k is an item set rated by u_k, where $t_i, t_k \subseteq I$. The association rule based on II measures is defined as $a \rightarrow b$, where a and b are two separate sets with $a \subseteq t_i, b \subseteq t_k$ and $a \cap b = \emptyset$. The rule is accepted with a threshold value of II measures $\varphi(a, b) \geq 1 - a$ with $0 \leq a \leq 1$, where $\varphi(a, b)$ is defined in Eq. (4).

Algorithm for Generating Association Rule Based on II Measures. According to the definition of association rules based on II measures presented in the previous section, we build the algorithm for generating association rules of the model, see Algorithm (1), from the training dataset including the following steps. In the first step, the algorithm generates sets from one element, two elements, ..., k elements. In the next step, for each set generates non-empty subsets of the candidate set. In the final step, the algorithm based on subsets to generate association rules with the threshold of II value. This threshold is determined based on the empirical results of specific datasets.

Algorithm 1. Generation of association rules based on II measures.

Input: Training dataset, threshold value of II measures.
Output: The association rule set based on II measures.

1: **Step 1:** Generating a candidate set from 1-item set to k-item set
2: **Step 2:** Generating nonempty subsets of candidate set
3: **for** each candidate set I **do**
4:    Generating nonempty subsets s of I
5: **end for**
6: **Step 3:** Generating association rule with threshold value of II
7: **for** each nonempty subsets s of I **do**
8:    Generating association rule $r : s \rightarrow (I - s)$
9:    **if** value of II of rule $r \geq a$ **then**
10:       Selecting association rule r for recommender model
11:    **end if**
12: **end for**

4.2 The CF Model Based on II Measures

Definition of Recommender Model Based on II Measures. Suppose that $U = \{u_1, u_2, \ldots, u_n\}$ is a user set; $I = \{i_1, i_2, \ldots, i_m\}$ is an item set; $R_{\text{train}} = \{r_{j,k}\}$ is a training set; $R_{\text{test}} = \{r_{i,l}\}$ is a test set; $RU = \{r_1, r_2, \ldots, r_t\}$ is an association rule set generated from the training set based on II measures; $\text{MATRIX}_{\text{Boolean}} = \{b_{t,i}\}$ is a boolean matrix with $b_{t,i}$ to be binary value of rule r_t and user u_i, if user u_i rated for items belong to left-hand side of rule r_t then $b_{t,i} = 1$, otherwise $b_{t,i} = 0$; $RU_{u_a} = \{r_{u_a1}, r_{u_a2}, \ldots, r_{u_ak}\}$ is an association rule set selected for u_a. The result of model is an item set belongs to the right-hand side of RU_{u_a} that u_a has not yet rated $I_{u_a} = \{i_{u_a1}, i_{u_a2}, \ldots, i_{u_aN}\}$. The process of determining the association rule set for each user in the testing set contains three steps. In the first step, we base on the number of rules generated from R_{train} and the number of users in R_{test} to create a boolean matrix including t rows and l columns (t the number of rules and l the number of users). The boolean matrix is created according to the rule: if user u_i rated for items belong to the left-hand side of rule r_t, then cell(t, i) is assigned value 1 ($b_{t,i} = 1$), otherwise cell(t, i) is assigned value 0 ($b_{t,i} = 0$). Next step, for each user i, if the value of cell$(t, i) = 1$, then select association rule r_t. The final step, sort the rule set of each user who needs recommendation in the descending order of II values and recommend to them N items belong to the right-hand side of the rules having the highest II value that they not yet rated.

Recommender Algorithm Based on II Measures. The algorithm generating results of CF model based on II is described in Algorithm (2).

5 The CF Model Based on SIS Measures (CFSISR Model)

5.1 Similarity Measures Based on II Measures

Similarity measures have an important role in User-based CF models because it directly affects the outcome of the recommender models [22]. Many solutions to determine the similarity value between two users in the CF models have been proposed, such as using statistical correlation [3], using cosine distance between two vectors, association rules, and other solutions [6,20]. In this section, we formulate a similarity measure using the asymmetric approach of the statistical implicative analysis method [19]. In particular, the similarity value between two users is determined based on the total statistical implicative distance of association rules that tend to support counter-examples generated from the rating data of the two users.

The SIS Measures Between Two Users The authors use SIS measures to determine the similarity value between two users based on association rules

Algorithm 2. Generation of recommendation results based on II measures.

Input: Set of association rules generated from the training set; test set.
Output: Recommended results for each user in the test set.

1: **Step 1:** Creating a boolean matrix based on association rule set and testing set
2: $\text{MATRIX}_{\text{boolean}} = \{b_{i,j}\}$, where $i = \{1, 2, \ldots, t\}$, $j = \{1, 2, \ldots, l\}$
3: **for** each association rule in *RU* **do**
4:    **for** each user in the test set **do**
5:       **if** (items on row i) $\cap$ (items rated by user on column j) $\neq \emptyset$ **then**
6:          $\text{MATRIX}_{\text{boolean}}[i, j] = 1$
7:       **else**
8:          $\text{MATRIX}_{\text{boolean}}[i, j] = 0$
9:       **end if**
10:    **end for**
11: **end for**
12: **Step 2:** Selecting association rules to recommend for each user
13: **for** each user in the test set **do**
14:    **for** each association rule selected **do**
15:       **if** $\text{MATRIX}_{\text{boolean}}[i, j] = 1$ **then**
16:          Selecting association rule on row i for user on column j
17:       **end if**
18:    **end for**
19: **end for**
20: **Step 3: Selecting items to recommend for each user**
21: **for** each user in the test set **do**
22:    Sorting descending rule set based on II values
23:    Selecting top N items that user j has not rated from the right hand side of rules to recommend for user j
24: **end for**

generated from the rating data of two users and the II measures. The SIS value between two users u_a and u_b is determined by Eq. (5).

$$\text{SIS}(u_a, u_b) = 1 - \frac{\sum_{i=1}^{k} \varphi(r_i)}{k}, \tag{5}$$

where $\text{SIS}(u_a, u_b)$ is the similarity value between two users u_a and u_b, $\varphi(r_i)$ is II value of the association rule r_i, and k is the number of association rule set that generated from rating data of two users u_a and u_b.

Algorithm to Measure SIS Value of Two Users. To determine the SIS value of two users using the II measures, we propose an algorithm, see Algorithm (3), based on calculating the statistical implicative distance of the association rules that tend to support counter-examples generated from the rating data of the two users. Specifically, the association rules chosen to determine the similarity value of two users u_a and u_b must meet two conditions. The left-hand side of the rule must have at least one item rated by the user u_a, and the right-hand side of the rule must have at least one item that has not been evaluated by user u_b. Based on the rules that tend to support the counter-examples of each pair of users, we calculate the average of the II value, e.g., statistical implicative distance, of the ruleset. Meanwhile, the SIS value between two users u_a and u_b is determined by $1-\bar{S}$ where $\bar{S}$ is the average value of the statistical implicative distance.

Algorithm 3. Measurement of SIS value between two users.

Input: Rating matrix, data rating of two users u_a and u_b.
Output: SIS value of two users u_a and u_b.

1: **Step 1:** Generating association rules from users' rating matrix
2: **Step 2:** Building association rule set of two users u_a and u_b
3: Selecting items rated by user $u_a : I_{u_a}$
4: Selecting items not rated by user $u_b : \bar{I}_{u_b}$
5: Selecting association rules of the form: $X \rightarrow Y$ with $X \in I_{u_a}$, $Y \in \bar{I}_{u_b}$ and $X \cap Y \neq \emptyset$
6: **Step 3:** Calculating SIS value of two users u_a and u_b : $\text{SIS}(u_a, u_b)$
7: Calculating mean of II values of rule set $\bar{S}$
8: Calculating $\text{SIS}(u_a, u_b) = 1 - \bar{S}$

5.2 CF Model Based on SIS Measures

Definition of Recommender Model Based on SIS Measures. The CF model based on SIS measures are defined as follows CFSISR $= <U, I, R, F>$, where $U = \{u_1, u_2, \ldots, u_n\}$ is a set of n users in the system. $I = \{i_1, i_2, \ldots, i_m\}$ is a set of m items in the system. $R = \{r_{j,k}\}$ is the rating matrix, $r_{j,k}$ is the rating value of user u_j for item i_k. $F = U \times I \times R \rightarrow I_{u_a}$ is a function to find out the items that need to recommend for user $u_a \in U$ (with $I_{u_a} = \{i_1, i_2, \ldots, i_N\}$).

Recommender Algorithm Based on SIS Measures. The collaborative filtering recommender algorithm based on SIS measures is presented in Algorithm (4).

6 Experiments

6.1 Experimental Data

In order to evaluate the effectiveness of the proposed recommender models, the empirical part is deployed on two internationally published standard datasets: MSWeb [4] and MovieLens [12]. The MSWeb dataset was created by sampling and processing the www.microsoft.com logs [4]. The dataset was randomly selected from 38000 anonymous users who visited the website for a week in February 1998. In this experiment, the binary rating matrix of the dataset has the

Algorithm 4. Recommender collaborative filtering based on SIS measures.

Input: User set U, item set I, rating matrix R, user recommendation result u_a
Output: Recommendation of items for user $u_a : I_{u_a} = \{i_1, i_2, \ldots, i_N\}$

1: **Step 1:** Identifying k user set similar to user u_a
2: **for** each user $u_i \in U$ **do**
3: Determining similarity value between u_a and u_i by SIS measures; $SIS(u_a, u_i)$
4: Sorting descending user list by similarity values
5: Selecting the first k users with highest similarity values $N(a)$
6: **end for**
7: **Step 2:** Calculating the rating value for items
8: Selecting items that user u_a has not rated yet ($r_{a,k} = \emptyset$)
9: Calculating the rating value for items: $\hat{r}_{a,k} = \frac{1}{\sum_{i \in N(a)} s_{a,i}} \sum_{i \in N(a)} S_{a,i} r_{i,k}$, where $s_{a,i}$ is the similarity value between user u_a and u_i. $r_{i,k}$ is the rating value of user u_i for item i_k.
10: **Step 3:** Selecting recommendation items for user u_a
11: Sorting descending item list by rating values
12: Selecting N items with the highest rating values to recommend to user u_a

structure 32710 × 285. The MovieLens dataset was created from available rating data on the MovieLens web site of GroupLens Research with ratings between 1 and 5 [12]. The dataset was collected over various periods, depending on its size. In this experiment, this dataset was collected from 09/19/1997 to 22/04/1998 and processed into a matrix with structure 943 × 1664. Two datasets are selected for this experiment because they have been used in many studies on CF models. In this study, we use the k-fold cross-validation technique (with k = 5) to process experimental datasets. This technique ensures that each user has at least one occurrence in the testing dataset corresponding to k evaluation models. In each evaluation, the models use one subset as a testing dataset, and the remaining k − 1 subset is a training dataset for training models. The effectiveness of the model is calculated based on the average value of k evaluation times.

6.2 The Tool for Experiments

In this experiment, ARQAT is developed on the R language by our research team [18]. This tool is implemented on ARQAT platform on Java language [13]. In this tool, we integrate the CF models from package tool recommender lab [11] and install additional functions: data processing; generating association rules based on II measures; recommender models are proposed in this study and measures used to evaluate these models.

6.3 The Experimental Results on Integer Rating Dataset

In the empirical part of the MovieLens data scenario, the models are evaluated five times (Cross-validation with k = 5) with the number of known ratings of each user by 6 (given = 6). In particular, the parameters when generating rules of the models: CFIIR model has Min_Supp = 0.01, Min_Conf = 0.3 and Min_Impli = 0.3; CFSISR model has Min_Supp = 0.01, Min_Conf = 0.3 and Min_Impli = 0.3; AR model has Min_Supp = 0.01, Min_Conf = 0.3.

Evaluating by Rating Values of Models on MovieLens. In this section, we conduct experiments to test the accuracy of the models on the integer rating dataset. The accuracy of the models is checked through error indicators: the square root of the mean square error between user rating value and model's predicted rating value (RMSE), the average square error between user rating value and model's predicted rating value (MSE), the average absolute error between user rating value and model's predicted rating value (MAE). Table (1) presents the comparative results of the above three indicators of two proposed models with traditional CF models: AR, UBCF, and IBCF. The results show that the index RMSE, MSE, MAE of our models have lower value than those indicators of the traditional CF models on the MovieLens dataset.

Table 1. Comparing RMSE, MSE, MAE of models on MovieLens dataset.

Models	RMSE	MSE	MAE
CFSISR	1.06481	1.13382	0.84720
CFIIR	1.06721	1.13893	0.84570
AR	1.06729	1.13912	0.85090
UBCF	1.08020	1.16684	0.86489
IBCF	1.07855	1.16328	0.85850

Evaluating by Recommendations of Models on MovieLens. We compare the accuracy between the proposed models and traditional CF models by examining the Precision/Recall (Pre/Rec) ratio of five models. In order to evaluate data accurately, we run models with the number of movies recommend to each user increasing from 1 to 15. Figure (1a) shows that the accuracy of two proposed models is better than the other three CF models. This chart also shows a higher Pre/Rec ratio of the CFSISR model than this ratio of CFIIR model. This implies that the proposed models are more accurate than the traditional CF models on the MovieLens dataset.

6.4 The Experimental Results on Binary Rating Dataset

In the empirical part of the MSWeb data scenario, the models are evaluated five times (cross-validation with k = 5) with the number of known ratings of each user by 4 (given = 4). In particular, the parameters when generating rules of the models: CFIIR model has Min_Supp = 0.01, Min_Conf = 0.1 and Min_Impli = 0.5; CFSISR model has Min_Supp = 0.01, Min_Conf = 0.1 and Min_Impli = 0.5; AR model has Min_Supp = 0.01, Min_Conf = 0.1.

To further compare the accuracy of the proposed models with traditional CF models on binary rating dataset, we built the model evaluation data with the number of websites recommend to each user increasing from 1 to 15 and draw a Pre/Rec ratio graph for five models, e.g., CFSISR, CFIIR, AR, UBCF, and IBCF. Figure (1b) shows that the two proposed models have better results than those of the three traditional CF models. In particular, the two models CFSISR and CFIIR have the Pre/Rec ratio almost similar, further confirming that the proposed models are more accurate than traditional CF models on the MSWeb dataset.

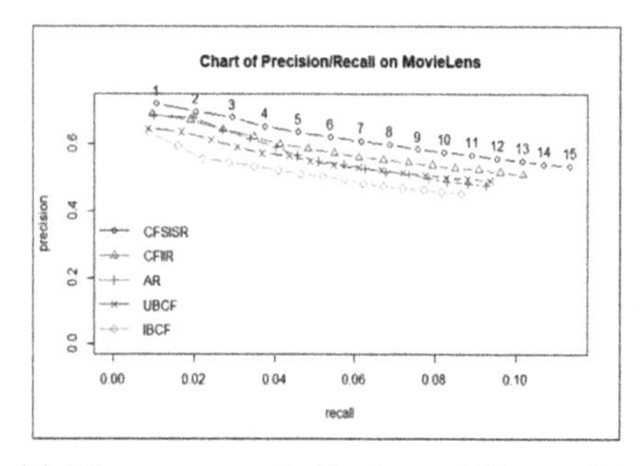

(a) The accuracy indicators of file models on MovieLens dataset.

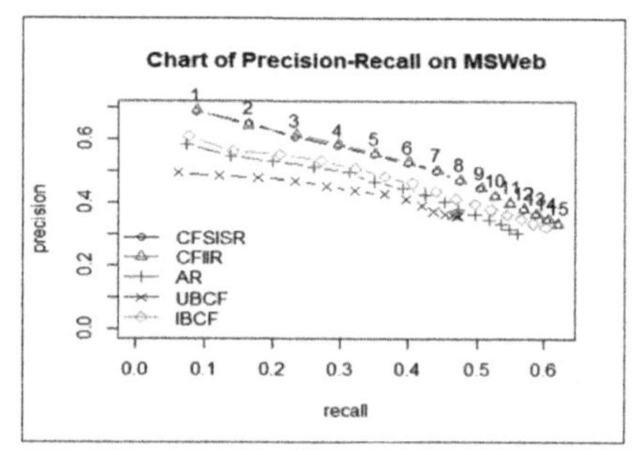

(b) The accuracy indicators of five models on MSWeb dataset.

Fig. 1. Precision-Recall of MovieLens and MSWeb datasets.

6.5 Comparative Results of Models on Datasets

Comparing the Accuracy of the CFSISR Model on Datasets. To evaluate the accuracy of the CFSISR model on two empirical datasets, we used the evaluation method based on recommendation results. In this method, we compare the three indicators: Precision, Recall and Fmeasure of the model on two empirical datasets. Figure (2a) compares the results of the above indicators of the model on two datasets. This result shows that the three indicators on MSWeb dataset have higher values than those on the MovieLens datasets. This shows that the CFSISR model give better results on binary rating dataset.

Comparing the Accuracy of CFIIR Model on Datasets. To compare the accuracy of CFIIR model, we conducted experiments on two datasets and compared the values of the indicators: Precision, Recall and Fmeasure. Figure (2b) presents the values of accuracy measurement of the models on the two empirical datasets. This result shows that the values of three indicators on MSWeb dataset are higher than those on the MovieLens datasets. This shows that the CFIIR model give better results on binary rating dataset.

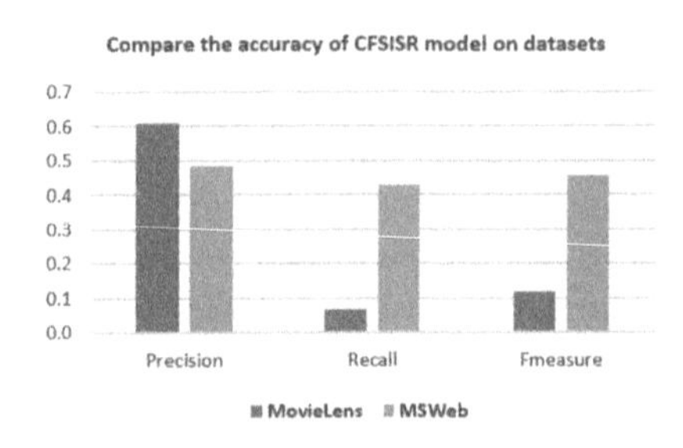

(a) The chart presents the accuracy indicators of the CFSISR model.

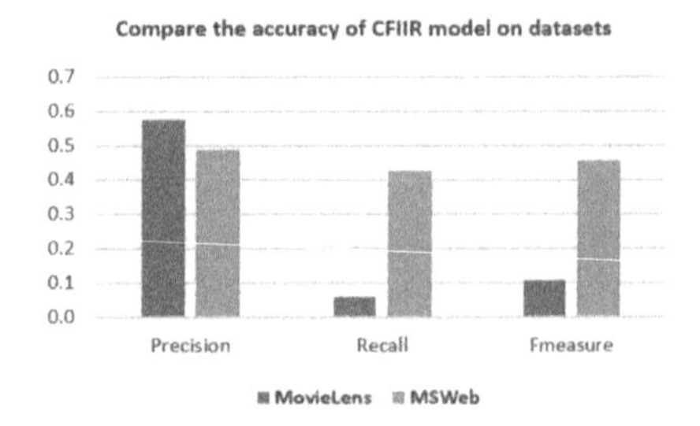

(b) The chart presents the accuracy indicators of the CFIIR model.

Fig. 2. Accuracy indicators of the CFSISR and CFIIR models.

7 Conclusions

In this paper, we propose two new models for CF models based on statistical implication analysis, e.g., CFIIR and CFSISR. The authors have proved that the asymmetric approach of the statistical implicative analysis empowers the association rules, which in turn improve the overall prediction performance of recommender models. These approaches recommend items to users based on a set of association rules built from measures that define the implicative relationship among users. An intensive evaluation and comparison have been conducted on several standard datasets using the ARQAT platform developed by our research team. The high accuracy results confirm the efficiency of the two proposed models.

References

1. Aggarwal, C.C., et al.: Recommender Systems, vol. 1. Springer, Heidelberg (2016). https://doi.org/10.1007/978-3-319-29659-3
2. Agrawal, R., Srikant, R., et al.: Fast algorithms for mining association rules. In: Proceedings of 20th International Conference Very Large Data Bases, VLDB, vol. 1215, pp. 487–499 (1994)
3. Avazpour, I., Pitakrat, T., Grunske, L., Grundy, J.: Recommendation systems in software engineering. Dimensions and Metrics for Evaluating Recommendation Systems, pp. 245–273 (2014)
4. Breese, J.S., Heckerman, D., Kadie, C.M.: Anonymous Web Data, pp. 98052–6399. Microsoft Research, Redmond (1998). www.microsoft.com
5. Ekstrand, M.D., Riedl, J.T., Konstan, J.A., et al.: Collaborative filtering recommender systems. Found. Trends® Hum.-Comput. Interaction **4**(2), 81–173 (2011)
6. Glass, D.H.: Confirmation measures of association rule interestingness. Knowl.-Based Syst. **44**, 65–77 (2013)
7. Gomez-Uribe, C.A., Hunt, N.: The netflix recommender system: algorithms, business value, and innovation. ACM Trans. Manag. Inf. Syst. (TMIS) **6**(4), 1–19 (2015)

8. Gras, R., Kuntz, P.: An overview of the statistical implicative analysis (SIA) development. In: Gras, R., Suzuki, E., Guillet, F., Spagnolo, F. (eds.) Statistical Implicative Analysis, pp. 11–40. Springer, Heidelberg (2008). https://doi.org/10.1007/978-3-540-78983-3_1
9. Gras, R., Kuntz, P., Greffard, N.: Notion de champ implicatif en analysis statistique implicative. In: The 8th International Meeting on Statistical Implicative Analysis, Tunisia, pp. 1–21 (2015)
10. Hahsler, M.: Lab for developing and testing recommender algorithms. Copyright (C) Michael Hahsler (PCA and SVD implementation (C) Saurabh Bathnagar) (2015). http://R-Forge.R-project.org/projects/recommenderlab
11. Hahsler, M.: Recommenderlab: a framework for developing and testing recommendation algorithms. Technical report (2015)
12. Harper, F.M., Konstan, J.A.: The movielens datasets: history and context. ACM Trans. Interact. Intell. Syst. (TIIS) **5**(4), 1–19 (2015)
13. Huynh, X.-H., Guillet, F., Briand, H.: A data analysis approach for evaluating the behavior of interestingness measures. In: Hoffmann, A., Motoda, H., Scheffer, T. (eds.) DS 2005. LNCS (LNAI), vol. 3735, pp. 330–337. Springer, Heidelberg (2005). https://doi.org/10.1007/11563983_28
14. Jooa, J., Bangb, S., Parka, G.: Implementation of a recommendation system using association rules and collaborative filtering. Procedia Comput. Sci. **91**, 944–952 (2016)
15. Li, X., Li, D.: An improved collaborative filtering recommendation algorithm and recommendation strategy. Mob. Inf. Syst. **2019** (2019)
16. Linden, G., Smith, B., York, J.: Amazon.com recommendations: item-to-item collaborative filtering. IEEE Internet Comput. **7**(1), 76–80 (2003)
17. Osadchiy, T., Poliakov, I., Olivier, P., Rowland, M., Foster, E.: Recommender system based on pairwise association rules. Expert Syst. Appl. **115**, 535–542 (2019)
18. Phan, L.P., Phan, N.Q., Nguyen, K.M., Huynh, H.H., Huynh, H.X., Guillet, F.: Interestingnesslab: a framework for developing and using objective interestingness measures. In: Akagi, M., Nguyen, T.-T., Vu, D.-T., Phung, T.-N., Huynh, V.-N. (eds.) ICTA 2016. AISC, vol. 538, pp. 302–311. Springer, Cham (2017). https://doi.org/10.1007/978-3-319-49073-1_33
19. Phan, N.Q., Dang, P.H., Huynh, H.X.: Statistical implicative similarity measures for user-based collaborative filtering recommender system. Int. J. Adv. Comput. Sci. Appl. **7**(11) (2016). https://doi.org/10.14569/IJACSA.2016.071118
20. Phan, N.Q., Dang, P.H., Huynh, H.X.: Collaborative recommendation based on statistical implication rules. J. Comput. Sci. Cybern. **33**(3), 247–262 (2017)
21. Spagnolo, F., Gras, R., Guillet, F., Suzuki, E.: Statistical Implicative Analysis, Theory and Applications. Springer, Heidelberg (2008). https://doi.org/10.1007/978-3-540-78983-3
22. Su, X., Khoshgoftaar, T.M.: A survey of collaborative filtering techniques. Ad. Artif. Intell. **2009** (2009)
23. Varzaneh, H.H., Neysiani, B.S., Ziafat, H., Soltani, N.: Recommendation systems based on association rule mining for a target object by evolutionary algorithms. Emerg. Sci. J. **2**(2), 100–107 (2018)
24. Xu, G., Tang, Z., Ma, C., Liu, Y., Daneshmand, M.: A collaborative filtering recommendation algorithm based on user confidence and time context. J. Electr. Comput. Eng. **2019** (2019)

Computer Vision Techniques

Object Searching on Video Using ORB Descriptor and Support Vector Machine

Faisal Dharma Adhinata[1], Agus Harjoko[2], and Wahyono[2](✉)

[1] Master Program of Computer Science, Universitas Gadjah Mada, Yogyakarta, Indonesia
faisaldharma@mail.ugm.ac.id

[2] Department of Computer Science and Electronics, Universitas Gadjah Mada, Yogyakarta, Indonesia
{aharjoko,wahyo}@ugm.ac.id

Abstract. One of the main stages in object searching on video is extracting object regions from video. Template matching is popular technique for performing a such task. However, the use of template matching has a limitation that requires a large object as a template. If the template size is too small, it would obtain few features. On the other hand, ORB descriptors are often used for representing the object with a good accuracy and fast processing time. Therefore, this research proposed to use machine learning method combining with ORB descriptor for object searching on video data. Processing video in all frames is inefficient. Thus, frames are selected into keyframes using mutual information entropy. The ORB descriptors are then extracted from selected frame in order to find candidate region of objects. To verify and classify the object regions, multiclass support vector machine was used to train ORB descriptor of regions. For evaluation, the use of ORB would be compared with other descriptor, such as SIFT and SURF for showing its superiority in both accuracy and processing time. In experiment, it is found that object searching with ORB descriptor performs faster processing time, which is 0.219 s, while SIFT 1.011 s and SURF 0.503 s. Meanwhile, it also achieves the best F_1 value, which is 0.9 compared to SIFT 0.63 and SURF 0.65.

Keywords: Object searching · Machine learning · ORB · SVM multiclass

1 Introduction

In recent years, an intelligent surveillance system has been developed to improve security in various public facilities, such as offices, universities, department stores, etc. Criminal theft requires forensic analysis as evidence in court. One of possible evidence is recorded video from CCTV camera which can be used to look for clues to the stolen item. Searching on evidence by looking at all of the entire video frames from beginning to end is inefficient because it requires a lot of time. Besides, the number of recorded videos is too large for manual analysis by human. Therefore, an automatic object searching system is needed for assisting this analysis.

M. Hernes et al. (Eds.): ICCCI 2020, CCIS 1287, pp. 239–251, 2020.
https://doi.org/10.1007/978-3-030-63119-2_20

The accuracy and speed of processing on video data are affected by pre-processing and feature extraction of a frame on video. The pre-processing stage is crucial in processing on video data because there are often frames that contain the same information. Thus, frames with same information should be only processed once. The utilization of keyframe selection could speed up processing on video data. Several keyframe selection methods have been studied by several researchers, including Ouyang et al. [1]. The results showed that the mutual information entropy method was the best among other methods. Then this method also produces a keyframe that matches the core of video content, and it is fast enough in its selection [2].

The feature extraction stage on video frame often uses the SIFT [3], SURF [4], and ORB [5] algorithms. The use of feature extraction algorithm usually uses a template matching technique for processing on video data [6, 7]. The use of a template matching technique with the ORB algorithm has several drawbacks, i.e. it cannot use a small template. This is because the features obtained from the small template would be few. Then, the accuracy is not very good compared to the other feature extraction methods. So, it is required use another technique to solve these problems.

The machine learning technique is often implemented for processing on image data. Research by Murugeswari and Veluchamy [8], studied the HMM, ANN, and SVM multiclass methods for classifying on hand movement in real-time video. The result is the SVM method resulting in the best accuracy. Li and Wang's research [9] also uses SVM multiclass to classify images on three types of vehicles, namely cars, trains, or planes. The classification result is 90% of accuracy. Therefore, this paper proposed a machine learning technique with the ORB algorithm as feature extraction on frame and SVM multiclass for object searching model.

This paper is organized as follows. Section 2 describes a brief explanation related to previous research in the field of object searching on video. Section 3 explains the proposed method, while Sect. 4 contains a discussion of the results and the evaluation of this system. In the end, Sect. 5 will conclude our work.

2 Related Works

The use of CCTV cameras for monitoring criminal acts has been widely implemented in public places. Recently, intelligent surveillance research has been developed through CCTV video. In general, the main stages in the processing of video data are pre-processing and feature extraction. Processing all video frames is inefficient in real-time video processing because many frames contain the same information. Therefore, the frame should be selected for avoiding multiple time process in same information, called keyframe. Research conducted by Ouyang et al. [1] made a comparison of keyframe extraction methods. The result is that the mutual information entropy method gives the best results for keyframe extraction, with 89.7% of recall and 69.9% of precision. Mutual information shows two frames that have different information, so it requires a comparison of one frame with another frame. The next stage is extracting features from the selected keyframes. Feature extraction using SIFT and SURF algorithms can search for household appliances in a video [10]. The results showed that the SIFT algorithm had an accuracy of 82% while the accuracy of the SURF algorithm was 18% in searching objects on video. Then, the time processing that needed for matching objects through a video frame, the SIFT algorithm takes 30 s while the SURF algorithm takes 9 s. The SIFT algorithm is resulting in better recall results than the SURF algorithm, but the processing time is inversely proportional to its accuracy.

The process of searching objects or detecting objects requires a long time to monitor CCTV cameras one by one manually. Therefore, an algorithm is needed to process features on a keyframe quickly and accurately. The results of the literature review from the research by Yu and Kong [11] and Toapanta et al. [12] obtained a fast and accurate algorithm in feature extraction, namely the ORB algorithm. The use of ORB is also used in an experiment using frames that have around 1000 features [5]. ORB spent processing time of 15.3 ms on each frame, whereas for SURF 217.3 ms and SIFT 5228.7 ms.

Two techniques are often used for searching objects in video data, namely, template matching and machine learning. The template matching technique matches the features generated by the query image directly with the keyframe. Some researchers use template matching with the ORB algorithm to process video data. This technique is used in some cases, which is background clutter, partial occlusion, and deformed objects [6]. There are drawbacks in using the ORB algorithm for template matching, namely the object template must be large. The use of ORB algorithm for template matching is also less optimal than the SIFT algorithm because of the few features that obtained [7]. In the other hand, machine learning does the modelling from query images that are used to match keyframes. Research by Murugeswari and Veluchamy [8] examined the HMM, ANN, and SVM multiclass methods for classifying hand movement in real-time video. The result is the best on the SVM method.

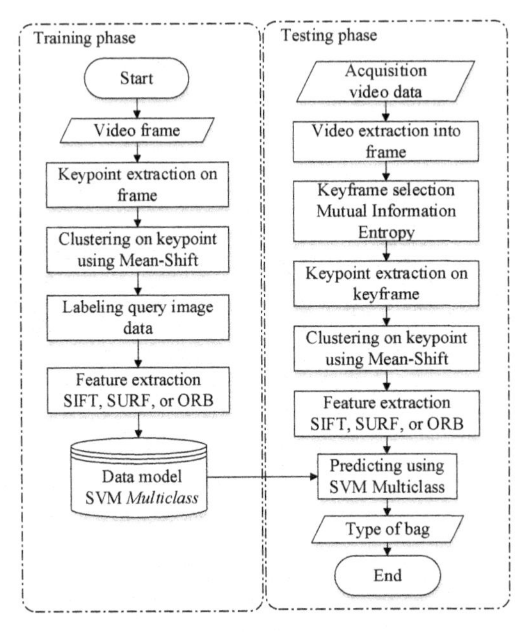

Fig. 1. The step of machine learning technique

3 The Proposed Method

The machine learning technique is divided into two stages: the training and testing phases. The training phase starts with the input frame on video, which was done by keypoint extraction based on ORB algorithm. These keypoints are then clustered using the Mean-Shift method. The clustering results produces several regions which are represented as bounding boxes, and they are stored and labelled according to object and non-object classes. For each obtained region, the ORB descriptors are extracted one more time, which will be used for the training process into a model using the one-vs-all on SVM multiclass method. The testing phase starts with a new video input which is then extracted into frames. These frames are selected into keyframes using the mutual information entropy method. Keypoint detection using the ORB algorithm is applied on the obtained keyframes. The keypoint obtained was also clustered to form the bounding box as same as the training stage. The result of the bounding box is done by feature extraction using the ORB algorithm to get the keypoint and descriptors. Descriptors that generated from each bounding box are used to predict bag motifs using SVM model. Figure 1 shows the illustration how our proposed method work for searching the object on the video.

3.1 Keyframe Selection

The change of scene video is detected by significant a decrease in the value of mutual information. The entropy value shows the randomness of gray-level distribution of an image. The more random on gray level distribution produce a higher entropy value. Image entropy is calculated based on image histogram information. The process of keyframe selection begins with that the first frame becomes a keyframe. Then, the second frame is calculated mutual information entropy value with the previous keyframe. The keyframe selection process is illustrated in Fig. 2. The mutual information entropy equation is shown in Eq. (1) [13].

Fig. 2. Illustration keyframe selection

$$I(X;Y) = \sum_{y \in Y} \sum_{x \in X} p(x,y) \log\left(\frac{p(x,y)}{p(x)p(y)}\right) \quad (1)$$

where $I(X; Y)$ is defined as Mutual information (number of information) need in a mean of frame x and y, while $p(x, y)$ is defined as a probability gray level frame x and y

If the value of mutual information entropy is getting closer to 0, so the two frames are mutually independent or become the next keyframe. Conversely, the higher mutual information entropy value means that the two frames have almost the same information [2].

3.2 Oriented FAST and Rotated BRIEF (ORB)

ORB is a combination of FAST keypoint detector and BRIEF descriptor [14]. Feature extraction using the ORB algorithm begins with transformation on the scale pyramid image. A FAST algorithm finds the keypoint that begin with a random point on the first check, whether the point is an angle by making a circle with 16 pixels and labelled using numbers 1 to 16 clockwise. After finding the keypoints, determining the best N points using Harris corner. After obtaining the keypoint, an orientation searching is performed using centroid intensity as shown in Eq. (2) [15].

$$C(\overline{x}, \overline{y}) = \left(\frac{m_{10}}{m_{00}}, \frac{m_{01}}{m_{00}}\right)$$

$$\theta = atan2(m_{01}, m_{10}) \tag{2}$$

where $C(\overline{x}, \overline{y})$ is centroid of object, m_{00} is moment level 0 (area of object), and m_{10}, m_{01} are moment level 1. The next stage after obtaining keypoints is extracting binary descriptors using the Binary Robust Independent Elementary Features (BRIEF) algorithm. BRIEF compares the values of all sampling pairs (the first pixel with the second pixel). If the first pixel is brighter than the second pixel, it has a value of 1; otherwise, it will be 0. This process is done with Eq. (3).

$$\tau(p; x, y) := f(x) = \begin{cases} 1, & p(x) < p(y) \\ 0, & p(x) \geq p(y) \end{cases} \tag{3}$$

where $p(x)$ is intensity value at pixel x and $p(y)$ is intensity value at pixel y. This process will be repeated until 256 pairs. Then, 256 bits are divided by 1 byte that was resulting in 32 dimensions of a binary descriptor. The result of ORB keypoints appears to be less than SIFT and SURF algorithms. Thus, it is expected to perform faster processing time in object searching. Figure 3 shows the results of keypoints on the SIFT, SURF, and ORB algorithms.

3.3 Features Clustering on Keyframe

Feature extraction using ORB algorithms on keyframes produces multiple keypoints and descriptors. Keypoints and descriptors on keyframes are clustered which aim to detect multiple objects on keyframes. The clustering algorithm used in this study is Mean-Shift. Mean-Shift algorithm is based on centroid, which always updates the candidate centroid by calculating the mean at all points according to the window area [16].

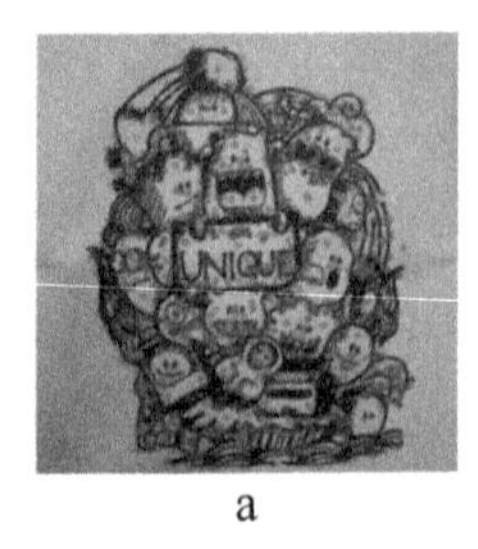

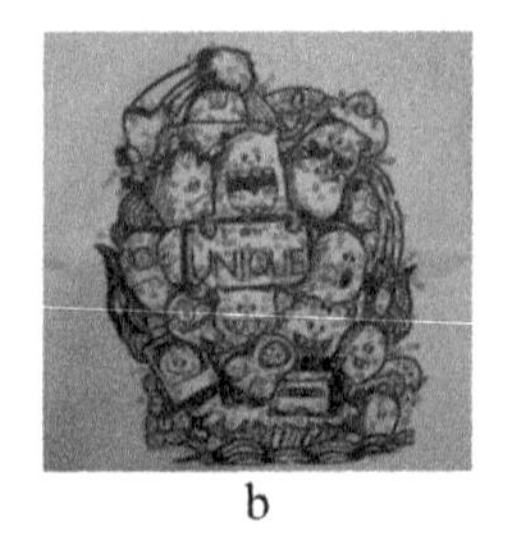

Fig. 3. The visualization (red point) of a) SIFT, b) SURF, and c) ORB keypoints (Color figure online)

The mean-shift algorithm begins with storing the coordinates of keypoint from the feature extraction of the ORB algorithm. All keypoints are cluster center, as shown in Fig. 4a. Furthermore, the window size (kernel bandwidth) is automatically determined by estimating the bandwidth. Then, the initial location of bandwidth estimation is determining that results window. According to the name of the algorithm, this algorithm calculates the mean of the cluster center on all points in the window. Initially, the cluster center is at the keypoint, as shown in Fig. 4b. Then, this algorithm makes a shift in a density area by updating the mean value of the cluster center, as shown in Fig. 4c. The algorithm stops when the cluster center position has not changed, as shown in Fig. 4d.

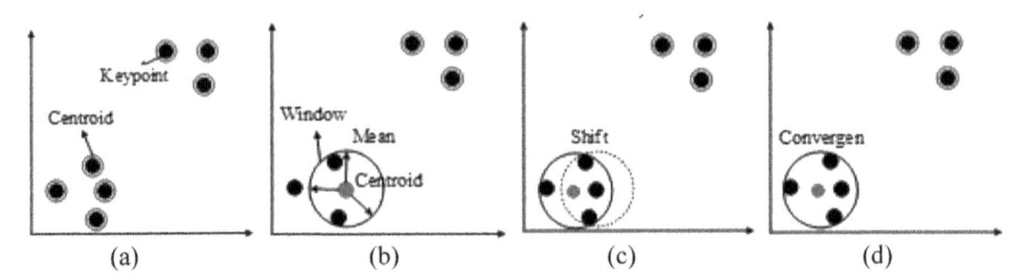

Fig. 4. The step of Mean-Shift algorithm

3.4 SVM Multiclass

The training process of making a model is using SVM (Support Vector Machine) multiclass one-vs-all. There are five classes used in this system, namely the class of textured bag (Batik), lettering patterned bag, black and white patterned bag, color pattern bag, and non-objects. This study uses several bag patterns because SIFT, SURF, and ORB algorithms are feature-based algorithms. In general, the pattern that contained in the bag are texture, lettering pattern, black and white pattern, and color pattern. Figure 5 represents the implementation of SVM multiclass.

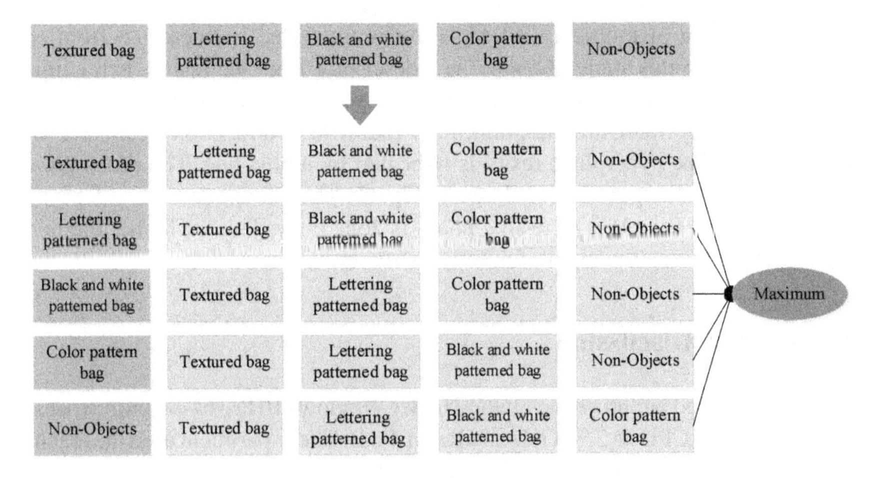

Fig. 5. The implementation of multiclass SVM

The SVM concept is simply an attempt to find the best hyperplane that functions as a separator of two classes in the input space. The best hyperplane separator between the two classes can be found by measuring the margin of the hyperplane and finding its maximum point. SVM multiclass in this study uses Eq. (4).

$$\begin{bmatrix} W_{11} & W_{12} & W_{13} & .. & W_{1D} \\ W_{21} & W_{22} & W_{23} & .. & W_{2D} \\ .. & .. & .. & .. & .. \\ W_{K1} & W_{K2} & W_{K3} & .. & W_{KD} \end{bmatrix} \begin{bmatrix} X_1 \\ X_2 \\ X_3 \\ .. \\ X_D \end{bmatrix} = \begin{bmatrix} s_1 \\ s_2 \\ s_3 \\ .. \\ s_K \end{bmatrix} \quad (4)$$

Where:
W_{KD} : Descriptor vector of the generated model
X_D : Descriptor vector of testing image
s_K : Class prediction results

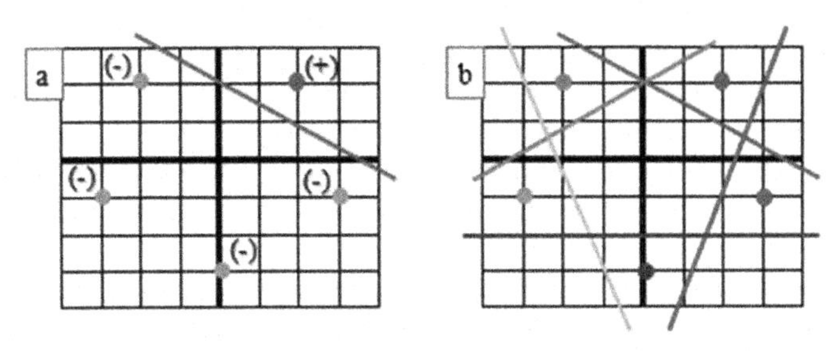

Fig. 6. The illustration of Support Vector Machine (left) Multiclass classification, (right) Result of training process.

In the prediction of multiclass SVM, a maximum value is calculated from each class comparison. In this study, the input is a descriptor from the feature extraction of SIFT, SURF, or ORB algorithm. For example, in Fig. 6a, a textured bag symbolized by a 2-dimensional descriptor in blue point, then fourth in yellow point is another class of patterned bag. The hyperplane result is indicated by a blue line. The illustration of hyperplane on the training process is shown in Fig. 6b. The training process will produce the value of W_{KD} in Eq. (4). The testing phase is done by finding the higher predictive value of s_K that entering the input of descriptor vector of X_D.

4 Results and Discussion

We have implemented our proposed method under Windows 10 Pro 64-bit with processor Intel Core i3-6100U CPU @ 2.30 GHz and RAM 4 GB. The experiment used the machine learning technique to see the speed and F_1 value of the ORB algorithm compared to SIFT and SURF algorithms.

4.1 Data Acquisition and Pre-processing

In this research, we collected the dataset using CCTV of 2.0 MP with 10 fps. The machine learning technique uses 100 query images on each motif and 150 on non-objects, so there are five classes. The amount of training data on non-objects is 150 because the amount of non-object data recorded by the camera at one frame is greater than the object. This experiment is using query image data with a resolution of 150 × 150 both on bag objects and non-objects.

Processing all frames causes delay in real-time processing, so frames are selected as keyframes. Keyframe selection is applied by comparison between frames only takes 6 ms. The mutual information entropy method is able to compensate for the frames that produced by CCTV camera every second. The use of keyframe selection is very helpful in real-time processing.

4.2 The Experiment of SVM Model Training

First, the proposed method was evaluated using data training for obtaining the best method for object searching. As shown Table 1, the ORB algorithm performs the fastest processing time in training dana, which is 9.93 s. The value of F_1 in the ORB algorithm is also the highest one compared to SIFT and SURF algorithms, which is 0.97. These results indicate that the fastest and most accurate is the ORB algorithm at the training stage. Thus, ORB is suitable to be used for real-time object searching which later will be evaluated using several scenarios.

In addition, it can be analyzed that the training process of the SIFT and SURF are slower than ORB. It is because both descriptors produce more dimension features such as 128 and 64 dimensions respectively, comparing to the ORB which only produces 32 dimensions. Furthermore, ORB utilized Hamming distance for matching one object to another object which is more faster than utilizing Euclidean distance.

Table 1. The results of query image modelling on data training.

Method	F-Measure	Processing time (s)
SIFT	0.95	47.33
SURF	0.93	17.40
ORB	0.97	9.93

4.3 Evaluation on Single Object

After obtaining the best model in the previous subsection, this model is then evaluated on Single Object Scenario. Figure 7 shows several experiments of data on four new video motifs. The experiment is carried out using 50 s of a new video with ten fps that were resulting in 500 frames on each motif.

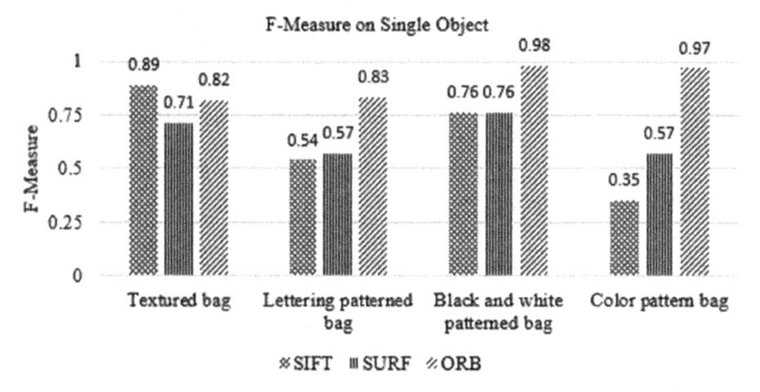

Fig. 7. The result of on several types of the bag toward F-Measure value

As shown in Fig. 7, the ORB algorithm achieves better F_1 value comparing to the SIFT algorithm except in texture motifs. For the textured motif, the F_1 value of the ORB algorithm is also not too different than the SIFT algorithm. Then, at the experiment of time processing on each frame, as shown in Fig. 8, the average processing time of the ORB algorithm is the fastest compared to SIFT and SURF algorithms. An example of single object results is shown in Fig. 11(a).

4.4 Evaluation on Multiple Objects

The object that recorded by CCTV camera is not only a single object. The Mean-Shift clustering method is used to detect multiple objects on keyframes. Without clustering, it can only detect a single object. The experiment is done using 50 s of a new video with ten fps, resulting in 500 frames on each motif. There are three people carrying bags and two people not carrying bags. Figure 9 shows that the ORB algorithm gains the best F_1 value comparing to SIFT and SURF algorithms. An example of multiple object results is shown in Fig. 11b.

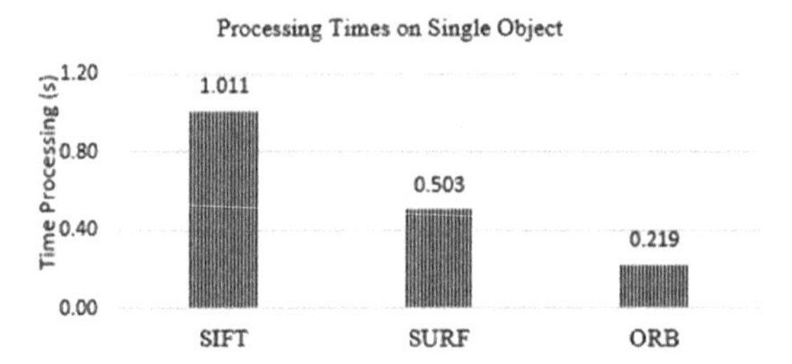

Fig. 8. The result of keyframe matching with respect to processing time

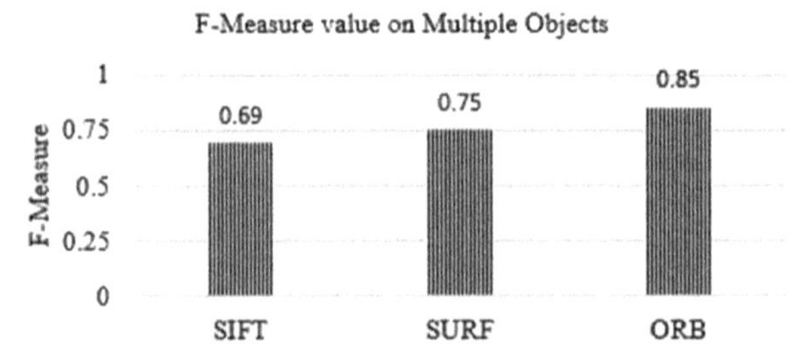

Fig. 9. The result of multiple object data experiment toward F-Measure value

4.5 Evaluation on Small Object

For evaluating small objects, the experiment was conducted using a single object data that recorded at a distance of 3 m. This experiment uses four bag motifs to see the resulting of F_1 value. Figure 10 shows that the ORB algorithm is almost the most superior in experiments except in the texture motifs. However, the F_1 value in textured motif of the ORB algorithm is not too different from the SIFT algorithm. In this experiment, the SURF algorithm is resulting in the worst results compared to other algorithms. An example of small object results is shown in Fig. 11c.

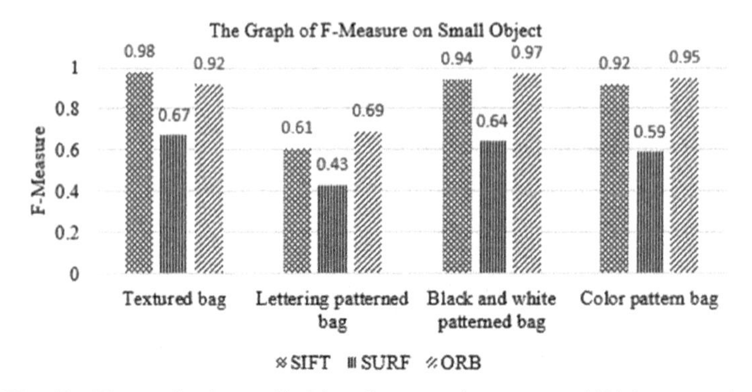

Fig. 10. The result of a small object data experiment toward F-Measure value

Fig. 11. The sample results in (left) single, (middle) multiple objects, (right) small objects.

4.6 Discussion

Figures 7 and 8 show that the ORB algorithm is the greatest of the SIFT and SURF algorithms. In template matching, the size of an object influences the result of matching because the size of an object that is too small produces fewer features [6]. Based on Fig. 10, the ORB algorithm can detect small bags well through machine learning technique. The use of template matching technique [7] also results in the SIFT better than ORB because the feature detection using SIFT is higher than ORB. However, in machine learning technique the ORB algorithm is the greatest in both speed and F-Measure values.

Feature extraction using SIFT, SURF, or ORB algorithms is a disadvantage when used on slightly or plain of featured object. This algorithm works to find feature interest on object. The feature interest can come from the angle of pattern, the color difference in pattern, or the change in color intensity of neighboring pixels. A plain patterned object without a pattern causes no features interest in the object.

5 Conclusion

This paper proposed a method for searching the object on the video. In our proposed method, the ORB algorithm was utilized as feature extraction, while the SVM method was utilized as a model of bag type classification. The mutual information was also applied for extraction keyframe from the video, which is expected to perform faster processing time. In the experiment, the use of the ORB algorithm was the most accurate and the fastest compared to the SIFT and SURF algorithms. Even the ORB algorithm can detect small objects in video data. ORB algorithm performed the fastest processing time, which is 0.219 s compared to the SIFT 1.011 s, SURF 0.503 s, and ORB is a superior in F_1 value of 0.9 compared to SIFT 0.63 and SURF 0.65. However, our method may not work well when the feature in texture-less object. Therefore, the next research can add a dataset for plain object, and researcher can modify the feature extraction method that can use in texture-less object.

Acknowledgment. This work was supported by the 2020 Final Assignment Recognition Program (RTA) of Universitas Gadjah Mada (No. 2488/UN1.P.III/DIT-LIT/PT/2020).

References

1. Ouyang, S.Z., Zhong, L., Luo, R.Q.: The comparison and analysis of extracting video key frame. IOP Conf. Ser. Mater. Sci. Eng. **359**(1), 012010 (2018)
2. Sun, L., Zhou, Y.: A key frame extraction method based on mutual information and image entropy. In: 2011 International Conference on Multimedia Technology, vol. 1, no. 6077, pp. 35–38 (2011)
3. Lowe, D.G.: Distinctive image features from scale-invariant keypoints. Int. J. Comput. Vis. **60**(2), 91–110 (2004)
4. Bay, H., Ess, A., Tuytelaars, T., Vangool, L.: Speeded-up robust features (SURF). Comput. Vis. Image Underst. **110**(3), 346–359 (2006)
5. Rublee, E., Rabaud, V., Konolige, K., Bradski, G.: ORB: an efficient alternative to SIFT or SURF. In: Proceedings of IEEE International Conference on Computer Vision, pp. 2564–2571 (2011)
6. Zhou, D., Tian, Y., Li, X., Wu, J.: ORB-based template matching through convolutional features map. In: Proceedings of 2019 Chinese Automation Congress, pp. 4695–4699 (2019)
7. Purohit, M., Yadav, A.R.: Comparison of feature extraction techniques to recognize traffic rule violations using low processing embedded system. In: 2018 5th International Conference on Signal Process. Integration Networks, pp. 154–158 (2018)
8. Murugeswari, M., Veluchamy, S.: Hand gesture recognition system for real-time application. In: 2014 IEEE International Conference on Advanced Communications, Control and Computing Technologies ICACCCT 2014, no. 978, pp. 1220–1225 (2014)
9. Li, Q., Wang, X.: Image classification based on SIFT and SVM. In: Proceedings of 17th IEEE/ACIS Conference on Computer and Information Science, vol. 1, no. 1, pp. 762–765 (2018)
10. Jabnoun, H., Benzarti, F., Amiri, H.: Object recognition for blind people based on features extraction. In: International Image Processing, Applications and Systems Conference, pp. 1–6 (2014)
11. Yu, H., Kong, L.: An optimization of video sequence stitching method. In: 2018 IEEE Conference on Mechatronics and Automation, pp. 387–391 (2018)
12. Toapanta, S.M.T., Cruz, A.A. C., Gallegos, L.E.M., Trejo, J.A.O.: Algorithms for efficient biometric systems to mitigate the integrity of a distributed database. In: CITS 2018 - 2018 International Conference on Computer, Information and Telecommunication Systems, pp. 1–5 (2018)

13. Sainui, J., Sugiyama, M.: Minimum dependency key frames selection via quadratic mutual information. In: 10th International Conference on Digital Information Management, pp. 148–153 (2015)
14. Calonder, M., Lepetit, V., Strecha, C., Fua, P.: BRIEF: binary robust independent elementary features. In: Daniilidis, K., Maragos, P., Paragios, N. (eds.) ECCV 2010. LNCS, vol. 6314, pp. 778–792. Springer, Heidelberg (2010). https://doi.org/10.1007/978-3-642-15561-1_56
15. Rao, T., Ikenaga, T.: Quadrant segmentation and ring-like searching based FPGA implementation of ORB matching system for full HD video. In: Proceedings of 15th IAPR International Conference on Machine Vision Applications MVA 2017, pp. 89–92 (2017)
16. Comaniciu, D., Meer, P.: Mean shift: a robust approach toward feature space analysis. IEEE Trans. Pattern Anal. Mach. Intell. **24**(5), 603–619 (2002)

An Improved of Joint Reversible Data Hiding Methods in Encrypted Remote Sensing Satellite Images

Ali Syahputra Nasution(✉) and Gunawan Wibisono

Department of Electrical Engineering, Faculty of Engineering, The University of Indonesia, Kampus Baru UI, 16424 Depok, Jakarta, Indonesia
ali.syahputra@ui.ac.id

Abstract. Data protection security is very necessary when distributing high resolution remote sensing satellite images from LAPAN to users via electronic media. Reversible data hiding and encryption are two very useful methods for protecting privacy and data security. This paper proposes an increase in the method of joint reversible data hiding on remote sensing satellite images based on the algorithm of Zhang's work, Hong et al.'s work, and Fatema et al.'s work. To evaluate the smoothness of the blocks, a modification of the fluctuation calculation function is presented. The experimental results show that the modified calculation function gives better estimation results. Then, the proposed method gives a lower extracted bit error rate and the visual quality of the image from the proposed method is better than the three references. For example, when the block size is 8×8, the extracted-bit error rate (EER) of the SPOT-6 test image of the proposed modified function was 8.40%, which is quite lower than the 14.14% EER of Zhang's function, 9.62% EER of Hong et al.'s function and 11.87% EER of Fatema's et al.'s method. Likewise, the quality of SPOT-6 image recovery represented by the peak signal-to-noise ratio (PSNR) of proposed modified function is 50.52 dB, which is slightly higher than the 48.23 dB PSNR of Zhang's function, 49.93 dB PSNR of Hong et al.'s function and 49.00 dB PSNR of Fatema's et al.'s function.

Keywords: Encrypted remote sensing satellite images · Joint reversible data hiding · Fluctuation function · EER · PSNR

1 Introduction

Data hiding and encryption are a combination of two approaches to privacy protection and data security that are popular this time. Encryption techniques convert plaintext content into unreadable chipper text. Data hiding techniques embed secret messages or bits of information into cover media such as images, images, audio or video by making a few modifications. Nowadays, joint reversible data hiding in encrypted images (RDHEI) is desired property where the cover image is encrypted by the content owner before forwarding it to the data hider for embedding data. Then, the receiver side can extract hidden additional information and recover the original cover image without loss

M. Hernes et al. (Eds.): ICCCI 2020, CCIS 1287, pp. 252–263, 2020.
https://doi.org/10.1007/978-3-030-63119-2_21

or distortion. In the joint RDHEI method, as shown in Fig. 1, embedded data can only be extracted after image decryption. In other words, additional data must be extracted from the plaintext domain, so that the main content is disclosed after data extraction.

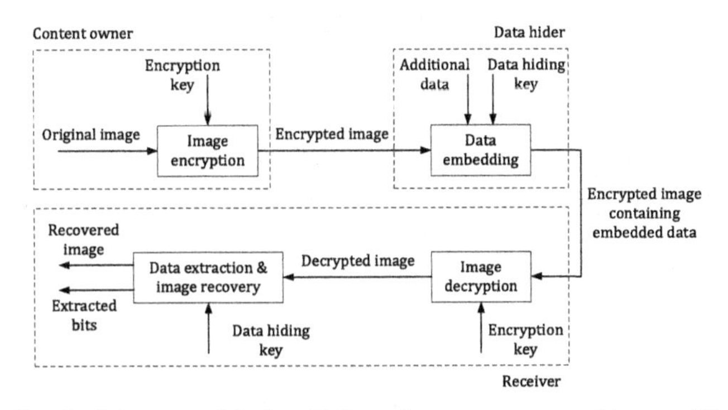

Fig. 1. Joint reversible data hiding scheme on encrypted images [3].

Some of the high-resolution remote sensing images such as SPOT-6, SPOT-7 and Pleiades are commercial and are limited by licensing in terms of data usage, where remote sensing satellite imageries are widely used by users/stakeholders to obtain information about natural resources, disasters, spatial planning. Therefore, the application of encryption and joint reversible data hiding techniques in high-resolution remote sensing satellite images is very useful for preserving privacy and data security when distributed over the internet network (electronic media). This activity also supports the role of LAPAN in Act Number 21 of 2013 concerning Space [1] and in Government Regulation Number 11 of 2018 regarding Procedures for Organizing Remote Sensing Activities [2] where LAPAN is required to collect, store, process, and distribute data through the National Remote Sensing Data Bank (BDPJN) as a remote sensing data network node in the spatial data network system national.

In recent years, there have been presented several methods of RDHEI, and can be classified into two categories. The first is to find space for confidential data after encrypting the image, referred to as the Vacating Room After Encryption (VRAE) [3–15]. The second is to reserve the amount of space needed for confidential data in a reversible manner before encrypting images. The additional data is hidden into the reserved space after encryption. This is referred to as 'Reserving Room Before Encryption' (RRBE) [16, 17].

Currently, many researchers have presented several studies related to joint RDHEI. In 2011, Zhang proposed joint RDHEI by dividing the encrypted image into several blocks and extracting the data as well as reproducing the image based on the smoothness of the image block [3]. In Zhang's method, the four pixel borders of each image block were avoided when calculating fluctuations in each block. Furthermore, in 2012, Hong et al. improve Zhang's algorithm by using a new method for calculating the smoothness of image blocks and exploiting side match techniques [4]. Hong et al. take border pixels from each block into the calculation, but only two adjacent pixels are used in evaluating the smoothness of each block. Fatema et al. also improved fluctuation function based on the Zhang's method by considering the actual value of four adjacent pixels in the calculation of fluctuations to reduce the extracted-bit error rate [10]. Therefore, this paper proposes a modified, more precise function to estimate the complexity of each image block by using two adjacent pixels to reduce extracted-bit error rate and increase peak signal-to-noise ratio of the recovered remote sensing satellite images.

This paper is organized as follows. Section 2, discusses several related studies such as the methods of Zhang [3], Hong et al. [4] and Fatema et al. [10] clearly. Section 3 explains the detailed procedure of the proposed method. Then experimental results and performance comparison with the methods of Zhang, Hong et al., Fatema et al. and proposed modified function is presented in Sect. 4. Conclusions are given in Sect. 5.

2 Related Works

In 2011, joint reversible data hiding in encrypted images was introduced by Zhang by proposing a new algorithm by dividing the image into blocks and using the LSB plane. Then, Hong et al. and Fatema et al. presented improvements in data extraction and image recovery.

In the Zhang method [3], the owner encrypts the image with an exclusive-or bitwise operation. Then the data hider will divide the image into several blocks of size s. After segmenting each block in two parts, he adds additional bits into each block by adopting 3 LSB planes. After sending the encrypted image containing additional data, the receiver will first decrypt it and divide decrypted image containing additional data into blocks of the same size s, then each block will be separated into two sets of the same size and extraction of data/image recovery will be carried out according to fluctuations in Eq. (1) from each block. The Zhang's fluctuation functions are as follows:

$$f_Z = \sum_{u=2}^{s-1}\sum_{v=2}^{s-1}\left|p_{u,v} - \frac{p_{u-1,v} + p_{u,v-1} + p_{u+1,v} + p_{u,v+1}}{4}\right| \tag{1}$$

Where $p_{u,v}$ is the pixel value located at (u, v).

In the Hong et al.'s method [4], they increase data extraction/image recovery based on the Zhang's method. First, they proposed a new function such as Eq. (2) where $p_{u,v}$ is the pixel value located at the position (u, v) of each image block with size s1 x s2 to estimate the smoothness of the block. They consider more pixels so that the extracted-bit error rate is reduced.

$$f_H = \sum_{u=1}^{s_2}\sum_{v=1}^{s_1-1}\left|p_{u,v} - p_{u,v+1}\right| + \sum_{u=1}^{s_2-1}\sum_{v=1}^{s_1}\left|p_{u,v} - p_{u+1,v}\right| \tag{2}$$

In the Fatema et al.'s method [10], they also improved data extraction/image recovery based on the Zhang's method by proposing new functions such as Eq. (3) where $p_{u,v}$ is the pixel value located at the position (u, v) of each image block with size s. They consider the actual value of four neighbouring pixels in the calculation of fluctuations to reduce the extracted-bit error rate.

$$f_F = \sum_{u=2}^{s-1}\sum_{v=2}^{s-1}\left|p_{u,v} - p_{u-1,v}\right| + \left|p_{u,v} - p_{u+1,v}\right| + \left|p_{u,v} - p_{u,v-1}\right| + \left|p_{u,v} - p_{u,v+1}\right| \tag{3}$$

3 Proposed Works

3.1 Modification of Fluctuation Function

The fluctuation function is very important when viewed from previous works. The accuracy of the data extraction is determined by the accuracy of the fluctuation function. In this study, the modification of the fluctuation function is used to calculate the fluctuation value, as shown in Eq. (4). This modified fluctuation function is similar to Hong, but especially when the block size is small, it has more accuracy than the Hong fluctuation function according to empirical results.

$$f_P = \sum_{u=1}^{s}\sum_{v=2}^{s-1}\left|2 * p_{u,v} - (p_{u,v-1} + p_{u,v+1})\right| + \sum_{u=2}^{s-1}\sum_{v=1}^{s}\left|2 * p_{u,v} - (p_{u-1,v} + p_{u+1,v})\right| \tag{4}$$

To further reduce the extracted-bit error rate, the fluctuation calculation of the image block can be estimated by calculating the absolute difference in horizontal and vertical neighbouring pixels. Figure 2 shows the distribution of neighbouring pixels from a given pixel. A grid circled in green in different positions means the pixel to be counted, and the colour marked is their neighbouring pixel. Three types of pixels according to their coordinates are shown in the figure. The marked in orange is Zhang's (f_Z) and Fatema et al.'s (f_F) fluctuation function calculation. The marked in blue second class is Hong et al.'s fluctuation function calculation (f_H). The marked in red is the proposed fluctuation function calculation (f_P).

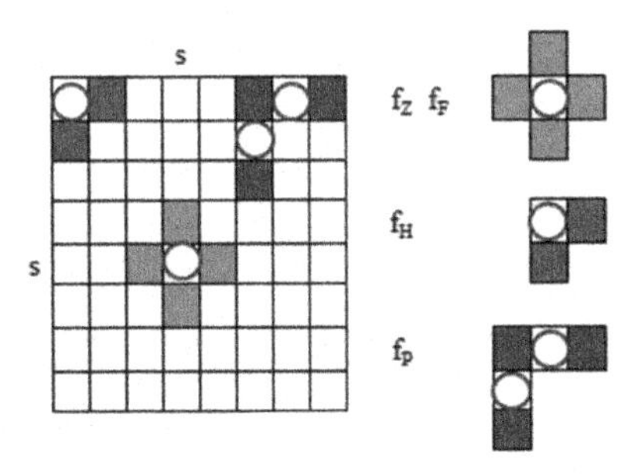

Fig. 2. Distribution of neighbouring pixels of a given pixel.

3.2 Procedures of the Proposed Method

Image Encryption. In the sender side, to begin image encryption phase, the original satellite image is loaded and resize it into a size of *MxN* pixels. Then, the color image is extracted into individual red, green, blue channels. After that, the original satellite image is encrypted by an encryption key by applying bitwise exclusive-or (XOR). Let P is an 8-bit uncompressed cover image of size MxN, and $p_{i,j}$ is the pixel value located at (i, j). Assume the pixel values $p_{i,j}$ range from 0 to 255 which can be represented by 8 bits $p_{i,j}^0, p_{i,j}^1, p_{i,j}^2, \ldots, p_{i,j}^7$. So, we have

$$p_{i,j}^k = \left\lfloor \frac{p_{i,j}}{2^k} \right\rfloor \bmod 2,\ k = 1, 2, \ldots, 7 \tag{5}$$

For encrypted images C, encrypted bits $C_{i,j}^k$ can be calculated with the exclusive-or operation as follows:

$$C_{i,j}^k = p_{i,j}^k \oplus r_{i,j}^k,\ k = 1, 2, \ldots, 7 \tag{6}$$

Where $r_{i,j}^k$ is generated by an encryption key using the standard stream cipher. Then $C_{i,j}$ encrypted data can be obtained as follows:

$$C_{i,j} = \sum_{k=0}^{7} C_{i,j}^k \times 2^k \tag{7}$$

Data Embedding. In data embedding stage (see Fig. 3), the block size value s is assumed. Then, the data hider segments the encrypted image into several non-overlapping blocks sized by sxs. Next, generate the data message to embed in the encrypted image by considering a matrix of 0 and 1. Get two sets S_0 and S_1 using the data hiding key. If data hiding key values at the pixel position is 0, then it goes into set S_0 otherwise set S_1. If the bit message to be embedded is '0' in each block of red channel image, flip the three least significant bits (LSB) of each encrypted pixel in set S_0 and pixel in set S_1 is not changed. On the other hand, If the bit message to be embedded is '1', flip the three LSB of each encrypted pixel in set S_1 and pixel in set S_0 is unchanged. After that, the resultant of red, green, blue channels is combined to get the encrypted image with additional data.

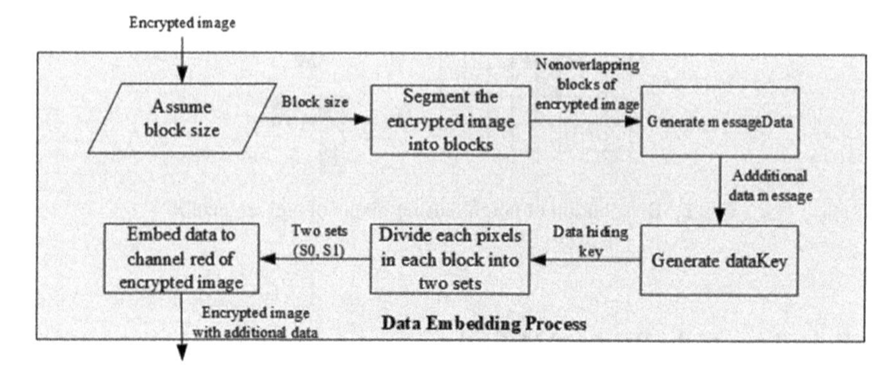

Fig. 3. Data embedding process scheme.

Image Decryption. In the receiver side, to begin image decryption phase, after receiving the encrypted image with additional data, the receiver decrypts it with a decryption key by applying bitwise XOR which is similar to image encryption procedures. For flipped encrypted bits $C'^{k}_{i,j}$, the marked decrypted bit can be calculated as follows:

$$C'^{k}_{i,j} \oplus r^{k}_{i,j} = \overline{C^{k}_{i,j}} \oplus r^{k}_{i,j} = \overline{p^{k}_{i,j} \oplus r^{k}_{i,j}} \oplus r^{k}_{i,j} = \overline{p^{k}_{i,j}} \tag{8}$$

In the same block, the bits that have not been flipped without the embedded bits will be the same as the original bits $p^{k}_{i,j}$.

Data Extraction and Image Recovery. In the data extraction and image recovery phase, the decrypted image with additional data is decomposed into red, green, blue channels. Then, the decrypted red channel image is segmented into several non-overlapping blocks sized by *sxs*. Next, each pixel in each block is divided into two sets $newS_0$ and $newS_1$ in the same way. If data hiding key values at the pixel position is 0, then it goes into set $newS_0$, otherwise set $newS_1$. Flip three LSB in set $newS_0$ and $newS_1$ to get two sets S_{00} and S_{11}. After that, make two sets H_0 and H_1. If data hiding key values at the pixel position is '0', then set S_{00} and $newS_0$ goes into set H_0 and H_1, respectively. Otherwise set $newS_1$ and S_{11} go into set H_0 and H_1, respectively. After that, calculate fluctuation H_0 and H_1 to determine which one is the original image.

Hereinafter, combine the red channel with the green and blue channels to give the recovered original image. Finally, calculate EER by comparing each pixel from the original matrix data message with a recovered data message and PSNR [6] to measure the quality of the final recovered image as follows:

$$PSNR = 10\log_{10}\frac{255^2}{\frac{1}{MN}\sum_{i=1}^{M}\sum_{j=1}^{N}\left(O_{i,j} - M_{i,j}\right)^2} \tag{9}$$

Where $O_{i,j}$ and $M_{i,j}$ are the original pixel values and the modified pixel values, respectively.

4 Results and Discussion

In this paper, three standard processing high-resolution remote sensing satellite images, i.e. SPOT-6, SPOT-7, and Pleiades-1A, as shown in Fig. 4, are considered. The test images have been resized to 512 × 512 where each pixel is represented by 8 bits. The range of the block size is from two to 32. In this research, two important performance parameters will be analyzed:

- Extracted-bit error rate (EER); this parameter shows the ratio of unrecovered bits to the total number of embedded bits.
- Peak signal-to-noise ratio (PSNR); this parameter is to show the differences between the original image and the recovered original image.

To compare the proposed joint data hiding method with the referenced methods, the error pattern of the data hiding system is shown in Fig. 5. For a SPOT-6 image and s = 14, error positions of Zhang's function [3], Hong's function [4], Fatema's function [10],

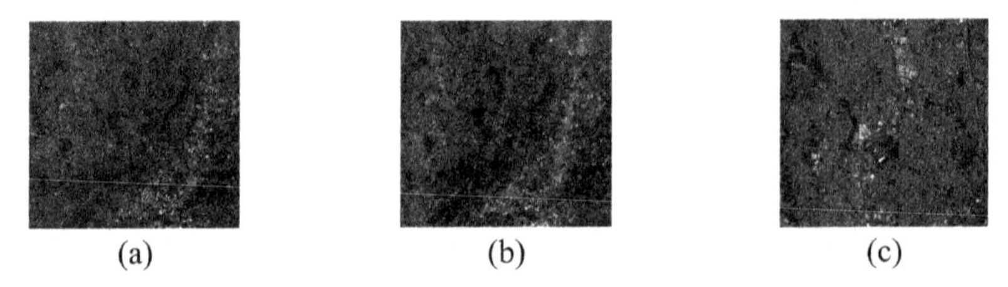

Fig. 4. Simulation test images. (a) SPOT-6; (b) SPOT-7; (c) Pleiades-1A.

and the proposed system with fluctuation functions in Eq. (4) are shown in (a), (b), (c) and (d) in Fig. 5, respectively. The small black square in a large square in Fig. 4 denotes *sxs* error pixels in a SPOT-6 image. From Fig. 5a to 5d, the recovery performance of the system with the proposed fluctuation functions in Eq. (4) is better than the ones of Zhang's method [3], Hong's method [4], and Fatema's method [10].

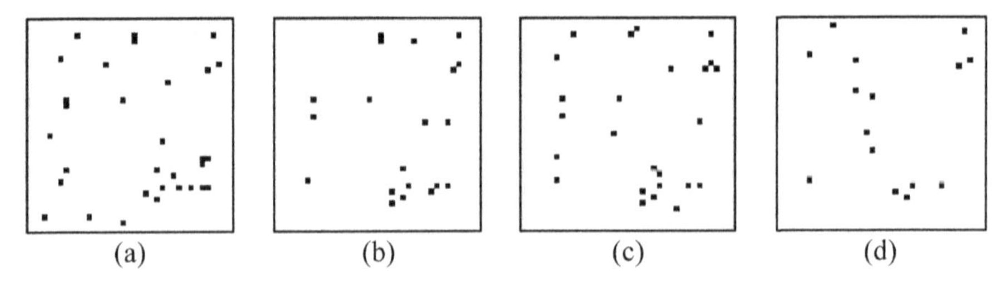

Fig. 5. Incorrect extracted-bit pattern of Zhang's method [3], Hong's method [4], Fatema's method [10], and proposed method in the SPOT-6 image for s = 14. (a) f_Z in (1); (b) f_H in (2); (c) f_F in (3), (d) f_P in (4).

In Fig. 6 to Fig. 8, the EER dan PSNR of image recovery performances of the referenced functions and proposed function are shown according to the block size of *s* when SPOT-6, SPOT-7 and Pleiades-1A images in Fig. 4 are considered. The 'Ref. Zhang [3] ', 'Ref. Hong et al. [4] ', and Ref. Fatema et al. [10] in these figures denote EERs for Zhang's function in (1), Hong's function in (2), and Fatema's function in (3) respectively. The 'Proposed Function' in these figures denotes for EERs for the proposed fluctuation functions in (4). The 'Inf' in these figures denotes infinite PSNR. Infinite PSNR cannot be illustrated in these figures. However, for convenience, it is located at 103.4 dB. For all test images, the EER of proposed function slightly lower than the other three references functions as the block size decreases. Likewise, the PSNR of proposed function is slightly higher than the other three references functions as the block size decreases.

In SPOT-6 test image, as shown in Fig. 6, when 4096 bits are embedded (s = 8 × 8), the EER of the proposed function is 8.4% which is 5.74% lower than Zhang's function of 14.14%, 1.22% lower than Hong's function of 9.62%, and 3.47% lower than Fatema et al.'s function of 11.87%. Likewise, when 4096 bits are embedded (s = 8 × 8), the PSNR of the proposed function is 50.52 dB which is 2.29 dB higher than Zhang's function of 48.23 dB, 0.59 dB higher than Hong's function of 49.93 dB, and 1.52 dB higher than Fatema et al.'s function of 49.00 dB. By using at least 784 bits (s = 18 × 18),

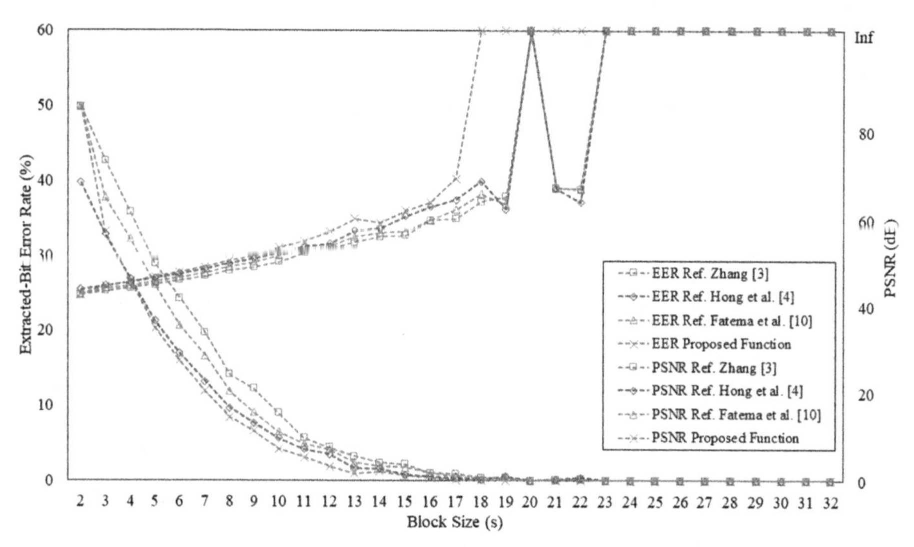

Fig. 6. EER and PSNR image recovery performances comparison of referenced and proposed function for the SPOT-6 test image.

error-free extracted bits and complete reversibility (PSNR = infinity) can be achieved by the proposed function meanwhile at least 625 bits (s = 20 × 20), error-free extracted bits and infinite PSNR can be achieved by the three references.

In SPOT-7 test image, as shown in Fig. 7, when 4096 bits are embedded (s = 8 × 8), the EER of the proposed function is 8.01% which is 5.61% lower than Zhang's function of 13.62%, 1.29% lower than Hong's function of 9.3%, and 3.03% lower than Fatema et al.'s function of 11.04%. Likewise, when 4096 bits are embedded (s = 8 × 8), the PSNR of the proposed function is 50.74 dB which is 2.35 dB higher than Zhang's function of 48.39 dB, 0.67 dB higher than Hong's function of 50.07 dB, and 1.44 dB higher than Fatema et al.'s function of 49.30 dB. By using at least 784 bits (s = 18 × 18), error-free extracted bits and complete reversibility (PSNR = infinity) can be achieved by the proposed function meanwhile at least 576 bits (s = 21 × 21), error-free extracted bits and infinite PSNR can be achieved by Hong and Fatema, and at least 484 bits (s = 23 × 23), error-free extracted bits and infinite PSNR can be achieved by Zhang.

In Pleiades-1A test image, as shown in Fig. 8, when 4096 bits are embedded (s = 8 × 8), the EER of the proposed function is 11.35% which is 7.13% lower than Zhang's function of 18.48%, 2.59% lower than Hong's function of 13.94%, and 5.67% lower than Fatema et al.'s function of 17.02%. Likewise, when 4096 bits are embedded (s = 8), the PSNR of the proposed method is 49.20 dB which is 2.15 dB higher than Zhang's function of 47.05 dB, 0.91 dB higher than Hong's function of 48.29 dB, and 1.79 dB higher than Fatema et al.'s function of 47.41 dB. By using at least 625 bits (s = 20 × 20), error-free extracted bits and complete reversibility (PSNR = infinity) can be achieved by the proposed function meanwhile at least 576 bits (s = 21 × 21), error-free extracted bits and infinite PSNR can be achieved by Zhang and Hong, and at least 529 bits (s = 22 × 22), error-free extracted bits and infinite PSNR can be achieved by Fatema.

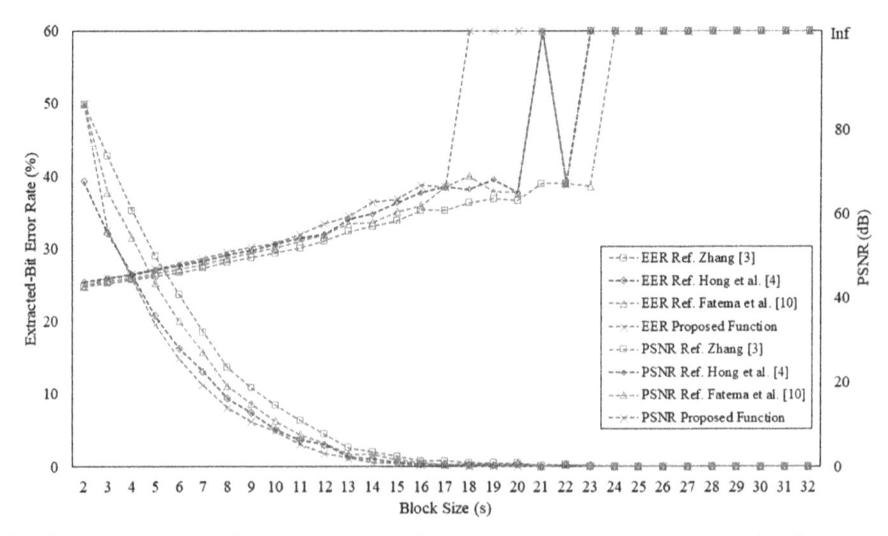

Fig. 7. EER and PSNR image recovery performances comparison of referenced and proposed function for the SPOT-7 test image.

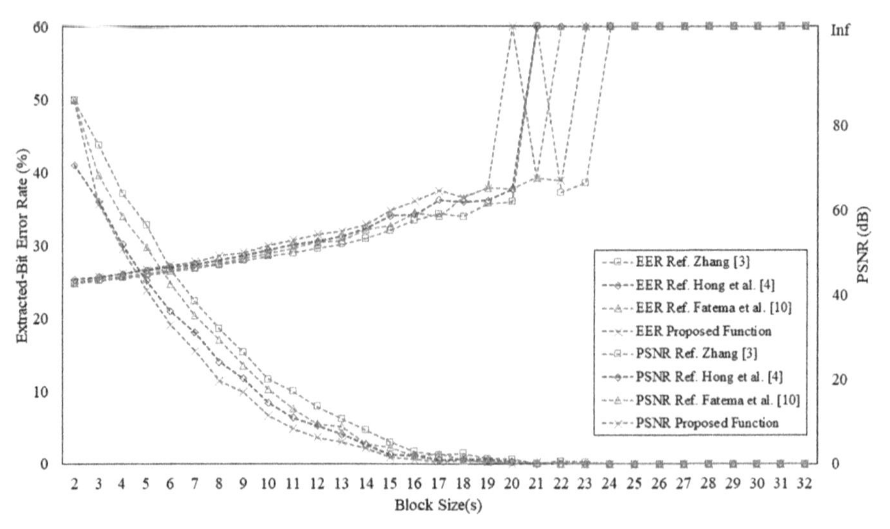

Fig. 8. EER and PSNR image recovery performances comparison of referenced and proposed function for the Pleiades-1A test image.

Table 1 through Table 3 show the comparison of the main results between the modification of the proposed function and the reference function Zhang [3], Hong et al. [4], and Fatema et al. [10] for error-free extracted-bits and infinite PSNR on the SPOT-6, SPOT-7, and Pleiades-1A test images. The "minimum block size (s)" in the Table is the minimum block size (s) that guarantees error-free extracted-bits and maximum PSNR (infinity). "Number of messages (bits)" in the Table is the actual number of bits that can be embedded that corresponds to the minimum block size (s). The "gain" in the Table states the ratio of the number of message bits of the proposed function to the number of message bits from the three reference functions Zhang, Hong et al., and Fatema et al. and written in percent (%).

Table 1. Comparison of minimum block size (s) to obtain error-free extracted-bits and infinite PSNR between proposed systems with references for SPOT-6 test image.

Methods	Minimum block size (s)	Number of Message (bits)	Gain (%) with Zhang [3]	Gain (%) with Hong et al. [4]	Gain (%) with Fatema et al. [10]
Ref. Zhang [3]	20 × 20	625	–	–	–
Ref. Hong et al. [4]	20 × 20	625	–	–	–
Ref. Fatema et al. [10]	20 × 20	625	–	–	–
Proposed	18 × 18	784	125,44	125,44	125,44

Table 2. Comparison of minimum block size (s) to obtain error-free extracted-bits and infinite PSNR between proposed systems with references for SPOT-7 test image.

Methods	Minimum block size (s)	Number of Message (bits)	Gain (%) with Zhang [3]	Gain (%) with Hong et al. [4]	Gain (%) with Fatema et al. [10]
Ref. Zhang [3]	23 × 23	484	–	–	–
Ref. Hong et al. [4]	21 × 21	576	119,01	–	–
Ref. Fatema et al. [10]	21 × 21	576	119,01	–	–
Proposed	18 × 18	784	161,98	136,11	136,11

As shown in Tables 1 through Table 3, the minimum block size (s) of the proposed system is always smaller than the reference systems. The number of message bits embedded in the proposed system is also greater than the reference system. In Table 1, the number of embedded message bits of the proposed system shows 125.44% greater than the three reference systems in the SPOT-6 test image. In Table 2, the number of embedded message bits of the proposed system shows 161.98% greater than the Zhang reference system, and 136.11% greater than the Hong and Fatema reference system in the SPOT-7 test image. In Table 3, the number of embedded message bits of the proposed

Table 3. Comparison of minimum block size (s) to obtain error-free extracted-bits and infinite PSNR between proposed systems with references for Pleiades-1A test image.

Methods	Minimum block size (s)	Number of Message (bits)	Gain (%) with Ref. Zhang [3]	Gain (%) with Hong et al. [4]	Gain (%) with Fatema et al. [10]
Ref. Zhang [3]	21 × 21	576	–	–	108,88
Ref. Hong et al. [4]	21 × 21	576	–	–	108,88
Ref. Fatema et al. [10]	22 × 22	529	–	–	–
Proposed	20 × 20	625	108,51	108,51	118,15

system shows 108.51% greater than the Zhang and Hong reference system, and 118.15% greater than the Fatema reference system in the Pleiades-1A test image.

5 Conclusion

In this paper, based on Zhang [3], Hong et al. [4] and Fatema et al. [10] functions, a modified fluctuation function to measure the smoothness of image block has been proposed. The experimental results show that the modified fluctuation function is more precise and can further reduce the extracted-bit error rate compared to the other three references functions. Furthermore, the visual quality of the recovered remote sensing satellite images, represented by PSNR, is quite higher than the other three references functions. In the future, error control coding can be used in the algorithm in order to improve the performance.

Acknowledgment. The authors would like to thank Universitas Indonesia for funding through PUTI Prosiding Universitas Indonesia (UI), under contract No. NKB-1183/UN2.RST/HKP.05.00/2020.

References

1. The Republic of Indonesia: Law Number 21 the Year 2013 Regarding Space. Ministry of Law and Human Rights of the Republic of Indonesia, Jakarta (2013)
2. The Republic of Indonesia: Government Regulation Number 11 the Year 2018 Regarding Procedures for Organizing Remote Sensing Activities. Ministry of Law and Human Rights of the Republic of Indonesia, Jakarta (2018)
3. Zhang, X.: Reversible data hiding in encrypted Image. IEEE Signal Process. Lett. **18**(4), 255–258 (2011)
4. Hong, W., Chen, T., Wu, H.: An improved reversible data hiding in encrypted images using side match. IEEE Signal Process. Lett. **19**(4), 199–202 (2012)

5. Li, M., Xiao, D., Peng, Z., Nan, H.: A modified reversible data hiding in encrypted images using random diffusion and accurate prediction. ETRI J. **36**(2), 325–328 (2014)
6. Wu, X., Sun, W.: High-capacity reversible data hiding in encrypted images by prediction error. Signal Process. **104**, 387–400 (2014)
7. Liao, X., Shu, C.: Reversible data hiding in encrypted images based on absolute mean difference of multiple neighboring pixels. J. Vis. Commun. Image R. **28**, 21–27 (2015)
8. Kim, Y.S., Kang, K., Lim, D.W.: New reversible data hiding scheme for encrypted images using lattices. Appl. Math. Inf. Sci. **9**(5), 2627–2636 (2015)
9. Pan, Z., Wang, L., Hu, S., Ma, X.: Reversible data hiding in encrypted image using new embedding pattern and multiple judgements. Multimedia Tools Appl. **75**(14), 8595–8607 (2016)
10. Fatema, K., Song, K.Y., Sunghwan, K.: A modified reversible data hiding in encrypted image using enhanced measurement functions. In: 2016 Eighth International Conference on Ubiquitous and Future Networks (ICUFN) 2016, pp. 869–872, Vienna (2016)
11. Zhang, X.: Separable reversible data hiding in encrypted image. IEEE Trans. Inf. Forensics Secur **7**(2), 826–832 (2012)
12. Zhang, X., Qian, Z., Feng, G., Ren, Y.: Efficient reversible data hiding in encrypted images. J. Bis. Commun. Image Represent. **25**(2), 322–328 (2014)
13. Qian, Z., Zhang, X.: Reversible data hiding in encrypted images with distributed source encoding. IEEE Trans. Circuits Syst. Video Technol. **26**(4), 636–646 (2016)
14. Xiao, D., Xiang, Y., Zheng, H., Wang, Y.: Separable reversible data hiding in encrypted image based on pixel value ordering and additive homomorphism. J. Vis. Commun. Image R. **45**, 1–10 (2017)
15. Chuan, Q., Zhihong, H., Xiangyang, L., Jing, D.: Reversible data hiding in encrypted image with separable capability and high embedding capacity. Inf. Sci. **465**, 285–304 (2018)
16. Ma, K., Zhang, W., Zhao, X., Yu, N., Li, F.: Reversible data hiding in encrypted images by reserving room before encryption. IEEE Trans. Inf. Forensics Secur. **8**(3), 553–562 (2013)
17. Zhang, W., Ma, K., Yu, N.: Reversibility improved data hiding in encrypted images. Signal Process. **94**, 118–127 (2014). ISSN 0165-1684

3D Kinematics of Upper Limb Functional Assessment Using HTC Vive in Unreal Engine 4

Kai Liang Lew(✉), Kok Swee Sim, Shing Chiang Tan, and Fazly Salleh Abas

Multimedia University, Jalan Ayeh Keroh Lama, 75450 Melaka, Malaysia
lewkailiang@gmail.com

Abstract. The purpose of research in this paper is to quantify the accuracy and precision of HTC Vive by making upper limb assessment measurements and performing functional tasks in the Unreal Engine 4. Thirty healthy males performed daily aim functional tasks, and arm length measurement and assessment were made. Each participant attended two testing sessions and one arm length measurement session. The upper limb length was measured using HTC Vive after making three types of hand posture exercises. The arm assessment included the minimum and maximum angle of shoulder adduction, abduction, flexion and extension. The experiment showed all the upper limb measurements collected from the functional tasks as well as the position and rotation of the upper limb could be estimated correctly.

Keywords: Unreal engine 4 · Arm assessment · HTC Vive

1 Introduction

The upper limb is used in most of the daily tasks such as grabbing, pushing, and pulling. There are a lot of studies that involve upper limb including motor control, rehabilitation and neurophysiology. Most of the quantitative measurements of upper limb functions are performed with Kinect. A motion capture system can predict the body joints position using 2D images with depth knowledge by using machine learning [1]. Although the Kinect is reliable in evaluating the upper limb, the system can detect only the user who is in front of it. The Kinect camera is sensitive to light which may not work well if it is exposed to light. Moreover, it cannot react well when the movement of the user is too fast. Thus, HTC Vive, a virtual reality (VR) and motion tracking system can be a possible alternative for evaluating the upper limbs. It works well in any place, can detect the front and back of the user and also can react with the movement of the user regardless of the speed.

HTC Vive has two base stations (sensor) along with two controllers and one head-mounted display (HMD) [4]. The base stations are used to track the position of the controller and HMD. The base station can track the movement of the controller and HMD correctly at any position. HTC Vive can be used as an upper limb performance assessment tool. The use of the HMD allows the user to perceive a 3D stereoscopic image.

M. Hernes et al. (Eds.): ICCCI 2020, CCIS 1287, pp. 264–275, 2020.
https://doi.org/10.1007/978-3-030-63119-2_22

The HTC Vive system combined with Unreal Engine 4 can provide real-time feedback for clinical use and rehabilitation training. Some researchers adopted similar devices such as Oculus Rift for clinical use and rehabilitation training [1]. Other researchers investigated the accuracy and precision of the VR systems [1].

The aim of this study is to quantify HTC Vive's accuracy and precision by performing measurements of the upper limbs and functional tasks in Unreal Engine 4. The experimental protocol is approved by the National Medical Research Register (NMRR-20-63-52503).

Section 2 of this paper presents a few work related to upper limb rehabilitation. The proposed upper limb rehabilitation system is described in Sect. 3. Section 4 explains a methodology designed for assessing the proposed system. Results are analysed in Sect. 5. Discussion is presented in Sect. 6 before concluding remarks are made in Sect. 7.

2 Related Work

Chen et al. [3] performed upper limb rehabilitation by using HTC Vive. They obtained consent from the patients to participate in rehabilitation training. The patients completed the training according to the game condition in the virtual environment. They analyzed the upper limb work by using the Denavit-Hartenberg parameter method to determine the safety of training. The user wore the HTC Vive HMD and held a controller to make movement in the rehabilitation training. All patients touched four different points in VR environment.

Cai et al. [1] performed an upper limb functional assessment using the Kinect V2 sensor. Their assessment involved 10 healthy males who repeated four tasks in two separate testing sessions. Records for the angle at points of target achieved and the range of motion were obtained using the Kinect V2 sensor. They found that Kinect could measure shoulder and elbow flexion/extension angle accurately but its accuracy in measuring shoulder internal/external angle was not high. Kinect could not detect the hand that did not appear in front of it.

Nilsson et al. [2] showed six different types of manipulation controls in VR which displayed different interactions in VR. 14 people participated in the test. All participants were required to sit, wear VR head-mounted displays and used two controllers to perform the movement in the test. They needed to move, scale and rotate a cube to fit it in a reference plane. The setup of the avatar was a master-slave system. Six manipulation controls were set in the test, which were thumbpad, menu, sphere, avatar, drag and pull. The thumbpad had the same size as the controller and it was used to rotate and move the cubes. The menu spawned in front of the participant with buttons and sliders which could rotate and move the cubes. The sphere's rotation and movement were based on the controller movement. When the controller in the sphere, it would rotate based on the rotation of the controller. If the controller was outside of the sphere, it would move based on the distance between the controller and sphere.

Chen et al. used HTC Vive to perform the arm assessment. They recorded the motion track of the upper limb wrist but they did not capture the angle at points of target achieved and the range of motion. Cai et al. used Kinect to assess the upper limb but it the system performed a poor measurement on shoulder internal/external angle and could not detect the hand when it was not shown in front of kinect. Nillson et al. required to use complete the training and analyzed the upper limb assessment. However, they did not measure the height of body and rely on wrist angle to evaluate the upper limb. The proposed method measured the angle at points of target achieved, range of motion, shoulder internal/external angle, shoulder and elbow flexion/extension angle, the height of body and length of arm.

3 Proposed Upper Limb Rehabilitation System

3.1 HTC Vive Overview

The HTC Vive devices consist of a head-mounted display (HMD), two controllers and two base stations (sensors). The HTC Vive HMD can virtualize the virtual content and it has per-eye displays in a 1080 × 1200 resolution and operates at 90 Hz; performs a 360-degree position tracking. The HMD is integrated with audio. The HTC Vive controller helps the user to contact the content in virtual world through physical movement in the real world and it consists of a haptic sensor. It utilizes the low-latency tracking technology to assess the headset's relative position. The controller is used to grab, hold, select options and so on in the virtual world. Moreover, it has a rumble motor to perform haptic feedback. HTC Vive uses two laser emitters (called Lighthouse or base station) to perform tracking. The base stations alternatingly send out horizontal and vertical infrared laser sweeps spanning 120° in each direction [3]. Two base stations are placed in front and behind of the user so that tracking can be made in 360°. The application is run on an RTX 2080Ti desktop computer. Two requirements shall be fulfilled before using VR; firstly, the area must be at least 1 m length and 2 m width, and secondly, no obstacle exists in the area. This is to avoid any unnecessary incidents during using the VR device.

3.2 Unreal Engine 4 (UE4)

The assessment program is developed using a game development tool called Unreal Engine 4 (UE4, version 4.24). The programming language used is C++. SteamVR software is used to connect the VR software to the VR device. When running the assessment in the Unreal Engine 4, it will create a log file to record d information for troubleshooting the program. The coordinated system used in the UE4 is a 3D Cartesian coordinate system in terms of x, y, and z [5]. The rotation used in UE4 is the Euler angle. The rotation axes are roll (φ), pitch (θ) and yaw (ψ) [9]. The position of any object in UE4 can be measured in real-time. Unreal Engine 4 has interchangeable codes with libraries in an object-oriented design framework. It is a computer-generated graphics system [6]. Moreover, it has a visual scripting system, Blueprint system [7]. The Blueprint systems are called as blueprints and act the same way as C++ classes. It has a color code which represents the type of function. The lines in Blueprint system is used to connect the pin. The red color function is header while the blue color is code function. The code function is represented as a node of a graph with pins for input and outputs.

3.3 Arm Length Measurement

The arm length measurement is based on the length of upper limb between each joint. This is one of the methods used to assess the performance of the HTC Vive. There are three joints namely Joint 1 that represents the joint at shoulder, Joint 2 that represents the joint at elbow and Joint 3 that represents the joint at hand. The position of the joints can be identified by performing three types of hand.

Table 1. The summary of the arm measurement coordinate.

Hand	Coordinate (LHC/RHC) (x, y, z)
First hand posture coordinate	(x_{1L}, y_{1L}, z_{1L}), (x_{1R}, y_{1R}, z_{1R})
Second hand posture coordinate	(x_{2L}, y_{2L}, z_{2L}), (x_{2R}, y_{2R}, z_{2R})
Third hand posture coordinate	(x_{3L}, y_{3L}, z_{3L}), (x_{3R}, y_{3R}, z_{3R})
Joint 1 coordinate	(x_{1L}, y_{1L}, z_{2L}), (x_{1R}, y_{1R}, z_{2R})
Joint 2 coordinate	(x_{3R}, y_{3R}, z_{2L}), (x_{3L}, y_{3L}, z_{2R})
Joint 3 coordinate	(x_{1L}, y_{1L}, z_{1L}), (x_{1R}, y_{1R}, z_{1R})

The first hand posture is shown in Fig. 1a. Figure 1a shows the hand posture that the hand is at a resting position. It has two coordinates namely, the left hand coordinate (LHC) (x_{1L}, y_{1L}, z_{1L}) and the right hand coordinate (RHC) (x_{1R}, y_{1R}, z_{1R}). The joint 3 has the same coordinate with the first hand posture. After performing the first hand posture, the user will continue with the second hand posture. Figure 1b shows the second hand posture that the hand is in a straighten position. The corresponding coordinates of second hand posture are the LHC (x_{2L}, y_{2L}, z_{2L}) and RHC (x_{2R}, y_{2R}, z_{2R}). Based on the two hand postures, the joint 1 coordinate can determine because the joint 1 has the height of Z_2 and the same position of x_1 and y_1. Therefore, the coordinates of joint 1 are LHC (x_{1L}, y_{1L}, z_{2L}) and RHC (x_{1R}, y_{1R}, z_{2R}).

The user will continue to perform the third hand posture. Figure 1c shows the hand posture that the joint 3 will place on the elbow. The coordinates are LHC (x_{3L}, y_{3L}, z_{3L}) and RHC (x_{3R}, y_{3R}, z_{3R}). After getting the coordinates, the joint 2 coordinates can be estimated. Joint 2 has the same height as Z_2 and the same position as x_3 and y_3. Therefore, the coordinates of joint 2 are LHC (x_{3R}, y_{3R}, z_{2L}) and RHC (x_{3L}, y_{3L}, z_{2R}). Figure 2 shows the position of the hand when it is lying down. Table 1 shows a summary of the arm measurement coordinate.

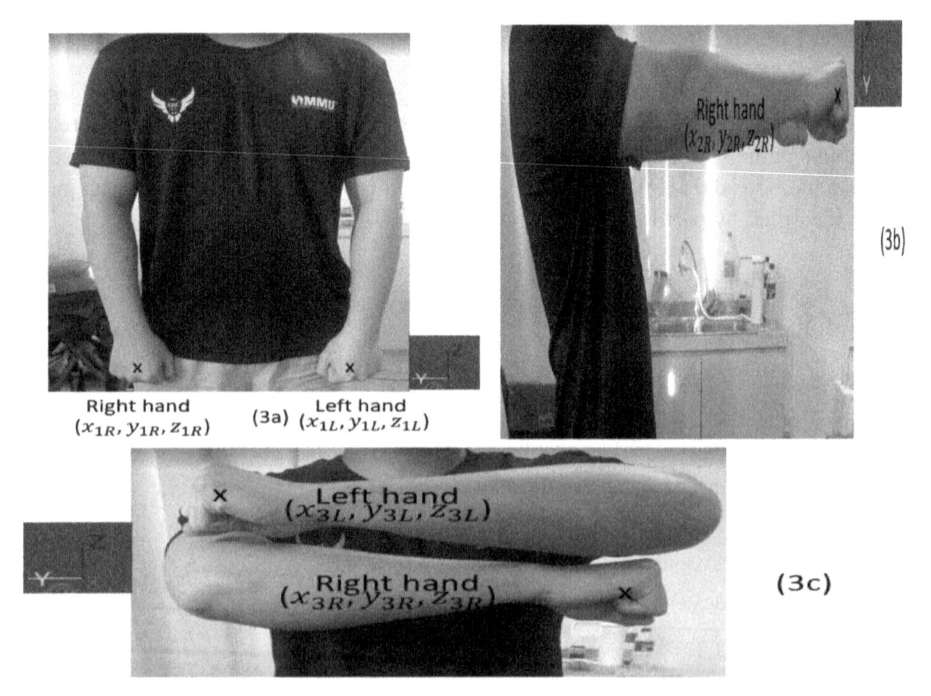

Fig. 1. a) the first hand-posture b) the second hand-posture c) the third hand-posture

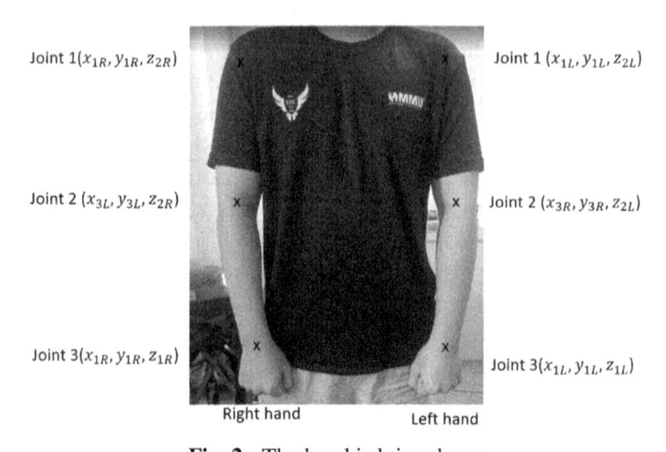

Fig. 2. The hand is lying down

3.4 Arm Assessment

Arm assessment was made in two sessions. The first session determined the minimum and maximum angle of the upper limb when doing adduction, abduction, flexion and extension. The participant moved the upper limb to a point where the upper limb could no longer move. The measurement would be recorded in the UE4. The default rotation in UE4 was (0, 0, 0).

The second session was to do a series of functional tasks [8]. Each task was repeated for at least three times. In this session, the position and rotation of the upper limb were determined using the HTC Vive's controller.

1. The first task was to move the hand to the contralateral shoulder. It defined the activities namely brushing axilla or zipping up a jacket. The participant would move his upper limb from a relaxed position (starting position) beside his body to touch the contralateral shoulder.
2. The second task was to move the hand to the back pocket. The task included activities such as reaching back and perineal care. The participant would move hand from a starting point to the back pocket.
3. The third task was to move the hand to comb hair. This task included the activities such as touching the back of the head and washing hair. The participant began to move a hand from the same starting point and lift the hand up to the back of head.
4. The fourth task was to move the hand to the mouth. Activities under this task included shaving face and brushing teeth. The participant began to move a hand from the same starting point and moved the upper limb towards his mouth.
5. The fifth task was to take an object on the table. This task included common activities in daily life. For example, the participant would start moving his hand from a starting point to the table to grab any object.

4 Methodology

4.1 Subject

Twenty healthy male and ten healthy female university students participated in the experiment. They do not have any upper limb injuries in the past neither taking any medication which can affect upper limb function. Before the test, all participants were informed of the basic procedure of the experiment and signed a consent form. Table 2 shows the information about the students.

Table 2. The student characteristic

Characteristic	Student (N = 30)
Sex (male/female)	20/10
Age (yr) (mean±SD)	25±2.5
Dominant (left/right)	11/19
Height (cm) (mean±SD)	167±5.2

4.2 Testing Procedure

The experiment was conducted at Multimedia University E-Health Laboratory. The upper limb information is recorded through unreal engine 4 platform (UE4, version

4.24) with HTC Vive. Two base stations were used to detect the position of the touch controller and integrate the data into UE4. The sensors were placed at the corner of an area of 1 m × 1 m. A participant was required to stand at the center of the area and holding the HTC Vive controllers during the testing session.

Each participant was requested to attend two testing sessions. In the first session, the arm length measurement was performed. Before that, the arm length was measured from shoulder to wrist once by using a measurement tape. The participant was required to hold the controller and does the following hand posture to get the length of the upper limb that have been described in Sect. 3.4

In the second testing session, the participants performed upper limb adduction, abduction, flexion and extension. The measurement was recorded in the UE4. After that, each of the participant performed a series of functional tasks that have been described in Sect. 3.5

5 Statistical Analysis

For the arm length measurement, the actual arm length was compared with the arm length in the VR. In the first session of arm assessment, all measurements were recorded in terms of mean, standard deviation, medium, minimum angle and maximum angle. In the second session of arm assessment, the range of motion and rotation were computed. The time movement was also recorded and normalized between 0% and 100%.

6 Results

6.1 Arm Length Assessment

The arm length of each participant was measured by UE4. The results of the arm length measurement are shown in Table 3. Table 3 shows the average value of the arm length measurement of 30 participants. The length between each joint was compared. The average accuracy is computed using Eq. 1 and Eq. 2. The length of arm measured by a tap is deemed as a reference. Both average accuracy between d_{1v} and d_1 and between d_{2v} and d_2 are 97%.

Table 3. The result of the arm length measurement.

	J1–J2* (cm)	J1–J2 MM (cm)	J2–J3* (cm)	J2–J3 MM (cm)	Accuracy (J1–J2, J2–J3) (%)
Measured by tape	31.5 ± 2.41	28, 35	26.6 ± 1.42	25, 29	100,100
Measure by VR	31 ± 2.18	27, 37	27.4 ± 3.13	24, 29	97,97

The value * is mean ± standard deviation, MM is minimum and maximum

$$average\ accuracy = \frac{\Sigma \frac{d_{nv}}{d_n}}{n} \times 100\%;\ d_{nv} < d_n \tag{1}$$

$$average\ accuracy = \frac{\Sigma \frac{d_n}{d_{nv}}}{n} \times 100\%;\ d_n < d_{nv} \tag{2}$$

6.2 Arm Assessment

The values for the first session of arm assessment estimated by the HTC Vive system are recorded in Table 3. The value shows the pitch, yaw values. The shoulder adduction, abduction, flexion and extension are observed. All values are mean, standard deviation, medium, minimum angle and maximum angle that can be obtained by the HTC Vive system in UE4. Figure 3 shows the average pitch and yaw in adduction, abduction, flexion and extension. In this assessment, only the roll value remains as zero; both pitch and yaw value are not 0. The pitch value varies between −90° and 90° and the yaw value varies between −180° and 180°. If the pitch value is negative, the position of upper limb is lower than upper body. If the yaw value is negative, the rotation of the upper limb is rotated in a counterclockwise direction. In other words, both pitch and yaw values indicate the movement of a participant's upper limb.

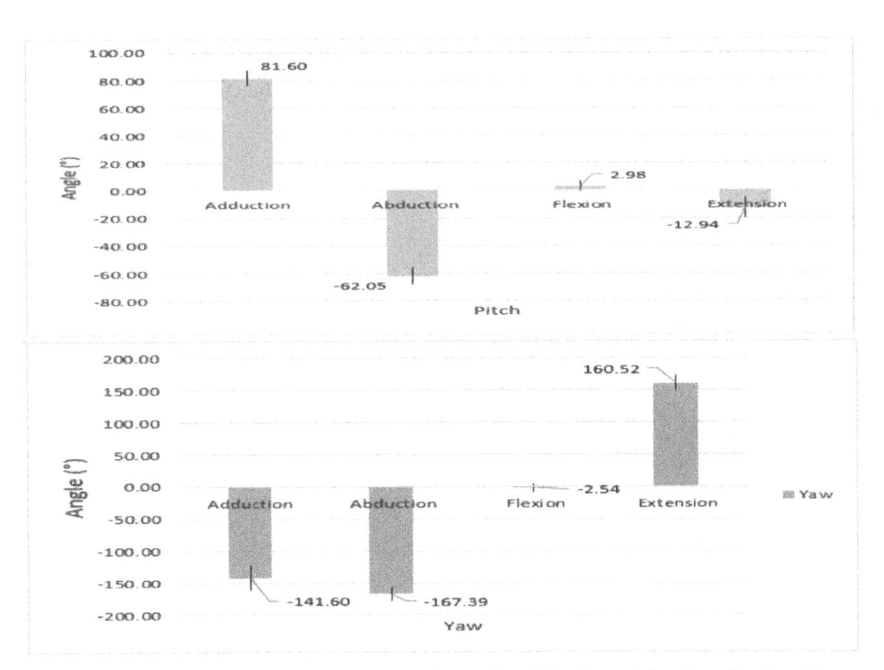

Fig. 3. Result of average pitch (up) and yaw (down) in adduction, abduction, flexion. The error bar shows the standard deviation.

The average range of motion for the upper limb functional tasks estimated by the HTC Vive system are shown in Fig. 4, Fig. 5 and Fig. 6. The z axis shows the position of the upper limb when moving up and moving down. The y axis shows the upper limb position when moving left and right. The x axis shows the position of the upper limb when. moving forward and backward. The records for average rotation of upper limb after completing functional tasks are depicted in Fig. 7. We observe that participants who did the third task would make more movements as compared to other tasks because the position in x, y and z change significantly. On the other hand, less movement are observed in the second task because the position in x, y and z did not change greatly. Thus, the third task is more complex than the other two.

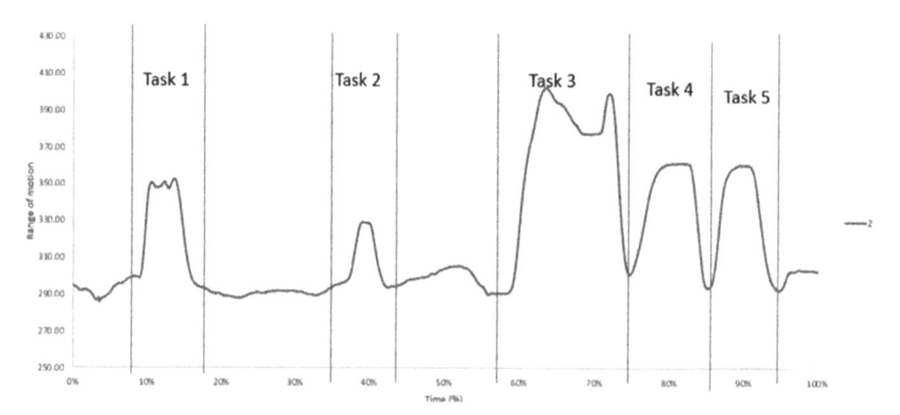

Fig. 4. The movement at the z-axis when doing the functional tasks

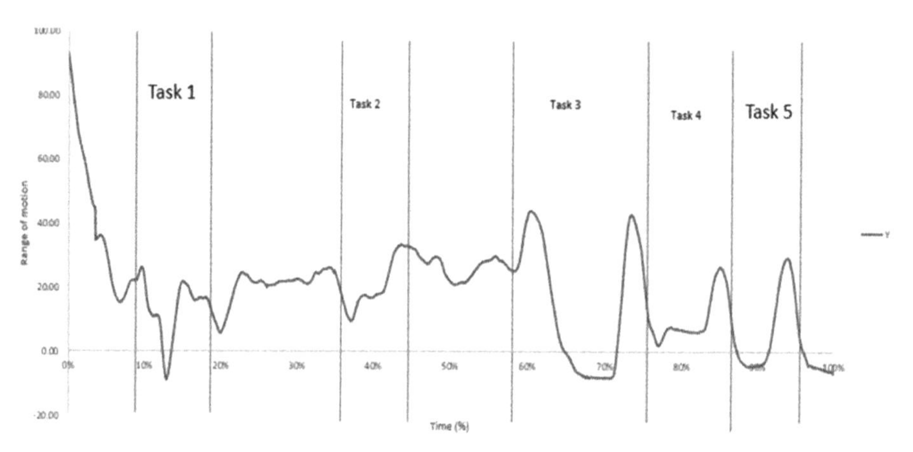

Fig. 5. The movement at the y-axis when doing the functional tasks

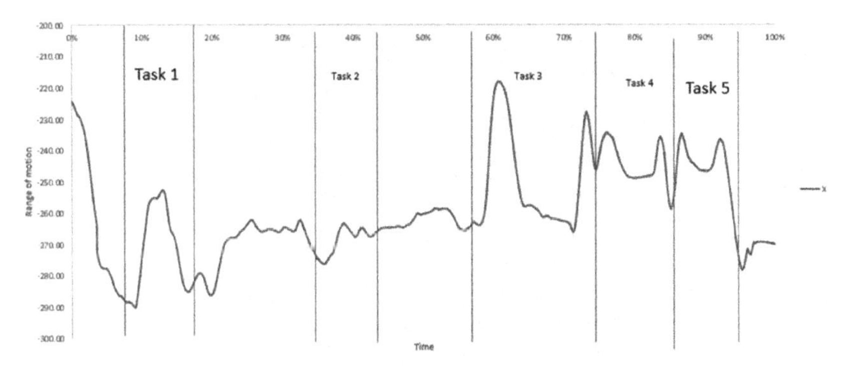

Fig. 6. The movement at the x-axis when doing the functional tasks

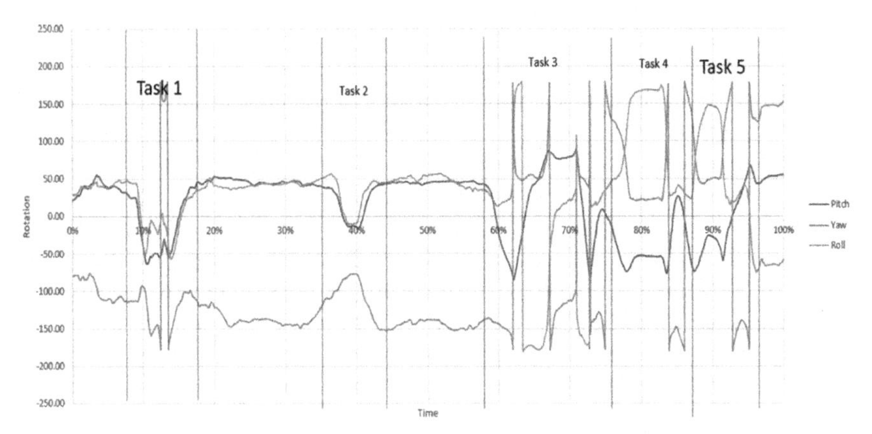

Fig. 7. The average rotation of the upper limb when doing functional tasks

7 Discussion

The study covers the test for the accuracy and precision of upper limb length measurement and upper limb assessment using HTC Vive. The HTC Vive has good accuracy in measuring shoulder flexion, extension, abduction and adduction, the upper limb length and the position and rotation of the upper limb when doing the functional task. It can detect the position of the controller when the controller is in the area of the lighthouse.

The default rotation for pitch (θ), yaw (ψ) and roll (φ) is (0, 0, 0). In the first testing session, the roll angle is 0 because the hand is not required to rotate when doing abduction, adduction, flexion and extension. The direction of participant always faces toward the x-axis. Figure 8 shows the default rotation and rotation direction in UE4 and. Roll is the rotation from x-axis, pitch is the rotation from y-axis, yaw is the rotation from z-axis. Thus, the angle in first testing session can be obtained by using the rotation.

During the second testing session, functional tasks, the position of the upper limb can be determined and recorded very well. It can detect the movement in 360° as long as the person is in the coverage area. With the UE4 recording features, it can playback

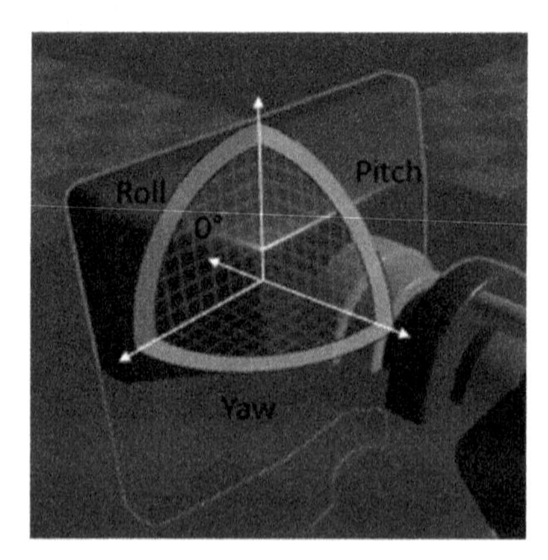

Fig. 8. The rotation direction for Pitch, Yaw and Roll in Unreal Engine 4

a visual log of the position of the user upper limb after the user finishes perform the functional tasks in the UE4.

The limitation of this study is the sample size is small. The system has not yet been tested involving other people such as upper limb stroke patients. The results from this study could be different if the evaluation is carried out in a clinical population. The healthy upper limb movements also have low variability which increases heterogeneity and can lead to high-reliability estimates. In the future, the system will be tested with a larger sample size, including stroke patients.

The movement of the elbow is not measurable if it depends on the HTC Vive controller alone. Therefore, the result only can measure the movement of joint 1 and joint 3. HTC Vive Company provides a tracker that can be attached on part of the body an act as a sensor. With the tracker, the movement of joint 2 can be measured. In the future, the system will be tested with the HTC Vive tracker and controller together to obtain the upper limb measurement.

Moreover, the precision of arm length measurement is not consistent when measured with HTC Vive. There are some inaccurate measurements when the patients are doing the arm length measurement alone. When a patient straightens his upper limb, the controller held at the hand is not parallel to the upper limb and both upper limb distances are closed to each other. Therefore, an assistant is required to help adjust the patient upper limb. In the future, the system will be implemented the angle detection and distance between two limbs detection for the controller so that accuracy of the arm length can be consistent.

8 Conclusion

The upper limb functional assessment system for HTC Vive is developed in this research to assess the upper limb length, adduction, abduction, flexion and extension. Moreover, it can measure the upper limb angle during the adduction, abduction, flexion and extension. It also can measure the range of motion when doing the functional task. All the functional tasks can be assessed accurately such as shoulder and elbow flexion/extension and shoulder adduction/abduction. The combination with HTC Vive and Unreal Engine

4 can record the data in real-time. The proposed system is potentially useful for assessing stroke rehabilitation in the hospital and rehabilitation center.

Conflicts of Interest. The authors declare that they have no conflicts of interest.

Acknowledgement. We would like to thank TM Research & Development Grant (MMUE/180026) for supporting this project financially.

References

1. Cai, L., Ma, Y., Xiong, S., Zhang, Y.: Validity and reliability of upper limb functional assessment using the microsoft kinect V2 sensor. Appl. Bionics Biomech. (2019). https://doi.org/10.1155/2019/7175240
2. Nilsson, F., Persson, T., Sadjadee, S., Berglund, E.: Upper body ergonomics in virtual reality an ergonomic assessment of the arms and neck in virtual (2017)
3. Chen, D., Liu, H., Ren, Z.: Application of wearable device HTC VIVE in upper limb rehabilitation training. In: Proceedings of the 2018 2nd IEEEAdvanced Information Management, Communicates, Electronic and Automation Control Conference IMCEC, (IMCEC) 2018, pp. 1460–1464 (2018). https://doi.org/10.1109/imcec.2018.8469540
4. Anthes, C., García-Hernández, R.J., Wiedemann, M., Kranzlmüller, D.: State of the art of virtual reality technology. In: IEEE Aerospace Conference Proceedings, June 2016–October 2017 (2016)
5. Li, C., Fahmy, A., Sienz, J.: An augmented reality based human-robot interaction interface using Kalman filter sensor fusion. Sensors **19**(20), 4586 (2019)
6. Torres-Ferreyros, C.M., Festini-Wendorff, M.A., Shiguihara-Juarez, P.N.: Developing a videogame using unreal engine based on a four stages methodology. In: 2016 Proceedings of the 2016 IEEE ANDESCON, ANDESCON (2017). https://doi.org/10.1109/andescon.2016.7836249
7. Drozina, A., Orehovacki, T.: Creating a tabletop game prototype in unreal engine, vol. 4, pp. 1568–1573 (2018)
8. van Andel, C.J., Wolterbeek, N., Doorenbosch, C.A., Veeger, D.H., Harlaar, J.: Complete 3D kinematics of upper extremity functional tasks. Gait Posture **27**(1), 120–127 (2008). https://doi.org/10.1016/j.gaitpost.2007.03.002
9. Ardakani, H.A., Bridges, T.J.: Review of the 3-2-1 Euler angles : a yaw – pitch – roll sequence map from E j to a j : the yaw rotation, pp. 1–9 (2010)

2D-CNN Based Segmentation of Ischemic Stroke Lesions in MRI Scans

Pir Masoom Shah[1], Hikmat Khan[1], Uferah Shafi[2], Saif ul Islam[3], Mohsin Raza[4], Tran The Son[5(✉)], and Hoa Le-Minh[6]

[1] COMSATS University Islamabad, Islamabad, Pakistan
[2] National University of Science and Technology (NUST), Islamabad, Pakistan
[3] Institute of Space Technology (IST), Islamabad, Pakistan
[4] Edge Hill University, Ormskirk, UK
[5] Vietnam – Korea University of Information and Communication Technology, Da Nang City, Vietnam
ttson@vku.udn.vn
[6] Northumbria University, Newcastle Upon Tyne, UK

Abstract. Stroke is the second overall driving reason for human death and disability. Strokes are categorized into Ischemic and Hemorrhagic strokes. Ischemic stroke is 85% of strokes while hemorrhagic is 15%. An exact automatic lesion segmentation of ischemic stroke remains a test to date. A few machine learning techniques are applied previously to beat manual human observers yet slacks to survive. In this paper, we propose a completely automatic lesion segmentation of ischemic stroke in view of the Convolutional Neural Network (CNN). The dataset used as a part of this study is obtained from ISLES 2015 challenge, included four MRI modalities DWI, T1, T1c, and FLAIR of 28 patients. The CNN model is trained on 25 patient's data while tested on the remaining 3 patients. As CNN is only used for classification, we convert segmentation to the pixel-by-pixel classification tasks. Dice Coefficient (DC) is used as a performance evaluation metric for assessing the performance of the model. The experimental results show that the proposed model achieves a comparatively higher DC rate from 4–5% than the considered state-of-the-art machine learning techniques.

Keywords: Stroke · MRI · Deep learning · Convolutional Neural Network

1 Introduction

Stroke is a medical condition of the brain, happens due to the short supply of oxygenated blood to the brain cells. Thus, cells begin dying. Sometimes in the brain, clots are produced in blood vessels and cause a limited supply of oxygenated blood to the brain that in turn causes a stroke. Stoke is the second leading death in humans [1] while the survivors are left with disability [2]. Generally, stroke-disability includes memory loss, paralysis, movement control, emotional disturbances [3], and so on. The primary drivers of strokes are

M. Hernes et al. (Eds.): ICCCI 2020, CCIS 1287, pp. 276–286, 2020.
https://doi.org/10.1007/978-3-030-63119-2_23

hypertension and high blood pressure [4]. Despite the fact that a stroke may happen at any age, most of the stroke patients are more than 65 years old [5]. Stroke is mainly categorized as Ischemic and Hemorrhagic Stroke. 85% of stroke patients are diagnosed with Ischemic type while the remaining 15% has Hemorrhagic-Stroke [6]. The Ischemic-Strokes is additionally divided into four stages, Hyper-Acute, Acute, Sub-Acute, and Chronic. All these four stages are categorized on the time-period basis. Hyper-acute refers to the first 4 h onset while acute is more than 6 and less than 24 h. In contrast, Sub-acute are one to seven days longer while chronic is longer than a whole week. This study focuses on the Sub-Acute Stage of Stroke. Techniques like computed tomography (CT), MRI (magnetic resonance imaging), and X-Ray are used to observe the detailed structure and condition of an organ, tissue, or cell in the human body. Each of these techniques has particular advantages and disadvantages associated with them and can be used for the specific application. Among all imaging techniques, MRI is found non-invasive which also offers intensive information in regards to ischemic stroke even in early stages [6]. Commonly after ischemic stroke, numerous changes happen in brain water content, while the MRI is extremely sensitive to detect alteration in tissue-free water content even after 1-h onset of ischemic stroke.

While providing treatment to patients of Ischemic-Stroke, a specialist is interested in discovering total volume, mass, and location of the lesion from MRI scans [7]. A conventional way of estimating the nature of the lesion is manual segmentation, in which physicians or radiologists manually assess the location and size of the lesion in all slices of MRI scans. Although this technique is effective but has several weaknesses. For instance, it is tedious and different observers may reach different conclusions. Due to the aforementioned problems, an extreme need for an automatic solution for localization and segmentation of lesions of ischemic-stroke exists. Although many CAD (Computer-Aided Diagnosis) based frameworks are proposed earlier for ischemic stroke lesions segmentation, they still lag to beat human observers.

Automatic lesions segmentation in MRI isn't a simple task, as the location and shapes rely upon several factors, such as occlusion site, time after symptom onset, and contrast in vessel anatomy of different patients. The presence of white matter hyper-intensities is another challenging issue as it intrudes on the accurate segmentation [8]. While dealing with these challenges, we implemented CNN for ischemic-stroke lesions segmentation and enhanced accuracy. The main contributions of this study are given below:

- A deep 2D-CNN technique is proposed for ischemic-stroke lesion segmentation.
- We achieved stat of the are results.
- Our proposed model is robust to overfitting issues on limited data.
- Additionally, batch normalization is applied before each LeakyReLU layer.

The organization of the paper is described as follows: detailed literature of ischemic-stroke lesion segmentation and classification is presented in Sect. 2.

Section 3 presents the methodology of the proposed approach. Section 4 contains the performance evaluation of the proposed scheme and includes experimental setup, performance measures, and evaluation of the proposed technique. Section 5 concludes the paper.

2 Literature Review

A considerable measure of work has already been done in the domain of medical image segmentation to achieve high accuracy and efficiency. It is one of the difficult but interesting research topics. Liang et al. [8] proposed a new technique for the segmentation of sub-acute stroke lesions. In this framework, the random forest was used as a classifier and intensities of the patches as features. The task was performed by the SISS data set, achieving a dice score of 0.55.

Maier et al. in [9] proposed an automatic sub-acute lesion segmentation, based on extra tree forest. They used a local dataset comprised of MRI scans of 37 patients with modalities of T1 weighted, T2 weighted, FLAIR, and DWI. Intensities based features were extracted and extra tree forest was used for segmentation. Their results show a DC value of 0.65.

Maier et al. [10] presented a Random Forest-based segmentation for stroke lesion. A total of 50 trees of the random forest were trained by 1,000,000 random samples. The technique was applied to ISLES 2015 challenge dataset. This framework achieved the mean DC, ASSD, and HD of 0.58, 7.91, and 34 respectively.

In this article [11] the authors extended their previous work from brain tumor segmentation to ischemic-stroke lesion segmentation. Initially, they extracted Local Texture Feature (LTF) from the MRI scans and subsequently performed Intensity Inhomogeneity correction. In order to get the local gradient, eigenvalues decomposition of the 2D structure tensor matrix was applied. Feature ranking techniques were further applied in order to select the top-ranking features. Only 19 of 35 features were selected for training the random forest. The experiments were performed on ISLES 2015 challenge dataset, their results show a mean dice score of 59%.

Mahmood et al. [12] presented a random forest-based technique to perform the automatic segmentation of ischemic-stroke lesions. The dataset was obtained from ISLES 2015 challenge. Bias field correction and normalization were applied in initial steps. Specific features were extracted by various techniques such as intensity, intensity difference, and gradient in the x-direction. They normalize all the selected features by zero mean and the resultant features were used by the random forest. The proposed model achieved the mean ASSD, Dice, Hausdor Distance, Precision and Recall of 10.30, 0.54, 82.78, 0.67, and 0.50 respectively.

2.1 CNN Based Approaches for Stroke Lesion Segmentation

In [13] the author proposed a CNN based approach for ischemic lesion segmentation in DWI MRI scans. The proposed framework consists of two convolutional neural networks, MUSCLE and EDD Nets. In the first phase, EDDNet was applied for the detection of lesions in scans while MUSCLE Net was used to evaluate the detection of EDD Net. The dataset used was obtained from a local hospital containing DWI scans of 746 patients. The model achieved a dice coefficient of 0.67.

Maier et al. performed a comparative study in [14] where several machine learning techniques such as Generalized Linear Models, Random Decision Forests, and CNN were applied to an MRI sequence of 37 sub acute patients of ischemic-stroke. The performance of these techniques along with the two human observers was compared and evaluated, but none of the machine learning techniques achieved an accuracy equivalent to the human observers. However, Random Decision Forest was found useful among all methods which outperformed all other techniques on the same dataset.

In [15] Dutil et al., used different Deep Neural Networks (DNN) architectures on ISLES and SPES challenge datasets. A two-pathway architecture was found useful, as it performed better than the rest of the network architectures. This model is a refined version of the network used in the MICCAI brain tumor segmentation (BRATS) challenge by the same author. In the CNN model, input images are passed to both path layers where each path layer is responsible to learn local or global details from the images. After passing images from these layers, the output features maps of these layers are rejoined at last convolutional layer, before the final probabilistic layer. The model has achieved the mean ASSD, Dice, Hausdor Distance, Precision, and Recall of 8.92, 0.6, 31.75, 0.72, and 0.67 respectively on SISS dataset. While on the SPES dataset the model achieved ASSD, Dice, Hausdor Distance, Precision and Recall of 1.76, 0.85, 23.28, 0.83, and 0.88 respectively.

Kamnitsas et al. in [16] presented CNN 11-layer deep 3D CNN for the segmentation of brain lesions. This model is the winner of the ISLES 2015 challenge and achieved promising results of 64% in terms of dice score. They also made it possible to train a CNN model on the smaller datasets, to avoid over-fitting problems. Table 1 summarizes the related work.

3 Methodology

The proposed methodology is described as follows:

3.1 Data Acquisition

To evaluate the proposed technique, we use ISLES challenge dataset [17]. This challenge aimed to detect tissue loss during ischemic-stroke in MRI. Worldwide experts were invited to take part in this challenge. A total of 28 patient's data was obtained and each patient contains four MRI modalities DWI, FLAIR T1, and T1c along with pixel-wise ground truth as shown in Fig. 1.

Table 1. Summary of literature review

Authors/Techniques	Dataset/type	DC	Flaws
Reza/local gradient and texture feature [11]	ISLES 2015 challenge/MRI scans	0.59	Trained on selected features
Liang/random forest [8]	SISS/MRI scans	0.55	Trained on selected features
Maier/extra tree forest [9]	Local Dataset/MRI scans	0.65	Not tested on benchmark dataset
Mahmood/Random Forest [12]	ISLES 2015 challenge/MRI scans	0.54	Trained on selected features
Liang chen et al./CNN (EDD and MUSCLE Nets.) [13]	Local Dataset/MRI (DWI scans)	0.67	Data class Imbalanced
Maier et al./Generalized Linear Models, Random Decision Forests, and Convolutional Neural Networks [14]	Local Dataset/MRI scans	0.67	Not tested on benchmark
Dutil et al./CNN (two-pathway architecture) [15]	SISS dataset/MRI scans	0.6	Data class Imbalanced
Kamnitsas et al./CNN [16]	ISLES 2015 challenge/MRI scans	0.64	Data class Imbalanced

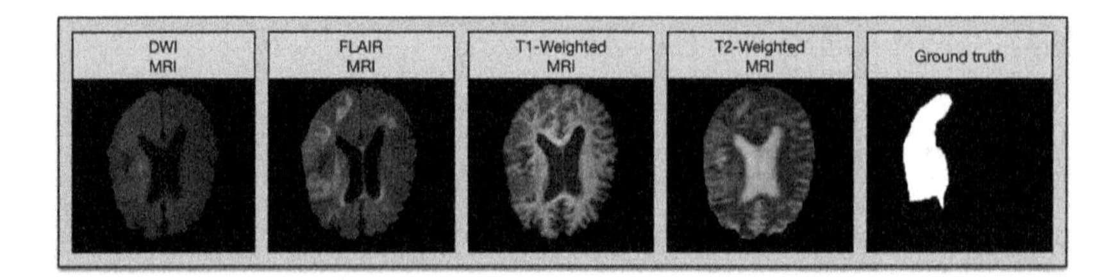

Fig. 1. Four different MRI modalities along with ground-truth against each patient.

3.2 Pre-processing

We use 25 patients' data for training and validation while the remaining three patient's data for testing the proposed model. The training and testing of patient sets are exclusive. The images are already skull-stripped and registered. As the convolutional neural network is only limited to image classification, we convert segmentation to the pixel classification task. Moreover, we extracted 50,000 patches of dimension 33×33 randomly while the number of strokes and health pixels are extracted in an equal ratio to avoid the imbalanced class distribution problem.

3.3 Proposed Model

Convolutional Neural Network (CNN) is a biologically inspired DNN. The self-feature learning capability of CNN enables it to produce better performance than other existing techniques for computer vision tasks. Due to the record-shattering performance of CNN, it is applied extensively to biological classification and segmentation problems during the last decade. Generally, a CNN model is composed of several layers including convolutional layers, pooling layer (max-min), and fully connected layers. Grouping these layers for a problem in an effective way is a key factor that subsequently produces the finest results. The upcoming sections will discuss the order and nature of layers used in the proposed model. Figure 2 shows the system model of the proposed scheme.

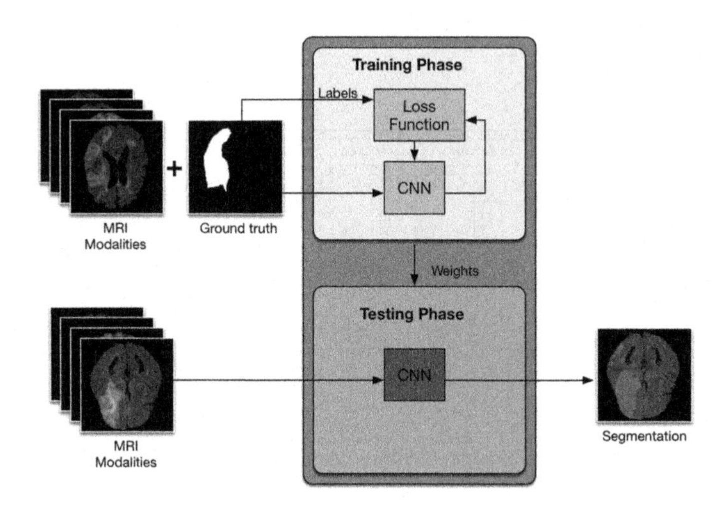

Fig. 2. Proposed framework for segmentation of Ischemic Stroke lesions.

3.4 Proposed Network Architecture

In this study, we propose 2D-CNN for Ischemic stork brain lesions segmentation. This model has the advantage of a small convolutional kernel of the size 3×3 for the entire network. Since we deal with limited data, therefore, the smaller conventional kernel with less parameter was a better option. However, on the other hand, a larger convolutional kernels demand a high amount of training data as the larger convolutional kernels have large parameters. In the proposed model we used advanced LeakyReLU action function and every convolutional kernel is followed by LeakyReLU. The LeakyReLU has the ability to sort out ReLU issues like dying ReLU, where LeakyReLU addresses this issue by adjusting negative gradients on backpropagation. We used batch normalization to enhance performance and training. This model takes the patches of the size 33×33 as input. Moreover, we used max-pooling layers to reduce learn-able parameters. Three max-pooling layers with the kernel size of 3×3 and stride 2×2 are placed between the 1st three convolutional layers, more precisely the first three convolutional layers are followed by max-pooling layers. The generated feature maps of the 1st three convolutional layers are input to the further three convolutional layers and thus the output of these layers is then forwarded to another max-pooling layer with the kernel size of 3×3 and stride 2×2. The final output (feature maps) of the pooling layer has the dimensions of 128 * 7 * 7. Further, these feather maps are then connected to the fully connected layers. In the proposed model we used only two fully connected layers the first fully-connected layer is consists of 512 while the second is consists of 256 neurons. In both fully connected layers, we used advanced regularization technique, dropout with 0.1 value, in order to reduce overfitting. The Soft-max layer is used to get the segmentation probabilities. The graphical illustration of the network is shown in Fig. 3.

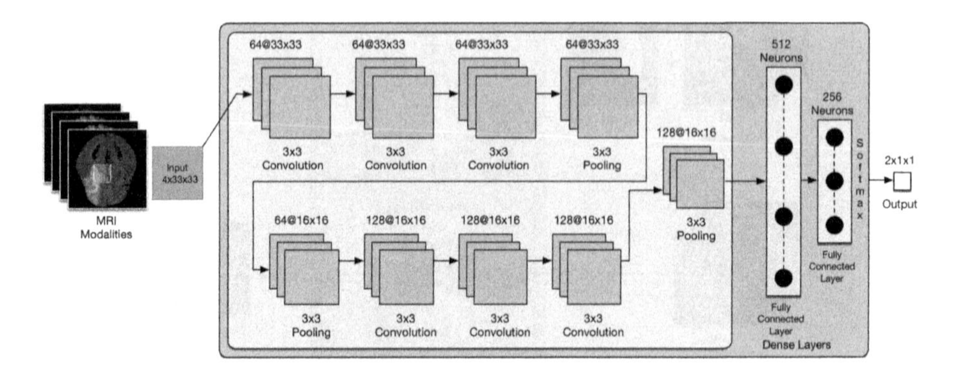

Fig. 3. Block diagram of CNN.

4 Performance Evaluation

As discussed earlier, in the pre-processing section, we divided the data into three sets, training, validation, and testing set. The training and testing of patient sets were exclusive. The training and validation progression of the model on the training data is shown in Fig. 4 while the respective minimization of the loss function can be seen in Fig. 5. We extracted 50,000 patches of dimension 33×33 randomly and in a balanced manner. (i.e. the count of stroke centered patch and a healthy brain are equal). This helped us to avoid the imbalanced class distribution problem. Our task is a binary classification. The CNN model has to classify the pixels as either stroke (i.e. lesion) or healthy. The coloring scheme is as follows, we used red color to refer to the stroked region. The pixel-wise classification of the proposed model results in the final lesion segmentation. The segmentation results are displayed without any post-processing. We used the Dice coefficient for the evaluation of the proposed model. It measures the overlap between the predicted outcome and provided ground truth. DC can be calculated as:

$$DC = \frac{2TP}{FP + 2TP + FN} \tag{1}$$

We used an exclusive set for testing the model after training on the training set. We reported DC for the middle 21 slices and for entire slices against each patient. Table 2 shows the DC obtained by the proposed model.

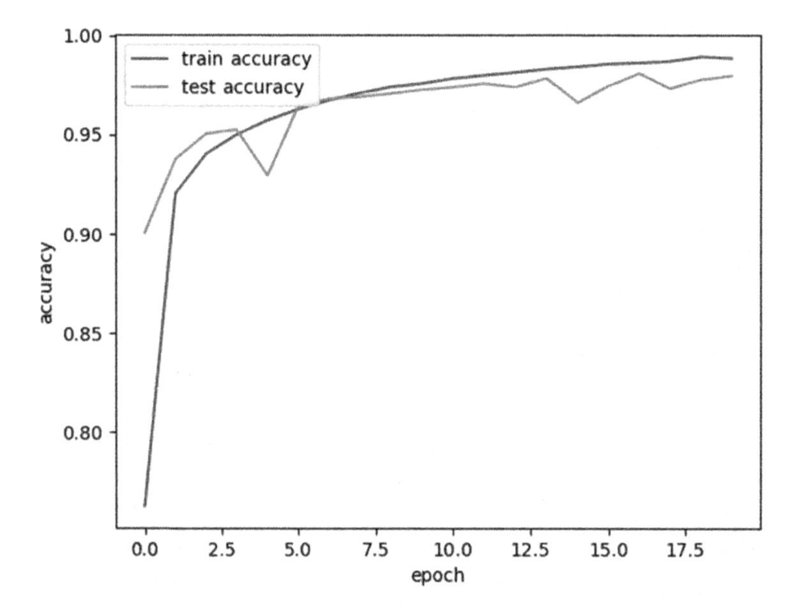

Fig. 4. This diagram shows the training and validation accuracy procession of the proposed CNN model.

Figure 6 shows the slice wise segmentation performed by the proposed model, where row 1 shows the slice of segmented MRI while row 2 shows the respective ground-truth of the slice. The proposed model achieved better accuracy than the random forest, extra tree forest, and other deep neural networks. The reason is the better handling of class imbalance problems and the learning of automated (respective to the problem) features. The detailed comparison of the proposed model along with the other existing techniques are presented in Table 3.

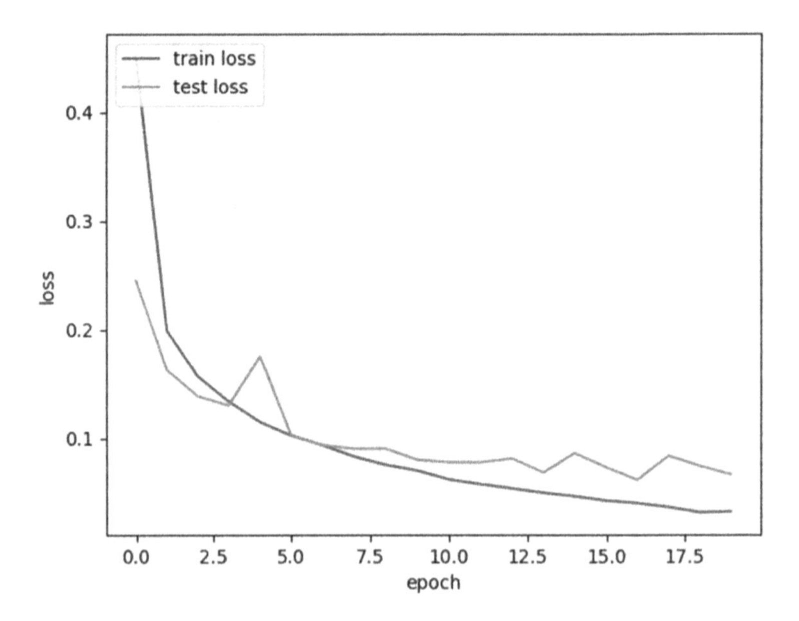

Fig. 5. This diagram shows the decrease in the loss (i.e. cost) function for the proposed CNN model.

Table 2. Model Performance on the test set

Test Patient	Mean dice coefficient over all slices	Mean dice coefficient over middle 21-Slices
Patient # 1	0.661320496	0.744979144
Patient # 2	0.822749173	0.869131752
Patient # 3	0.662616723	0.808938034
Mean Dice Coefficient	**0.715562131**	**0.807682977**

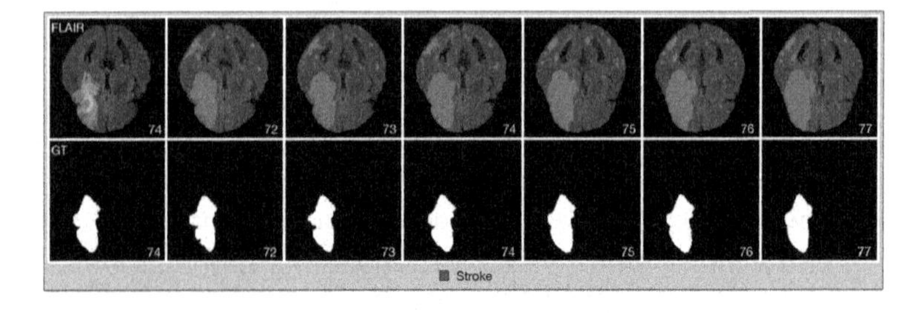

Fig. 6. Slice wise segmentation performed by the proposed model.

Table 3. Comparison of the proposed model with state-of-the-art machine learning techniques

Technique	Dataset/type	Mean dice coefficient
Random Forest [12]	ISLES 2015 challenge/MRI scans	0.54
Random Forest [8]	SISS/MRI scans	0.55
Extra tree Forest [9]	Local Dataset/MRI scans	0.05
CNN [16]	ISLES 2015 challenge/MRI scans	0.64
Proposed CNN	**ISLES 2015 challenge/MRI scans**	**0.71**

5 Conclusion

In this study, we proposed an entirely automatic brain lesion segmentation system based on CNN architecture. The proposed model with six convolutional layers is capable to learn the complex patterns of the lesions. The experiments are carried out on the benchmark ISLES 2015 challenge dataset. We extracted 50,000 patches of dimension 33×33 randomly and in a balanced manner. These patches were then incorporated into CNN. Since it is a binary classification task, CNN has to classify it as a lesion or a healthy pixel. During training, we noticed that limited samples of data are an issue for CNN that results in the model overfitting. However, the proper placement of dropout and the other network layers suppress the overfitting problem and efficiently improved the accuracy up to 10% when compared to other techniques.

References

1. Monteiro, M., et al.: Using machine learning to improve the prediction of functional outcome in ischemic stroke patients. IEEE/ACM Trans. Comput. Biol. Bioinform. **15**(6), 1953–1959 (2018)
2. Rajinikanth, V., Satapathy, S.C.: Segmentation of ischemic stroke lesion in brain MRI based on social group optimization and fuzzy-Tsallis entropy. Arab. J. Sci. Eng. **43**(8), 4365–4378 (2018). https://doi.org/10.1007/s13369-017-3053-6
3. Ghotra, S.K., Johnson, J.A., Qiu, W., Newton, A.S., Rasmussen, C., Yager, J.Y.: Health-related quality of life and its determinants in paediatric arterial ischaemic stroke survivors. Arch. Disease child. **103**(10), 930–936 (2018)
4. Wang, S., Head, B.P.: Caveolin-1 in stroke neuropathology and neuroprotection: a novel molecular therapeutic target for ischemic-related injury. Curr. Vas. Pharmacol. **17**(1), 41–49 (2019)
5. Gopalratnam, K., Woodson, K.A., Rangunwala, J., Sena, K., Gupta, M.: A rare case of stroke secondary to iron deficiency anemia in a young female patient. Case Rep. Med. 2017 (2017)
6. Meadows, K.L., Silver, G.M.: The effects of various weather conditions as a potential ischemic stroke trigger in dogs. Vet. Sci. **4**(4), 56 (2017)
7. Mitra, J.: Lesion segmentation from multimodal MRI using random forest following ischemic stroke. NeuroImage **98**, 324–335 (2014)

8. Chen, L., Bentley, P., Rueckert, D.: A novel framework for sub-acute stroke lesion segmentation based on random forest. Ischemic Stroke Lesion Segment. 9 (2015)
9. Maier, O., Wilms, M., von der Gablentz, J., Krämer, U.M., Münte, T.F., Handels, H.: Extra tree forests for sub-acute ischemic stroke lesion segmentation in MR sequences. J. Neurosci. Methods **240**, 89–100 (2015)
10. Maier, O., Wilms, M., Handels, H.: Image features for brain lesion segmentation using random forests. In: Crimi, A., Menze, B., Maier, O., Reyes, M., Handels, H. (eds.) BrainLes 2015. LNCS, vol. 9556, pp. 119–130. Springer, Cham (2016). https://doi.org/10.1007/978-3-319-30858-6_11
11. Reza, S.M., Pei, L., Iftekharuddin, K.: Ischemic stroke lesion segmentation using local gradient and texture features. Ischemic Stroke Lesion Segment. 23 (2015)
12. Mahmood, Q., Basit, A.: Automatic ischemic stroke lesion segmentation in multi-spectral MRI images using random forests classifier. In: Crimi, A., Menze, B., Maier, O., Reyes, M., Handels, H. (eds.) BrainLes 2015. LNCS, vol. 9556, pp. 266–274. Springer, Cham (2016). https://doi.org/10.1007/978-3-319-30858-6_23
13. Chen, L., Bentley, P., Rueckert, D.: Fully automatic acute ischemic lesion segmentation in DWI using convolutional neural networks. NeuroImage: Clin. **15**, 633–643 (2017)
14. Maier, O., Schröder, C., Forkert, N.D., Martinetz, T., Handels, H.: Classifiers for ischemic stroke lesion segmentation: a comparison study. PLoS ONE **10**(12), e0145118 (2015)
15. Havaei, M., Dutil, F., Pal, C., Larochelle, H., Jodoin, P.-M.: A convolutional neural network approach to brain tumor segmentation. In: Crimi, A., Menze, B., Maier, O., Reyes, M., Handels, H. (eds.) BrainLes 2015. LNCS, vol. 9556, pp. 195–208. Springer, Cham (2016). https://doi.org/10.1007/978-3-319-30858-6_17
16. Kamnitsas, K., Chen, L., Ledig, C., Rueckert, D., Glocker, B.: Multi-scale 3D convolutional neural networks for lesion segmentation in brain MRI. Ischemic Stroke Lesion Segment. **13**, 46 (2015)
17. Isles challenge dataset. http://www.isles-challenge.org/ISLES2015/. Accessed 22 Mar 2020

Melanoma Skin Cancer Classification Using Transfer Learning

Verosha Pillay, Divyan Hirasen, Serestina Viriri(✉), and Mandlenkosi Gwetu

School of Mathematics, Statistics and Computer Science,
University of KwaZulu-Natal, Durban, South Africa
{214539347,215018696,viriris,gwetum}@ukzn.ac.za

Abstract. Melanoma is one of the most aggressive types of skin cancer as it rapidly spreads to various areas of the body. With the increase and fatal nature of melanoma, it is of utmost importance to establish computer assisted diagnostic support systems to aid physicians in diagnosing skin cancer. In this paper, we make use of deep learning and transfer learning by testing 14 pre-trained models for the classification and detection of skin cancer. Historically, the data in which Deep Convolutional Neural Networks are fed and trained on comes predominantly from European datasets resulting in biased data. To overcome this issue, we first determine the differences of melanoma that lie within people of different skin tones. Thereafter, we make use of the GrabCut segmentation technique to accurately segment the lesion from the surrounding skin tone in order to solely focus on the lesion. The pre-trained CNN, Squeezenet1-1, achieved the best experimental results with an accuracy rate of 93.42%, sensitivity of 92.11% and specificity of 94.74%. The experimental results achieved indicate that there is a possible solution to the underrepresented data of dark-skinned people.

Keywords: Melanoma · Deep learning · Transfer learning · Dark-skinned

1 Introduction

Skin cancer is the out of control growth of irregular skin cells causing the skin to rapidly multiply and form malignant tumours [2]. Between 2008 and 2019 the number of new invasive melanoma cases rapidly increased by 54% [3]. With such a substantial escalation, it is of great importance for us to inaugurate and build on techniques to assist health-care providers in diagnosing skin cancer at its preliminary stages [4]. Although, melanoma is one of the most deadly types of skin cancer, an early diagnosis can significantly increase the survival rate of patients [5].

A key concern with skin cancer detection using Machine Learning (ML) and Deep Learning (DL) is research is primarily focused on fair-skinned populations which include Australia, Europe and the United States [6]. Consequently, little to no research has been conducted on skin cancer detection for a diverse

M. Hernes et al. (Eds.): ICCCI 2020, CCIS 1287, pp. 287–297, 2020.
https://doi.org/10.1007/978-3-030-63119-2_24

range of skin tones, resulting in a collection of biased data. As far as we know, no previous research has investigated skin cancer detection on a wide range of skin tones. However, Azehoun-Pazou et al. [1] proposed a study which focused on macroscopic image segmentation of black skin lesions. The authors conclude that their study was able to identify the significant differences that exist between dark and fair skin tones which include colour and anatomical structures. Over time, research work has developed a focus on training Artificial Intelligence (AI) algorithms with light skin types mainly due to the unavailable or underrepresented skin cancer datasets for darker skin tones. This poses some problems when researching skin cancer detection for all skin tones.

In order to rectify the problem of the lack of available datasets for all skin tones, this research work will focus on using a segmentation technique to accurately segment the suspected cancerous lesion from the background. Thereafter training a DL model by testing 14 pre-trained Convolutional Neural Networks (CNN) with these segmented images in order to determine the best model for this scenario. This allows us to focus on the lesion and removes skin colour from the overall problem. Thereby, allowing us to train the model with the dominant European datasets until future datasets incorporating all skin tones become available. Previous research showed DL with the use of CNNs achieved an error rate of 16.4% while other popular ML classifiers achieved an error rate between 26–30% [7,8]. Traditional ML models become progressively effective at their required function; however, it is highly dependent on human assigned features requiring prior domain knowledge. On the other hand, DL structures algorithms in layers to develop an "Artificial Neural Network" that can learn and determine features necessary to make intelligent decisions on its own [9]. DL is discovered and proves to have the best techniques with state-of-the-art performances [10,11]. This research work makes use of macroscopic images to distinguish melanoma and benign skin lesions. The use of conventional digital cameras are the most commonly utilized equipment to examine the features of pigmented skin lesions. These images are known as macroscopic or clinical images. The digital revolution is beginning to enhance standard dermoscopic procedures; hence macroscopic images are becoming widely popular due to its accessibility [12].

The main objective of this work is to demonstrate the feasibility of accurately detecting skin cancer by solely focusing on the skin lesion and not the skin tone surrounding it. Furthermore, to model a framework for accurate detection of skin cancer from macroscopic images.

2 Skin Cancer in People of Colour

Skin cancers are less prevalent in people of colour. On the other hand, when it develops or emerges it tends to be diagnosed at the last stages resulting in an enigmatic prognosis. In an article posted by the American Cancer Society [13], they estimated an average 5 year melanoma survival rate of only 65% in black people versus 91% in white people. Furthermore, in another paper [14] proposed, the authors found that late stage melanoma diagnosis are more frequent among Hispanic and black patients than non-Hispanic white patients. Furthermore, 52% of non-Hispanic black patients and 26% of Hispanic patients receive an initial diagnosis of advanced stage melanoma, versus 16% of non-Hispanic white patients. From the viewpoint of a medical professional, often there is negligible index of suspicion for skin cancer in patients of colour due to the small chances of occurrence [15]. Hence, it is likely that these patients have reduced full body skin examinations. With such staggering statistics, there is an essential need to improve or develop new non-invasive Computer Assisted Diagnostic (CAD) support systems which incorporate people of all colours in aiding physicians with the prognosis of melanoma.

It is of paramount importance for us to understand the difference between melanoma found in fair-toned people and dark skinned people. Azehoun-Pazou et al. [1] established the differences that exist between the various skin tones can be observed in terms of colour, anatomical structure and the part of the body the lesion develops. Generally, melanoma skin cancer tends to form on areas of the body which are sun-exposed; however, for people of colour melanoma tends to develop in more out of the way areas resulting in detection difficulties [15]. Acral Lentiginous melanoma (ALM) is a type of cancer that appears on the palms, nailbeds and soles of a person's foot). ALM has a much higher percentage of occurring in people of colour than Caucasians [16]. It is interesting to note that, although, ALM is less prevalent in fair-skinned people, it is often misdiagnosed or mistaken as an injury or nail fungus. Due to the common misdiagnosed, ALM is deeply invasive when it is finally identified [17]. Light-skinned people indicate the appearance of spreading macules or patch from the suspected lesion while darker skin tones share the same changes but the presence of a changing pigmented band on the nail.

3 Segmentation

In this paper we make use of the iterative GrabCut image segmentation method which is based on graph cuts. In this technique the user indicates a bounding box by drawing a rectangle around the suspected lesion to be segmented and also allows the user to do corrective editing by selecting background and foreground pixels. The GrabCut algorithm then estimates the colour distribution of the target (lesion within the rectangle) and the background by making use of a Gaussian Mixture Model (GMM) shown in calculation (1). This is used to construct a Markov random field over the pixel labels, with an energy function (3) that prefers connected regions having the same label, and running a graph cut based optimization to infer their values.

$$p(x) = \sum_{i=1}^{M} \pi_i N_i(x|\mu_i C_i) \tag{1}$$

Where $N_i(x|\mu_i C_i)$ denotes a bivariate or trivariate normal distribution with mean vector μ_i and a covariance matrix $C_i \times \pi$ is a mixture proportion for each group and it is regarded as the ith prior probability of Gaussian distribution that data sample produces. These prior probabilities should satisfy (2) :

$$\sum_{i=1}^{M} \pi_i \; and \; 0 \leq \pi_i \leq 1 \tag{2}$$

$$E = (I_{i,j}, S, C, \lambda) \tag{3}$$

Where S takes values 0 for background and 1 for foreground to perform hard segmentation (algorithms that produce a binary map, i.e. a pixel belongs to either foreground or background). C denotes the colour parameter and λ represents the coherence parameter [18].

The GrabCut technique allows the segmentation of images in colour as it makes use of an energy minimization approach by taking colour and contrast information into account [19]. This method is suitable for highly textured and noisy images as macroscopic images indicate the presence of noise such as hair, illumination variation etc. Furthermore, this method assists in reducing the user interaction by using mechanisms called "iterative estimation" and "incomplete labelling" [20].

4 Methods and Materials

4.1 Datasets

In this paper, a variety of popular pre-trained models were applied on a combination of three datasets i.e.: MED-NODE [24], DermIS [25] and DermQuest [26] in order to determine the best performing model for skin cancer detection. The sample data consists of a total of 376 macroscopic images. The sample size in this study may not be considered large enough for Deep Learning as large datasets proved to be very powerful especially for medical analysis [21]. Thus, additional data will be added from non-public datasets as we are still awaiting images to be sent which will be then be tested and added to the final paper. However, datasets of comparable size are not unusual for macroscopic images [22], due to the challenging nature (manual labour) of acquiring labelled data from medical experts which can be tedious and result in vacillations in the medical diagnosis [23]. Moreover, zipped files of the selected macroscopic images from DermQuest and DermIs can be found on [22] which is provided by the Vision and Image Processing Lab research group of the University of Waterloo.

4.2 Transfer Learning

Transfer learning is a riveting paradigm, most popularly employed in DL due to the massive amounts of data required to train models. Transfer learning trains a base network on a large dataset and task in which the learned features are transferred to a second network to be trained on a target dataset (small and different from the base dataset) and task [29]. Due to the scarcity of sufficiently large datasets being available, a small number of people opt to train entire CNNs from scratch. Thus, it is common for people to pre-train a CNN on a large dataset such as ImageNet [27]; thereafter, use the CNN to jump start the development process on the new task [28,29]. Traditionally, the concept of transfer learning and pre-training are conceptually fairly similar; however, in pre-training, the network architecture is defined then trained on a large dataset; whereas in transfer learning, the network architecture such as Resnet, Alexnet, VGG etc. must be transferred as well as the weights [21]. Transfer learning helps decrease the required time necessary to develop and train a model which in turn accelerates results [28]. Furthermore, by making use of the source model knowledge, this can assist in fully learning the target task rather than it learning from scratch.

4.3 Data Argumentation

In order to ensure the deep learning models achieve good performance with the small amount of training data we currently have attained, we have opted to make use of data argumentation. Data augmentation artificially generates new training data from existing training data through various ways of processing/transformations. These transformations usually include random rotations, flips, zoom, translations, shifts etc. Essentially employing data augmentation does not make any changes inside the image (to the human eye) but rather change its pixel values [31]. Data augmentation is one of many methods developed to reduce the problem of overfitting by converging to the root of the problem i.e.: the training data [21]. Moreover, models trained with data augmented data will tend to generalize better.

4.4 Batch Normalization

Batch normalization is a technique employed to increase the stability of a neural network by normalizing the output of an antecedent activation layer. This is completed by subtracting the batch mean and dividing the standard deviation [32], depicted in (4). Moreover, this technique enables each layer to learn by itself independently of the other layers in the network.

$$B(x) = \gamma \frac{(x - \mu)}{\sigma} + \beta \tag{4}$$

Where μ and β represent the mean and standard deviation of the batch. γ and β are learned parameters. In an article posted by Jason Brownlee [33], he states batch normalization stabilizes the learning process and ideally cuts back on the number of epochs used for training. Using batch normalization helps reduce the dependency of gradients on the scale of parameters and makes training more resilient to the parameter scale [30]. Batch normalization allows us to make use of significantly higher learning rates resulting in the speed at which a network trains to be increased.

4.5 Evaluation Metrics

The performance of the proposed model makes use of the standard medical imaging evaluation metrics, Sensitivity, Specificity and Accuracy. However, the evaluation metrics employed for segmentation include Matthew Correlation Coefficient (MCC), Dice (DSC) and Jaccard (JAC) index.

5 Discussion

Table 1 lists the results of various techniques implemented in literature. This table suggests the use of transfer learning models are becoming a widely popular choice for the classification of skin cancer. Furthermore, the results highlight that large amounts of labelled data is necessary to obtain good results or to further improve results. However, large datasets for macroscopic images are challenging to acquire due public unavailability or difficulties obtaining these images by medical experts. Contrary, to publically available datasets, to our knowledge there are no macroscopic datasets consisting of all skin tones for skin cancer detection. One concern regarding results in literature is due to the lack of fair and dark toned people which leads to biased results for skin cancer detection. This is a major source of limitation and can only be accounted for once datasets consisting of a variety of skin tones are constructed and made publicly available.

There are four main types of melanoma skin cancer - superficial spreading, nodular, lentigo maligna and acral lentiginous. Each of these types can develop on different parts of the body e.g.: Lentigo maligna forms on areas of the skin such as the face, ears and arms while superficial spreading melanoma develops on the central part of the body which includes the back, arms and legs. Research and investigation of these four types of melanoma for CAD systems are rarely analysed in literature as datasets provided don't present labelled data by the type of melanoma but rather the type of skin cancer i.e.: basal cell carcinoma (BCC), squamous cell carcinoma (SCC), and melanoma. This is important because each of the four types of melanoma differ in terms of colour and structure, which are two highly important features when diagnosing melanoma. Furthermore, the dissimilitude of features can greatly impact the number of misclassifications between malignant and benign lesions. Moreover, identifying the correct type of melanoma can assist in early detection as some of the four types of melanoma are hard to detect or misclassified as a regular mole. Thus, physicians can identify the stage of the melanoma and correctly treat the patient. Figure 1 shows the four types of melanoma with their associated features.

Taking a closer look at the literature Table 1, we note that the results demonstrated in paper [41], proposed by Hosny et al. achieved an accuracy rate of 98.61% which is 5.19% higher than the proposed method. The authors make use of the Ph2 dataset which consists of a total of 200 dermoscopic images [45]. Although, Honsy et al. achieved a higher accuracy, sensitivity and specificity, the proposed method is trained and tested using datasets that consist of images which contain all four types of melanoma. Whereas, the Ph2 dataset contains only 40 melanoma images which may or may not incorporate the different

Table 1. Comparative analysis of various methods and techniques in literature

Paper	Proposed	Datasets	Image total	Classification model	Image type	Sr %	Sp%	Acc%
[36] 2018	Classification of 2 different types of skin cancer i.e: Malignant melanoma ad Basal cell carinoma (12 lesions) using Deep Learning	Med-Node, Edinburg and Atlas	111069	Resnet152	Macroscopic	96 AUC	95 AUC	–
[37] 2018	Deep learning algorithm to classify the clinical images of 12 skin diseases of skin cancer	Asan, MED-NODE and Alas site images	19398	Resnet152	Macroscopic	88 AUC	94 AUC	57.3
[38] 2017	Classification of skin lesions using a single CNN	–	129450	InceptionV3	Macroscopic	96 AUC	–	72.1
[10] 2018	Diagnostic classification of dermoscopic images of melanocytic origin	–	–	InceptionV4	Deromscopic	–	82.5	–
[39] 2014	Distinguish between benign and malignant pigmented lesions	DermAtlas, Dermnet, DermQuest, DermIs and Atlas	282	SVM	Macroscopic	77	86.93	82.2
[40] 2018	A system for the melanoma skin cancer detection to differentiate benign and malignant lesions	MED-NODE	170	Decision Tree	Macroscopic	–	–	82.35
[41] 2018	An automated skin lesion classification method is proposed to classify melanoma and nevus	Ph2	200	alexnet	Dermoscopic	98.33	97.73	98.61
[42] 2016	A proposed system distinguishing between melanoma and benign cases	MED-NODE	6120 original and synthesised	CNN	Macroscopic	81	80	81
[43] 2018	A model proposed to distinguish between melanoma and benign lesions to generate mobile compatible models	ISIC archive, Dermnet NZ, MED-NODE and Ph2	–	CNN	Macroscopic and Dermoscopic	88.23	83.82	86
[44] 2016	Track and classify suspicious skin lesions as benign or malignant	ISIC	1280	VGG16	Macroscopic	–	–	78
Proposed	Differentiate between Melanoma and Benign lesions	MED-NODE, DermQuest and DermIs	376	$Squeezenet1_1$	Macroscopic	92.11	94.74	93.42

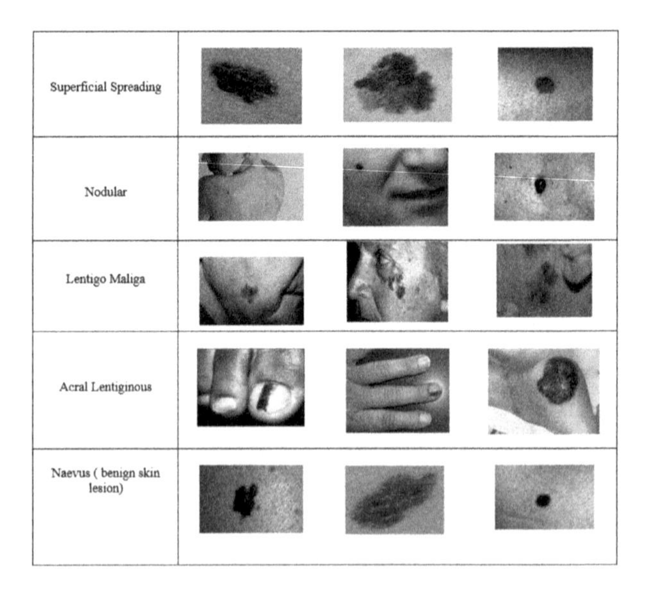

Fig. 1. Examples of the 4 types of Melanoma

melanoma types resulting in a less adaptable model when given other types of melanoma. This could possibly result in a handful of misclassifications. Hence, the proposed model is better equipped to identify variability among the four types of melanoma. Contrary to the findings of paper [41], it is not necessarily a fair comparison to make as Honsy et al. make use of dermoscopic images while the proposed method uses macroscopic images, each image type has its own set of differences, limitations and benefits.

6 Conclusion

This research paper investigates the precise segmentation of lesions from surrounding skin in order to overcome the issue of biased data. Furthermore, we make use of deep learning and transfer learning by testing 14 pre-trained models to determine which of these models are best suited for melanoma detection. We perform quantitative evaluation of our work by employing three traditionally used metrics for classification problems which include: accuracy, sensitivity and specificity. Moreover, we conduct an extensive comparative analysis of various methods and techniques implemented in literature with use of machine learning and deep learning. Generally there are two types of images used for skin cancer detection mainly, dermoscopic and macroscopic images. For this work, we make use of macroscopic images which are images acquired from conventional cameras. In conclusion, the Squeezenet1-1 model achieved the highest results with an accuracy rate of 93.42%, sensitivity 92.11% and specificity 94.74%. By focusing on the skin lesion and not the surrounding skin tone, we are able to remove any confusion with the amount of brown shades from the skin tone and lesion to correctly differentiation between a benign and malignant lesion. This indicates a possible solution to overcoming the lack of data available in order to create systems that are able to detect melanoma on all skin tones. In future research, more research is needed to apply and test methods on people of colour. However, this can only be accomplished once datasets consisting of a range of skin tones are presented. In addition, future research could look into synthesizing data until data from all ethnic groups are represented.

References

1. Assogba, M., Vianou, A., Azehoun-Pazou, G.: A method of automatic black skin lesion's macroscopic image analysis Géraud Azehoun-Pazou Letia Epac 01 BP 2009 Cotonou Rep BENIN, **4**, 2278–7720 (2014)
2. Skin Cancer Foundation. https://www.skincancer.org/. Accessed 16 July 2019
3. Skin Cancer Facts. www.skincancer.org/skin-cancer-information/skin-cancer-fact. Accessed 18 Apr 2019
4. Zhang, Y.J., Tavares, J.M.R.S.: Computational Modeling of Objects Presented in Images: Fundamentals, Methods, and Applications. Springer, Heidelberg (2014). https://doi.org/10.1007/978-3-319-09994-1
5. Brinker, T., et al.: Skin cancer classification using convolutional neural networks: systematic review (2018)
6. AI-Driven dermatology could leave dark-skinned patients behind. https://www.theatlantic.com/health/archive/2018/08/machine-learning-dermatology-skin-color/567619/. Accessed 21 July 2019
7. Fujisawa, Y., et al.: Deep learning-based, computer-aided classifier developed with a small dataset of clinical images surpasses board-certified dermatologists in skin tumor diagnosis. Br. J. Dermatol. (2018). https://doi.org/10.1111/bjd.16924
8. Russakovsky, O., et al.: ImageNet large scale visual recognition challenge. Int. J. Comput. Vis. **115**(3), 211–252 (2015). https://doi.org/10.1007/s11263-015-0816-y
9. A simple way to understand machine learning vs deep learning. https://www.zendesk.com/blog/machine-learning-and-deep-learning/. Accessed 23 July 2019
10. Haenssle, H.A., et al.: Man against machine: diagnostic performance of a deep learning convolutional neural network for dermoscopic melanoma recognition in comparison to 58 dermatologists. Ann. Oncol. **29**, 1836–1842 (2018)
11. What's new in digital health. https://www.dermengine.com/blog/artificial-intelligence-in-dermatology-diagnosis-deep-learning. Accessed 17 July 2019
12. Pillay, V., Viriri, S.: Skin cancer detection from macroscopic images, pp. 1–9 (2019). https://doi.org/10.1109/ICTAS.2019.8703611
13. Cancer facts and figures. https://www.cancer.org/research/cancer-facts-statistics/all-cancer-facts-figures/cancer-facts-figures-2019.html. Accessed 11 Sept 2019
14. Hu, S., Soza-Vento, R.M., Parker, D.F., et al.: Comparison of stage at diagnosis of melanoma among Hispanic, black, and white patients in Miami-Dade County, Florida. Arch Dermatol. **142**(6), 704–8 (2006)
15. Skin Cancer Foundation. https://www.skincancer.org/blog/ask-the-expert-is-there-a-skin-cancer-crisis-in-people-of-color. Accessed 11 Sept 2019
16. Skin Cancer Foundation. Skin Cancer Facts and Statistics. https://www.skincancer.org/skin-cancer-information/skin-cancer-facts/. Accessed 11 Sept 2019

17. AIM at Melanoma Foundation. Melanoma in People of Colour. https://www.aimatmelanoma.org/melanoma-risk-factors/melanoma-in-people-of-color/. Accessed 11 Sept 2019
18. Parvathy, R., Livingston, J.: A review on various graph cut based image segmentation schemes (2013)
19. Franke, M.: Color image segmentation based on an iterative graph cut algorithm using time-of-flight cameras. In: Mester, R., Felsberg, M. (eds.) DAGM 2011. LNCS, vol. 6835, pp. 462–467. Springer, Heidelberg (2011). https://doi.org/10.1007/978-3-642-23123-0_49
20. Kulkarni, M., Nicolls, F.: Interactive image segmentation using graph cuts. In: Proceedings of 20th Annual Symposium of the Pattern Recognition Association of South Africa (PRASA 2009), Stellenbosch, South Africa (2009)
21. Shorten, C., Khoshgoftaar, T.M.: A survey on image data augmentation for deep learning. J. Big Data **6**(1), 1–48 (2019). https://doi.org/10.1186/s40537-019-0197-0
22. Abbas, Q., Emre Celebi, M., Garcia, I.F., Ahmad, W.: Melanoma recognition framework based on expert definition of ABCD for dermoscopic images. Skin Res. Technol. **19**, e93–e102 (2013). https://doi.org/10.1111/j.1600-0846.2012.00614.x
23. Giotis, I., Molders, N., Land, S., Biehl, M., Jonkman, M.F., Petkov, N.: MED-NODE: a computer-assisted melanoma diagnosis system using non-dermoscopic images. Expert Syst. Appl. **42**(19), 6578–6585 (2015). https://doi.org/10.1016/j.eswa.2015.04.034
24. Dermatology database used in MED-NODE. http://www.cs.rug.nl/~imaging/databases/melanoma_naevi/. Accessed 9 Aug 2019
25. Derm101. http://www.dermquest.com. Accessed 9 Aug 2019
26. DermIs. http://www.dermis.net. Accessed 9 Aug 2019
27. Jia, D., Wei, D., Richard, S., Li-Jia, L., Kai, L., Li, F.-F.: ImageNet: a large-scale hierarchical image database. In: CVPR 2009 (2009)
28. An introduction to transfer learning in machine Learning. https://medium.com/kansas-city-machine-learning-artificial-intelligen/an-introduction-to-transfer-learning-in-machine-learning-7efd104b6026. Accessed 20 Aug 2019
29. Transfer learning introduction. https://www.hackerearth.com/practice/machine-learning/transfer-learning/transfer-learning-intro/tutorial/. Accessed 20 Aug 2019
30. Ioffe, S., Szegedy, C.: Batch normalization: accelerating deep network training by reducing internal covariate shift. In: ICML (2015)
31. Vision Transform. https://docs.fast.ai/vision.transform.html. Accessed 20 Aug 2019
32. Batch normalization in neural networks. https://towardsdatascience.com/batch-normalization-in-neural-networks-1ac91516821c. Accessed 20 Aug 2019
33. Accelerate the training of deep neural networks with batch normalization. https://machinelearningmastery.com/batch-normalization-for-training-of-deep-neural-networks/. Accessed 20 Aug 2019
34. Matthews correlation coefficient. https://en.wikipedia.org/wiki/Matthews_correlation_coefficient. Accessed 20 Aug 2019
35. Goyal, M., Yap, M.H.: Multi-class semantic segmentation of skin lesions via fully convolutional networks (2017)
36. Mendes, D.B., Silva, N.C.: Skin lesions classification using convolutional neural networks in clinical images. CoRR, abs/1812.02316 (2018)
37. Han, S.S., et al.: Classification of the clinical images for benign and malignant cutaneous tumors using a deep learning algorithm. J. Invest. Dermatol. **138**, 1529–1538 (2018)

38. Esteva, A., et al.: Dermatologist-level classification of skin cancer with deep neural networks. Nature **542**, 115–118 (2017)
39. Ramezani, M., Karimian, A., Moallem, P.: Automatic detection of malignant melanoma using macroscopic images. J. Med. Sig. Sens. **4**, 281 (2014)
40. Shalu, Kamboj, A.: A color-based approach for melanoma skin cancer detection. In: First International Conference on Secure Cyber Computing and Communication (ICSCCC), Jalandhar, India, pp. 508–513 (2018). https://doi.org/10.1109/ICSCCC.2018.8703309
41. Hosny, K., Kassem, M., Fouad, M.: Skin cancer classification using deep learning and transfer learning (2018). https://doi.org/10.1109/CIBEC.2018.8641762
42. Nasr-Esfahani, E., et al.: Melanoma detection by analysis of clinical images using convolutional neural network. In: 38th Annual International Conference of the IEEE Engineering in Medicine and Biology Society (EMBC), Orlando, FL, 2016, pp. 1373–1376 (2016) https://doi.org/10.1109/EMBC.2016.7590963
43. Ly, P., Bein, D., Verma, A.: New compact deep learning model for skin cancer recognition. In: 9th IEEE Annual Ubiquitous Computing, Electronics & Mobile Communication Conference (UEMCON), pp. 255–261 (2018)
44. Kalouche, S.: Vision-based classification of skin cancer using deep learning (2016). https://www.semanticscholar.org/paper/Vision-Based-Classification-ofSkin-Cancer-usingKalouche/b57ba909756462d812dc20fca157b3972bc1f533
45. Mendonça, T., Ferreira, P., Marques, J., Marçal, A., Rozeira, J.: PH2 - A dermoscopic image database for research and benchmarking. In: Conference proceedings: Annual International Conference of the IEEE Engineering in Medicine and Biology Society. IEEE Engineering in Medicine and Biology Society. Conference, pp. 5437–5440 (2013). https://doi.org/10.1109/EMBC.2013.6610779

Decision Support and Control Systems

Design a Neural Controller to Control Rescue Quadcopter in Hang Status

Nguyen Hoang Mai[1], Le Quoc Huy[1], and Tran The Son[2](✉)

[1] Danang University of Science and Technology, Da Nang, Vietnam
{nhmai,lqhuy}@dut.udn.vn

[2] Vietnam – Korea University of Information and Communication Technology, Da Nang, Vietnam
ttson@vku.udn.vn

Abstract. Quadcopters can be used for various applications in many fields including service, life, security, military… This is a research of oriented application development prospects in the near future because of the quadcopters ability to move flexibly regardless of the terrain and can support human issues that quadcopter on the ground do not. However, due to the motion characteristics in the air, so the quadcopter has certain limitations, including moving the static problem, the suspend state of the quadcopter. This is a complex subject and many scientists are interested in studying the development of large size quadcopters that can carry both people and heavy equipment. This paper analyzes some problems associated with quadcopter motion in a static state and designs a neural controller as the basis for developing more advanced applications in practice to manufacture the big quadcopter for load. The simulation results illustrate the problem and explain the relevance of the theory.

Keywords: Quadcopter · Helicopter · Body frame · Inertial frame · Control of quadcopter · Aerodynamics · Flight

1 Problem

It is well know that small quadcopters are in an environment with very strong interferences and the classic controller cannot meet desired characteristics, especially for controlling in hanging mode [1], the anti disturbance has canceled by using artificial neural network with a stabilization in fix region. Some study used ANN to help a signal process from camera or sensors [2]. This is a static state operation that requires dynamic balancing of many factors. Unlike the robot motion on hard surface or floating on the water, which we can break motion or anchor central, with quadcopter, only solve coordination problems in 4 engines and 4 rotors [5].

Indeed, driven by four propellers should quadcopter better stability than conventional helicopter with one rotor [9–11]. Therefore, if one can exploit this advantage of quadcopters to fabricate larger ones it will be possible to carry big loads more flexibly and more safely than current helicopters.

M. Hernes et al. (Eds.): ICCCI 2020, CCIS 1287, pp. 301–313, 2020.
https://doi.org/10.1007/978-3-030-63119-2_25

However, whereas the working conditions will change drastically as poor air lift, the focus asymmetry. If quadcopter carrying arms would have manipulated strong nonlinearity [2]. The studies used ANN with PID controller to control drone fly in a region for variable payloads [3–6, 12, 14]. The using of RBF combined with an image based visual servo can tract a moving target by mean of a camera. In the studies before [1–3, 5, 10], they used successfully predict the steering angles required to move the drone along a predetermined path [13]. The using of ANN and CNN to control quadcopter have focused to process information from camera [15, 16]. A robust neural network-backstepping sliding mode to monitor path of the drone's flight has a good convergence of the position without suspension [18]. The using of a spherical coordinates to describe aero dynamics with uncertain parameters [19]. So the hang status or suspense status only used GPS to fix position. That is an application to view drone in both attitude and position system [17]. To keep a position fixing to do things in air environment then we must control angles well. This a difficult problem for quadcopter. The results in [12, 13, 16] give good trajectory results when there is no air change. But it does not describe the suspended state, when it is necessary to perform the operation, so the stability characteristics in this state have not been seen. So, one should design sustainable large controller to improve stability for quadcopter in hang mode, the ability of quadcopter activities will be increased. In fact, the rescue robot only needs to identify the correct stopping position and hanging in that position rather than moving exactly in orbit. The inclined state of the robot is necessary to ensure safety in the event of an aerial collision such as near high-voltage lines or antenna poles or high-rise buildings. In this case, fixing the angle of body is necessary to perform the operation. We consider the ideal model quadcopter to solve the problem posed in this study.

The rest of the paper is structured as follow. Section 2 discusses the quadcopter dynamics. Then Sect. 3 and 4 propose the design of signal analyzer, and H^{-1} controller. Simulation results and analysis is presented in Sect. 5. Finally, the Sect. 6 concludes the paper.

2 Quadcopter Dynamics

The ideal quadcopter will satisfy the following basic conditions:

- Symmetric construction
- The propellers have not change tilt angle and elastic properties.
- The body of quadcopter has not deformation in the motion.
- The space of air cushion is not change at proximity surface of the quadcopter.

We can see a kinematic system of the quadcopter as Fig. 1. Quadcopter kinetics is determined by 6 coordinates of **InFr**:

$$\eta = [\phi\ \theta\ \psi]^T;\ \boldsymbol{\xi} = [x\ y\ z]^T;\ \mathbf{q} = [\xi\ \eta]^T \tag{1}$$

Among them, three coordinates $\boldsymbol{\eta}$ to describe Euler angle with 3 axis of the quadcopter body coordinates $O_B x_B y_B z_B$ of **BoFr**, and three coordinates $\boldsymbol{\xi}$ to describe related position in the inertial frame system [7].

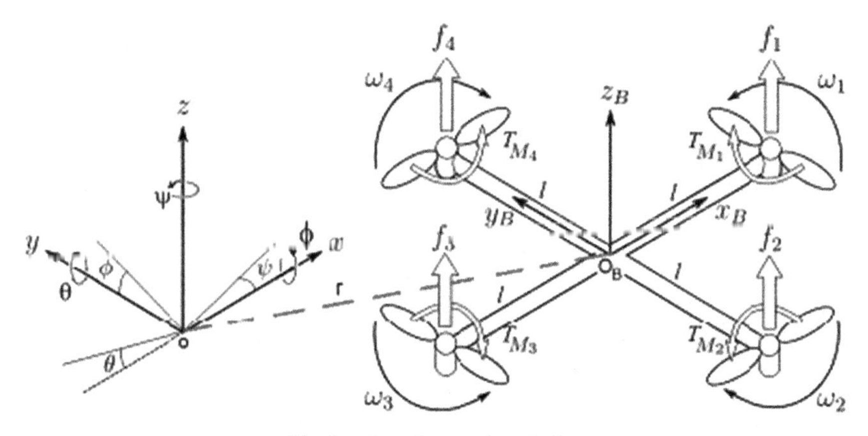

Fig. 1. Coordinates description

The transformation from the **BoFr** (Body Frame: Body Frame Coordinates) coordinate to the **InFr** (Inertial Frame: Inertial Frame Coordinates) coordinate is done by the rotation matrix **R** which is defined as:

$$\mathbf{R} = \begin{bmatrix} C_\psi C_\theta & C_\psi S_\theta S_\phi - C_\phi S_\psi & C_\psi S_\theta C_\phi + S_\psi S_\phi \\ S_\psi C_\theta & S_\psi S_\theta S_\phi + C_\psi C_\phi & S_\psi S_\theta C_\phi - S_\phi C_\psi \\ -S_\theta & S_\phi C_\theta & C_\theta C_\phi \end{bmatrix} \quad (2)$$

where:

$S_x = \sin(x)$; $C_x = \text{cosin}(x)$; $T_x = \tan(x)$; $cT_x = \cot an(x)$. **R** is an orthonormal rotation matrix so $\mathbf{R}^{-1} = \mathbf{R}^{\mathrm{T}}$ is used to describe the inverting transformation from the **InFr** to **BoFr**.

The transformation matrix for angular velocity **G** from **InFr** to **BoFr**, and $\mathbf{G}^{-1}$ from **BoFr** to **InFr** are defined as [8, 20]:

$$\dot{\eta} = \mathbf{G}^{-1}\boldsymbol{\gamma}, \quad \begin{bmatrix} \dot{\varphi} \\ \dot{\theta} \\ \dot{\psi} \end{bmatrix} = \begin{bmatrix} 1 & S_\phi T_\theta & C_\phi T_\theta \\ 0 & C_\phi & -S_\phi \\ 0 & S_\phi / C_\theta & C_\phi / C_\theta \end{bmatrix} \begin{bmatrix} p \\ q \\ r \end{bmatrix} \quad (3)$$

$$\boldsymbol{\gamma} = \begin{bmatrix} p \\ q \\ r \end{bmatrix} = \mathbf{E}\dot{\boldsymbol{\eta}} = \begin{bmatrix} 1 & 0 & -S_\phi \\ 0 & C_\phi & S_\phi C_\theta \\ 0 & -S_\phi & C_\phi C_\theta \end{bmatrix} \begin{bmatrix} \dot{\varphi} \\ \dot{\theta} \\ \dot{\psi} \end{bmatrix} \quad (4)$$

The matrix **E** is invertible only if $\theta \neq (2\,\mathrm{k} - 1)\phi/2$, ($k \in \mathrm{Z}$).

Supposed that the quadcopter structure is totally symmetrical, then its center of gravity will be its body center. Hence the moment of inertia is determined as:

$$\mathbf{I} = \begin{bmatrix} I_{xx} & 0 & 0 \\ 0 & I_{yy} & 0 \\ 0 & 0 & I_{zz} \end{bmatrix}; \ \mathbf{I}_{xx} = \mathbf{I}_{yy}. \quad (5)$$

According to [6], the thrust f_i and the torque τ_i for each quadcopter motor are determined:

$$f_i = k\omega_i^2; \; T_{M_i} = b\omega_i^2 + I_M\dot{\omega}_i \tag{6}$$

In fact k and b are nonlinear variables [6], but in most of cases for controller designing we can consider them constants. $\mathbf{I}_M$ is the moment of inertia of the motor's rotor.

From (6) we can deduce the total thrust L and the total torque τ:

$$L = \sum_{i=1}^{4} f_i = k\sum_{i=1}^{4}\omega_i^2; \; \mathbf{L}^B = [0\ 0\ L]^T \tag{7}$$

$$\mathbf{T}^B = \begin{bmatrix} T_\phi \\ T_\theta \\ T_\psi \end{bmatrix} = \begin{bmatrix} lk\left(-\omega_2^2 + \omega_4^2\right) \\ lk\left(-\omega_1^2 + \omega_3^2\right) \\ \sum_{i=1}^{4} T_{M_i} \end{bmatrix} \tag{8}$$

where l is the distance from the rotor's center to the quadcopter's center of gravity. From (8) it is shown that the forward rotation requires reducing the second rotor's angular velocity and while increasing the fourth rotor's angular velocity. Similarly, the first rotor's angular velocity must be reduced while the third rotor's angular velocity must increase.

From these analyses one can deduce the Lagrange equation:

$$L(q, \dot{q}) = E_{trans} + E_{rot} - E_{pot} = \frac{1}{2}m\dot{\xi}^T\dot{\xi} + \frac{1}{2}\boldsymbol{\gamma}^T\mathbf{I}\gamma - mgz \tag{9}$$

where E_{trans}, E_{rot}, E_{pot} is the kinetic energy of the translational movement, the kinetic energy of the rotational movement, and the potential energy, respectively, of the quadcopter. One can then obtain the Lagrange II equation:

$$\mathbf{F} = \left[\frac{\mathbf{f}}{\mathbf{T}}\right] = \frac{d}{dt}\left(\frac{\partial\mathbf{L}}{\partial\dot{\mathbf{q}}}\right) - \frac{\partial\mathbf{L}}{\partial\mathbf{q}} \tag{10}$$

The force that induces the translational movement is the rotors' total thrust:

$$\mathbf{f} = \mathbf{RT}_B = m\ddot{\boldsymbol{\xi}} + mg[0\ 0\ 1]^T \tag{11}$$

The rotational movement energy is:

$$E_{rot} = \frac{1}{2}\boldsymbol{\gamma}^T\mathbf{I}\gamma = \frac{1}{2}\ddot{\eta}^T\mathbf{J}\eta \tag{12}$$

With the Jacobian matrix $\mathbf{J}(\eta)$ in the form:

$$\mathbf{J}(\boldsymbol{\eta}) = \mathbf{J} = \mathbf{G}^T\mathbf{I}\mathbf{G} = \begin{bmatrix} I_{xx} & 0 & -I_{xx}S_\theta \\ 0 & I_{yy}C_\phi^2 + I_{zz}S_\phi^2 & (I_{yy} - I_{zz})C_\phi S_\phi C_\theta \\ -I_{xx}S_\theta & (I_{yy} - I_{zz})C_\phi S_\phi C_\theta & I_{xx}S_\theta^2 + I_{yy}S_\phi^2 C_\theta^2 + I_{zz}C_\phi^2 C_\theta^2 \end{bmatrix} \tag{13}$$

Then:

$$\mathbf{T} = \mathbf{T}_B = \mathbf{J}\ddot{\boldsymbol{\eta}} + \frac{d}{dt}(\mathbf{J})\dot{\boldsymbol{\eta}} - \frac{1}{2}\frac{\partial}{\partial\boldsymbol{\eta}}\left(\dot{\boldsymbol{\eta}}^T\mathbf{J}\dot{\boldsymbol{\eta}}\right) = \mathbf{J}\ddot{\boldsymbol{\eta}} + \mathbf{C}(\eta, \dot{\boldsymbol{\eta}})\dot{\boldsymbol{\eta}} \tag{14}$$

where the matrix $\mathbf{C}(\eta, \dot{\boldsymbol{\eta}})$, describing the Coriolis force and the centrifugal force, whose components are [4]:

$$\mathbf{C}(\boldsymbol{\eta}, \dot{\boldsymbol{\eta}}) = \begin{bmatrix} c_{11} & c_{12} & c_{13} \\ c_{21} & c_{22} & c_{23} \\ c_{31} & c_{32} & c_{33} \end{bmatrix} \tag{15}$$

$$c_{11} = 0;\ c_{12} = (I_{yy} - I_{zz})\left(\dot{\theta}C_\phi S_\phi + \dot{\psi}S_\phi^2 C_\theta\right) + (I_{zz} - I_{yy})\dot{\psi}C_\phi^2 C_\theta - I_{xx}\dot{\psi}C_\theta$$

$$c_{13} = (I_{yy} - I_{zz})\dot{\psi}C_\phi S_\phi C_\theta^2;$$

$$c_{21} = (I_{zz} - I_{yy})\left(\dot{\theta}C_\phi S_\phi + \dot{\psi}S_\phi C_\theta\right) + (I_{yy} - I_{zz})\dot{\psi}C_\theta C_\phi^2 + I_{xx}\dot{\psi}C_\theta \quad c_{22} = (I_{zz} - I_{yy})\dot{\phi}C_\phi S_\phi$$

$$c_{23} = -I_{xx}\dot{\psi}S_\theta C_\theta + I_{yy}\dot{\psi}S_\phi^2 S_\theta C_\theta + I_{zz}\dot{\psi}C_\phi^2 S_\theta C_\theta \quad c_{31} = (I_{yy} - I_{zz})\dot{\psi}C_\theta^2 S_\phi C_\phi - I_{xx}\dot{\theta}C_\theta$$

$$c_{32} = (I_{zz} - I_{yy})\left(\dot{\theta}S_\theta C_\phi S_\phi + \dot{\phi}S_\phi^2 C_\theta\right) + (I_{yy} - I_{zz})\dot{\phi}C_\phi^2 C_\theta + I_{xx}\dot{\psi}S_\theta C_\theta - I_{yy}\dot{\psi}S_\phi^2 S_\theta C_\theta - I_{zz}\dot{\psi}C_\phi^2 S_\theta C_\theta$$

$$c_{33} = (I_{yy} - I_{zz})\dot{\phi}C_\phi S_\phi C_\theta^2 - I_{yy}\dot{\theta}S_\phi^2 S_\theta C_\theta - I_{zz}\dot{\theta}C_\phi^2 S_\theta C_\theta + I_{xx}\dot{\theta}S_\theta C_\theta$$

3 Design of Signal Analyzer

Based on abovementioned analysis, this study proposed the model analysis method from 4 control signals for 4 motors. Unlike industrial robots that have independently controlled joints, quadcopter has 4 motors that run simultaneously producing 4 different forces to move the quadcopter [8, 9]. Therefore we can devide these forces into force components for translational movement $\mathbf{f} = [f_X\, f_Y\, f_Z]^T$ and torque $\mathbf{T} = [T_R\, T_P\, T_Y]^T$ for orientation [9]. **M** is the block including 4 motors and blades. The whole model is shown in Fig. 2.

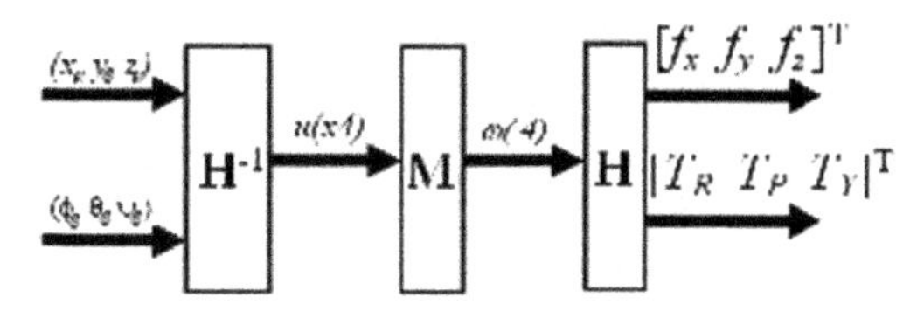

Fig. 2. Diagram of the analysis block.

Designing **H**:
In Fig. 2 the **H** block is the analyzer that converts 4 motors velocity signals to 2 translational vectors of force and torque. The $\mathbf{H}^{-1}$ block makes the inverting conversion of

2 input vectors in **InFr** to 4 controlling signals for motors. **M** is the 4 DC motors block and $\mathbf{H}^{-1}$ is the controller.

If α is the angle between the force vector f with the Ox axis in **InFr** and β is the angle between the projection off on the xOy plan with the Ox axis in **InFr**, then:

$$f_z^2 = \mathbf{f}^T\mathbf{f}C_\alpha;\ f_y^2 = \mathbf{f}^T\mathbf{f}S_\alpha S_\beta;\ f_x^2 = \mathbf{f}^T\mathbf{f}S_\alpha C_\beta \tag{16}$$

We consider the case when the plan of 4 blades is always in parallel with quadcopter body plan ($x_B O_B y_B$) as depicted in Fig. 1. Supposed that in initial state the quadcopter is in position in parallel with **InFr** and is at a certain altitude z corresponding to the angular velocity $\omega_i(0)$ for all 4 motors. Then the flip torque is determined:

$$T_{l1\to3} = 2l.(f_1 - f_3) = -T_{l3\to1} :\ \theta \text{ control}$$

$$T_{l2\to4} = 2l.(f_2 - f_4) = -T_{l4\to2} :\ \phi \text{ control}$$

It can be seen that ϕ and θ vary from values less than $-90^0 \to$ less than 90^0, i.e. the quadcopter can not be in the position perpendicular to the ground. Then when one controls the flip angular θ it will attain the value corresponding to the new balance position:

$$(mgl + T_{l1\to3}).\cos\theta = mgl \Rightarrow \theta = \arccos\left(\frac{mgl}{T_{l1\to3} + mgl}\right) \tag{17}$$

Similarly, if one controls the flip angular ϕ its new balance value is:

$$(mgl + T_{l2\to4}).\cos\phi = mgl \Rightarrow \phi = \arccos\left(\frac{mgl}{T_{l2\to4} + mgl}\right) \tag{18}$$

where m is the weight converted to the quadcopter center of mass.

4 Design of $\mathbf{H}^{-1}$ Controller

The models is done from the mathematical structures analyzed in (1), (3), (8), (10) to be structured object part as shown in Fig. 3. In particular, the dynamic built directly. In practice the simplest model for motors is the first-order model. We will thus use this 1st-order model for controller design:

$$\omega(s) = \frac{K_D}{Ts + 1}U(s) \tag{19}$$

Where $U(s) \to u(t)$ is the control signal. However, the goal of the research is to focus on the hang state, so we consider the quadcopter as a rigid moving object with 3 position coordinates and 3 direction coordinates, so we use the 6 DoF model to describe a robot is subjected to 3 translational forces and 3 torques. Transforming a quadcopter into a 6DoF model will transform the control of the rotor blades into the robot body and will not be subject to physical violations.

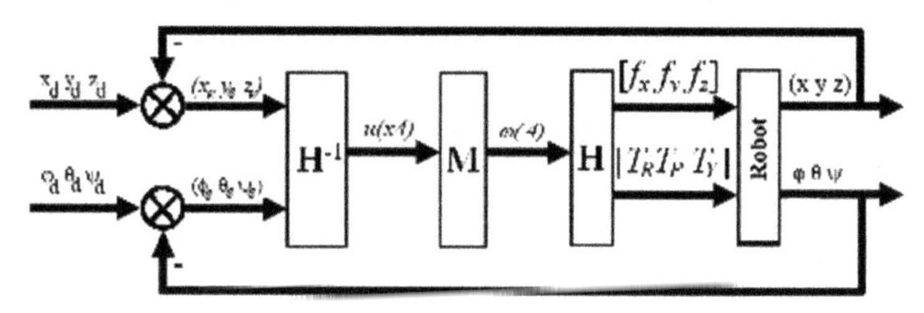

Fig. 3. Controller's model

For translational part we have:

$$\ddot{\boldsymbol{\xi}} = \left[\ddot{x}\,\ddot{y}\,\ddot{z}\right]^T = (1/m)\mathbf{f} - g[0\,0\,1]^T \tag{20}$$

For the rotational part RPY:

$$\ddot{\boldsymbol{\eta}} = [\ddot{\phi}\ddot{\theta}\ddot{\psi}]^r = \mathbf{J}^{-1}(\mathbf{T}_B - \mathbf{C}(\boldsymbol{\eta}\,\dot{\boldsymbol{\eta}})\dot{\boldsymbol{\eta}}) \tag{21}$$

$\mathbf{H}^{-1}$ receives 6 error signals as input and controls 4 combinational output signals. At the same time, for practice requirements, there are some more input signals such as density of air, vortex velocity of air, quadcopter body vibration, dynamic load etc. Then there can be many input signals. Figure 3 shows the system block diagram.

Note that in Fig. 3, **H** is a hidden structure inside the quadcopter body, **M** is a convertor $\mathbf{u}(t) \rightarrow \boldsymbol{\omega}(t)$. Measured signals for quadcopter position and real Euler angle are obtained by positional inertial sensors and direct inertial sensors.

In order to design a controller suitable for inputs expanding, this study proposed to use the linear neural controller having structure containing 15 hidden layers and 4 output layers as shown in Fig. 4.

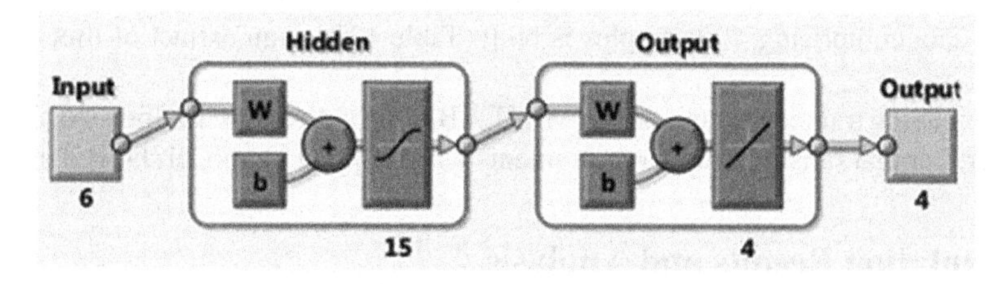

Fig. 4. Neural controller's structure.

We define an input weights matrix $\mathbf{W} = [w_{ij}]$, with $i = 1,..4$ and $j = 1,..6$; input vector of error $\mathbf{q}_e = [x_e\ y_e\ z_e\ \phi_e\ \theta_e\ \psi_e]^T$; output vector $\mathbf{u} = [u_1\ u_2\ u_3\ u_4]^T$ are voltages applied to motors; vector of slope $b = [b_1\ b_2\ b_3\ b_4]^T$. We define the relation:

$$\mathbf{u} = h(\mathbf{a}) = h(\mathbf{W}.\mathbf{q}_e + \mathbf{b}) \tag{22}$$

The function f is chosen having the form *tanh*(.) to describe a strong error variation in the range under consideration and saturated for large error so that it will not induce large voltage variation during controlling. Then layer's weight components are determined:

$$w_{ij}(n+1) = w_{ij}(n) + \Delta w_{ij}(n) \tag{23}$$

We can define the sum of squares of deviations:

$$E(n) = \frac{1}{2}\sum_{j=1}^{k} e_j^2(n) = \frac{1}{2}\sum_{j=1}^{k}\left[t_j(n) - u_j(n)\right]^2 = \frac{1}{2}\sum_{j=1}^{k}\left[t_j(n) - h_j(a_j(n))\right]^2$$
$$= \frac{1}{2}\sum_{j=1}^{k}\left\{t_j(n) - h_j\left[\sum_{i=0}^{6}\left(\mathrm{w_{ij}}.q_j(n) + b_j\right)\right]\right\}^2 \tag{24}$$

where u_{j} is the outputs of the *j*-th intermediate layer, t_{j} is the desired output of the *j*-th intermediate layer *j*. Then u_{k} is the last output voltages applied to motors.

From (23) we have:

$$\Delta \mathrm{w_{ij}} = -\eta\delta_j(n)x_i(n) \tag{25}$$

With hidden nodes:

$$\delta_j(n) = -\frac{\partial E(n)}{\partial \mathrm{w_{ij}}} = -\frac{\partial E}{\partial y_i(n)}\frac{\partial h(n)}{\partial u_j(n)} \tag{26}$$

With output nodes:

$$\delta_j(n) = e_j(n)\frac{\partial h_j(n)}{\partial u_j(n)} \tag{27}$$

Based on the quadcopter parameters and the Inertial Frame Coordinate, the network training data comprising 400 samples is built. Table 1 show an extract of this training data.

After having trained the network by MATLAB Neuron Network Toolbox we obtained the controller that satisfies output requirements when applied for model showed in Fig. 5.

5 Simulation Results and Analysis

Previous quadcopter research focused on controlling three position coordinates [12–14], controlling the robot in orbit. Today's robots often use GPS to navigate. But in some conditions without GPS such as deep caves, indoors or loss of GPS, the estimation of location and direction becomes necessary. System simulation is carried-out for duration of 100 s with quadcopter initially in the Inertia Frame Coordinate origin, then moves to a pre-determined position and keeps stable at this position. Simulation results are shown in Figs. 6 7, 8, 9, Fig. 10.

Quadcopter parameters are: mass $m = 3$ kg; arm length $l = 0, 25$ m; 24 V DC brushless motor with 2 A nominal current; symmetrical center of mass.

Table 1. Input-output data for training.

x_e	y_e	z_e	ϕe	θe	ψe	u_1	u_2	u_3	u_4
1.50	0.40	1.00	1.00	2.00	3.00	1.93	1.98	1.99	1.94
1.13	0.29	0.76	0.54	1.19	1.73	1.86	1.95	1.98	1.89
0.65	0.14	0.46	-0.02	0.60	0.47	1.80	1.93	1.97	1.84
0.15	-0.04	0.16	-0.52	0.30	-0.61	1.71	1.90	1.96	1.79
-0.34	-0.18	-0.11	-0.81	0.10	-1.45	1.69	1.88	1.95	1.74
-0.87	-0.22	-0.37	-0.93	-0.20	-2.07	1.64	1.86	1.94	1.70
-1.51	-0.13	-0.67	-0.84	-0.71	-2.41	1.59	1.84	1.93	1.66
-2.29	0.13	-1.13	0.69	-1.13	-1.18	1.55	1.82	1.92	1.62
-3.08	0.70	-1.87	1.42	-0.65	-0.53	1.51	1.80	1.91	1.58
-3.69	1.78	-2.74	1.35	-0.05	-0.41	1.47	1.78	1.90	1.55
-4.00	3.38	-3.23	0.99	0.51	-0.11	1.44	1.76	1.90	1.52

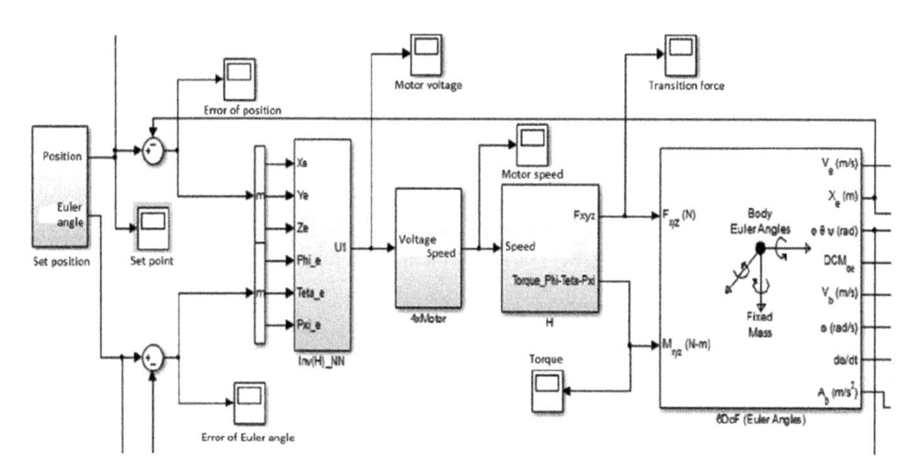

Fig. 5. System's SIMULINK model

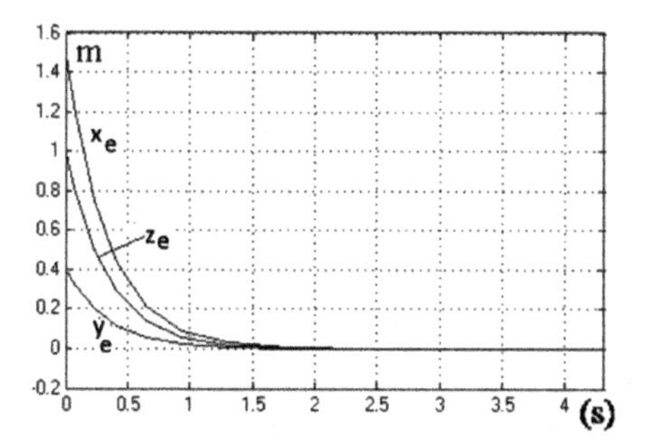

Fig. 6. Translational movement error

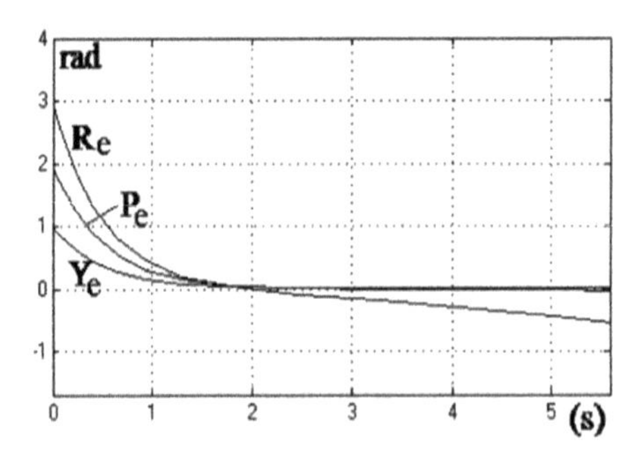

Fig. 7. Euler angle error

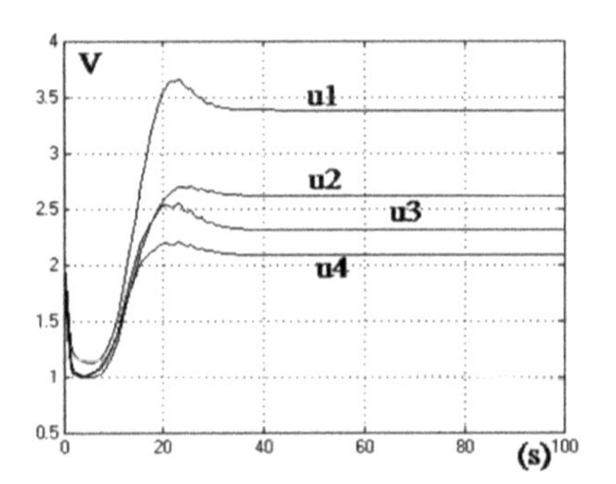

Fig. 8. Voltage of the four motors

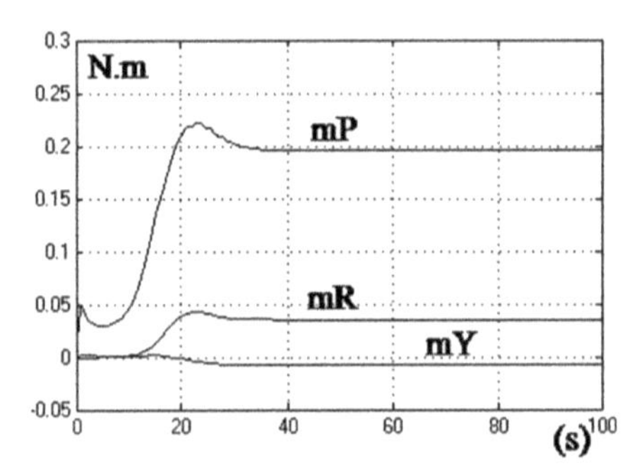

Fig. 9. Euler rotational torque

According to the Fig. 6 and Fig. 7 one can see that after 2 s the quadcopter moves to the predetermined position and errors approach to zero. Similarly, Euler angles approaches to zero, while Pitch angle deviates 0.5 rad causing a 28° declination of the quadcopter.

This can be the case when we choose voltage signals u_1 and u_4. It is shown more clearly in voltages form in Fig. 8.

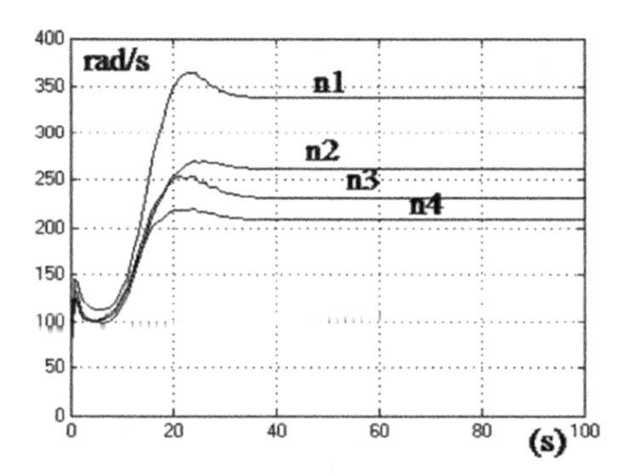

Fig. 10. Speed of the four motors.

Figure 8 shows that the quadcopter transient takes about first 40 s. Although the position error is stable from the 5th second, the quadcopter continues to rotate for a short duration due to the fact that the moment of inertia change is not yet stable.

The Euler rotational torque change shown in Fig. 9 shows that the quadcopter still moves in the inclined plan. This inclination is intended because the quadcopter position set point is not parallel to any plan in the Inertial Frame Coordinate.

Simulation result shown in Fig. 10 indicates that motors speed on attain the stable value after 40 s and then sustain this stable speed, which means that the thrust is stable. Then the quadcopter sustains its inclined position corresponding to the position set-point. Figure 11 shows the experimental model. We have some difficult to record data in real time to present. We hope will present experimental results in next studies.

Fig. 11. Experimental model

6 Conclusion and Discussion

In this study we have designed a multi-layer structure neural controller which is applied for autonomous quadcopter. Simulation results have shown that transient response is stable and satisfy required characteristics.

The neural controller can show its advantage for expanding more inputs to process more input information so that the quadcopter can move totally automatically.

The density of the surrounding air is continuously varied and unpredictable at the points that quadcopter passes. It is then really necessary to study the system robustness. The proposed controller in this work can be promising to deal with nonlinear conditions and MIMO cases.

The description of the air cushion, which is the quadcopter movement environment, is not yet considered in this paper. This important problem will be dealt in another study.

The paper has not considered the rotor's nonlinear stiffness encountered in heavy load quadcopters. In this study we have supposed that quadcopter's rotors are inelastic and experience no strain. Further ongoing study will evaluate these problems.

References

1. Jafari, M., Xu, H.: Intelligent control for unmanned aerial systems with system uncertain and disturbances using artificial neural network. In: MDPI2018, vol. 89557, University of Nevada, Reno (2018)
2. Amer, K., Samy, M., ElHakim, R., Shaker, M., ElHelw, M.: Convolutional neural network-based deep urban signatures with application to drone localization. In: IEEE International Conference on Computer Vision Workshops (ICCVW), pp. 2138–2145. IEEE (2017)
3. Amer, K., Samy, M., Shaker, M., ElHelw, M.: Deep convolutional neural network-based autonomous drone navigation, arXiv:1801.08214,Center for Informatics Science Nile University Giza, Egypt (2018)
4. Padhy, R.P., Verma, S., Ahmad, S., Choudhury, S.K.: Deep neural network for autonomous UAV navigation in indoor corridor environments. In: International Conference on Robotics and Smart Manufacturing (RoSMa2018), pp. 643–650, Elsevier (2018)
5. De Mel, D.H., Stol, K.A., Mills, J.A., Eastwood, B.R.: Vision-based object path following on a quadcopter for GPS-denied environments. In: IEEE International Conference on Unmanned Aircraft Systems (ICUAS), pp. 456–461 (2017)
6. Johnson, W.: Calculated hovering helicopter flight dynamics with a circulation-controlled rotor, Ames research center NASA (1977)
7. Ribera, M., Celi, R.: Helicopter Flight Dynamics Simulation, Doctor of Philosophy, Department of Aerospace Engineering-University of Maryland (2007)
8. Houghton, Aerodynamics for engineering students. 5th edn. Buttrworth-Heinemann, England (2003). ISBN 0750651113
9. Caldera, H.S.M.M., Anuradha, B.W.S.: A self-balancing quadcopter design with autonomous. In: SAITM Research Symposium on Engineering Advancements 2014 (SAITM – RSEA 2014) (2014)
10. Luukkonen, T.: Modelling and control of quadcopter, Aalto University School of Science, Mat-2.4108, Independent Research Project In Applied Mathematics Espoo, 22 August 2011
11. Amin, M.Y.: Modeling and neural control for quadcopter helicopter. Yanbu J. Eng. Sci. (2011). ISSN 1658-5321
12. Teng, Y.-F., Hu, B., Liu, Z-W., Huang, J., Guan, Z.-H.: Adaptive neural network control for quadrotor unmanned aerial vehicles. In: The 11th Asian Control Conference (ASCC), pp. 988–992, Australia (2017)
13. Shirzadeh, M., Asl, H.J., Amirkhani, A., Jalali, A.A.: Vision-based control of a quadrotor utilizing artificial neural networks for tracking of moving targets. J. Eng. Appl. Artif. Intell. **58**, 34–48 (2017)

14. Lower, M., Tarnawski, W.: Quadrotor navigation using the PID and neural network controller. In: Zamojski, W., Mazurkiewicz, J., Sugier, J., Walkowiak, T., Kacprzyk, J. (eds.) Theory and Engineering of Complex Systems and Dependability. AISC, vol. 365, pp. 265–274. Springer, Cham (2015). https://doi.org/10.1007/978-3-319-19216-1_25
15. Padhy, R.P., Verma, S., Ahmad, S., Choudhury, S.K.: Deep neural network for autonomous UAV navigation in indoor corridor environments. In: International Conference on Robotics and Smart Manufacturing (RoSMa2018), Procedia Computer Science, pp. 643–650, Elsevier (2018)
16. Luo, C., Du, Z., Yu, L.: Neural network control design for an unmanned aerial vehicle with a suspended payload. Electronics **8**(9), 931 (2019)
17. Muliadi, J., Kusumoputro, B.: Neural network control system of UAV altitude dynamics and its comparison with the PID control system. Hindawi J. Adv. Transp. Article ID 3823201 (2018)
18. Celen, B., Oniz, Y.: Trajectory tracking of a quadcopter using fuzzy logic and neural network controllers. In: The 6th IEEE International Conference on Control Engineering and Information Technology (CEIT), Istanbul (2018)
19. Dierks, T., Jagannathan, S.: Neural network control of quadrotor UAV formations. In: American Control Conference, Hyatt Regency Riverfront, pp. 2990–2996, St. Louis (2009)
20. Martin, R.S., Barrientos, A., Gutierrez, P., del Cerro, J.: Unmanned Aerial Vehicle (UAV) modelling based on supervised neural networks. In: Proceedings of the 2006 IEEE International Conference on Robotics and Automation, pp. 2497–2502, Orlando (2006)

Multidimensional Analysis of SCADA Stream Data for Estimating the Energy Efficiency of Mining Transport

Paweł Stefaniak[1(✉)], Paweł Śliwiński[2], Natalia Duda[1], and Bartosz Jachnik[1]

[1] KGHM Cuprum Research and Development Centre, gen. W. Sikorskiego 2-8, 53-659 Wrocław, Poland
pkstefaniak@cuprum.wroc.pl

[2] KGHM Polska Miedź S.A., M. Skłodowskiej-Curie 48, 59-301 Lubin, Poland

Abstract. This paper outlines the recommendation of analytical tools likely to be derived from the data recorded within the industrial automation system. The means might facilitate optimization of process efficiency, especially in terms of energy efficiency. Basically, each electromechanical device is electrically charged and controlled by the industrial automation system. A kind of the signal usually depends on various operational modes of the given device which are classified by its load. Available signal segmentation and statistical methods lead to the automatic identification of these modes and working patterns or abnormal performances caused by poor technical condition. Therefore, simple electrical signal allows to count the real device performance time and utilities usage, to identify its working modes, to recognize process losses, to specify KPI factors and to develop diagnostics. This paper describes multidimensional processing of conveyor stream data along with their exemplary use in real-time data. The algorithm of identifying operational regimes is characterized based on machine learning and further in-context analyses paired with visualisations.

Keywords: Belt conveyor · Energy efficiency · SCADA · Mining · Big data · Mining transportation

1 Introduction

Basically, energy efficiency means the ratio of utility-efficiency result of the plant, device and installation under the ordinary exploitation conditions to their energy consumption [1].

The global trend of energy policy has been currently observed in the mining industry. It forges balancing the costs of energy management according to ISO 50001 standard for Energy Management Systems. Regarding this issue, firstly, energy streams and their supervision should be considered. Next, the strategy for reducing energy consumption and its losses are to be defined and implemented. Supervising and constantly improving energy receivers and their recycling are yet other significant issues. The reliability of

M. Hernes et al. (Eds.): ICCCI 2020, CCIS 1287, pp. 314–325, 2020.
https://doi.org/10.1007/978-3-030-63119-2_26

mining plant's assets and the organizational aspect of the whole mining infrastructure should be also taken into consideration.

For example, in Polish underground copper ore mines there are three most energy consuming processes, namely ventilation, belt conveyor transportation system and air-conditioning. Energy consumption is often subject to seasonal changes, mainly in terms of air-conditioning, where it decreases sharply in winter months and increases in summer months as the result of natural temperature changes and the performance of free-cooling system. These areas have been classified as the main energy consuming ones also in other industries [2].

The domain of analytic improvement in the belt conveyor transportation system in mines is considered in this paper. In the literature there are a number of publications considering the issue of conveyor transport in terms of efficiency estimation [3, 4], tracking material flow [5–7] energy consumption [8], movement limitations [9], predictive maintenance [10] and operational management [11, 12]. The majority of these areas require professional measuring equipment, improving complex infrastructures and IT devices.

The authors highlight the analytical capacity of the signals recorded with the industrial automation systems. Currently the leverage of this data is severely limited. SCADA signals serve highlighting the factors, which might significantly facilitate the optimization of operational and energy efficiency. Basically, each electromechanical device is charged by electricity and controlled by the industrial automation system. Therefore, the access to the information provided by SCADA is inclusive and is not charged extra. The form of the signal often depends on various working modes of the given device which are subject to load, organizational scheme of a mine and climate conditions in case of air-conditioning. Available signal segmentation and statistical methods lead to the automatic identification of these modes and working patterns or abnormal performances caused by poor technical condition. Therefore, simple electrical signal allows to specify the real device performance time, utilities usage, working modes of the given device, process losses, efficiency parameters and to develop technical diagnostics methods (Fig. 1).

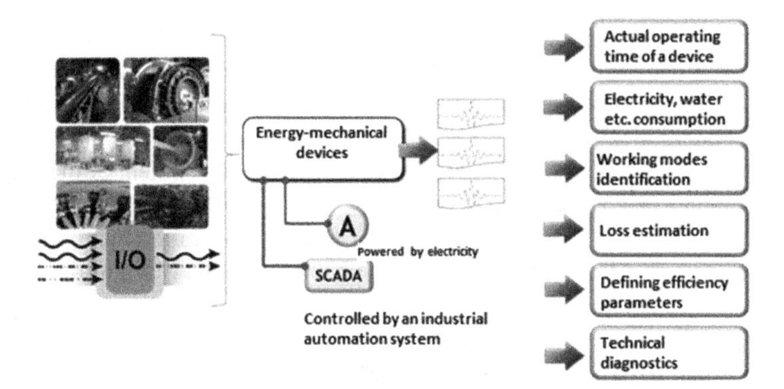

Fig. 1. How to use tools to optimize energy and operational efficiency based on data recorded by SCADA.

For instance, in Polish KGHM copper ore mines almost all underground infrastructure devices are operationally connected with a few other devices constructing wide, networked technical plants. Each point of the plant has a specific function assigned. The points are ordered hierarchically and by their dependency rules. Such "chains" occur in conveying system, air-conditioning and drainage. Many of the chained machines generate real-time operational data and periodically provide the information to enterprise resource planning records (*enterprise resource planning software*, SAP-ERP).

The article covers the notion of implementing simple analytical tools in multidimensional energy efficiency analysis as researched on a section of the belt conveyor transport system in one of KGHM underground copper ore mines. General method of data analysis derived from industrial automation systems is presented herein. It is thought that the presented method might facilitate the process of curbing the energy consumption in the industrial infrastructure.

The paper is divided into 5 main parts. First, the description of the issue researched has been outlined. In the second section the subject belt conveyor transportation system is described. Next section covers input data, the dependencies among them and their analytical capacity. An applied methodology has been explained in the fourth section. It includes the description of a key algorithm for the identification of the belt conveyor engine working modes and further multidimensional analysis and its leverage in real-time data. Finally, there is the summary of the research.

2 Description of the Subject Object

In KGHM mines conveyor system is a continuous, horizontal system of material transportation from mining departments to mining shafts with which material is carried to the surface. The system is an example of networked technical points including several dozen of conveyors, which carry the material on the belt (Fig. 2a). Transportation system consists of conveyors, bunkers, loading and reloading points. Loading of a conveyor is conducted either from another conveyor with which the conveyor is connected, or from a grid to which the material is carried directly from mining faces with self-propelled haul trucks.

The conveyor transport is a perfect example of the "chain". Each conveyor is operationally linked with at least one conveyor, constituting the integral part of a bigger system. Each object in such a system fulfils its role and is ordered hierarchically. There are main and department transport (Fig. 2b).

Nevertheless, the scale of the problem regarding the development of the analysis is even greater than it might seem due to the size of a machinery park, number of technical objects, their dispersion and variety of design features, operations and technologies used. Possibilities and limitations of the analytical development in multi-plant mining enterprise are considered in [13].

3 Input Data and Its Analytical Capacity

3.1 Input Data - Basic Information

Depending on the number of engines mounted, each conveyor generates from 1 to 4 current signals. The more significant ones provide also the signal on the weight of

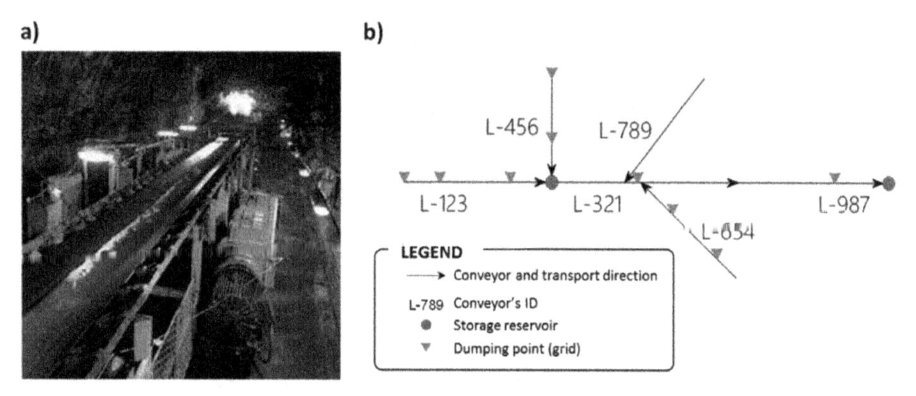

Fig. 2. a) Belt conveyor which is transporting material in the underground mine, b) sample section of a belt conveyor network depicting the example of transportation system and its main points.

carried material. Data sampling period is not fixed. It is recorded once the amperage raises by the minimum of Δ (1A). Alike, measuring weight signal is triggered when the weight of carried material increases by the minimum of Δ (min. 1 Mg). The signal is reset after the preset period of time, usually defined on a specific basis (by a shift, a day, a week) for each conveyor. The signal is recorded incrementally. Due to its diversification, an instantaneous material weight on a belt can be measured. The examples of input data are presented in Fig. 3.

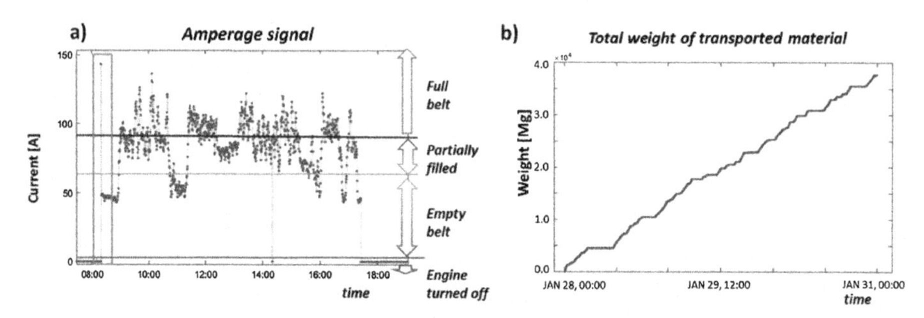

Fig. 3. Input data: current signal (left) and weight signal (right).

Furthermore, department conveyors under the grids also generate data [0/1] about conveyors including the specific times when material is transported by haul trucks (Fig. 4).

3.2 Identified Connections and Analytical Capacity

Fluctuating amperage reflects different operational modes of the conveyor depending on external load (turned off, start-up, empty belt operation, material transportation) - in particular cases it is possible to recognize full and partial material load on the belt. This way, start-up might be discerned and classified as light (empty belt) and heavy (full

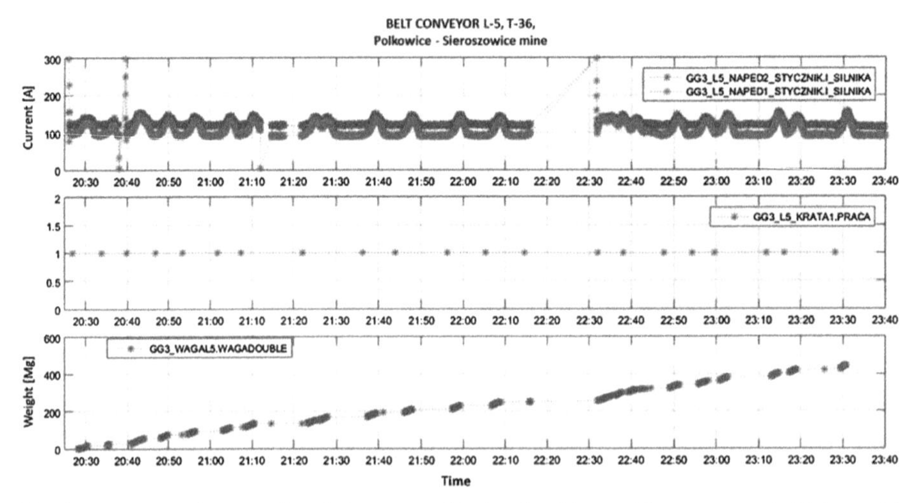

Fig. 4. Examples of current signals (upper graph), signals of the working conveyor (middle graph), incremental weight signal (lower graph).

belt). This classification is vital for the analysis of the belt robustness. The identification of operation modes leads to the insight into the conveyor operation measured by a time unit. Additionally, energy consumption can be estimated in context for current of the empty (idle) and full (load operation) belt; so can be both energy efficiency and process losses of transportation system (lost capacity). Such information can be presented as a set - for the whole system - or divided by specific conveyors. It is vital to consider the threshold above which current values indicates the operation with external load (carrying of material). Idle level might differ depending on each device and each engine used, even in the same conveyor. Moreover, current of idle operation of a single engine is subject to change in time because of changing movement limitations. Tracking daily idle levels can be further applied in diagnostics. The following factors are simply specified: (a) tripping operation of conveyor's engines (for example when the oil level in clutch is improper) or (b) abnormal operation of the conveyor's parts.

Current signal depends on the weight of material carried on the conveyor belt. Dependency between the total weight of material and total amperage is approximately linear. Therefore, the weight of material can be forecast based on amperage. Currently there is a limited access to weight data. Weight is commonly recorded only at the key conveyors (output) of the particular areas. Also, defects in a scale sensor result in loss of basic information on the mining transport efficiency. The studied method has been described in [3]. Firstly, this method is leveraged in validation processes. It can also be applied to conveyors without a scale, which are connected to the ones with scales. In this case, the time shift has to be administered. This way, the access to weight data is boosted facilitating the process of tracking material flow in the transportation network [5]. At this point, the integration of data from the conveyors is of the utmost importance. The authors of this paper have developed an algorithm which traces the load on grid and presents the percentage rate of particular areas in loading to the area conveyor and of particular underground machines at a given area dump point.

4 Methodology

The algorithm-based method of identifying the working modes of an engine of a conveyor belt is described. The results acquired might be indicative of the working mode of the whole conveyor. Considering this issue, it is crucial to prepare the data correctly so that k-means clustering can be applied to identify the process.

The identification of the idle mode constitutes the key point for the effective use of infrastructure. Further in this paper, there is also presented the following equation to calculate energy consumed by amperage data.

$$Energy\ consumed\ = \sum current \cdot U \cdot \sqrt{3} \cdot 0.86 \cdot \frac{1}{1000} \cdot \frac{1}{3600},$$

where U is amperage, and the results from the preset time period are totalled.

4.1 Identified Connections and Analytical Capacity

Current measurements from the engines of the conveyor belt constitute input data. The data is not raw - before the algorithm which identifies the working modes is triggered, the data is cleaned and interpolated. Cleaning stands for deleting incorrect read-outs from a sample. Such incorrect read-outs include minus amperage and temporary value drops to zero (interpreted as the shut-down of the engine, which appeared to be incorrect read-outs). In order to interpolate the missing values, a repetitive method was used. This method multiplies previous read-outs as the amperage recording follows the change in amperage.

Cleaned and interpolated data are divided by a date and an engine number. The algorithm is triggered for each section of data in order to identify the working modes. The picture (Fig. 5) below depicts the sample input data.

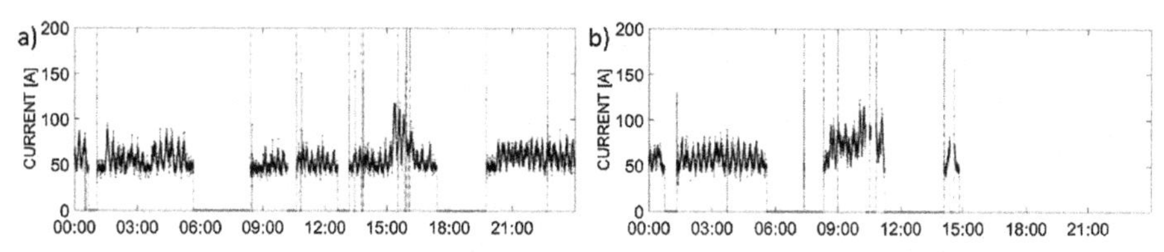

Fig. 5. Sample amperage read-outs from the engine of the belt conveyor: a) conveyor A, b) conveyor B.

Working modes are specified with k-means clustering. This method is chosen because of its low computational complexity as compared to other algorithms and it is easy to implement.

So that k-means clustering is reasonable, daily amperage at each point should meet the following criteria:

1. Non-zero and non-NaN values are at least 2/3 of all read-outs as per one working shift[1] or the whole day.

[1] Working shift - for the workers it usually means 7.5 or 6 h under the ground.

2. Quantile difference of 0.95 and 0.05 (calculated for non-zero and non-NaN amperage) is higher than the preset threshold of L. The threshold means low alternating current.

The threshold of $L = 13A$ is characteristic of low alternating current as calculated with the historical data.

There are the following conveyor working modes: shut down, idle operation, light load operation, heavy load operation, start-ups. Similar modes can be outlined for the engines of the conveyor. Various amperage is applied to the given modes. Thus, 5 clusters are extracted from a daily signal. Therefore, k-means clustering is conducted. A cluster with the lowest centroid value stands for the engine shut-down. Idle operation means the maximum amperage of the second cluster (i.e. the cluster with the second lowest centroid value). The consecutive clusters follow incrementally by centroid values and they stand for: light load operation, heavy load operation and start-ups. This approach is shown in the below block scheme (Fig. 6).

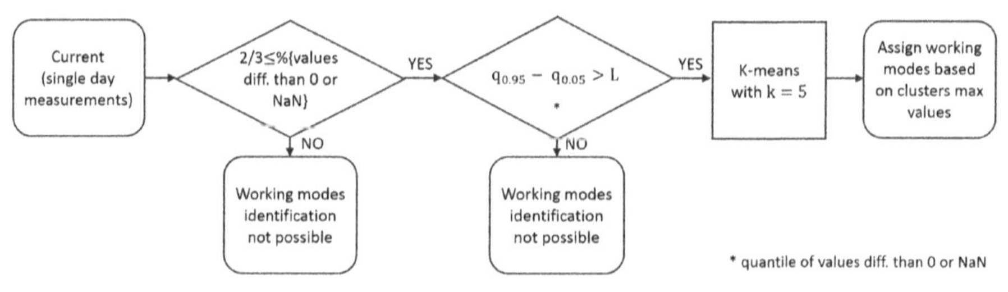

Fig. 6. Block scheme of the algorithm used to indicate the working modes of the engine of the belt conveyor.

There are no initial centroid values in the applied input parameters of k-mean clustering; thus, the k-means clustering++ has been applied to predetermine them. The process is listed below.

1. Randomly choose one point out of the whole X set (discrete uniform distribution). The chosen data is the first centroid, marked with c_1.
2. Specify the distance from each point to c_1. Let $d(x_m, c_j)$ be a distance between c_j and m.
3. Randomly choose the next c_2 centroid from X dataset, with the $d^2(x_m, c_1)/\sum_{j=1}^{n} d^2(x_j, c_1)$ probability.
4. In order to calculate j-center of the cluster:
   a. Determine the distances from each point to each centroid and assign each point to the nearest centroid.
   b. For $m = 1, \ldots, n$ and $p = 1, \ldots, j - 1$, randomly choose j centroid with $d^2(x_m, c_p)/\sum_{\{h:x_h \in c_p\}} d^2(x_h, c_p)$ probability, where C_p is the set of the points nearest to c_p centroid and x_m belongs to C_p set. Then, choose the next cluster

central point with the probability relative to its distance from the nearest centroid (already chosen one).

5. Repeat the step number 4 until k centroids is selected.

K-means clustering++ application increases the distances among the preset centroids, which are commonly dispersed in a random way. The initial parameters which are determined this way lead to better, more coherent results as compared to the traditional method, in which the ordinary k-means clustering algorithm is restarted multiple times and the model of the lowest SSE value is applied. Moreover, the innovative approach boosts the pace of the algorithm by simplifying the computing method.

Due to the fact that k-means clustering is non-deterministic, this method of identifying the working modes might be enhanced by Monte Carlo method. The groups respective to the given working modes are then determined multiple times. The maximum values of the clusters - the working mode thresholds - are samples. The samples along with statistics can be used to set the correct threshold of the given mode.

4.2 Application of the Algorithm to Real Data

The graphs below (Fig. 7) show the results of k-mean clustering used to identify the idle state of the engine of the belt conveyor and performance regimes.

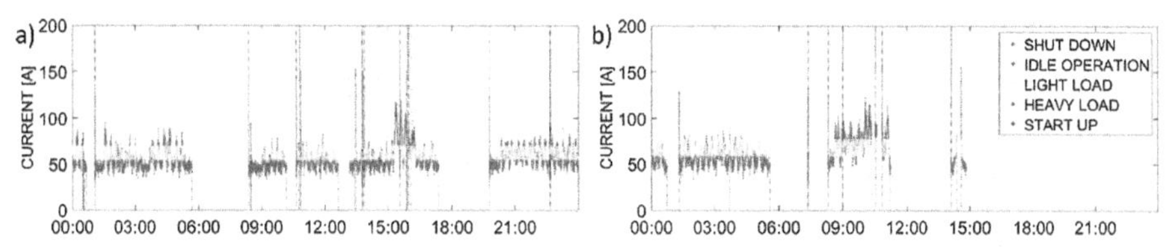

Fig. 7. Working modes determined with k-means clustering for the above examples (see Fig. 6): a) conveyor A, b) conveyor B.

Graphs in Fig. 8a present the frequency at which the modes of engines change while operating the conveyor; particularly in case of empty belt operation and partial load operation for the majority of time.

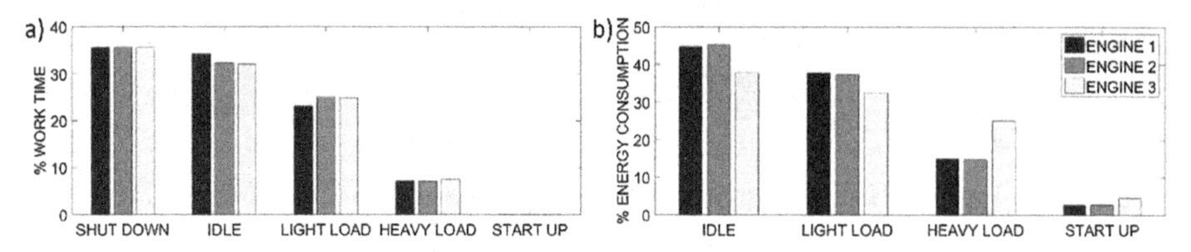

Fig. 8. a) Daily-based working rate of the engines of the subject conveyor per the working modes, b) percentage compilation of energy consumption by the engines per the working modes.

The algorithm applied facilitates tracking the total time of conveyors performance per the mode (Fig. 8a). On a daily basis start-ups, due to their specification, contribute a little to the overall result as compared to the remaining modes. The management should perform agile process planning in order to shorten the time of idle operation and effectively exploit the infrastructure. The above graph can be also completed with the information on the total energy consumption per the various working modes (Fig. 8b).

Additionally, the algorithm can be used to assess the technical conditions of particular engines. For example, assuming that the above graphs depict the identification results for the engines of the same type, while working with the same conveyor, operational time share in the particular modes should be similar in each case. Substantially bigger operational time share of one engine under heavy load might indicate, for instance, an unequal load of material to specific engines.

Figure 9, which depicts the amperage box plots, indicates that engines 1 and 2 show the similar statistics in each working modes. Average amperage is higher for engine 3, regardless of its working mode. Such a state might imply that, in order to balance odd number of drive units, the third engine has to compensate the excess mechanical resistance (Fig. 10).

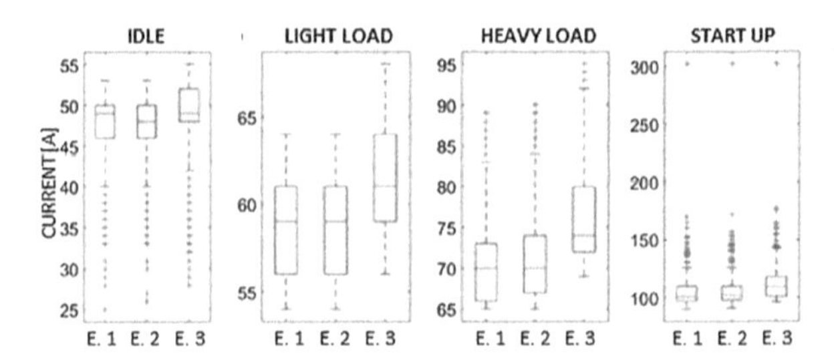

Fig. 9. Compilation of box plots showing amperage of the engines as per the working modes

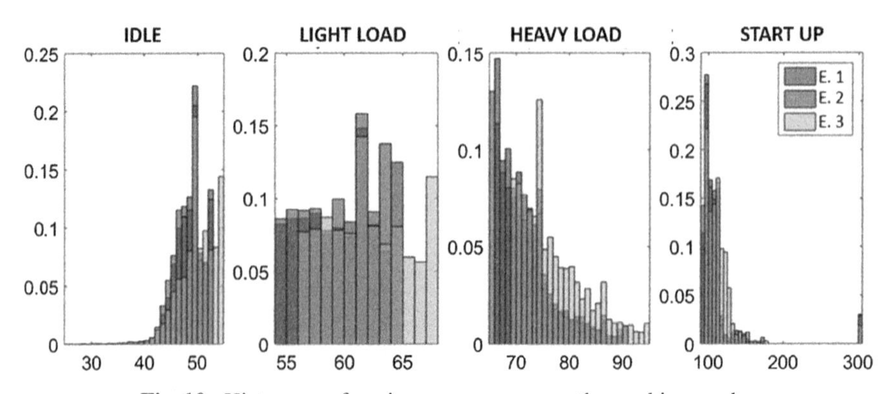

Fig. 10. Histogram of engine amperage as per the working modes

4.3 Sample Multidimensional Analyses

Multidimensional analyses improve the assessment of infrastructure efficiency. They foster the process of compiling the identified working mode and another device information. For instance, the information concerning energy consumption in the given working mode (Fig. 11a) or position of the researched point (Fig. 11b).

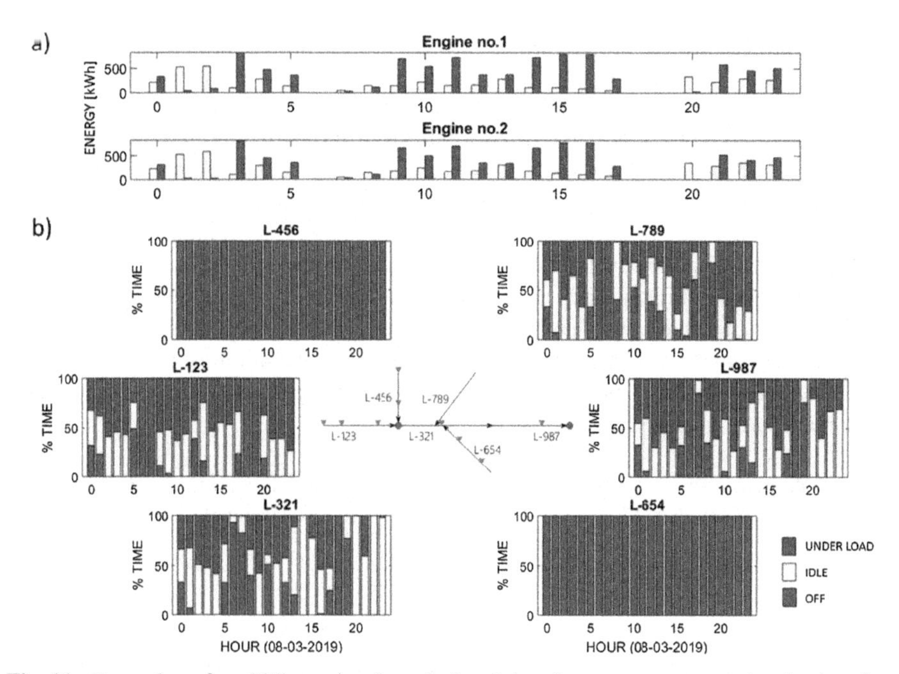

Fig. 11. Examples of multidimensional analysis: a) hourly energy consumption in the given working mode, b) compilation of pictures with the conveyor position in a working day.

Dimensional approach with the carrying direction of material provides the insight into the performance of the whole system. Besides, it facilitates the identification of "bottle necks" of the network and understanding the reasons of the overload at specific points. This approach provides significant information for further development of the conveyor network. As already indicated in the paper, the working modes of conveyor are directly related to the working modes of its engines. Accordingly, it is possible to determine the load of a conveyor with the load of an engine with the intersection method.

5 Summary

The paper covers the analytical capacity of the industrial automation data on the example of the belt conveyors. The basic multidimensional algorithm was applied to process the amperage data in order to enhance KPI efficiency model, regardless of time and spatial aggregation of the indicators. It enables the KPI model to be expanded with organizational, performance and technical aspects based on widely available SCADA data as well as automation of the reporting process and data validation. The recommended solution is the identification of the characteristic working modes (external load level) of the engine of the conveyor and the conveyor itself per the segmentation of the amperage signal. The data identification with k-means clustering has been recommended. Low computational complexity of the method, as well as its popularity are definite advantages of the solution, which allow real-time application. Next, the amperage signal is used to estimate consumed energy and multidimensional computing is conducted to calculate the indicators of the conveyor operation under various conditions. The method is adapted to a large number (~200) of various (in terms of technical parameters, purpose, construction of the route) conveyors, and its applicability is estimated at 95% of cases. After minor modifications, it could be used in other areas of industrial infrastructure like fans, crushers, mills etc. Unfortunately, when large number of conveyors is taken into consideration, the necessity to integrate large amount of data implies the use of BIG DATA tools. In that case also, due to the large number of objects, continuous validation of data and KPIs is required. Further development of analytics is required to better support production, planning and optimization of conveyors operation. In this paper the method of data visualisation is presented, which explicitly traces the conveyors operation, indicates the process losses (empty belt energy) and identifies tripping operations of the engines. The possibility to divide start-ups into light and heavy (belt robustness) and trace grids and points loads have been also discussed in this paper.

Acknowledgment. This work is supported by EIT RawMaterials GmbH under Framework Partnership Agreement No. 17031 (MaMMa-Maintained Mine & Machine).

References

1. A Energy Efficiency Act of 15 April 2011 r. (Dz. U. no. 94, item 551 as amended). http://prawo.sejm.gov.pl/isap.nsf/download.xsp/WDU20110940551/T/D20110551L.pdf
2. Efremenko, V., Belyaevsky, R., Skrebneva, E.: The increase of power efficiency of underground coal mining by the forecasting of electric power consumption. In: E3S Web of Conferences, vol. 21, EDP Sciences (2017)
3. Stefaniak, P., Kruczek, P., Śliwiński, P., Gomolla, N., Wyłomańska, A., Zimroz, R.: Bulk material volume evaluation and tracking in belt conveyor network based on data from SCADA. In: Widzyk-Capehart, E., Hekmat, A., Singhal, R. (eds.) Proceedings of the 27th International Symposium on Mine Planning and Equipment Selection - MPES 2018, pp. 335–344. Springer, Cham (2019). https://doi.org/10.1007/978-3-319-99220-4_27
4. Stefaniak, P., Zimroz, R., Obuchowski, J., Sliwinski, P., Andrzejewski, M.: An effectiveness indicator for a mining loader based on the pressure signal measured at a bucket's hydraulic cylinder. Procedia Earth Planet. Sci. **15**, 797–805 (2015)
5. Bardziński, P.J., Król, R., Jurdziak, L.: Empirical model of discretized copper ore flow within the underground mine transport system. Int. J. Simul. Modell. **18**, 279–289 (2019)
6. Jansen, W., Morrison, R., Wortley, M., Rivett, T.: Tracer-based mine-mill ore tracking via process hold-ups at Northparkes mine. In: Tenth Mill Operators' Conference, Adelaide, October 2009
7. Kruczek, P., Polak, M., Wyłomanska, A., Kawalec, W., Zimroz, R.: Application of compound Poisson process for modelling of ore flow in a belt conveyor system with cyclic loading. Int. J. Min. Reclam. Environ. **32**(6), 1–16 (2017)

8. Zhang, S., Xia, X.: Optimal control of operation efficiency of belt conveyor systems. Appl. Energy **87**, 1929–1937 (2010)
9. Bukowski, J., Gładysiewicz, L., Król, R.: Tests of belt conveyor resistance to motion. Maint. Reliab. **3** (2011)
10. Wodecki, J., et al.: Automatic calculation of thresholds for load dependent condition indicators by modelling of probability distribution functions–maintenance of gearboxes used in mining conveying system. Vibroeng. Procedia **13**, 67–72 (2017)
11. Stefaniak, P., Wodecki, J., Zimroz, R.: Maintenance management of mining belt conveyor system based on data fusion and advanced analytics. In: Timofiejczuk, A., Łazarz, B.E., Chaari, F., Burdzik, R. (eds.) ICDT 2016. ACM, vol. 10, pp. 465–476. Springer, Cham (2018). https://doi.org/10.1007/978-3-319-62042-8_42
12. Galar, D., Gustafson, A., Martínez, B.V.T., Berges, L.: Maintenance decision making based on different types of data fusion. In: Eksploatacja i Niezawodnosc-Maintenance and Reliability, vol. 14, pp. 135–144. Polish Maintenance Society (2012)
13. Dudycz, H., Stefaniak, P., Pyda, P.: Advanced data analysis in multi-site enterprises. basic problems and challenges related to the IT infrastructure. In: Nguyen, N.T., Chbeir, R., Exposito, E., Aniorté, P., Trawiński, B. (eds.) ICCCI 2019. LNCS (LNAI), vol. 11684, pp. 383–393. Springer, Cham (2019). https://doi.org/10.1007/978-3-030-28374-2_33

A Simple Method of the Haulage Cycles Detection for LHD Machine

Wioletta Koperska, Artur Skoczylas(✉), and Paweł Stefaniak

KGHM Cuprum Research and Development Centre Ltd., gen. W. Sikorskiego 2-8, 53-659 Wroclaw, Poland

askoczylas@cuprum.wroc.pl

Abstract. In underground mining of metal ores, horizontal transport of material is performed using self-propelled machines, especially Load-Haul-Dump machines. For example, in KGHM underground mines, where room-and-pillar system is used to deposit exploitation, the haulage process is provided by wheel loaders and haul trucks with suitably adjusted operation configuration. In case of shorter haulage routes, only wheel loaders take part in haulage process. Currently, there is observed a global tendency reliant on develop predictive maintenance as well as navigation or production optimization using Industrial Internet of Things (IIOT). Unfortunately, analytics development in this domain requires full insight into machine's workflow in mining excavations and multivariate analysis in order widely understanding of machine operating contexts. In this article, a quick method to haulage cycle identification on example of wheel loader has been proposed. Developed algorithm is based on hydraulic pressure signal segmentation which provides to recognize loading operation, haulage and return of machine to mining face after unloading material in dumping point. The method is based on smooth hydraulic pressure signal in order to reduce signal interference but introduce to apply a convolution of smoothed signal with inverted step function.

Keywords: Performance assessment · Load-Haul-Dump machine · Mining transport · Signal segmentation

1 Introduction

Current challenges of modern mining enterprises are associated, among others, with the implementation of modern IT tools ensuring the improvement of planning processes efficiency. To maintain competitiveness on the raw material market, it is therefore necessary to constantly develop production planning methods that adapt to the individual and time-varying needs of the mining company. By the way, the planning process begins all subsequent mining processes. In this regard, it is necessary, inter alia, to develop tools to support the management in the area of production chain optimization, which

The original version of this chapter was revised: The order of the first names and surnames of the authors has been corrected. The correction to this chapter is available at https://doi.org/10.1007/978-3-030-63119-2_67

M. Hernes et al. (Eds.): ICCCI 2020, CCIS 1287, pp. 326–337, 2020.
https://doi.org/10.1007/978-3-030-63119-2_27

consists of many operational and operational-related threads. For example, in the underground mining of copper ores, the key among them are primarily: planning the location of mining fronts, choosing the right assets to select the deposit or spoil disposal, as well as the selection of auxiliary machines, including their management. The very issue of spoil disposal optimization is closely related to geological modelling, planning the deposit exploitation process or information flow in the company's internal logistics system. Therefore, when developing analytics to support planning processes, a systematic and fully integrated approach is key. One of the basic goals is, among others, defining the analytical assumptions for spoil logistics.

The main task is, among others, estimation of the shift amount of hauling cycles for operators of loaders and haulage trucks in individual mining departments. These values are calculated in conditions of considerable uncertainty, based on mining practice, catalogue data without actual readings of length and slope of haulage roads, indicators of road quality assessment and operational parameters of machines. Therefore, the key in this respect is to create tools to track the level of resource utilization in various mining conditions, including operational and quality indicators of transport means, and to assess the effectiveness of work organization methods. Another task is the reduction of machine downtime due to irregularities in the condition, as well as resulting from the wrong selection of the configuration of self-propelled machinery for haulage operation leading to disruption of the production process.

It is therefore important to have access to reliable sources of information on the efficiency of spoil haulage, from each individual machine, obtained from real industrial data, preferably from an on-board monitoring system. This article is about the proposition of an algorithm for automatic identification of haulage cycles and their component operations for a self-propelled wheel loader. In practice, the settlement of the production of loaders is usually carried out by means of verbal declarations by operators. In [10], the authors mention the use of video recordings, without providing specific information as to whether the method was automated, which was not the subject of their research. However, in the literature, we can find several similar works carried out for KGHM. The article [12] presents the use of Kalman filter for smoothing the pressure signal as well as a simple method based on the if-then-else rule to identify cycles using a constant pressure threshold. The algorithm allows one to identify the loader ride with a full bucket and empty drive backward after unloading. The disadvantage of the method is, above all, counting the spoil in the loading mode of the loading and unloading process. In addition, the work lacks the presentation of the results of testing the algorithm on a wider population of machines that would show the algorithm's resistance to different signal behaviours resulting from different haulage routes or operators' driving styles. Article [9] presents the identification of labour regimes on the basis of their statistical properties, which should be different for each of them. The described method is based mainly on the empirical second moment, whose main advantage is the assumption of the lack of distribution in the analysed time series. The development of signal segmentation based on its statistical properties is presented in publication [6]. The method described is used to find transition points between processes observed in the signal based on the non-parametric statistical regression method (MARSplines). In [13], the authors reviewed and tested methods of smoothing the pressure signal for the purpose of identifying cycles. However, there is no reference to the method from previous work using the

Kalman filter as well as the review is based only on simple methods of smoothing the signals. Continuation of work is an article [7] in which a method was proposed extending previous versions of the algorithms with additional identification of loader bucket loads based on pressure signal segmentation using regime variance. The concept of using the classification algorithm to recognize different loader modes based on vibrational data and the cardan shaft speed signal has been presented in [8]. An interesting solution was presented in [11], where the authors developed a simple method for estimating the motion trajectory for loading and haulage machines implementing the haulage process in underground excavations. The algorithm is based on the speed signal and inertial data. The method allows to identify the route of the machine in time, as well as track the level of vibration recorded on various parts of the route, which gives an insight into the dynamic overload of the machine and the condition of the road surface.

This work presents a new algorithm for automatic identification of self-propelled wheel loader offload cycles as well as component operations based on a pressure signal. The method is based on smoothing the pressure signal to eliminate interference and apply the convolution to the smoothed signal with inverted Heaviside step function. The advantage of the algorithm is its simplicity, high accuracy, resistance and low algorithmic complexity. Moreover allows to perform full identification of cycles, along with each component operation. Additional goal of the algorithm is to prepare a learning sample for building a more complex data-based algorithm: engine speed, fuel consumption and travel speed, which are more accessible from the point of view of the entire population of loaders operated in KGHM mines.

The article is divided into 5 main parts. In the first step, an introduction to research issues was made along with a brief analysis of the literature. Next, the input data and the relationships appearing in this data, as well as the basic problems of analysis were presented. Section 3 describes methodologies, including the process of smoothing the pressure signal, recognition of jumps, detection of haulage cycles and component operations of these cycles, especially loading, driving with a full bucket and returning to the ancestor after unloading. The next section evaluates the algorithm using measures of sensitivity, precision and accuracy. The article ends with a summary.

2 Input Data

The algorithm has been developed based on data acquired by on-board monitoring system called SYNAPSA common used in Polish underground copper ore mine. A vehicle monitoring system operates on a large scale of self-propelled machines (wheel loaders, haul trucks, drilling and bolting machines) in the KGHM mines. This recorder in real time, with 100 Hz sampling, reads selected machine parameters. These data are averaged to 1 s before being sent to the database that collects them. Access to densely sampled data is possible, however, it requires operator intervention and connection of a properly prepared portable flash memory (pendrive) which will initiate data dump. This option has limited time availability because densely sampled data is constantly overwritten. Therefore, despite the possibility of using more accurate data, one has decided to use rare-sampled signals to create the possibility of implementing the algorithm on a large scale.

As a result of the analysis performed together with the mining staff supervising and operating vehicles, it has been provided that the best suited variable for the purpose of identifying the operational cycles of LHD (Load-Haul-Dump) machines is HYDOILP variable.

HYDOILP contains information about the pressure in the bucket tilting hydraulic system. Even unprocessed variable allows the observation of operational cycles. Each cycle can be divided into 3 main operations: loading, driving with full bucket (haulage) and driving with the empty bucket (return). Despite the fact that individual cycles are clearly visible, they are not identical. In some cycles there occur deviations such as e.g. longer and more complicated loading, or sudden increases or decreases of the variable values during driving. For these and similar reasons, automatic detection cannot be performed using basic methods like threshold values, but requires the use of a more complex algorithm. The variable with reference to the above mentioned operations are presented and marked on Fig. 1.

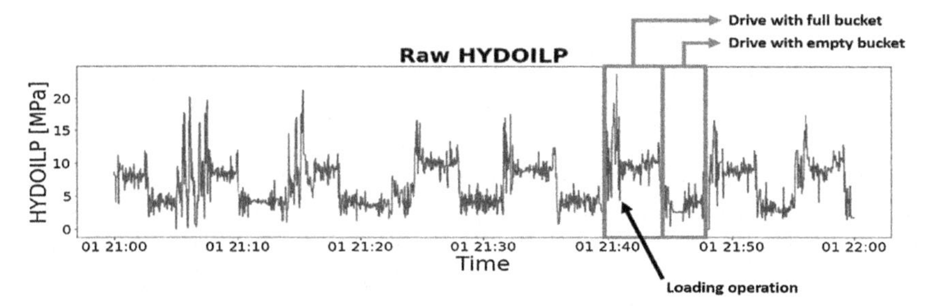

Fig. 1. HYDOILP hydraulic pressure graph from 7 examples of spoil haulage cycles.

3 Methodology

The presented algorithm for identifying operational cycles of haulage process can be divided into 3 main tasks: data preparation, data analysis and finally proper detection. The algorithm diagram is presented on Fig. 2. The first iteration of the algorithm is aimed at cycles detection, while the next iteration detects loading moments using information about cycles. The actions performed during the first start are: smoothing the entire variable signal, then its convolution with the inversion of the unit jump and the detection of cycles after simple signal cleaning. The second run performs very similar actions, however, instead of using the whole signal. Now, algorithm uses only segments of signal containing driving with full bucket, and cuts out using information about the cycles obtained from the first iteration. Despite the fact that the algorithm roughly performs the same actions in both cycles, it was decided to present them as two different procedures, so as to indicate that the functions differ from each other. Those involved in the second iteration are more suited to work with smaller fragments, so that less visible on a wide scale changes in the signal are more highlighted.

3.1 Data Smoothing

In order to correctly detect the LHD cycles, the HYDOILP signal should be smoothed. In the algorithm design process, the following methods of data smoothing have been

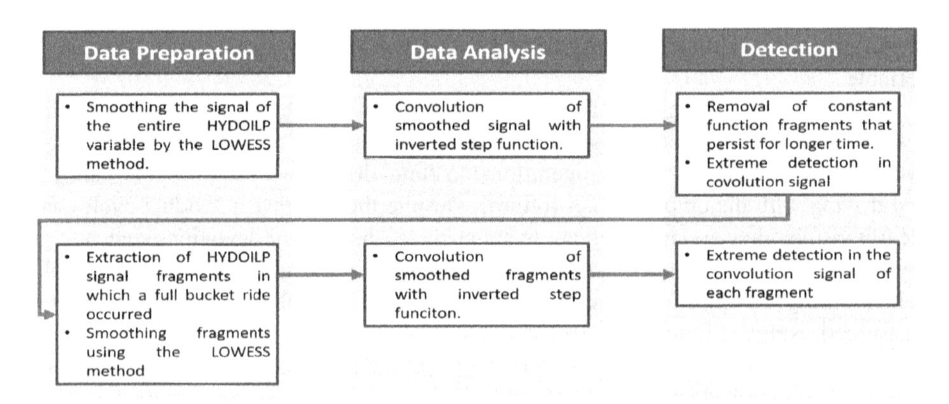

Fig. 2. Diagram of the algorithm for detecting work cycles of LHD machine.

tested: LOWESS, MA, EMA, DEMA, WMA, HMA, ALMA, LR, QR and Kalman filter. The experimental method found that the LOWESS method is the relatively best for the purpose of the described task.

LOWESS (Locally Weighted Scatterplot Smoothing) is a method of fitting a curve to data based on the method of least squares presented in [3] by W. S. Cleveland, later further developed in [4]. The main idea of this algorithm is to divide the data into smaller fragments, and then to each of these fragments fit a straight line using the least squares method. More accurately describing the algorithm, we can divide it into 5 main steps. The first step determines the width of the window being analysed. An increase in the width of the analysed window increases the degree of smoothing of the function. The second step involves assigning weighted local regression weights to individual points. The weight allocation function is determined by the formula:

$$w(x_k) = \left(1 - \left|\frac{x_i - x_k}{d_i}\right|^3\right)^3, \text{ for k} = 1, \ldots, \text{N} \tag{1}$$

where: d_i describes distance from x_i to Nth neighbour. After assigning the weights, the third step follows, i.e. linear regression based on the least squares method. The linear function determined by the formula is fitted to the points:

$$\hat{y}_k = a + bx_k \text{, for k} = 1, \ldots, \text{N} \tag{2}$$

Because of the window, linear regression adjusts the linear function based on a small number of points. This action significantly increases the impact of outliers on the target function. For this reason, the fourth stage of the algorithm reduces the impact of outliers by establishing new weights. These weights are assigned based on formula:

$$G(x_k) = \begin{cases} \left(1 - \left(\frac{|y_i - \hat{y}_i|}{6median(|y_i - \hat{y}_i|)}\right)^2\right)^2, & \left|\frac{|y_i - \hat{y}_i|}{6median(|y_i - \hat{y}_i|)}\right| < 0 \\ 0, & \left|\frac{|y_i - \hat{y}_i|}{6median(|y_i - \hat{y}_i|)}\right| \geq 0 \end{cases} \tag{3}$$

The last fifth step assumes the re-estimation of the linear regression function, this time using both weights calculated in step two and those from step four. This action can be determined by the formula:

$$\sum_{k=1}^{N} w(x_k)G(x_k)(y_k - a - bx_k)^2 \quad (4)$$

In order to increase the precision of the curve fitted to the data, the procedure can be repeated several times, but with each iteration this improvement decreases drastically. As a result of using the LOWESS smoothing method, the signal shown in Fig. 3 was obtained.

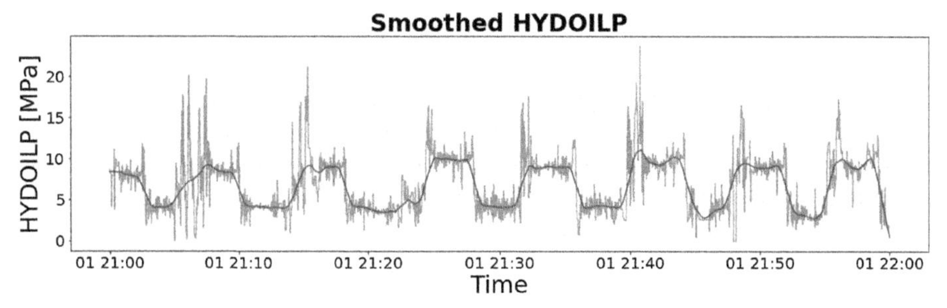

Fig. 3. An example of the result of smoothing operation (red) of a raw signal (blue) by the LOWESS method. (Color figure online)

3.2 Step Detection

The smoothed variable signal with a waveform close to a quadratic function shown in Fig. 3 has clearly visible cycles. Each cycle is divided into 2 basic segments, driving with a full (upper segment > 6 MPa) and an empty bucket (lower segment < 6 MPa). Driving with a full bucket is characterized by values oscillating within 8 MPa, while for driving with empty it is about 4 MPa. Oscillations when the value is changed mean optional loading/unloading. In order to detect similar jumps in the signal, it is possible to use a wide range of algorithms that operate in real time or with delay [1]. To detect individual cycles it was decided to use the convolution operation [5]. Convolution is a mathematical operation of two functions creating a third function. This action can be determined by the formula:

$$(f * g)(t) = \int_{-\infty}^{\infty} f(\tau)g(t - \tau)d\tau \quad (5)$$

This action can be understood as the use of a mirror image of the function $g(\tau)$, or $g(-\tau)$, and then moving it along the τ axis. Moving the function $g(\tau)$ will create regions that can be divided into two categories, the functions will overlap (i.e. both functions will have non-zero values at the same time) or one of the functions will be equal to 0, in which case the convolution resulting value will also be 0 in that moment. In each part of this displacement there is a multiplication between the functions $g(\tau)$ and $f(\tau)$, followed by integration of the multiplication result within $(-\infty, \infty)$. As a result of these actions convolution value for time t is calculated. The function obtained by the convolution has values that depend mainly on the overlap of functions, with the increase of overlapping non-zero fragments of both functions at the moment τ, the value of the convolution function increases.

However, for the convolution to be used to detect cycles, first function $g(\tau)$ need to be created, assuming that $f(\tau)$ is the signal of the smoothed HYDOILP variable. For this purpose, the reverse unit jump function (Heaviside step function [2]) was created, which can be described by the following Eq. (6):

$$g(\tau) = \begin{cases} 1, & \tau \leq 0 \\ 0, & \tau > 0 \end{cases} \tag{6}$$

Signal which is the result of convolution between these two functions is shown in Fig. 4. This signal has clearly marked not only the start of subsequent cycles, but also differentiated driving moments with a full and empty bucket. At the beginning of loading and further driving with a full bucket, the convolution signal increases to create a local maximum, in the case of driving with an empty bucket this signal decreases to create a local minimum.

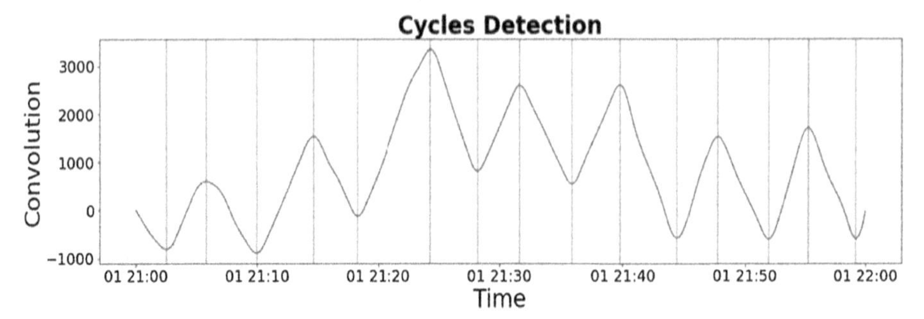

Fig. 4. The result of the HYDOILP signal convolution with reverse jump function together with the detection of the moments of the beginning of loading (red lines) and unloading (green lines). (Color figure online)

3.3 Cycles Detection

In order to detect individual cycles, the function extrema should be found, for this purpose it was decided to use the *argrelextrema()* function from the scipy library (scipy.signal.argrelextrema) available in Python. This function in this algorithm was implemented with two input arguments: a variable containing the result signal of the convolution operation and a function accepting two variables, used to compare them (comparator). Extrema locations are found by searching for places for which the following condition is true:

```
Comparator(data[n], data[n: n + x])
```

In order to find both the local minimum and maximum, the *argrelextrema()* function was run twice with different comparators. In both cases, the numpy library function became the comparator. In order to find local maxima, the *greater(x1, x2)* function was used, which returns the true value when x1 > x2. For local minima, the *less(x1, x2)* function was used, which returns the true value when the x1 < x2 condition is met.

As a result of the presented actions, selected positions of signal moments were received, responsible for the beginning of driving cycles with an empty and full bucket. In some cases, however, such convolution signal analysis may not be sufficient. Values that stay at a relatively low level for a long time can cause incorrect detection of extrema. The solution to this problem is to remove these fragments from analysis. This operation is performed by means of a movable window with a width of 1000 samples, which is moved by 100 samples. The maximum signal value is checked in each window and if it is less than the set limit value, in this case the signal values in the given window are replaced with NaN values.

In this way, cycles are distinguished from running with an empty and full bucket. The detection results are shown in Fig. 5. As one can see, the main parts of each of the visible cycles are correctly classified. There are also some misclassified border samples, however, their impact on the detection of the cycle state is minimal (1 Hz sampled signal). Furthermore it should be highlighted that monitoring of heavy-duty vehicles operating in harsh condition of underground mine is related to presence of many source of interference. Moreover, the machines can change the haulage routes as well as road conditions (surface condition, road shape, slope, traffic, length of road sections) during one work-shift. Each operator also has a different style of driving and operating the machine. Thus, we should except that signal, especially fast-changing can fluctuate abnormally, patterns characteristic of a given operating regime may also sometimes differ from classic cases. Therefore, sometimes the algorithm detects changes in operating regimes too early or too late. However, these are errors of a few seconds and are negligible in terms of mining stuff needs.

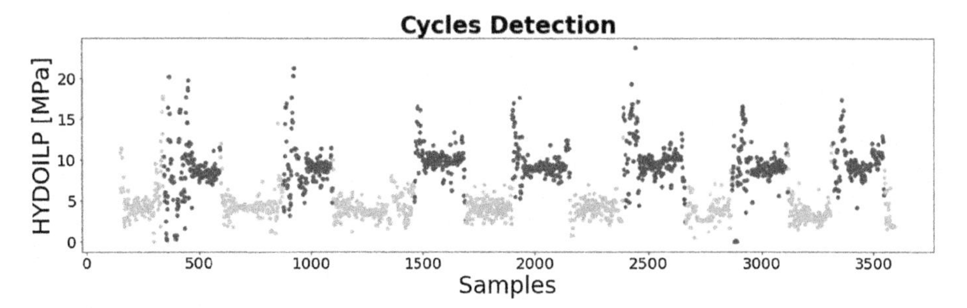

Fig. 5. Loader cycles detection result.

3.4 Detection of Loading Operation

Load detection is based on finding the initial excitation that occurs at the beginning of each cycle. In order to detect these moments, the methods described earlier in the article and information about cycles have been proposed. To detect loads, one must first detect cycles. After obtaining information about the cycles, segments that correspond to driving with a full bucket are cut out from the raw HYDOILP signal, because they are the ones that begin the individual cycles. In the next step, these segments are smoothed and subjected to convolution with the inverse of the step function. This time the process is repeated that the values are adjusted to suitably smaller time intervals, so that more subtle changes occurring in the signal can be better detected. The signal from the convolution also does not require additional processing as in the case of cycle detection, where values that remained constant for too long have been removed. This is due to the short segment of the signal, and the fact that the segment contains only such operational regimes as the loading and driving with a full bucket.

As a result of the convolution, a signal should be created in which the end of loading determines the local minimum. This is due to the fact that the loading has much higher values than driving with a full bucket, and when it ends, these values fall and stabilize. In some cases, however, before stabilization occurs, the values increase again for a while (which is probably the result of operator actions), as a result of which the local maximum instead of the minimum is observed in the convolution signal. In both cases, the extreme associated with the end of loading is the only one observed, which is why the algorithm looks for any extreme and marks the moment of its occurrence as the end of loading.

The effects of load detection are shown on Fig. 6. The colours assigned to each from three identified sub processes of single haulage cycle: loading, driving with a full bucket and driving with an empty bucket.

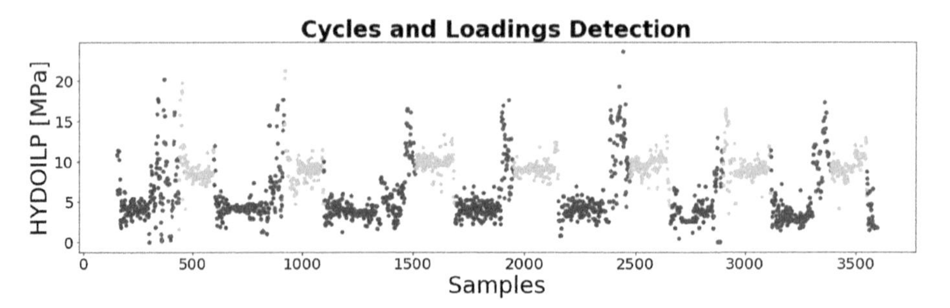

Fig. 6. Result of loadings detection procedure.

4 Algorithm Efficiency

In order to obtain a measure of the algorithm's effectiveness, the confusion matrix was used. This matrix classifies the detection in terms of their compliance with reality, in this case manually checked. Each detection was determined by one of four labels:

true positive - detection consistent with the real state, true negative – lack of detection consistent with the real state (not included in the analysis), false positive – detection not consistent with the real state and false negative – lack of detection not consistent with the real state.

The signals from 5 full working days of the machine were analysed, on which over 300 cycles were recorded. The entire analysis was carried out separately for the detection of parts with full bucket driving, empty bucket driving and loads. The analysis does not include truly negative values, because these occur only when the machine is not in the middle of any of the cycles, i.e. it is turned off, and such fragments were not analysed. The analysis was not performed for each sample separately, instead the focus was on the entire stages of each cycle. The result is that errors due to incorrect classification of border samples are not taken into account. However, given the sampling rate of 1 Hz, the impact of several misclassified samples on the whole algorithm is minimal (for full/empty bucket driving detection, this effect is usually less than 5% of the stage).

During the analysis, it was found that single detection errors occur mainly at the beginnings and ends of entire signal sections, which can be explained by strange signal behaviour resulting from the preparation of the operator and machine for work, performing other side tasks. The validation algorithm for signal fragments during which the machine has not or did not perform the spoil disposal process is not the subject of this article. Obtained confusion matrix has been presented in Table 1.

Table 1. Confusion Matrix – Driving with full/empty bucket and loading operation detection

Date	Driving with full bucket			Driving with empty bucket			Loading operation		
	TP	FP	FN	TP	FP	FN	TP	FP	FN
01.03.2017	40	2	2	39	2	2	28	0	14
02.03.2017	87	2	2	85	2	2	50	0	39
03.03.2017	78	3	0	77	3	0	74	1	4
05.03.2017	26	0	0	25	0	0	13	0	13
06.03.2017	83	1	0	83	1	0	50	0	33
Sum	314	8	4	309	8	4	215	1	103

From the values derived from the error matrix, it is possible to determine a wide range of coefficients that relate to the quality of detection. However, the focus is on 3 basic ones that describe the algorithm well. They are: sensitivity, precision and accuracy.

Sensitivity (TPR - True Positive Rate) - describes the ability of the algorithm to detect the studied phenomenon. If this parameter reaches 100%, it will mean that all occurring manifestations of the examined feature (cycle starts) have been correctly detected. Sensitivity can be determined by the formula:

$$TPR = \frac{TP}{TP + FN} \tag{7}$$

Precision (PPV - Positive Predictive Value) - defines the degree of veracity of detections. If the parameter reach 100%, it will mean that each positive detection was consistent with real state (according to the standard sample). The precision is given by the formula:

$$PPV = \frac{TP}{TP + FP} \tag{8}$$

Accuracy (ACC - Accuracy) - defines the degree of compliance of real detections (TP and TN) in relation to all possible detections (TP, TN, FP, FN). The maximum value of this parameter means that during operation the algorithm did not make a mistake in the detection, thus all manifestations of the variable (start of cycles) were detected, and there was no other detection. Accuracy is determined by the formula:

$$ACC = \frac{TP + TN}{TP + TN + FP + FN} \tag{9}$$

The calculated measures are presented in Table 2. The highest factor in the case of cycles with empty and full bucket is characterized by sensitivity, which means that the majority of cycle start moments have been detected. All parameters for both variants are at a high level of over 96%, which means good operation of the algorithm. Slightly better results are achieved by full-bucket driving detection, but the increase in accuracy is at the level of a decimal or even a hundredth percent, so it has no noticeable effect. In the case of loading, the algorithm no longer achieves such good results. Its accuracy is at the level of 67%, but it is characterized by very high sensitivity (above 99%).

Table 2. Values of basic algorithm quality measures

Cycle	Sensitivity (TPR)	Precision (PPV)	Accuracy (ACC)
Driving with full bucket	98,742%	97,515%	96,319%
Driving with empty bucket	98,722%	97,476%	96,261%
Loading operations	67,610%	99,537%	67,398%

5 Conclusions

Recognition of machine operating cycles as well as parameterization of each cycle is used in current production accounting, but above all is key from the point of view of assessing the effectiveness of machine work in various contexts as well as supporting the implementation of planning tasks. In practice, the number of completed cycles is still recorded by means of verbal declarations by operators, which in many cases differs from the actual performance of the machines. This paper presents a simple method for identifying loader haulage cycles with their component operations, i.e. loading, driving with a full bucket and empty return after unloading to the operating face. The input data used in the work is a pressure signal in the hydraulic system of the loader bucket tilting cylinder. The developed algorithm is based on smoothing the pressure signal to eliminate interference and apply the convolution to the smoothed signal with inverted unit pitch. The advantage of the algorithm is its simplicity, high accuracy, resistance and low algorithmic complexity. The accuracy of the developed algorithm was estimated at 96%. Due to the noticeable problems with the accessibility of the pressure signal, the developed algorithm will be used to develop a learning sample to build a neural network classifier model for detecting withdrawal cycles based on other variables and unknown relationships hidden in them.

Acknowledgment. This work is supported by EIT RawMaterials GmbH under Framework Partnership Agreement No. 17031 (MaMMa-Maintained Mine & Machine).

References

1. Basseville, M., Nikiforov, I.V.: Detection of Abrupt Changes: Theory and Application, vol. 104. Prentice Hall, Englewood Cliffs (1993)
2. Bracewell, R.N., Bracewell, R.N.: The Fourier Transform and Its Applications, vol. 31999. McGraw-Hill, New York (1986)
3. Cleveland, W.S.: Robust locally weighted regression and smoothing scatterplots. J. Am. Stat. Assoc. **74**(368), 829–836 (1979)
4. Cleveland, W.S., Devlin, S.J.: Locally weighted regression: an approach to regression analysis by local fitting. J. Am. Stat. Assoc. **83**(403), 596–610 (1988)
5. Hirschman, I.I., Widder, D.V.: The Convolution Transform. Courier Corporation (2012)
6. Kucharczyk, D., Wyłomańska, A., Zimroz, R.: Structural break detection method based on the adaptive regression splines technique. Phys. A **471**, 499–511 (2017)
7. Polak, M., Stefaniak, P., Zimroz, R., Wyłomańska, A., Śliwiński, P., Andrzejewski, M.: Identification of loading process based on hydraulic pressure signal. In: The Conference Proceedings of 16th International Multidisciplinary Scientific Geoconference SGEM 2016, pp. 459–466 (2016)
8. Saari, J., Odelius, J.: Detecting operation regimes using unsupervised clustering with infected group labelling to improve machine diagnostics and prognostics. Oper. Res. Perspect. **5**, 232–244 (2018)
9. Sikora, G., Wyłomańska, A.: Regime variance testing-a quantile approach. arXiv preprint arXiv:1203.1144 (2012)
10. Skawina, B., Greberg, J., Salama, A., Gustafson, A.: The effects of orepass loss on loading, hauling, and dumping operations and production rates in a sublevel caving mine. J. South Afr. Inst. Min. Metall. **118**(4), 409–418 (2018)
11. Stefaniak, P., Gawelski, D., Anufriiev, S., Śliwiński, P.: Road-quality classification and motion tracking with inertial sensors in the deep underground mine. In: Sitek, P., Pietranik, M., Krótkiewicz, M., Srinilta, C. (eds.) ACIIDS 2020. CCIS, vol. 1178, pp. 168–178. Springer, Singapore (2020). https://doi.org/10.1007/978-981-15-3380-8_15
12. Stefaniak, P., Zimroz, R., Obuchowski, J., Sliwinski, P., Andrzejewski, M.: An effectiveness indicator for a mining loader based on the pressure signal measured at a bucket's hydraulic cylinder. Procedia Earth Planet. Sci. **15**, 797–805 (2015)
13. Wodecki, J., Michalak, A., Stefaniak, P.: Review of smoothing methods for enhancement of noisy data from heavy-duty LHD mining machines. In: E3S Web of Conferences, vol. 29, p. 00011. EDP Sciences (2018)

Haul Truck Cycle Identification Using Support Vector Machine and DBSCAN Models

Dawid Gawelski, Bartosz Jachnik(✉), Pawel Stefaniak, and Artur Skoczylas

KGHM Cuprum Research and Development Centre, Wrocław, Poland
{bjachnik,pkstefaniak}@cuprum.wroc.pl

Abstract. The haul trucks are one of the most often used assets in horizontal transport in underground copper ore mining. This haulage process has a cyclic form and in simple terms the machine drives from point A to point B, where its cargo box is respectively loaded and dumped. What is most important in its basic performance assessment is to identify each cycle and its parametrization in terms of total duration, idling, fuel consumption, and driving speed. In the literature, we can find a few similar works but the majority of them is based on a poorly available hydraulic pressure signal of the actuator in cargo box unloading system or braking system pressure signal. Unfortunately, all of them are not robust and unreliable in real, noisy signals. For this reason, searching for an innovative new concept of the solving problem seems right in this state. This paper describes the new method of the operation cycles identification for underground haul trucks, which is based on multidimensional techniques of operational data analysis using machine learning. The leading idea consists of three parts: searching characteristic non-hydraulic values in signals which correspond to cycles, identifying distinctive periods in haulage process and splitting signal into cycles accordingly. In the first step, the data mining techniques are used to find significant variables, afterwards SVM classifying model identifies unloadings, which are then applied to cluster data by DBSCAN algorithm. The whole process is presented on haul trucks real data from KGHM Lubin mine.

Keywords: Load-Haul-Dumping machines · Underground mining · Mining transport · Performance assessment · Signal segmentation · Support vector machines · Density-based spatial clustering

1 Introduction

Performance assessment issue is widely investigated in the literature. We can find basic approaches related to the simple assessment of effectiveness of dedicated mining assets with multivariate parameter analysis, methods for diagnosing the entire mining processes, as well as mining enterprise performance assessment. It is worth to highlight that these methods have a wide range of applications in mining and that there is significant supply of this type of information in various business areas. The assessment of the efficiency of mining infrastructure is crucial from the point of view of production/working time settlement, its planning or optimization.

M. Hernes et al. (Eds.): ICCCI 2020, CCIS 1287, pp. 338–350, 2020.
https://doi.org/10.1007/978-3-030-63119-2_28

In [1] authors described the model of time series classification in order to identify operation regimes. Model is based on polynomial regression and the concept of hidden Markov chains. In paper [6] concerns mitigation of fuel consumption, reduction of costs, as well as efficiency improvement issues for operations by introducing automatic support systems for operators in the mines. Other paper [7] presents the result of an analysis of the working time of automatic LHD machines in an underground mine in Sweden. In this paper much attention was paid to the impact of the total length of downtime in machine operation. We can also find the application of the FTA method to total work time analysis. The method of estimation of a residual life-time of a machine, taking into account the changes in operating modes based on homogenous Markov process, has been proposed in [17]. The review of most popular KPIs used to optimize maintenance strategy in relation to behavior of assets NPV has been shown in [10]. Similar subjects were considered in [1], which presents a method for evaluating the effectiveness of work in open-cast mines, especially for transport aspects. Analysis of the effectiveness of Turkish coal mines in 2003–2010 according to DEA (Data Envelopment Analysis) methodology was presented in [2]. The authors of work [11] conducted a literature review concerning measures for effectiveness of deposit extraction, loading and transport in the mining industry. The procedure to improvement of effectiveness of equipment usage by the elimination of non-productive working time of the machines has been presented in [8].

It is obvious, that the assessment of the effectiveness of mining processes should be carried out in a holistic approach and cover a longer period of time. However, access to reliable information on the performance of individual machines is crucial and should be based on real industrial data. In this paper we consider the subject of automatic identification of machine operation cycles based on real-time acquired operational parameters by on-board monitoring system on example of haul trucks used in one of the Polish underground copper mines. The number of the recognized operation cycles and statistics for each cycle can be used in the settlement of production, but also as an input data to all the efficiency assessment methods described above. In the literature, we can find several similar works. The papers [13, 20] shows method based on hydraulic pressure signal for recognizing the operation modes for underground wheel loader. The proposed approach enables to segment the hydraulic pressure signal and identify such subprocesses like: loading, driving with a full bucket and driving with an empty bucket. The concept of usage of classification algorithm to recognize the loader operation modes based on vibration data and speed of the Cardan axle has been shown in [14]. In paper [19] the authors proposed a simple method for estimating trajectory of a haul truck motion in mining excavations during the haulage process based on inertial data. In case of underground haul trucks, selection procedures of variables acquired by the on-board monitoring system to determine operational cycles for vehicle has been proposed in [18]. The author in paper [22] proposed the use of blind source separation technique and statistical analysis for better recognition of haul truck modes using three signals: engine rotations per minutes, fuel consumption and vehicle motion speed. Other similar algorithm based on simple if-then-else rules is presented in [9]. This algorithm uses such parameters as engine rotations per minute, vehicle motion speed, braking system pressure and gear selection by operator. Algorithm evaluation was performed on three test data sets with 97 events of truck unloading (no more than 30 standard work-shifts) and reliability of event detection is about 90%.

In summary, most of the available methods for automatic detection of haulage cycles and operation modes of LHD machines are based on hydraulic pressure signals of the actuator in cargo box unloading system, vibration data, or braking system pressure signal. In the case of many vehicles, this data is not available in standard on-board monitoring systems. Parameters registered in a wide range of underground vehicles are engine rotations per minute, vehicle motion speed, or fuel consumption. These parameters have also hidden information on the haulage process but are heavily dependent on many time-varying factors existing in the underground mine condition. In case when whole haulage process is homogeneous in time like in [22] (cycles follows a set pattern) it seems that cycle recognition problem will be a trivial task for a data analyst. In practice, the procedure of cycles recognition is much more sophisticated. What is really important here, is algorithm robustness to varying mining conditions. During a single shift, the machine can change the haulage route even several times. Route change does not only affect the distance but also road conditions (bump, slope, traffic) and route shape (turns, maneuvers, lengths of individual road segments). It is necessary to fit the model by testing it on a huge population of data obtained from different mining areas, which very often is omitted in the literature. For this reason, the amount of performed haulage cycles in practice is still recorded based on the operator's verbal statement, which often deviates from actual performance. In this paper, we propose a novel method to identify operation cycles for underground haul trucks based on basic parameters of the machine driving system. The algorithm is based on a multidimensional analysis of time series using machine learning techniques like support vector machine and DBSCAN algorithms.

This paper is organized as follows: firstly the state-of-the-art and problem definition was defined. Next, investigated underground haul truck and haulage process have been described. The following methodology, which is the main part of this article was split into five subsections: preliminary expectation, exploring data and data mining, pre-processing data, identification of haul truck unloading process via SVM, clustering results with DBSCAN.

2 Description of Mining Machine and Cyclic Haulage Process

2.1 Investigated Machine - Haul Truck

Haul truck (Fig. 1) is a basic mining machine used in Polish underground copper mines in cyclic horizontal haulage processes from mining faces to dumping points secured with grids, where extracted ore is preloaded into the conveyor system. Haul trucks are specially designed to operate under harsh mining conditions in order to ensure that they are able to achieve satisfactory performance even on roads with a slope up to 8°. In this respect, the key design features which are the subject of optimization are machine dimensions, reliability, stability and drive unit parameters. The load capacity of the cargo box is 20 Mg. The operator uses the hydraulic unit to control the trailer. Transported copper ore is dumped through the removable slide wall.

Fig. 1. Investigated machine - haul truck.

2.2 Characteristics of Load-Haul Procedure

As mentioned above, haul truck is intended to haul extracted ore from mining faces to dumping point with grid and hydraulic hammer to break oversized rocks. In a simple sense, it performs cyclic work transporting material from point A to point B. We can distinguish four basic operational modes of a haul truck: (a) loading of the cargo box at the mining face, (b) hauling material to the dumping point, (c) unloading material onto the grid, (d) returning to the loading zone at the mining face. It usually takes three buckets of a wheel loader to fill the haul truck's cargo box. This is due to a lack of space at the mining face, which is dictated by the technology used in mines. The total length of the haulage route should not exceed 1 500 m. The driving speed during passing the haulage route should not exceed 12 km/h. The haulage process is usually short and lasts less than a minute (Fig. 2).

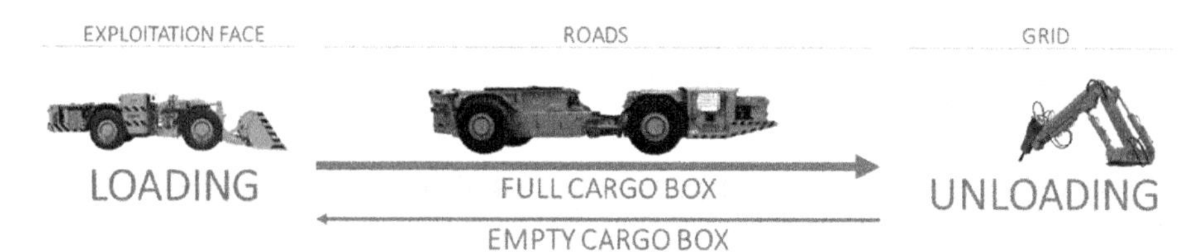

Fig. 2. The course of load – haul – dump process using LHD machines in room and pillar exploitation system.

2.3 Source Data from Haul Truck

In the study, real data from the mining machine registered in the SYNAPSA system was used. As mentioned earlier, the whole process is based on the movement of the haul-age vehicle from point A to B, where loading and unloading take place. Due to the distinction of the forward and return route, the whole process consists of 4 modes (operational regimes). Considering possessed knowledge, the most important variables for the operational modes identification problem are BREAKP, GROILP, HYDOILP, SELGEAR, SPEED, ENGRPM, FUELS. The first three of them are the operational parameter of the following pressure: breaks, gear oil, and hydraulic oil. The SELGEAR informs about gear and motion direction selection by the operator. SPEED is vehicle motion speed in kilometers per hour. Engine rotation per minute is described by ENGRPM. FUELUS informs about fuel consumption per hour. The shortest time dimension that continuous measurements can have is one mine's shift. Then we are sure that the sample contains only complete cycles. However, samples were usually taken from a couple of days (end of day overlapped with the end of shift). Two samples with few days were used in the algorithm. One for teaching the model and the other for testing. The signal from the haul truck consists of characteristic patterns related to the machine's operational regimes. They are visible in Fig. 3, which shows the haul truck's signals registered during one shift.

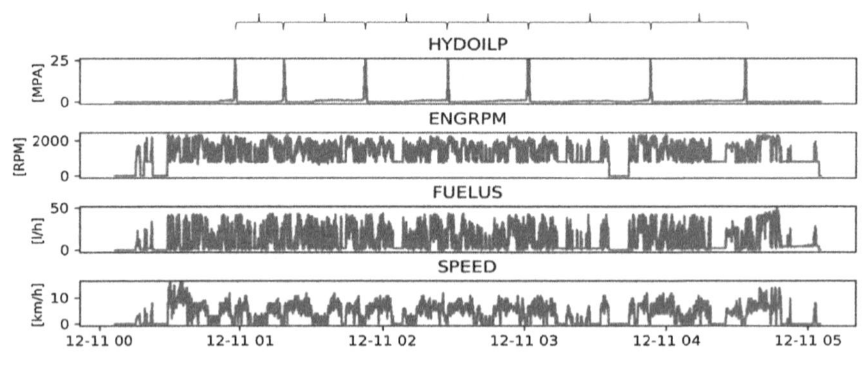

Fig. 3. Time series from one work-shift segmented regard to haul truck operational cycles.

3 Methodology

3.1 Analysis of Relationships in the Data and Identification of Patterns

Before choosing the correct model, we need to define initial assumptions. In order to identify cycles, firstly we decide to find unloadings of the machine. It is the most distinctive operational mode in the whole cycle and additionally, it is considered to be the last one. Therefore, it allows us to identify the end of a cycle. Other modes in the cycle are not as characteristic. Loading is taking place during the stoppage which is no different from any other stoppage of the machine. Similarly, driving with a full-filled box and an empty one are difficult to distinguish when considering just the SYNAPSA signal. All the unloading identifications methods developed so far were based on hydraulic signal analysis. But due to a low-quality of received signals, we decided to find a correlation in other variables that would allow us to use this data for the same purpose.

Before training and accurate models selection, it necessary to explore the data. It helps to find a relationship between variables and to understand which of them can be useful for cycle identification purposes. For preparing the training data set, another algorithm has been used, which was based on the identification of HYDOILP peaks, which can be associated with unloadings (Fig. 4). It turns out that SPEED and ENGRPM variables are the most valuable for the prediction of unloadings. SPEED gets only low values in that mode and ENGRPM was grouped in three separate clusters, with center values of 800, 1500, 2200.

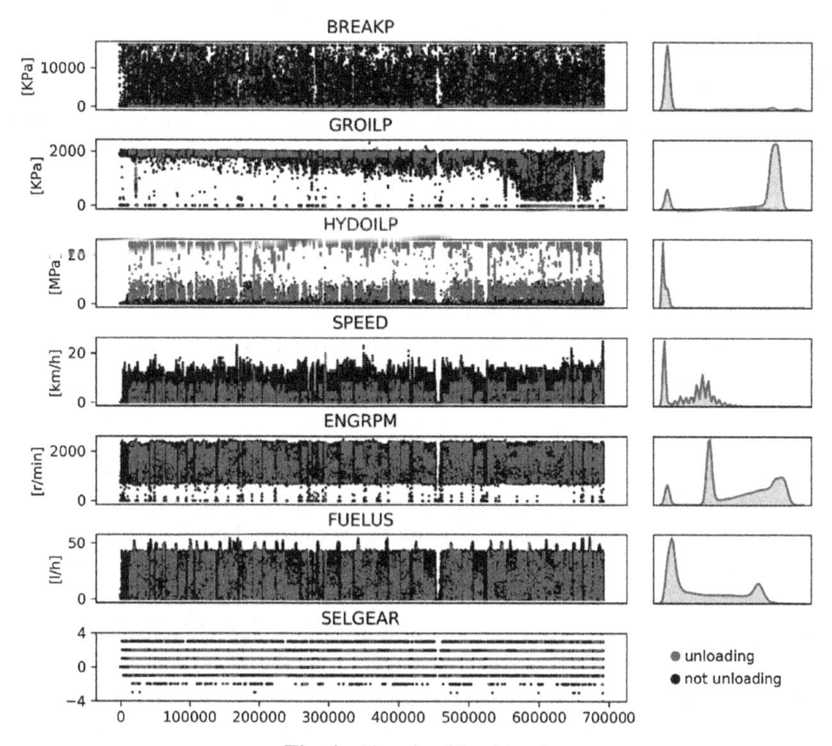

Fig. 4. Signals of haul truck

For the next steps of analysis, we decided to find a correlation between SPEED and ENGRPM in reference to HYDOILP. For this reason in Fig. 5, different values of HYDOILP were marked with different colors. It can be noticed, that high values of pressure can be associated with specific values of other variables. It allowed us to choose this data for unloadings identification.

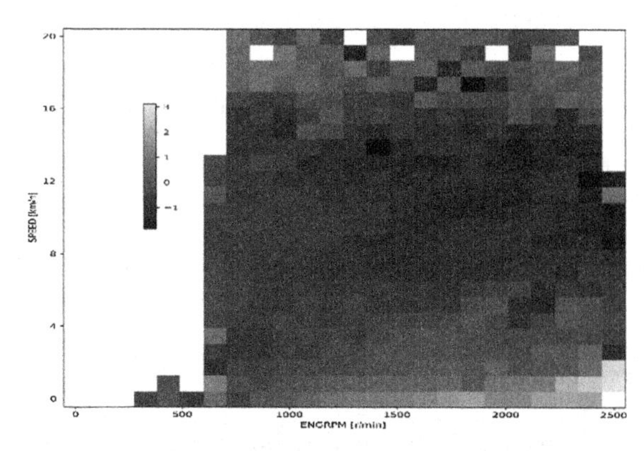

Fig. 5. Relations between ENGRPM, SPEED and HYDOILP.

3.2 Pre-processing Data

Until implementation data into the model, we decided to analyze in detail signals from HYDOILP, SPEED, and ENGRPM during the unloading of a haul truck. HYDOILP could be subject to a significant error that could influence unloadings identification. In some specific cases, this variable reaches its top values when the machine is moving, which is impossible in reality. To identify cycles correctly, we choose only the one which machine was stopped. It helps to highlight the typical values of ENGRPM, which depends on a specific cycle. Furthermore, filtered data allows us to notice that the value of ENGRPM near 800 could be not connected with unloading. After mining, exploring, and pre-processing data for the algorithm, we can predict unloading without HYDOILP, based on the following observations:

- During unloading the vehicle is not moving.
- The hydraulic cylinder is moved by a hydraulic pump, which is moved by the engine. Due to this fact, the engine speed is rising. The various characteristics of the signal could depend on machine or operator working, but it shouldn't have an influence on cycle prediction.

Because of that, only parts of the signal with SPEED = 0 and ENGRPM in the range between 1200 and 2500 (speed engine above idle level) should be considered. The data implemented in the model was prepared according to the above restrictions. Besides, the speed variable is smoothed by using a moving average with a 5-second window. The whole process of cycle identification is as follows:

1. Select SPEED, ENGRPM (attributes), and labels whether the sample meets the criteria (speed, engrpm).
2. Use the moving average for SPEED variable.
3. Train Support Vector Machine Classifier.
4. Identify unloadings using SVM.
5. Cluster unloading samples.
6. Fill empty values with the following number of cycle.

3.3 Identification of Haul Truck Unloading Process via SVM

For unloadings identification using variables SPEED and ENGRPM, we decided to use the Support Vector Machine model [4, 15]. SVM is a supervised learning model, which works particularly well for small sets of data. Three signals were used as input data: the variable SPEED filtered with the moving average, the variable ENGRPM and the new signal describing whether a given sample meets the values described earlier (SPEED = 0, ENGRPM < 2500, ENGRPM > 1200). As a result of the SVM algorithm's operation on the sample, the signal indicating the places of unloading occurrence is returned (signal is of the Boolean type).

SVM was invented by Vladymir Vapniik in the 60' of the XX century. It has been considered to be one of the best models for classification before the expansion of neural networks. The model doesn't calculate the probability of belonging to category (such as most of them), but it tries to find border which will separate two categories of data. The base idea of the SVM is creating hyperplane, which separates two classes. The formula to define hyperplane is as follows:

$$h(\boldsymbol{x}) = \boldsymbol{w}^T * \boldsymbol{x} + b \tag{1}$$

The slope of hyperplane is determined by the norm $||\boldsymbol{w}||$. The distances between classes and the plane (called margin) are defined by points where decision function (1) is equal to 1 or −1. When the slope decreases, the margin increases. It has been shown on the one dimension data at the plot bellow [Fig. 6].

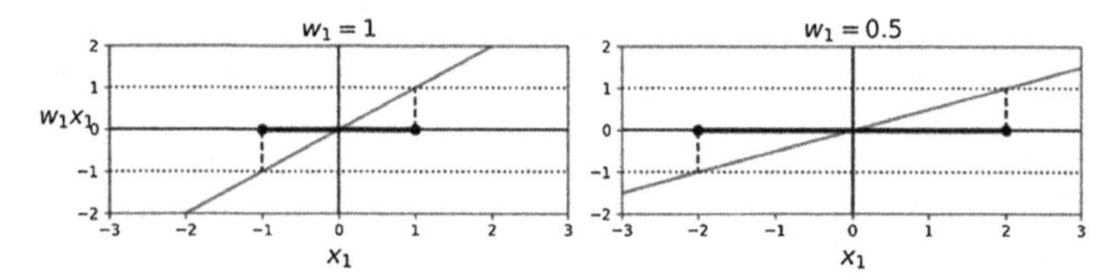

Fig. 6. Influence of norm to slope of hyperplane in one dimensional example.

The goal of SVM classifier algorithm, which is used in this article, is to find hyperplane by minimalizing the norm $||\boldsymbol{w}||$, so that it separates two classes of data (the value of classifying function has to be always above 1 for the positive class, and below −1 for the negative class). In this case the problem is described by the following formula:

$$\min\ (||\boldsymbol{w}||)$$

subject to

$$t^{(i)}\left(\boldsymbol{w}^T * \boldsymbol{x}^{(i)} + b\right) \geq 1\ for\ i = 1, 2, \ldots, n$$

where $t^{(i)} = 1$ for positive class and $t^{(i)} = -1$ for negative class.

As mentioned above, the formula tries to maximize the distance between the two classes. To define the hyperplane (in our 2-dimensional case it is a line) only adjacent samples are needed. These points are called support vectors. For this reason, the addition or subtraction of points that are not in the boundary has no impact on the model. However, the case where it is not possible to linearly divide the data, and therefore a separating hyperplane cannot be defined is easy to imagine. Then, we can add a counter condition, by implementing a slack variable that describes the value of permission to abuse the border. Because the $||\boldsymbol{w}||$ has no derivative in 0, additionally we decided to change optimization formula as follows:

$$\min\ (\frac{1}{2}\boldsymbol{w}^T \cdot \boldsymbol{w} + C\sum\nolimits_{i=1}^{m} \vartheta^{(i)})$$

subject to

$$t^{(i)}\left(w^T * x^{(i)} + b\right) \geq 1 \text{ and } \vartheta^{(i)} \geq 0 \text{ for } i = 1, 2, \ldots, n$$

where:

$\boldsymbol{x}^{(i)}$: vector of i-item attributes,
w: vector of parameters,
b: bias term,
$\vartheta^{(i)}$: value of slack variable for i-item,
C: ratio of abusing border.

The compromise between maximizing border and permission can be defined by ratio C. We are able to adjust the model by using many hyperparameters to find a compromise between the number of crossing boundaries and the effectiveness of the model. The above formula allows splitting sets linear. In some cases, it is impossible to split the data using this approach. The clever way to solve this problem is to use a polynomial feature. The easiest way is to raise to power each of the variables. It creates additional features, which could be possible to divide by a line. It should be done after the classification of data quality because this method creates next dimensions, which can lead to the curse of dimensionality. The example of this approach is presented in Fig. 7. More about SVM can be found in the articles [12, 15, 21].

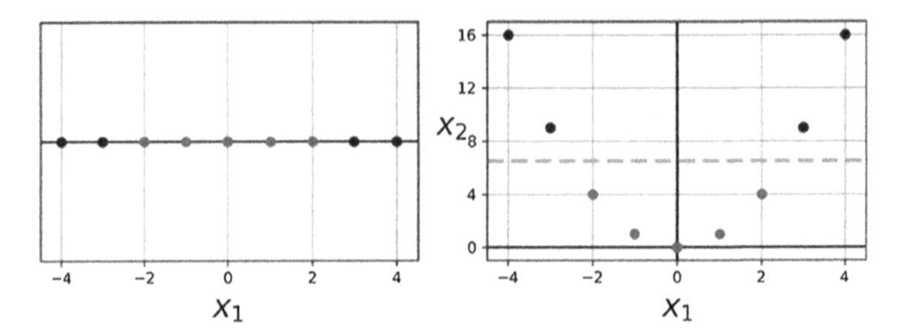

Fig. 7. Influence of norm to slope of hyperplane in one dimensional example.

3.4 Clustering Results with DBSCAN

As previously explained, the results of the SVM-based model gives only information about unloadings and they are not so useful for others mode identification yet. For the task of detecting other work regimes signal need to be segmented using that information. The fragments between the two unloadings are selected and cut out. The next step of the algorithm is focused on grouping those fragments. Each of them has a different duration and depends on varying mining conditions or operator. In our case for identification cycles, we use clusters of unloadings which are close in time. For this reason, the best model to solve this problem is the usage of the DBSCAN clustering algorithm [5, 16]. It is based on searching clusters areas with high density separated by areas of low density. In general, we define a core sample as being a sample in the dataset such that there exist other samples within a minimum distance of eps, which are defined as neighbors of the core sample. The procedure is looped until all points will be assigned to one of the groups. If the group consists of a too small number of elements, it is removed. In our case, the algorithm works with parameters, which are presented below:

1. One cycle has a minimum of 3 samples with unloading.
2. The minimum distance between neighbors is 30 s.

It should remove anomalies and solve the problem with double identification of one cycle. As mentioned before, the unloading is the last sub-process of the whole haulage cycle. For this reason, after clustering unloading segment of data, all the empty values should be filled by a given cycle number.

4 Testing on the Industrial Data

For testing of the algorithm haul truck data from the period between 01.12.2019-31.12.2019 and 01.02.2020-15.12.2020 was used. Data from December will be used for training of the model and from February – for testing purposes. Due to the low informational character of the data from records, it is difficult to establish the actual effectiveness of the new model. The only way to compare the results with reality is to manually count the cycles that occur or through records kept by operators. Therefore, testing results can now be reduced to comparing the occurring cycles without comparing the times and lengths of the cycles. The effect of the detection process is compared with other approaches (HYDOILP based algorithm and manually recording method) in Table 1.

As can be seen, the algorithm mostly finds results similar to the HYDOILP-based algorithm. It should be pointed out, that the new approach is not based on the HYDOILP variable. Tests on real data prove that it is possible to identify cycles by the method of increasing hydraulic pressure, without measurements of this variable. By comparing the results of the new approach to data recorded by operators it can be seen that the results are similar. It should be emphasized that the signal from the change February 4, 2020, 18:00 was damaged, because of this it was not possible to correctly reproduce the number of cycles by the algorithm. Despite this, it is considered that the operation of the algorithm is in the limits of correctness, whereas the number of cycles identified by the operator is not equal to the number of pressure increases. Besides actual changes, there are also cycles recorded during an unregistered change. These are mostly periods when repairs or maintenance work on the machine was carried out.

Table 1. Comparison of results between algorithms and records kept by operators.

Date	Peaks of hydraulic pressure	SVM + DBSCAN	Manual record entered by operator
01.02.2020 06:00	4	3	–
03.02.2020 00:00	9	7	8
03.02.2020 06:00	0	1	–
03.02.2020 12:00	7	8	8
03.02.2020 18:00	7	9	8
04.02.2020 12:00	8	8	8
04.02.2020 18:00	3	3	9
05.02.2020 00:00	9	9	8
05.02.2020 06:00	1	1	–
05.02.2020 12:00	7	6	–
05.02.2020 18:00	12	13	13

5 Summary

The work presents an effective method of determining cycles based on the identification of pressure increases by means of engine revolutions and machine speed. The proposed algorithm is based on the new approach of recording only valuable variables, excluding those whose quality was low (e.g. HYDOILP). Through this approach, the algorithm is more effective and more resistant to signal quality. The tests carried out in the second part took part in an underground mine on the data from the haul-age vehicle. They allowed verifying the assumed method. As a result, a summary of the number of cycles per shift was compared with manual records. Studies have shown great effectiveness in identifying pressure increases when analyzing other variables. Minimal differences between the results may be due to a lack of accuracy between the number of increases and the recorded number of cycles. In further work suggests that the algorithm be developed to validate results for the occurrence of the change.

Acknowledgment. This work is supported by EIT RawMaterials GmbH under Framework Partnership Agreement No. 17031 (MaMMa-Maintained Mine & Machine).

References

1. Assumpção, M.R.P., de Medeiros, C.A.: Estudo sobre eficiência técnica na extração de nióbio. Revista de Ciência & Tecnologia, **17**(35), 115–128
2. Bakirci, F., Yakut, E., Demirci, A., Gündüz, M.: Efficiency measurement in Turkish coal enterprises using data envelopment analysis and data mining. Can. Soc. Sci. **10**(1), 103–110 (2014)

3. Chamroukhi, F., Samé, A., Aknin, P., Govaert, G.: Model-based clustering with Hidden Markov Model regression for time series with regime changes. In: The 2011 International Joint Conference on Neural Networks, pp. 2814–2821. IEEE (2011)
4. Chang, C.-C., Lin, C.-J.: LIBSVM: a library for support vector machines. ACM Trans. Intell. Syst. Technol. (TIST) **2**(3), 1–27 (2011)
5. Ester, M., et al.: A density-based algorithm for discovering clusters in large spatial databases with noise. In: KDD, vol. 96. no. 34 (1996)
6. Frank, B., Skogh, L., Filla, R., Fröberg, A., Alaküla, M.: On increasing fuel efficiency by operator assistant systems in a wheel loader. In: International Conference on Advanced Vehicle Technologies and Integration (VTI 2012), Changchun, China (2012)
7. Gustafson, A., Schunnesson, H., Galar, D., Kumar, U.: The influence of the operating environment on manual and automated load-haul-dump machines: a fault tree analysis. Int. J. Min. Reclam. Environ. **27**(2), 75–87 (2013)
8. Jakkula, B., Mandela, G., et al.: Improvement of overall equipment performance of underground mining machines- a case study. Model. Measur. Control C, **79**(1) (2018). https://doi.org/10.18280/mmc_c.790102
9. Krot, P., Sliwinski, P., Zimroz, R., Gomolla, N.: The identification of operational cycles in the monitoring systems of underground vehicles. Measurement **151**, 107111 (2020)
10. Kumar, U., Parida, A., Duffuaa, S.O., Stenström, C., Galar, D.: Performance indicators and terminology for value driven maintenance. J. Qual. Maint. Eng. (2013)
11. Mohammadi, M., Rai, P., Gupta, S.: Performance measurement of mining equipment. Int. J. Emerg. Technol. Adv. Eng. **5**(7), 240–248 (2015)
12. Noble, W.S.: What is a support vector machine? Nat. Biotechnol. **24**(12), 1565–1567 (2006)
13. Polak, M., Stefaniak, P., Zimroz, R., Wyłomańska, A., Śliwiński, P., Andrzejewski, M.: Identification of loading process based on hydraulic pressure signal. In: The Conference Proceedings of 16th International Multidisciplinary Scientific Geoconference SGEM 2016, pp. 459–466 (2016)
14. Saari, J., Odelius, J.: Detecting operation regimes using unsupervised clustering with infected group labelling to improve machine diagnostics and prognostics. Oper. Res. Perspect. **5**, 232–244 (2018)
15. Schölkopf, B., et al.: New support vector algorithms. Neural Comput. **12**(5), 1207–1245 (2000)
16. Schubert, E., et al.: DBSCAN revisited, revisited: why and how you should (still) use DBSCAN. ACM Trans. Database Syst. (TODS) **42**(3), 1–21 (2017)
17. Si, X.S., Wang, W., Hu, C.H., Zhou, D.H., Pecht, M.G.: Remaining useful life estimation based on a nonlinear diffusion degradation process. IEEE Trans. Reliab. **61**(1), 50–67 (2012)
18. Śliwiński, P., Andrzejewski, M., et al.: Selection of variables acquired by the on-board monitoring system to determine operational cycles for haul truck vehicle. In: Mueller, C., et al. (eds.) Mining Goes Digital © 2019. Taylor & Francis Group, London, ISBN 978-0-367-33604-2 (2019)
19. Stefaniak, P., Gawelski, D., Anufriiev, S., Śliwiński, P.: Road-quality classification and motion tracking with inertial sensors in the deep underground mine. In: Sitek, P., Pietranik, M., Krótkiewicz, M., Srinilta, C. (eds.) ACIIDS 2020. CCIS, vol. 1178, pp. 168–178. Springer, Singapore (2020). https://doi.org/10.1007/978-981-15-3380-8_15

20. Stefaniak, P., Zimroz, R., Obuchowski, J., Sliwinski, P., Andrzejewski, M.: An effectiveness indicator for a mining loader based on the pressure signal measured at a bucket's hydraulic cylinder. Procedia Earth Planet. Sci. **15**, 797–805 (2015)
21. Widodo, A., Yang, B.-S.: Support vector machine in machine condition monitoring and fault diagnosis. Mech. Syst. Signal Process. **21**(6), 2560–2574 (2007)
22. Wodecki, J., Stefaniak, P., Śliwiński, P., Zimroz, R.: Multidimensional data segmentation based on blind source separation and statistical analysis. In: Timofiejczuk, A., Chaari, F., Zimroz, R., Bartelmus, W., Haddar, M. (eds.) Advances in Condition Monitoring of Machinery in Non-Stationary Operations, pp. 353–360. Springer, Cham (2018). https://doi.org/10.1007/978-3-319-61927-9_33

Intelligent Management Information Systems

Data Quality Management in ERP Systems – Accounting Case

Marcin Hernes[1(✉)], Andrzej Bytniewski[1], Karolina Mateńczuk[1], Artur Rot[1], Szymon Dziuba[1], Marcin Fojcik[2], Tran Luong Nguyet[1], Paweł Golec[1], and Agata Kozina[1]

[1] Wrocław University of Economics and Business, Komandorska 118/120, 53-345 Wrocław, Poland
{marcin.hernes,andrzej.bytniewski,karolina.matenczuk,artur.rot, szymon.dziuba,tran.loung.nguyet,pawel.golec, agata.kozina}@ue.wroc.pl

[2] Western Norway University of Applied Sciences, Førde, Norway
marcin.fojcik@hvl.no

Abstract. ERP systems process data obtained from heterogeneous sources and therefore the data are characterized by different quality. In order to effectively support management, ERP systems must be based on high-quality data. This is a prerequisite for making decisions within the company. The aim of this paper is to analyse the problems of data quality management in ERP systems and its main contribution is to develop procedures for data quality management in data processing, especially accounting data. The basic outcome of the conducted research is the systematization of the knowledge concerning the data allowing for making quick and effective economic decisions. Based on the results of the research it was concluded that a large number of data control procedures embedded in the ERP allows for ensuring appropriate data quality.

Keywords: Data quality · Data quality management · ERP systems

1 Introduction

Data quality is an important element of the proper functioning of business organizations. Only based on the data of appropriate quality can managers make correct decisions. Data quality in ERP systems is of particular importance. These systems collect data from many internal and external sources around the world, so the data may be of different quality [17]. Quality is measured in ERP by using the utility function [1]. An important problem is that real utility is only known when the effects of a decision made using this data are known. Important features of the data include urgency and topicality, because the right decision can only be made if the system is capable of quick collecting the updated data needed to make that decision [18]. Other properties of the data are undeniableness, comprehensibility and reliability. Data are often considered reliable if they can be confirmed in many sources.

M. Hernes et al. (Eds.): ICCCI 2020, CCIS 1287, pp. 353–362, 2020.
https://doi.org/10.1007/978-3-030-63119-2_29

The motivation for presented research rise from need reported by participants of scientific conferences. After our presentations related to using artificial intelligence in ERP, there were many similar comments from scientists: "If the data quality is insufficient then artificial intelligence methods will be not effective". Therefore, we decided to explain this problem in this paper.

The aim of this paper is to systematize the problems related to data quality and its management in ERP systems. The main contribution is developing procedures for data quality management in data processing, especially accounting data, by ERP. The first part presents the concept of data quality. Next, the issues of data quality management in ERP were characterized. The last part of the paper is an analysis of practical aspects of data quality management.

2 Research Methodology

In our researches we use the Design Science Research Process (DSRP) methodology [2]. This methodology involves the realization of research works as process-dependent steps responsible for the implementation of basic tasks and the delivery of basic work artefacts. This methodology is based on five basic steps. The first, "Awareness of problem", focuses on the preparation of a preliminary proposal for a new research topic, which usually results from the analysis of already available achievements and results in related disciplines or disciplines, as well as in industry. This step is performed in "Introduction" section. The second, development of the proposal "Suggestion" – the "Related works section" – follows the initial proposal, and is a creative step aimed at defining new models, methods, functionalities based on new configurations of existing or new elements. The third, development of the proposal "Development", is the stage where the initial research is developed and extended. We perform "Development" stage in Data quality management section. The fourth, "Evaluation" – "Discussion" section in this paper - is the stage where the prepared artefacts are evaluated according to the adopted criteria. The fifth, "Conclusion" section in this paper – can be the end of a given research cycle or is the end of the entire research undertaking. As outlined, each stage ends with a specific result, which then forms the basis and allows for the next step in the process. In addition, this methodology assumes that it is possible to return to the previous steps under the observed restrictions.

We use mainly the literature of subject analysis, as research method in the Awareness of problem and the Suggestion stages. In the development stage, we use such research methods, as modelling, design and implementation and observation of phenomena's.

The Evaluation and Conclusion phases we based on synthesis, analysis, induction and deduction research methods.

3 Related Works

Adequate data quality is a prerequisite for the proper functioning of ERP systems. In the literature on the subject, the concept of data quality is defined in many ways. For example, in the study conducted by Batini et al. [1], data quality is defined as the qualification of the correctness of data and their utility.

Maślanowski [3] argues that data quality is defined by such criteria as:

- accuracy,
- usefulness, meaning that they should be relevant to the recipient,
- timeliness[1], i.e. they should be supplied at a specific time.

According to the European statistical system, data quality is determined by the following components [4]:

- usefulness,
- accuracy,
- timeliness[2] and punctuality[3],
- accessibility and transparency,
- comparability,
- consistency.

The quality features of the data are influenced by the complexity of many problems, which may include [5]:

- data arriving too late, and concerning events that previously occurred and one cannot control the processes that have occurred any more,
- excessively detailed data, although in the case of IT systems this should not be viewed as a major drawback, as these data can be easily aggregated automatically
- too extensive data, i.e. they cover a fairly large period of time
- future-oriented data,
- data of only quantifiable character,
- mutually exclusive data,
- unclear data for strategic purposes.

In business analytics, data quality is identified with completeness and consistency being central dimensions of data quality in analytical systems [8]. The quality of business data should be considered in three steps [9]:

- at the input,
- during processing,
- at the output

Lee et al. [10] define data quality as a hierarchy of attributes (Fig. 1).

[1] A description of the problems concerning the increasing the timeliness of data was presented in the paper [6].

[2] Timely means having a deadline, a precisely defined moment of completion of a given action [6].

[3] Punctual means that the entity's actions are synchronized with those of other entities. Punctuality applies to each phase of the action [7].

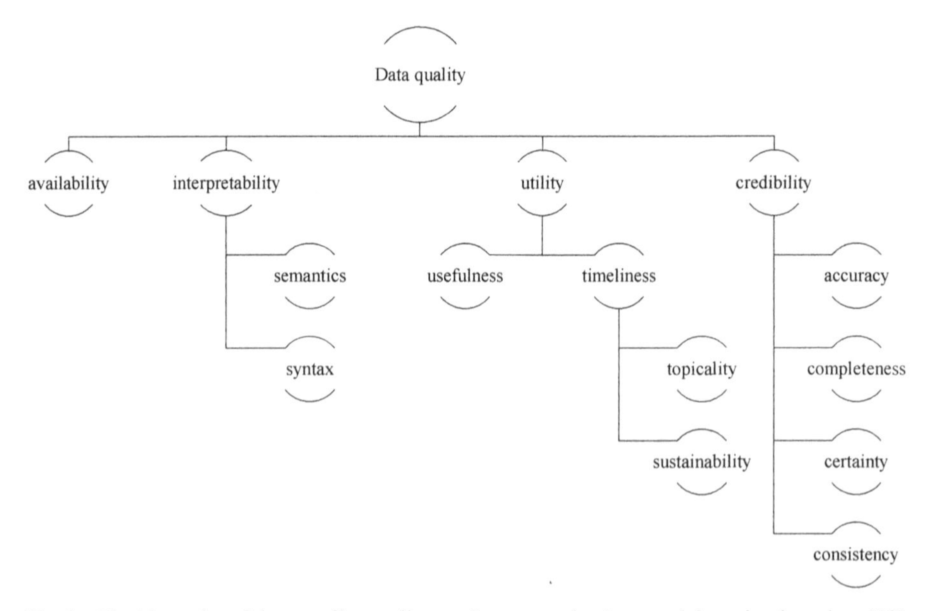

Fig. 1. The hierarchy of data quality attributes. Source: author's own elaboration based on [12].

The authors emphasized that in this relation the superior attribute is true (fulfilled) if all the subordinate attributes (logical product) are true. For example, data are interpretable if they have the correct syntax and semantics.

Therefore, it can be concluded that data quality is a very broad concept. The definition of attributes (and their hierarchy) that have to be fulfilled when analysing data quality is most often based on the nature of the decision-making situation. For example, the data used to make decisions on marketing activity do not have to be fully consistent or complete, while with regard to production control data, inconsistency or incompleteness is unacceptable as it could lead to damages to machinery/equipment on the production line or even to accidents.

There are many works related to data quality in ERP. The work [19] presents issues related to data quality in the implementation phase of ERP life cycle. The authors presents several factors, which influent on the data quality. The article [20] proposes the assessment of quality criteria method based on dimensions analysis. The Normal Accident Theory is used for data quality assessment in [21]. Paper [22] provides guidance for ERP system use in DQM in the insurance sector. Work [23] presents the ERP data quality control method based on schema defined in XML. The authors of paper [24] identified the most important factors for accounting information quality in ERP and their impact on data quality outcomes. The work [25] employ data mining to accounting transactions in order to discover patterns of data handling in accounting system.

However existing works are related mainly to the data quality management methods and procedures, but they not explain effects of implementation of the developed solution in the sufficient way. The managers and also employees of business organizations are very confused reading these works – they often think, that quality management is performed not by ERP but by separate software. This problem is related to the accounting managers and employees in a high degree. In addition there is small number of publications related to data quality in accounting systems. Therefore in this paper we provide knowledge for managers and employees and develop the procedure for data quality management designed mainly to accounting area in ERP system.

4 Data Quality Management

The process of data quality management consists of the following phases [11]:

1. Data profiling – this is a process of analysis and understanding of existing data. Data that already exist in the system must be correctly interpreted and used. The data profiling phase is therefore the starting point for the subsequent phases. As a result of this phase, the reply is obtained whether the data are complete and accurate, i.e.:

    - have an appropriate range (e.g. the age of a person 200 years is outside the permissible range),
    - are standardized (for example, if the height of one person is recorded in centimetres and the height of another person is input in metres, the data are not standardized),
    - have correct values (e.g. if the sex can only take the value "M" or "F" and the data contain the value "L", their value is incorrect),
    - are compatible with local settings and adopted formats (for example, different regions of the world have different date formats and therefore there may be incompatibility; different currencies may also exist),
    - there is a consistency of the links between the data (e.g. if the data contain an incorrect postal code of the town, the links are inconsistent).

2. Assigning the quality of acquired data. Based on the results obtained in the previous phase, rules of correctness with regard to the acquired data are construed. Next, it is checked whether the data obtained comply with these rules. As a result of this phase, the following actions are undertaken:

    - data exclusion: if the error in the data is too big, the best approach may be to delete the data,
    - data acceptance, even if it is known that there are errors in the data, but they are within tolerance; sometimes the best solution is to accept data with an error,
    - correcting data, if it is known how this can be done,
    - inserting a default value; sometimes it is important that the field has a value (e.g. unknown).

3. Data integration. Data on the same object or real-world phenomenon stored in different databases may differ. Different methods may be used to integrate data, for example [13]

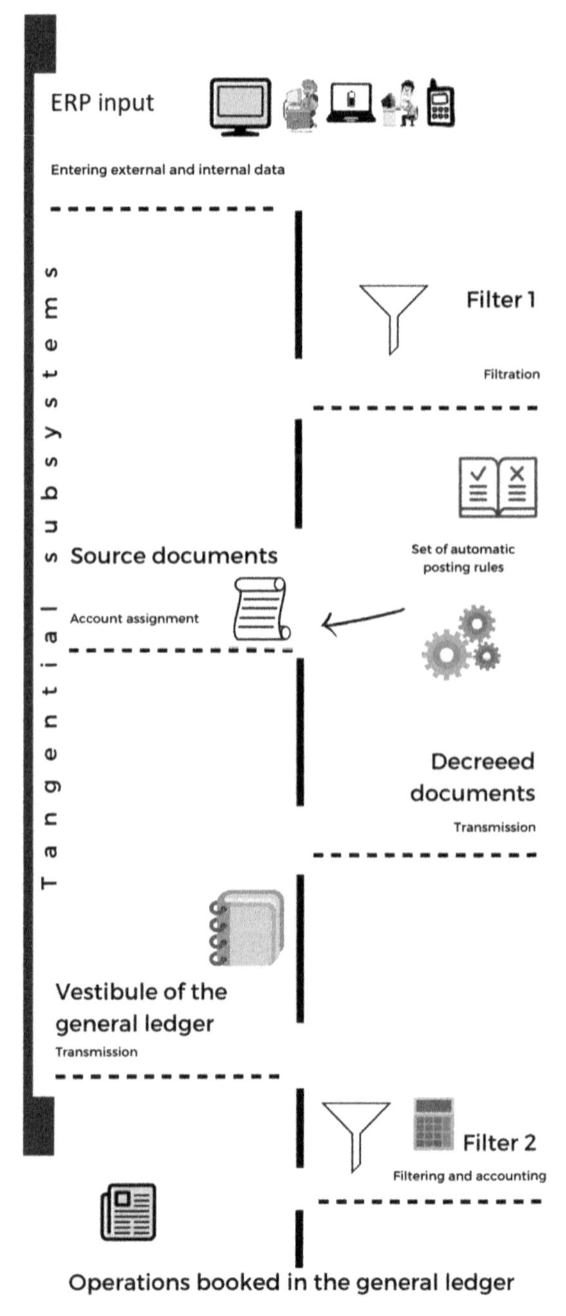

Fig. 2. Procedures for data quality management in accounting data processing.

- choice,
- negotiations,
- consensus.

4. Data augmentation (increasing the analytical capabilities of data). This is the last phase of data quality management. Data are converted to a form that enables deeper business analysis (for example, OLAP cubes are created to enable multidimensional data analysis).

ERP systems have built-in procedures for data quality management, which are implemented in the process of data processing (Fig. 2). The data for the tangential subsystems can be input by users or from electronic devices (Internet of Things). They can be internal or external. The data entered are filtered (Filter 1). For example, the limit values of the entered data are checked. After filtering, the data become source documents. Then an automatic assignment of these documents is performed based on the accounting principles[4] and these documents are transferred to the so-called vestibule of the general ledger. These documents are then filtered (Filter 2) for the correctness of the synthetic entries in the general ledger (GL) and booked.

5 Discussion

In practice, data quality management is carried out primarily by means of procedures/conditions performed automatically by ERP [15]. Systems operating in large companies and medium-sized enterprises may contain even about 200 such procedures/conditions (apart from the conditions implemented in the production control module, whose number can reach thousands or even hundreds of thousands depending on the level of complexity of the technological line).

Most data errors are removed during the data acquisition process. It should be clearly emphasised that data obtained from the company's internal environment (e.g. entered by employees) are already checked and assigned quality during the data entry process. The system allows only standardised data to be entered (for example, if the unit of measure is a piece, the system allows this unit to be entered as "pcs" and the employee is not allowed to enter "piece" etc.).

Data exclusion in ERP can only occur in relation to external data, e.g. from cyberspace. For example, one can take into account customer feedback, obtained from social networking sites, about the product offered by the company. If 100 opinions are automatically analysed by the system and 99 are found to be positive and 1 is negative, the system can automatically exclude this negative opinion [16].

In the production management subsystem, data errors are inadmissible. For example, in the process of production control, the error of even a single data bit can have very negative consequences not only in business and technological aspects, but also in terms of human health and life (for example, incorrect machine control can lead to an accident

[4] Detailed solutions for accounting rules were described in the studies [14].

at work). Very often, a change in the state of the controller's binary output requires checking dozens of conditions (sensor status, the current state of the production line, etc.).

Data errors in ERP can only occur for a short period of time after they have been acquired, until they are checked automatically or by a person. An example is financial-accounting subsystem, where the amounts of all documents to be booked must correspond to the facts. Even if an input error occurs that cannot be detected at the "input" by the system, it will be identified at a later stage of processing. For example, the input of an erroneous value of a VAT invoice for the purchase of material will be revealed when confronted with a material acceptance document.

The control of data obtained from the tangential subsystems (fixed assets, logistics, human resources management) with the financial-accounting subsystem, i.e. the general ledger, becomes particularly important. Data obtained from these subsystems are automatically controlled in various ways according to predefined rules. Examples of such rules are, in terms of fixed asset depreciation data, symbols of synthetic and analytical accounts and the locations of their use. The above symbols must be consistent with the general ledger account symbols, otherwise the automatic booking process is stopped and requires human intervention to correct an incorrectly declared account in the fixed assets subsystem, or it may be a coincidence that the error occurred in the general ledger because there was no such account created before.

Another example of control of data acquisition (booking) from the logistics subsystem to the general ledger is checking the correctness of assigning the synthetic and analytical account symbol to the items of the goods catalogue, since the data defining the booking pattern contained in this catalogue is the basis for the automatic creation of the goods receipt assignment. For example, if a certain group of goods (e.g. juices) is booked on a specific goods analytical account and the beer symbol was incorrectly entered in the catalogue template during its configuration, the previously defined rules in the profiling phase that analyse this assignment (at the entry to the general ledger) will show that an error occurred in the automatically generated assignment in the logistics subsystem. In such a situation, the booking of the document of goods receipt in the general ledger is withheld and requires correction. This correction can be made by the program, but only if such an error has already occurred and is already covered by the rules of correctness, human intervention is required to rectify the error.

Another example of testing the correctness of data from tangential subsystems booked in the general ledger is the control of documents against limit values. If they are zero or exceed the minimum or maximum values, they are not booked and need to be corrected or eliminated.

This research analyse the data quality management in ERP problem from the users (managers and employees) point of view. This has been done only in a small degree in existing papers. We take the attempt of explanation, why data in ERP are characterized the proper quality and why on the basis of this data the more deeper analysis (e.g. Using artificial intelligence tools) can be performed.

6 Conclusion

Data quality is a key element of the proper functioning of ERP systems. There are different definitions of the concept of data quality and different approaches to data quality management. Their common feature is the final effect of the data processing, i.e. the system delivery of data allowing for making quick and effective economic decisions. It should be clearly emphasized that a large number of data control procedures embedded in the ERP allows for ensuring appropriate data quality.

The main limitations of the research are related to the failure to consider the assessment of selected ERP systems in terms of data quality management. The specificity of individual industries, which also has an impact on the quality management methodology in ERP systems, has not been taken into account.

Further research work should focus on, for example, the analysis of data quality in systems processing incomplete, uncertain and inconsistent data, such as multi-agent systems, as well as the analysis of the value of information and knowledge stored in ERP.

Acknowledgement. The project is financed by the Ministry of Science and Higher Education in Poland under the programme "Regional Initiative of Excellence" 2019 - 2022 project number 015/RID/2018/19 total funding amount 10 721 040,00 PLN.

References

1. Batini, C., Rula, A., Scannapieco, M., Viscusi, G.: From data quality to big data quality. In: Big Data: Concepts, Methodologies, Tools, and Applications, pp. 1934–1956. IGI Global (2016)
2. Vaishnavi, V., Kuechler, W., Petter, S. (eds.): Design science research in information systems, 20 January 20 2004. (created in 2004 and updated until 2015 by Vaishnavi, V. and Kuechler, W.). http://www.desrist.org/design-research-in-information-systems/. Accessed 30 June 2019
3. El Sibai, R., Chabchoub, Y., Chiky, R., Demerjian, J., Barbar, K.: Assessing and improving sensors data quality in streaming context. In: Nguyen, N.T., Papadopoulos, George A., Jędrzejowicz, P., Trawiński, B., Vossen, G. (eds.) ICCCI 2017. LNCS (LNAI), vol. 10449, pp. 590–599. Springer, Cham (2017). https://doi.org/10.1007/978-3-319-67077-5_57
4. Shekhar, S., Xiong, H.: Data quality. In: Shekhar, S., Xiong, H. (eds.) Encyclopedia of GIS, p. 220. Springer, Boston (2008). https://doi.org/10.1007/978-0-387-35973-1_248
5. Hernes, M., Bytniewski, A.: Knowledge integration in a manufacturing planning module of a cognitive integrated management information system. In: Nguyen, N.T., Papadopoulos, George A., Jędrzejowicz, P., Trawiński, B., Vossen, G. (eds.) ICCCI 2017. LNCS (LNAI), vol. 10448, pp. 34–43. Springer, Cham (2017). https://doi.org/10.1007/978-3-319-67074-4_4
6. Otto, B., Österle, H.: Corporate data quality. Springer, Heidelberg (2013)
7. Popescu, M.A.M., Ge, M., Helfert, M.: The social media perception and reality-possible data quality deficiencies between social media and ERP. In: ICEIS, no. 1, pp. 198–204 (2018)
8. Kwon, O., Lee, N., Shin, B.: Data quality management, data usage experience and acquisition intention of big data analytics. Int. J. Inf. Manag. **34**(3), 387–394 (2014). https://doi.org/10.1016/j.ijinfomgt.2014.02.002
9. Haug, A., Arlbjorn, J.S., Zachariassen, F., Schlichter, J.: Master data quality barriers: an empirical investigation. Ind. Manag. Data Syst. **113**(2), 234–249 (2014). https://doi.org/10.1108/02635571311303550

10. Owoc, M., Weichbroth, P., Żuralski, K.: Towards better understanding of context-aware knowledge transformation. In: 2017 Federated Conference on Computer Science and Information Systems (FedCSIS), pp. 1123–1126. IEEE (2017)
11. Lee, Y.W., Pipino, L., Funk, J.D., Wang, R.Y.: Journey to Data Quality. MIT Press, Cambridge (2006)
12. Geiger, J.: Data quality management: the most critical initiative you can implement. In: The Twenty-Ninth Annual SAS User Group International Conference (2004)
13. Hernes, M., Sobieska-Karpińska, J.: Application of the consensus method in a multiagent financial decision support system. IseB **14**(1), 167–185 (2015). https://doi.org/10.1007/s10257-015-0280-9
14. Gill, A.A., Shahzad, A., Ramalu, S.S.: Examine the influence of enterprise resource planning quality dimensions on organizational performance mediated through business process change capability. Glob. Bus. Manag. Rev **10**(2), 41–57 (2018)
15. Hernes, M., Sobieska-Karpińska, J.: Reduction of a forrester effect in a supply chain management system. J. Intell. Fuzzy Syst. **37**(6), 7325–7335 (2019). https://doi.org/10.3233/JIFS-179342
16. Rot, A., Olszewski, B.: Advanced persistent threats attacks in cyberspace. Threats, vulnerabilities, methods of protection. In: FedCSIS Position Papers, pp. 113–117 (2017)
17. Pondel, M.: A concept of enterprise big data and BI workflow driven platform. In: Ganzha, W.M., Maciaszek, L., Paprzycki, M., Ganzha, M., Maciaszek, L., Paprzycki, M. (ed.) Proceedings of the 2015 Federated Conference on Computer Science and Information Systems, 13–16 September 2015, Łódź, Poland (T. 5, ss. 1699–1704). Polskie Towarzystwo Informatyczne ; Institute of Electrical and Electronics Engineers (2015). http://doi.org/10.15439/2015F343
18. Becker, J., Jankowski, J., Wątróbski, J.: Transformations of standardized MLP models and linguistic data in the computerized decision support system. In: Różewski, P., Novikov, D., Bakhtadze, N., Zaikin, O. (eds.) New Frontiers in Information and Production Systems Modelling and Analysis. ISRL, vol. 98, pp. 213–231. Springer, Cham (2016). https://doi.org/10.1007/978-3-319-23338-3_10
19. Xu, H., Nord, J.H., Brown, N., Nord, G.D.: Data quality issues in implementing an ERP. Ind. Manag. Data Syst. **102**(1), 47–58 (2002)
20. Haug, A., Arlbjørn, J.S., Pedersen, A.: A classification model of ERP-system data quality. Ind. Manag. Data Syst. **109**(8), 1053–1068 (2009). https://doi.org/10.1108/02635570910991292
21. Cao, L., Zhu, H.: Normal accidents: data quality problems in ERP-enabled manufacturing. J. Data Inf. Qual. (JDIQ) **4**(3), 1–26 (2013)
22. Glowalla, P., Sunyaev, A.: Managing data quality with ERP systems-insights from the insurance sector. In: ECIS, p. 184 (2013)
23. Sieniawski, P., Trawiński, B.: An open platform of data quality monitoring for ERP information systems. In: Sacha, K. (ed.) Software Engineering Techniques: Design for Quality. IIFIP, vol. 227, pp. 289–299. Springer, Boston, MA (2006). https://doi.org/10.1007/978-0-387-39388-9_28
24. Xu, H.: What are the most important factors for accounting information quality and their impact on ais data quality outcomes? J. Data Inf. Qual. (JDIQ) **5**(4), 1–22 (2015)
25. Alpar, P., Winkelsträter, S.: Assessment of data quality in accounting data with association rules. Expert Syst. Appl. **41**(5), 2259–2268 (2014)

A Model of Enterprise Analytical Platform for Supply Chain Management

Paweł Pyda[1(✉)], Helena Dudycz[2], and Paweł Stefaniak[3]

[1] KGHM Polish Copper S.A. O/COPI, KGHM Polska Miedź S.A., M. Skłodowskiej-Curie 45b, 59-301 Lubin, Poland
pawel.pyda@kghm.com
[2] Department of Information Technology, Wroclaw University of Economics and Business, Komandorska 118/120, 53-345 Wrocław, Poland
helena.dudycz@ue.wroc.pl
[3] KGHM Cuprum Research and Development Centre, Gen. Wł. Sikorskiego 2-8, 53-659 Wrocław, Poland
pkstefaniak@cuprum.wroc.pl

Abstract. The article discusses an analytical platform for executives in the context of the needs for integrated supply chain management in large multi-site production companies. The first part discusses the development of methods, techniques, and technologies to improve supply chain management, while the second part presents an enterprise analytical platform model along with its basic assumptions and requirements. The basic elements of the platform and their functional description were discussed and groups of users who apply them to perform particular business roles in the company were identified. The third part presents a case study of selected business processes performed using on the enterprise analytical platform as part of integrated supply chain management. The contribution attempted with this article is to propose a model of an enterprise analytical platform taking into account an integrated supply chain for large multi-site production companies. Conclusions presented in this article may be significant in the following areas: the development of an enterprise analytical platform in terms of selecting individual components, understanding the process of developing executive information, making key decisions related to the development of an analytical platform, as well as developing the company's strategy.

Keywords: Advanced data analysis · Master data · Analytical platform · Supply chain management · Production company

1 Introduction

In order to be able to operate on the international market and maintain their leading position, multi-site production companies must follow the current trends of business processes' global digitisation and development in line with the idea of Industry 4.0 [1–3]. As a result of these companies' all activities (i.e. direct and indirect ones), cost reduction is expected in any critical area. Each action is accompanied by decision-making

M. Hernes et al. (Eds.): ICCCI 2020, CCIS 1287, pp. 363–375, 2020.
https://doi.org/10.1007/978-3-030-63119-2_30

based on information available at a given place and time. The information is generated on the basis of data most often coming from the company's IT systems. Several years ago, reports from a single system, generally of the Enterprise Resource Planning (ERP) type, were sufficient for the management staff. Nowadays, information from a few or sometimes more than a dozen systems from various business areas, i.e. finance, purchasing, personnel and payroll, material management, as well as production and industrial automation, is required at the same time to enable decision-making. The expectations of the management, resulting from the need to have the necessary information so as to make the best decisions, also include the integration of data coming from the Internet (e.g. websites, quotations, services, or cloud applications). This issue was presented in more details in [4, 5]. The results of data processing, as well as access to data itself, should be available to managers on an ongoing basis, from anywhere in the world. These objectives can be achieved through the implementation of a dedicated analytical platform.

Challenges of the concept of Industry 4.0 faced by multinational manufacturing companies make it indispensable to introduce relevant organisational, legal, and technological changes [2, 6]. One of these consists in providing the company with a way to efficiently handle business processes [7]. As part of its integrated supply chain, which concerns the flow of information, products, and services inside as well as outside the company. The internal flow includes issues related to sourcing, production, and distribution, while the external one integrates the company with its suppliers and customers.

Supply chain management is the decision-making process associated with the synchronisation of physical information-related, and financial supply and demand streams flowing between its participants in order to enable them to achieve a competitive advantage and create added value through synergies for the benefit of all its beneficiaries [8]. Supply chain management aims to help companies meet their customers' needs in the most efficient and cost-effective way [9]. The following basic elements of supply chain management can be distinguished:

- Supply chain business processes - activities carried out by participants to attain specific objectives (e.g. maintaining production levels, ensuring machinery availability for use, product development and commercialisation).
- Supply chain management components - designed to support business process management. Two groups can be distinguished here. The first concerns physical and technical elements (e.g. planning and supervision methods), while the second concerns management and behavioural elements (management methods and leadership structure).
- The network structure of the supply chain - created by all participants of the supply chain, i.e. companies cooperating with each other.

The primary purpose of supply chain analysis is to facilitate the understanding of all data produced by different elements of the supply chain [10]. These analyses result from the management's needs for making the right decisions [4]. Immense amounts of data are collected and displayed in a clear manner, accompanied by visualisations in the form of charts and graphs, and then used for managers' planning and decision-making activity [11]. The analyses make it possible to extract certain hidden patterns from the data as well as acquire valuable insights. The issue is not an easy one, as supply chains have a very complex structure within different industries and their operation requires well-qualified staff. The subject of supply chain analysis is constantly developing, with new technologies and methods coming into being to increase the effectiveness of forecasts, help detect inefficient areas, better respond to customer needs, and simulate and implement innovative solutions [12, 13].

Having complete product information (ready capital-intensive goods, spare parts, consumables, etc.) is a key issue of advanced data analysis within an integrated supply chain [14]. The aim of this article is to present an enterprise analytical platform taking into account an integrated supply chain for managers in large multi-site production companies. This article presents the results of the research carried out according to the following procedure: (1) literature research including advanced data analysis and supply chain management; (2) identifying the needs for the implementation of advanced data analysis on the supply chain management in multi-site production companies; (3) study of IT infrastructure in production companies; (4) analysis of software platforms in companies such as Microsoft, Facebook, Amazon, Oracle, SAP; (5) developing an enterprise analytical platform model; (6) a validation of the proposed model in a production company. The article is structured as follows. Firstly, the impact of IT tools and new technologies on business processes within the integrated supply chain is presented. The subsequent section presents an enterprise analytical platform model, by describing its basic components and user groups and discussing the identified data sources and the data integration process. Next, selected business processes implemented on the basis of the proposed an enterprise analytical platform model are presented. The article ends with a summary.

2 The Development of Methods, Techniques, and Technologies to Improve the Efficiency of Supply Chain Management

The methods of the analysis, collection, and acquisition of data change over the years, which also applies to the supply chain [15]. In the past, supply chain analysis was limited to simple statistics and measurable performance indicators which only allowed planning and forecasting demands [16]. Data was stored in spreadsheets or in the form of flat files, which did not allow for more advanced analysis. Frazzon et al. [17] show in their work a digital age evolution.

Later, in the 1990s, enterprises started to introduce systems such as EDI (electronic data exchange systems) and ERP (enterprise resource management systems). This opened up many new possibilities for supply chain analysis. These systems made it possible to create a network of interconnections enabling the exchange of information between all participants in the supply chain, which improved forecasting and planning [18].

The first decade of the 21st century saw the emergence of software for intelligent data analysis and predictive analysis [17, 19], which contributed to a better understanding and use of data, thus improving decision-making and decision optimisation by businesses [20].

Today we are facing the challenge of using a very large amount of data generated by the consecutive links in the supply chain. The amount of data obtained has increased dramatically in recent years; in 2017, a typical supply chain gained access to a data set 50 times larger than it was 5 years earlier [21]. In the era of a multitude of available technological solutions and immense amounts of acquired and stored data, the development of analytics for SCM should take into account developments including machine learning tools, Big Data [22, 23], Business Intelligence, or cloud computing solutions [24]. Machine learning enables the detection of patterns in data from the supply chain. Using machine learning algorithms that can continuously learn in the supply chain process enables one to quickly identify the most important factors for supply chain success. The algorithms iteratively process data, and many of them use constraint-based modelling to find the basic set of factors offering the highest predictive accuracy.

One of the most difficult aspects of supply chain management is predicting future production demands. Various approaches are used in practice: from basic statistical analysis techniques (including moving averages) to advanced simulation modelling. Machine learning proves to be very effective as a solution allowing for factors of existing methods that do not enable tracing or quantifying over time [25].

In the era of global challenges, enterprises, especially large, multi-site, and production companies, have to take up the challenges they face in connection with the need for advanced data analysis (this issue is discussed in more detail in [26]. The rapidly changing markets force almost real-time decision-making. Companies that will have appropriate tools (IT solutions supporting supply chain management) will have a chance for survival and competitive advantage.

3 A Proposal of an Enterprise Analytical Platform Model for the Data Integration of SCM

Nowadays, large multinational enterprises, in order to operate efficiently, make quick and agile decisions in a rapidly changing market, as well as remaining competitive, must have quick access to data reflecting company operation. In addition to the access to up-to-date information, companies must have access to information, tools, and analyses that allow them to forecast future events, while the existing IT infrastructure in multinational companies causes many problems with the implementation of advanced data analysis [27].

In particular, the implementation of predictive analyses is important for decision making as part of integrated supply chain management. The number and range of business processes inside and outside the organisation that need to be analysed translate into the amount and size of data. These data need to be analysed by building proper analytical and predictive models and reports or key performance indicators [28]. Without access to appropriate tools, such operations are impossible. The fulfilment of these conditions involves designing appropriate infrastructure, implementing suitable tools, as well as introducing organisational changes. The only possible solution to this problem is to implement an enterprise analytical platform. An example of such analytics platform for emergency and security purposes was presented in [29].

The analytical platform should be a combination of appropriate ICT (Information and Communication Technologies) infrastructure together with a set of IT systems and tools enabling transfer, storage, processing, management, presentation and sharing of data. The platform should be designed in such a way that it is scalable, easy to integrate, easy to maintain and configure, and built from components that will not lead to dependence on a single supplier (so-called vendor lock-in). This issue has been widely described in [30].

The main expectation of the platform is fast and easy access to up-to-date, consistent, standardised, and reliable data for presentation or use for reprocessing. In order to meet this requirement, it is necessary to perform a number of activities, starting from the identification of business data sources, through the development of a standard for their acquisition, to the processing and storage in a data warehouse. Next, one needs to start developing business data models. They contain physically recorded data in tables, linked by means of relationships. Relationships should be built in such a way that they reflect the reality of the business area concerned to the highest possible extent. In the case of an area related to supply chain management, these can be tables containing data on, for example, products, suppliers, orders, or customers. In addition, data models may contain tables with economic and financial indicators, calculated on the basis of implemented algorithms. They can also be fed with the results of calculations performed using external tools. Depending on the needs, the data model can be extended by adding data from another business area or it can be connected with another business model using a relation. Having at their disposal a wide set of coherent data located in one place, an analyst can create any combination of relations and data processing (the so-called logical data model) providing information for a specific business process.

An enterprise analytical platform is a set of interconnected systems and IT tools between which data is transmitted. In principle, three groups of tools that can create a platform are distinguished. The first one includes basic software that enables storing, processing, and acquiring data:

- Databases – used for auxiliary purposes, e.g. as a database consolidating data from multiple systems;
- Data warehouses – the main element of the platform, one where temporary data obtained from sources is stored, read-only business data, and thematic data warehouses (Data marts) also referred to as business models;
- Data storage and processing platforms based on computer clusters (the most popular of which is Apache Hadoop);
- ETL (Extract Transform Load) tools for creating data flow processes.

The second group includes supporting tools, i.e.:

- Data Governance tools
- Tools ensuring data quality and consistency, sometimes included in ETL tools; also available as independent modules;
- Tools for developing data models, drawing up documentation, and developing data dictionaries, often referred to as metadata management tools, may be parts of a larger Data Governance software package;

- Data security tools (authorisation systems, logging, and audit);
- Tools for operational basic data management as the only source of master data concerning e.g. customer and product data consolidated using a number of domain systems.
- Tools supporting the use of algorithms and methods of machine learning and artificial intelligence.

The third group includes tools designed to support business processes, i.e.:

- Tools for creating logical data models for the users' own needs, so-called Self-service BI tools;
- Tools for reporting and presenting data (managerial cockpits and KPIs);
- Tools for statistics, analytical model development, prediction, forecasting, and what-if analysis;
- Tools for planning, budgeting and controlling;
- Analytical master data management system;
- Other dedicated tools catering to specific company needs.

Depending on specific business requirements, the platform may consist of any combination of the above-mentioned components. In the development phase, it usually includes tools belonging to the first group; along with the growth and complexity of processes, tools from the second group are added. With the development of business needs, further components from the third group may be added.

While creating an enterprise analytical platform to support an enterprise's decision-making, it is necessary to identify user groups to be associated with the platform. Each user group has relevant business roles assigned.

The RACI Matrix is a powerful tool to assist in the identification of roles and assigning of cross-functional responsibilities to a project deliverable or activity. RACI represents: R - Responsibility, A - Accountable, C - Consulted, and I - Informed. Table 1 represents RACI Matrix of activities within the Enterprise Analytical Platform Model.

In order to start any analytical work using dedicated tools or create the necessary reports, data from the source system must go through a long process involving the various stages of data acquisition, purification, transformation, integration, and recording in the target data warehouse tables. Figure 1 shows a data integration model from source systems into a data warehouse. Data can come from different sources. Most often, the platform is primarily fed with data from ERP systems and domain-specific systems such as CRM, contract management system, project management system, etc. Access to this data is implemented most often through a dedicated API (Application Programming Interface) or direct reading from the database. Another source of data is various types of files (CSV, XML, txt, etc.). A common case of importing data from files is that of older industrial devices, for which it is the only possibility of exporting data.

An analytical platform can also be powered by an enterprise service bus that acquires data from both the company's internal and third-party systems (websites and cloud services). In the case of very large volumes of data, especially unstructured data, the source is the Hadoop platform using computer clusters, often referred to as Big Data. The data stored in the warehouse can be used by external applications dedicated to

Table 1. RACI Matrix of activities within the Enterprise Analytical Platform Model.

Activity \ Role	System administrator	System architect	Business domain software administrator	Data security administrator	Data governance tool administrator	System analyst	Data flow developer	Data architect	Developer (SQL, R, Python)	System operator	Data Scientist	Business user	SCM analyst	Branch manager	CEO Member
First group of tools															
Administration of systems including databases as well as the data warehouse and Big Data platform	A	C	I	C	I	I	I	I	I						
Supervision the integration components within the analytical platform and its development	R	A	R	C	I	I	I	I	I						
Second group of tools															
Administration of business specific systems (silos)	C	C	A	C	I	I	I	I		I		C	I		
Creation of governance security rules, permission management, creating safety guidelines	R	C	I	A	R	I	C	C	C	I	I	C			
Administration of data governance tools (dictionary of business terms, data quality, data stage, security)	R	C	R	R	A	C	C	C	I	I	I	I			
Analyzing, designing and implementing IT systems	R	C	R	C	C	A	I	I	C	I	I	I	I		
Design and implementation of data flows (i.e. ETL processes)		C	C	C	R	I	A	R	R	I	I	I	I		
Creating data models and structures		I		I	I	C	C	A	R		I				
Designing, creating and implementing database systems and algorithms based on end user requirements	R	I		C	I	C	C	C	A		I	C	I		
Third group of tools															
Operational tasks e.g. creating purchasing procedures, configuring contracts etc.			C	I						A		C	C		
Designing data modeling processes, creating algorithms and predictive models to extract the data needed in business.				I	I		C	C	R	I	A	C	C	I	I
Business process support (e.g. plannig, budgeting and finance controlling).			C	I	I			C	C	I	I	A	C	I	I
Monitoring the entire process of SCM using available reports			C	C	I	C	C	C	C		C	C	A	I	I
Monitoring and managing subordinate business processes		I			I			I			C	C	R	A	I
Making strategic decisions based on available performance indicators.											C	C	C	R	A

R (Responsibility) – person or role responsible for ensuring that the item is completed

A (Accountable) – person or role responsible for actually doing or completing the item

C (Consulted) – person or role whose subject matter expertise is required in order to complete the item

I (Informed) – person or role that needs to be kept informed of the status of item completion

specific business tasks (statistical tools, tools for building analytical models, planning and budgeting tools, machine learning, artificial intelligence, etc.). The results of the work from these applications can also be a source of data for the platform (so-called feedback).

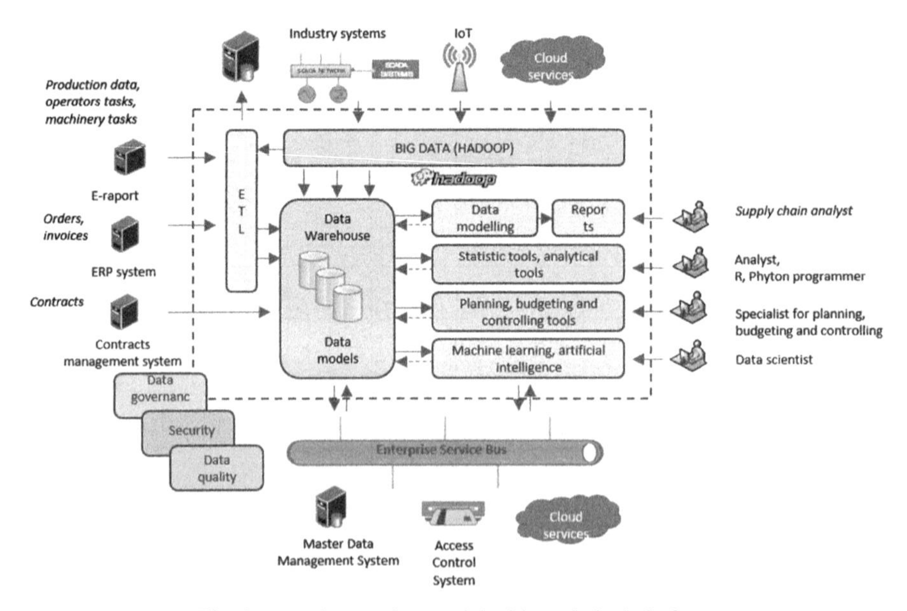

Fig. 1. Data integration model with analytical platform

4 Implementation of Selected Business Processes - A Case Study

In large manufacturing companies, material management plays an enormous role due to the high cost of materials. In the field of material index analysis, the key element is the master data management system, which should meet the following criteria:

- It should be a single source of information about material indices for the analytical platform for the purpose of analytics, as well as for other IT systems for operational purposes;
- It should consolidate and maintain a standardised material index database for all branches of a given company as well as its external collaborators;
- It should enable the enrichment of the information on material indices by describing additional attributes, defining substitutes, adding overview photos, descriptive product sheets, etc.;
- It should allow creating references to databases of suppliers, material producers (substitute list, producers' price lists, and market data).

The master data management system, coupled with a data warehouse as part of the enterprise analytical platform, can also enable other, more sophisticated analytical capabilities.

By expanding the data model for the integrated supply chain by feeding it with data from external cooperating entities (suppliers and buyers of goods) or websites (e.g. cloud services, databases of manufacturers and suppliers of parts, and market data) we obtain a strong foundation for building further analytical algorithms (prediction, machine learning, and artificial intelligence). The extension of the scope of data and advanced tools for data analysis makes it possible to implement business processes including the following examples:

- Generating a list of the best materials in terms of quality;
- Generating consumption lists for individual materials by assortment groups;
- Evaluating suppliers in terms of delivery dates;
- Evaluating suppliers in terms of the quality of materials supplied by them;
- Preparing procurement plans for suppliers;
- Examining material price increases in particular assortment groups;
- Comparing material prices by suppliers;
- Preparing material demand plans;
- Automatically generating orders for suppliers;
- Automatically generating proposals for purchase budget plans.

Described in previous chapter the analytical platform model was used in a big multi-site company to support business processes in materials management area. In the data warehouse the data from the ERP system (mainly from MM module) and from the main materials supplier's ERP system are collected. Additionally, some algorithms have been implemented, which prepare aggregate data, prepare some KPIs and store the data in suitable data structures. Analytics can explore the data with dedicated analytics tools (like Excel, Statistica or SPSS Modeler) for their own needs. Additionally for the management needs several reports has been created which let you monitor actual company status and support your strategic decisions making.

Manager dashboards have been divided into four main areas: own inventory, stocks of deposit, number of inventory items and unrealized order items. In each of these dashboards you can carry out data mining operations according to different perspectives and layouts.

The own inventory dashboard presents a table and charts data regarding to own inventory for whole company with limit and percentage and value deviation by the type of index: CAPEX and OPEX. These data relate to the previous day. In addition, stock history for the last year is presented. After performing the drill-down operation on the table we receive inventory data broken down by industry. The next drill-down operation displays data in the table broken down into a given branch of the company. The next drill down operation allows for a valuable presentation of own inventory divided into warehouses: rotary warehouse, investment warehouse, etc. At the same time, it is possible to go to the report presenting the state of own inventories according to the hierarchy of indexes. By default, the report displays the values for all departments, but the user can select the data view only for the selected department with the index hierarchy tree along with the sums of values for individual groups.

The inventory of deposit stocks dashboard presents a table and charts containing data on the status of deposit stocks together with the expected level as well as a percentage and value deviation. These data relate to the previous day. In addition, a history of inventory levels for the last year is presented along with a curve representing the expected level. Data mining operations allow us to perform analysis in two directions. In the first one, after performing the drill-down operation in the table, we receive data on deposit stocks with an additional breakdown into specific manufacturers and suppliers. The next drill down operation allows you to receive a more detailed report containing the indexes of materials of a given supplier presented according to the hierarchy of product groups. The second direction is that after performing the drill down operation, the value of deposits is displayed, broken down into individual branches of the enterprise. From this report you can go to the report as part of the first analysis, but limited to indexes by a given branch of the company.

The number of inventory items dashboard is presented in the table containing data on the number of inventory items in three categories: inventory below the warning level, lack of inventory and all index items. These data relate to the previous day. From this level, after performing the drill-down operation, you can go to the statement showing the position of the inventory, broken down into individual departments of the company. By performing the next drill-down operation, data already filtered for the given department at the material index level is displayed. This report, in turn, allows you to filter the list of material indexes by specific warehouses by showing indexes on one tab with the zero stock, on the other with the stock below the warning level and the tab with all items, i.e. items from the previous two tabs and additionally the indexes with the stock above warning level. The warning level is set by the warehouse manager. From the main level of the dashboard, by performing drill-down operations, you can get a list of material indexes taking into account three categories, broken down into specific suppliers. After performing the drill-down operation, a statement is created containing additional division into a given department.

The fourth cockpit presents unrealized orders for the entire enterprise divided into three categories due to the exceeded time of unrealized orders: up to 14 days, up to 30 days and over 30 days. The next drill-down operation allows you to display a detailed report showing, among others: order number, material index, material description, time the order was exceeded, purchase order number, date of sending the order, etc.

The above examples of manager dashboards implementation show how great potential is provided by the analytical platform by providing key data for the management from the SCM area as well as for the management board itself. Key data and indicators of the company's economic efficiency are very easily available in one place for the managers.

The examples of dashboards in this section show only a fraction of the possibilities offered by the application of an analytical platform in the area of supply chain management.

5 Conclusions and Future Work

This article presents a proposal for an enterprise analytical platform model taking into account the requirements and needs of managers dealing with supply chain management. The key elements of the platform are indicated, i.e. the data warehouse together with the ETL toolkit (support for data downloading, processing, and loading processes), as well as the Data Governance toolkit and the analytical and reporting toolkit. Examples of supply chain management business processes that can be supported by the proposed enterprise analytical platform are discussed, based on the example of a real multi-site production company. The following elements of the enterprise analytical platform were implemented in the company under study: a data warehouse in the mass parallel processing mode, a set of tools for conducting analyses, building data models, and developing reports and tools to support Data Governance processes. The company's further challenges with respect to the platform are related, among other things, to the launch of the Hadoop environment (a platform designed for distributed storage and processing of large data sets using computer clusters). This is a natural direction in the face of the requirements related to analytics not only associated with supply chain management, but also more broadly related to production and industrial automation. The result of these works will be a comprehensive and scalable environment enabling efficient processing and management of large data sets in one place for the whole multi-departmental production company operating on the international market and cooperating with contractors from all over the world.

The proposed enterprise analytical platform model has the following advantages. First of all, it is characterized by flexibility, which means that it allows to connect to compatible commercial and open source systems. Thus, it is possible to integrate software needed from various suppliers in the company, without limiting to systems from a particular software vendor. Secondly, it allows you to become independent of a particular supplier, i.e. it does not have to be based on the platform solution proposed by only one supplier (the so-called vendor lock). An enterprise can develop its own and unique processes and data flow models. Thirdly, small companies, especially those without a large budget, can develop their own business processes, build a data model and validate this solution at relatively low costs. However, a thorough study of the cost-benefit analysis (CBA) should be carried out to implement the proposed analytical platform model. Such research should identify both the expense related to the implementation and maintenance of this platform and the countable and uncountable effects.

Future works will be related to the implementation of the Master Data Management System (MDMS) in the area of material indices. There is a need to unify and standardise the index database for all branches in the production company under study, as a result of which it will be possible to create collective plans related to supply, storage, production, and transport of manufactured goods for the whole company. The implementation of tools for planning and budgeting material needs is being considered.

Further works will also concern the construction of a model of production data to be ultimately associated with the integrated supply chain model. It is assumed that this will allow the identification and analysis of the correlation between production plans and orders for materials and services.

Acknowledgement. The project is financed by the Ministry of Science and Higher Education in Poland under the programme "Regional Initiative of Excellence" 2019–2022 project number 015/RID/2018/19 total funding amount 10 721 040,00 PLN.

References

1. Kosacka-Olejnik, M., Pitakaso, R.: Industry 4.0: state of the art and research implications. Logforum, **15**(4), 475–485 (2019). https://doi.org/10.17270/J.LOG.2019.363
2. Xu, L.D., Xu, E.L., Li, L.: Industry 4.0: state of the art and future trends. International Journal of Production Research, **56**(8), 2941–2962 (2018). https://doi.org/10.1080/00207543.2018.1444806

3. Resman, M., Pipan, M., Simic, M., Herakovic, N.: A new architecture model for smart manufacturing: a performance analysis and comparison with the RAMI 4.0 reference model. Adv. Prod. Eng. Manag. **14**(2), 153–165 (2019). https://doi.org/10.14743/apem2019.2.318
4. Lai, Y., Sun, H., Ren, J.: Understanding the determinants of big data analytics (BDA) adoption in logistics and supply chain management. Int. J. Logist. Manag. **29**(2), 676–703 (2018). https://doi.org/10.1108/IJLM-06-2017-0153
5. Kruczek, P., et al.: Predictive maintenance of mining machines using advanced data analysis system based on the cloud technology. In: Widzyk-Capehart, E., Hekmat, A., Singhal, R. (eds.) Proceedings of the 27th International Symposium on Mine Planning and Equipment Selection - MPES 2018, pp. 459–470. Springer, Cham (2019). https://doi.org/10.1007/978-3-319-99220-4_38
6. Pyda, P., Stefaniak, P., Dudycz, H.: Development assumptions of a data and service management centre at KGHM S.A. In: Mueller, C., et al. (eds.) Mining goes Digital Proceedings of the 39th International Symposium 'Application of Computers and Operations Research in the Mineral Industry' (APCOM 2019), pp. 569–577 (2019)
7. Brinch, M.: Understanding the value of big data in supply chain management and its business processes. Int. J. Oper. Prod. Manag. **38**(7), 1589–1614 (2018). https://doi.org/10.1108/IJOPM-05-2017-0268
8. Cai, J., Liu, X., Xiao, Z., Liu, J.: Improving supply chain performance management: a systematic approach to analyzing iterative KPI accomplishment. Decis. Support Syst. **46**(2), 512–521 (2009). https://doi.org/10.1016/j.dss.2008.09.004
9. Naraharisetti, P.K., Karimi, I.A., Srinivasan, R.: Supply chain redesign through optimal asset management and capital budgeting. Comput. Chem. Eng. **32**(12), 3153–3169 (2008). https://doi.org/10.1016/j.compchemeng.2008.05.008
10. Wong, C.W.Y., Lai, K., Cheng, T.C.E.: Value of information integration to supply chain management: roles of internal and external contingencies. J. Manag. Inf. Syst. **28**(3), 161–200 (2011). https://doi.org/10.2753/MIS0742-1222280305
11. Tan, K.H., Zhan, Y., Ji, G., Ye, F., Chang, C.: Harvesting big data to enhance supply chain innovation capabilities: an analytic infrastructure based on deduction graph. Int. J. Prod. Econ. **165**, 223–233 (2015). https://doi.org/10.1016/j.ijpe.2014.12.034
12. Chen, D.Q., Preston, D.S., Swink, M.: How the use of big data analytics affects value creation in supply chain management. J. Manag. Inf. Syst. **32**(4), 4–39 (2015). https://doi.org/10.1080/07421222.2015.1138364
13. Shafiq, M., Savino, M.M.: Supply chain coordination to optimize manufacturer's capacity procurement decisions through a new commitment-based model with penalty and revenue-sharing. Int. J. Prod. Econ. **208**, 512–528 (2019). https://doi.org/10.1016/j.ijpe.2018.12.006
14. Vilminko-Heikkinen, R., Pekkola, S.: Master data management and its organizational implementation. J. Enterp. Inf. Manag. **30**(3), 454–475 (2017). https://doi.org/10.1108/JEIM-07-2015-0070
15. Tiwari, S., Wee, H.M., Daryanto, Y.: Big data analytics in supply chain management between 2010 and 2016: insights to industries. Comput. Ind. Eng. **115**, 319–330 (2018). https://doi.org/10.1016/j.cie.2017.11.017
16. Fosso Wamba, S., Gunasekaran, A., Papadopoulos, T., Ngai, E.: Big data analytics in logistics and supply chain management. Int. J. Logist. Manag. **29**(2), 478–484 (2018). https://doi.org/10.1108/IJLM-02-2018-0026
17. Frazzon, E.M., Rodriguez, C.M.T., Pereira, M.M., Pires, M.C., Uhlmann, I.: Towards supply chain Management 4.0. Braz. J. Oper. Prod. Manag. **16**(2), 180–191 (2019) https://doi.org/10.14488/BJOPM.2019.v16.n2.a2
18. Gunasekaran, A., Kumar Tiwari, M., Dubey, R., Fosso Wamba, S.: Big data and predictive analytics applications in supply chain management. Comput. Ind. Eng. **101**, 525–527 (2016). https://doi.org/10.1016/j.cie.2016.10.020

19. Waller, M.A., Fawcett, S.E.: Click here for a data scientist: big data, predictive analytics, and theory development in the era of a maker movement supply chain. J. Bus. Logist. **34**(4), 249–252 (2013). https://doi.org/10.1111/jbl.12024
20. Trkman, P., Budler, M., Groznik, A.: A business model approach to supply chain management. Supp. Chain Manag.: Int. J. **20**(6), 587–602 (2015). https://doi.org/10.1108/SCM-06-2015-0219
21. Sanders, N.R., Ganeshan, R.: Big data in supply chain management. Prod. Oper. Manag. **27**(10), 1745–1748 (2018). https://doi.org/10.1111/poms.12892
22. Schoenherr, T., Speier-Pero, C.: Data science, predictive analytics, and big data in supply chain management: current state and future potential. J. Bus. Logist. **36**(1), 120–132 (2015). https://doi.org/10.1111/jbl.12082
23. Yablonsky, S.A.: Data and analytics innovation platforms. In: the Proceedings of XXIX ISPIM Innovation Conference, Stockholm, Sweden (2018)
24. Guanochanga, B., et al.: Real-time air pollution monitoring systems using wireless sensor networks connected in a cloud-computing, wrapped up web services. In: Arai, K., Bhatia, R., Kapoor, S. (eds.) FTC 2018. AISC, vol. 880, pp. 171–184. Springer, Cham (2019). https://doi.org/10.1007/978-3-030-02686-8_14
25. Waller, M.A., Fawcett, S.E.: Data science, predictive analytics, and big data: a revolution that will transform supply chain design and management. J. Bus. Logist. **34**(2), 77–84 (2013). https://doi.org/10.1111/jbl.12010
26. Dudycz, H., Stefaniak, P., Pyda, P.: Problems and challenges related to advanced data analysis in multi-site enterprises. Vietnam J. Comput. Sci. (2020, in print)
27. Dudycz, H., Stefaniak, P., Pyda, P.: Advanced data analysis in multi-site enterprises. basic problems and challenges related to the IT infrastructure. In: Nguyen, N.T., Chbeir, R., Exposito, E., Aniorté, P., Trawiński, B. (eds.) ICCCI 2019. LNCS (LNAI), vol. 11684, pp. 383–393. Springer, Cham (2019). https://doi.org/10.1007/978-3-030-28374-2_33
28. Govindan, K., Cheng, T.C.E., Mishra, N., Shukla, N.: Big data analytics and application for logistics and supply chain management. Transp. Res. Part E: Logist. Transp. Rev. **114**, 343–349 (2018). https://doi.org/10.1016/j.tre.2018.03.011
29. Pérez-González, C.J., Colebrook, M., Roda-García, J.L., Rosa-Remedios, C.B.: Developing a data analytics platform to support decision making in emergency and security management. Expert Syst. Appl. **120**, 167–184 (2019). https://doi.org/10.1016/j.eswa.2018.11.023
30. Woo, J., Shin, S.-J., Seo, W., Meilanitasari, P.: Developing a big data analytics platform for manufacturing systems: architecture, method, and implementation. Int. J. Adv. Manuf. Technol. **99**(9–12), 2193–2217 (2018). https://doi.org/10.1007/s00170-018-2416-9

Identification the Determinants of Pre-revenue Young Enterprises Value

Robert Golej(✉)

Wroclaw University of Economics and Business, Wrocław, Poland
robert.golej@ue.wroc.pl

Abstract. Valuation of a companies at the start-up stage is an important factor in its development. Like determining the risk of its success. Valuation process support by machine learning systems can significantly simplify investment decision making, identify risks and start-up value. Hence the problem of how to estimate the value of an enterprise and what factors determine it. The purpose of the article is to identify independent variables (attribute) for the enterprise value in the pre-revenue phase. The practical result of the research is the possibility of using machine learning solutions defining the dependence of values on independent variables in internal, spatial and temporal systems. For the identification of variables', a methodical review of the start-up's valuation was used. An analysis of the innovation evaluation proposals and the significant concept of the Business Model were carried. The analysis of independent variables' in various methods was performed by means of comparisons method in the form of a table. As a result of research, main independent variables were identified. Variables are both internal and external to the organization. Limitations: access to data to build a model with the right amount of data.

Keywords: Valuate start-up · Machine learning · Expert algorithms

1 Introduction

The application problem of predictive models, expert algorithms and machine learning in the perspective of property values deal with Talaga et al. [1] Piwowarczyk et al. [2]. Obtaining reliable information about the value of a start-up often determines the possibility of financing it and further development. Its value should be determined despite the lack or low level of achieved revenues. Commonly used valuation methods are not always applicable to start-ups [3]. While assets or DCF (discount cash flow) methods work well in valuing traditional enterprises, they can distort the value of a start-up. Despite the difficulties in valuating start-ups, especially those that are in the early stages of the life cycle, there are investors who see growth potential. At the earliest stage of development, they are business angels, and at a later stage private equity/venture capital funds. Lack or a small level of revenues achieved in the early stage of start-up development and a significant level of costs incurred result in a negative cash flow (CF). Deferring revenues over time, and thus positive cash flow (CF), is a typical phenomenon. Many

M. Hernes et al. (Eds.): ICCCI 2020, CCIS 1287, pp. 376–388, 2020.
https://doi.org/10.1007/978-3-030-63119-2_31

enterprises survived the early phase and generated significant revenues in the mature phase. This means that early-stage start-ups can be valued with some probability (risk). Identification of factors affecting the success of a start-up, which is not yet confirmed by revenues, but only potential and expected is the key to making good management and investment decisions. First, I will briefly characterize the valuation entity which is the start-up in the context of its valuation. Next, I will present selected methods for valuing start-ups. In this case, however, the project uses widely understood resources of organization to which facilitates and simplifies the process of innovation development Selection methods for new projects (innovations) in established organizations can also be an inspiration [4]. Business Model Canvas is another tool for identifying and structuring start-up activities focused on success According to the author, there is a research gap regarding the identification of factors affecting the value of start-ups in the pre-revenue phase of their measurement and evaluation. Information on factors is dispersed in such areas as: Value Based Management, start-up theory, start-up valuation, Business Model Canvas, Balanced Scorecard and innovation assessment.

2 Specificity of Entities Start-Up in the Early Phase

2.1 Start-Up Concept

According to Calopa et al. [5] start-ups are defined as newly created enterprises that are in the phase of product development and market research in the area of new technologies. Blank [6] radically defines a start-up as "a temporary organization that is looking for a profitable, scalable and repeatable business model." By nature, start-ups are a temporary organization that is looking for a business model. Once he finds his model, he ceases to be a start-up. This statement is radical and identifies the start-up as a phase of incubation, testing hypotheses and answering the questions: What? Who? How? And how to scale? The answers to these questions form the business model. Scalability, i.e. the ability to quickly acquire many customers and sales is an important feature of start-ups also emphasized by Robehmed [7]. Łuczak [8] also refers to the business model, he defines a start-up as an enterprise at an early stage of development that does not have a stable and sustainable business model.

Gemzik-Salwach [9] indicates distinguishing features start-ups: the fact of starting a business, offering innovative goods or services for which there is demand or will only be created. In turn, Ries [10] claims that a start-up is an institution created with a view to building new products and services in conditions of high uncertainty. Researchers clearly point to the innovative nature of the start-up offer, which gives rise to pay attention to the innovation evaluation system.

To sum up, it is an enterprise: small (micro or small), newly created, often operating in the area of new technologies, without a shaped business model, whose activity is subject to a significant risk of failure, with the potential to increase value in the long run. The simplest model envisages four basic business development phases: the seed or start stage, the start or startup stage (early stage), the growth stage and the expansion stage. In this phase, the company bears operational losses because the initial phase is very costly. In start and early phase, the costs of remuneration, expertise, market research and research and development are incurred in order to achieve the expected revenues in the next phase. The lack or low level of revenues achieved in the initial stage of the company's activity is characteristic for many start-ups. Regardless of the number of separate stages of the start-up life cycle, a significant share of the initial stages in the total life cycle of this type of enterprises can be observed - the conceptual, initial, and early stages. Such a life cycle of a startup with a "stretched" initial phase determines its low profitability (and even negative) during the first periods of activity. This allows you to identify the first variable, which is the period and expected duration of the initial phases, it also means the need for knowledge, innovation (usable creativity), efficiency, entrepreneurship. We can indicate pre-revenue period in early stage of development of a young enterprise. It can also be divided into pre-money periods and post-money periods (before and after investing).

As emphasized by Armstrong [11], high level of start-up elimination rate is not due to internal reasons but to external factors. It is therefore necessary the appropriate reaction of the entrepreneur and management of the start-up to a change in external conditions in the context of the transition from one phase of the life cycle to another is crucial. So the transition, in the marketing perspective, to the clients of innovators/and further from the clients of precursors to the early majority (this transition is called the valley of death) requires appropriate experience. This requires start-ups' managers to be active in the area of marketing-sales and operational management-costs in seed stage and start-up stage - identifying necessary resources in a broad sense; revenues and costs; profitability and sources of financing of activities; need to define a business model on which assumptions a business plan (BP) will be based is also emphasized; testing and refining business model assumptions.

2.2 Problems of Company Valuation in the Start-Up Phase

The most important and most frequently used method of valuation of an enterprise is the DCF (Discount Cash Flow) method. For further considerations, let us recall the DCF calculation formula. Damodaran [12] Wrzosek et al. [13], state that:

$$V = \sum_{t=1}^{n} CF_t * a_t + RV * a_n, \tag{1}$$

$$where\colon a_t = \frac{1}{(1 + WACC)^t}, \tag{2}$$

$$and \quad CF_t = (P_t - K_t) * (1 - TK) + Am_t - I_t - \Delta WC, \tag{3}$$

$$WACC = \frac{E}{E + D} r_E + \frac{E}{E + D} r_D(1 - TK), \tag{4}$$

$$according\ to\ the\ model\ CAPM \quad r_E = r_{RF} + \beta(r_M + r_{RF}), \tag{5}$$

Where: ***V*** *- enterprise value calculated using the DCF method;* ***CF*** *- Enterprise Cash Flow (NCFF);* ***P*** *- operating income;* ***K*** *- operating costs;* ***Am*** *depreciation as an element of K;* ***I*** *investments; delta WC change in working capital;* $\boldsymbol{r_E}$ *- cost of equity/expected rate of return for risk class described by;* $\boldsymbol{r_D}$ *- cost of debt;* $\boldsymbol{r_M}$ *- rate of return on the market* *portfolio;* r_{RF} *- rate of return on safe instruments; E - value of equity;* ***D*** *- debt value;* ***TK*** *- effective CIT rate;* ***BETA*** *- this beta ratio indicates which direction the movements of the share price of the selected company are correlated with the movements of the entire stock exchange or the selected block of shares, a factor determining the specific risk of the entity;* ***RV*** *- Residual Value calculated for CF from period n, last year of the detailed forecast;* ***WACC*** *- weighted average cost of capital, expected rate of return for capital, discount rate.*

The main problems in valuation of strart-ups are:

- the problem of estimating forecast revenues as a key variable, the problem of revaluation of forecast revenues,
- the problem of calculating the discount rate representing the risk associated with the implementation of the project. It is the discount rate that is subject to the most frequent modifications in the calculation of the start-up value.

One of the most difficult issues when valuing is determining the growth rate of start-up revenues in the initial phase, which in turn form the basis for calculating free cash flow. The more accurate the forecast of the revenue growth rate, the more reliable the company's valuation. It is relatively easy in enterprises with a stable market position and operating in a stable environment. In this case, the revenue growth rate is determined based on historical data. In the case of start-ups in the pre-revenue phase, revenue estimation is very difficult and burdened with high unpredictability. The uncertainty of revenue forecasts by start-up concerns the following issues: Is there a market? When will the revenues be achieved? What will be the revenues and what will be their volatility? Degree of New Product Development? What influences the achieved revenues? Do we have the resources necessary to enter the market effectively?

Accepting too high projected revenues will result in an inflated young enterprises valuation. In the literature, the phenomenon of excessive certainty in the context of start-up valuation has been described, among others Wickham [14], pointing out that entrepreneurs should make it easier for investors to assess risk, and not just convince them that the investment generates a certain level of risk. With regard to the rate of revenue growth, it should be noted that this does not apply to start-ups in the conceptual and initial stages. Estimating revenue forecasts applies to start-ups in the initial phase, but already generating revenue.

The purpose of start-up valuation from the investor's perspective is often determine the likely ROI (Return on Investment). The level of expected return depends on many factors, for example investor preferences, investment risk or the strategy adopted. Valuation, especially in the initial and development phases, must take into account the growth potential on the one hand, despite the lack of revenues and a positive balance of free cash, and on the other the risk of failure. Determining the expected rate of return (ROI, ke) reflecting the estimated risk is also a problem. The cost of capital is often taken as the average value, as if all start-ups in the pre-revenue phase were in the same risk class, it is possible that this is the case, but it is doubtful that all start-ups are identical. What happens at use the average value of start-ups. Opposition to this approach are valuation methods focusing on determining the risk of an individual start-up.

When valuing such entities, it is therefore necessary to focus on factors that are able to substantiate the future market success of the idea and the expected future cash flow, for a given risk class (often not quantitative). We cannot examine the stability of revenues because they do not exist, neither the margin, because there is no production, etc. there are only some guesses. Instead, we have an idea, a team and an environment (sometimes rudimentary, because a start-up can create a new need and a new market). Therefore, you can examine the relationship of the organization's attributes (assets) and its environment at time t (pre-income phase) and compare them with the value from the period t + n (durability of the enterprise (bankruptcy), the value calculated using the DCF method, or information on sales transactions start- ups). Therefore, with attribute markings and an algorithm formulated (machine learning), we are able to determine the value. However, the beginning of the considerations is to determine the leading attributes of the organization (its assets) and its environment that determine the value of start-ups. There may be a situation in which the idea itself is interesting, while its implementation is weak, due to a weak team or lack of sufficient funding.

It is important that the works carried out in the pre-revenue phase lead to the creation of assets that have the ability to generate revenues (patents, know-how, knowledge), which means that the start-up already has assets that can be developed, sold, licensed etc. (commercialize), thus gaining additional decision-making possibilities (flexibility).

3 Research Method

The question was asked in the study: What variables determine the value of start-up at an early stage of development? As previously stated, the start-up value in the pre-revenue phase is difficult to determine. How to determine the value determinants of such enterprises. For this purpose, a method of comparisons of valuation methods, assessment models of innovative projects and business modeling was used, which, according to modern strategic management, is an important tool for strategic analysis. As a result of the research, independent variables affecting the start-up value were identified. Each time the value is marked by a relationship.

$$V = f\,(1, \ldots, n);\ \textit{where: } V \textit{ - start-up value};\ 1, \ldots, n \textit{ - independent variables}, \quad (6)$$

The study focused on identifying independent variables (1, ..., n). Next, it was identified how often the variable appears in the models, which proves its impact on the value. To conduct the comparison, it was necessary to conduct extensive literature studies. Criteria were identified, then adequate measure should be established for the criteria and the entire base subjected to machine learning. The scope of research is related to the multitude of value variables indicated by various valuation models. It is important not to omit any, and in the course of further research the real relationship between dependent and independent variables (attributes) will be established. A summary of the results of comparative research is presented in Table 1.

4 Measurement of the Value of Start-Up in the Pre-revenue Phase

4.1 Value Based Management

The situation of shaping the value of entities with an established position is fundamentally different than start-ups' in the pre-revenue phase. However, analyzing the factors determining the value of start-ups, one cannot ignore the concept of Value Based Management (VBM), which aims to increase value by controlling the factors that shape value. One of the first to address this issue was Rappaport [15], who proposed the following key value factors: period of value increase, revenue growth rate, sales operating profitability, income tax rate, investments in net working capital, investments in net fixed assets, period of competitive advantage (the period of creating the added value of the company's operational value), structure and cost of capital. It should be noted that the theory of VBM value management has gained a lot of research, publications, studies, concepts of identifying and controlling value factors (tools). Among many factors determining value, both financial and non-financial ones were mentioned and examined. These factors were divided into: financial, operational and strategic, available and used factors. These factors are expected to significantly affect the increase in cash flow, volatility of cash flows and optimization of the capital structure in the long term (factors increasing profitability and reducing risk). Skoczylas et al. [16] discusses these concepts more widely. Further discussion revealed that value is generated from tangible and intangible assets (intellectual capital). As part of intellectual capital, Guthrie [17] lists: structural, human and customer capital. And the whole range of measures of intellectual capital refers to the entity that already earns revenues. Only some of them are adequate to the start-up situation. According to the author, there is a research gap regarding the identification of factors affecting the value of start-ups in the pre-revenue phase of their measurement and evaluation. Copland Koller and Murrin [18] hold a similar view, claiming that the factors that shape the value of an enterprise are not universal and should be identified individually for each company. The article refers to the method of measuring the value of start-ups based on valuation models, which are often of a scoring nature, and thus referring to broadly understood factors that shape value, but typical for start-ups in the pre-revenue phase.

4.2 Scoring Methods

The Scorecard Valuation method, also called the Bill Payne valuation method [19], consists in comparing a start-up to other start-ups, taking into account in the valuation factors such as: economy, location, development phase of the enterprise, sector to which the enterprise belongs, management team, intellectual capital, market size, competition and possible revenues. Within the Scorecard Valuation method, three stages are distinguished - determining the average value of pre-revenue start-ups; comparing the measured start-up with the accepted standards (determining the correction factor) and correcting the average value of similar start-ups by the factor for the valued enterprise. The accepted standards: entrepreneur, team, management 30%; opportunity to develop 25%; product/technology 15%; competition 10%; sales/ marketing 10%; the need for additional funding 5%, others 5%. Each of the factors for the start-up is assigned a value from 0–150%. The value of the factors is multiplied standard (eg 100% * 30% = 0.3). Then the results obtained for the factors are added together, and the value obtained is multiplied the average value of start-ups.

The Berkus method [9, 20] was developed by David Berkus, a well-known business angel. Estimating the start-up value in this method requires taking into account the following carriers of its value: idea (idea, product risk, pre-design phase); prototype (reduces technological risk, design phase); quality of the management team (reduces the risk of failure, each phase); strategic partnership and cooperation (reduces market and competition risk); product offer and sales (reduces financial risk, post-design phase). In this method, the start-ups' value is calculated as the sum of the values of the factors. The value of the factors is set in the range of 0 to $ 500. The pre-money value cannot be greater than $ 2M.

Cayenne Consulting [21] has developed a startup value calculator. In the introduction, the company points out that traditional valuation methods, such as the DCF method or the P/E ratio are inadequate for unlisted start-ups. That is why - according to the company - the valuation should be based more on qualitative data. The Cayenne calculator contains 25 single-choice closed questions.

4.3 Methods Derived from the DCF Concept

Wong and Schootbrugge [22] draw attention to the fact that the DCF method is not suitable for start-ups, listing four main limitations: the problem of time; the problem of including the risk in the discount rate; applying too high discount rate by venture capital funds; unable to include - stop and change option. The discount rate in the DCF method reduces the value of future cash flows to their present value. Most often, the discount rate is the weighted average cost of capital from the CAPM model. This leads to risk estimation at a high and constant level, which means a high discount rate and can lead to project rejection using the traditional DCF method. However, the risk is different at every stage of start-up development. After positive market research or implementation of the technology, it is much lower than in the design phase. Okafor [23] undertakes the problem of valuation of start-up by financial methods. The disadvantages of the traditional DCF method have resulted in numerous concepts for its modification corresponding to the information needs of investors and the essence of start-ups.

The Venture Capital Method. In the venture capital (VC) method [24], there are two concepts related to start-up valuation: post-money valuation means the value of a start-up after external financing and the value of pre-money, i.e. the value of a start-up before obtaining financing. In other words, the post-money value is the term value that we can get as a result of selling a start-up. The relationship between post-money and pre-money can be written as follows:

$$POST\text{-}M = PRE\text{-}M + I \quad (7)$$

$$POST\text{-}M = RV/expected\ ROI \quad (8)$$

Where: ***POST-M*** *- post-money value; PRE-M - pre-money value;* ***I*** *- investment;* ***expected ROI*** *- rate of return required by investors (including their risk assessment);* ***RV*** *- Residual Value, Terminal Value, Exit Value.*

For start-ups from the later development phase, the Gordon model used in the DCF method is used, according to which the forward value is the value of free cash flow from the last year of the detailed forecast with a constant growth rate.

$$RV = \frac{CF_t(1+g)}{r-g} \quad (9)$$

*Where: **g** - constant growth rate after the detailed forecast period; **r** - discount rate.*

Scenario Method. One of the modifications is the so called scenario DCF, according to which DCF is calculated for three possible variants: optimistic, realistic and pessimistic.

The Multi-step Damodaran Method. Young enterprises have a dispersed ownership structure. Therefore, market risk is not the only type of risk that should be included in the estimation of the cost of equity. Specific risk should be taken into account, which is not possible using the classic Beta formula and the cost of capital because, as a rule, start-ups in the initial phase are not listed on the stock exchange [3]. For start-ups, a different approach should be used to estimate the discount rate. A. Damodaran divides the start-up life cycle into several phases, calculating the Beta coefficient for each one separately. The value of the BETA ratio depends on the sources of financing that the start-up uses. This can be represented as follows: determination of the market BetaM coefficient; BetaM correction in the period before financing from the VC fund; Market BetaM correction during the period of financing from the VC fund; Market BetaM for the period after IPO.

$$AV\ \beta L = AV\ \beta u\,(1 + (1 - TR)\,(Av.D/E)) \quad (10)$$

*Where: **Av. β_L** average Beta value for listed companies of the sector; **TR** - income tax rate; **Av.D/E** average value of the Debt/Equity ratio for the sector/ industry; **AV β_U** average Beta value for unlisted listed companies.*

For the period of startup's activity before financing with VC funds, the Beta market coefficient is adjusted by the correlation of companies from a given sector or industry with the market, which increases the value of the coefficient.

$$Total\ Beta = (Market\ Beta/market\ correlation) + / - (VC\ financing\ effect) \quad (11)$$

$$Cost\ of\ equity = r_{RF} + Total\ Beta * (PR) \quad (12)$$

*Where: **PR** - Risk Premium, assumed depending on the industry; **r_{RF}** - return on safe investment (risk free).*

DCF Correction According to Wong and Schootbrugge. KM. Wong and E. Schootbrugge presented a different approach to valuation [22]. They propose so-called multi-stage valuation for start-ups and technology projects, based on discounted cash flows. The calculation of the start-up value in this method is as follows:

$$V_i = PVTV_i - \sum\nolimits_{t=i}^{n} (1/s)I_t \quad (13)$$

Where: $\boldsymbol{V_i}$ *- enterprise value at i-moment;* $\boldsymbol{PVTV_i}$ *the company's forward value at i-moment, with a discount rate equal to the interest rate on the bonds, present value terminal value;* $\boldsymbol{s}$ *- probability of success in a given phase,* $\boldsymbol{I_t}$ *- the necessary investment outlay at i-moment to go to the phase.*

Capital Cost Valuation According to Morningstar. An interesting method of determining the cost of capital is presented in the Morningstar valuation guide. It included the following elements: revenue repeatability, operating leverage, financial leverage and the level of systematic risk as well as the adjustment resulting from the premium for the given country [25]. The equity cost determined in this way should still be adjusted by the premium for the investment risk in the given country.

Correction of the BETA Coefficient According to Festel, Wuermseher and Cattanco. Festel, Wuermseher and Cattanco [26] created a model divided into five categories - technology, products, implementation, organization and finances - and 4 sub categories. Every time it gives 20 independent variables. Each subcategory contains five elements that have been assigned a Beta correction of $+1$ to -1. The sum of corrections for all subcategories gives the final correction of the Beta coefficient (up or down). They indicated the following areas:

- Technology: Maturity of technology, Advantages compared to competitive technologies, Reputation of scientist, Patent protection.
- Products: Product benefits, Unique selling proposition, Scalability, Competition.
- Implementation: Business plan, Technical development plan, Marketing plan, Business development plan.
- Organization: Competences of the management team, Headquarters location, Competences of advisory board, Process efficiency.
- Finances: Sales plan, Costs plan, Profitability, Liquidity plan.

Real Options. The traditional DCF method takes into account the value of investment outlays and projected free cash flow for one specific moment. Thus, this method eliminates the use of the "stop and change" option that every manager has. Start-ups are flexible entities that are able to quickly respond to signals coming from the environment and change the adopted strategy. The method that allows taking into account current market and internal conditions as well as the decision-making of managers is the method of real options. More about real options in reference items e.g. Amram and Kulatilaka [27]; Brach [28]; Copeland and Antikarov [29].

4.4 Other Areas of Research Exploration Related to the Value of Start-Ups

As part of Strategic Management, a tool called Business Model Canvas was developed. This is the start template for shaping the business model. It consists of nine fields. These are elements describing the value of the product for the client, resources, customer segment and relationships with them, distribution channels, revenues, costs, key partners [30, 31]. Business Model Canvas was proposed by Osterwalder based on his previous studies on business model [32]. From a start-up value perspective, it can be said that these areas determine the value of a start-up.

Another important area in which a relationship with start-up value can be sought is the innovation assessment system. From this perspective, the start-up is an organization set up to implement innovation. For this reason, the assessment used to assess innovative projects in existing entities can be partially transferred to the start-up valuation ground. Of course, it should be borne in mind in this case that the assessment of start-up values must take into account variables not taken into account in entities that have already been operating for a long time or vice versa (e.g. strategic convergence, use of existing resources, or on the other hand investment capital demand). In this approach, market factors, customer and product identification dominate the assessment. The assessment

Table 1. Analysis of factors influencing the value of pre-money start-up.

METHOD VARIABLES	Scorecard Valuation Bill Payne method	Berkus method	Cayenne method	Assessment of innovations	Business model canvas	Festel, Wuermseher and Cattanco	Morningstar	Scenerio method	Wong and Schootbrugge	The multi-step Damodaran method	Frequency
Market (Opportunity to develop)	YES		Customer identification, market growth rate, market size	Is the market growing, market size	Customer identification and customer relationship	Systematic risk	Systematic risk	Systematic risk	Systematic risk	Market Beta / market correlation	9
Management	YES	YES	intellectual property and corporate attorney, experience	Experience	Key activities	Competences of advisory board					6
Product	YES	Idea	Maturity of product (idea, prototype, sales) Product benefits, Unique selling proposition,	Product benefits, Unique selling proposition,	Product benefits, Unique selling proposition,	Product benefits, Unique selling proposition,					6
Technology	YES	Prototype	Maturity of technology, Patent protection	Maturity of technology	Resource Availability	Maturity of technology, Advantages compared to competitive technologies, Patent protection					6
Team	YES		PhD, marketing experience	Resource Availability	Resource Availability	Reputation of scientist					5
Sales plan			YES		YES	YES		YES	YES		5
Cost plan					YES	YES		YES	YES		4
Competition	YES	-	YES	Are there similar entities, products		YES					4
Sales/Marketing	YES	YES	Is there a sale? Sales Potential.		Sales channels						4
Probability of success				YES				YES	YES	YES	4
Average start-up value	YES	YES	YES								3
Opportunity to develop	YES		Is the market growing,	Is the market growing							3
Additional funding	YES	-							YES	YES	3
Investing						YES		YES	YES		3
Beta correction, measure of start-up risk						YES	YES			YES	3
Partnership and cooperation		YES	Ability to cooperate		Key partners						3
Planned return/margin			YES	YES				YES			3
Scalability					**YES**	**YES**					2
Business plan			YES			YES					2
Business model				**YES**	**YES**						2
Entrepreneur/owners	YES		invested money and time, determination								2
Marketing plan			YES			YES					2
Revenue variability							YES	YES			2
Degree of Financial Leverage							YES			YES	2
VC financing effect										YES	1
Technical development plan						YES					1
Business development plan						YES					1
Degree of Operating Leverage							YES				1

is scoring, however, with a positive assessment of financial efficiency. Cooper et al. [33] addresses this issue more broadly.

5 Discussion and Conclusion

The research identifies many independent variables that determine the value of an enterprise in the pre-revenue phase. Many factors are repeated, testifying to their importance. In all models, the team, product, technology maturity and growing market are decisive. External factors can often affect the value with some delay. They are an opportunity for possible use. Whether it will be used will be determined especially by marketing competences. The importance of business modeling skills, including business scaling, is increasingly being drawn, which is not yet widely identified in valuation models, but noticeable in the strategic management trend. Some methods do not identify factors, but rather provide tips on how to negotiate with owners, shares and their price. Valuation methods from the group of discount methods indicate the importance of risk (systematic and specific), investment outlays and free cash flows. Part of the information is very credible, but many can be accused of wishfulness and illusiveness. This applies especially to sales, cost and investment forecasts. Even small revenues confirming the assumptions make the forecasts more credible and the valuations. Particularly important for the identification of independent variables are scoring type models and those that correct the DCF model, shaping BETA based on extensive knowledge of the entity and the market in which it operates. The review of methods of pre-revenue start-ups valuation leads to the conclusion that the main dependent variables of the model are: probability of bankruptcy/failure; risk measured with beta and discount rate, and the value of the start-up itself. The group of independent variables includes:

Market: does the market exist; expected market share; industry growth rate; market size; B2B, B2C market type; competition offer; competition reaction; number of bankruptcies; transaction value in the case of sale of start-ups; location; market and safe rates of return; systematic industry risk; are there similar solutions on the market; idea of customer relationships; market research; identification of customer needs.

Product/Technology: the complexity of the product; product uniqueness; product benefits; technology maturity (are all problems solved); interesting idea; product maturity.

Scalability: the possibility of multiplication of the business model, rapid customer growth and revenues.

Management and management activity: management team (structure, competences, experience, marketing experience, industry experience and successes with start-up companies); who is attorney of company; are the implementation/sales/marketing/distribution plans defined (Business Plan defined, Business Model); the quantity and type of key resources; defined key activity and distribution channels.

Owner involvement: financial outlays; dedicated time; determination; experience.

Support: whether a business angel or experienced industry managers are involved in the start-up; whether, if necessary, the start-up has a relationship with the scientific.

Decision flexibility - are start-up development other options possible?

Participation in networks/cooperation: strategic alliances; clusters; cooperation (participation in the production, value or logistics chain), degree of formalization.

Low: legal restrictions, product, technology, market.

Assets protection - protection of intellectual values and key resources; patent protection; who provides legal services for intellectual property; threat of appropriation; market value of developed solutions.

Time: what is the expected period until the first sale (period of deferred expected revenues); how long does it take to work on the idea; development phase (concept, prototype, tests, trial series, trial sale, …).

Financial management: investment outlays (incurred, must incurred); certainty of further financing; expected revenues/costs; expected ROI; operating leverage; financial leverage; average start-up value.

Determining the model and significance of individual variables requires collecting a large amount of the above-mentioned data at various times of the entities' operations. This issue seems very interesting especially in the perspective of external changes determining the valuation. This will allow the use of deep learning and machine learning. Decision support by machine learning systems will allow for real valuation and approximation of the positions of the owner and investor. Thus, it will limit the area of information asymmetry and may allow for informed negotiations. It can lead to the acceleration of investment transactions. On the one hand, it can lead to better risk identification, on the other, it can reduce investors' expected profitability. It has been noticed that many valuation models do not take into account the important aspect of scalability.

The assessment of the usefulness of independent variables to describe the dependent variable is the goal of the next stage of research. The further step is to collect data and propose valuation models using a machine learning, test them and provide machine learning, building experts algorithm.

Acknowledgment. "The project is financed by the Ministry of Science and Higher Education in Poland under the programme "Regional Initiative of Excellence" 2019–2022 project number 015/RID/2018/19 total funding amount 10 721 040,00 PLN".

References

1. Talaga, M., Piwowarczyk, M., Kutrzyński, M., Lasota, T., Telec, Z., Trawiński, B.: Apartment valuation models for a big city using selected spatial attributes. In: Nguyen, N.T., Chbeir, R., Exposito, E., Aniorté, P., Trawiński, B. (eds.) ICCCI 2019. LNCS (LNAI), vol. 11683, pp. 363–376. Springer, Cham (2019). https://doi.org/10.1007/978-3-030-28377-3_30
2. Piwowarczyk, M., Lasota, T., Telec, Z., Trawiński, B.: Valuation of building plots in a rural area using machine learning approach. In: Nguyen, N.T., Chbeir, R., Exposito, E., Aniorté, P., Trawiński, B. (eds.) ICCCI 2019. LNCS (LNAI), vol. 11683, pp. 377–389. Springer, Cham (2019). https://doi.org/10.1007/978-3-030-28377-3_31
3. Damodaran, A.: Valuing Young, Start-up and Growth Companies: Estimation Issues and Valuation Challenge (2009). http://people.stern.nyu.edu/adamodar/pdfiles/papers/younggrowth.pdf. Accessed 14 Apr 2020
4. Golej, R.: Selekcja projektów nowych produktów w controllingu innowacji, PN UE we Wrocławiu, nr 291, pp. 147–159 (2013)

5. Calopa, M.K., Horvat, J., Lalic, M.: Analysis of financing sources for start-up companies. Management **19**(2), 19–44 (2014)
6. Blank, S., Dorf, B.: Podręcznik startupu, Helion (2012)
7. Robehmed, N.: What Is A Startup? Forbes, 16 December 2013. Accessed 16 Apr 2020
8. Łuczak, K.: Rachunkowość innowacji na przykładzie przedsiębiorstw określanych mianem start-up, Finanse, RFU, ZNUSz No **830**(70), 79–87 (2014)
9. Gemzik-Salwach, A.: Wykorzystanie metody Dave'a Berkusa do analizy potencjału rozwojowego firm start-up w Polsce, AUL Folia Oeconomica **2**(300), 111–122 (2014)
10. Ries, E.: Metoda Lean Startup, Helion, Gliwice (2011)
11. Armstrong, A.: Key risks of the 4 stages of a startup's life cycle (2016). www.itproportal.com/2016/02/11/key-risks-of-the-4-stages-of-a-startups-life-cycle. Accessed 03 Dec 2019
12. Damodaran, A.: The Dark Side of Valuation: Valuing Young, Distressed, and Complex Businesses, 3rd edn. Pearson FT Press (2018)
13. Ocena efektywności inwestycji. Wrzosek S. (ed.) Wyd. UE we Wrocławiu, Wrocław (2008)
14. Wickham, P.A.: Overconfidence in new start-up success probability judgement. Int. J. Entrepreneurial Behav. Res. **12**(4), 210–227 (2006)
15. Rappaport, A.: Wartość dla akcjonariuszy, Wig-Press Warszawa (1999)
16. Determinanty i modele wartości przedsiębiorstw, Skoczylas W. (ed.) PWE Warszawa (2007)
17. Guthrie, J.: The management, measurement and the reporting of intellectual capital. J. Intell. Capital **2**(1), 27–41 (2001)
18. Copeland, T., Koller, T., Murrin, J.: Wycena mierzenie i kształtowanie wartości firmy. WIG Press, Warszawa (1997)
19. Payne, B.: Scorecard Valuation Methodology (Rev 2019): Establishing the Valuation of Pre-revenue, Start-up Companies, Frontier Angels, 21 October 2019. https://www.angelcapitalassociation.org/blog/scorecard-valuation-methodology-rev-2019-establishing-the-valuation-of-pre-revenue-start-up-companies/. Accessed 16 Apr 2020
20. Berkus, D.: After 20 years: Updating the Berkus Method of valuation, 04 November 2016. https://berkonomics.com/?p=2752. Accessed 14 Apr 2020
21. Cayenne Calculator. https://www.caycon.com/valuation. Accessed 16 Apr 2020
22. Wong, K.M., Schootbrugge, E.: Multi-stage valuation for start-up high tech projects and companies. J. Account. Finan. **13**(2), 45–56 (2013). http://www.na-businesspress.com/jafopen.html
23. Okafor, A.: The impact of valuation methods on likelihood of mergers and acquisitions of high-tech startup companies in Nigeria, Walden University (2018)
24. How does the venture capital method value a business, Venionaire Capital. https://www.venionaire.com/venture-capital-method/. Accessed 14 Apr 2020
25. Valuation Methodology Handbook (General Model v6), Morningstar (2012)
26. Festel, G., Wuermseher, M., Cattanco, G.: Valuation of early stage high-tech start-up companies. Int. J. Bus. **18**(3), 216–231 (2013)
27. Amram, M., Kulatilaka, N.: Real Options: Managing Strategic Investment in an Uncertain World. Harvard Business School Press, Boston (1999)
28. Brach, M.A.: Real Options in Practice. Wiley, Hoboken (2003)
29. Copeland, T.E., Antikarov, V.: Real Options: A Practitioner's Guide. Texere, NY (2001)
30. Osterwalder, A., Pigneur, Y.: Tworzenie modeli biznesowych – Podręcznik wizjonera, Wydawnictwo Helion, Warszawa (2012)
31. Business Model Canvas – myślenie modelem biznesowym. www.pi.gov.pl. Accessed 16 Apr 2020
32. Osterwalder, A., Pigneur, Y.: An ontology for e-Business models. In: Currie, W. (ed.) Value Creation from E-Business Model, pp. 65–97. Elsevier (2004)
33. Cooper, R., Edgett, S., Kleinschmidt, E.: Portfolio Management-Fundamental to New Product Success. The PDMA Toolbook for New Product Development (2002)

Blockchain Platform Taxonomy

Andrew A. Varnavskiy(✉), Ulia M. Gruzina, and Anastasiya O. Buryakova

Financial University under the Government of the Russian Federation,
Leningradsky Prospekt 49, Moscow, Russia
avarnavskii@gmail.com

Abstract. Now blockchain technology continues to develop – lots of new projects and platforms are being created. Some of these technological solutions are based on approved platforms, but other developers use their own new model types. The statistics show that blockchain technology is one of the future-proof technologies like artificial intelligence, Internet of Things and nanotechnology. However, the lack of standardisation in the field of the modern technologies can cause negative consequences for developers and financial institutes. The purpose of this study is to create the common open classification of existing blockchain platforms. Classification is a hierarchical structure consisting of six key characteristics: token, transaction, block, framework, network and communication. Each characteristic consists of groups (qualitative features), including sets. Such approaches allow to standardise the process of forming the blockchain platform architecture. Moreover, it is possible to create various scenarios of the network changes.

Keywords: Blockchain technology · Blockchain platforms · Token · Transaction · Block · Framework · Network

1 Introduction

More than ten years have passed since the first article about the blockchain platform «Bitcoin: A Peer-to-Peer Electronic Cash System» [1] was published. There was a rapid development of the blockchain technology – lots of new different platforms like Hyperledger, Corda, EOS were developed. Each of these platforms has their own technological characteristics and frameworks. The increase of numerous solutions causes the necessity to create universal, global classification of blockchain platforms. This classification can be used to regulate or ease the selection of solutions for various business processes.

So, blockchain is one of the types of distributed ledger technologies, that consists of the chain of the related and encrypted blocks. Key characteristics of the blockchain technologies are «distributed», chronological data record, inability to replace data in the past. Moreover, using the blockchain, it is possible to automate some transactions through smart contracts. These characteristics are similar for all blockchain platforms. But experts have discovered others important features. These features are the subject of our research.

The lack of standardisation in the field of the modern technologies can cause negative consequences for developers and financial institutes, which are improving operational

M. Hernes et al. (Eds.): ICCCI 2020, CCIS 1287, pp. 389–401, 2020.
https://doi.org/10.1007/978-3-030-63119-2_32

processes. Such uncertainty also spreads on the legislative authorities, which are not able to distinguish all existing possibilities. As a consequence, the appropriate legislation can't be provided. Medium term absence of the standards will induce risks of confidentiality, interoperability, security for all market parties.

The purpose of this study is to create a common open classification of existing blockchain platforms. The developed classification is based on the accumulated theoretical and practical experience. Our proposed classification is the hierarchical structure of sets of technological elements of platform solutions. Due to rapid development of technology the tree structure will most likely to expand. We are sure, that new branches will appear, but the key characteristics should remain the same.

2 Review of Existing Blockchain Classification

Various classifications of blockchain platforms were presented by both domestic and foreign authors. They often distinguish public and private, open and closed blockchain platforms. [2–5] However, it seems to us that such a classification does not give a complete view of all the characteristics. Moreover, various authors consider different «public/private» perspectives, without taking into account the research of their colleagues. As a result, different platforms with fundamentally different characteristics can be called, for example, «open platform», «public platform» and so on.

For example, M. Swan in «Blockchain: Blueprint for a New Economy» distinguishes: blockchain 1.0 («settlement» blockchain), blockchain 2.0 (smart-contracts), blockchain 3.0 (decentralized applications and organizations). Alternatively, blockchain-platform are classified by consensus: Practical Byzantine Fault Tolerance Algorithm, Proof-of-Work, Proof-of-Stake, and others. [6] Earlier we noted such classification features as «access to the network», «restriction of actions with information», «payment for transactions». [7] However, we find that these characteristics will not cover all the technological aspects.

In our opinion, the classification of the authors Tasca P. and Tessone C. J. [8] is most coherent. The authors presented a taxonomy of the blockchain technology formed in the systematic, hierarchical groups. A reverse-engineering approach was applied. Taxonomy includes eight components (levels), which are divided into sub-levels.

The first component is «Consensus». The authors use an extended interpretation of this term. Consensus is a set of rules and mechanisms that allows to maintain and update a distributed ledger, ensure the reliability of the record in it.

The second component is the «Transactional capabilities», characteristics associated with transactions and blocks. In particular, the authors use such characteristics as the «transaction model», «limits to scalability», «data structure in the blockheader» and others.

The third component is the «native currency/tokenization», which includes three subgroups: native asset, tokenization, asset supply management.

The fourth component is the possibility of developing a blockchain network (internal and external capabilities, management model, scripting languages).

The fifth group is security and privacy. The feature includes two components: data encryption and data privacy.

The sixth is the codebase. The seventh component is identity management. There are several levels in this group: «access and control Layer» and «Identify Layer».

The eighth component is called «Charging and rewarding system». In fact, the characteristics of this group are based on the economic component - the motivation for the activity of blockchain agents.

Despite all the advantages of the Tasca P. and Tessone C.J. classification, there are some controversial points. Is it necessary to combine the block and the transaction on the same level (component)? Is it advisable to relate organizational details (open/closed code, separate project/consortium) and scripting languages to the characteristics of the network development prospects? Should we allocate "pseudo-anonymity" as one of the sub-levels of «identity management»? And other moments. We will try to show our answers in the developed classification.

An analysis of other classifications also needs to be taken into consideration. Tasca P. and Tessone C. J. present the blockchain classification, although there are many studies on distributed ledger technology in general. For example, in an article by Swiss experts «Decrypting Distributed Ledger Design-Taxonomy, Classification and Blockchain Community Evaluation» [9], the attribute «DL Type (blockchain, DAG and other)» is just one of the distinguishing characteristics. In general, the authors consider two branches of taxonomy:

- DLT (distributed ledger technology) – attributes directly related to technological aspects (consensus, Turing complete language, verification, reward system, etc.).
- CED (cryptoeconomic design) - attributes related to tokenisation and design of a distributed system (public/private networks, the process of creating new base units, the purpose of tokenisation).

Another difference between the classification developed by the authors and the classification of Ballandies M. C., Dapp M. M., Pournaras E. is the interpretation of the consensus algorithm [9]. The authors say that «consensus» consists of many different characteristics: the type of consensus, block confirmation algorithm, permission to write and read the registry, reward. We believe that these characteristics should be divided into different levels. Another aspect in which we hold a slightly different opinion is the division of platforms into private and public. For example, authors distinguish two attributes - «Permission to Write» and «Permission to Validate», but one platform can be assigned to both sets - the platform can be public and limited at the same time (Byteball, IOTA). It causes a serious terminological confusion.

The following attributes of the work of Ballandies M. C., Dapp M. M., Pournaras E. seemed interesting to us:

- Transaction structure. The attribute is responsible for the possibility or inability to create transactions with arbitrary content. In the blockchain such an opportunity exists. For example, the portrait of Nelson Mandela is placed on the Bitcoin blockchain, but this is more of an experiment than a systematic use of the network, because its main function is to transmit information about the movement of Bitcoin as a unit of account. In 2013, OP_RETURN was adopted as a valid type of transaction, providing users with 80 bytes of space for embedding arbitrary data. OP_RETURN allows users to

mark transaction output as non-disposable, removing this output from the UTXO set while maintaining its data on the blockchain. This allows users to validate a data point without forcing nodes to support that data point.
- Ability to transfer token. An example is the Akasha blockchain, where tokens are assigned to specific users. A similar attribute can be reflected in our classification at the «Token» level, however, there are some doubts about the expediency of creating a system with «non-transferring» units of accountin general.
- Reward system. The authors emphasize that various options are possible: a reward for mining, a reward for mixing transactions or no reward at all.
- Tokenisation goals - providing access to the platform, renting computing power, providing services, the token can be a form of security, stock, coupon, etc. This attribute is not technological and is very closely related to business models of projects. These points were undoubtedly included in our classification.

Next classification of distributed ledgers is the article «A taxonomy of blockchain-based systems for architecture design» [10] by Australian researchers. In our opinion, it is necessary to pay attention to the fundamental point - is it necessary to include in the taxonomy aspects that go beyond the technological framework, the so-called «off-chain» characteristics? For example, in the article, researchers highlight the feature «Item data» - the source of data acquisition and transaction initiation («embedded» transactions, smart contracts - on-chain; user transaction - off-chain and others). The «Scalability» subset of the «New blockchain» set (which includes, for example, sidechains) is controversial. We believe that sidechains will complement existing networks and they should not be distinguished within the classification of blockchain platforms. Another aspect is the «Data Structure», which includes «Blockchain», «GHOST», «BlockDAG», and «Segregated witness». Different approaches are possible, but within the framework of our taxonomy, we stick to broader view and disclose only two characteristics - «Block header» and «Block structure», without focusing on the type of distributed registry.

And the final feature that was reviewed is the taxonomy published in 2018 in the article «Distributed ledger technology systems: a conceptual framework» [11]. The authors, Martino Recanatini and Kathryn Vagneur, show five technological levels: formed transactions (transaction); unconfirmed transactions (log); confirmed transactions (record); the set of records held by a node, although not necessarily consistent with the consensus of other nodes (journal); transactions, accepted into a block and distributed between nodes, but with an «intermediate» status (ledger). The authors' taxonomy in this paper is similarly formed. It is noteworthy that Martino Recanatini and Kathryn Vagneur do not consider the classification features in terms of levels. The authors distinguish three blocks: «Protocol», «Network» and «Data».

A number of aspects of this classification were taken into account in our proposed version, but in a different modified form. For example, the presence or absence of the parent network is reflected at the «Communications» level - the division of accounting units into tokens and coins, the period between the initiation of the transaction and its confirmation are reflected in two sets - the «Latency» and the «Gossiping».

The literature review of classification of decisions based on blockchain technology made it possible, on the one hand, to identify a number of terminological inconsistencies.

and on the other hand, to see completely different approaches to the construction of taxonomies and hierarchical structures, which allowed us to expand our classification.

3 Methodology

The methodological approach is composed of the following steps:

1. Analysis of existing blockchain classifications, identification their strengths and weaknesses;
2. Blockchain platform research. Selection of key technical characteristics of the platforms;
3. Taxonomy setting. Taxonomy has been populated by main elements and their components;
4. Description of the technical characteristics of blockchain classification. Identification of comparability of characteristics of different directions/branches.

4 Developed Blockchain Taxonomy

Identifying the relationships between different characteristics helps to identify different alternatives in the work of blockchain platforms. This allows to develop the optimal architecture for each specific task. The basis of the developed classification consists of six levels of blockchain platforms functioning: token, transaction, block, framework, network and communication.

4.1 Token

Platforms differ in the possibility of the creation and operation of units of account. At this level, we consider two distinctive characteristics (see Fig. 1).

a) Existence of a token. If there are no tokens on the platform, blockchain is a «documentary blockchain». Token (unit of account) indicates a «settlement blockchain».
b) Emission. Issue of tokens takes place when the platform starts or during its life. Also emission may be limited (like Bitcoin) or not limited.

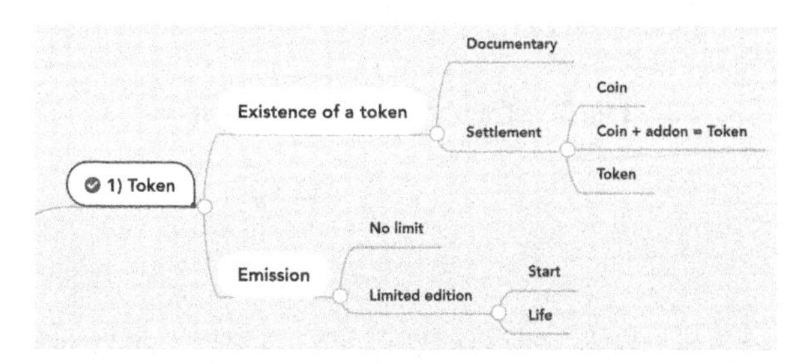

Fig. 1. «Token» classification.

4.2 Transaction

The basic component «Transaction» describes how nodes, connected to a P2P network, store and update information in a distributed ledger. The task of the transactional model is to prevent the appearance of data on the network that should not be trusted, for example, entering «double costs» into the ledger. We also distinguish two subsets of blockchain platforms, based on transactional characteristics (see Fig. 2).

a) Structure of transaction:

  - The Unspent Transaction Output (UTXO). UTXO implies that in each transaction the output amount at the output cannot exceed the input amount. Network members are not able to use the same output in different transactions.
  - Traditional Ledger. Compared to UTXO, various blockchains, such as Stellar and Ripple, use a more traditional model for recording transactions. In particular, Stellar lists every single transaction in the history of the distributed Stellar leger. Ripple also uses the traditional transaction model to register an increase/decrease in balance and clear all balances on accounts. In Ethereum, some transactions are used to perform actions in the smart contracts defined in specific atomic records in the blockchain. These transactions are the execution of orders of interested parties that perform actions from the specified smart contracts.

b) Transaction Composition:

  - Restricted. In some blockchains the composition of the transaction is predetermined. For example, restrictions can be set in the form of «fields» of records, limited volume, etc. An example is the IOTA blockchain where recording of any additional information is impossible.
  - Free. For example, a transaction can be made of text of arbitrary length, and special functions can be provided to add any other information for the standardised record.

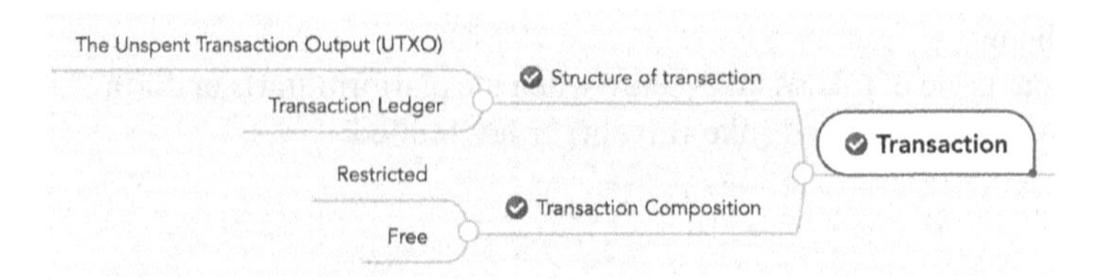

Fig. 2. «Transaction» classification.

4.3 Block

The «Block" component assumes that the rules for creating a block vary depending on the platform. We distinguish two distinctive characteristics (see Fig. 3).

a) Block Storage. Depending on what kind of information is stored in the blockchain, the scalability of the system is determined. In this regard, we distinguish two sets based on the "Blockheader":

   – Set of the blockchain platforms in which information is stored exclusively on transactions (systems such as Bitcoin). These transactions contain a set of inputs and outputs that help identify the sender and receiver in a particular transaction. This approach is also used in blockchains such as IOTA, which uses the principle of DAG (oriented acyclic graph) to store each individual transaction.
   – Balance is a set of platforms, where the User Balance is also stored in the block. For example, in Ripple, the decentralized repository contains information about user balance. This approach may limit the system's storage requirements, but at the same time reduces the possibility of transaction rollback.

b) Blockheader. Depending on the data structure in the block header, it can perform various functions. On the one hand, it includes transaction hashes for building a chain of blocks, on the other hand, it contains additional information for various application levels or technological platforms of the blockchain. The data structure in the block header describes the ability of the system to store information about transactions. We have identified two possible types of data structures in the block header:

   – Binary Merkle Tree. For example, a Binary Merkle Tree is used to store transactions in Bitcoin. This hash is included in the block header. The information in the block header contains the hash of the previous header, timestamp, mining difficulty, nonce and root hash of the Merkle tree containing all the transactions of this block.
   – Patricia Merkle Tree. On the one hand, Patricia Merkle Tree (a practical algorithm for extracting information encoded in alphanumeric form) allows you to perform operations such as insertion, editing information related to the balance and nonce. This means that it is possible to check transactions faster than in the model of a simple Merkle tree. [12] An important advantage is the availability of checking specific branches of the tree, which also saves time. Ethereum uses the Patricia Merkle Tree for the block header to store more information than it is possible in the Binary Merkle Tree.

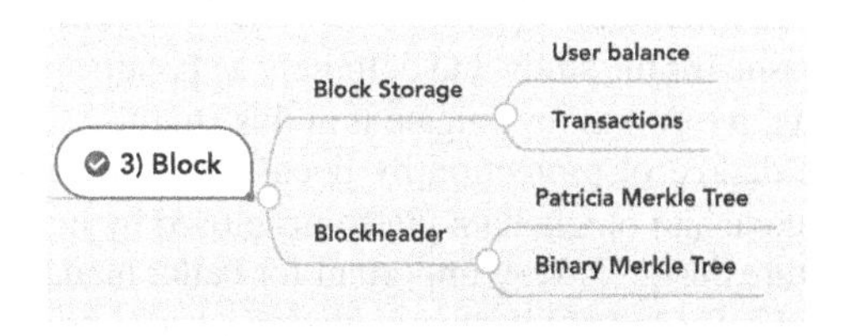

Fig. 3. «Block» classification.

4.4 Framework

Component "Framework" is the most variable in the whole classification. The main characteristics include the following (see Fig. 4).

a) Cording language. This characteristic illustrates the interconnection of basic programming languages in a specific blockchain platform. Blockchain platforms rarely use a single language for all tasks. Cording languages can be divided into two levels: the blockchain «the language in which it was created» - the basic level, and the languages in which the «applications» were created – the script languages. The «Cording language» component shows the interconnection of programming languages in a particular blockchain platform. Two options are possible:

   - Single cording language. For example, Bitcoin Core uses C ++ as its programming language.
   - Several cording languages - several languages are used. For example, Ethereum uses C ++, Ethereum Virtual Machine Language and Go, which allows to better interact with other languages. Stellar supports JavaScript, Java, and Go. There are also community-supported SDKs for Ruby, Python, and C-Sharp.

b) Script language. Widespread cording languages are Turing complete. This means that you can implement an algorithm for modeling any Turing machine. Thus, these are general purpose languages in which undefined calculations can be performed. Languages may be incomplete in Turing due to constructive considerations — to prevent some ways of executing code (for example, indefinitely). The scripting language used in some blockchain platforms allows changing the conditions under which certain information (for example, transactions) will be included in the public record. These conditions must be determined algorithmically (smart contracts).
c) Finality. In various platforms, an agreement on when a block is considered confirmed, is applied or not. The following options are possible: "deterministic" or "non-deterministic." For example, in Bitcoin the protocol is not deterministic. "Waiting to add 6 confirmed blocks to the chain" reduces the probability that the transaction will be redefined. But completely exclude the possibility that a previously verified block will be removed or «removed» from the blockchain in the future is impossible. Accordingly, when we talk about a deterministic blockchain, the situation will be reversed.
d) Consensus: Proof-of-Work, Proof-of-Stake, Proof-of-Authority, Proof-of-Capacity, Proof-of-Burn, Byzantine Fault Tolerance consensus, hybrid consensus and other. This is a relevant issue for the study and definitely a classification feature. However, due to its versatility, we do not investigate it in this study.
e) Depending on the degree of protection by licenses, we are considering an option with open source code and blockchain platforms closed by licenses.
f) Software architecture design is very important for better management and development. The choice of software architecture includes specific structural options among the features available for software development. The following design options can be distinguished: monolithic (single-level software application without modularity, like Bitcoin) and polylithic (Tendermint, Hyperledger Fabric).

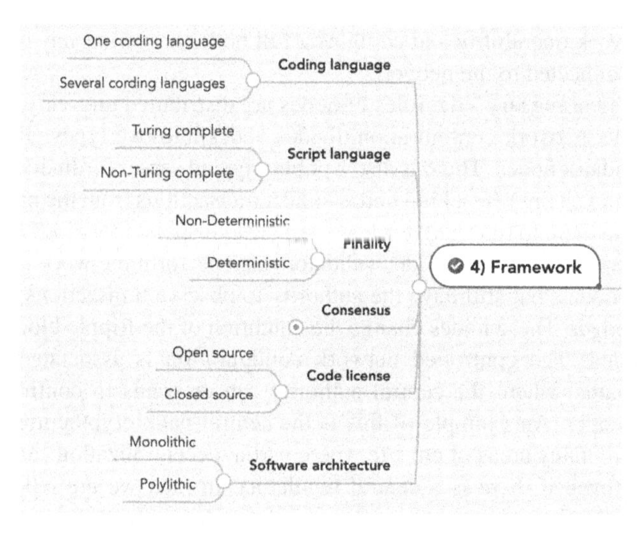

Fig. 4. «Framework» classification.

4.5 Network

«Network» component answers the question of how network continuity is ensured, how nodes and chains interact, and what roles nodes play (see Fig. 5).

a) The basis of all blockchain systems is their decentralized nature. It is necessary that the nodes connected to the network are indistinguishable from each other. This concept, however, can not be fully implemented. There are several reasons for this: limited computing capacity or limited network bandwidth. Under these conditions, different nodes have access to different levels of information. Those that do not store information completely are «thin nodes» connected to a peer-to-peer network. For this reason, in the "Network" we distinguish the following types of blockchain platforms: with Full Nodes (all participants are equal) and with Thin Nodes Capabilities.
b) Latency is a subcomponent that describes the rule for distributing messages in networks. This set consists of two subsets: synchronous communications (systems that set upper bounds for the "process speed interval" and "communication delays") and asynchronous communications (the block will not be closed until all members of the network have confirmed it).
c) Network topology. The essence of the characteristic is to determine the type of relationship between network nodes and the type of information flow between them: the inclusion of a transaction in the ledger and/or only the verification of the transaction. Based on this characteristic, three sets can be determined:

  – The decentralized ledger is a distributed P2P network that allows direct transactions to each node. Everyone can check the validity, like any miner who maintains the network operability and contains a full node, as well as any other full node that is connected to the network.

- Hierarchical registry - the roles of nodes are distributed unevenly. For example, in Ripple, network configuration divides nodes into two types: Tracking nodes and Validator nodes. There is also a public rippled server (ordinary nodes, like in other blockchains). Tracking nodes collect transactions from the participants and distribute them to the validator node, they also respond to various requests about the status of the ledger. Nodes validators also perform the work of the previous type of nodes, but still have the authority to observe transactions and add them to the ledger. These nodes change the condition of the Ripple blockchain.
- Controlled. The controlled network configuration is associated with private blockchains, where the central authority can or wants to control information in the ledger. An example of this is the central bank cryptocurrency (CBCC). There are many areas of our life where partial decentralization can give positive results (even if there is a central regulator). In fact, we are talking about the decentralization of certain interactions under the control of a single or several administrators.

d) Gossiping is a characteristic of the process of passing information through servers. In the absence of centralized routing control (as in traditional electronic payment systems), nodes must transmit the available information. When information about a new block appears in a node, it distributes it to all peers:

- Local exchange - messaging occurs first locally, until consensus is reached. This is also called the «federal consensus» used, for example, in Ripple. In Ripple, nodes can exchange transaction records with other nodes and reach consensus

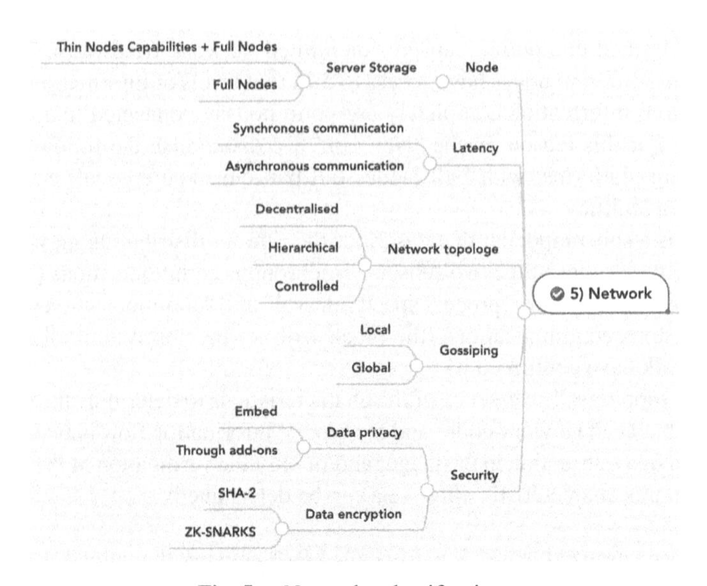

Fig. 5. «Network» classification.

without communicating with other nodes in the network. Most of the information is transmitted «locally» - in terms of the P2P network. Only then the information is sent through all the other nodes.

- Global information sharing. In most blockchains (Bitcoin, Ethereum and others) the exchange of messages occurs with a specific list of peers (equal nodes) that are considered reliable. In Bitcoin, they are called fallback nodes. These nodes maintain a list of all peers on the network.

e) Security. New implementations of blockchain technology carry both technical and operational risks associated with security and privacy. Two sets can be distinguished: data encryption (SHA-2, ZK-SNARKS) and data privacy (embedded confidentiality and confidentiality through add-ons).

4.6 Communication

The platforms differ depending on the motivation of the economic agents for certain aspects (see Fig. 6). Consider the following options:

a) Access and management. Management is the definition of «interaction rules». Summarizing, some rules give access to the platform resources or determine the order of their use. This component consists of four sublevels:

- Public. Basic characteristics for the network to be considered public will be open code, free ability to read the ledger and verify blocks. Distinctive characteristics are «unlimited freedom» (the user himself determines the level of access) and «rules» (anyone can set and change the rules).
- Limited public. Basic features can also be «positive» for private platforms. Distinctive characteristics must be negative. For example, everyone can read, but only a limited number of participants can offer transactions to the ledger. The «read» characteristic will be positive, the «offer» - negative. Therefore, «unlimited freedom» will be negative. That is, the rights to "read" and "offer" are provided only centrally. Full (unlimited) freedom is already absent.
- Private – This is a registry for which the basic and distinctive characteristics of a public registry are inverse.
- Extended private. The platform may provide for certain freedom for a limited number of persons. Such platform can not be public, but it cannot be called absolutely private either. We call these platforms «extended private». Expansion here may occur «in any direction». For example, «read everything».

b) Identification. We define three possible levels: KYC/ALM (Know Your Customer/Anti-Money Laundering), pseudo-anonymous (Bitcoin, Ethereum), anonymous (Monero, Zcash).

c) Tokenisation. Two features are considered:

- Rewarding System: Lump-sum Reward (persons involved in storage or verification are rewarded by the system.), no reward (for private blockchains), additional payments (Block + Security Reward). This set illustrates the mechanism that is set automatically to reward participants.

- Fee System is payments made on the platform. The parties to the payment are participants of the platform. For example, a fee for a network request, a request for storage, data retrieval, computational operations, etc. In this regard, we can distinguish 2 subsets: the fee reward and the fee structure.

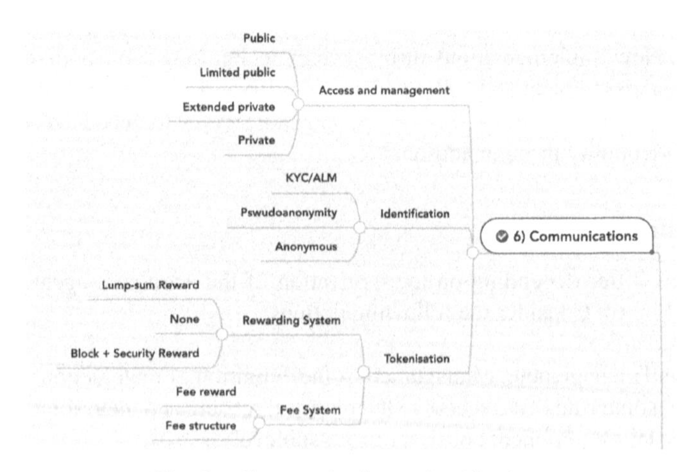

Fig. 6. «Communications» classification.

5 Conclusion

The developed classification is a hierarchical structure consisting of six main components. Each component includes characteristics and sets. The proposed approach includes both technological and economic (organizational) characteristics. Advantages of taxonomies are flexibility (the ability to add new branches) and unambiguity (exact characteristics for the proposed elements). At the same time, taxonomy takes into account the strengths of existing classifications, detailing them.

Classification criteria can be considered as a set of alternative rules for the operation of large-scale systems. During the development, possible alternatives will be selected depending on the requirements for the system. Such approaches allow to standardize the process of forming the blockchain platform architecture. Moreover, it is possible to create various scenarios of the network changes. For example, local gossiping and asynchronous communications will cause a transition from a public network to a private network. Also for the controlled ledger the «platform reward» characteristic will not be fundamental for improving the network health.

References

1. Nakamoto, S.: Bitcoin: a peer-to-peer electronic cash system (2008)
2. Zhdanov, I., Fedorov, A., Balashov, A.: Introduction to blockchain 2.0. Sci. J. **10**(23) (2017)
3. Peskova, O., Polovko, I., Zakharchenko, A.: Use of blockchain technologies in electronic document management systems: analysis and software implementation. Inzhenernyj vestnik Dona **5**, 3–6 (2019)
4. Zavarzin, A.: The prospects of the blockchain technology in the context of the growth of the welfare of society. Bull. North Caucasus Federal Univ. **3**(66), 76–84 (2018)
5. Benos, E., Garratt, R., Gurrola-Perez, P.: The economics of distributed ledger technology for securities settlement. Staff Working Paper Staff Working Paper. Bank of England (2017)
6. Puthal, D., et al.: Everything you wanted to know about the blockchain: its promise, components, processes, and problems. IEEE Consum. Electron. Mag. **4**, 6–14 (2018)
7. Varnavskiy, A., Buryakova, A., Sebechenko, E.: Blockchain for the government, KNORUS, Moscow (2020)
8. Tasca, P., Tessone, C.J.: Taxonomy of blockchain technologies. Principles of identification and classification, arXiv preprint arXiv:1708.04872 (2017)
9. Ballandies, M.C., Dapp, M.M., Pournaras, E.: Decrypting distributed ledger design-taxonomy, classification and blockchain community evaluation. arXiv preprint arXiv:1811.03419 (2018)
10. Xu, X., et al.: A taxonomy of blockchain-based systems for architecture design. In: IEEE International Conference on Software Architecture (ICSA), Sweden, pp. 243–252 (2017)
11. Rauchs, M., et al.: Distributed ledger technology systems: a conceptual framework. SSRN 3230013 (2018)
12. Morrison, D.R.: PATRICIA - practical algorithm to retrieve information coded in alphanumeric. J. ACM (JACM) **4**, 514–534 (1968)
13. Morabito, V.: Business Innovation Through Blockchain, 188 p. Springer, Heidelberg (2017)
14. Sompolinsky, Y., Zohar, A.: PHANTOM: A Scalable BlockDAG Protocol. IACR Cryptology ePrint Archive (2018)

Brain Tumor Medical Diagnosis. How to Assess the Quality of Projection Model?

Pawel Siarka(✉)

Wroclaw University of Economics and Business, Wrocław, Poland
pawel.siarka@ue.wroc.pl

Abstract. One of the critical problems of decision models used for medical purposes is to assess its projection quality. The paper refers to the most popular approaches (e.g., SKALE, GSS, CRGS) used to estimate postoperative outcomes in neurosurgery i.e., before the surgery begins. The methods were examined to select the best one in terms of model performance i.e., predictive power.

The analysis revealed that the PS SCORE approach gave the most reliable forecasts. For the purpose of model performance assessment, the authors used AUROC statistics. It was shown that AUROC could be interpreted as the probability of correct classification. As the clinical data are usually scarce, the authors decided to enrich the model performance analysis by incorporating the confidence interval approach. This gave additional information regarding model stability.

Keywords: Brain tumor · Model validation · Artificial intelligence

1 Introduction

One of the founders of modern computers and AI was Alan Turing (1950). His "Turing test" addressed the problem of achieving human-level performance in cognition related tasks (Mintz and Brodie 2019). During the last two decades, artificial intelligence methods became popular and widely implemented in the health care industry. There are multiple solutions regarding artificial neural networks, Bayesian networks, or fuzzy expert systems. Most of the methods aim to identify clinical correlations between symptoms and correct diagnosis. This problem was originally addressed with traditional statistical models, however, AI methods are being implemented due to large databases with historical observations.

A large amount of data required for AI solutions is a substantial limitation, and therefore in many cases, only traditional statistical methods can be leveraged. The challenge for model developers is to validate the models in terms of their performance. Various assumptions made at the stage of model development make the comparative analysis complex.

One of the most common problems in medicine comes from patient classification into one of two possible groups. This is the classical forecasting problem of a dichotomous variable based on multiple characteristics. In this case, the traditional AUROC analysis can be leveraged, which allows us to assess the probability of correct classification.

M. Hernes et al. (Eds.): ICCCI 2020, CCIS 1287, pp. 402–410, 2020.
https://doi.org/10.1007/978-3-030-63119-2_33

Furthermore, simulations approach e.g., bootstrapping, can also be used for assessment of the model stability i.e., AUROC volatility analysis.

Unlike the traditional dichotomous regression methods, a deep learning or pattern recognition solutions involve teaching computers with repetitive algorithms. The model learns to classify the observations based on specific clinical images and known results.

In the literature, there are multiple implementations of AI methods in medicine. There are studies presenting online scheduling of appointments or online check-ins in medical centers, digitization of medical records, reminder calls, adverse effect warnings while prescribing multidrug combinations. Also, radiology was approached by new technology (Mayo and Leung 2018). Computer analysis of clinical imaging is now a vital component of the diagnosis process. So-called Computer Assisted Diagnosis is widely used for screening mammography analysis or brain tumor assessment. AI may provide substantial help (Mayo and Leung 2018) in radiology or magnetic resonance images analysis, especially in hospitals with scarce highly qualified human resources.

An example of an intelligent system is Germwatcher, which was developed by the University of Washington to detect hospital-acquired infections (Kahn et al. 1993). Another system is Babylon, which helps patients to consult the doctor online in the UK. Its functionality covers the examination of reported symptoms and health monitoring. Another successful implementation of AI methods is Da Vinci robotic surgical system. It has substantially changed the field of surgery. The robotic arms mimic a surgeon's hand movements to perform. It provides better precision by eliminating hand tremors and gives 3D images and necessary magnifications (Hamlet and Tremblay 2017). Technically advance Boston children's hospital is working since 2018 on an AI system that provides help to parents for their ill children. This solution is based on a web interface where parents can report symptoms and get immediate advice.

Another field where AI is intensively implemented is smartphones and smartwatches industry. Apple, Fitbit, or Xiaomi offer trackers that monitor heart rate, sleep levels, and ECG tracings. It can alert the user or his family and even call for an emergency. The National Institute of Health (NIH) presented AiCure App, which helps people regularly take medications (Labovitz et al. 2017).

IBM's Watson Health is developing similar systems to identify symptoms of heart disease. Stanford University is also working on an AI system called PAC. PAC is a solution addressed to older people living alone (Bianconi et al. 2019). Furthermore, several projects are being under development regarding conversational healthcare analysis. Siri and Google Assitant are testing such solutions.

Most of the econometric models are developed based on stringent assumptions impacting the final distribution function of a model quality indicator. For this reason, it is hard to compare outcomes with results achieved using AI methods. As the distributions for quality measures are not known, it is difficult to estimate its confidence intervals and asses model stability. The paper addresses this problem by applying a bootstrap simulation analysis under the model validation procedure. The purpose of this paper is to present an innovative approach that can be leveraged for AI and traditional statistical models.

The method is presented for models developed using clinical data regarding patients with brain tumors. The authors examined popular medical systems like SKALE, GSS, CRGS, PS Score, and assessed their ability to form an accurate diagnosis.

This paper starts with an introduction where the problem and literature are presented. The next chapter includes a description of a statistical approach that can be used for the assessment of model quality. The last part of this paper contains a summary and principal conclusions.

2 Modeling Postoperative Outcomes

Meningiomas are the most common primary brain tumors. They represent almost 25% of all primary intracranial tumors (Wiemels et al. 2010). The incidence within the general population is 2.3:100 000. However, when the data from autopsies are included, it rises to 5.5:100 000 (Cohen-Inbar et al. 2010). This ratio shows that half of the tumors remain clinically silent. They are usually benign and curable after surgical resection. However, some complications within elderly patients are quite common. As the population is aging, the morbidity is expected to rise as well. It was observed that within the population between 60–70 years old, the incidence is 8.5:100 000. The statistical report prepared by the Central Brain Tumor Registry of the United States says that the incidence rate increases progressively with age and increases significantly after age 65. So, it should be expected that a longer life span and better health care will result in a much higher incidence rate in the nearest future.

When meningioma is diagnosed, the neurosurgeon has to face the dilemma of whether to operate it. It is necessary to assess the potential risk associated with the surgery resulting in deterioration of the patient's health or even decease. The results of the surgery are commonly presented based on the GOS (Glasgow Outcome Scale). This scale allows the objective assessment of the patient recovery by dividing them into five classes. The first class (1) means death without recovery of consciousness. Class number two refers to a persistent vegetative state, a prolonged state of unresponsiveness and lack of higher mental functions. Number three includes severe disability resulting in a permanent need for help with daily living. Another fourth class incorporates patients with moderate disability. These people do not need assistance in everyday life. Some of them can be even professionally active but may require some special equipment. The last fifth group relates to patients with low disability. They are characterized by light damages with minor neurological deficits.

For the first time, the quantitative model for brain tumors assessment was developed by Arienta et al. (1990). The authors created the CRGS (Clinical-Radiological Grading System) system for elderly patients with intracranial meningiomas. The model incorporated six criteria: tumor location, size, peritumoral edema presence, the neurological status of the patient, concomitant diseases, and preoperative Karnofsky Performance Scale (KPS) score.

The next attempt was made by Sacko et al. (2007). They leveraged the following characteristics: sex, Kanofsky Scale, American Society of Anesthesiology (ASA) score, tumor location, edema. Their model is known as the SKALE system. In 2010 the GSS model (Geriatric Scoring System) was presented by (Cohen-Inbar et al. 2010). The purpose of the model was to identify the elderly subpopulation, which is likely to benefit from surgical intervention. It was developed based on tumor size, location, peritumoral edema, neurological deficits, Karnofsky score, and associated diabetes, hypertension, or lung disease.

Any projection model requires validation to assess the goodness of the classification. This problem was addressed by Anderson (2007) and Řezáč (2009). Engelman (2006) made a substantial contribution by leveraging AUROC statistic, implementing statistical tests and confidence intervals. Sobehart and Keenan (2001) also referred to that problem. Siarka (2012) presented his research, where he leveraged a simulation approach by using bootstrap analysis in order to assess the volatility of AUROC measure. He also measured the impact of the size of a sample on the model stability. This issue is particularly important due to the problem of scarce data in medical applications.

3 The Assessment of the Model Quality

One of the most popular methods for decision-making models validation is AUROC analysis (Areas Under ROC Curve). The approach can be used for models where each observation is classified into one of two populations. The ROC curve is created by pairs of Hit Rate and False Alarm Rate calculated for given observation outcome e.g., score. Hit Rate shows the percentage of correctly classified observations from the first population. At the same time, False Alarm Rate shows the fraction of observations selected from other population which were misclassified. The surface area under the ROC curve ranges from zero to one, where one indicates a perfect model. The surface area equal to 0.5 means that the model outcomes are equivalent to a coin toss experiment.

AUROC statistic also has other interpretation. It may be considered on the ground of probability theory as:

$$AUROC = P\left(\hat{S}_1 < \hat{S}_2\right) + \frac{1}{2}P\left(\hat{S}_1 = \hat{S}_2\right)$$

where $\hat{S}_1$ and $\hat{S}_2$ are random variables taking score values for observations derived from the first and the second populations, respectively. This measure corresponds to Mann-Whitney statistics (U) which can be calculated as follows:

$$U = \frac{1}{N_1 N_2} \sum_{1,2} u_{1,2}$$

where N_1 and N_2 are the numbers of observations in sample one and two. The above equation summing up is conducted for all possible pairs of observations derived from first and second populations. Statistics $u_{1,2}$ is calculated based on the following formula:

$$u_{1,2} = \begin{cases} 1 \; when \; \hat{s}_1 < \hat{s}_2 \\ \frac{1}{2} \; when \; \hat{s}_1 = \hat{s}_2 \\ 0 \; when \; \hat{s}_1 > \hat{s}_2 \end{cases}$$

where $\hat{s}_1$ is the empirical realization of variable $\hat{S}_1$, and $\hat{s}_2$ is empirical realization of variable $\hat{S}_2$. Mann Whitney statistics (U) can be interpreted as the probability of correct classification, providing that there are two observations (each of them is derived from a distinct population), and the classification is made based on score value derived from the examined model.

As U statistics is calculated based on the given sample, it can take various values for the same model i.e., depending on the sample. So, its volatility may provide significant information regarding the stability of the projection model. If, for example, the U statistics equals to 0.7, but its standard deviation is 0.3, then the model cannot be deemed as reliable due to its large span of the quality measure. The too-wide confidence interval shows that the real value of AUROC is expected to be somewhere between 0.4 and 1. This is why it is important to estimate the distribution function of U statistics and calculate the confidence interval.

The variance of U statistics can be calculated based on the following formula:

$$\hat{\sigma}_U^2 = \frac{1}{4(N_1-1)(N_2-1)}[\hat{P}_{1\neq 2} + (N_1-1)\hat{P}_{1,1,2} + (N_2-1)\hat{P}_{2,2,1} \\ -4(N_1+N_2-1)(U-0.5)^2]$$

where $\hat{P}_{1\neq 2}, \hat{P}_{1,1,2}, \hat{P}_{2,2,1}$ are the estimators of following probabilities:

$$\hat{P}_{1\neq 2} = P(S_1 \neq S_2),$$

$$\hat{P}_{1,1,2} = P(S_{1,1}, S_{1,2} < S_2) - P(S_{1,1} < S_2 < S_{1,2}) - P(S_{1,2} < S_2 < S_{1,1}) + P(S_2 < S_{1,1}, S_{1,2}),$$

$$\hat{P}_{2,2,1} = P(S_{2,1}, S_{2,2} < S_1) - P(S_{2,1} < S_1 < S_{2,2}) - P(S_{2,2} < S_1 < S_{2,1}) + P(S_1 < S_{2,1}, S_{2,2}),$$

where $S_{1,1}$ and $S_{1,2}$ are two independent observations of a random variable with a distribution function of scoring points for the first population. Respectively, $S_{2,1}$ and $S_{2,2}$ refer to two independent observations from the second population.

The knowledge regarding the expected value and standard deviation of U statistics is sufficient to estimate its distribution function due to the following convergence:

$$\frac{U - E(U)}{\hat{\sigma}_U} \xrightarrow{N_1, N_{2n\to\infty}} N(0,1)$$

Hence, the confidence interval can be calculated for given significance level α using the following formula:

$$\left[U - \hat{\sigma}_U \Phi^{-1}\left(1 - \frac{\alpha}{2}\right), U + \hat{\sigma}_U \Phi^{-1}\left(1 - \frac{\alpha}{2}\right)\right]$$

The confidence interval provides essential information concerning the range of model performance measure. The narrower is the interval, the more certainty regarding the model quality we have.

The knowledge regarding confidence intervals is relevant, although it is not sufficient to identify the best model. The confidence intervals may overlap each other making the final decision difficult. Thus, it is convenient to test the statistical hypothesis concerning the difference between U statistics calculated for two distinct models. The null hypothesis says that both models are equally good. The alternative hypothesis states that the difference between the given U statistics is statistically significant. The test statistics (T) has χ^2 distribution function with one degree of freedom:

$$T = \frac{(U_1 - U_2)}{\hat{\sigma}^2_{U_1} + \hat{\sigma}^2_{U_2} - 2\hat{\sigma}_{U_1,U_2}}$$

where $\hat{\sigma}_{U_1,U_2}$ is a covariance value calculated for U_1 and U_2 statistics. It can be derived based on the following formula:

$$\hat{\sigma}_{U_1,U_2} = \frac{1}{4(N_1 - 1)(N_2 - 1)}[\hat{P}^{1,2}_{1,1,2,2} + (N_1 - 1)\hat{P}^{1,2}_{1,1,2} + (N_2 - 1)\hat{P}^{1,2}_{2,2,1} - 4(N_1 + N_2 - 1)(U_1 - 0,5)(U_2 - 0,5)]$$

where $\hat{P}^{1,2}_{1,1,2,2}$, $\hat{P}^{1,2}_{1,1,2}$ and $\hat{P}^{1,2}_{2,2,1}$ are calculated as follows:

$$\hat{P}^{1,2}_{1,1,2,2} = P\left(S^1_1 > S^1_2, S^2_1 > S^2_2\right) + P\left(S^1_1 > S^1_2, S^2_1 < S^2_2\right) - P\left(S^1_1 < S^1_2, S^2_1 < S^2_2\right) - P\left(S^1_1 < S^1_2, S^2_1 > S^2_2\right)$$

$$\hat{P}^{1,2}_{1,1,2} = P\left(S^1_{1,1} > S^1_2, S^2_{1,2} > S^2_2\right) + P\left(S^1_{1,1} > S^1_2, S^2_{1,2} < S^2_2\right) - P\left(S^1_{1,1} < S^1_2, S^2_{1,2} > S^2_2\right) - P\left(S^1_{1,1} < S^1_2, S^2_{1,2} < S^2_2\right)$$

$$\hat{P}^{1,2}_{2,2,1} = P\left(S^1_1 > S^1_{2,1}, S^2_1 > S^2_{2,2}\right) + P\left(S^1_1 > S^1_{2,1}, S^2_1 < S^2_{2,2}\right) - P\left(S^1_1 < S^1_{2,1}, S^2_1 > S^2_{2,2}\right) - P\left(S^1_1 < S^1_{2,1}, S^2_1 < S^2_{2,2}\right)$$

4 Empirical Results

The study was conducted based on clinical data regarding operated intracranial meningioma in 2009–2014. All patients were over 65 years old and didn't have undergone resection before. The average age in a sample was equal to 71.02. There were selected two classes based on the early surgical outcome at discharge (GOS). First of them incorporated deaths cases and vegetative states observed after the operation (1 and 2 in the GOS scale). The second class included all other outcomes (3–5 in the GOS scale). This way, two distinct populations were identified, determining the dichotomous variable. Each available neurosurgery model (CRGS, GSS, SKALE, PSscore) was assessed in terms of projection quality.

The AUROC analysis was performed to assess model quality. As was emphasized earlier, this measure shows the probability of correct classification i.e., for two randomly chosen patients where one of them gets 1 or 2 points in the GOS scale, and the second one has 3, 4, or 5 points in the GOS scale. Table 1 presents the results of the analysis for selected neurosurgery systems:

Table 1. AUROC results for neurosurgery systems.

Method	AUROC	Confidence interval $\alpha = 0{,}1$	
PS Score	73,1%	61,1%	84,6%
SKALE	65,3%	52,8%	76,6%
CRGS	64,5%	52,8%	76,6%
GSS	66,0%	53,7%	78,5%

Source: author's work.

The outcomes presented in Table 1 show that the highest performance was observed for PS Score. The AUROC value was equal to 73.1%, which is higher than the second-best result achieved for the GSS method (66%). It can be noticed that GSS CRGS and SKALE gave very similar results ranging from 64.5% to 66%. Hence, the PS Score system gave significantly better results than other methods.

For each model, there were calculated confidence intervals of AUROC statistics. As there were various models developed based on multiple assumptions examined, the authors decided to leverage boot bootstrap analysis. This approach allows to estimate the actual distribution function of AUROC instead of relying on questionable normal distribution assumption. For each method, one thousand samples were drawn from a given population, and AUROC statistics were calculated. These results allowed to estimate actual distribution functions. The graphical presentation of confidence intervals is shown on Fig. 1.

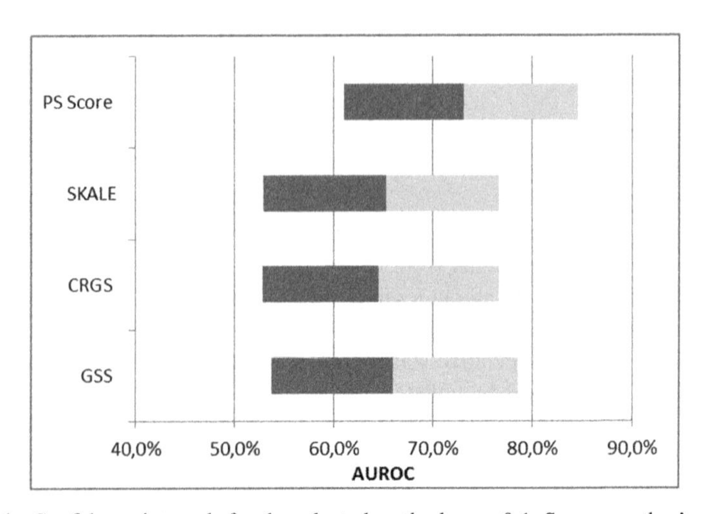

Fig. 1. Confidence intervals for the selected methods $\alpha = 0.1$. Source: author's work.

Figure 2 presents the distribution functions of AUROCs calculated for selected neurosurgery systems. SKALE system and CRGS gave similar outcomes, while GSS appeared to be slightly better. AUROC distribution function of PS Score is shifted to the left, which confirms its better performance.

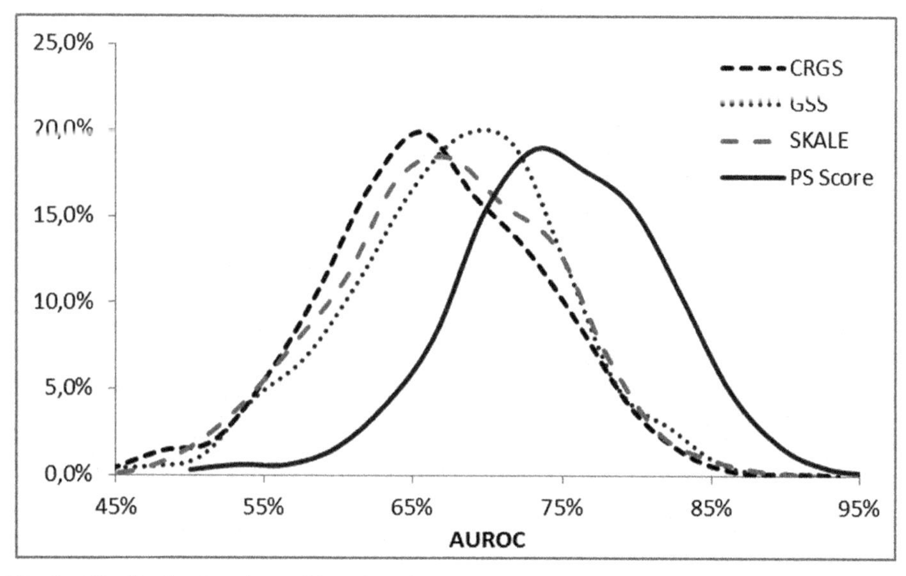

Fig. 2. Distribution functions of AUROCs for the selected neurosurgery systems. Source: author's work.

5 Summary

The study presents several approaches that are being leveraged in modeling meningioma postoperative outcomes. The theoretical part of this paper includes a description of AUROC statistics, which allows to assess the quality of the models. According to this method, the PS Score system appeared to be most accurate.

A variety of projection methods do not solve the critical problem, which is a selection of the best one with the best performance. The scarce data is a fundamental issue that needs to be addressed by model developers. Furthermore, the availability of various characteristics is also limited. Some studies are showing that tumor chromosomal deletions are correlated with the tumor size. Therefore, it is likely that some additional variables fed with DNA test outcomes would improve model accuracy. Therefore it is necessary to collect data by integrating existing data sources and expand the scope of potential characteristics.

The implementation of decision models for medical purposes requires further research of projection quality assessment techniques. Only sound and reliable methods should be leveraged in such an important area as health care. AI methods give new opportunities but also require a detailed examination to prove their forecasting ability.

References

Anderson, R.: The Credit Scoring Toolkit: Theory and Practice for Retail Credit Risk Management and Decision Automation. Oxford University Press, New York (2007)

Arienta, C., Caroli, M., Crotti, F., Villani, R.: Treatment of intracranial menigiomas in patiens over 70 years old. Acta Neurochir (Wien) **107**(1–2), 47–55 (1990)

Bianconi, G.M., et al.: Vision-based prediction of ICU mobility care activities using recurrent neural networks. In: Machine Learning for Health Workshop, Neural Information Processing Systems (NIPS), December 2017 (2019). https://med.stanford.edu/cerc/research/new-pac.html. Accessed 27 Mar 2019

Cohen-Inbar, O., Soustiel, J.F., Zaaroor, M.: Menigiomas in the elderly, the surgical benefit and a new scoring system. Acta Neurochir. **152**, 87–97 (2010)

Engelman, B.: The Basel II Risk Parameters: Estimation, Validation, and Stress Testing. Springer, Heidelberg (2006). https://doi.org/10.1007/978-3-642-16114-8

Hamlet, P., Tremblay, J.: Artificial intelligence in medicine. Metabolism **69S**, S36–S40 (2017)

Kahn, M.G., Steib, S.A., Fraser, V.J., Dunagan, W.C.: An expert system for culture-based infection control surveillance. In: Proceedings of the Annual Symposium on Computer Application in Medical Care, p. 171-5, PMID: 8130456 (1993)

Labovitz, D.L., Shafner, L., Reyes Gil, M., Virmani, D., Hanina, A.: Using artificial intelligence to reduce the risk of nonadherence in patients on anticoagulation therapy. Stroke **48**, 1416–1419 (2017)

Mayo, R.C., Leung, J.: Artificial intelligence and deep learning-Radology's next frontier? Clin. Imaging **49**, 87–88 (2018)

Mintz, Y., Brodie, R.: Introduction to artificial intelligence inmedicine. Minim. Invasive Ther. Allied Technol. **28**, 73–81 (2019)

Řezáč, M., Řezáč, F.: Measuring the Quality of Credit Scoring Models, Credit Research Conferences 2009, The University of Edinburgh (2009)

Sacko, O., Sesay, M., Rous, F.-E., et al.: Intracranial meningiomas surgery in the ninth decade of life. Neruosurgery **61**(5), 950–954 (2007)

Siarka, P.: Quality measures of scoring models. J. Risk Manag. Finan. Inst. Henry Steward **5**(4), 432–446 (2012)

Sobehart, J., Keenan, S.: Measuring default accurately. Risk, pp. S31–S33 (2001)

Wiemels, J., Wrensch, M., Claus, E.B.: Epidemilogy and etiology of meningioma. J. Neurooncol. **99**, 307–314 (2010)

Meta-learning Process Analytics for Adaptive Tutoring Systems

Gracja Niesler(✉) and Andrzej Niesler

Wroclaw University of Economics and Business, ul. Komandorska 118/120, 53-345 Wrocław, Poland
{Gracja.Niesler,Andrzej.Niesler}@ue.wroc.pl

Abstract. The effectiveness of the learning process depends to a large extent on the provision of an adequate instruction program. Individual approach to learners' needs posits the necessity to comply with the characteristics of their preferences and predispositions for learning, which ought to be considered in tutoring content delivery. Adapting the instruction material to a particular learner requires intelligent tutoring system (ITS). In turn, the adaptation mechanism itself is based on comprehensive data analytics derived from the learning environment and the learner's behaviour as well. The paper presents a new approach in which intelligent tutor can deliver a support on meta level of learning process through analytics and adaptability. Most commonly, the analytics and adaptation in ITS are targeted mainly at the instruction program, but since the individual approach assumes that each learner should be treated individually, we must consider replacing active role of a human teacher, by the functionality of the intelligent tutor. Therefore, the role of supervising the learning process depends on the user themselves. It means that the meta-learning process has to be the subject of analysis and adjustment. Users usually do not have enough knowledge about how to learn, and for that reason the task of monitoring learner's performance and adjusting all activities should be ensured by the system.

Keywords: Meta-learning · Learning process analytics · Intelligent tutoring systems (ITS) · Adaptability · Adaptive systems

1 Introduction

Musashi Miyamoto, a great Japanese swordsman and philosopher, wrote '*each man practices as he feels inclined*'. It means that developing perfection requires following one's own, unique way of acting and choosing the means to succeed. The need for an individual approach to learning has become a broad research topic, which inevitably requires the application of intelligent technologies. Over the past few decades, the educational environment has developed into a computer-aided teaching and learning area. Numerous studies have been presented since the early 1980s and a plethora of problems have not yet been solved. The identified research issues incline evolutionary approach according to the expanding knowledge about human's psychological and cognitive processes, technological development, and general social awareness and maturity.

M. Hernes et al. (Eds.): ICCCI 2020, CCIS 1287, pp. 411–423, 2020.
https://doi.org/10.1007/978-3-030-63119-2_34

The aim of this paper is to propose an extended approach to adaptability in intelligent tutoring systems, based on meta-learning analytics. The observations and conclusions derive from many years of research and examination of students' performance according to their innate learning abilities. The educational data analytics can bring benefits in terms of providing help in the selection of teaching methods, creation of conceptual tutoring materials, defining the way courses are conducted, identification of students' needs, as well as monitoring of learning outcomes and the adaptation of relevant tools and methods accordingly. Essentially, the approach contributes a theoretical foundation for gaining a better understanding of the individual nature of the learning phenomenon.

The paper is structured as follows. Section 2 addresses the issues of the general scope of learning analytics, educational data mining, and the meta-learning approach. Section 3 presents the basis of tutoring systems and discusses conditions for adaptability. The main perspectives of regarding adaptation have been analysed. In Sect. 4 the significance of meta-learning process have been introduced, featuring three perspectives: educational path, instruction program, and mental model. Specific proficiency indicators for meta-learning analytics have been proposed as a tool for evaluation and adaptation. Section 5 refers to the typical structure of a tutoring course and applies the adaptation layers of meta-learning elements in the analytical cycle. The concept of phenomena relationship is demonstrated in the form of a model. Because the issues discussed are considerably complex in nature, there is a lot of areas and directions worth considering for further research and solutions. Both the aspects have been discussed in Sects. 6 and 7. The paper closes with the conclusion and further research outlook.

2 Background and Related Work

There are many applications in the field of educational data analysis. Its purpose of use is mainly global or rather institutional. The feedback from the analysis process can be useful to authors or managers of didactic courses, materials, and curricula. Technologies transposed to educational field, such as data analytics and data mining, are called *Learning Analytics* and *Educational Data Mining*. In Romero and Ventura's review, there appear more of related terminology: *Academic Analytics*, *Institutional Analytics*, *Teaching Analytics*, *Data-Driven Decision Making in Education*, *Big Data in Education* and *Educational Data Science* [1]; all dependent on the analysis perspective.

Learning Analytics comprises visual data analysis techniques, social network analysis, semantic and educational data mining including prediction, clustering, relationship mining, discovery with models and separation of data for human judgment [2]. Generally, all the methods rest on examining the educational process to improve learning. Educational data mining tends to be more focused on automated methods for data discovery in the learning context, meanwhile the learning analytics revolves around human-led methods for data exploration [3]. While the analytical data are time-sensitive, the capturing should be conducted just as they occur. Hence, the technology should be applied in the form of a tutoring system, constantly monitoring students' performance based on all their activities, such as giving responses, using discussion board, accessing reading materials, completing assignments or other assessment activities, etc.

The learning process analytics can be perceived then as 'the measurement and collection of learner actions and learner productions, organized to provide feedback to learners, groups of learners and teachers during a teaching/learning situation. This information can be presented in various forms, e.g., a browsable analytics website or a dashboard and should engage participants in reflection with respect to their different goals, roles, tasks, productions, and so forth' [4].

The issue reveals differently if the main beneficiaries of analytics are the learners themselves. The use of general data cannot result automatically in greater performance, effectiveness, or affect achieving goals. The learner alone is not able to use broad, statistical data for their own use. Moreover, such data are implicit enough to exclude the necessity of analysis, especially as the range in case of a singular user is not significant.

Therefore, taking the learner's perspective requires both learning analytics and educational data mining in a process-driven approach, but on the meta-learning level. The point of reference is the learning process and data that can be extracted from it. In this context the useful and relevant information is tacit, even for the learner themselves, because it relates to meta learning methods. For this aim, we adapt Maudsley's definition of **meta-learning**, as 'the process by which learners become aware of and increasingly in control of habits of perception, inquiry, learning, and growth that they have internalized' [5] and introduce it into the area of intelligent tutoring systems. Because the learners differ by their ability to learn, we should consider adapting tutoring system functionality to the individuals and on cognitive and psychometric level, as a support for the limitations in their ability to learn and awareness of the self-learning process.

3 Adaptability in Intelligent Tutoring Systems

The technological advancements deliver a great support in the field of the teaching-learning process assistance, inasmuch as for effective tutoring each learner ought to be treated individually. That kind of personalization can be provided only by an intelligent tutor, which is able to adapt the teaching process to the learner's needs. There are many solutions in terms of electronic-aided learning tools. Romero and Ventura distinguish some of such solutions in the field of Computer Based Education (CBE) [1]:

- Adaptive and Intelligent Hypermedia System (AIHS) – based on the goals, preferences and individual knowledge and adaptation arising from the interaction analysis;
- Intelligent Tutoring System (ITS) – providing customized instruction and feedback by adapting the interactions accordingly to the domain, student, pedagogical models;
- Learning Management System (LMS) – delivers most common requisites for teaching and learning (group and individual), focused on recording all learners' activities and provides the tools for managing a didactic course;
- Massive Open Online Course (MOOC) – web-based system with the functionality of LMS, but for the large number of learners.

They also mention 'test and quiz systems' and other types, such as: '*wearable learning systems, learning object repositories, concept maps, social networks, WIKIs, forums, educational and serious games, virtual and augmented reality systems*' [1], which generally represent just technological or methodological aspects of the types mentioned above. The AIHS and ITS principally refer to the human-system interactions while the

LMS implies mainly the active role of a human tutor (commonly used for blended-learning). The MOOCs can comprise intelligent elements providing adaptability as well [6]. The most extensive is the term Intelligent Tutoring System, therefore we will treat every kind of educational systems with the significant role of intelligent solutions in ITS. The adaptability of the tutoring systems requires applying the intelligent factor, so every adaptable tutoring system will be treated as ITS as well.

Adaptability of tutoring systems is the ability to react to the learner's behaviour and adjust their actions accordingly to the user's expectations, in particular to the individual needs, especially the tacit ones. We can distinguish two types of adaptation: active and passive. The active one is constantly monitoring both, the learner's activities, and the changes in learning environment as well, and is validating and redefining patterns and predicting user's further activities. The passive adaptation is rather being executed on demand (of a user or prerequisites and pre-set conditions). The ITS with both passive and active adaptation fits the term of Self-Improvable Adaptive Instructional System (SIAIS). The SIAIS generally consist of the following components: human learners (constantly self-improving), self-improvable learning resources (human and digital, capable of improving), learning environments and learning processes (as instructional sequences for specific domain for particular learners group) [7].

In this paper, the approach to ITS analytical functionality is beyond the instruction adaptation. From the learning analytics perspective, we can distinguish three process elements of adaptation in tutoring systems, in terms of a source of educational data: instruction adaptation, interaction adaptation and cognitive adaptation (see Fig. 1). The instruction is the process initiated and controlled by the ITS.

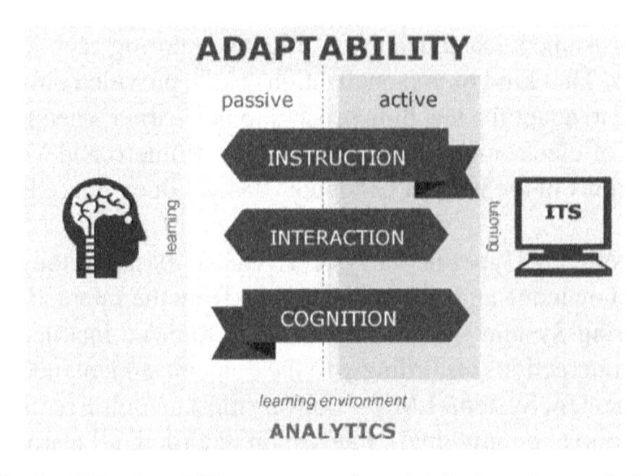

Fig. 1. ITS adaptability conceptual model for learning analytics

The process of instruction adaptation includes typically the following steps: diagnosing or assessing learner's knowledge state of all or part of the target area of learning; recommending actions to the learner; interacting and teaching the learner to understand what they have not learned to improve understanding and learning; providing novel practice opportunities with assessments and scaffolding to re-diagnose the learner's knowledge state post-instructional interaction. The **instruction adaptation** can be delivered on *macro-level* (outer loop, on the tasks level) and *micro-level adaptivity* (inner loop, on activities' level within task) [8]. The **cognition adaptation** process starts from the learner's perspective and requires constructing a **mental model**, which is an axiomatic structure, created consciously or subconsciously in the user's mind. This is particularly important when modelling the interaction layer for tutoring process, as it determines the level of understanding of information, tasks, or commands by the learner.

The foundation of creating a mental model is perception, as experiencing the phenomena of objective reality, followed by the proper process of creating the model, namely the cognitive process (in the sense of understanding), allowing mental representation for objects of reality. Hence, it follows that the mental model is expressed in natural language, which makes it difficult to transfer the concept to systemic ground.

There is also an additional aspect related to the mental model. A human-made image of reality can objectively be a false representation. This poses further challenges by the manner and form of representation and presentation of information and knowledge to ensure the perceptual uniqueness of perceived objects. For that purpose, we can use the instruments for creating adaptive tutoring content [9], based on learning determinants as memory, understanding and content associations, relating respectively to tutoring outcomes, such as: knowledge, skills, and competences [10]. The natural learning predispositions meet the following adaptivity instruments: frequency and 'time unit' – for the memory determinant; analytics (deduction) and synthetics (induction) representation – for understanding determinant; perceptiveness, language (keywords and semantic relations), logic (causal-result relationship) and numbers (formula representation) – for content association determinant. The combination of adaptation instruments should be delivered according to interaction history analysis and results of instruction process.

The **interaction** is being initiated both ways and the **adaptation** is based both on information about instruction and cognition process. For interaction adaptability we use the differences derived from the individual personality types. The type categorization brings inaccuracies with personality identification, so we have treated each characteristic individually, which resulted in acquiring 16,384 types instead of 16, according to Myers-Briggs Type Indicator [11]. The adaptability of interaction requires conducting a personalization process based on the preference-oriented tutoring delivery. Individual preferences depend on orientation (towards outer world) and processing (acquired knowledge) mental functions [10]. According to those, we can distinguish the following personalization processes: tutoring organization, content presentation, and content acquiring. In this article we focus mainly on interaction adaptation, which provides the meta-level of the learning process, including instruction and cognition process results.

4 Elements of the Meta-learning Process

The tutoring outcomes indirectly depend on learning style. The analysis of current research indicate what elements of the ITS are being adapted according to the learning style adaptation, which is: learning material/learner characteristics, learning media, contents and resource format, learner knowledge, recommended tutoring materials, intelligent game, learner evaluation and practice [12]. We can say that the learning style is an expression of meta-learning in execution. The adaptability of the style by the ITS substitutes the meta-learning process on the learner's side. *Ipso facto*, we reduce the user's cognitive overload, who can focus on performing the task and learning.

We can distinguish the three perspectives of learning process and that gives a broad spectrum of analytical meta-learning data for further adaptation: **educational path**, expressed in educational levels, according to the ISCE standard elaborated by UNESCO Institute for Statistics [13]; **tutoring course/instruction program**, as the set of activities, conducted in a computer-aided learning; **mental model**, which bases on Cartesian reality division and epistemological states of human's mind, as perceptive and reflective modes, allowing identification and encoding the information for the further processing and usage [10]. Levels on the educational path differ with how the instruction is organised (as the characteristics of teaching process and applied assessment methods) according to related learning objectives expressed as the set of outcomes (knowledge, skills, and competences) derived from the European Qualification Framework [14].

According to these levels, we can gradate the learning outcomes for evaluation of tutoring content delivery and adapting learning style to the learner's needs. In addition to learning preferences and predispositions analysis, the evaluation of learner's **proficiencies** is required, expressed by knowledge, skills, and competences **indicators**.

Table 1 presents the indicators based on the proficiency level. They can be used to precisely describe the level of expected learning outcomes, and treat it as a foundation for students' assessment. It brings more flexibility to the educational courses creation and conducting, and to design content for didactic units. Delivery of expected outcomes allows to examine not only the level of learning outcomes, expressed in grades, but also to identify what exact kind of knowledge, skill, or competence needs improvement.

We identified seven kinds of knowledge indicators: type, genre, complexity, abstraction level, range, scope, and required way of thinking. For example, for some didactic content the expected outcome can be knowledge which is theoretical, general, specialised and limited to one field, and, therefore, needs only imitative way of thinking and processing. The skills are described by the following indicators: attitude, advancement, difficulty level, intricacy, recurrence, problem solving, and application, and the competences by: adaptation, management, context level, self-reliance, involvement, decision making, and responsibility. To describe one course or specific didactic unit we need to combine all three objectives' specifications. The indicators sets are different according to the educational level. It is possible that the lack of proficiency on one level restrains from acquiring the knowledge and skills which are required on the specific level in the learning process. For example, if the learner is not able to fulfil autonomic tasks or absorb the knowledge using critical understanding, the system-assisted meta-learning helps to diagnose the problem and adjust the next steps in instruction process.

To achieve the meta-learning scope on learning process, we should apply **mental model** to the course perspective, where the learner is using innate preferences and predispositions to acquire knowledge, skills, and competences. The most common is the Kirkpatrick's training evaluation model [15], that consists of four levels: reaction, learning, behaviour, and results.

For meta-learning analytics, we will treat *reaction* – as the learner's attitude towards delivered instruction content; *learning* – as evaluation of meeting the objectives level; *behaviour* – as reactions and results to tutoring delivery changes; *results* – verification

Table 1. Proficiency level indicators for meta-learning objectives

Knowledge		Skills		Competences	
Indicator	Value	Indicator	Value	Indicator	Value
Type	- factual - theoretical	Context level	- conditions - situation - circumstances	Adaptation	- scheduled - flexible
Genre	- academic - professional	Advancement	- basic - advanced - expert	Management	- individual - group - leading
Complexity	- homogeneity - heterogeneity	Difficulty level	- elementary - professional - master	Attitude	- cognitive - emotional - behavioural
Abstraction	- general - detailed	Intricacy	- simple - complex	Self-reliance	- confident - insecure
Range	- specialized - comprehensive	Recurrence	- unrepeatable - recurrent - extrapolatable	Involvement	- proactive - receptive
Scope	- field - domain - interdisciplinary	Problem solving	- deductive - inductive - abductive	Decision making	- logic-based - principle-based
Thinking	- imitative - constructive - critical - originative	Application	- projection - creativity	Responsibility	- duty - authority - support

of adopted learning style. The mental perspective focuses on cognitive and functional aspects of the meta-learning process, in which we can propound the following modes:

- Empirical – based on individual perception according to the content delivery and presentation. Corresponds with reaction evaluation;
- Reflective – a result of experiencing the interaction with learning content, the actual learning process;
- Operational – the external action according to the processing results. In the context of evaluation corresponds with the behaviour to the results obtained.

In order to apply meta-learning aspects to analytics we must relate to the tutoring constructs in ITS. The analytic process retrieves the data from the instruction activities and interactions, and builds a foundation for adaptability on the meta-learning level.

5 Analytic Cycle for Meta-learning Adaptation

Structuration of analytic activities in educational environment into macro-level process of collecting and analysing the data is called **learning analytics process**. In a period of one semester it generates thousands of transactions per student [16]. According to

Campbell and Oblinger, the process involves gathering and organizing information in varied forms and from different sources, then analysing and manipulating data, functionally exceeding the traditional reporting by providing decision-support capabilities [17]. It should rather be a cycle, where collecting data from the learners and processing them as suitable metrics transpose into the intervention means that affect students' behaviour, and collecting the additional data continues until the next round [18]. In case of the meta-learning process, we operate on the micro-level and the collected data concern all the activities and interactions initiated by a user. We can identify the following steps in the analytical process:

- Capturing – selecting and organizing data;
- Reporting on patterns, trends, and exceptions, using descriptive statistics;
- Model prediction, based on statistical regression data;
- Acting as informing or intervening to improve learning performance;
- Refining the model, as self-improvement process.

As the proficiencies appoint the educational view, the tutoring course/instruction program establishes the second perspective for the learning process. All the activities and events connected with conducting the course concern the user and, arranged chronologically, indicate the learning process. The typical structure for tutoring course consists of the following elements [19]: course introduction – basic information about the course, learning objectives, and main instructions on the course materials; module – consists of module introduction, menu, lessons containing the topics, which correspond to the objectives and summary; lesson – consists of group of topics, which should correspond with learning objectives from the module introduction, each lesson should provide the content and knowledge checks; topic – as elementary unit of each lesson.

For the scope of meta-learning analytics, we reduced all the activities to the lesson or module level, which naturally extrapolates to the whole course and then to group of courses, study, major or educational path. As a result, we have acquired general elements, which indicate the stages in the tutoring process: introduction, instruction, learning objectives, content, practice activities, knowledge checks, assessment and supporting resources. All those elements relate to content delivery. Based on learner's preferences for learning, we have to focus on course organization as well. According to most of tutoring systems, the course consists of the following elements [15]: interface – visual framework of a course, the way the content is delivered and displayed; text; navigation – all the elements that enable the user sequential fulfilling all the stages of learning process; interactions – any events and activities that need user's or system's reaction, especially when the learner is doing knowledge checks, or is following the didactic material; tests – e.g. multiple choice, drag and drop, true/false, fill in the blank, short answer, essay, simulation; media – audio, video, graphics and animations, the form content is presented; collaboration – concerning the group work and knowledge sharing for task accomplishing; discussion forums (including also social media plug-ins).

Figure 2 presents the content- and organization-oriented scopes on elements of learning process, from the tutoring course perspective in the scope of meta-learning adaptation and analytics. For the instruction adaptation, the elements on the left concern the content-based activities that represent the following steps in the learning process. The elements we have to focus on, while designing learning process analytics, are: content, practice and evaluation. Those elements present the actual learner's work with knowledge acquiring, practicing skills, and evaluating own results. Based on the results or independently the learner can use additional resources provided for the course or from the outside. Every knowledge check provides the feedback for additional practice. The assessment is conducted by the tutor or the system itself to provide the scores for learning goals achieving.

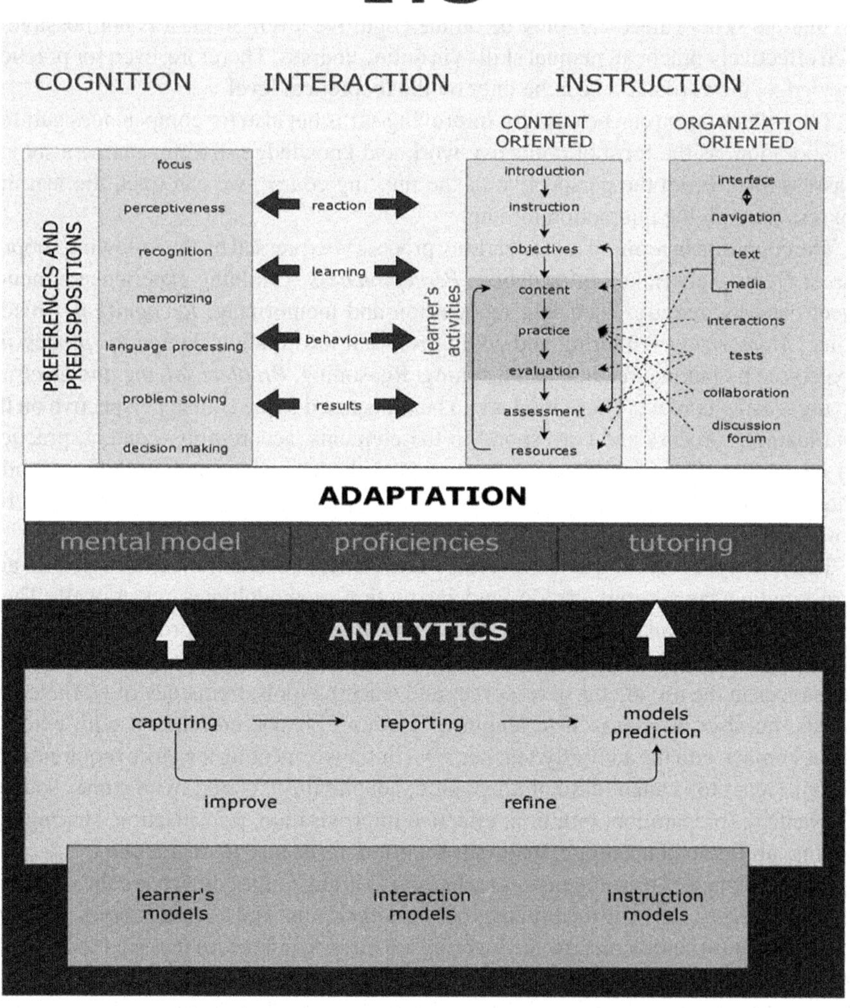

Fig. 2. Analytic model for meta-learning adaptation

On the right side there are elements which represent organisation of instruction course. Interface and navigation represent the static and dynamic aspects of display.

They have no value for learning process analytics, apart from being the framework for elements that user interacts with. Text and media are the elements that represent the means to deliver learning content. On the basis of the knowledge representation and methods effectiveness measured by the semi or final results we can do the analytics proposing the adjustments of didactic method to the particular needs of individual student. Nevertheless, visualization concerning learning outcomes supports achievement goal orientation and, therefore, motivation itself [20].

Interaction, tests, and collaboration indicate different ways of practicing. Interaction with lesson content are designed for knowledge acquiring. Test can be used for knowledge checks, but also as skills practicing. In case of typical courses, we must point out, that the skills gained can only be on the cognitive level. So far it is not possible to teach effectively practical, manual skills via online courses. Therefore, even for practical knowledge, the verification can be only on the theoretical level.

The collaboration can be used for improving skills but also for competences gaining. Collaboration, as the form of collective work and knowledge sharing, can be a subject of assessment. From the perspective of the tutoring course, we can track the learning progress through the interaction mining.

The cognition in terms of meta-learning process is expressed by the following subprocesses: *Focus* – goals, key information; *Perceptiveness* – building experience – mental object patterns, requires focusing, recognition and memorizing; *Recognition* – media choice; *Memorizing* – filtering and coding relevant information; *Language processing* – expressed by talking, reading and listening; Reasoning, *Problem solving and Decision making* – using context. Those modes can be transposed to the course perspective on the meta-learning process and correspond to the elements, accordingly: content, practice, and evaluation. Empirical mode corresponds with the senses, and further with adequate content presentation. Reflective mode is responsible for acquiring the delivered knowledge.

The intelligent tutoring system creates and verifies constantly learner's profile and corresponding interactions profile, and instruction methodologies set as well. There are many levels of analytics. The learner's analytics data source profile may include: patterns of communication and interaction with others in the group, group role, order and choices in the use of course resources and teaching tools, frequency of work, length of tests, number of approaches, length of learning breaks, correlation with external events, contact with the authority/teacher, own initiative, meeting the basic requirements, ability to react to change (ease of adaptation, adaptability), coping with stress, sources of advantage (preparation, intuition, effective improvisation, prioritisation, sharing and winning, ability to concentrate, maintain attention, resistance to distraction.

The efficiency of learning process relates to learning curves, which are the representation of data sets, showing the progress of learning results. The learning curves analytics can give the information on how the learning experience influences the objective results. The monitoring of the learning process, divided into many subprocesses connected with different disciplines and levels, results in the big amount of different data. The analysis may include calculating how much time does it take to learn a particular type of material and predicting how much the learner will need for material with similar characteristics – this is helpful for planning the learning, e.g. for an examination.

6 Discussion and Future Work

The learning process analytics can and should be applied and performed on each of the presented perspectives. The main objective of the process analysis is the feedback information, which should be the answer to the problem of meta-learning. Nevertheless, we should not burden the learner with additional cognitive chore for identifying the results and applying the suitable methods for learning improvement. We have to remember that we cannot expect from the student to be a professor skilled in didactics and psychology. Therefore, the part of meta-learning responsibility should be taken by the tutoring system itself. The user ought to be equipped only with the relevant data, which can help to make the right decision about the training conduction.

We can distinguish two teaching-learning process modes in the tutoring systems (which differ in learning process supervision): self-tutoring and assisted-learning. Self-tutoring is the process, in which the learner takes an active role in the training. Assisted learning is dedicated specially to the lower educational levels, in which the teacher decides about the training conduction and designs the plan and rhythm of material repetitions. Self-tutoring is limited to the part of practice activities and knowledge checks. The adaptation from the learner's perspective is also limited to this area (content, practice, evaluation). The more data are collected the more precise results of analytics and prediction can be achieved. The learning process analysis should fulfil the adaptation needs. The measurements and indicators can be extracted from the user's actions, recorded in tutoring system's logs. It can work as an early warning system, identifying learner's manifestations of discouragement and providing actions to increase motivation [21]. The juxtaposition of meta-learning process and teaching methods is based on preferences, predispositions, and proficiencies results with emerging the analytical data. Combining the learner's chosen path and the result allows for measuring learning efficiency. Continuous recordings in the field can be used to draw the learning curve, which can then be used as a source of input data and reference for further analysis.

Since testing of preferences and aptitudes can be inaccurate and is always subject to measurement errors, further research into the analysis of interactions and behaviour is advisable. The learner often has no knowledge of his or her mental processes. Only an analysis of the selection and effectiveness of tutoring instruments can give measurable results (experiential analytics). Experiential Analysis approach consists of research based on experience from individual and group analyses of learners' interactions and behaviour. This is the valid approach to extracting data from tutoring systems.

There are also many aspects for future research. A very important issue is semantic analytics of the learning process – Semantic Process Mining, i.e. discovery and improvement of recurrent behaviours in order to determine the presence of different learning patterns in process models [22]. It requires defining ontologies for the elements of meta-learning as representation of knowledge in the ITS [23]. The meta-learning feedback and (re)presentation of results indicate the use of big data technologies and artificial intelligence algorithms, which still require further development due to many limitations and difficulties in designing versatile solutions [24]. Apart from cognitive, mental models, another important issue is affective analytics, which can help with emotional engagement and adjustments to the needs in that area [25].

7 Conclusion

To carry out an effective analysis, it is necessary to determine very precisely the teaching material (another analysis). The didactic material is a reference point for the verification of the effects achieved by the student. Through the prism of achieving educational results, determining the methods used and precisely defining the components of the teaching material, we are able to draw conclusions about the individual needs of the learner expressed through preferences, predispositions and preparation (proficiency), and then adjust the behaviour of the system (analytical results) to them, giving him or her the opportunity to choose to improve his or her skills through appropriate teaching methods. Also feedback on weaknesses that need some improvement. The predispositions mainly relate to the way the didactic material is presented, preferences for the organisation and selected teaching methods, and proficiencies provide feedback on the current knowledge, skills, and competences. The predispositions are inborn, preferences may change over time, and proficiency is the most dynamic and practically changes after each of the didactic content.

The conclusion is based on determining the values of the defined indicators, which are then transferred into the result (in the form of learning outcomes) and correlated with the replacement of a given indicator. That kind of detailed analytics cannot be done without technological support of intelligent tutoring systems [26]. Hence, the meta-learning process should be automated and conducted by the intelligent tutor. The main condition here is the adaptability, as the sensitivity towards the learner, their needs and learning progress.

References

1. Romero, C., Ventura, S.: Educational data mining and learning analytics: an updated survey. WIREs. Data Min. Knowl. Discov. **10**(3), 16 (2020)
2. Avella, J.T., Kebritchi, M., Nunn, S.G., Kanai, T.: Learning analytics methods, benefits, and challenges in higher education: a systematic literature review. J. Interact. Online Learn. **20**(2), 1–17 (2016)
3. Baker, R., Siemens, G.: Educational data mining and learning analytics. In: Sawyer, K. (ed.) Cambridge Handbook of the Learning Sciences, 2nd edn, pp. 253–274 (2014)
4. Schneider, D.K., Class, B., Benetos, K., Da Costa, J., Follonier, V.: Learning process analytics for a self-study class in a Semantic Mediawiki. In: Proceedings of the International Symposium on Open Collaboration, New York (2014)
5. Maudsley, D.B.: A Theory of Meta-Learning and Principles of Facilitation: An Organismic Perspective, University of Toronto (1979)
6. Yousef, A.M.F., Chatti, M.A., Schroeder, U., Wosnitza, M.: An evaluation of learning analytics in a blended MOOC environment. Int. Rev. Res. Open Distance Learn. **16**(2), 69–93 (2015)
7. Hu, X., Tong, R., Cai, Z., Cockroft, J.L., Kim, J.W.: Self-improvable adaptive instructional systems (SIAISs): a proposed model. In: Sinatra, A., Graesser, A., Hu, X., Brawner, K., Rus, V. (eds.) Design Recommendations for Intelligent Tutoring Systems: Self-improving Systems, vol. 7. Army Research Laboratory, Orlando (2019)
8. Tong, R., Rowe, J., Goldberg, B.: Architecture implications for building macro and micro level self-improving AISs. In: Sinatra, A., Graesser, A., Hu, X., Brawner, K., Rus, V. (eds.) Design Recommendations for Intelligent Tutoring Systems: Self-improving Systems, vol. 7. Army Research Laboratory, Orlando (2019)

9. Niesler, A., Wydmuch, G.: Predisposition-based intelligent tutoring system: adaptive user profiling in human-computer interaction. In: Proceedings of the Fourth International Conference on Web Information Systems and Technologies, Funchal, Portugal (2008)
10. Niesler, G., Niesler, A.: Methodological apparatus and instruments for personalization in adaptive tutoring systems. Bus. Inform. **4**(54), 60–73 (2019)
11. Niesler, A., Wydmuch, G.: User profiling in intelligent tutoring systems based on myers-briggs personality types. In: Proceedings of the International MultiConference of Engineers and Computer Scientists, Hong Kong (2009)
12. Kumar, A., Singh, N., Ahuja, N.: Learning styles based adaptive intelligent tutoring systems: document analysis of articles published between 2001 and 2016. Int. J. Cogn. Res. Sci. Eng. Educ. **5**, 83–98 (2017)
13. UNESCO: International Standard Classification of Education ISCED 2011. UNESCO Institute for Statistics, Montreal, Quebec (2012)
14. European Commission: The European Qualification Framework for Lifelong Learning. Education and Culture DG (2012)
15. Elkins, D., Pinder, D.: E-Learning Fundamentals. ATD, Alexandria (2015)
16. Picciano, A.G.: The evolution of big data and learning analytics in American higher education. J. Asynchronous Learn. Netw. **16**(3), 9–20 (2012)
17. Campbell, J.P., Oblinger, D.G.: Academic analytics: a new tool for a new era. EDUCAUSE Rev. **42**(4), 40–57 (2007)
18. Clow, D.: The learning analytics cycle: closing the loop effectively. In: Proceedings of the 2nd International Conference on Learning Analytics and Knowledge, New York (2012)
19. DOL Learning Link: Best Practices in Instructional Design for Web-based Training. U.S. Department of Labor, Washington, D.C. (2011)
20. Shirazi, S., Hatala, M., Gasevic, D., Joksimovic, S.: The role of achievement goal orientations when studying effect of learning analytics visualizations. In: Proceedings of the Sixth International Conference on Learning Analytics & Knowledge (2016)
21. Cabrera-Loayza, M.C., Cadme, E., Elizalde, R., Piedra, N.: Learning analytics as a tool to support teaching. In: Botto-Tobar, M., Zambrano Vizuete, M., Torres-Carrión, P., Montes León, S., Pizarro Vásquez, G., Durakovic, B. (eds.) ICAT 2019. CCIS, vol. 1195, pp. 415–425. Springer, Cham (2020). https://doi.org/10.1007/978-3-030-42531-9_33
22. Okoye, K., Tawil, A.-R., Naeem, U., Lamine, E.: Discovery and enhancement of learning model analysis through semantic process mining. Int. J. Comput. Inf. Syst. Ind. Manag. Appl. **8**, 93–114 (2016)
23. Dermeval, D., Albuquerque, J., Bittencourt, I., Isotani, S., Silva, A., Vassileva, J.: GaTO: an ontological model to apply gamification in intelligent tutoring systems. Front. Artif. Intell. **2**, 13 (2019)
24. Winer, A., Geri, N.: Learning analytics performance improvement design (LAPID) in higher education: framework and concerns. Online J. Appl. Knowl. Manag. **7**(2), 41–55 (2019)
25. Arroyo, I., Woolf, B., Burleson, W., Muldner, K., Rai, D., Tai, M.: Correction to: a multimedia adaptive tutoring system for mathematics that addresses cognition, meta-cognition and affect. Int. J. Artif. Intell. Educ. **28**(10), 470 (2018)
26. Sinatra, A.M., Ososky, S., Sottilare, R., Moss, J.: Recommendations for use of adaptive tutoring systems in the classroom and in educational research. In: Schmorrow, D.D., Fidopiastis, C.M. (eds.) AC 2017. LNCS (LNAI), vol. 10285, pp. 223–236. Springer, Cham (2017). https://doi.org/10.1007/978-3-319-58625-0_16

Innovations in Intelligent Systems

Visualization of Structural Dependencies Hidden in a Large Data Set

Bogumila Hnatkowska(✉)

Wroclaw University of Science and Technology, Wyb. Wyspianskiego 27, 50-370 Wroclaw, Poland

Bogumila.hnatkowska@pwr.edu.pl

Abstract. Conceptual models are very often used for visualization of complex domains. Such models help, for example, in understanding the relationships between entities. The paper presents valuable extensions to a method proposed by the author that creates a domain model in the form of a UML class diagram from raw data included in a data set. The formally defined extensions address the main limitations of previously presented algorithms, significantly increasing the readability of generated diagram. The new elements include inferring data types and class names from additional information in data.

Keywords: Conceptual models · Data retrieval · Class diagram · UML · Raw data

1 Introduction

The number of electronically available data on the Web is overwhelming. The data can be both structured and unstructured and relating to different domains. The relationships between data can be difficult to understand, especially when the only accessible asset is a text file without any additional description (e.g., csv file). To enable a better understanding of data and to increase their potential applications (e.g., training models, data models, testing data), one can require to extract a meta-data and visualize the dependencies between them graphically, e.g. with the use of a conceptual model [1].

Retrieving information about the entities and the relationships among them from raw data can be challenging and time-consuming. It is why it should be at least partially automatized. The first attempts to such automation were presented in [2, 3]. This paper is a continuation of the above-mentioned works. It presents valuable extensions to the existing method, addressing its most critical known limitations.

The data within a specific data source can be related to each other by different types of dependencies, e.g., functional, inclusion, or conditional [4]. In the proposed approach, the data mining process is based on functional dependencies read from the data. Its implementation was based on the Fun algorithm [5]. The result is presented in the form of the UML class diagram, nowadays, one of the modeling standards [6].

M. Hernes et al. (Eds.): ICCCI 2020, CCIS 1287, pp. 427–439, 2020.
https://doi.org/10.1007/978-3-030-63119-2_35

The mining method presented in the previous version assumed that input data match a specific pattern, i.e. the vertical table. Its extension allows the data to follow another pattern, called header composition [7], however adapted to the csv format.

The rest of the paper is structured as followed. Section 2 shortly describes the competitive solutions to the presented problem. Section 3 presents (at a very general level) the existing method with its drawbacks. Section 4 defines proposed extensions and demonstrates how they work by simple examples. The last Sect. 5 concludes the paper.

2 Related Works

There is not much research addressing the problem of inferring a structural diagram from a data frame. The closest is the TANGO method (Table ANalysis for Generating Ontologies) [8]. The method is very complex and covers 4 main steps: (1) retrieving data from different sources and construction of a so-called normalized table, (2) construction of mini-ontologies from normalized tables, (3) discovering of inter-ontologies mapping, (4) merging mini-ontologies into one ontology. Our method covers only the 2nd step from TANGO method. Similarly to [8], we look for functional dependencies in data and use them to find out if a relationship is mandatory or optional. Authors of [8, 9] also use additional sources of knowledge (e.g., lexicons/data frames) to find hypernyms/hyponyms. Unfortunately, the data mining process is described very vaguely (no algorithms or formal translations), and presented by a few examples. The other main difference between TANGO and our approach is that TANGO represents mini-ontologies with the use of Object-oriented Systems Model (OSM) notation, not used very often nowadays while we use the UML class diagram for that purpose.

Another interesting approach to the extracting of class diagrams from spreadsheets is presented in [10]. Instead of analyzing functional dependencies in data, authors try to identify a matching pattern (the library of patterns was also proposed). Depending on how data are written, they are translated into classes (groups of data), fields or methods. A known limitation of this method is the impossibility to infer the inheritance relationship between classes. Unfortunately, neither a pattern library or the prototype tool (Gyro) is still available under the given link (http://www.st.ewi.tudelft.nl/~hermans/Gyro). The main difference between the Gyro tool and the proposed approach is that the proposed method processes raw data (no specific meta-information from the spreadsheet is available, e.g., colors, types) trying to identify entities based on functional dependencies hidden in data.

3 Generation Algorithm

3.1 Basic Notions

A data frame, which is the input of the generation algorithm, is defined as a tuple $DF = <H, B>$, where H is the data frame header, and B represents the body.

The header H is a set of attribute names with types assigned (option). Formally, it can be defined as $H = \{<attr_i: T_{attri}>\}$, where $attr_i$ – the name of an attribute, T_{attri}

– the attribute type (it is assumed that it is a primitive type). For a given attribute of name *attr*, its type is denoted by *type*(*attr*).

The body B is a set of items (records). Each item of the data frame is a partial function from attribute names into the respective data types (or value of unknown type). Notation <*attr*, *v*> represents a value v assigned to the attribute *attr* (v can be empty ($\perp$)) in a specific item. The projection of an item t into a subset of attribute names $A \subseteq H$ is defined as $t[A] = \{<attr, v> \mid attr \in H\}$ [4]

Body analysis allows to find functional dependencies between frame attributes. Such relationships form the core of the generation mechanism.

Let X, $Y \subseteq H$. The subset X is said to define functionally the subset Y (denoted by $X \to Y$) when for any t_1, $t_2 \in B$, if $t_1[X] = t_2[X]$ then $t_1[Y] = t_2[Y]$. A functional dependency $X \to Y$ is minimal if removing an attribute from X makes the dependency invalid [4]. A functional dependency $X \to Z$ is transitive, if there exists such X, Y, $Z \subseteq H$ that $X \to Y$ and $Y \to Z$. An attribute A is partially functionally dependent on a set of attributes X, if there exists such $X' \subset X$ that $X' \to A$ [4, 11].

A subset X ($X \subseteq Y \subseteq H$) is called a candidate key with respect to Y, if $X \to Y$ and it doesn't exist such X' ($X' \subset X$) that $X' \to Y$ [11].

3.2 Procedure

The generation algorithm goes through three main stages, presented graphically in Fig. 1. The details are given in [2, 3]. Here only a general idea of the algorithm is presented together with the known limitations, some of which are addressed in this research. The detail description of the algorithm is too long to be presented here. The algorithm complexity is O(m * n!), where n is the number of attributes in the header H, and m is the number of entities in the data frame. It is why some heuristics are applied to avoid the complete search of a space of solutions.

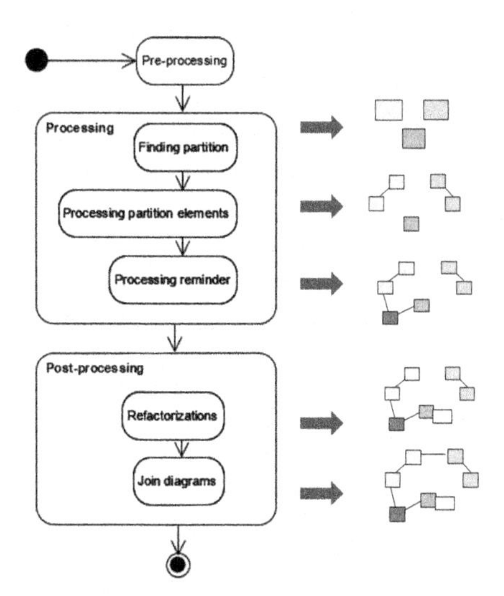

Fig. 1. Main stages of the generation algorithm

During the pre-processing stage, a data frame is read, and functional dependencies are found based on the frame body. From practical reasons, the algorithm considers functional dependencies of a set containing from 1 to 4 attributes. The algorithm processes a singular source of data, e.g. a csv file.

Processing is the main stage. It is split into three sub-stages. The first sub-stage aims to divide the attributes into functionally independent subsets (partition elements) and so-called reminder. They are marked in Fig. 1 symbolically with different gray colors. Each partition element is a source for a single class diagram, initially generated during the *Processing partition elements* activity. The activity distributes the attributes of any partition element – by the normalization process (adapted to object-oriented paradigm) – among one or more classes connected with associations or generalizations. The first generated class is called a root class. These classes play important role in the later phases. The last sub-stage processes the remainder, i.e., the attributes that somehow are dependent on more than one partition element. In general, the remainder is processed in the same way the partition elements are, but, in the end, the resulting class diagram is connected to existing diagrams.

The post-processing stage is divided into two sub-stages. The first sub-stage aims in refactoring of generated diagrams mostly by the introduction of self-associations and, in some cases – generalizations. Refactorizations are performed based on additional information delivered by an expert about the intended semantics of attributes. Before the last activity, the generated graph still can be inconsistent. Thus, the *Join diagrams* activity links root classes of diagrams generated for partition elements with associations.

3.3 Known Limitations

Figure 2 presents an example of the algorithm application for real data [2, 3]. The diagram was generated automatically on the basis of data from university documents filled in by teachers at the end of the semester. Teachers are obliged to give grades (attributes *Grade*, *Date*) for each student (*Album*, *Surname*, *FirstNames*) enrolled in any group (*GroupId*) being a part of the course (*CourseId*, *CourseName*), and assigned to the lecturer (*EmployeeId*, *EmployeeData*, *Time*). The group is conducted within a specific winter or summer semester (*SemType*), which occupies a specific place in the study plan (*Semester*, *YearOfStudy*). All the above-mentioned entities are considered within a specific academic year.

As one can observe, the diagram lacks some basic information, e.g., types for attributes. Class names are artificial (*DummyX*), and they don't represent domain entities. In all cases, the generation of self-associations (not presented in Fig. 2) requires expert support. The expert is asked if the semantics of two (or more) attributes is the same, and – on that basis – self-associations (and generalizations) are introduced. What is more, two classes can be linked with at most one association.

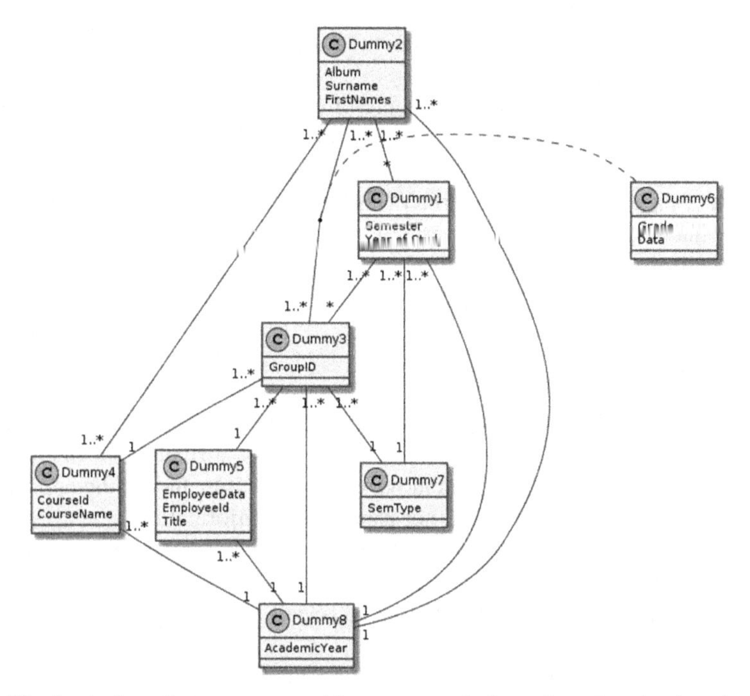

Fig. 2. A class diagram generated for a case-study from the university domain.

4 Proposed Extensions

4.1 Working Modes

The algorithm described in the previous section requires a decision made by a user to perform specific re-factorizations. The user has to decide if the semantic of some attributes is the same. The decision influences a transformation rule to be run.

The proposed improvement allows us to change the mode in which the application is working from manual (an expert is necessary) to automatic. In the latter mode, a default answer of the expert is assumed (negative answer), what runs a default transformation.

4.2 Data Types

As mentioned above, the previous version of the algorithm has extracted only the attribute names from the data frame (DF). However, according to basic notions, each value should belong to a (primitive) data type. More formally, each attribute name $attr \in H$ has a data type T_{attr} assigned, noted as $attr$: T_{attr}. Information about T_{attr} can be extracted from data and presented on the diagram, increasing its readability.

The existing implementation extracts type information, but uses it to determine whether to communicate with an expert. The types were not shown in the diagram, and the method used to determine them is not documented.

UML introduces only a few primitive types, i.e., Boolean, String, UnlimitedNatural, Integer, and Real [6]. The difference between UnlimitedNatural and Integer is slight - the first includes only non-negative values plus the value *unlimited*, represented by '*'.

As the *unlimited* value has no representation in data, UnlimitedNatural is not used on the generated class diagram.

Attribute types are retrieved in a new pre-processing stage, run at the beginning of the algorithm. The entire content (data) of the specific attribute is checked, whether it belongs to a specific type or is empty (empty values are skipped). It is assumed that Boolean values are represented by 'true'/'false' or 'yes'/'no' literals. Belonging of the attribute to Integer/Real numbers is verified by values conversion to Long, and Double types, respectively. If, for any value, the above-defined checks fail, the String data type is inferred.

4.3 Class Names

The existing algorithm generates a class diagram with anonymous class names (e.g., *Dummy1*, *Dummy2*). Finding proper names for groups of attributes is a challenge. Different scenarios of solving that problem were considered, including the application of external data sources (e.g., ontologies) or using a meta-information.

The proposed extension tires to re-use additional data (if they are present) as the source for class names. The solution resembles the approach to data generation described in [10], where authors assume that data in spreadsheets are organized according to some pattern.

To formally describe the extension, the header definition is changed to:

$$H = \{<id, attr: T_{attr}, l_{attr}>\},$$

where *id* is a unique identifier (a subsequent natural number), *attr* – is the attribute name, T_{attr} – is the attribute type (inferred by the tool), l_{attr} is a textual label assigned to the attribute. The label can be empty. It is assumed that labels are given directly above the row with attribute names, as it is shown in Table 1 (the arrows in the bottom represent existing functional dependencies in data.). The labels don't need to be unique (for readability they spread over more than one column in the figures). The labels assignment to particular attributes is set during the pre-processing stage (when data are read).

Table 1. Exemplary data frame for a car fleet.

	Car			Owner			Co-owner	
Reg no	Brand	Model	Owner license	Owner name	Owner surname	Co-owner license	Co-owner name	Co-owner sur-name

Additionally, the following definitions are proposed. Let:

- *D* represent a generated diagram,

- *attr*(*c*) – a set of attributes in the *c* class (*attr*: *Class* → Set(*Attribute*)),
- *labels*(*c*) – a set of labels assigned to the attributes of *c* class (can be empty) (*labels*: *Class* → Set(*String*)),
- *label*(*a*) – a label assigned to the *a* attribute (can be null) (*label*: *Attribute* → *String*),
- *attrLab*(*c*, *l*) – a set of attributes of the class *c* with the *l* label assigned (*attrLab*: *Class* × *String* → Set(*Attribute*)),
- *name*(*c*) – a name of the *c* class (*name*: *Class* → *String*), and
- *len*(*s*) – the number of elements in the *s* set (len: *Set*(*Type*) → *Integer*).

Diagram *D* is a container of classes (with attributes) and their relationships. A class may have many so-called candidate keys. A subset of class attributes is considered as a candidate key if it functionally defines all other attributes (including another candidate keys). Assume, that the access to any class key in the class is given by the function with the signature: *aKey*: *Class* → Set(*Attribute*).

Labels, if they are present, are considered as potential class names in the post-processing stage. When possible, the automatically generated names (*Dummy*) will be replaced by the proper labels. The substitution algorithm is presented below (see Algorithm 1). It has to solve serval potential conflicts, e.g., the same label is assigned to attributes of more than one generated classes (Fig. 3, c); one label covers the class attributes only partially (Fig. 3, b), or the attributes of one class have different labels assigned (see Fig. 3, d).

Algorithm 1: Diagram refactoring - class names substitution with labels

---

Input: *D* – generated diagram; *c* – class (diagram element)
begin
  for $\forall_{c \in D}$ **do**
  begin
   if $\forall_{l_i, l_j \in labels(c)}(l_i = l_j)$ and $l_i \neq null$ and $\nexists_{c_k \in D}\, l_i \in labels(c_k)$ **then**
    name(*c*) ← *l*   // *case 1*
   else if $\forall_{l_i, l_j \in labels(c)}(l_i = l_j \text{ or } l_i = null \text{ or } l_j = null)$ and $\nexists_{c_k \in D}\, (l_i \neq null) \in labels(c_k)$
     and $\forall_{a_k \in aKey(c)} label(a_k) = l_i$ **then**
    name(*c*) ← *l* // *case 2a*
   else if $\forall_{l_i, l_j \in labels(c)}(l_i = l_j \text{ or } l_i = null \text{ or } l_j = null)$ and $\nexists_{c_k \in D}\, (l_i \neq null) \in labels(c_k)$
     and $\exists_{a_k \in aKey(c)} label(a_k) \neq l_i$ and $\frac{len(attrLab(c, l_i))}{len(attr(c))} > 0.5$ **then**
    name(*c*) ← *l* // *case 2b*
   else if $\exists_{c_k \in D}(labels(c) \cap labels(c_k) = \emptyset)$
     and $\exists_{l \in labels(c)} \forall_{a_m \in aKey(c)}\, label(a_m) = l \text{ and } l \neq null$
     and $(\nexists_{a_n \in aKey(c_k)}\, label(a_n) = l \text{ or } aKey(c) \longrightarrow aKey(c_k))$ **then**
    name(*c*) ← *l* // *case 3*
   else if $\exists_{l_i, l_j \in labels(c)} l_i \neq l_j \neq null$ and $\forall_{a_m \in aKey(c)}\, label(a_m) = l_i$ **then**
    name(*c*) ← l_i // *case 4a*
   else if $\exists_{l \in labels(c)} \frac{len(attrLab(c, l))}{len(attr(c))} > 0.5$ and $\nexists_{c_k \in D}\, l \in labels(c_k)$ **then**
    name(*c*) ← *l* // *case 4b*
  end
end

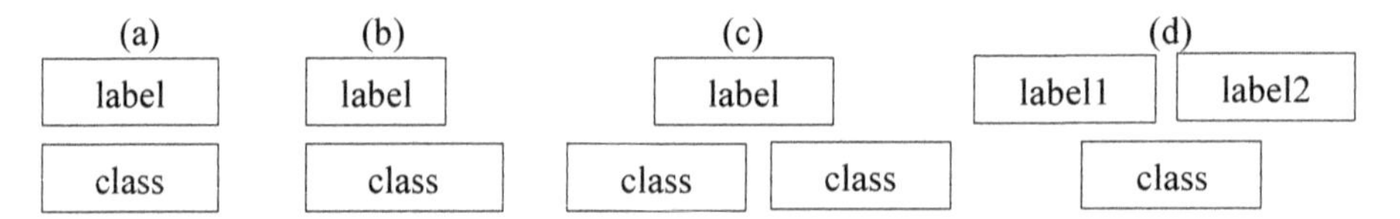

Fig. 3. Possible relations between the scope of labels and generated classes.

For the first case (Fig. 3, a) – all class attributes share the same label – the label is used as the class name.

For the second case (Fig. 3, b) – the class attributes refer to only one label, but it covers not all attributes – the decision whether to use the label as the name of the class is determined from the fact if the key column in the class has the label assigned. Note: the key column (columns) in the class functionally defines the rest of the columns. If the answer is positive, the label name replaces the actual name of the class. Otherwise, it is checked if at least 50% of class attributes have the label assigned. In such a case, the label is used instead of the *Dummy* name.

The third case (Fig. 3, c) – the same label is spread over more than one class – is served differently. The label is used as a class name for that class, whose candidate key has the label assigned. If more than one class satisfies this condition, the label replaces the name of that class, which candidate key defines functionally other keys. If none of the classes have a key (very rare), then the label is not used. Please note, that the presented algorithm is written in a simplified manner which suggest existence of only one candidate key in the class. In fact, the conditions are checked for all candidate keys.

In the last case (Fig. 3, d) – more than one label refers to the class attributes – the decision about which of the labels use is taken based on the label assignments to a candidate key. If such a label defines keys in more than one class, the 3rd rule is applied. If the label assigned to the key of the class can't be used (rule 3), but at least 50% of attributes are described by another label *l*, which is not assigned to another class, then the label *l* replaces the class name.

The algorithm assures the uniqueness of class names.

Figure 4 presents a class diagram resulting from Table 1: (a) without labels, (b) with labels.

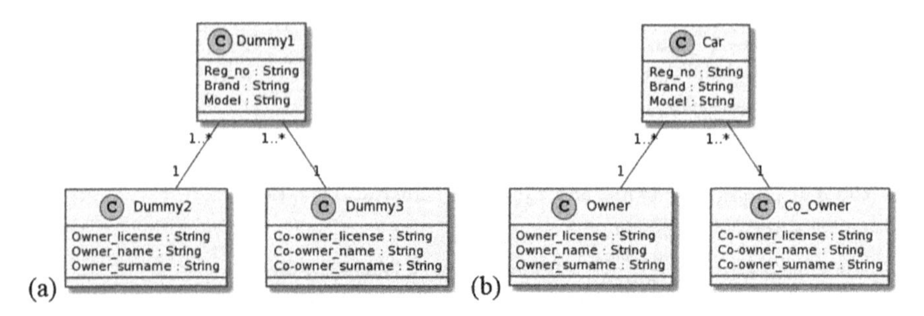

Fig. 4. A class diagram generated from exemplary data frame (car fleet).

Another example of class name generation is shown in Fig. 5 (case 3). The left part presents the existing functional dependencies between attributes and the labels assigned. The right part visualizes the generated diagram. As one can observe, the *Student* label is not used for the *Grade* attribute. This is because that attribute is contained in a separate class without a key.

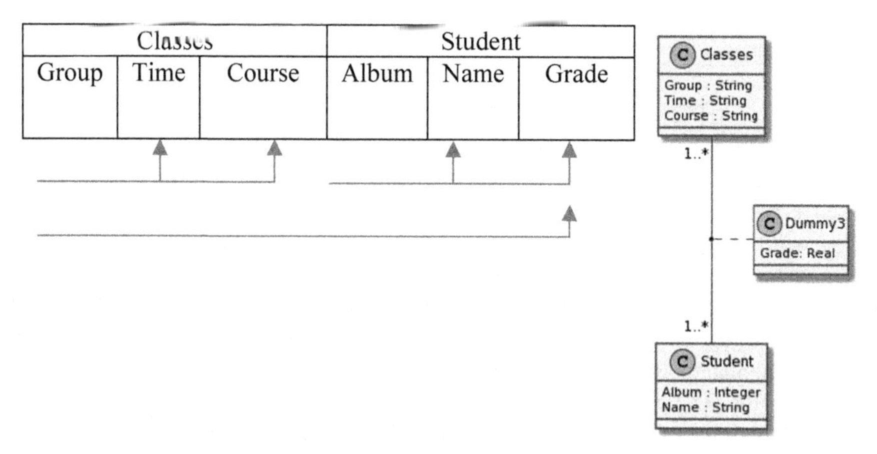

Fig. 5. An example of Algorithm 1 application (case 3).

4.4 Many Relationships Between the Same Classes

The generation algorithm can process a single data frame only. In consequence, two entities (represented by their instances in DF), can be connected by at most one relationship (association/composition). So, when one row contains data representing the same type of entity more than once, the algorithm cannot recognize this.

Let us demonstrate the problem by example. Let's assume that we would like to process data about a car fleet along with information about the car owner and co-owner (if any) for any car – see Table 1. One car (identified by registration number (*Reg no*)) must have exactly one owner assigned (identified by owner license) and may have at most one co-owner assigned (identified by co-owner license).

The previous version of the generation algorithm would identify two separate classes representing a person playing the role of a car owner/co-owner – see Fig. 4a. The better version would produce only two classes: one representing cars, and one representing people somehow connected with cars.

The solution to the problem mentioned above (see algorithm 2) assumes that the names of attributes can be repeated (such replication would not influence the shape of the generated diagram). Additionally, labels can describe the role the entity plays – see an example in Table 2.

Table 2. Data frame for a car fleet with repeated attribute names.

Car			Owner			Co-owner		
Reg no	Brand	Model	License	Name	Surname	License	Name	Surname
1	Fiat	Fiesta	L1	John	Nowak	L2	Eva	Nowak
2	Opel	Corsa	L2	Steve	Smith			
3	Kia	Ceed	L3	Ann	Turner	L4	Trevor	Frank

Algorithm 2 is run in the post-processing stage, after the algorithm 1. The algorithm requires the introduction of a few formal elements. First of all, it is run for classes not marked as root by the generating algorithm. A class is a root class when it was generated as first within a specific partition element, and its functional dependencies cause the generation of other classes [3]. The refactoring merges two classes into one by deletion of one class (possibly with *Dummy* name), and changing one of association ends the deleting class has. The class can only be deleted if it doesn't have connections to other classes. In manual mode, a user can tell the name of the class, which is to be used instead generated name. The previously used class names become names of association roles.

The following auxiliary functions are used within the algorithm:

root: *Class* → *Boolean* – returns true if the class (parameter) is a root class

attrS: *Class* → Set(Tuple(*String*, *String*)) – returns a set of tuples containing (for each class attribute) the attribute names and types (a: T_a)

ends: *Association* → Set(*Class*) – returns a set of classes constituting ends of the association

ends: *Association*, *Integer* → *Class* – gives an access to the association n-th end

roles: *Association*, *Integer* → *String* – gives an access to the association n-th role name

Algorithm 2: Diagram refactoring – more than one association between the same classes

Input: D – generated diagram, m – mode (automatic or manual),
c_i, c_j – classes (diagram elements), ass_k, ass_l – associations (diagram elements)

begin
for $\forall_{c_i, c_j \in D}$ $(i \neq j \; and \; not \; root(c_i) \; and \; not \; root(c_j))$ **do**
begin
if $m = automatic \; and \; attrS(c_i) = attrS(c_j) \; and \; (name(c_i) \neq "Dumm*")$
$or \; (name(c_i) = "Dumm*" \; and \; name(c_j) = "Dumm*")$
$and \; \exists_{ass_k, ass_l \in D}(c_i \in ends(ass_k) \; and \; c_j \in ends(ass_l) \; and \; ends(ass_k)$
$\cap \; ends(ass_l) = c_n) \; and \; \nexists_{ass_m \in D}(c_i \in ends(ass_m) \; or \; c_j \in ends(ass_m))$ **then**
$ends(ass_l, j) \leftarrow c_i$;
$D.delete(c_j)$;
$roles(ass_l, roles) \leftarrow name(c_i)$
else if $m = manual \; and \; attrS(c_i) = attrS(c_j) \; and \; (name(c_i) \neq "Dumm*")$
$or \; (name(c_i) = "Dumm*" \; and \; name(c_j) = "Dumm*")$
$and \; \exists_{ass_k, ass_l \in D}(c_i \in ends(ass_k) \; and \; c_j \in ends(ass_l) \; and \; ends(ass_k)$
$\cap \; ends(ass_l) = c_n) \; and \; \nexists_{ass_m \in D}(c_i \in ends(ass_m) \; and \; c_j \in ends(ass_m)$ **then**
$name \leftarrow read()$;
$ends(ass_l, j) \leftarrow c_i$;
$D.delete(c_j)$;
$roles(ass_l, j) \leftarrow name(c_i)$;
$roles(ass_l, i) \leftarrow name(c_i)$;
$name(c_i) \leftarrow name$
end
end

Figure 6 presents an application of Algorithm 2 for Table 2 (functional dependencies are the same as in Table 1): (a) for automatic mode, (b) for manual mode (assume that an expert used the name *Person* for a common class).

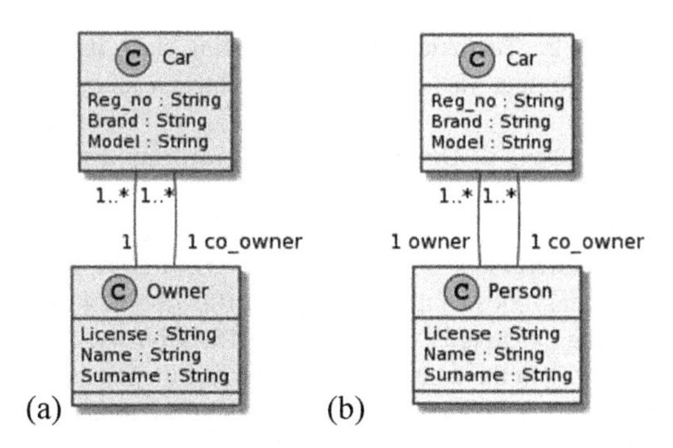

Fig. 6. An example of Algorithm 2 application.

5 Summary

The paper presents valuable extensions to the method of generating a class diagram from raw data. Extensions address, among others, derivation of data types and class names.

The latter was possible with the assumption that data are organized according to one of the supported patterns. At that moment only basic types are inferred from data. The list is to be extended with, e.g., dates.

The extensions were thoroughly tested and also checked for randomly selected spreadsheets satisfying the input criteria (the proper format) taken from existing corpora: Integer [12], and modified Euses [13]. In most cases, data in the spreadsheets had to be cleared (e.g., summary row deleted) before translating it to a csv file. In all cases, the obtained results were consistent with expectations, i.e. the generated class diagram matched the diagram prepared in advance.

The method is still under development. Now, it is able to process a single csv file with non-normalized data. In the future, the method is to be extended with the possibility of processing a set of files containing connected data. It is also considered to include external data sources, e.g., ontologies or lexicons, to identify class names and generalization relationships among them.

The potential application of the presented method is not limited to the visualization of complex dependencies among data. It can also be used as a tool for accessing the quality of data against a known domain represented by a class diagram (or any notation that could be translated to a class diagram, OWL2 for example).

References

1. Embley, D.W., Liddle, S.W.: Big data—conceptual modeling to the rescue. In: Ng, W., Storey, V.C., Trujillo, J.C. (eds.) ER 2013. LNCS, vol. 8217, pp. 1–8. Springer, Heidelberg (2013). https://doi.org/10.1007/978-3-642-41924-9_1
2. Hnatkowska, B., Huzar, Z., Tuzinkiewicz, L.: A data-driven conceptual modeling. In: Jarzabek, S., Poniszewska-Marańda, A., Madeyski, L. (eds.) Integrating Research and Practice in Software Engineering. SCI, vol. 851, pp. 97–109. Springer, Cham (2020). https://doi.org/10.1007/978-3-030-26574-8_8
3. Hnatkowska, B., Huzar, Z., Tuzinkiewicz, L.: Extracting class diagram from hidden dependencies in data set. Comput. Sci. **20**(2) (2020). https://doi.org/10.7494/csci.2020.21.2.3483
4. Liu, J., Li, J., Liu, C., Chen, Y.: Discover dependencies from data – a review. IEEE Trans. Knowl. Data Eng. **24**, 251–264 (2012)
5. Novelli, N., Cicchetti, R.: FUN: an efficient algorithm for mining functional and embedded dependencies. In: Van den Bussche, J., Vianu, V. (eds.) ICDT 2001. LNCS, vol. 1973, pp. 189–203. Springer, Heidelberg (2001). https://doi.org/10.1007/3-540-44503-X_13
6. OMG Unified Modeling Language 2.5, 01 March 2015. http://www.omg.org/spec/UML/2.5/
7. Teixeira, R., Amaral, V.: On the emergence of patterns for spreadsheets data arrangements. In: Milazzo, P., Varró, D., Wimmer, M. (eds.) STAF 2016. LNCS, vol. 9946, pp. 333–345. Springer, Cham (2016). https://doi.org/10.1007/978-3-319-50230-4_25
8. Tijerino, Y., Embley, D., Lonsdale, D., et al.: Towards ontology generation from tables. World Wide Web **8**, 261–285 (2005)
9. Embley, D.W., et al.: Conceptual-model-based data extraction from multiple-record web pages. Data Knowl. Eng. **31**(3), 227–251 (1999)
10. Hermans, F., Pinzger, M., van Deursen, A.: Automatically extracting class diagrams from spreadsheets. In: D'Hondt, T. (ed.) ECOOP 2010. LNCS, vol. 6183, pp. 52–75. Springer, Heidelberg (2010). https://doi.org/10.1007/978-3-642-14107-2_4

11. Kedar, S.: Database Management System. Technical Publications, USA (2011)
12. Ausserlechner, S., et al.: The right choice matters! Solving substantially improves model-based debugging of spreadsheets. In: 13th International Conference on Quality Software (QSIC), pp. 139–148. IEEE (2013)
13. Hofer, B., Riboira, A., Wotawa, F., Abreu, R., Getzner, E.: On the empirical evaluation of fault localization techniques for spreadsheets. In: Cortellessa, V., Varró, D. (eds.) FASE 2013. LNCS, vol. 7793, pp. 68–82. Springer, Heidelberg (2013). https://doi.org/10.1007/978-3-642-37057-1_6

Internet Advertising Strategy Based on Information Growth in the Zettabyte Era

Amadeusz Lisiecki[1] and Dariusz Król[2(✉)]

[1] Faculty of Computer Science and Management, Wrocław University of Science and Technology, Wrocław, Poland
[2] Department of Applied Informatics, Wrocław University of Science and Technology, Wrocław, Poland
Dariusz.Krol@pwr.edu.pl
http://krol.ksi.pwr.edu.pl

Abstract. This research introduces a new method to evaluate and make most practical use of the growth of information on the Internet. The method is based on the "Internet in Real Time" statistics delivered by WebFX and Worldometer tools, and develops a combination of these two options. A method-based application is deployed with the following five live counters: *Internet traffic*, *Tweets sent*, *Google searches*, *Emails sent* and *Tumblr posts*, to measure the dynamics of the overall trend in user activity. Four parts of a day and three categories of days are examined as corresponding discrete inputs. In the search for the most effective web advertisement, the displaying strategy will vary directly with the level of users' activity in the form of live counters over 12 months of 2018. Two competent surveys consolidate the outcome of our work and demonstrate how the company can identify the best web advertising strategy closer to his business needs and interests. We conclude with the discussion whether the considered method can lead to superior efficiency level of all other communication activities.

Keywords: Advertising efficiency · Internet metrics · Knowledge integration · User activity

1 Introduction

Today, the increase of information in the modern, digital world becomes more complex issue and much more difficult than it used to be in the past. By 2022, annual global IP traffic will reach 4.8 ZB with tidy CAGR of roughly 24%, which translates to quadrillions in MB and smoothed rate of compound annual growth to consider when making investment decisions [1]. On average, the digital universe is doubling in size every two years, and will probably reach the level of 175 ZB by 2025 [2].

However, for the average Internet user, this phenomenon may seem insignificant. An easy way for a user to log into a social network, write a post, put a photo gives the impression of a simple and fast action. The standard operation

M. Hernes et al. (Eds.): ICCCI 2020, CCIS 1287, pp. 440–452, 2020.
https://doi.org/10.1007/978-3-030-63119-2_36

on a computer takes less than a minute [3]. However, all these operations, in millions, are most often unevenly distributed and concurrently executed. There is also a problem with the additional workload caused by variable network traffic flow. The other traffic volume occurs over business hours on weekday, and totally different on Saturday or Sunday evening. Thus, the increase in information in these two cases will be significantly different [4].

There are two fundamental topics to study the phenomenon of rapid information growth in the zettabyte era [5]. The first is the quality of information [6]. However, this is not the topic discussed here. The second is to focus on the amount of information conveyed by each communication activity [7]. One may approach this issue in an algorithmic way. Particular attention should be paid to the reliability of such algorithm. It should be based on the analysis of the real factors that have a bearing on this phenomenon. It should be noted that due to the complexity and size of the problem, it is not possible to determine the very exact procedure. You can only estimate it with the best possible accuracy, how to count the increase in information [8].

The aim of this work is to investigate the problem of information growth on the Internet and propose a new method to evaluate the increase of information based on real-time evolving statistics, as described in detail in Sect. 2 and 3. The specific objectives are: (1) to select the relevant counters based on Internet live stats to adhere to current trends in users' overall activity, and (2) to identify the most effective web advertisement strategy, aligned with the activity of users [9], and convergent on specific business needs and interests. As an illustrative example discussed in Sect. 4, we apply our method in two practical surveys of using five counters to obtain insights for decision-making in web advertisement policy. The first case concerns with the setting the best moment in time for displaying the advertisement. The second case shows the best way of selecting the right market strategy most closely match the expected goals. A properly running ads strategy can be an effective way to increase company's gross and deliver operating profit [10].

Table 1. World Internet usage and population statistics estimates for 30–June–2019. Percentages may not add up to 100% due to rounding error (Source: Internet World Stats)

Regions	Population	Pop. % of world	Internet users	Penetration rate (% Pop.)	Internet world %
Africa	1,320,038,716	17.1	522,809,480	39.6	11.5
Asia	4,241,972,790	55.0	2,300,469,859	54.2	50.7
Europe	829,173,007	10.7	727,559,682	87.7	16.0
Latin America	658,345,826	8.5	453,702,292	68.9	10.0
Middle East	258,356,867	3.3	175,502,589	67.9	3.9
North America	366,496,802	4.7	327,568,628	89.4	7.2
Australia	41,839,201	0.5	28,636,278	68.4	0.6
TOTAL	7,716,223,209	100.0	4,536,248,808	58.8	100.0

2 Internet Statistics

A number of different statistics aimed at digital data measuring were suggested over the years [11]. At present the most popular set is Internet World Stats[1], an international website that features Internet usage, population stats, social media stats, travel stats, and market research data, for over 243 individual countries and world regions, cf. Table 1 with non-live statistics. On the other hand, based on Internet Live Stats[2] (ILS) captured on 30–May–2019 we are informed that:

- More than 4,240,065,773 people were active Internet users in the world.
- Total number of Websites discovered so far was 1,689,910,748 and over 75% are inactive.
- About 37,495,249,892,071 valid emails were sent.
- Approximately 109,477,293,627 tweets were sent, part of what boosts the rate of growth to 30% per year.

In addition, they include, the numbers of sent emails, Google searches, articles and blog posts shared by, short messages posted on Twitter, videos shown on YouTube, photos posted on Instagram, conversations on Skype, attacked websites, sold computers and mobile devices, data sent via the Internet, electricity use and carbon footprint of the Internet. ILS website contains counters based on the Real Time Statistics (RTS) algorithm. On some of them you can see the substantial increase when compared to their levels from a previous year.

In this paper, to provide a reference for the growth rate of information we use two others tools: WebFX The Internet in Real Time[3] and Worldometer[4]. The first one contains several internet metrics counters, which show change from the moment you enter the site every 0.2 s. The latter service uses an advanced RTS algorithm for each live counter. Details on the RTS algorithm are not publicly available due to the commercial activities of Worldometer. You can only observe the current state in a limited scope. It is not possible to make forecasts or to study trends in the past. Worldometer' algorithm includes several advanced features such as: formulas compounded every second, weekday/weekend/holiday sensitive counters, time zone and worldwide consistent counters. Worldometer provides live counters that are associated with: the population of people in the world, politics, economics, media, environment, energy, food, drinks and health including seasonal flu or most recent Wuhan coronovirus causing COVID-19, now named SARS-CoV-2. These real-time statistics show how vast the Internet is, not to mention how much it's growing every day.

---

[1] *Internet World Stats.* Retrieved 30–May–2019 from http://www.internetworldstats.com.

[2] *Internet Live Stats.* Retrieved 30–May–2019 from https://www.internetlivestats.com.

[3] *WebFX The Internet in Real Time.* Retrieved 30–May–2019 from https://www.webfx.com/internet-real-time.

[4] *Worldometer.* Retrieved 30–May–2019 from https://www.worldometers.info.

For the purpose of our research, we selected common live counters between WebFX and Worldometer. These are: *Internet traffic*, *Tweets sent*, *Google searches*, *Emails sent* and *Tumblr posts*. All these stats refer to one day.

3 Measuring the Information Growth Related to User Activity

Our method is based on the combination of two types of Internet real-time stats. The first one is provided by WebFX. The second is based on the RTS algorithm provided by Worldometer. As it has been presented in the previous section, there are 5 common live counters for each type of source. The WebFX page does not show the value of real-time but presents averaged values. Consequently, RTS-based counters from Worldometer are more preferred. They attain much better accuracy. For this reason, values from WebFX will be scaled to the corresponding Worldometer and ILS stats. The exact determination of changing values from counters based on the RTS algorithm will take place through two single measurements. The time difference between them will be approximately one hour. In this way, we can compensate for the effect of taking a screen shot with the current state of counters. It is essential that all measurements are made in the same time zone, therefore, the time convertion and calculation of changes during 1 day were implemented.

The proposed method is designed to be accurate, robust and reliable. Worldometer distinguishes the difference between workdays, weekends and holidays in the context of increments in each metric. On a day off, people have more time available. They can use it for various activities. These include, for example, a meeting with family, going out to a party, or carrying out renovation works at home. People generate a lot of changes while using Internet. They often use the Google search engine, communicate with clients or colleagues via email. They also send files. In this way, they increase global network traffic. Among others days, one can also distinguish a specific day associated with the current events. It can be either planned, for example, championships in sports, talks of politicians, music festivals or unplanned, for instance, floods, hurricanes, tornadoes, volcanic eruptions, earthquakes, tsunamis, storms, and other geologic processes. The larger the area affected by this event, the greater the increase in users' statistics. Moreover, every day you can distinguish four parts of a day: night, morning, afternoon and evening. For each of them, there is other change in metrics. For further examinations, in terms of clock time night is considered to start at midnight, morning at 6 a.m., afternoon at noon, and evening at 6 p.m.

4 Results

The goal of advertising is to deliver information that positively influence current or prospective customers. Therefore, we deliver two competent surveys: (1) finding the best time for ads to be displayed and (2) deploying the right market strategy for decision-making. The key idea is based on the observation that the faster the rate of live counter change, the more preferential moment in time is to choose when the ad should appear. This study is limited to Poland, because additionally, we would like to use the statistics from Public Opinion Research Center[5] to include their reports focused on different demographic parameters like age, education, place of residence and material standing. However, before these estimations are possible, stratified sampling needs to be done, for example, the selection of proper set of counters, vast measurements according to parts of a day and categories of days. Relevant to our purpose, we conducted detailed thorough analysis of the gathered data over the predefined time interval, that is 12 months of 2018, to determine the optimal settings. As expected, the findings are reasonable, fair and meet requirements, however, we found some discrepancies between WebFX and Worldometer live stats. To correct this problem, elementary mismatching coefficient was introduced. Generally we confirmed, that the assumptions made are empirically testable and considered valid, both remained on a satisfactory level. These results are not reported here due to space limit. In short, this preliminary research allows us to continue further examination.

4.1 What Are the Best Times for Ads to Be Displayed?

The first survey aims to determine the definitive ranking of the months, from best time to worst, resulting from 5 live stats. It covers 12 calendar months in 2018 (January to December). The change in metrics will be calculated for each calendar month. A target audience span users in the age range ⟨18–24⟩.

The results of this survey, depicted in Fig. 1 and Table 2 suggest as follows:

- For each metric, the minimum monthly result belongs to February. Based on the corresponding waterfall charts, it can be concluded that the median line is very close to the top quartile. Therefore, many results are above the middle value. For *Tweet sent*, *Google searches* and *Emails sent*, the maximum value is close to the median, the rankings look very similar.
- For most metrics, the highest increase occurs in July. This is due to the fact that July has 31 days and contains several days with the special events, for example, FIFA World Cup.
- Two rankings look identical in terms of month breakdown with counters: *Google searches* and *Emails sent*.

The results shed light on a number of important issues in advertising. First, more preferred in the context of changes to web stats is the number of days per month than the number of days with events. The choice of the date for displaying the ad should consider the dynamics of growth between individual months and their position in the ranking. In the case of the first criterion, March is the best choice, and according to the second, July.

[5] *Public Opinion Research Center*. Retrieved 30–May–2019 from https://www.cbos.pl/EN/home/home.php.

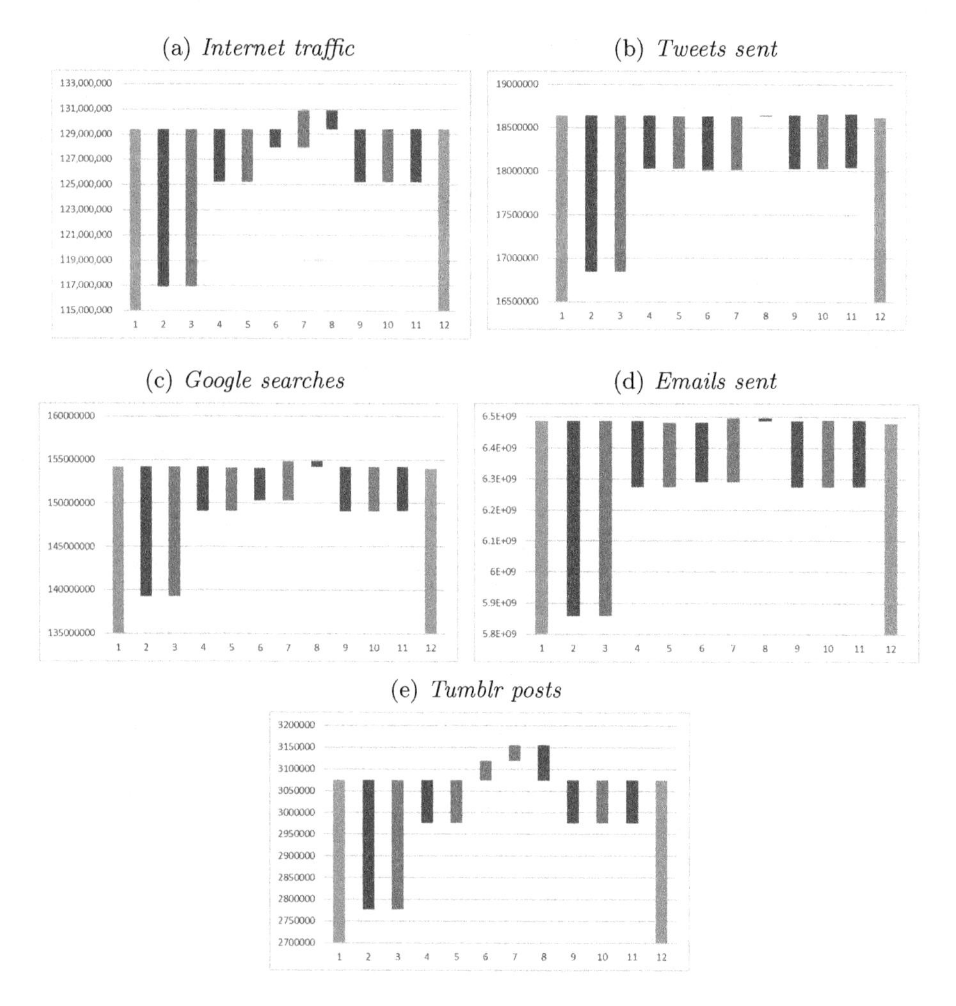

Fig. 1. The month-based distribution of (a)–(e) counters: *Internet traffic*, *Tweets sent*, *Google searches*, *Emails sent*, *Tumblr posts* for the population aged ⟨18–24⟩. The column type chart shows all pertinent user activity changes over 12 months plotted on the x-axis. The scale of the y-axis reflects the values for the associated criterion variable

4.2 What Should Be the Right Market Strategy?

The second survey consists of three questions: (1) variant-based of which variant to advertise, (2) language-base in which language to advertise and (3) branched-based of which branch to advertise.

The answer to the first question explains by example how the right strategy for online selling of skin care and hair care products can be deployed. The company has identified two target groups of customers. The first includes population aged ⟨45–54⟩, the second aged ⟨55–64⟩. The company plans to produce 138,000 cosmetics of the first type (skin care) and 117,300 cosmetics of the second (hair

Table 2. The month-based user activity ranking for changes in (a)–(e) counters: *Internet traffic*, *Tweets sent*, *Google searches*, *Emails sent*, *Tumblr posts* for the population aged ⟨18–24⟩. The ranking is dependent on the changes over 12 months

(a) *Internet traffic*

Rank	Month	Counter
1	Jul	130,927,134
2	Jan, Mar, May, Aug, Oct, Dec	129,405,696
8	Jun	127,936,098
9	Apr, Sep, Nov	125,231,318
12	Feb	116,882,564

(b) *Tweets sent*

Rank	Month	Counter
1	Oct	18,653,967
2	Jan, Mar, Aug	18,644,951
5	Jul	18,631,427
6	May	18,626,919
7	Dec	18,617,903
8	Nov	18,040,883
9	Apr, Sep	18,031,867
11	Jun	18,013,835
12	Feb	16,841,764

(c) *Google searches*

Rank	Month	Counter
1	Jul	154,837,390
2	Oct	154,199,513
3	Jan, Mar, Aug	154,154,433
6	May	154,064,274
7	Dec	154,019,194
8	Jun	150,367,741
9	Nov	149,168,622
10	Apr, Sep	149,123,543
12	Feb	139,242,079

(d) *Emails sent*

Rank	Month	Counter
1	Jul	6,495,743,524
2	Oct	6,489,477,450
3	Jan, Mar, Aug	6,486,871,845
6	May	6,481,660,635
7	Dec	6,479,055,030
8	Jun	6,291,764,787
9	Nov	6,276,861,448
10	Apr, Sep	6,274,255,844
12	Feb	5,859,446,260

(e) *Tumblr posts*

Rank	Month	Counter
1	Jul	3,155,577
2	Jun	3,119,513
3	Jan, Mar, May, Aug, Oct, Dec	3,074,433
9	Apr, Sep, Nov	2,975,258
12	Feb	2,776,908

care), but only one of these products will be advertised. Based on previous sales experience, the company has determined the popularity of products: 30% of population aged ⟨45–54⟩ and 70% of people aged ⟨55–64⟩ for skin product. The hair product was equally popular in both age groups. The ad will be displayed during the second week of September 2018. The questions is variant-based of which product should be advertised.

To find the answer, we identify the increase in web metrics for the given time of ad display. Then, on the base of the population pyramids we evaluate how

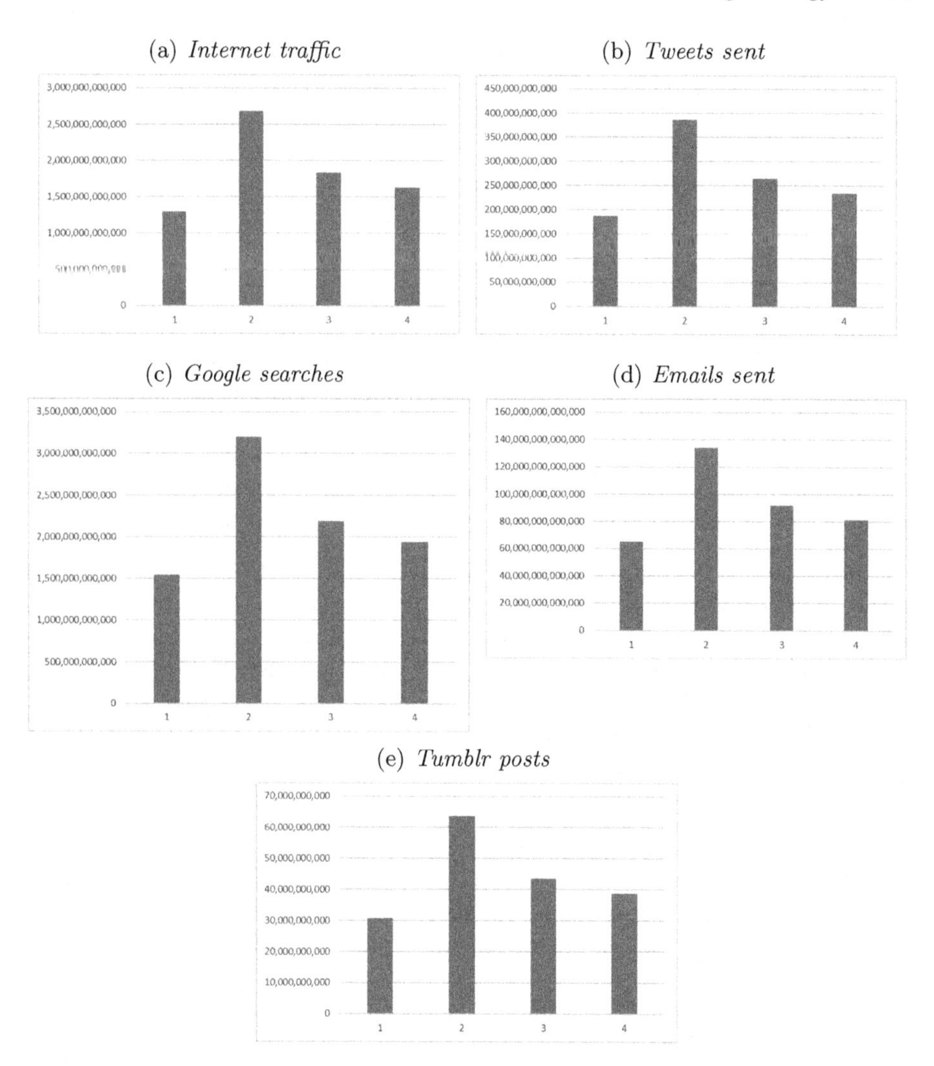

Fig. 2. The variants-based distribution of (a)-(e) counters: *Internet traffic*, *Tweets sent*, *Google searches*, *Emails sent*, *Tumblr posts* for two products and the population aged ⟨45–54⟩ and ⟨55–64⟩. The column type chart shows all pertinent user activity changes over four variants as pairs (product, age range). The scale of the y-axis reflects the values for the associated criterion variable

many Poles has representatives in each age group. The next step is to determine the percentage of active users in defined age categories. Then we calculate the increase in each of the metrics. To be more precise, we calculate the increase in the second week of September for five metrics. The next step is to read from the statistical data the population in two age ranges: ⟨45–54⟩ and ⟨55–64⟩. The following step consists in calculating the number of Internet users in two age ranges based on statistical data: $\langle 45\text{–}54\rangle$ and $\langle 55\text{–}64\rangle$ in relation to the number of all Internet users. Then we specify four variants V_1, V_2, V_3, and V_4 as pairs (product, age range). The next one is determining the increase in each of the metrics for each age group. Then the largest score which defines the winning variant is picked. Finally, we return the product: skin care or hair care for which there is an advantage in at least 3 metrics.

The results of this part, depicted in Fig. 2 suggest as follows:

- For each metric, the distribution between four variants looks similar. The highest score was obtained for V_2 (skin care, $\langle 55\text{–}64\rangle$), then V_3 (hair care, $\langle 45\text{–}54\rangle$), V_4 (hair care, $\langle 55\text{–}64\rangle$), and the lowest for V_1 (skin care, $\langle 45\text{–}54\rangle$).
- The age group $\langle 45\text{–}54\rangle$ despite a smaller number of population than the group $\langle 55\text{-}64\rangle$, has a larger number of Internet users. In the case of the same popularity of a hair product in both age groups, only the live counter affects the value of the variant. The result is intuitive, the product with the highest popularity should be advertised first, variant V_3 (hair care, $\langle 45\text{–}54\rangle$).

The answer to the second question aims to determine the best language used in the ad. The company wants to advertise its products on the Internet, which will be distributed to various different speaking countries. The research stems from the fact that people are more likely to buy products if they watch the advertisement in their native language. The following languages are considered: German, French and Spanish. The advertisement would be aired during the last two weeks of August 2018. The language with the highest increment in stats should be used for advertising.

The results of this part, depicted in Fig. 3 suggest as follows:

- The largest increase in all metrics is for Spanish. It is more than twice as large as for German. It is similar for short messages on Twitter, although the difference between German and French is smaller than between Spanish and German.
- With a similar number of countries for French-speaking countries (21) and for Spanish-speaking countries (20), Spanish has a clear, more than twice the advantage in online metrics. With the significant difference between the number of French-speaking (21) and German-speaking countries (4) German has a slight advantage in stats.

The answer to the third question aims to determine the best ad display strategy out of three options. For example, the company has 3 separate branches and only one product ad is considered. The first branch deals with courses at the university. The advertisement would be displayed during the second and third week of June 2018 in the evening. The second branch distributes waistcoats in Sweden. The advertisement would be displayed from 11:10 a.m. to 2:37 p.m. over September. The third branch distributes the memorabilia. The ad would be displayed during the last three weeks of August from 9 a.m. to 9 p.m. on

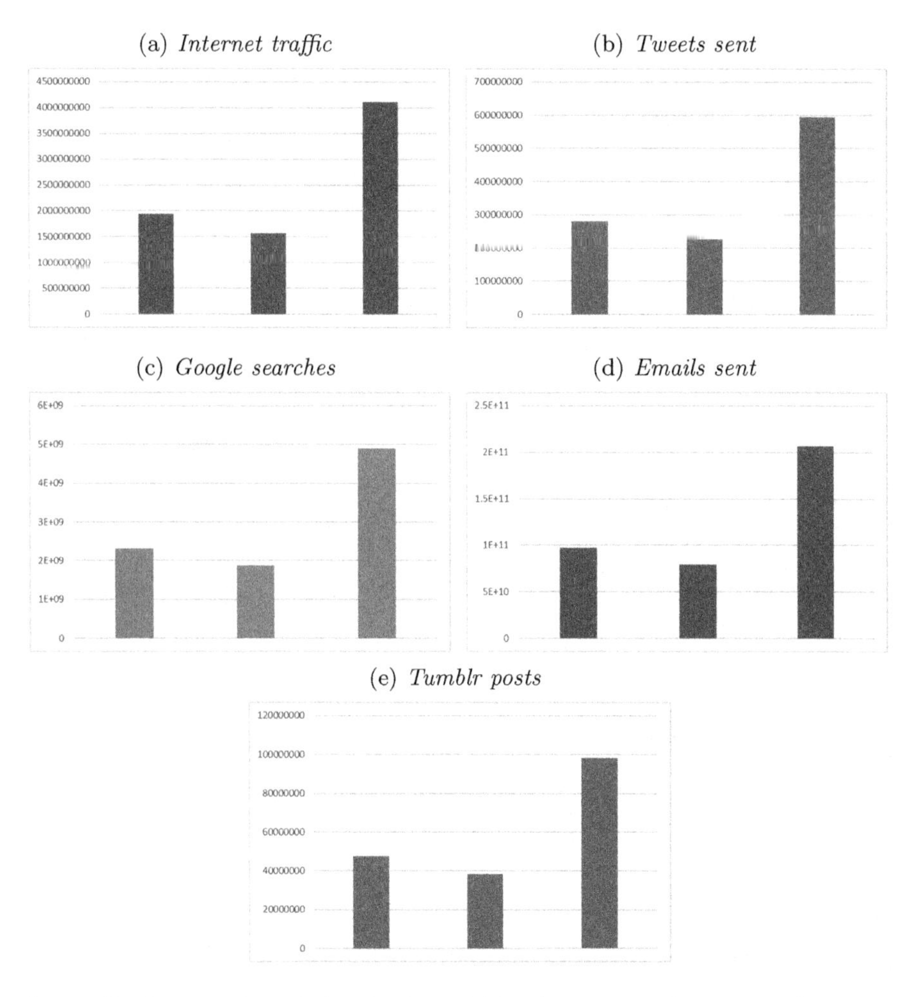

Fig. 3. The language-based distribution of (a)–(e) counters: *Internet traffic*, *Tweets sent*, *Google searches*, *Emails sent*, *Tumblr posts* for German, French and Spanish plotted on the x-axis. The column type chart shows all pertinent user activity changes over three languages. The scale of the y-axis reflects the values for the associated criterion variable

Saturdays and Sundays. The increase in web metrics for the first, second and third branch with given restrictions should be evaluated.

The results of this part, depicted in Fig. 4 suggest as follows:

- In all metrics the biggest change is for the second branch. The first one has a higher rate than the third. The ad should be displayed for the second branch.

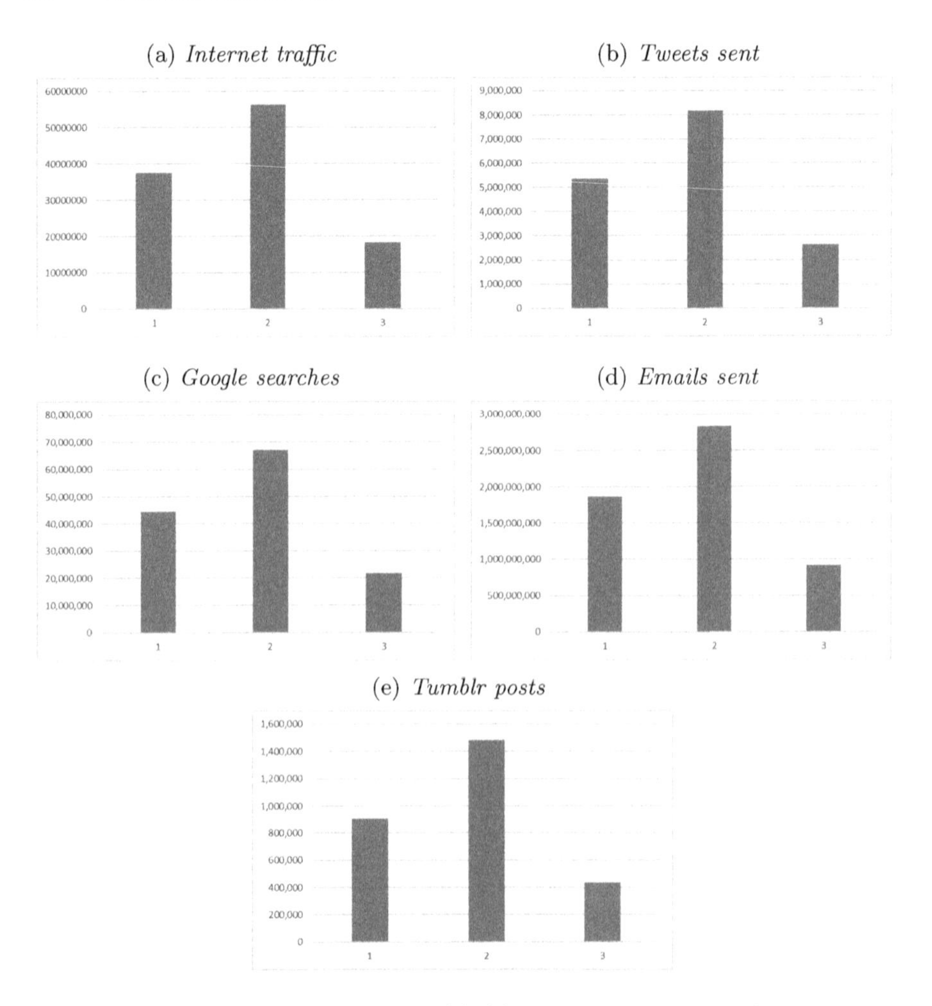

Fig. 4. The branch-based distribution of (a)–(e) counters: *Internet traffic*, *Tweets sent*, *Google searches*, *Emails sent*, *Tumblr posts* for three branches plotted on the x-axis. The column type chart shows all pertinent user activity changes over three branches. The scale of the y-axis reflects the values for the associated criterion variable

5 Conclusion

In this study, we provide a new method to measure the information growth by combining different live counters, and we model the relevant online advertisement strategy using the observed changes to five stats. Two competent types of research demonstrate how the company can identify the best web advertising strategy closer to his business needs and interests. In the first case, the best month for displaying ads is selected. The best possible marketing strategy is analysed in the second case. At this point, this might prove that a proper live counter can be identified in any single trial.

It is equally clear that our research brings potential economic benefits and can help increase overall advertising efficiency but also impose certain limitations. First, the set of live counters could be extended, in the first row considering the other indices based on RTS algorithm. From experience, it is possible that recalculations and alignment of the results for a specific region will be refreshed automatically. Moreover, a limitation of our data is that it comprises the period starting on 1–Jan–2018 and ending on 31–Dec 2018, small numbers of measurements – only four parts of a day, and thus cannot account for the best effect on advertising efficiency in the next year or later. Finally, the focus of this study was solely on Internet advertising. The larger issue, which we may well address in future work, is whether the proposed method leads to superior efficiency level in other important communication activities including direct marketing, public relations and publicity, sales promotions, and sponsorships. This is the overarching question that up-and-coming research should ultimately address and go beyond. Consequently, we believe that our initial work lays a foundation for addressing this intriguing question.

Acknowledgments. This research received financial support from the statutory funds at the Wrocław University of Science and Technology.

References

1. White paper: The Zettabyte Era: Trends and analysis. Technical report C11–739110-00, Cisco, June 2017
2. Reinsel, D., Gantz, J., Rydning, J.: The digitization of the world. from edge to core. Technical report US44413318, IDC, November 2018
3. Saganowski, S.: Predicting community evolution in social networks. In: Proceedings of the IEEE/ACM International Conference on Advances in Social Networks Analysis and Mining, pp. 924–925. ACM (2015). https://doi.org/10.1145/2808797.2809353
4. Aleshkin, A., Lesko, S.: Predicting the growth of total number of users, devices and epidemics of malware in internet based on analysis of statistics with the detection of near-periodic growth features. In: XXI International Conference Complex Systems: Control and Modeling Problems, pp. 347–352 (2019). https://doi.org/10.1109/CSCMP45713.2019.8976674
5. Jin, E.M., Girvan, M., Newman, M.E.J.: Structure of growing social networks. Phys. Rev. E **64**, 046,132 (2001). https://doi.org/10.1103/PhysRevE.64.046132
6. Król, D.: Measuring propagation phenomena in social networks: promising directions and open issues. In: Nguyen, N.T., Trawiński, B., Fujita, H., Hong, T.-P. (eds.) ACIIDS 2016. LNCS (LNAI), vol. 9621, pp. 86–94. Springer, Heidelberg (2016). https://doi.org/10.1007/978-3-662-49381-6_9
7. Coffman, K.G., Odlyzko, A.M.: Internet Growth: Is There a "Moore's Law" for Data Traffic?, pp. 47–93. Kluwer Academic Publishers, Dordrecht (2002)
8. Dobrescu, L., Obreja, S., Vochin, M., Dobrescu, D., Halichidis, S.: New approaches for quantifying internet activity. In: E-Health and Bioengineering Conference, pp. 1–4 (2019). https://doi.org/10.1109/EHB47216.2019.8969998

9. Boudreau, M., Watson, R.: Internet advertising strategy alignment. Internet Res. **16**, 23–37 (2006). https://doi.org/10.1108/10662240610642523
10. Pergelova, A., Prior, D., Rialp, J.: Assessing advertising efficiency. J. Adv. **39**(3), 39–54 (2010). https://doi.org/10.2753/JOA0091-3367390303
11. Huberman, B.A., Adamic, L.A.: Growth dynamics of the world-wide web. Nature **401** (1999). https://doi.org/10.1038/43604

An Approach Using Linked Data for Open Tourist Data Exploration of Thua Thien Hue Province

Hoang Bao Hung[1(✉)], Hanh Huu Hoang[2], and Le Vinh Chien[1]

[1] Thua Thien Hue Center of Information Technology, Hue, Viet Nam
hbhung@viethanit.edu.vn
[2] Posts and Telecommunications Institute of Technology, Ha Noi, Viet Nam
hhhanh@ptit.edu.vn

Abstract. In five domains of Digital Transformation, data plays an important role as a key tangible asset for value-creation. Nowadays, data is continuously generated everywhere, even the unstructured data is increasingly becoming usable and valuable. The challenge of data is turning it into valuable information and unleash its potentials for Digital Transformation. Motivated by Open Data initiatives in over the world which took the lead for publishing government data for public use, Thua Thien Hue Open Data Initiative (TTHODI) (Thua Thien Hue Open Data Portal: http://opendata.thuathienhue.gov.vn/) took first steps in public its local government data to create data-driven benefits of community and enterprise in the region. On this trend, Linked Data is naturally coming at the right time for us as an ideal standard for representing our open datasets thanks to its expressiveness in form of triples. Furthermore, with huge potentials for linking our open datasets to existing datasets in Linked Open Data Cloud for enriching our data cloud for an added value to TTHODI. This paper presents our approach using Linked Data for Tourist Open Data as a case study and also shows the first results of using Linked Open Data for TTHODI in Tourism.

Keywords: Digital transformation · Open data · Linked data

1 Motivation

"If I had to express my views about digital future – that of Europe or indeed, the whole world – I could do it in one word: data," addressed Andrus Ansip, Vice President Digital Single Market of EU [1].

This statement expresses the vital requirement or new oil for digital age is *data*, and data has been an integral one in Digital Transformation's five domains. Data Transformation is making breakthrough in global scope thanks to the rapid development and convergence of technologies such as cloud computing, mobile computing, internet of things (IoT), Big Data, block chain and artificial intelligence. Data transformation or Industry 4.0 has led the industry breakthrough in developed countries and making huge potentials of development for new emerging economies. This trend is also changing

M. Hernes et al. (Eds.): ICCCI 2020, CCIS 1287, pp. 453–464, 2020.
https://doi.org/10.1007/978-3-030-63119-2_37

the way we communicate, interact, study, go shopping and entertain. In national-level, digital transformation is making changes in governments' state management, improving e-government (e-gov) services for new demands from their citizens and enterprises.

In five domains of Digital Transformation, data is a key tangible asset for value creation in governments, enterprises and people in their own digitally transformation [2]. Eventually, data analytics become vital for the value-chain generation with the processing capacity in turning data into valuable information. Furthermore, in order to make maximum usage of the big data analytics, the vital assets – data of governments and enterprises are needed to be open for. Open Data will play a vital part of the whole digital transformation process of governments in e-gov services as they are as the new oil for the digital transformation mechanism.

Nowadays, wide practical use of big data means that data transformation is more important for businesses than ever before. An ever-increasing number of programs, applications, and devices continually produce massive volumes of data. And with so much disparate data streaming in from a variety of sources, data compatibility is always at risk [3]. Therefore, in order to making data of companies and organizations ready as open data for sharing and re-using, efforts to convert data from any source into a format that can be integrated, stored, analyzed, and ultimately mined for actionable business intelligence have been made, but we are still struggle for open data format or standard for easy to use with their original meanings.

In this paper, based on previous studies [4] and the emerging of Semantic Web technologies for new paradigm namely Web of Data, we focus on Linked Data design principles with the core of RDF data model for representing open datasets for the better expressiveness and data relationship formulation. This paper is structured as follows, Sect. 2 introduce the role of open data in Digital Transformation, then Sect. 3 briefly explain why Linked Data and Linked Open Data is the choice our approach. A practical result will be described in Sect. 4 with Linked Open Data for tourism in Thua Thien Hue. Finally, the paper is concluded with outlook for future work.

2 Open Data as the Core of the Digital Transformation

Open Data is the idea that some data should be freely available to everyone to use and republish as they wish, without restrictions from copyright, patents or other mechanisms of control [3, 4]. The philosophy behind Open Data has been long established, but the term "Open Data" itself is recent, gaining popularity with the rise of the Internet and World Wide Web and, especially, with the launch of Open Government Data (OGD) initiatives such as *data.gov* in the United States., *data.gov.uk* in the United Kingdom and *data.europa.eu* in the European Union.

We are witnessing the trend of opening governmental data crossing the world first for the transparency of the governments and later for the digital transformation revolution in the industrial revolution 4.0 wave. As mentioned, developed countries or bloc like the US, the UK and lately EU with its "Digital Agenda for Europe" have been putting the Open Data is the core of their digitalization strategy. And now, we can see what similarly happening in ASEAN countries as well [4].

Open data goes beyond data from the government alone: non-governmental organizations (NGOs), universities, companies, or individuals can post open data publicly for people to analyze.

Official portals with OGD where government agencies and national statistics offices of countries feed open data portals are attracted the attention in the field. In this paper, we studied two of the main definition frameworks for open data, including the Open Definition and the Open Government Data Principles [1], and the "Open data" term implies data that conform to both. There are properties for open data that we can share as common:

- anyone can access, use and share.
- becomes usable when made available in a common, machine-readable format.
- must be licensed. Its license must permit people to use the data in any way they want, including transforming, combining and sharing it with others, even commercially.

3 Linked Data – a Choice for Open Data Format

3.1 Principles of Linked Data

Open Data analysis is urgently demanded for a new approach in which the datasets must be taken into account in an integrated view by exploring and discovery new knowledge and information from explicit inter-lined data items from open datasets. Linked Data is based on RDF for designing a new paradigm of linking the data items on the Web. Linked Data is one of the core pillars of the Semantic Web, also known as the Web of Data [5]. Linked Data is a set of design principles for sharing machine-readable interlinked data on the Web. The four design principles of Linked Data proposed by its inventor Tim Berners-Lee as early as in 2006 described as follows [6]:

1. Use URIs as names for things. The URI is a single global identification system used for giving unique names to anything – from digital content available on the Web to real-world objects and abstract concepts.
2. Use HTTP URIs so that people can look up these names. As the HTTP protocol provides a simple mechanism for retrieving resources, when things can be identified by URIs in conjunction with this protocol, they become easier to find.
3. When someone looks up a URI, provide useful information, using the standards (RDF, SPARQL).

The RDF is a graph-based representation format for data publishing and interchange on the Web developed by the W3C. It is also used in semantic graph databases (also known as RDF triplestores) – a technology developed for storing inter-linked data and inferring new facts out of existing ones. In Fig. 1 shows a sample data items published in Linked Data (RDF format).

SPARQL, on the other hand, is the W3C-standardized query language for extracting, retrieving and manipulating data stored in RDF format. As such, it allows us to search the Web of Data and discover relationships.

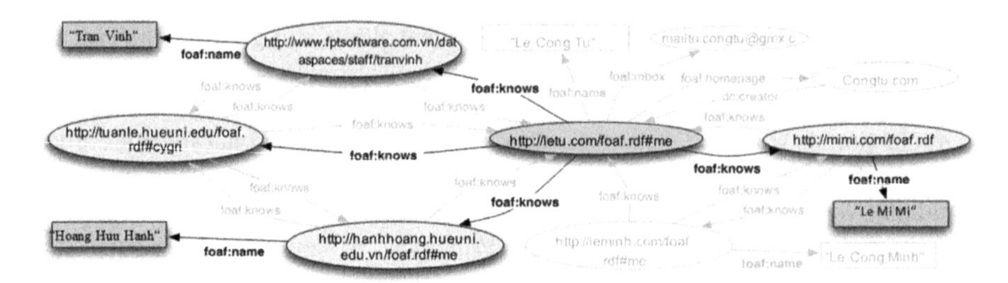

Fig. 1. An example of data represented in RDF/Linked Data

4. Include links to other URIs so that they can discover more things By interlinking new information with existing resources, we maximise the reuse and inter-linking among existing data and create a richly interconnected network of machine-processable meaning.

In order to achieve and create Linked Data, data should be available for a common format (RDF) to make access to existing databases. Provided by W3C, technological standards RDF, RDFa, R2RML, RIF, and SPARQL are for us to get access to the machine-readable datasets. Linked Data breaks down the information silos existing between various formats and puts down the fences between various valuablessources.

3.2 5-Star Linked Open Data Scheme

Linked Open Data (LOD) is a powerful blend of Linked Data and Open Data: it is both linked and uses open sources. LOD is Linked Data that is released under an open license and does not impede its reuse for free. One notable example of an LOD set is DBpedia [7] – a crowd-sourced community effort to extract structured information from Wikipedia and make it available on the Web.

In 2010, Tim Berners-Lee suggested a 5-star deployment scheme as Table 1 for Open Data [6]. The rating begins at one-star and the more proprietary formats are removed and links added, the more stars data gets.

3.3 Linked Open Data Cloud

The LOD Cloud [8] is a knowledge graph that manifests as a Semantic Web of Linked Data. It is the natural product of several ingredients:

- Open Standards—such as URI, URL, HTTP, HTML, RDF (plus other RDF Notations), SPARQL;
- Data curators across industry and academia;
- A modern DBMS platform: Virtuoso from OpenRDF (RDF4J for now);
- Seed Databases—initially, DBpedia and Bio2RDF[1] formed the core; significant contributions have come from the Wikidata project and the Schema.org-dominated SEO

[1] bio2rdf.org.

Table 1. Linked Open Data quality scheme [6]

Quality	Description	Format / Example
★	Available on the web *but with an open license, to be Open Data*	whatever format
★★	Available as machine-readable structured data	Excel instead of image scan of a table
★★★	As (2) plus non-proprietary format	CSV or JSON instead of Excel
★★★★	All the above, plus: use open standards to identify things	RDF
★★★★★	All the above, plus: link your data to other people's data to provide context	Linked Data / RDF

and SSEO axis supported by Search Engine vendors (Google, Bing, Yandex, and others).

The LOD Cloud provides a loosely coupled collection of data, information, and knowledge that is accessible by human or machine using Internet with the abstraction layer provided by the Web. It permits both basic and sophisticated lookup-oriented access using either the SPARQL or SQL queries.

3.4 Thua Thien Hue Open Data Initiative Goes for Linked Open Data

TTHODI is an initiative of Thua Thien Hue for laying down a data-centric foundation and infrastructure for their e-gov services. In order to realize this goal, HueCIT[2] is tasked to building an Open Data Agenda for the provincial authorities for publishing OGD datasets. Considering several approaches for OGD publishing for the official data portal of Thua Thien Hue, we have found the "Digital Agenda for Europe" have systematic and shared guidelines for building a data-driven foundation for the digital societies. According guidelines of the EU Open data portal[3], characteristics of new paradigm of open data should contain at least following criteria:

- Accessibility: open data should be accessed by
  - URL, URI
  - Datasets
- Machine Readability: open data should be serialized in easily machine-readable formats of:
  - XML

[2] www.huecit.vn.

[3] data.europa.eu.

  - RDF
  - Linked Data

- DCAT Compliance: data categories should be well defined in: Catalogue
- License: government license for using the open data published by authorities.

We have decided to develop the TTODI using the approach of using 5-star Linked Open Data for open data quality and planned to build the catalogues with high-level ontologies or vocabularies which are inter-linked to the LOD Cloud[4] focusing on datasets of DBpedia [6] and Geonames [7].

Hue, the capital city of Thua Thien Hue, is a heritage city and a hotspot for tourists. The smart city strategy of Thua Thien Hue stating with smart tourism, that is the reason we are choosing tourist open data as case study. Our objective is aimed at demonstrating the ideas of having as much Linked Open Datasets then as much benefits we can achieve in perspectives of authorities, its citizen and businesses. In this paper, we start with the focal area of publishing data into Linked Data and a prototype of demonstrating the advantages of using Linked Data for exploring information and also helping for smart information querying.

4 Linked Open Data Mashup for Open Tourist Data Exploration in Thua Thien Hue

4.1 Transforming Raw Data to RDF/Linked Data

The first step of the data transformation of the existing data sets in TTHODI legacy systems is about converting the data into RDF format and then attach them to the LOD cloud. We have used the tools provided by W3C and trusted service providers[5] for accomplishing this task.

Platform for Linked Open Data Publishing

- D2R (Database to RDF) Server: a platform for transform relational databases into linked data formats
- Talis Platform: a platform for storing published linked data
- Pubby: a platform supports querying the Linked Data using SPARQL
- Paget: a an application platform for Linked Data application development
- Linked Media Framework: a framework for supporting of publishing, querying and semantic searching datasets in Linked Data.
- PublishMyData: and this platform for publishing our data sets in form of Linked Data into the LOD cloud by associating data items with existing vocabularies.

Validating the Published Linked Data Sets

- Hyena: we have used this tool for as RDF editing framework.

[4] lod-cloud.net.

[5] linkeddata.org/tools.

- Vapour: then we have used this tool as RDF validator for validating syntax of the RDF/Linked Data sets.

A segment of resulted RDF datasets is shown below (Fig. 2 (a) and (b)) in which, the RDF/XML serialisation of the mashuped data is about a set of hotels and its relevant relationship information, such as < `dbpedia-owwl:isPartOf` > to a resources described in DBpedia dataset[6]. This complies with Linked Data principles above.

Using GeoNames

When converting the collected data into Linked Data, we have taken four design principles of Linked Data into account by having our data links to be inter-connected to the data items indatasets in LOD cloud, such as DBpedia and GeoNames[7].

GeoNames [9] is a geographic dataset published in LOD cloud and also provide the services for access the data. By associating our tourist dataset to GeoNames data items, we have established the linkage between the two datasets and from one we can navigate other by RDF links (Fig. 3).

Extracting RDF Resources by Associating Local to Global Data Sources

Sole datasets cannot provide more valuable information as data are inter-linked each other in many ways. For instance, a people working in a university in France looks for a hotel in Hue for his coming conference in the region; he also would like to do some social activities during his stay. With these information, sole datasets could not provide him an appropriate answer. By associating datasets from different domains by "cross-references" using RDF links based on URI, we can easily retrieve the valuable information, then mashup them up as solution for the request.

The RDF datasets were formulated from different sources using different corresponding services: Relational database (D2R) for legacy systems, websites (Web spider) for provincial offices which do not have a database systems, together with text mining for data items analysis techniques (NLP and Open Calais), other data sources using Freebase or DBpedia.

While we have data from legacy databases and systems, we still need to crawl the information which is available in heterogenous data sources on the Internet to enrich or complete the missing data attributes in the datasets. With LOD cloud, we have a huge semantic database containing billions of triples that enable the associating our local data items to the global ones. This will unleash the potentials of accessing data in over the world for our contextualized solution which is likely the way we have been using World Wide Web for our content sharing platform and changing our lives and how we live today.

4.2 Tourist Service Support Systems Using the Tourist Linked Open Dataset

Motivation

Building smart applications on top of the Linked Open datasets for smart support businesses and peoples is the second objective of the TTHODI. This objective is aimed

[6] dbpedia.org.

[7] geonames.org.

```
10 <rdf:Description rdf:about="http://tss.huecit.vn/Romance+Hotel">
11     <rdf:type rdf:resource="http://tss.huecit.vn/location"/>
12     <tss:id>1</tss:id>
13     <tss:city>Hue</tss:city>
14     <tss:name>Romance Hotel</tss:name>
15     <tss:type>Hotel</tss:type>
16     <tss:address>16 Nguyễn Thái Học, Phú Hội, Hue, Huế</tss:address>
17     <tss:star>4.0</tss:star>
18     <tss:price>34.0</tss:price>
19     <tss:rate>4.5</tss:rate>
20     <tss:image>https://d2zsw2b3bfsikz.cloudfront.net/hotelpictures/VNTLII/2RC390000.jpg</tss:image>
21     <tss:review>In the heart of Hue, Romance Hotel is close to Dong Ba Market and Truongtien Bridge.
        City</tss:review>
22     <dbpedia-owl:isPartOf rdf:resource="http://dbpedia.org/page/Hue,_Vietnam"/>
23 </rdf:Description>
24
25 <rdf:Description rdf:about="http://tss.huecit.vn/Vina+Hotel+Hue">
26     <rdf:type rdf:resource="http://tss.huecit.vn/location"/>
27     <tss:id>2</tss:id>
28     <tss:city>Hue</tss:city>
29     <tss:name>Vina Hotel Hue</tss:name>
30     <tss:type>Hotel</tss:type>
31     <tss:address> 57 Nguyễn Công Trứ, Phú Hội, Hue, Huế</tss:address>
32     <tss:star>4.0</tss:star>
33     <tss:price>18.0</tss:price>
34     <tss:rate>5.0</tss:rate>
35     <tss:image>https://d1pa4et5htdsls.cloudfront.net/images/88/4822/111742/111742-2977162_29_b-origi
36     <tss:review>A stay at Vina Hotel Hue places you in the heart of Hue, convenient to Dong Ba Marke
        Imperial City.</tss:review>
37     <dbpedia-owl:isPartOf rdf:resource="http://dbpedia.org/page/Hue,_Vietnam"/>
38 </rdf:Description>
```

(a)

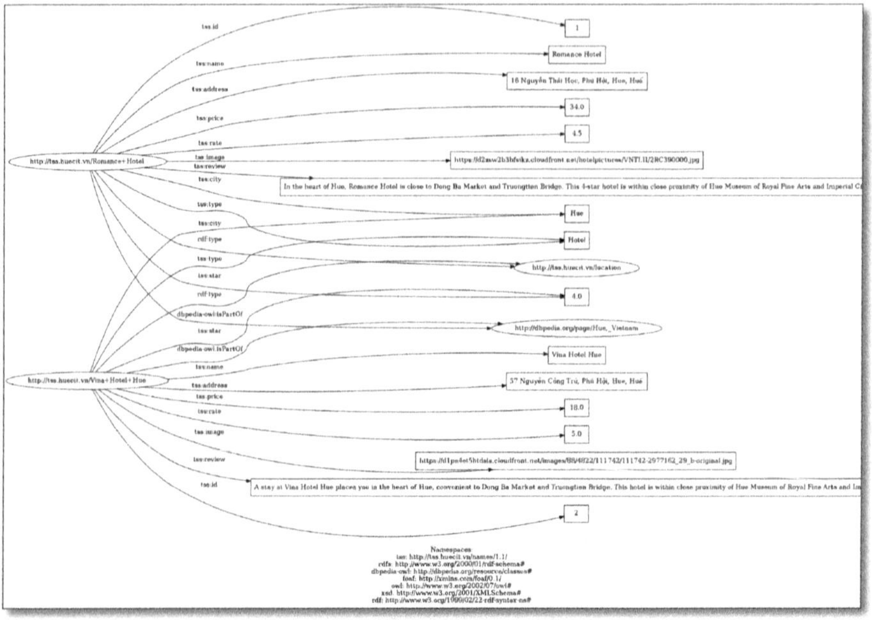

(b)

Fig. 2. A segment of TTHODI's Linked open data with two resources {"Romance Hotel" and "Vina Hotel Hue"} in form of (a) RDF/XML and (b) RDF graph visualization

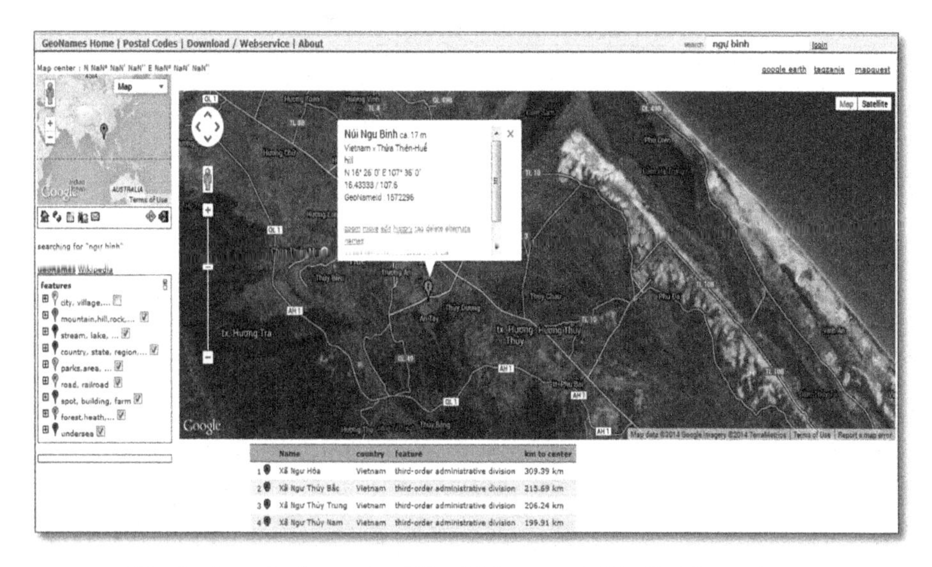

Fig. 3. A result of accessing "Ngu Binh mountain" in GeoNames dataset

make the data "living" in smart applications so that they are widely used by different stakeholders.

Tourist Service Support system (TSS) is an in-house application that we have developed to demonstrate the new abilities with the published Linked Open Datasets of hotels and places in the region such as linking relevant data items, smart querying. The dataset used in TSS is a part of Thua Thien Hue tourism open database and published using Lined Data design issues. The dataset could be found in Web address of http://huecit.vn/lods/hotels_places.rdf. TSS is designed and developed to provide a smart city service (smart tourist services) aligning to the Thua Thien Hue Smart city strategy.

TSS is a Linked Data-based Web application allowing its users to search for tourist information in Thua Thien Hue area and able to explore further with the advantages of Linked Data. TSS also plays as a recommender system based on the RDF links within Linked Data sets and also on the search results. The recommendations range from accommodation, tours, monuments, restaurants and also special local areas for tourists.

TSS System Architecture

As described in Fig. 4, TSS system architecture is designed in layered model with lower layers providing services for the uppers. Layers are defined as follows:

- Presentation Layer: front-end layer of the TSS: the UI/UX of the systems.
- Mashup Layer: this is the most important layer of the TSS for mashing up Linked Data sources within the TSS and other services such as Google Maps (Google location service).
- Data Access, Integration & Storage Layer: this core layer provides services for converting data to RDF/Linked Data format: The collected and crawled data will be process and transformed into Linked Data. During this process, the data links have

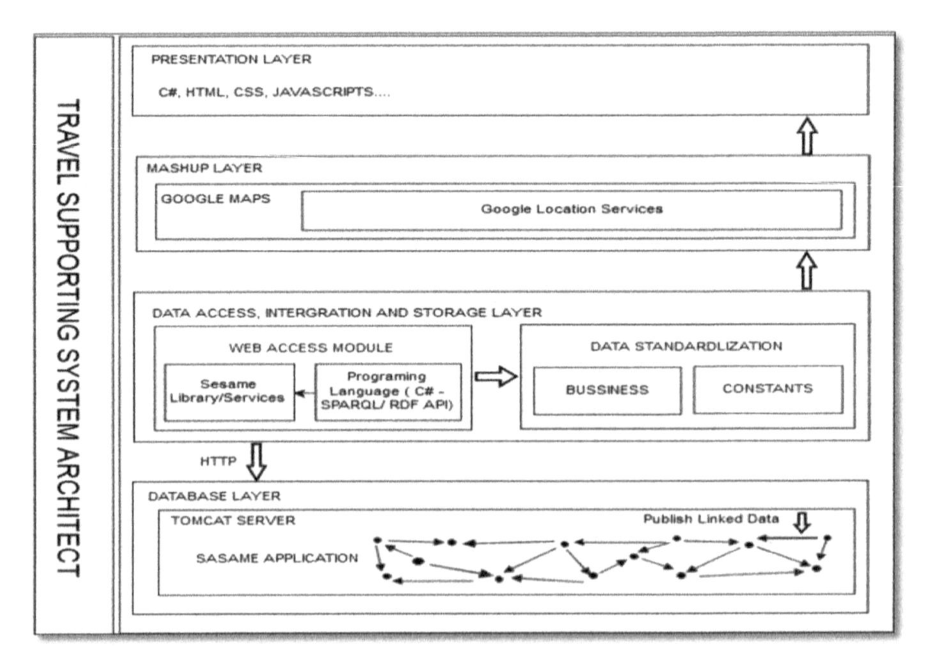

Fig. 4. Proposed TSS system architecture

been created and validated. SPARL queries for data and information querying and processing also provided in this layer.
- Database Layer: storing tourist linked datasets and crawled datasets in triple store for local and remote use.

Preliminary Result
TSS is developed using framework such as Tomcat server[8], Sesame Framework[9] and .NET framework[10]. TSS creates and publishes the data in format of RDF/XML into the local servers. With support by C# Visual Studio, dotNetRDF[11] to access to Sesame-server for information updating and retrieval. We also using Google Maps services and Google location services for showing Linked Data items with Geonames [7] information on the map for increasing visually and interactively.

TSS provides a semantic search for location, hotels, restaurants, tours and monuments and results will be shown on the interactive map (Fig. 5). Based on the limited data on the legacy databases and also crawled data from Web, beside the tourist information for Hue, TSS also provides a limited portion of tourist information for Saigon (Ho Chi Minh city), Hanoi, Danang and Quy Nhon.

[8] http://tomcat.apache.org/.
[9] http://www.openrdf.org (rdf4j.org).
[10] https://dotnet.microsoft.com/.
[11] http://www.dotnetrdf.org.

The semantic search has been also prototyped in TSS by allowing users entering a sentence for querying. The querying text will be parsed using NLP and Open Calais to extract the "objects" from the query. Extracted objects are then mapped to data items that are stored in the Linked Data sets of TSS system. For example:

```
String content = "Mr. Hung would like to book a hotel with rating
                  over 4.5 rate for Hue conference";
Map<String, List<String>> results = new OpenCalaisClient().
getPropertyNamesAndValues(content);
for (String key : results.keySet()) {
System.out.println(" " + key + ": " + results.get(key));}
```

The above code segment is executed and produces the following data objects:

```
Person: [Hung]
Relations: [HotelBooking]
City: [Hue]
Rate: [> 4.5]
Country: [Vietnam]
```

Based on the parsed objects, we then map the object to data items and formulate the corresponding SPARQL queries against the dataset.

Fig. 5. Results for best hotel in Hue based on the linked datasets

Figure 5 shows the found hotel for the posted query for Hotel in Hue. TSS also provides the navigation ability to LOD cloud data items thanks to the established RDF links between TSS data and the LOD. The geographic data retrieved from DBpedia and GeoNames could also be presented into TSS application but in English.

5 Conclusion and Outlook

This paper presented the approach using Linked Data as the data model and format representation for open datasets and make it as data-centric infrastructure for smart services and applications in local e-government services. This approach is aligned to the initiative of TTHODI in Digital Transformation agenda of Thua Thien Hue province. Based the survey and practical approached from official open data portals of developed countries in over the world, our decision to choose Linked Data for publishing local OGD to make a breakthrough in the trend of unleased government data to public and also motivate open data from enterprise as well.

This is the first step of the long journey of TTHODI and a case study showing that if we have a lot of data in the form of Linked Data, we can have and build smart applications using those data for benefits of community and business.

In the near future, HueCIT will expend this research results for all legacy and local government data which are licensed for opening and sharing in to the Linked Data format and launching a program for promoting and motivating developer and start-up communities to build up smart applications based on the linked open data platform of TTHODI.

References

1. Carrara, W., Cha, W.S., Fischer, S., van Steenbergen, E.: Creating value through open data: study on the impact of re-use of public data resources. Eur. Commission (2015). https://doi.org/10.2759/328101
2. Rogers, D.L.: Digital Transformation Playbook - Rethink Your Business for the Digital Age. Columbia Business School Publishing (2016)
3. Zaidi, E., Thoo, E., Heudecker, N.: Magic Quadrant for Data Integration Tools. Gartner Inc. (2019)
4. Stagars, M.: Open Data in Southeast Asia – Towards Economic Prosperity, Government Transparency, and Citizen Participation in the ASEAN. Palgrave Macmillan (2016)
5. Heath, T., Bizer, C.: Linked data: evolving the web into a global data space. Synth. Lect. Semant. Web: Theory Technol. **1**(1), 1–136 (2011)
6. Tim Berners-Lee: Linked Data – Design Issues. https://www.w3.org/DesignIssues/LinkedData.html
7. Auer, S., Bizer, C., Kobilarov, G., Lehmann, J., Cyganiak, R., Ives, Z.: DBpedia: a nucleus for a web of open data. In: Aberer, K., et al. (eds.) ASWC/ISWC -2007. LNCS, vol. 4825, pp. 722–735. Springer, Heidelberg (2007). https://doi.org/10.1007/978-3-540-76298-0_52
8. Linked Open Data Cloud. https://lod-cloud.net/
9. Wick, M., Vatant, B.: GeoNames Geographical Database. https://geonames.org/

A Literature Review on Dynamic Pricing - State of Current Research and New Directions

Karol Stasinski(✉)

Wroclaw University of Science and Technology, 27 Wybrzeże Wyspiańskiego Street, 50-370 Wrocław, Poland
karol.stasinski@pwr.edu.pl

Abstract. In recent years, the topic of dynamic pricing has drawn a significant attention. Many different scientific communities have studied demand estimation techniques and price policies, committing greatly to its development. This paper surveys latest literature streams, providing a brief overview of research background and state of current works, identifying most active sub-fields and emerging directions. It focuses mostly on computer science and operations literature, however it also discusses contributions from management, economics and marketing fields. This study goal is to provide scientific community, as well as business sector with brief summary of latest research efforts.

Keywords: Dynamic pricing · Algorithmic pricing · Pricing strategy · Demand estimation

1 Introduction

1.1 Motivation

Online shopping is one of the most popular online activities worldwide with global e-retail being 14.1% of all sales and reaching 3.5 trillion U.S. dollars in 2019 [13]. Together with the advancement of information technologies and growing transportation networks, the general availability of goods worldwide is increasing. This phenomenon forces local supplier to compete on global, allowing sellers not only to expand their operations, but also requiring them to adjust for rival actions. Together with big-data revolution, and the ever growing improvement in processing power (thanks to cloud computing boom [38]), companies enthusiasm in use of data analytics to support, or even automate decision making process is rising. The trend of identifying market information, as a key aspect in best product placement and pricing rather than relying on seller's "gut instinct", is developing. As a result, organizations start to process extensive data to gain customer insights, propagate knowledge to their consumers and examine current trends to compete for customer's share of wallet [19]. This data revolution

M. Hernes et al. (Eds.): ICCCI 2020, CCIS 1287, pp. 465–477, 2020.
https://doi.org/10.1007/978-3-030-63119-2_38

is driving intent of more frequent and targeted price adjustments to maximize revenue, while preserving clients' attitude towards the product.

Similar phenomenon can be spotted on energy supply market. In 2017 total world's electricity consumption reached 23,696 PWh and has been steadily growing 2.3% rate year-to-year [33]. The on-going urbanization process causes regular increase in baseload energy demand, however uneven adaptation of power-hungry technologies such as AC units, entertainment devices in the past and today's electric vehicles, results in local demand spikes. The increasing and uncoordinated electrical load imposes significant challenges for the stable operation of the power grid [42], and most often needs to be handled by "peaker plants" or energy storage facilities [12]. That eventuality forces power producers to adequately plan the variability of demand, creating the market for peak-demand suppliers that activate on specific conditions, such as amount of power needed or price per watt-hour.

Those issues are the main motivation of this paper. Dynamic pricing is a study of determining optimal pricing policy for sold products or services and can be seen as special form of demand response, allowing better optimization of resource utilization. Furthermore, for goods and services with price-dependant nature of consumption it can even be utilized as demand shaping mechanism. This applies to vendors selling their products through variety of channels, as long as they allow them to quick adjust for changing environment. Whether it comes to e-commerce or, mentioned before power grid, dynamic pricing addresses greater part of issues concerned. Generally, it can be divided into two, strictly connected problems: demand estimation and understanding demand-price relation, allowing for price forecasting and shaping. This characteristic makes dynamic pricing quite popular scientific topic, as well as thanks to digital technology, this technique is nowadays widely utilized in various businesses, sometimes considered to be an indispensable part of pricing policies.

Dynamic pricing approaches are already discussed in existing literature reviews. The power gird an EV charging concerns have been widely discussed by Limmer [42], Siano [56], Vardakas et al. [61], Deng et al. [17], and for retail markets, den Boer [8]. This article aims at being a dynamic pricing guide, discussed as a concept, without further narrowing into either of subfields. The main goal is to identify what builds DP, what problems researchers solved in this area, what obstacles require further research, as well as summarise to what other fields dynamic pricing can be or is applied.

1.2 Methodology

In order to find all relevant references to include in this article, three science knowledge aggregation websites were used: Google Scholar, Springer Link, and ResearchGate. Summary is build upon papers published since October 2016. There were included overlapping works, such as extended conference papers published in journals, as well as double versions or duplicates. The aim of this work is to be an ad-hoc review of relevant publications in dynamic pricing literature. In other sections, the scope is restricted only to key papers and reviews for clarity.

1.3 Composition of the Paper

The following survey is organized as follows: Sect. 2 consists of review of most valuable historical work and past reviews in the field of algorithmic price calculation, and addresses initial difficulties, as well as practical application problems of reviewed topic. Its goal is to provide a scientifically sound context for current work analysis. Section 3 focuses on review of most recent (since 2016) literature published in the field of dynamic pricing, grouped by inventory restrictions considered, demand estimation approach and application area. Its goal is to clearly summarize current state of research in DP field. This section is the core of the paper, while others are supporting. Section 4 aims to highlight new directions in scientific research that have emerged on the basis of papers discussed in Sect. 3, as well as to identify possible extensions to those studies. Section 5 is meant to summarize all discussed topics.

2 Background

Dynamic pricing as a subject of study takes its roots in economics. As mentioned previously, it generally breaks into two problems: demand estimation and understanding of demand-price relation, allowing price forecasting. As such, both these fields are not anything new and date back to more than a century. Only recently, those works have been repurposed to work in virtual environment as core decision-aiding or decision-making mechanisms in computer science. This section covers fundamental breakthroughs and concepts that emerged in the past, allowing shaping up nowadays algorithmic pricing techniques.

Demand Estimation. It is practically impossible to make an arbitrary decision upon the optimal pricing without identification of customer demand. As discussed by Meghir and Robin [46], the analysis based on actual number of purchases needs to be concluded and fed to theoretical mathematical model in order to draw probable demand estimation. First acknowledged work identifying demand curves was done by Charles Davenant [16] more than 3 centuries ago, which relates the supply to price of corn (mentioned further in [8]). With the advancement of statistical analysis and increase in population (therefore inclining demand) in the 20th century, it can be seen that more advanced research in market and demand understanding took off.

Price Forecasting. First acknowledged source that described price as mathematical function, encompassing multiple features was "Researches into the Mathematical Principles of the Theory of Wealth" [15]. Cournot defined demand-price relation of products as mathematical function, paving the way to determine the optimal price, stated as mathematical problem. This definition was picked by Fisher [21], who studied application of mathematical methods in economic problem. As stated in the abstract of his work in 1898, "Cournot's genius must give a new mental activity to everyone who passes through his hands." [21]. The mathematical description of demand-price relation was groundbreaking concept of its time. Cournot showed that, by assuming demand as a function of price, it is possible to find unique point being the best price (this is described in Chapter IV of Cournot's book [15] and more broadly covered in Sect. 2 of den Boer review [8]). Even though, Cournot was the first to solve "static pricing" problem, as den Boer mentions [8] his work was neglected for several decades.

Practical Appliance Issues. Initially, the goal of estimating demand curves was not to maximize profits, but used to support macro-economic theories on price, supply, and demand [8]. Application of algorithmic price adjustments was rejected as a concept, due to complexity and difficulty of estimating the demand curves for products [31]. Moreover, lack of or (later on) not sufficient automatisation meant that frequency of price adjustment would be too low to accommodate for sudden demand bursts [5].

3 Dynamic Pricing

As mentioned previously, dynamic pricing is composition of two problems: demand estimation and price forecasting. Furthermore, as price forecasting depends on identified demand and inventory restriction (i.e. vendor cannot sell wares that are not present in inventory), it is distinguishable that analysed papers can be characterized on the basis of demand estimation technique proposed and inventory restrictions considered. In addition to that, dynamic pricing approach can also be characterized by the field to which it is applicable. This section aims to cover these selected areas, presenting current state of research and identifying key contributions in recently published works.

3.1 Inventory Restrictions

It is crucial to identify inventory restrictions, when defining pricing policy for dynamic pricing solution. That issue is one of major features characterizing dynamic pricing concerned papers. There are two important research streams that focus on that matter. The first covers inventory-procurement type of problems and is connected to inventory size limitation in time. The second is revenue-quality management type of problems, where inventory can deteriorate its quality over time.

Inventory-Procurement Problems. This stream focuses on analysis of limitations concerning inventory size and expected sell-out time. It is assumed that a vendor has some amount of products, which need to be sold in a finite amount of time. Depending on the inventory replenishment strategy and supply type, within time-fixed selling horizon, it can be simplified that considered limitations split further into two problems: selling a fixed number of products during a finite time period and selling maximum (sustainable) number of products during a finite time period.

Dynamic pricing with fixed inventory has been discussed by Fisher et al. [22] paper. A competition-based dynamic pricing model is developed to adjust to rapidly-changing market - for e-commerce market. The work focuses on strategic consumer and on choice of model identification. It aims to thoroughly analyse the price and sensitivity of target audience in order to optimize products' price in selling time window. In comparison, most studies in that area [6,9,10,23,40,48,63] treat product-specific price sensitivity as constant, Fisher et al. conduct several experiments that allow them to measure the extent to which price sensitivities vary across products, proving that consumer willingness to buy specific product can be additionally amplified, as well as product price endogeneity concerns can be addressed in randomized real-life environment. The findings show that a best-response pricing algorithm that takes into account consumer behaviour, competitor actions, and supply parameters results in significant revenue improvement, and in case of team's specific product category final calculated improvement was 11%.

Dynamic pricing under strictly limited supply has been described by Chekired et al. [11] and Luo et al. [45] regarding electric vehicles charging infrastructure and power grid. In this case, dynamic pricing is used as smoothing agent for uneven demand management. Additionally, similar practise can also be observed in Lei and Ouyang work [39] regarding urban parking management. In both scenarios, the studies identify limited supply (being electrical power or number of parking spots available respectively), and devise algorithmic price estimation as a way to manage sustainable resource distribution and counteract sudden demand bursts. The approach to supply limitation presented in cited papers ranges from most general demand-procurement description and optimization done by Chekired's team; through multi-period multi-level decision models used by Lei and Ouyang to stochastic local optimizations done by Luo's team. Despite different methods of inventory-procurement management in all of the studies, dynamic pricing played key role as optimization agent, yielding significant difference when comparing static price definitions.

Revenue-Quality Management. Compared to main assumptions of literature discussed above, revenue-quality management stream focuses mostly on situations in which inventory value may decrease in time due to its quality or incurred utilization costs. Therefore, it is assumed that a vendor has some amount of perishable products, that needs to be sold with the maximum revenue in time-bounded selling horizon. In literature such situation is described as inventory effect, and has been extensively studied in the past (see [14,28,30,51,62]).

Decaying client interest has been discussed by Herbon and Khmelnitsky [29]. In their work, the researchers addressed the problem of determining the optimal dynamic pricing policy and replenishment schedule for a perishable product, with the aim of maximizing the retailer's profit. Effectively, the result of their study is an algorithm for the best replenishment schedule, allowing to fine-tune client's decay of interest and profit management.

Another notable paper has been written by Diabat et al. [18]. It covers distribution problems of perishable items. This work inspired further study done by Avinadav et al. [3], which analyses dynamic pricing together with promotion expenditure model to find an optimal way to adjust to decaying product interest. Additionally, the paper shows that, in the proposed EOQ model, the general demand function is separable into multiplicative components of selling price, products' age and promotion expenditure. That allows these parameters to be fine-tuned accordingly to products' need independently.

3.2 Demand Estimation Technique

Another, important feature that characterizes the reviewed papers is how the authors approach the problem of demand identification and estimation. Most noticeable streams regarding that matter are: action-set models, stochastic models, fuzzy models, learning models.

Action-Set Model, called also event based approach or model, relies on set of predefined reactions to market changes. That model has been partially used by Avramidis [4] to evaluate purchase probability based on historical data, as well as Nayak and Patil patent [49] describing dynamic pricing eventing system.

Stochastic Model, called also probabilistic models (in dynamic pricing context), are meant to adjust to random consumer or competition variations causing price fluctuations, allowing to estimate probability distribution of potential decision(s) outcomes. This type of demand modelling is by far the most popular, and has been used in: energy consumption management [45,55], routing [27,52,53], deteriorating stock management [20,37,41,47], as well as in general dynamic pricing model evaluation [7,24].

Luo et al. [45] and Reka and Ramesh [55] discussed the application of stochastic dynamic optimization of demand curve by application of dynamic pricing. Luo's [45] team focuses on EV charging infrastructure, showing that the proposed model can achieve up to 7% better efficiency in energy consumption management. Reka and Ramesh [55] paper, on the other hand, focuses on residential energy market and attempts to solve the smart-grid design problem.

Prokhorchuk et al. [52], Rambha et al. [53] and Han et al. [27] discussed routing and delivery cost optimization, by application of dynamic pricing with stochastic demand prediction. Prokhorchuk et al. [52] work discusses the problem of same-day delivery in e-commerce space. Rambha et al. [53] studies the costs of day-to-day routing with several choice models. Han et al. [27] paper analyses dynamic road congestion pricing with the use of stochastic user optimal principle.

Li et al. [41], Moustafa et al. [47], Kitaeva et al. [37], Dye [20] studied perishable products management in e-commerce stores with stochastic inventory system. Li et al. [41] studied the dynamic pricing and periodic ordering with stochastic stock analysis for inventory with deteriorating items. The work [47] by Moustafa et al. focuses on stochastic demand forecasting for sustainable pricing of food items. The work [37] by Kitaeva et al. analyses dynamic pricing model under stochastic demand for zero ending inventory, with fixed lifetime and order quality. The paper [20] by Dye addresses the joint problem of optimal pricing of items and marketing efforts in perishable-items-inventory management.

Bi et al. [7] studied dynamic pricing model under the assumption that consumers' memories are limited and their recall of previous prices obeys a first-order Markov stochastic process. The paper shows that with the increasing loss-averse of consumers, the steady-state price range widens, concluding on the memory window importance in profit forecasting.

Fuzzy Logic Model, called also fuzzy sets, are meant to process vague and imprecise information (coming from market, i.e. competitor price change without cause context). This type of demand modelling is substantially useful to deal with limited information coming from consumers, markets or competitors.

Tsao et al. [60] studied the problem of the sustainable smart-grid design in three-dimensional sustainability dimensions: economy, environment, and society. Authors apply multi-objective robust fuzzy stochastic programming in order to minimize the network costs under proposed scenarios. The paper concludes with further application of derived model in Vietnamese residential sector, reducing environmental costs by 3%. Future research work identified by Tsao's team covers further scaling of paper's concept and parameter optimization.

Similar work has also been published by Yu et al. [64]. Their paper studies interval-fuzzy chance-constrained programming as the means of demand forecasting for most optimal peak electricity management planning. The study uncovers several findings regarding peak power consumption and electricity-generation patterns. The team identifies that further investigation of trade and supply of energy is needed to mitigate limitations of the proposed model.

Learning Model, called also learning methods, have been mentioned many times over the years in dynamic pricing context. However most dominant term in recent papers is reinforcement learning (RL). It is used to solve decision making problem and may utilize several environment definitions, though typically Markov decision process is used.

Lu et al. [44] investigated the application of RL in hierarchical, smart-grid concept. Q-learning is employed to solve the decision making problem. The team proposes auto-learning model that does not require pre-specified input and is taught through dynamic on-line interaction. The authors concluded that future extension of their model is an investigation of optimal values of weighting factor between service provider's profit and customers' final cost.

Arango [2] described car-sharing platform optimization by utilizing dynamic pricing with reinforcement of learning demand estimation. The author proposed a Markov decision process, and sets pre-defined environment setup in order to smooth-out demand curve of the proposed platform. Arango identified that most valuable improvement to his work would come from reward-definition investigation as the current design ignores bias of user occurrence distribution.

Zhang et al. [65] reviewed the current state of research and practice of deep learning and reinforcement learning in smart grid. Their paper compares three learning techniques, namely: deep learning, reinforcement learning and their combination - deep reinforcement learning. It introduces the concept of status quo on these methods, and summarizes their potential in smart-grid application.

3.3 Application Area

The last and the most important feature characterizing reviewed papers are the industries to which these works are applicable.

Power Management. Dynamic pricing in power management, for example of EV charging sustainability management, has already been mentioned in works of Chekired et al. [11] and Luo et al. [45]. However, in recent years a large number of papers have been published in this field covering several sub-areas such as: smart grid management (Nguyen et al. [50], Almahmoud et al. [1], Tang et al. [59], Srinivasan et al. [57]), distributed energy management (Ji and Tong [34], Chekired et al. [11], Liu et al. [43]), load forecasting (Khan et al. [36], Rasool et al. [54]), showing inclining scientific interest in energy management and sustainability.

E-Commerce and Services. Papers published on dynamic pricing in e-commerce field have covered several sub-areas such as: same-day delivery (see the work [52] by Prokhorchuk et al.), deteriorating stock management (see Li et al. [41], Moustafa et al. [47], Kitaeva et al. [37], Dye [20]), ticket sales (see Dutta [19], Xu et al. [35], Gupta et al. [26]), car sharing (see Arango [2]), hotels and apartments (see Gibbs et al. [25], Zhuang et al. [66]).

4 New Directions in Dynamic Pricing

Most of the literature discussed above have been covered in a very general form. This section aims to summarize new areas of research that emerged in recently published papers, additionally identifying potential extensions to mentioned works in terms of research or application.

New Approach to Reinforcement Learning. The learning techniques in dynamic pricing have been discussed over the years [8], however only recently the computational power enabled researchers of deep and reinforcement learning, to cover full nature of algorithmic pricing. As discussed by den Boer [8], learning techniques have been mostly used in demand estimation. Nevertheless, recent studies discussed by Lu et al. [44] assume that reinforcement learning can also be used for pricing policy to learn market dynamics. This allow the model to encompass both demand estimation and price calculation into single package. This may allow to create a solution for more natural market management with greater autonomy than dual-model approach and is worth further study.

Negative Demand Markets. Electric energy supply market has noted a booming scientific interest in last four years, as can be seen by the works summarized in Sect. 3.3 of this paper. The discussed papers identify new unprecedented challenges that dynamic pricing is facing. The most important being demand management and identification modifications, that must take under consideration the fact that consumer can be treated as a power-storage, or even temporal producer (see work [58] by Subramanian and Das and [45] by Lue et al.). These new extensions open new section in dynamic pricing history, as they can be used to shape market demand from perspective of a consumer, contrary to most of the papers discussed in this work that study dynamic pricing from the perspective of monopolist producer.

Used Goods Market. Identified in e-commerce market related articles (see Sect. 3.3 of this paper) used goods market have received little scientific interest, while some of its branches such as used books market noted huge boom of commercial interest, according to BBC reports [32]. This trend may be the result of relaxed market restrictions on certain markets (such as Central European Market), and will develop as goods flow between countries increases. Further understanding of the challenges of this branch can significantly commit to dynamic pricing due to high competitive nature of those markets. Moreover, low single-item profit forces competing sellers to focus greatly on price optimization of volume sales, suggesting that dynamic pricing can be the best technique used for revenue management.

5 Conclusion

Dynamic pricing, in general, has received considerable research attention in recent years. Several different scientific communities have studied demand estimation techniques and price policies, most often with different goals. It can be noted that the most active research has been done in the area of electrical energy supply, especially in smart-grid sustainability. In spite of those different fields, many of papers overlapped in the means of the most general approach. This survey aims to provide ad-hoc overview on those studies, and facilitate further research on the broad topic of dynamic pricing.

References

1. Almahmoud, Z., Crandall, J., Elbassioni, K., Nguyen, T.T., Roozbehani, M.: Dynamic pricing in smart grids under thresholding policies. IEEE Trans. Smart Grid **10**(3), 3415–3429 (2018)

2. Arango, M.: Toll road with dynamic congestion pricing using reinforcement learning (2019)
3. Avinadav, T., Chernonog, T., Lahav, Y., Spiegel, U.: Dynamic pricing and promotion expenditures in an EOQ model of perishable products. Ann. Oper. Res. **248**(1–2), 75–91 (2017)
4. Avramidis, A.: Dynamic pricing with finite price sets and unknown price sensitivity. Author's Original (2018)
5. Bauer, D., Reiss, M.C.: Dynamic pricing: some thoughts and analysis. J. Accounting Finance **19**(3) (2019). https://doi.org/10.33423/jaf.v19i3.2029, https://www.articlegateway.com/index.php/JAF/article/view/2029
6. Besbes, O., Zeevi, A.: Dynamic pricing without knowing the demand function: risk bounds and near-optimal algorithms. Oper. Res. **57**(6), 1407–1420 (2009)
7. Bi, W., Li, G., Liu, M.: Dynamic pricing with stochastic reference effects based on a finite memory window. Int. J. Prod. Res. **55**(12), 3331–3348 (2017)
8. den Boer, A.V.: Dynamic pricing and learning: historical origins, current research, and new directions. Surv. Oper. Res. Manag. Sci. **20**(1), 1–18 (2015)
9. Boyd, E.A., Bilegan, I.C.: Revenue management and e-commerce. Manag. Sci. **49**(10), 1363–1386 (2003)
10. Caro, F., Gallien, J.: Clearance pricing optimization for a fast-fashion retailer. Oper. Res. **60**(6), 1404–1422 (2012)
11. Chekired, D.A., Khoukhi, L., Mouftah, H.T.: Decentralized cloud-SDN architecture in smart grid: a dynamic pricing model. IEEE Trans. Ind. Inform. **14**(3), 1220–1231 (2017)
12. Clarke Energy: Electricity peaking stations using gas engines. https://www.clarke-energy.com/natural-gas/peaking-station-peak-lopping-plants/. Accessed 17 Jan 2020
13. Clement, J.: Global e-commerce share of retail sales 2023, August 2019. https://www.statista.com/statistics/534123/e-commerce-share-of-retail-sales-worldwide/. Accessed 16 Jan 2020
14. Costa, A.M., dos Santos, L.M.R., Alem, D.J., Santos, R.H.: Sustainable vegetable crop supply problem with perishable stocks. Ann. Oper. Res. **219**(1), 265–283 (2014)
15. Cournot, A.A.: Researches into the Mathematical Principles of the Theory of Wealth. Macmillan (1897)
16. Davenant, C.: An essay upon the probable methods of making a people gainers in the ballance of trade. James Knapton, at the Crown in St. Paul's Church-yard (1700)
17. Deng, R., Yang, Z., Chow, M.Y., Chen, J.: A survey on demand response in smart grids: mathematical models and approaches. IEEE Trans. Industr. Inf. **11**(3), 570–582 (2015)
18. Diabat, A., Abdallah, T., Le, T.: A hybrid tabu search based heuristic for the periodic distribution inventory problem with perishable goods. Ann. Oper. Res. **242**(2), 373–398 (2016)
19. Dutta, A.: Capacity allocation of game tickets using dynamic pricing. Data **4**(4), 141 (2019)
20. Dye, C.Y.: Optimal joint dynamic pricing, advertising and inventory control model for perishable items with psychic stock effect. Eur. J. Oper. Res. **283**(2), 576–587 (2020). https://doi.org/10.1016/j.ejor.2019.11.008, http://www.sciencedirect.com/science/article/pii/S0377221719309191

21. Fisher, I.: Cournot and mathematical economics. Q. J. Econ. **12**(2), 119–138 (1898). https://doi.org/10.2307/1882115, https://academic.oup.com/qje/article-pdf/12/2/119/5465583/12-2-119.pdf
22. Fisher, M., Gallino, S., Li, J.: Competition-based dynamic pricing in online retailing: a methodology validated with field experiments. Manag. Sci. **64**(6), 2496–2514 (2018)
23. Gallego, G., Van Ryzin, G.: Optimal dynamic pricing of inventories with stochastic demand over finite horizons. Manag. Sci. **40**(8), 999–1020 (1994)
24. Ghoreishi, M.: Testing the robustness of deterministic models of optimal dynamic pricing and lot-sizing for deteriorating items under stochastic conditions. Central Eur. J. Oper. Res. **27**(4), 1131–1152 (2019). https://doi.org/10.1007/s10100-018-0538-7
25. Gibbs, C., Guttentag, D., Gretzel, U., Yao, L., Morton, J.: Use of dynamic pricing strategies by airbnb hosts. Int. J. Contemp. Hospital. Manag. (2018)
26. Gupta, V., Singh, K., Arjaria, S.K., Biswas, B.: Dynamic pricing in movie tickets using regression techniques. In: 2018 International Conference on Advanced Computation and Telecommunication (ICACAT), pp. 1–4. IEEE (2018)
27. Han, L., Zhu, C., Wang, D.Z., Sun, H., Tan, Z., Meng, M.: Discrete-time dynamic road congestion pricing under stochastic user optimal principle. Transp. Res. Part E: Logist. Transp. Rev. **131**, 24–36 (2019)
28. Herbon, A.: Dynamic pricing vs. acquiring information on consumers' heterogeneous sensitivity to product freshness. Int. J. Prod. Res. **52**(3), 918–933 (2014)
29. Herbon, A., Khmelnitsky, E.: Optimal dynamic pricing and ordering of a perishable product under additive effects of price and time on demand. Eur. J. Oper. Res. **260**(2), 546–556 (2017)
30. Herbon, A., Levner, E., Cheng, T.: Perishable inventory management with dynamic pricing using time-temperature indicators linked to automatic detecting devices. Int. J. Prod. Econ. **147**, 605–613 (2014)
31. Hicks, J.R.: Annual survey of economic theory: the theory of monopoly. Econometrica **3**(1), 1–20 (1935). http://www.jstor.org/stable/1907343
32. Hooker, L.: The booming trade in second-hand books, December 2018. https://www.bbc.com/news/business-46386557. Accessed 17 Jan 2020
33. International Energy Agency: World Energy Outlook 2019. OECD Publishing, Paris (2019). https://doi.org/10.1787/caf32f3b-en, https://www.oecd-ilibrary.org/content/publication/caf32f3b-en
34. Jia, L., Tong, L.: Dynamic pricing and distributed energy management for demand response. IEEE Trans. Smart Grid **7**(2), 1128–1136 (2016)
35. Jiaqi Xu, J., Fader, P.S., Veeraraghavan, S.: Designing and evaluating dynamic pricing policies for major league baseball tickets. Manuf. Serv. Oper. Manag. **21**(1), 121–138 (2019)
36. Khan, A.R., Mahmood, A., Safdar, A., Khan, Z.A., Khan, N.A.: Load forecasting, dynamic pricing and dsm in smart grid: a review. Renew. Sustain. Energy Rev. **54**, 1311–1322 (2016)
37. Kitaeva, A.V., Stepanova, N.V., Zhukovskaya, A.O.: Zero ending inventory dynamic pricing model under stochastic demand, fixed lifetime product, and fixed order quantity. IFAC-PapersOnLine **52**(13), 2482–2487 (2019)
38. Klebnikov, S.: Microsoft is winning the 'cloud war' against amazon: Report, January 2020. https://www.forbes.com/sites/sergeiklebnikov/2020/01/07/microsoft-is-winning-the-cloud-war-against-amazon-report/. Accessed 16 Jan 2020
39. Lei, C., Ouyang, Y.: Dynamic pricing and reservation for intelligent urban parking management. Transp. Res. Part C: Emerg. Technol. **77**, 226–244 (2017)

40. Levin, Y., McGill, J., Nediak, M.: Dynamic pricing in the presence of strategic consumers and oligopolistic competition. Manag. Sci. **55**(1), 32–46 (2009)
41. Li, Y., Zhang, S., Han, J.: Dynamic pricing and periodic ordering for a stochastic inventory system with deteriorating items. Automatica **76**, 200–213 (2017)
42. Limmer, S.: Dynamic pricing for electric vehicle charging–a literature review. Energies **12**(18), 3574 (2019)
43. Liu, Y., Zuo, K., Liu, X.A., Liu, J., Kennedy, J.M.: Dynamic pricing for decentralized energy trading in micro-grids. Appl. Energy **228**, 689–699 (2018)
44. Lu, R., Hong, S.H., Zhang, X.: A dynamic pricing demand response algorithm for smart grid: reinforcement learning approach. Appl. Energy **220**, 220–230 (2018)
45. Luo, C., Huang, Y.F., Gupta, V.: Stochastic dynamic pricing for EV charging stations with renewable integration and energy storage. IEEE Trans. Smart Grid **9**(2), 1494–1505 (2017)
46. Meghir, C., Robin, J.M.: Frequency of purchase and the estimation of demand systems. J. Econ. **53**(1), 53–85 (1992). https://doi.org/10.1016/0304-4076(92)90080-B
47. Moustafa, G.Y., Galal, N.M., El-Kilany, K.S.: Sustainable dynamic pricing for perishable food with stochastic demand. In: 2018 IEEE International Conference on Industrial Engineering and Engineering Management (IEEM), pp. 961–965. IEEE (2018)
48. Nair, H.: Intertemporal price discrimination with forward-looking consumers: application to the US market for console video-games. Quantitative Market. Econ. **5**(3), 239–292 (2007)
49. Nayak, A., Patil, V.: Dynamic pricing systems and methods (Sep 4 2018), US Patent 10,068,241
50. Nguyen, D.T., Nguyen, H.T., Le, L.B.: Dynamic pricing design for demand response integration in power distribution networks. IEEE Trans. Power Syst. **31**(5), 3457–3472 (2016)
51. Olsson, F.: Analysis of inventory policies for perishable items with fixed leadtimes and lifetimes. Ann. Oper. Res. **217**(1), 399–423 (2014)
52. Prokhorchuk, A., Dauwels, J., Jaillet, P.: Stochastic dynamic pricing for same-day delivery routing. arXiv preprint arXiv:1912.02946 (2019)
53. Rambha, T., Boyles, S.D.: Dynamic pricing in discrete time stochastic day-to-day route choice models. Transp. Res. Part B: Methodol. **92**, 104–118 (2016)
54. Tahir, A., Khan, Z.A., Javaid, N., Hussain, Z., Rasool, A., Aimal, S.: Load and price forecasting based on enhanced logistic regression in smart grid. In: Barolli, L., Xhafa, F., Khan, Z.A., Odhabi, H. (eds.) EIDWT 2019. LNDECT, vol. 29, pp. 221–233. Springer, Cham (2019). https://doi.org/10.1007/978-3-030-12839-5_21
55. Reka, S.S., Ramesh, V.: Demand response scheme with electricity market prices for residential sector using stochastic dynamic optimization. In: 2016 Biennial International Conference on Power and Energy Systems: Towards Sustainable Energy (PESTSE), pp. 1–6. IEEE (2016)
56. Siano, P.: Demand response and smart grids–a survey. Renew. Sustain. Energy Rev. **30**, 461–478 (2014)
57. Srinivasan, D., Rajgarhia, S., Radhakrishnan, B.M., Sharma, A., Khincha, H.: Game-theory based dynamic pricing strategies for demand side management in smart grids. Energy **126**, 132–143 (2017)
58. Subramanian, V., Das, T.K.: A two-layer model for dynamic pricing of electricity and optimal charging of electric vehicles under price spikes. Energy **167**, 1266–1277 (2019)

59. Tang, R., Wang, S., Li, H.: Game theory based interactive demand side management responding to dynamic pricing in price-based demand response of smart grids. Appl. Energy **250**, 118–130 (2019)
60. Tsao, Y.C., Thanh, V.V., Lu, J.C.: Multiobjective robust fuzzy stochastic approach for sustainable smart grid design. Energy **176**, 929–939 (2019)
61. Vardakas, J.S., Zorba, N., Verikoukis, C.V.: A survey on demand response programs in smart grids: pricing methods and optimization algorithms. IEEE Commun. Surv. Tutorials **17**(1), 152–178 (2014)
62. Yadavalli, V.S., Sundar, D.K., Udayabaskaran, S.: Two substitutable perishable product disaster inventory systems. Ann. Oper. Res. **233**(1), 517–534 (2015)
63. Ye, S., Aydin, G., Hu, S.: Sponsored search marketing: dynamic pricing and advertising for an online retailer. Manag. Sci. **61**(6), 1255–1274 (2015)
64. Yu, L., Li, Y., Huang, G.H., Shan, B.: A hybrid fuzzy-stochastic technique for planning peak electricity management under multiple uncertainties. Eng. Appl. Artif. Intell. **62**, 252–264 (2017)
65. Zhang, D., Han, X., Deng, C.: Review on the research and practice of deep learning and reinforcement learning in smart grids. CSEE J. Power Energy Syst. **4**(3), 362–370 (2018)
66. Zhuang, W., Chen, J., Fu, X.: Joint dynamic pricing and capacity control for hotels and rentals with advanced demand information. Oper. Res. Lett. **45**(5), 397–402 (2017)

Intelligent Modeling and Simulation Approaches for Games and Real World Systems

Sentiment Analysis by Using Supervised Machine Learning and Deep Learning Approaches

Saud Naeem, Doina Logofătu(✉), and Fitore Muharemi

Frankfurt University of Applied Sciences, Frankfurt, Germany
logofatu@fb2.fra-uas.de

Abstract. With the growth of online review sites, we can use the opportunity to find out what other people think. Sentiment analysis is a popular text classification task in the data mining domain. In this research, we develop a text classification machine learning solution that can classify customer reviews into positive or negative class by predicting the overall sentiment of the review text on a document level. For this purpose, we carry out the experiment with two approaches, i.e. traditional machine learning approach and deep learning. In the first approach, we utilize four traditional machine learning algorithms with TF-IDF model using n-gram approach. These classifiers are multinomial naive Bayes, logistic regression, k-nearest neighbour and random forest. Out of these four classifiers, logistic regression achieves the highest accuracy. The second approach is to utilize deep learning methodologies with the word2vec approach, for which we develop a sequential deep learning neural network. The accuracy we achieve with deep learning is much lower than our traditional machine learning approach. After finding out the best performing approach and the classifier, the next step of the work is to build our final model with logistic regression using some advanced machine learning methodologies, i.e. Synthetic Minority Over-sampling for data balancing issues, Shuffle Split cross-validation. The accuracy of the final logistic regression model is approximately 87% which is 3% higher from the initial experimentation. Our finding in this research work is that, in smaller dataset scenarios, traditional machine learning would outperform deep learning models in terms of accuracy and other evaluation metrics. Another finding in this work is, by addressing data balancing issues in the dataset the accuracy of the model can be improved.

Keywords: Sentiment analysis · Text classification · Supervised learning · Machine learning · Deep learning

1 Introduction

Nowadays one of the important and typical task in supervised machine learning in the field of sentiment analysis is a text classification. Sentiment analysis

M. Hernes et al. (Eds.): ICCCI 2020, CCIS 1287, pp. 481–491, 2020.
https://doi.org/10.1007/978-3-030-63119-2_39

is a field dedicated to extracting subjective emotions and sentiments from the text. A common use of sentiment analysis is to find out whether a text expresses negative, positive or neutral sentiment. Sentiment analysis provides the comprehension information related to public views, as it analyze different tweets and reviews. It is a verified tool for the prediction of many significant events such as box office performance of movies and general elections [15]. The purpose of sentiment analysis is to automatically determine the expressive direction of user reviews. The demand of sentiment analysis is raised due to increase requirement of analyzing and structuring hidden information which comes from the social media in the form of unstructured data [16]. Machine learning based techniques are implemented by extracting the sentences and aspect levels. The features consist of Parts of Speech (POS) tags, n-grams, bi-grams, uni-grams and bag-of-words. Machine learning contains three flavors at sentence and aspect, i.e., Naive Bayes, Support Vector Machine (SVM) and Maximum Entropy. Machine learning approach has three categories: i) supervised; ii) semi supervised; and iii) unsupervised. This approach is capable of automation and can handle huge amount of data therefore these are very suitable for sentiment analysis [17].

Deep Learning was firstly proposed by G.E. Hinton in 2006 and is the part of machine learning process which refers to Deep Neural Network[18]. Neural network is influenced by human brain and it contains several neurons that make an impressive network. Deep learning networks are capable for providing training to both supervised and unsupervised categories [19]. Deep learning includes many networks such as CNN (Convolutional Neural Networks), RNN (Recurrent Neural Networks), Recursive Neural Networks, DBN (Deep Belief Networks) and many more. Neural networks are very beneficial in text generation, vector representation, word representation estimation, sentence classification, sentence modeling and feature presentation [20]. For the accurate classification of sentiments, many researchers have made efforts to combine deep learning and machine learning concepts in the recent years, here in this work we make a comparison between the two using a review dataset.

2 Problem Statement

In this work we want to predict the positive or negative sentiment on a document level using two different approaches for a provided review text. The steps we follow in this work are:

- Features extraction from a review text e.g.:

  " *I was a happy customer for over ten years, but now I want to change my opinion and want to say that it is a pathetic company.*"

  Utilizing data preprocessing techniques to remove those words which will not help the machine learning solution to predict the polarity of a review, e.g. was, for, but etc.
- By utilizing features, e.g. "happy", "pathetic" to create a vocabulary space.
- From the vocabulary extracted from the reviews predict the polarity of a review text, i.e. positive for "happy" or negative for "pathetic".

There are several ML approaches available for performing sentiment analysis, however, finding out what approach and algorithm would perform better for this dataset is our concern in this research work. For this purpose, we will carry out experimental work with a small labelled dataset containing customer reviews with positive and negative labels.

3 Related Work

In [5], Sawakoshi et al. proposed a system by using support vector machine as a machine learning model, where they experimented whether "opinion" and "fact" sentences influence on the classification results. After conducting experiments on "Hotel reviews dataset", they establish in the first experiment that "opinion" and "fact" sentences were not affecting the classification of reviews into negative or positive if only opinion sentences were used. In another experiment, they concluded from test data that it has a significant influence on "recall" value for negative reviews. In future work, they want to utilize deep learning techniques to classify positive and negative reviews more efficiently, method which we applied here.

In [7], Kumar et al. proposed a system for opinion mining using three different machine learning algorithms such as Naïve Bayes, Logistic Regression and SentiWordNet. This system is capable of extracting reviews from amazon.com, preprocessing the text, select a particular product to obtain reviews for that product, classify those reviews and display the results. Different experiments were done for different products to find out the best performing algorithm. The performance was based on various parameters such as recall, precision and F-measure. The best results were generated with Naïve Bayes classifier. In future work, they want to use other algorithms on other different datasets to analyze their efficiency.

In this study [14], author has proposed a model for sentiment analysis considering both visual and textual contents of social networks. This new scheme used deep neural network model such as Denoising auto encoders and skip gram. The base of the scheme was CBOW (Continuous Bag-Of-Words) model. The proposed model consisted of two parts CBOW-LR (logistic regression) for textual contents and was expanded as the CBOW-DA-LR. The classification was done according to the polarity of visual and textual information. Four datasets were evaluated, i.e., Sanders Corpus, Sentiment140, SemEval2013 and SentiBank Twitter dataset. The proposed model outperformed the CBOWS+SVM and FSLM (fully supervised probabilistic language model).

An interesting work is the one of Prabowo et al.[13], they combine rule-based classification, supervised learning and machine learning into a new combined method. This method is tested on movie reviews, product reviews and MySpace comments. The results show that a hybrid classification can improve the classification effectiveness in terms of micro- and macro-averaged F1 score. We have not used any hybrid systems in our work, but we will test it in the future work.

4 Approaches

This section explains what approaches we will be utilizing for sentiment analysis work. Our first approach is with traditional machine learning techniques, which includes algorithms like multinomial naive bayes, logistic regression, k-nearest neighbour and random forest. Our second approach is to experiment with a deep learning model utilizing Keras API. For both approaches, we are using the accuracy evaluation metric score as our comparison metric.

Multinomial Naive Bayes uses probabilistic method of learning. This algorithm is widely used in NLP problems and used mostly for text classification where a label of a text is predicted. Naive Bayes is a family of probabilistic algorithms based on applying Bayes theorem with the naive assumption of conditional independence between every pair of a feature [8].

Logistic Regression is a statistical ML algorithm that classifies the data by taking out resulting variables on the maximum ends and try to create a logarithmic line that differentiates between them. It is also one of the popular algorithms used for text classification problems. The name Logistic derived from Logit Function which is used in this classification algorithm to calculate the relationship between dependent and independent variable/variables, this function is also called sigmoid-Function. The second half of the name Regression is derived from Linear Regression because it uses the same technique as Linear Regression does for problem solving [9].

K-Nearest Neighbor, also known as KNN is a supervised data classification algorithm that estimates how probable a data point can be a part of one class or depending on what class the data points nearest to it are in. It is also known as lazy-learner algorithm because it will not build a model utilizing the training dataset until unless a query of the data set is executed [10].

Random Forest also falls into the category for supervised machine learning algorithms. As the name suggests, it creates a forest and makes it somehow random. The larger the number of trees in the forest, the accurate the results gets. In other words, *it creates multiple decision trees and merge them together to get a more precise and stable prediction.* It is somehow related to the decision trees algorithm, but there is a twist, and the twist is it's randomness, the splitting and finding the root node in the random forest will run randomly [11].

Deep Learning Networks, In order to train the model we are going to use a type of Recurrent Neural Network, know as LSTM (Long Short Term Memory). This network is able to remember the sequence of past data i.e. words in our case in order to make a decision on the sentiment of the word. We are going to create the network using Keras. Keras is built on tensorflow and can be used to build most types of deep learning models. In order to estimate the parameters such as dropout, no of cells etc I have performed a grid search with different parameter values and chose the parameters with best performance. I have used ReLU as the activation function, and adam optimizer.

5 Implementation

This section provides a detailed insight into the experimental work carried out during this work utilizing traditional machine learning and deep learning approaches. The dataset used during the work is a labelled customer reviews dataset from an automotive company, which consists of approximately three thousand reviews. Python was used as a programming language with Spyder IDE. Table 1 below provides us with the dataset information:

Table 1. Dataset columns and their datatypes

Column name	Data type	Description
Review_Source	String	Tells us the source of the review
Reviewer_Name	String	Who was the reviewer
Customer_Reviews	String	Actual Review text about the product or service
Label	String	is it a positive or negative review
Year	Integer	What year when the review was posted

5.1 Traditional Machine Learning Approach

Following *no free lunch theorem* we selected four classification algorithm for our experimentation work in traditional machine learning approach. Starting with data preprocessing steps we performed following steps:

- Dropping irrelevant columns from our dataset.
- Removing null values from the data.
- Removing punctuation and unnecessary spaces.
- Normalization of normal words and negation words in the dataset.
- Removing duplicate entries in the dataset.
- Removing stopwords from the dataset.

After data preprocessing, we performed some initial data visualization steps to understand our dataset; this includes wordcloud for positive and negative words in the dataset, bar-plot for the classes distribution. After that, we performed feature engineering step in which we first mapped our class labels into a numerical representation. In the next step, we created tf-idf representation of the text data. After features extraction step, we fit and predicted with our four algorithms/classifiers, after cross-validating our classifiers, logistic regression came up with the highest accuracy. In the next step, the best parameters for the logistic regression was extracted, and with those parameters, we developed our final logistic regression classifier in the machine learning approach. With shuffle split cross-validation on our final logistic regression classifier, we achieved a good level of accuracy as an evaluation metric.

5.2 Deep Neural Networks

In our second approach, we utilized the Keras API to build our deep learning network for our text classification problem. Same data preprocessing steps were following from our traditional machine learning approach. For features extraction step, we utilized word2vec model, where a numerical form of a word is represented in a vector. We selected sequential type of a model. Our model consisted of an input layer, followed by a layer which helps the model to deal with the input of the variable length, output from the previous layer is piped through fully-connected layer with 16 hidden units which is followed by the layer which represents a single output node which is densely connected and in the end an output layer. The accuracy achieved with this approach was much lower than our previous machine learning approach. The loss amount was also much higher than expected.

6 Experimental Results

In this section the results of our experimentation work from the mentioned two approaches for sentiment analysis will be presented.

6.1 Machine Learning Approach

In the initial stage of this work, we selected four popular classification algorithms for our problem. In the first stage, we found out which classifier is performing well. The training dataset was 75% of the dataset, and the validation contained 25% of the data. Different evaluation metrics were considered before selecting the final machine learning classifier. Let's have a look at the results from the first step:

In Table 2, we can see different evaluation metrics generated from the algorithms we chose for the experimentation. Here CV represents cross-validation. As we can see that logistic regression is the winner here, we are keeping accuracy and ROC-AUC metric as a measuring parameter to choose the algorithm for further process. In the next step, we performed randomized search cross-validation to find our the best parameters for logistic regression. After finding out the best parameters, we developed our final machine learning classifier with some advanced machine learning techniques, i.e. Synthetic Minority Over-sampling for data balancing and shuffle split cross-validation. 25% of the validation data was reserved during the training process. The evaluation metrics from our final machine learning model are shown in Table 3 below. In Fig. 1, we can see the performance of our classifier in the form of the ROC-AUC curve, as we can see that our classifier has achieved 93% of ROC-AUC. This means that there is 93% chance that the model will be able to differentiate between positive and negative class, so the True Positive and True Negative rate is high in this case. The ideal situation in the figure will be if the curve is at the top left corner, which means that there is no False Negative and False Positive predicted by the model.

Table 2. Experimentation with four algorithms

	Naive Bayes	Log. Regression	KNN	Random Forest
Accuracy	85.50	84.84	82.31	79.52
Precision	86.11	84.06	82.27	79.57
Recall	85.50	84.84	82.31	79.52
F1 score	85.24	84.69	82.28	79.28
CV with Accuracy	85.34	85.47	81.15	80.55
CV with ROC-AUC	91.37	91.39	87.24	87.30

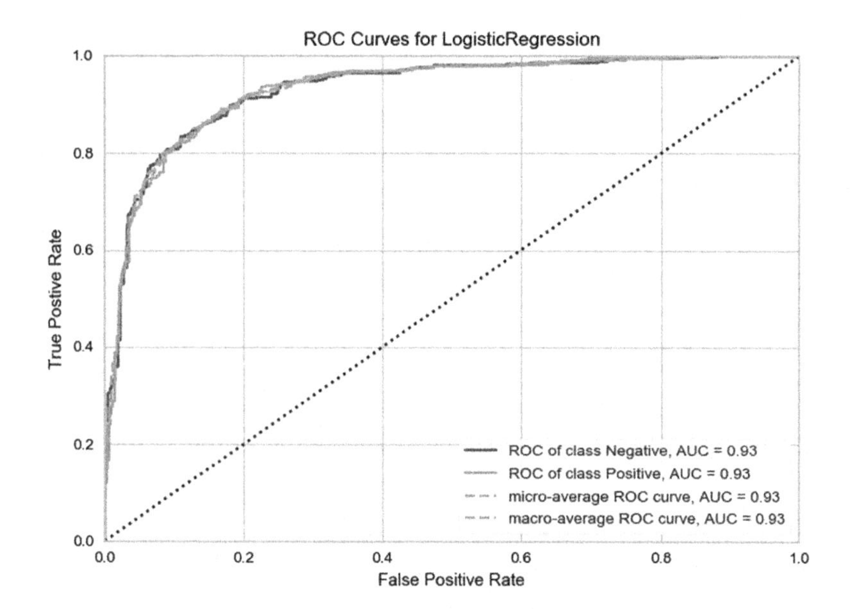

Fig. 1. ROCAUC curve from final ML model

Table 3. Evaluation metrics from final ML model

Accuracy	86.96%
Precision	87.006
Recall	86.96
F1 Score	86.96
ROC-AUC	93.00

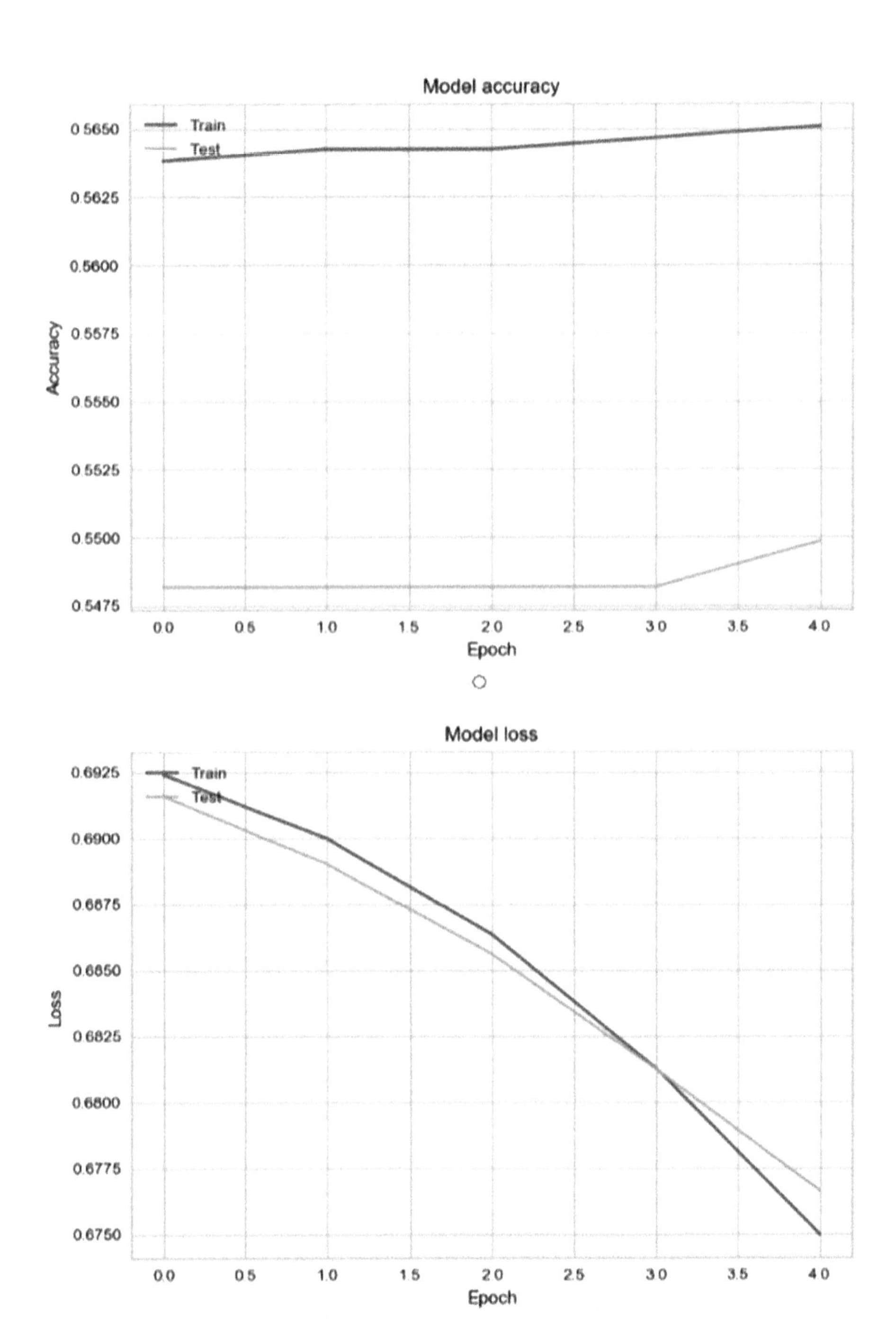

Fig. 2. Accuracy and loss with deep learning model

6.2 Deep Learning Model

We trained our deep learning model with 80% of training data and then validated with 20% of the remaining data, the accuracy we achieved was approximately 55%. In Fig. 2, we can see the accuracy in the training and validation process, which remained constant through the epochs. The second vital metric we considered for evaluation is the loss of our deep learning model, which was approximately 09%. Here loss means how much the model predicted away from an actual value. The higher the loss, the model is performing worse.

7 Conclusion and Future Work

In this research work, we have utilised traditional machine learning with the supervised approach and deep learning techniques to develop a sentiment analysis model that would predict the overall sentiment of text into positive or negative class. In our two different approaches, we carried out experimentation work with different algorithms to analyse the performance on the dataset we provided. Various evaluation metrics were considered to choose the right algorithm for further proceeding in our work. After several processes and considering various evaluation metrics in concern, the traditional machine learning approach with logistic regression provided us the best of the results for our research problem, with the logistic regression we achieved the accuracy of approximately 87% after applying a couple of advanced machine learning techniques, i.e. synthetic minority oversampling for data balancing issues and shuffle split cross-validation to verify the results with different splits of the data. The model was also applied on an unseen small dataset, and the accuracy of approximately 93% was achieved with it. The possible future work would be:

- Other approaches such as word level, sentence level and comparative sentence analysis can be utilized to check if a better accuracy can be achieved.
- A lexicon-based approach can be followed instead of machine learning, the advantage would be there is no training data required.
- Further data preprocessing techniques such a stemming, lemmatization and part-of-speech tagging to improve the model's accuracy.
- In features engineering step, increasing the number of n-grams in the tf-idf model to check whether if it has any impact on the model's accuracy.
- Multi-label classification for predicting more sentiments, i.e. neutral reviews.
- Handling smileys in the text to predict the sentiment.
- Experiment with techniques to handle multi-polarity within a text.
- Further work to process ambiguous wording in the text.

References

1. Gupta, S.: Applications of sentiment analysis in business. https://blog.paralleldots.com/product/applications-of-sentiment-analysis-in-business. Accessed 25 Jan 2019

2. Bhanot, K.: Why and what of machine learning? https://towardsdatascience.com/why-and-what-of-machine-learning-c0eda6ebe5b0. Accessed 26 Jan 2019
3. Grossfeld, B.: A simple way to understand machine learning vs deep learning. https://www.zendesk.com/blog/machine-learning-and-deep-learning/. Accessed 27 Jan 2019
4. Garbade, M.J.: A simple introduction to natural language processing. https://becominghuman.ai/a-simple-introduction-to-natural-language-processing-ea66a1747b32. Accessed 26 Jan 2019
5. Sawakoshi, Y., Okada, M., Hashimoto, K.: An investigation of effectiveness of "opinion" and "fact" sentences for sentiment analysis of customer reviews. In: 2015 International Conference on Computer Application Technologies, pp. 98–102, August 2015. https://doi.org/10.1109/CCATS.2015.33
6. Soni, S., Sharaff, A.: Sentiment analysis of customer reviews based on hidden Markov model. In: Proceedings of the 2015 International Conference on Advanced Research in Computer Science Engineering 38, Technology (ICARCSET 2015), ICARCSET 2015, pp. 12:1–12:5. ACM, New York (2015). ISBN 978-1-4503-3441-9. https://doi.org/10.1145/2743065.2743077
7. Kumar, K.L.S., Desai, J., Majumdar, J.: Opinion mining and sentiment analysis on online customer review. In: 2016 IEEE International Conference on Computational Intelligence and Computing Research (ICCIC), pp. 1–4, December 2016. https://doi.org/10.1109/ICCIC.2016.7919584
8. Sanghi, D.: Applying multinomial Naive Bayes to NLP problems. https://www.geeksforgeeks.org/applying-multinomial-naive-bayes-to-nlp-problems/. Accessed 29 Jan 2019
9. Narkhede, S.: Understanding logistic regression. https://towardsdatascience.com/understanding-logistic-regression-9b02c2aec102. Accessed 30 Jan 2019
10. Technopedia. K-nearest neighbor (K-NN). https://www.techopedia.com/definition/32066/k-nearest-neighbor-k-nn. Accessed 30 Jan 2019
11. Donges, N.: The random forest algorithm. https://towardsdatascience.com/the-random-forest-algorithm-d457d499ffcd. Accessed 30 Jan 2019
12. Heller, M.: What is keras? The deep neural network API explained. https://www.infoworld.com/article/3336192/what-is-keras-the-deep-neural-network-api-explained.html. Accessed 1 Feb 2019
13. Prabowo, R., Thelwall, M.: Sentiment analysis: a combined approach. J. Informetrics **3**(2) (2009)
14. Claudio, B., Tiberio, U., Marco, B., Del Bimbo, A.: A multimodal feature learning approach for sentiment analysis of social network multimedia. Multimedia Tools Appl. **75**(5), 2507–2525 (2015). https://doi.org/10.1007/s11042-015-2646-x
15. Heredia, B., Khoshgoftaar, T.M., Prusa, J., Crawford, M.: Cross-domain sentiment analysis: an empirical investigation. In: 2016 IEEE 17th International Conference on Information Reuse and Integration, p. 160165 (2016)
16. Haenlein, M., Kaplan, A.M.: An empirical analysis of attitudinal and behavioral reactions toward the abandonment of unprofitable customer relationships. J. Relatsh. Mark. **9**(4), 200228 (2010)
17. Aydogan, E., Akcayol, M.A.: A comprehensive survey for sentiment analysis tasks using machine learning techniques. In: 2016 International Symposium on Innovations in Intelligent Systems and Applications, p. 17 (2016)

18. Day, M., Lee, C.: Deep Learning for Financial Sentiment Analysis on Finance News Providers, no. 1, pp. 1127–1134 (2016)
19. Vateekul, P., Koomsubha, T.: A Study of Sentiment Analysis Using Deep Learning Techniques on Thai Twitter Data (2016)
20. Zhang, Y., Er, M.J., Wang, N., Pratama, M., Venkatesan, R.: Sentiment Classification Using Comprehensive Attention Recurrent Models, pp. 1562–1569 (2016)

EEG Based Source Localization and Functional Connectivity Analysis

Soe Myat Thu(✉) and Khin Pa Pa Aung

University of Information Technology, Yangon, Myanmar
{soemyatthu,khinpapaaung}@uit.edu.mm

Abstract. Nowadays Electroencephalography (EEG) is one of the most attractive Brain-Computer Interfaces (BCI) models to analyze brain signals source localization and connectivity estimation. Unlike other neuroimaging modalities such as fMRI, MEG, and PET; EEG has its higher temporal resolution that senses EEG is currently interesting area for many BCI researchers. However, the precise results of source localization and connectivity are challenging problems that mostly depend on the head models and inverse modeling methods. This paper focused on source localization and functional connectivity analysis using EEG signal over single-trial movement imaginary (MI) tasks by using brainstorm toolbox. Data obtained from the nature dataset that was recorded from 12 subjects, 29 recordings with 19 channels EEG device and MATLAB software utilized for the task.

Keywords: EEG data · Brainstorm toolbox · EEGLab · Biomedical imaging

1 Introduction

The source localization and the functional conductivity modeled can be determined in the volume conduction. Boundary Element Method (BEM) volume conduction model of the head (head model) is conducted from the subject's MRI (scalp, inner skull, outer skull) and source space (cortical surface). The Brainstorm can generate approximations based on the subject's cortex, head surfaces, ICBM152' inner and outer skull surface. The anatomical MRI data can be available from the Brainstorm Toolbox. Inhomogeneous, anisotropic and no spherical features of human heads affect EEG signals more than MEG. Thus, it is important to define more realistic head models for EEG data. To beat these problems, other imaging modalities, such as computerized tomography or magnetic resonance imaging are applied for extraction of the skull, brain, and scalp surface boundaries. Accurate source localization is highly dependent on the electric forward and inverse modelling methods.

Our main goal is to carry the attention of the neuroimaging researchers to the problems of neuronal sources estimation on the basis of the multichannel EEG recordings using OpenMEEG BEM and dipole modelling in Brainstrom Toolbox. Also, the problem of brain functional connectivity was analyzed by using Phase Locking Values (PLV) algorithm.

M. Hernes et al. (Eds.): ICCCI 2020, CCIS 1287, pp. 492–502, 2020.
https://doi.org/10.1007/978-3-030-63119-2_40

The proposed model includes four main parts. In the first section, data preprocessing over raw EEG data are discussed and then the literature reviews are described in next section. In the third section, Source Localization and Functional Connectivity are described with demonstrations. Finally, better experimental results are shown on the Motor Imagery (HaLT) dataset. These data are recorded Electro-Caps - 19 Channel EEG as shown in Fig. 1.

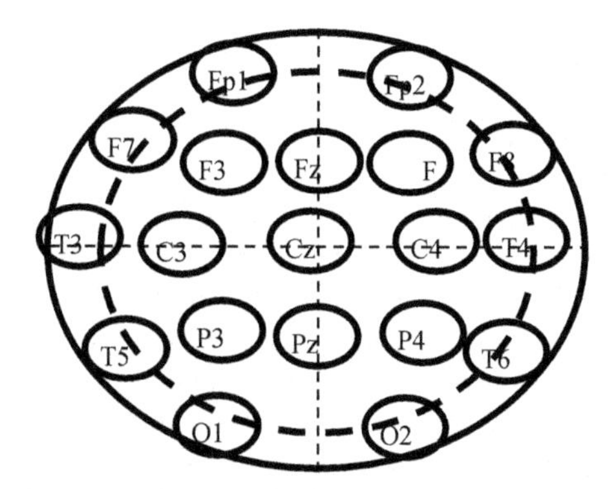

Fig. 1. Electro-Caps - 19 Channel EEG

2 Related Works

Source localization of signal information from distinct directions was been solved through researchers in many ways. Auditory signals are decoded in the brain and interpreted. The auditory cortex or temporal lobe in the brain is involved for processing auditory stimuli. The HRTF method was used to produce sound in distinct directions [1]. Experiments were conducted with auditory stimulation of human subjects and EEG was recorded from the subjects during the exposure for the sound. The cyclic short time Fourier transform and a set of time frequency analysis operators consisting of the continuous wavelet transform is applied to pre-processed EEG signal and the classifier was trained with time-frequency power from training data. The source localization of the sound is been used the SVM classifier.

Sound localization has been analyzed by different signal processing and statistical methods. In [2], Support Vector Machines (SVMs) classified virtual auditory stimuli in 6 directions with headphones to Human Subjects, recorded EEG data. Binaural and Monaural auditory stimulus was classified using ERPs in A Bednar et al. [3]. In Letawski [4], Interaural Time Difference (ITD) was used for auditory localization. It presents to analyze single-trial Movement Imagination (MI) tasks source localization using EEG signal. Wadsworth physio bank dataset for right hand versus left hand movement imagination is considered for 109 subjects performing. Forward modeling was based on 3 layered head geometry is co-registered with ICBM 152 template anatomy, that is a non-linear average of fMRI scans of 152 subjects. Inverse modeling was done with the assist of Standardized Low Resolution Electromagnetic Tomography (sLORETA) [5].

In [6], it represents the technique to cope with the high-dimensionality of source connectivity analysis, without requiring a priori input from the investigator. Brainstorm features cortical source estimation techniques integrated with anatomical MRI data volumes with a simple and intuitive graphical user interface, combined with powerful graphical pipeline builders and scripting abilities. In [7], it discusses several approaches of Bayesian framework, inverse problem and forward problem have been explored to overcome and address the uncertainties and issues of localization for the neural activities incurring in the brain.

3 Dataset Description

In this work, motor imagery (MI) BCI Nature datasets which are published as a scientific data was used [8]. This dataset was collected during the development of a slow cortical potentials-based EEG BCI. It contains 60 h of EEG BCI recordings across 75 recording sessions of 13 participants, 60,000 mental imageries, and 4 BCI interaction paradigms, with multiple recording sessions and paradigms of the same individuals. The participants included 8 males (61.5%) and 5 females (38.5%) who are between the ages of 20 and 35 participated in the study. BCI interactions involving up to 6 mental imagery states was considered. This dataset is one of the largest EEG BCI datasets published to date and presents a significant step from existing datasets in terms of uniformity, longitudinal and lateral coverage, and interaction complexity. These four (4) BCI interaction paradigms are 5F, CLA, HaLT, FREEFORM. The system focused on HaLT paradigm, which are recorded from 12 subjects, 29 recordings and the eGUI that are displayed six icons symbolizing right hand, left hand, tongue, left leg and right leg motor imageries together with a passive imagery indicated by a round circle as shown in Fig. 2. In the center of the screen the fixation point was shown. Action signals selecting the imagery to be implemented were shown as a red rectangle around the respective motor imagery icon. The advantage of this paradigm is contained all the movements of the participants. The abbreviation used in the data record naming convention is shown in Table 1.

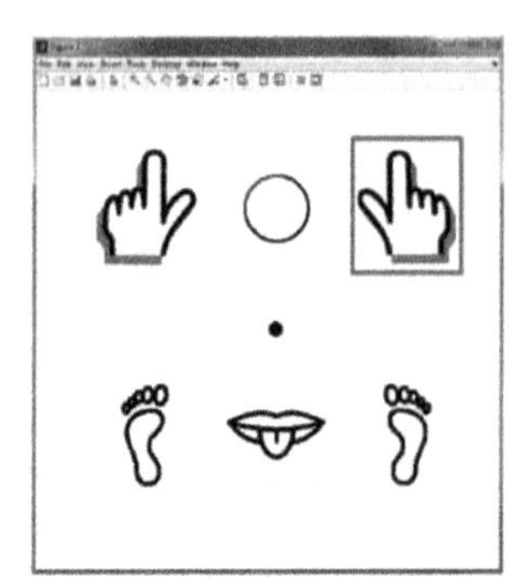

Fig. 2. The graphical user interfaces (eGUI) used for BCI interactions for HaLT Paradigm

Table 1. Abbreviate naming convention used to record data.

HaLT Paradigm	
"marker" code	Meaning
1	Left hand
2	Right hand
3	Passive/neural
4	Left leg
5	Tongue
6	Right leg
91	Session break
92	Experimental end
99	Initial relaxation

4 Data Preprocessing

EEG raw data are influenced the artifacts from non-physiological and physiological such as bad electrode contacts, broken electrodes, muscle activity and eye movements etc.

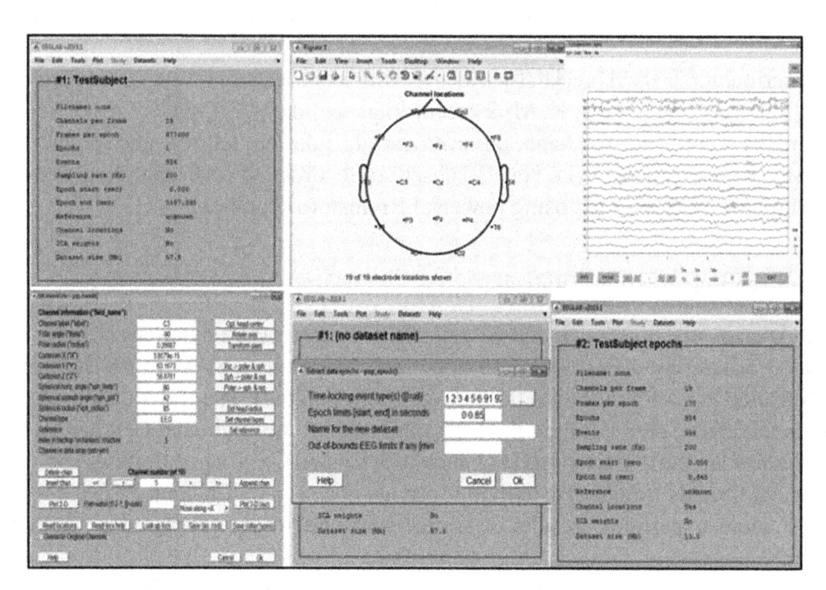

Fig. 3. Artifacts removal and epoch extraction steps in EEGlab toolbox

In the preprocessing step, the proposed method used EEGLab toolbox for artifacts removal and epoch extraction. EEGLab is an attractive Matlab toolbox for event-related EEG data. Moreover, each recording has six events (i.e. Right and Left hands, Right and

Left legs, passive and tongue) and each event has multiple trials in the same recording. These all trials are extracted by using epoch extraction option in EEGlab. Moreover, the proposed model is removed easily non required trials or events by using EEGlab for further experiment. In our experienced EEGlab is the one of the valuable research tool for brain signals. Some artifacts removal and epoch extraction steps are shown in Fig. 3.

5 Source Localization and Functional Connectivity

There are many available methods to solve the forward modeling such as Single sphere, Overlapping and Open MEEG BEM model. The proposed model is used Open MEEG BEM model for forward problem.

In the literatures, there are many solutions to solve the inverse problems, based on different assumptions on the way the brain works and depending on the amount of information. There are many methods that are available in Brainstorm, however, it represents three widely use approaches to the inverse problem in MEG/EEG source imaging: such as minimum-norm solutions, dipole modeling and beamformers. The proposed source localization model is used minimum-norm imaging method and is measured Current density map algorithm. Then the Functional connectivity was analyzed by using Phase Locking Values (PLV) algorithm among many algorithms such as coherence, Granger C, Correlation and so on.

6 Experimental Results of the Proposed Model

The experiments focused on HaLT paradigm, which are recorded from 12 subjects (Subject A, B, C, E, F, G, H, I, J, K, M) 29 recordings and the eGUI that are displayed with six symbols right hand, left hand, passive/neutral, right leg, left leg, and tongue motor imageries well-organized. The HaLT dataset are recorded 200 Hz sampling frequency rate. These results are tested using powerful Brainstorm Toolbox.

6.1 Source Localization Analysis

In the experiments, it is concluded that all of the Subjects (participation) clearly visible in marker 3(neutral), 4 (left leg), 5 (tongue) and 6(right leg). Some of the Subjects are also activated in Cz for mark 1(left hand) and 2 (right hand). The HaLT data is recorded 19 channel device. According to the Literature review, the Right and Left hands, Right and Left legs are mostly activated in channel C3, C4 and Cz. Tongue motions can visible on channel Cz. Neutral/Passive activities can also be seen on Channel Pz [7].

The detail experiment results of source localization are expressed as follows in Figures.

In Fig. 4, the experiments of the Left-Hand Activation on all of the Subjects (A, B, C, E, F, G, H, I, J, K, M, L) are expressed on the Human brain with source activation. Subjects C, G, I, J, L and M are clearly visible their right source activation in channel location C4. Otherwise, Subjects A, B, E, F, H and K are clearly visible in Cz cortical region. According to the results, all of the Subjects are well working their activities with brainstorm toolbox.

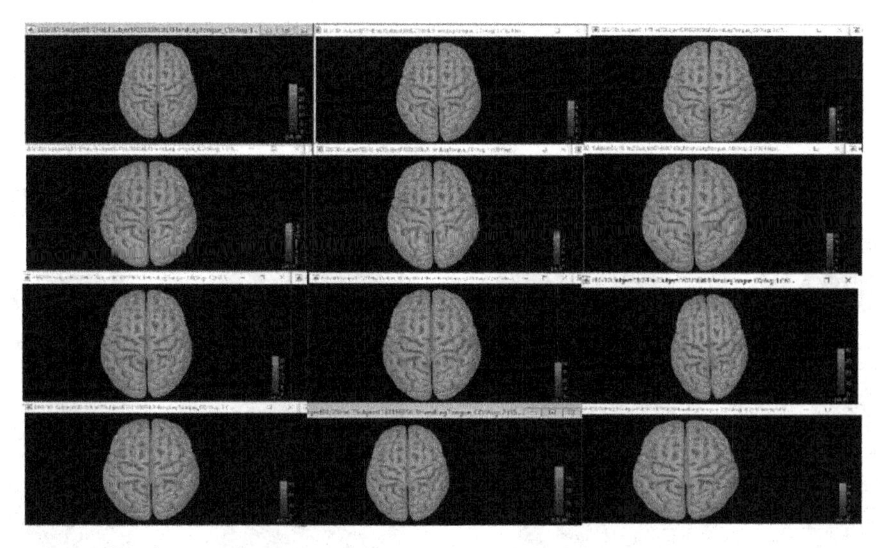

Fig. 4. Experiments results of Left-Hand Activation on Twelve Subjects

In this Fig. 5, the experiment results of Right-Hand Activation on Twelve Subjects are shown. Most of the subjects are activated in the required cortical C3 parts/region. Some of the subjects can be seen in Cz cortical region. Only Subject E are wrongly activated in Channel Pz cortical part.

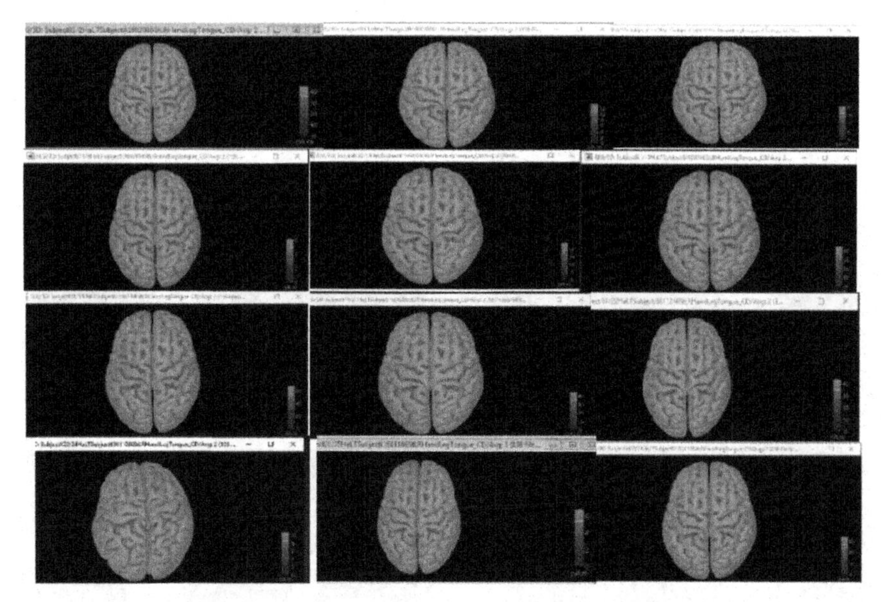

Fig. 5. Experiments results of Right-Hand Activation on Twelve Subjects

In this Fig. 6, all events are nearly located Cz. Most of the Subjects are clear visible activation for those cortical parts because of location in Cz. But Subject F, L and M are not clear because these events are located nearly Pz.

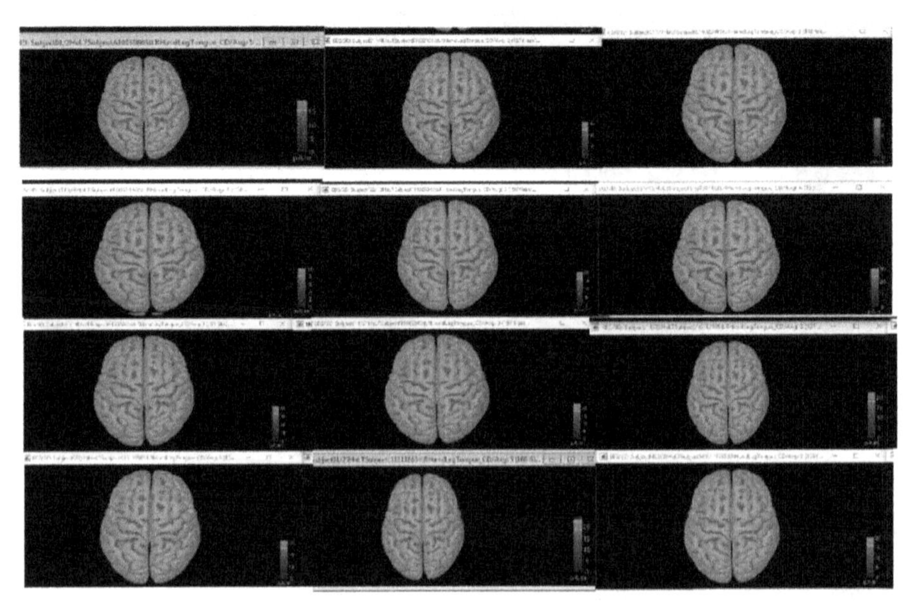

Fig. 6. Experiments results of Neutral Activation on Twelve Subjects

In this Fig. 7, all of the Subjects are clear visible for those cortical parts because of location in C4 and Cz. But Subject C is slightly not clear because of relating on Channel location C4.

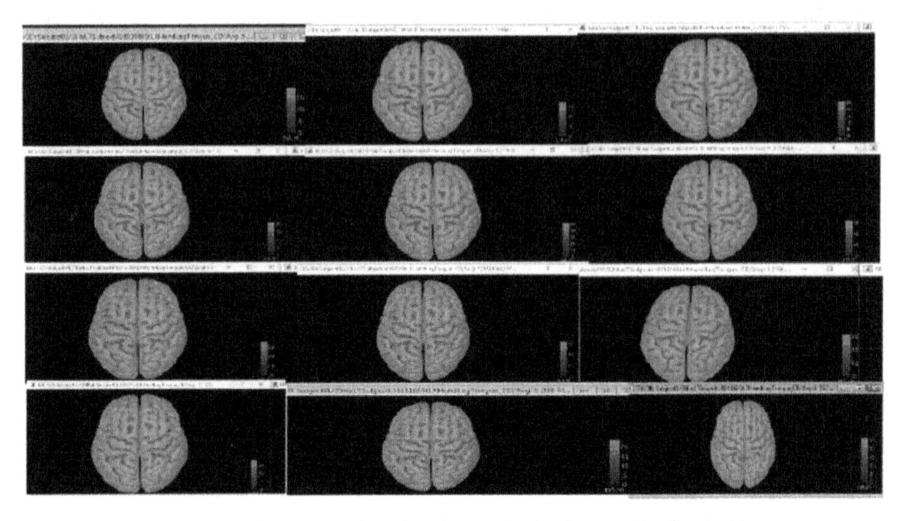

Fig. 7. Experiments results of Left Leg Activation on Twelve Subjects

In Fig. 8, most of the Subjects (tongue) are clearly visible these source activations that it is located in Cz and Pz. Only Subject L is not clear that it is activated on location C3.

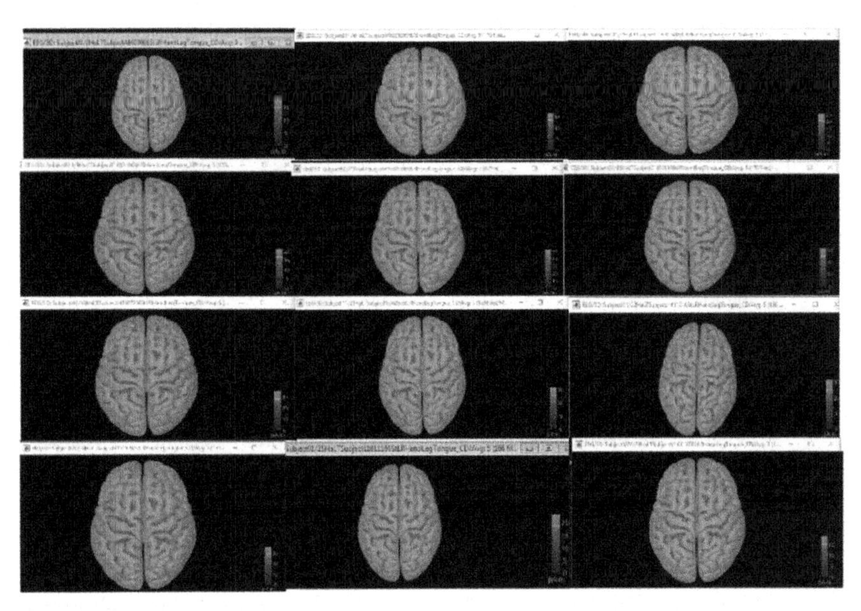

Fig. 8. Experiments results of Tongue Activation on Twelve Subjects

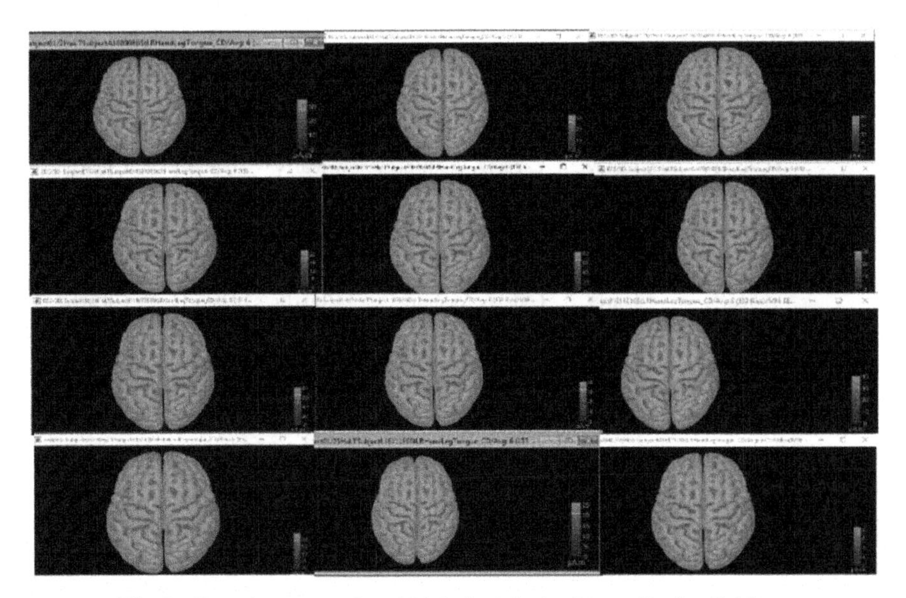

Fig. 9. Experiments results of Right Leg Activation on Twelve Subjects

In this Fig. 9, although all required cortical parts are activated, all events are not clearly related visible. Almost, the Subjects are clear visible for those cortical parts because of location in C3 and Cz in these Right Leg Events. But only one Subject G is activated at Location Pz so it is not accurate region.

Table 2 and Fig. 10 show all the results of the EEG signal Source Localization based on twelve (12) subjects. In Table 2, the accuracy of Maker 1 (Left hand), 2-Right hand, 4-Left leg and 5-Tongue is above 90%. The accuracy of 3-Passive/Neutral is 75%

Table 2. Source localization results

Subject	Maker 1 (Left hand)	2-Right hand	3-Passive/Neutral	4-Left leg	6-Right leg	5-Tongue
A	Clearly Visible	Clearly Visible	Clearly Visible	Clearly Visible	Clearly Visible	Clearly Visible
B	Clearly Visible	Clearly Visible	Clearly Visible le	Clearly Visible	Clearly Visible	Clearly Visible
C	Clearly Visible	Clearly Visible	Clearly Visible able	Not Clearly Visible	Not Clearly Visible	Not Clearly Visible
E	Clearly Visible	Clearly Visible le	Clearly Visible	Clearly Visible	Clearly Visible	Clearly Visible
F	Clearly Visible	Not Clearly Visible	Not Clearly Visible	Clearly Visible	Clearly Visible	Clearly Visible
G	Clearly Visible	Clearly Visible e	Clearly Visible	Clearly Visible	Not Clearly Visible	Clearly Visible
H	Clearly Visible	Clearly Visible	Clearly Visible	Clearly Visible	Clearly Visible	Clearly Visible
I	Clearly Visible le	Clearly Visible	Clearly Visible	Clearly Visible	Clearly Visible	Clearly Visible
J	Clearly Visible	Clearly Visible	Clearly Visible	Clearly Visible	Clearly Visible	Clearly Visible
K	Clearly Visible able	Clearly Visible le	Clearly Visible	Clearly Visible	Clearly Visible	Clearly Visible
L	Clearly Visible	Clearly Visible	Not Clearly Visible	Clearly Visible	Clearly Visible	Clearly Visible
M	Clearly Visible le	Clearly Visible	Not Clearly Visible	Clearly Visible	Clearly Visible	Clearly Visible
Average Accuracy	100%	92%	75%	92%	83%	92%

and 6-Right leg is 83% respectively. The proposed model is able to classify into 1-Left hand, 2-Right hand, 4-Left leg, 5-Tongue, 3-Passive/Neutral and 6-Right leg using the minimum-norm imaging method with the current density map algorithm with the average accuracy of 89%.

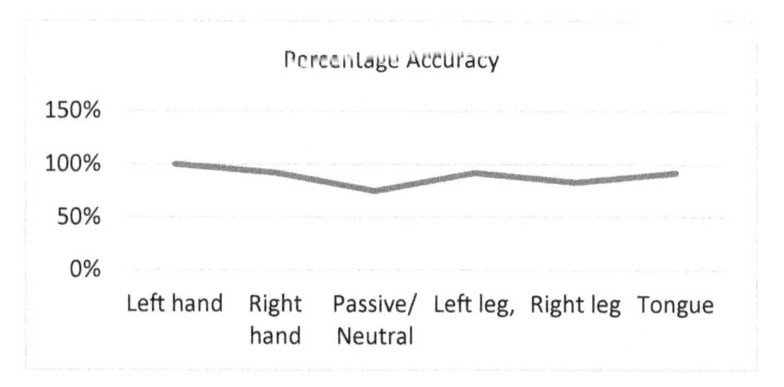

Fig. 10. Percentage Accuracy of the Human Activities on HaLT Paradigm Datasets

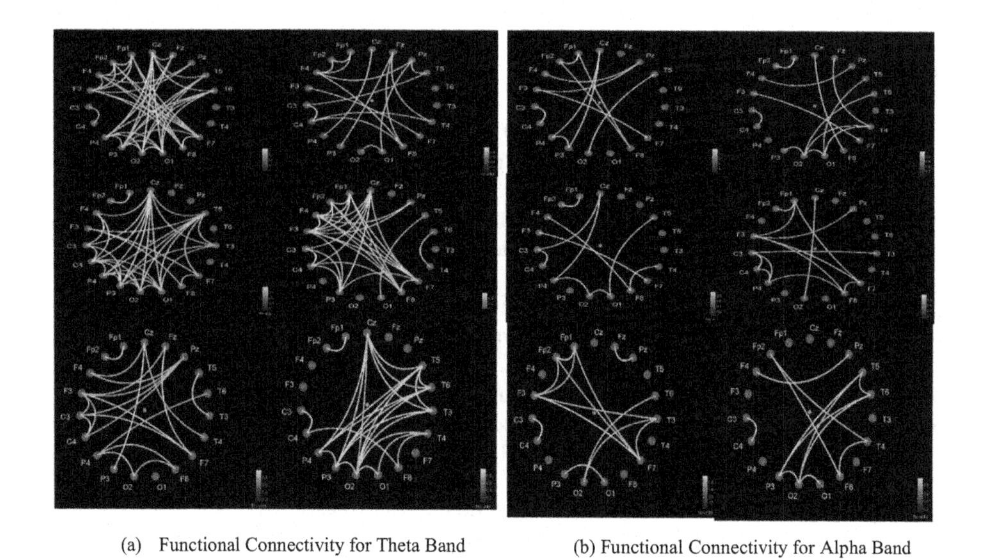

(a) Functional Connectivity for Theta Band (b) Functional Connectivity for Alpha Band

Fig. 11. Functional Connectivity of the Human Activities on HaLT Paradigm Datasets

6.2 Functional Connectivity Analysis

The analysis of functional connectivity faces multiple major challenges in terms of (1) high-dimensional, computationally-demanding data processing (2) visualization of large volumes of information and (3) efficient implementation of a diversity of connectivity metrics.

The proposed system is conducted large-scale data-driven cortical connectivity analysis, with large data dimensions using Brainstorm. Brainstorm supports four metrics such as Correlation, Coherence, Phase Locking Values and Ganger Causality. It uses Phase Locking Values algorithm to compute metrics. PLV algorithm is stability of two signals phase difference at a specified frequency band. In this Fig. 11, the proposed model is tested alpha (8–13 Hz) and theta (4–8 Hz) frequency band using PLV algorithm for each activity on Subject J.

7 Conclusion

The proposed system mainly analyzed source localization and functional connectivity of EEG signal over movement imaginary (MI) tasks by using brainstorm toolbox. Proposed system proves that brainstorm-based source localization and functional connectivity is the effective model with the average accuracy of 89%. The variation among the dataset and lack of the certain dataset are highly challenges for EEG researcher to analyze. Our future research will be focused on our own collected dataset and classify the human activities to get better result than this work.

References

1. Morenol, C.A.A., Manian, V., Rodriguez, D., Valera, J.: Auditory source localization by time frequency analysis and classification of electroencephalogram signals. J. Sci. Tech. Res. **19**(5) (2019). ISSN 2574-1241
2. Nambu, Ebisawa, M., Kogure, M., Yano, S., Hokari, H.: Estimating the intended sound direction of the user: toward an auditory brain computer interface using out of head sound localization. Plos One **8**(2) (2013)
3. Bednar, A., Boland, F.M., Lalor, E.C.: Different spatio-temporal electroencephalography features drive the successful decoding of binaural and monaural cues for sound localization. Eur. J. Neurosci. **45**(5), 678–689 (2017)
4. Letowski, T.R., Letowski, S.T.: Auditory spatial perception. ARL-TR-6016 (2012)
5. Handiru, V.S., Vinod, A.P., Guan, C.: Cortical source localization for analyzing single-trial motor imagery EEG. In: IEEE International Conference on Systems, Man, and Cybernetics (2015)
6. Dery, S., et al.: Functional Connectivity using Brainstorm, January 2014
7. Gaho, A.A., Musavi, S.H.A., Jatoi, M., Shafiq, M.: EEG signals based brain source localization approaches. (IJACSA) Int. J. Adv. Comput. Sci. Appl. **9**(9) (2018)
8. Kaya, M., Binli, M.K., Ozbay, E., Yanar, H., Mishchenko, Y.: A large electroencephalogram graphic motor imagery dataset for electroencephalogram graphics brain computer interfaces, 16 October 2018

Fitness Function Design for Neuroevolution in Goal-Finding Game Environments

K. Vignesh Kumar, R. Sourav, C. Shunmuga Velayutham, and Vidhya Balasubramanian(✉)

Department of Computer Science and Engineering, Amrita School of Engineering, Coimbatore, Amrita Vishwa Vidyapeetham, Coimbatore, India
{b_vidhya,cs_velayutham}@cb.amrita.edu

Abstract. Recently, games like Pac-Man have been hotbeds for neuroevolution research and NEAT has emerged as one of the leading techniques in the game playing domain [14]. In the context of the snake game, the goal of this paper is to enhance neuroevolution strategies with better fitness functions for effective goal finding. We develop greedy and non-greedy fitness functions, and demonstrate the effectiveness of these functions in both environments with and without dynamic obstacles. We then present an alternate implementation using the NEAT algorithm combined with Novelty Search to increase the genetic diversity of the agent population and explore the problem space without specifying direct objectives. These conclusions suggest that even with a low number of simple inputs, and simple fitness functions, agents are quickly able to achieve a novice amount of expertise in the Snake game using NEAT.

Keywords: Neuroevolution · Fitness function · NEAT · Novelty search

1 Introduction

Recent advancements in artificial intelligence and deep learning have opened up opportunities for disruptive solutions in various areas. However, the effectiveness of these solutions depends on the neural network design which has to be tweaked over time to suit the application. Neuroevolution (NE) provides an automated solution to address this issue through which a neural network can be trained without explicit training data—by evolving an initial population of randomized neural networks over several generations by assessing each individual's fitness [4]. Neuroevolution approaches like NEAT [16] and its variants have been designed to address several gaps perceived in fixed topology NE. Games have been the primary source of design and testing for developing and refining these approaches since they offer complexity in decision making, the ability to introduce uncertainty, and a platform with varying levels of complexity.

M. Hernes et al. (Eds.): ICCCI 2020, CCIS 1287, pp. 503–515, 2020.
https://doi.org/10.1007/978-3-030-63119-2_41

One of the primary challenges in designing neural networks using neuroevolution is understanding the effect of the fitness function on the effectiveness of the network. While several studies have focused on the neuroevolution approach itself, the design of fitness function is not made clear. It is important to analyze the design of fitness functions and their impact on the resultant neural network performance. In this paper, we explore an incremental design of fitness functions for a snake game to analyze the performance of neuroevolution. The snake game was chosen for its simplicity in the experiment implementation, target finding, and obstacle avoidance features. An effective neuroevolution strategy for such a game shows opportunities for designing solutions for real-time target finding, and navigation applications.

Interestingly the challenges posed by the objective-based approach, predominantly in neuroevolution, have led to *Quality diversity* - a recent family of evolutionary algorithms proposed to find a balance between both convergence and diversity during the search. These include novelty-based searches [8], behavioral diversity approach [11] and Curiosity Search [17]. These algorithms emerged due to the result of bringing a new perspective to the fitness function i.e. focus on the novelty of solutions rather than the mere quality of solutions. Hence, involving these algorithms in our fitness function design experiments is justified. Consequently, this paper also provides a comprehensive insight into the implementation of the novelty-based approach, including analysis of its many flavors and augmentations, and the drawbacks present in each. We then formulate an augmented Novelty Search of our own, which provides a performance comparable to that of the highest performing existing methods.

The primary contributions of this paper are as follows:

- An incremental design of experiments and a comprehensive analysis (especially the fitness function) while employing neuroevolution for games
- Adapting Novelty Search, and exploring its variants for the snake game and evaluating its performance over the proposed fitness functions.

The rest of the paper is organized as follows: Sect. 2 reviews the state of the art in neuroevolution with an emphasis on games, while Sect. 3 explains our architecture and design of the fitness functions. Section 4 analyses the results of our simulations and incremental experiments. Further, in Sect. 5 we discuss various novelty-based approaches with their implementation, analysis, and fine-tuning. Finally, we conclude in Sect. 6 with our observations.

2 Related Work

Previous work related to playing the Snake game have utilized various methods such as Q-learning [10,19] and genetic programming [2]. Our implementation builds upon these insights, with an emphasis on recent advancements in evolutionary algorithms such as Neuroevolution [4], NEAT [16] and Novelty Search [8].

Neuroevolution aims to find an efficient neural network for the given task through the evolution of neural network topology. The synaptic weights and connection between neurons are taken as genetic traits that are encoded. These traits are then subjected to evolutionary methods such as selection, crossover, and mutation. The gaps in the formulation of neuroevolution include the assorted encoding schemes in offer and their varying efficiency [13], the domination of mature features over newly evolved features in the population, and the need to maintain a simple network topology throughout the evolution process [16].

The *NeuroEvolution of Augmenting Topologies* (NEAT) algorithm advocates for starting the evolutionary process with a minimal structure with no hidden nodes and then subsequently evolving the neural network structure along with their connecting weights [16]. Stanley et al. went on to implement a real-time version of the NEAT algorithm to play the NERO video game, showcasing the diverse behavior and adaptability of game agents evolved through NEAT [15]. Games are found to be great test-beds for neuroevolution and more suited for the task than robotics; noting their excellent performance and numerous means of implementation [12].

Recent insights into evolutionary algorithms have called for conducting a *novel behavior* driven search replacing the traditional objective-based measure of fitness [8]. This Novelty Search approach exhibits adaptive behavior in deceptive environments and avoids local optima in the learning process, as evident in its success in navigating the maze experiment. It is shown that in complex environments, where the behavioral space is vast, the performance of Novelty Search is lacking [7]. Studies have tweaked the parameters of Novelty Search such as the archive size and the value of k [5], but there remains an inherent gap in the understanding of Novelty Search especially the modifications that lead to better results or are more suited for larger behavioral spaces.

While several studies have focused on the neuroevolution approach itself, the design of the fitness function is not made clear. This paper attempts to fill that gap, as well as provide insights into designing fitness functions and provides analysis into their performance in game environments. We then test these objective-based functions in increasingly complex scenarios to test their adaptability. We proceed to analyze modifications to Novelty Search that provide a better performance in large behavioral spaces with resource constraints. The merits and demerits of each approach are then analyzed and presented.

3 Architecture and Design of Experiments

This paper uses the classic snake game as the testbed for the neuroevolution experiments. Considering the fact that the focus of the paper is on the incremental design of experiments (especially the fitness function) while employing neuroevolution for games, the choice of a simple game as a testbed is well justified. As Fig. 3 shows, the goal of the game is to make a snake moving in a 2D space eat as many food blocks as possible without biting itself, both in the presence and absence of obstacles. Figure 1 shows the neuroevolution workflow adopted towards evolving a neural network controller for the snake game. The first step involves creating a population of randomized neural networks that will be subsequently evolved for the task at hand. The fitness function evaluates each neural network by simulating a game with that corresponding neural network as that snake's *brain*. The fitness function essentially awards good moves the snake makes (like eating the food block) and penalizes bad moves (like the snake biting itself or colliding with the wall) throughout the snake's lifetime. Snakes that do not show progress over a period (1600 steps) are killed, so as to not pollute the gene pool. Once the fitness evaluation has been completed, the top-performing candidates are deemed as "elites" and chosen to be a part of the next generation directly (after mutation), while the rest of the population is subjected to crossover proportionate to fitness values, and then randomly mutated (with a fixed probability for mutation selection and mutation rate) thereby potentially yielding a new generation of *better* neural networks. Each neural network of this newly created generation is put through the gaming simulation, assessed, and the evolution process repeats until we reach a fitness criterion that has been met. The evolution engine i.e. the neuroevolution algorithm adopted in the above workflow is NEAT with parameters given in Fig. 2.

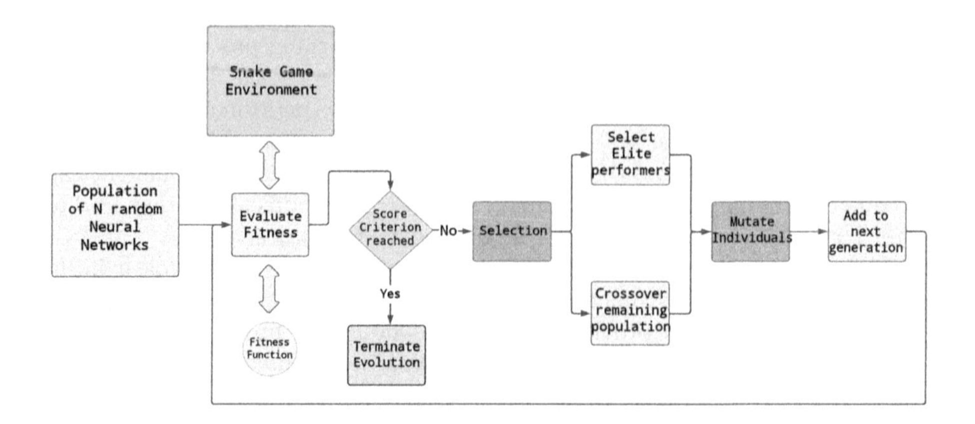

Fig. 1. Our proposed neuroevolution architecture

Each candidate neural network, which is represented by a direct encoding scheme, begins with no hidden nodes and with fully connected input and output nodes. A hidden layer architecture, as shown in Fig. 4, is formed with mutations of new hidden nodes with their connections and weights evolved by determining the fitness in subsequent generations which will give the controller its very intelligence in playing the snake game effectively, of course, guided by the fitness function we have been designing. Crossover of candidate neural networks is performed by tracking historical markers and the introduction of speciation [16], whereby innovations are protected and differing network topologies are merged.

Parameter	Value
Population size	100
Crossover rate	75%
Mutation rate	50%
Crossover method	Uniform
Elite selection rate	25%

Fig. 2. Table: parameter settings for NEAT workflow

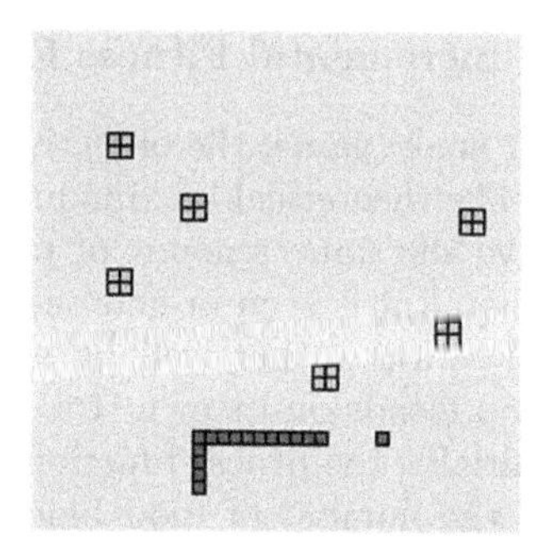

Fig. 3. The Snake game environment, with the addition of dynamic obstacles

3.1 Choice of Inputs and Output

In game playing environments, the input given to the neural network is often raw image data and previous studies have used screenshots of the snake game environment as input [10]. However, NEAT with direct encoded topologies perform well on pre-processed representations of the game state [6]. In our implementation, the coordinates of every element involved like the position of the food object and the coordinates of the snake's own body are known at every tick of the game clock and can be provided as input features.

Figure 4 shows the features which were given as input to the neural network. Initially, the distance in units to the food particle, and the distance to the nearest obstacle in three directions: *in front of the snake*, *to the left*, and *to the right* were used. The directional distance units were then replaced with an approach proposed by [2] where binary inputs indicated only the presence of food or obstacles in that particular direction. This modification helped to evolve neural networks with simpler topologies and with less overhead.

Interestingly, it has been seen that humans perform complex maneuvers by their knowledge (or memory). In the snake game, knowledge of the previously made moves dictates the next one. The performance of Q-learning in the snake game could be attributed to the previous state information found in the Markov decision process [10]. To emulate this human behavior, we propose a memory unit that contains the history of moves the snake has previously made. This has been implemented as a set of inputs to the neural network that specifies a set of previous turns made by the snake, the size of which was varied from one to ten. These inputs were binary states that specified whether the turn made previously was a left or right turn.

The outputs for each candidate neural network are defined to be the two actions that can be taken by the snake viz. to turn left or to turn right, which are again binary representations. It can decide to do neither and continue forward (neither of the neural network's outputs gets activated).

3.2 Incremental Fitness Function Design

In our snake game, the objective is *to have the maximum score when our snake dies.* The theoretical maximum score in the snake game is a fixed objective and we give the same amount of points for eating food at any point in the game. In general, it is seen in neuroevolution, and confirmed in our experiments, that NEAT works rather well by specifying general instructions and by letting it develop trends on its own [16].

Initially, the fitness function has been designed as a greedy strategy to maximize the number of food blocks eaten by the snake (henceforth representing candidate neural network controllers). The flip side of such a greedy approach is that the fitness function wrongly encourages the candidate neural networks to always maneuver the snake toward food even at the cost of biting itself thereby ending the game. Consequently, the fitness function should reward the candidate solutions that let the snake live longer while devouring more food blocks. Lockhart [9] proposes a reward for *staying alive.* Accordingly, the fitness function has been redesigned to reward those candidate controllers that let the snake live longer. The relative lifespan of a snake for each candidate controller is computed using each tick of the game clock. In addition, the rewarding and penalizing of the controllers for respectively steering the snake towards and away from the food is weighted to smoothen the greedy approach for eating more food blocks. Figure 4 shows the template of the neural network controller with the activation of input and outputs for the scenario in Fig. 3. In this scenario, none of the output nodes are activated, thus the agent continues in the current direction.

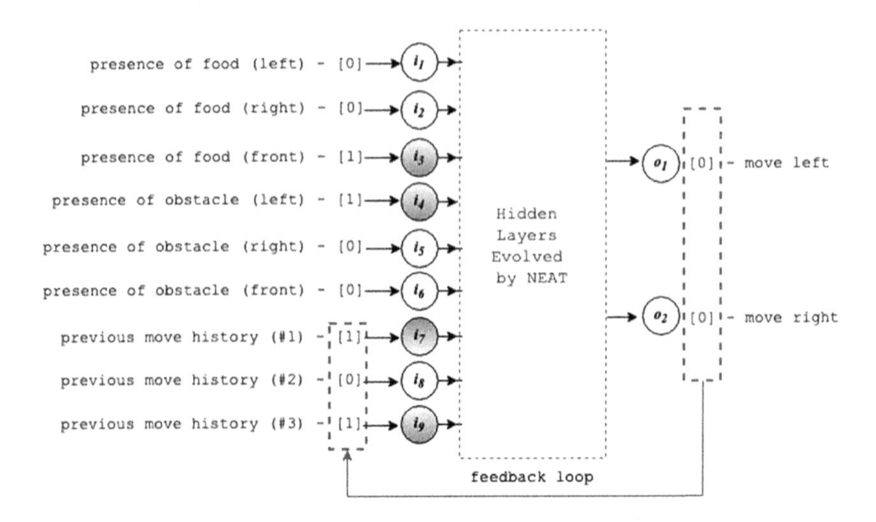

Fig. 4. Template of neural network controller for snake with sample activation

4 Simulation Results and Analysis

4.1 Objective Fitness Function Designs

The game is designed over a 40×40 grid and the learning process took place over 200 generations, by which time the agent's performance generally became saturated. The greedy fitness function rewarded candidate solutions with 1 point for making the snake move in the direction of the food, -1 for moving away from the food and 2 points for eating the food, analogous to the *distance reward* in Wei et al. [19]. With this fitness function (labelled version a in Fig. 5) snakes learnt to move toward food as early as the 2nd generation, by virtue of the penalty for avoiding food. By the 10th generation, some snakes obtained a good score of 66 food particles eaten while avoiding walls swiftly. A high score of 75 was attained by a snake in the 13th generation. Later generations were saturated with no improvement. The snakes always killed themselves around the time when its length reached twice the size of the board.

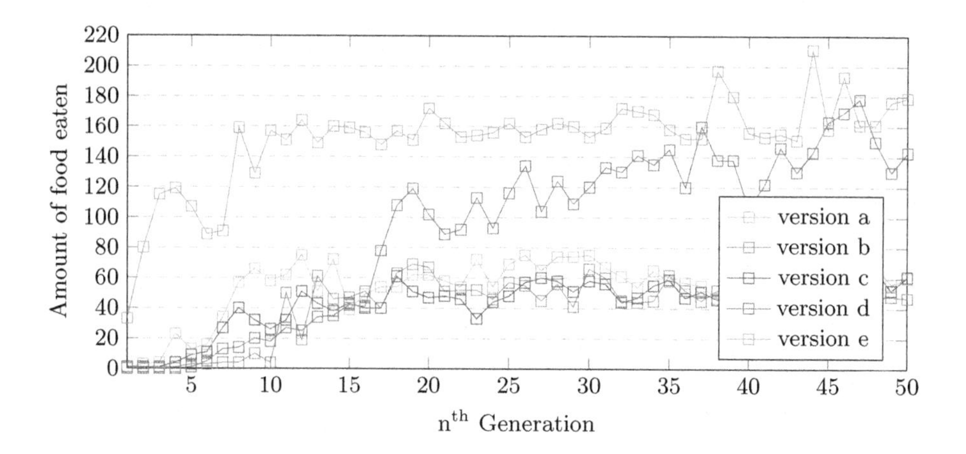

Fig. 5. Observations of controller scores with varying fitness metrics

We then reduced the penalty for moving against the food from -1 to -0.2, hoping the snakes can now learn not to kill themselves in pursuit of food. With this weighted greedy fitness function, snakes have learnt to go after food after around 2 generations, and by 10 generations, they learn to properly avoid walls. Subsequent generations did not show any improvement in strategy. A high score of 69 was attained by a snake of version b at the 20th generation. In both fitness functions, almost all snakes died by biting themselves. However, in the weighted greedy version, we observed that the snakes tried to stick to the game area boundaries thereby keeping the area predominantly empty to facilitate eating food.

Since the very objective of the game is to keep the snake alive while eating as much food as possible, the new greedy fitness function (version c) has been designed to additionally reward snakes 0.01 points for staying alive for each game tick. With this fitness function, a high score of 66 has been obtained in 15 generations. As has been observed earlier, in as early as 3 generations, the snakes have learnt to avoid walls and go after food. However, the evolution stagnated after 24 generations with no new strategies. Interestingly, as the fitness function now rewards the snakes for their longevity, the snakes have now learnt to choose the longest paths (often zig-zag) thereby delaying their end. This greedy strategy helps snakes to maximize fitness. However, it does not allow the snakes to learn to not eat themselves.

The reward for survival led snakes to learn counterintuitive tactics and strategies. Additionally, the greedy approach in the form of penalty for moving away from food also hinders the progress of the snake. So the non-greedy fitness function (version d) awards 1 point for moving towards food and 2 points for eating food. With this fitness function, the simulation resulted in a high score of 178 within 50 generations. As there is no negative reinforcement when the snake decides to move away from the food, it takes a while (observed to be around 8 generations) before it learns to go after it. By virtue of the non-greedy approach, snakes try various maneuvers instead of going only for food thus contributing to diverse strategies. By generation 21, it has been observed that a snake has learned the strategy of following its tail to stay alive. With this non-greedy version, snakes do not stick to the wall, they prefer to eat the food in the shortest distance if possible, and learn to avoid spiraling deaths even after 20 generations.

To achieve human-like maneuvers, a feedback loop in the neural network architecture has been implemented in version e which enables the snake to keep track of its previous moves along with the fitness function without a greedy approach. The simulation with this feedback controller has achieved a high score of 211 within 50 generations. After 20 generations, the knowledge of one's history of paths has resulted in highly sentient moves matching human performance. In fact, the optimal size of the memory (i.e. the number of previous moves) was found to be three. Beyond this limit, there was no noticeable increase in the performance of the snake.

4.2 Snake Game Environment with Obstacles

This section intends to investigate the potential of the designed non-greedy fitness function for a custom-designed version of the snake game to reflect real-world complexity. Accordingly, the environment has been modified to dynamically place and remove random obstacle blocks as shown in Fig. 3. This modification has high relevance to real-world environments where an autonomous agent has to navigate through space towards a target while constantly avoiding random moving obstacles. The snake's performance was promising as it learned that the new obstacles were like walls and it stayed clear of them, thus showcasing the generalizing ability of NEAT powered with our fitness function design. Figure 6 shows the performance of the non-greedy fitness function for the snake game with and without obstacles, and with a memory supported network. While the

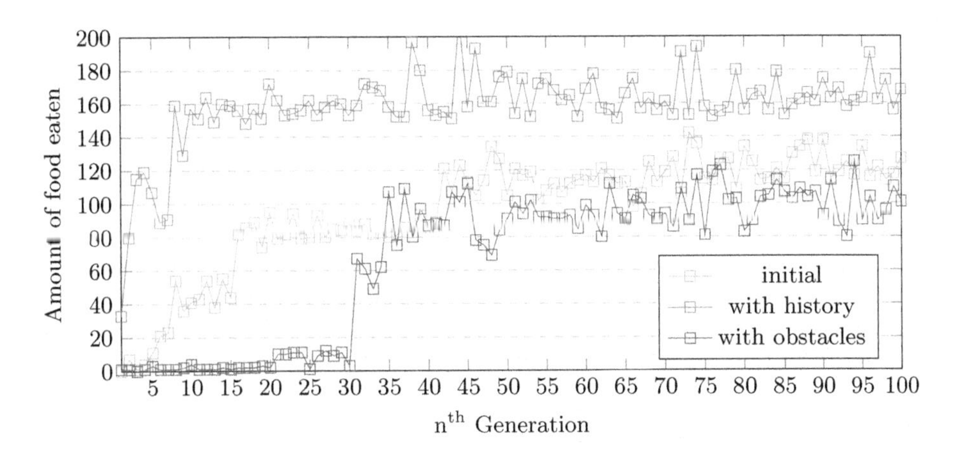

Fig. 6. Performance of non-greedy approach in different cases

memory supported network performs best for the case of no obstacles, the performance with obstacles shows improvement as the generations increase demonstrating the robustness of the fitness function.

5 Augmenting Fitness Using Novelty Search

The foundation of novelty search (NS) is based upon defining the *behavioral space*. In the snake game environment, we choose the behavior of the snake agent to be evaluated as the path traversed by the snake during its lifetime. However, the selection of this behavior as a novelty metric poses a challenge. Consider a game instance G_1 with the spawning of food at positional coordinates $(x_1,\ y_1), (x_2,\ y_2), ..., (x_n,\ y_n)$ to be considered as F_1 and the path required by the snake to perfectly consume these food to be $P1$.

Let us suppose instead that this snake takes a path P_2 slightly different to the perfect path P_1 during the game. In the next generation, let us consider that for a game instance G_2, food is spawned at places corresponding to F_2 and the perfect path required to be P_2 which the agent ends up traversing. Since the path P_2 has been previously traversed by a past instance, the novelty score given to the agent of game G_2 will be low, even though it was the perfect path required. To address this issue, during the learning stage of novelty search, a constraint is imposed on the positions of the food spawns, such that their positions will be static across all generations.

A new agent seen with high dissimilarity to previous agents can be deemed to be novel. The need arises for a function that is a measure of similarity between two paths. We propose the use of discrete Fréchet distance [3] between the paths traversed by two agents to be the similarity function. Fréchet distance for two polygonal curves P and Q is given as the minimum length required to join any two points between P and Q when moving forward along both the curves. These curves, which exist in metric space, are analogous to the path traversed by our agent in its lifetime. The inputs for the distance function are the coordinates of the ordered points of the curves.

The Fréchet distance of an agent's path is calculated to every other agent's path present in the generation and the novel archive. The k-nearest neighbors ($k = 3$) found are averaged to derive a novelty score. This novelty metric was then used as a replacement for the fitness function in our experiment. Since computing the different snake paths and their similarity in a large game environment is expensive, the experiments are run on a 10×10 game dimension with a theoretical maximum score of 100 with a 3000 generation limit. The earlier experiments have also been run in the new game conditions for comparison.

Table 1. Comparison of different algorithms as a measure of food eaten

Method	Static Env			Dynamic Env		
	Max	Mean	SD	Max	Mean	SD
Greedy Fitness	57	29.6	15.91	51	24.72	9.34
Non-Greedy Fitness	67	47.27	8.90	66	36.58	7.58
Novelty Search	18	9.93	2.29	26	11.72	3.91
Modified NS	32	24.86	4.27	39	29.34	5.28
Modified NS + Non-Greedy Fitness	72	47	9.37	55	**38.61**	10.07
Modified NS + Guiding Metric	**87**	**50.47**	14.98	**70**	37.98	8.10

An assortment of distinct flavors of the NS algorithms has been tested over the years. In our complex problem space, an approach similar to [18] where a novel archive is omitted and behavioral differences are calculated only among the k-nearest neighbors (kNN) is a plausible enhancement. This approach is a mix of the classic NS and a behavioral diversity based implementation [11], thus formulating a modified NS which seeks out an ideal behavior that exists in a search space between that of the traditional NS and Curiosity Search [17]. This implementation of NS is analogous to *intra-generational novelty search*, as it calculates the novelty of the agents among the current generation of the population alone, forgoing the presence of a novel archive found in [16]. It is seen that striking a balance between the exploratory nature of NS and the goal-drivenness of objective fitness functions leads to good performance on deceptive tasks [1]. Consequently, the modified NS has been augmented with the non-greedy fitness function as follows:

$$f(x) = modifiedNovelty(x) + (0.5 * modifiedNovelty(x) * nongreedyFitness(x)) \quad (1)$$

This augmented novelty metric calculation helps guide the NEAT algorithm to escape the local optima. While this function is comprehensive, the complexity of the guiding objective is in question and a simple metric guiding the search may well be sufficient. Therefore, we propose the use of a *guiding metric* analogous to a fitness metric, but one that is loosely defined based on the general goal of the game ie., maximizing the food eaten. Thus, we modify the previous fitness function to include a guiding metric (GM) as follows:

$$f(x) = modifiedNovelty(x) + (0.5 * modifiedNovelty(x) * GM(x)) \quad (2)$$

Table 1 shows the performance of the proposed greedy and non-greedy fitness functions on the constrained environment along with the modified NS functions ($n = 15$). The usage of the fitness metric of an agent as a parameter in calculating the novelty score shows an increase in the maximum score over the usual NS approach. It is also observed that the simple guiding metric yields comparable results for the maximum and average scores to a full-fledged fitness function as reported in Table 1. However, its reliability needs improvement. The fitness-based approaches are less computationally intensive while providing comparable performance, but their demerit lies in the fact that the manual and iterative design of the fitness function is laborious. On the other hand, a novelty based approach involves more computing and memory overhead (especially when a novel archive is used) but the implementation, particularly with a simple guiding metric, is significantly less complex.

The constraint of static food positions used in the training of the novelty algorithm arises a question of whether these agents would be able to maintain their performance when this restriction is removed. High performing agents which were trained by each strategy were evaluated on instances of game environments with dynamic food positions as given in Table 1 with $n = 100$. The novelty algorithms perform well in dynamic environments even when their learning was performed with environmental constraints. This validates the adaptability of NEAT and novelty based approaches.

6 Conclusion and Future Work

In this paper, we have provided an in-depth look into the implementation of NEAT using the Snake game as a testbed. Both greedy and non-greedy fitness functions were designed and the performance of the latter showed promise. It also performs well in the presence of obstacles and adding memory to the Neural Network controller led to the evolution of better strategies leading to higher scores. Augmenting this fitness function with a novelty search drastically improved the game scores obtained. By simplifying the fitness function over novelty search using a guiding metric, we demonstrate that a comparable result is achievable with lesser overhead. By proposing an incremental approach to fitness function design for a goal-finding game, this paper has demonstrated the impact of both the quality and novelty of the fitness function. This opens up opportunities to evolve similar strategies using recurrent neural networks and adapt them for dynamic environments using neuromodulation.

Recent explorations have been made into automating the formation of the ideal fitness function, particularly through the use of a neural network as a fitness function. The use of a similarly trained neural network which could approximate novelty score might prove to be advantageous in NS as it would eliminate the storage and computational costs associated with a novel archive, which was a major limiting factor in our experiments. The results of our experiment indicate that the large behavioral spaces of complex game environments can be reduced with constraints to simpler problems for the implementation of NS. This method can be implemented for other complex environments to conduct further research.

References

1. Cuccu, G., Gomez, F.: When novelty is not enough. In: Di Chio, C., et al. (eds.) EvoApplications 2011. LNCS, vol. 6624, pp. 234–243. Springer, Heidelberg (2011). https://doi.org/10.1007/978-3-642-20525-5_24
2. Ehlis, T., Hattan, J., Sikora, D.: Application of genetic programming to the snake game. Gamedev. Net **175** (2000)
3. Eiter, T., Mannila, H.: Computing discrete Frechet Distance. Technical Report CD-TR 94/64, p. 8 (1994)
4. Floreano, D., Dürr, P., Mattiussi, C.: Neuroevolution: from architectures to learning. Evol. Intell. **1**(1), 47–62 (2008)
5. Gomes, J., Mariano, P., Christensen, A.L.: Devising effective novelty search algorithms: a comprehensive empirical study. In: Proceedings of the 2015 Annual Conference on Genetic and Evolutionary Computation, pp. 943–950 (2015)
6. Hausknecht, M., Lehman, J., Miikkulainen, R., Stone, P.: A neuroevolution approach to general Atari game playing. IEEE Trans. Comput. Intell. AI Games **6**(4), 355–366 (2014)
7. Kistemaker, S., Whiteson, S.: Critical factors in the performance of novelty search. In: Proceedings of the 13th Annual Conference on Genetic and Evolutionary Computation, pp. 965–972 (2011)
8. Lehman, J., Stanley, K.O.: Abandoning objectives: evolution through the search for novelty alone. Evol. Comput. **19**(2), 189–223 (2011)
9. Lockhart, C.: Application of temporal difference learning to the game of Snake. Ph.D. thesis, University of Louisville (2010)
10. Ma, B., Tang, M., Zhang, J.: Exploration of reinforcement learning to snake (2016)
11. Ollion, C., Doncieux, S.: Why and how to measure exploration in behavioral space. In: Proceedings of the 13th Annual Conference on Genetic and Evolutionary Computation - GECCO 2011, p. 267. ACM Press, Dublin (2011)
12. Risi, S., Togelius, J.: Neuroevolution in games: state of the art and open challenges. IEEE Trans. Comput. Intell. AI Games **9**(1), 25–41 (2017)
13. Rodzin, S., Rodzina, O., Rodzina, L.: Neuroevolution: problems, algorithms, and experiments. In: 2016 IEEE 10th International Conference on Application of Information and Communication Technologies (AICT), pp. 1–4. IEEE (2016)
14. Samothrakis, S., Perez-Liebana, D., Lucas, S.M., Fasli, M.: Neuroevolution for general video game playing. In: 2015 IEEE Conference on Computational Intelligence and Games (CIG) (2015)
15. Stanley, K.O., Bryant, B.D., Miikkulainen, R.: Real-time neuroevolution in the nero video game. IEEE Trans. Evol. Comput. **9**(6), 653–668 (2005)
16. Stanley, K.O., Miikkulainen, R.: Evolving neural networks through augmenting topologies. Evol. Comput. **10**(2), 99–127 (2002)

17. Stanton, C., Clune, J.: Curiosity search: producing generalists by encouraging individuals to continually explore and acquire skills throughout their lifetime. PLOS ONE **11**(9), 1–20 (2016)
18. Urbano, P., Georgiou, L.: Improving grammatical evolution in Santa Fe Trail using novelty search. In: Advances in Artificial Life, ECAL 2013, pp. 917–924. MIT Press, September 2013
19. Wei, Z., Wang, D., Zhang, M., Tan, A.H., Miao, C., Zhou, Y.: Autonomous agents in snake game via deep reinforcement learning. In: 2018 IEEE International Conference on Agents (ICA), pp. 20–25. IEEE, Singapore, July 2018

An Application of Machine Learning and Image Processing to Automatically Detect Teachers' Gestures

Josefina Hernández Correa(✉), Danyal Farsani, and Roberto Araya

Center for Advanced Research in Education, Institute of Education (IE) of Universidad de Chile, Santiago, Chile
josefina.mhc@gmail.com, danyal.farsani@ciae.uchile.cl, roberto.araya.schulz@gmail.com

Abstract. Providing teachers with detailed feedback about their gesticulation in class requires either one-on-one expert coaching, or highly trained observers to hand code classroom recordings. These methods are time consuming, expensive and require considerable human expertise, making them very difficult to scale to large numbers of teachers. Applying Machine Learning and Image processing we develop a non-invasive detector of teachers' gestures. We use a multi-stage approach for the spotting task. Lessons recorded with a standard camera are processed offline with the OpenPose software. Next, using a gesture classifier trained on a previous training set with Machine Learning, we found that on new lessons the precision rate is between 54 and 78%. The accuracy depends on the training and testing datasets that are used. Thus, we found that using an accessible, non-invasive and inexpensive automatic gesture recognition methodology, an automatic lesson observation tool can be implemented that will detect possible teachers' gestures. Combined with other technologies, like speech recognition and text mining of the teacher discourse, a powerful and practical tool can be offered to provide private and timely feedback to teachers about communication features of their teaching practices.

Keywords: Automatic teacher´s gesture detection · Open pose · Machine learning · Pattern recognition · Intelligent image processing

1 Introduction

The applications of gesture-based computing is an emerging technology in practice and a research topic of great interest [1]. Initially, gesture-based computing received great attention due to its numerous applications in domains like gaming, computer vision, animation, surveillance, man-machine interaction, robotics and mobile devices [2], but recently its potential for learning purposes has generated a great interest among educators, learning technology specialists, and researchers [1]. Gesture-based devices offer opportunities to experiment and innovate in teaching in many areas, such as special education, physics, mathematics, physical therapy, arts, music, science and literacy [1, 3].

M. Hernes et al. (Eds.): ICCCI 2020, CCIS 1287, pp. 516–528, 2020.
https://doi.org/10.1007/978-3-030-63119-2_42

Researchers and developers have realized that gesture-based computing within education requires "interdisciplinary collaborations and innovative thinking about the very nature of teaching, learning and communicating" [4]. Hence, silent and nonverbal messages that are communicated through teachers' gestures are of high value in education and educational research, not only in terms of teaching, but also when assessing students' understandings [5]. Teachers' gestures can attract students' visual attention [6] and convey pertinent mathematical information [7]. However, previous studies have relied on tedious time-consuming methods to analyze the nonverbal language in different social contexts such as psychotherapy [8], medical education [9] and classroom settings [10]. Measuring and analyzing teachers' gestures is very difficult to achieve without computational support that could allow monitoring a teacher's gestures at a large scale.

The term 'gesture' has many definitions depending on the context of study. For example, [11] defines the term gesture as "movements of the arm(s) and hands…closely synchronized with the flow of speech". [12] refers to the spontaneous hand movements produced while talking as 'gesticulation'. [13] defines a gesture as a "body movement fulfilling communicational function". Also, a gesture can be defined as a physical movement of a part of the body (e.g., hands, arms, eyes and face) [14] in a communicative situation [15].

In this work we define the term gesture as "synchronous movements of hands or arms with an intended educational meaning". This definition includes regulatory gestures such as those for 'be quiet' or 'sit down' and excludes "preening gestures" such as those used to put the hair behind the back of the head/ear, or biting the nails, since they are not considered of educational meaning.

Gestures not only help constitute thought [11], but also help the speaker access a wide range of possible resources to convey meaning [16]. Studies conducted in noneducational settings show that listeners gain information from speakers' gestures [17], facilitating listeners' comprehension of the accompanying speech, particularly when the verbal message is ambiguous [18], highly complex [19, 20], or degraded in some way [21]. Therefore, gestures are particularly important in classroom settings because a teacher's instructional discourse presents new concepts and usually uses unfamiliar terms [22].

Even though coding videos in search of gestures is overly burdensome and ultimately counter-productive [23], previous studies have analyzed and compared teachers that gesticulate more or in different forms and the effects on student's learning outcomes based on video-lesson recordings. For example, [24] analyzed 10 eighth-grade lessons from different teachers that were videotaped in Hong Kong, Japan and the United States, randomly sampled from the TIMSS 1999 Video database [25]. The study looked for hand or arm gestures that signaled an intended comparison (e.g., pointing back and forth between a scale and an equation), among other cognitive supports, and concluded that Chinese and Japanese teachers used far more gestures that emphasized comparison than did U.S. teachers, as a strategy to reduce processing demands on their students. Hong Kong and Japan were selected for comparison to the United States because their students consistently outperform U.S. students on the TIMSS International achievement tests.

Also, [26] experimented with two fifth grades classes, where both groups were instructed on division problems, but one received a "high cuing" instructional lesson and the other received a "low cuing" instructional lesson. The results show that students in the high cuing condition outperformed those in the low cuing condition on problems that required an extension from the instruction.

[27] investigated whether teachers' gestures influence student's learning with preschool children. The students viewed one of two videotaped lessons about the concept of symmetry: a verbal-plus-gesture lesson, where the teacher produced pointing and tracing gestures as she explained the concept, or a verbal-only lesson, where the teacher did not produce any gestures. On the posttest, children who saw the verbal-plus-gesture lesson scored higher than children who saw the verbal-only lesson, concluding that gestures may play an important role in instructional communication.

Finally, [22] examined a teacher's use of verbal and gestural forms of communication using video analysis techniques. The video recording was a sixth-grade mathematics lesson that focused on algebraic relations, and all observable gestures that the teacher produced along with her speech were identified. The study focused on the use of gestures to "ground" their instructional language, that is, to link their words with real-world, physical referents such as objects, actions, diagrams, or other inscriptions, making the information conveyed in the verbal channel more accessible to students. The results concluded that gesture is pervasive in instructional communication, that gestures serve a scaffolding function, and that gestures were used most frequently for new material, for referents that were highly abstract, and in response to students' questions and comments. Thus, gesture appears to be one means by which this teacher attempted to scaffold student comprehension.

Therefore, if we were able to automatically detect when a teacher gesticulates in a video class recording, we could extend current results by studying the effects on student's engagement, learning outcomes, motivation, behavior and the teacher-student relationship, to name a few, through a large-scale analysis.

This paper presents the application of an automatic gesture recognition methodology in a K-12 classroom, with the teacher as the target. It contributes with a novel, low cost and easy to apply, automatic gesture spotting method that stores a reduced version of the video files and can analyze automatically and detect gestures, as a first step. The proposed method is then tested in a real educational scenario, with promising results.

2 Related Work

Even though information retrieval in the multimedia-based learning domain is an active and multidisciplinary research area [28], in current literature, studies on automatic gesture recognition are not applied to a teacher's gestures in a classroom. Also, we found studies on gesture-based teaching and learning with the purpose of teaching students through gesture-based devices such as the Nintendo Wii or the Microsoft Kinect technologies.

Most of the current work in automatic gesture recognition is limited to hand gestures. For example, [29] proposes a tool for foreign students to learn English and its corresponding cultures online, by enabling students to interact with relevant and authentic cultural content by using natural hand gestures. This tool uses computer vision technology to recognize hand gestures from live images to promote immersive and interactive language learning. By wearing a pair of mobile computing glasses, users can interact with virtual contents in a three-dimensional space by using intuitive hand gestures.

[30] uses videos to automatically recognize Arabic sign language, reporting a word recognition rate of up to 98.13%. Also, [31] uses wireless surface electromyography for gesture number recognition, achieving a real-time recognition accuracy higher than 90%. In hand-gesture recognition, the accuracy rates are very high, but these methods are not applicable to whole body movements and therefore, are not useful in a school context to study a teacher's movements.

Regarding full body movements there is very little research, and it is generally highly invasive for the subjects whose gestures are being detected, and/or uses very expensive technology to obtain high accuracy rates.

Regarding invasive technologies, [32] shows the results of automatic body gesture detection with body-worn inertial sensors, with a precision rate of approximately 70%.

There are also investigators trying to automatically detect corporal body movements, not gestures. For example, [33] detected 14 gymnastics movements, with an accuracy rate of 91%. Also, [2] presents a motion analysis through elbow, knee, and other joint movements, successfully detecting through video frames when the subject bends, performs a side-step, a lateral lunge, and other movements of this sort, with an accuracy rate of around 90%.

When looking into expensive technologies, [34] presents three applications of a natural human movement detection interface based on a flexible Wireless Body Area Network solution, where sensors are represented by tri-axial integrated MEMS accelerometers to detect postures. This type of study is only interested in a predetermined set of postures. Therefore, they do not present accuracy results because they are only interested in determining the cross-point between two different postures.

Current research on gesture-based teaching focusses on students' gestures to enhance learning through gesture-based devices, not on recognizing the gestures themselves. Most studies are in the domain of special education, followed by science and math, and they focus mainly on motor skills and behavior skills training. These topics are alike in the sense that the primary goal or target skills to learn were kinetic-centered [1].

The most popular gesture-based devices are Nintendo Wii, including Wii remote, Wii balance board, Wii Fit, Wii console and Wii Nunchuk, and Microsoft Xbox Kinect. There are also studies that use their own designs, combining multiple gesture-based devices [1, 35].

For example, Ching-Hsiang Shih conducted many studies using a Wii and related products to support learners with both physical and cognitive difficulties to learn daily life skills (motor skills) and specific job tasks (job training), physical rehabilitation, and other behavior skills training [36–44].

[45] presents a virtual reality training application integrated with motion capture technology for dance training. The users wear a motion capture suit and receive feedback on how to improve their dance movements. The experimental results show that the system better assists students in learning comparing to the "watching video" approach. The students also think the learning process is fun and motivates them to learn.

At Nottingham Trent University, a Ph.D. researcher experimented on the effects of the Nintendo Wii and Xbox Kinect in helping college students with learning difficulties. The participants were 16 to 24 years old, with disabilities ranging from Down's Syndrome to autism spectrum disorders. The students played tennis using a Wii tennis game and/or bowling using a Kinect game, and the results show that with five weeks of computer game training, the participants improve significantly comparing the pre and post test scores in their corresponding real activities [46].

[35] lists several more specific projects, such as DepthJS and 3Gear System. DepthJS is a Kinect project that allows users to interact with web browsers by gestures, and 3Gear System uses gloves and computer cameras to trace the movements of hands.

Finally, [47] proposes the Presentation Trainer, that supports the practice of nonverbal communication skills for public speaking through real-time feedback. The tool tracks the user's voice and body and recognizes whether the learner uses gestures while speaking, using the Microsoft Kinect sensor V2 input of the coordinates of the learner's joints, keeping track of the angles between forearms and arms, and between arms and shoulder blades. The current version of the Presentation Trainer supports the training of basic public speaking skills by providing learners with feedback about their use of pauses, voice volume, body posture, use of gestures, use of phonetic pauses, and steadiness in body posture. However, this tool is only programmed to identify mistakes in this set of factors and does not present accuracy rates in the detection of the factors.

3 The Study

Being able to recognize a teacher's gestures is crucial to better understand students' behavior, learning outcomes, and engagement [5]. However, so far there are no studies that have tried to automatically recognize a teacher's gestures in a K-12 classroom, and the only available mechanisms to study teachers' gestures is to manually detect them, which is a very tedious and slow job. Also, if the intent is to analyze hundreds of hours of recordings for a large-scale analysis, then the computational requirements could also be a grand limitation. The contribution of this paper is to respond to the following research questions: (1) Is it possible to automatically detect a person's gestures from a normal-quality recorded video? This question aims at defining a method with a high accuracy rate for automatically detecting when a subject gesticulates through a normal-quality video camera recording; and (2) Is this method applicable to a K-12 classroom with the teacher as the target? This question aims at successfully applying the automatic gesture detection methodology in a school classroom without interfering either the teacher or the students, and at the same time, maintaining the accuracy rates.

By automatically recognizing a person's gestures through a video recording, we can objectively measure the extent to which the subject (classroom teacher) employs gestures during his/her instructional talk to keep the task moving forward and engage with his/her students. Using this method enables us to open the black box of nonverbal classroom interaction, and eventually we could compare teachers' nonverbal practices across different subjects, schools, grades and/or countries

3.1 Methods

In this study, we worked with two different types of video recordings. First, an exact gesture recording was created, where different persons were recorded acting many gestures.

Then a teacher was recorded giving a lecture in a normal elementary school classroom, where her gestures were motioned naturally throughout the course of the class.

Exact Gesture Recording. The gestures that were to be detected have many ways of being gesticulated. For example, a pointing gesture can be gesticulated by pointing with one finger, with the entire hand, by raising the arm up to your shoulder, by raising the arm just a bit, etc. Therefore, to detect these gestures and distinguish them from the non-gestures, different people were recorded acting each gesture in all possible angles. Five actors represented 10 different forms of emblem gestures [48], 18 forms of metaphoric gestures [11], 9 forms of pointing gestures [49], and 37 forms of non-gestures. Seven cameras were placed in front of the actors as can be seen in Fig. 1, capturing the actor's entire body.

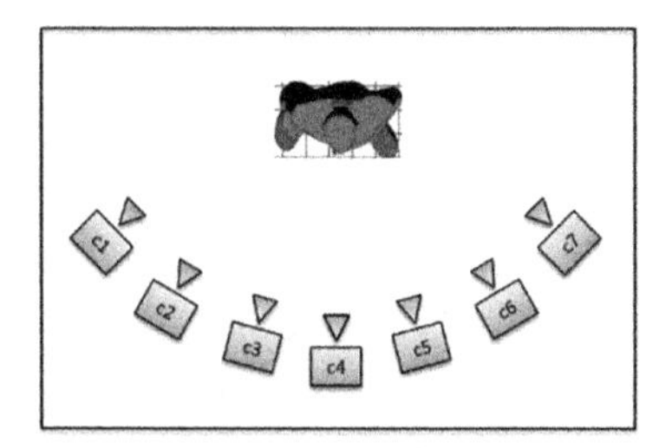

Fig. 1. Cameras in 7 different angles as seen from the ceiling

The five actors, three women and two men, represented different gestures a varied number of times. Also, the same actors represented what we define as "non-gestures", which are movements of the body that are not intended as having any meaning in the classroom context, therefore, not categorized as a gesture.

Real-Class Recording. To detect a teacher's natural gestures in a classroom, we recorded a 70 min 3rd grade science class. Of the entire recording, 22 min were only of the teacher, without students appearing in the image, therefore these were the parts of the class we worked with for now, to avoid confusion between subjects.

3.2 The Automatic Gesture Recognition Procedure

The automatic gesture recognition procedure has many stages. After obtaining a recording of a person gesticulating in a determined context, the recording is processed with OpenPose (from now on OP), and finally, run through a Machine Learning Software.

An advantage of this procedure is that normal video recordings weigh at least twice as their corresponding OP processed output. As an example, for a 132 MB class recording, the output would only be 70 MB. This is a good starting point for designing a practical system that could monitor teacher's gesture at a large scale. With the possibility of storing a reduced version of the video files, the system can run in a regular home or office computer, with no complicated computational requirements. Also, with this work procedure, there are no privacy issues. If children are recorded, OP does not identify them because it eliminates their faces from the outputs and only works with their skeletons, leaving their identities unknown.

OP [50, 51] is a real-time multi-person system that jointly detects human body, hand, facial, and foot keypoints on single images. To do this, the software separates a video recording into its frames, and processes frame by frame, returning a text file for each processed frame with the X and Y coordinates of the previously mentioned body key points. The body keypoints can be seen in Fig. 2(a) and the hand keypoints can be seen in Fig. 2(b).

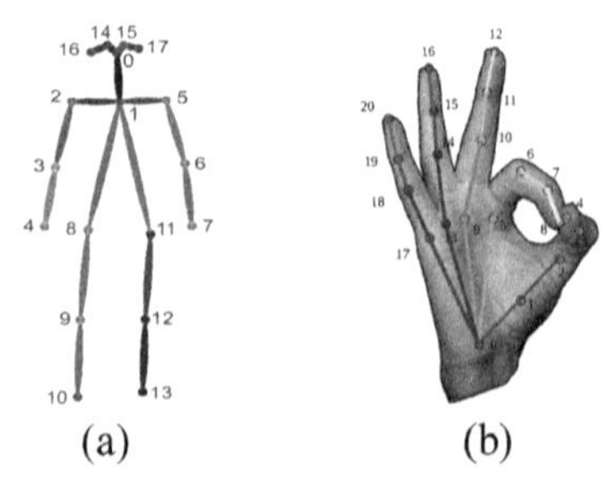

Fig. 2. (a) Set of 18 body points generated by OP [52]. (b) Set of 21 hand points generated by OP [51]

In this case, we only used the 18 body key points, 21 left-hand key points and 21 right-hand key points (60 key points in total). Figure 3(a) shows an example of the gesture frame outputs, and Fig. 3(b) shows an example of non-gesture frame outputs. The recordings had a resolution of 10FPS (frames per second), therefore, we had almost 60.000 frames of data to work with from the exact gesture recordings, and 13.000 frames from the classroom recording.

OpenPose's X and Y output coordinates consider the point (0,0) as the top right corner of an image. Therefore, there was a need to normalize each person's body coordinates to not depend on where the person was standing in the image, and only consider how the person was positioned regarding him/herself. The 18 body key points were normalized according to the neck's coordinates (body position 0) and the hands were each normalized regarding the wrist's key point (hand position 0).

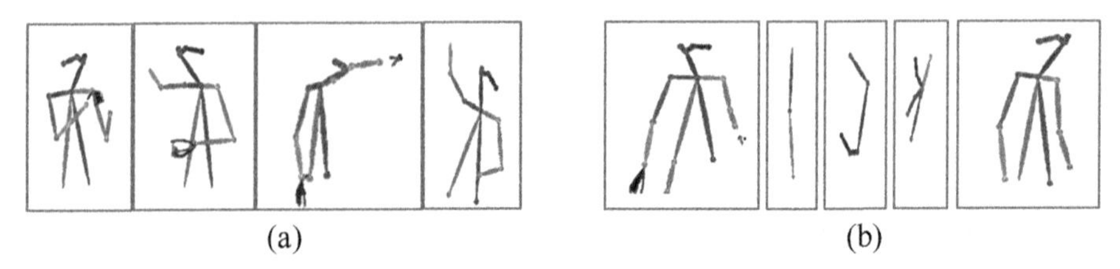

Fig. 3. (a) Gesture frame examples. (b) Non-gesture frame examples. Frames b.1 and b.5 show a person that is not gesticulating. Frames b.2, b.3 and b.4 are examples of when OpenPose didn't recognize a person or the person was in a position that made the body unrecognizable

Finally, the OP outputs must be run through an artificial intelligence algorithm to recognize from the subject's position (represented by the body's and hand's X and Y coordinates) if the subject is gesticulating or not. In this work we used Microsoft Azure Machine Learning Studio, which is a GUI-based integrated development environment for constructing and operationalizing Machine Learning workflow [53]. It allows the user to train and test a model with different classifiers. The models were trained with a binary classifier using a neural network algorithm, that was to determine, for each frame, if the subject was gesturing or not. As input, the machine used the information of the 120 coordinates of each frame, and a classification criterion: "G" for gesture and "X" for non-gesture. All frames were classified semi-automatically, and then some frames were also classified manually by gesture experts.

The semi-automatic classification consisted in watching the corresponding videos, determining the parts of the video where the subject gesticulated, and then calculating which frames were in the gesture-segments. The manual classification consisted in observing each OP processed frame and determining if the skeleton was gesticulating or not.

4 Results and Discussion

The training and testing methods combine the different available datasets to obtain the best possible results in automatically recognizing the teacher's gestures in the real-class recording dataset. The available training and testing datasets are the acted recordings and the real-class recordings; in their semi-automatic classification versions and manual classification versions. First, we trained the model with the acted gesture dataset and tested with the real class video recording, considering the semi-automatic classifications for both. Then, we trained the model with the acted gesture dataset and tested with the real class video recording, considering the manual classifications for both. Finally, we trained the model with the first half of the real class video recording dataset and tested with the second half, considering the manual classifications.

Regarding RQ1, in Table 1 we observe that by training the machine learning software with the exact gesture recording frames, and then testing the algorithm with the real class recording frames, considering the semi-automatically classified frames, the overall accuracy rate is 54%. This preliminary result was not as expected, which led us to manually review the frame's classification and reclassify some of them manually. By doing this, we were able to improve the gesture-prediction accuracy to 65% and to 49% when predicting non-gesture frames as can be observed in Table 2. The overall accuracy improves to 59%, which is much higher than the baseline accuracy of declaring all frames as non-gesture for this case (41%), and a bit higher than the accuracy baseline of declaring all frames as gesture (58%). This is a clear signal that if we were to have more manually classified frames, and hopefully more real-class recordings, this accuracy could improve significantly. Regarding RQ2, in Table 3, we observe that when training the machine learning software with 75% of the frames of the manually classified frames of a real-class and testing the algorithm with the other 25%, we obtain an accuracy rate of almost 55% for gestures, 86% for non-gesture frames, and an overall accuracy rate of 78%. This is much higher than what related works have obtained when trying to identify complete body movements, and higher than the accuracy baseline of declaring all frames non-gesture (73.5%) and if declaring all frames as gesture (26%).

Table 1. Training: recorded gestures. Testing: real-class rec. & semi-automatic classifications

Overall Accuracy: 54%		Predicted Class frames	
		G	X
Class frames	G	1.614 (47%)	1.794 (53%)
	X	4.268 (44%)	5.395 (56%)

Table 2. Training: recorded gestures. Testing: real-class recording & manual classifications

Overall Accuracy: 59%		Predicted Class frames	
		G	X
Class frames	G	467 (65%)	234 (35%)
	X	244 (51%)	252 (49%)

Table 3. Training and testing with the real-class recording and manual frame classifications

Overall Accuracy: 78%		Predicted Class frames	
		G	X
Class frames	G	73 (55%)	60 (45%)
	X	53 (14%)	317 (86%)

5 Conclusions and Future Work

We can conclude that Machine Learning and Image Processing algorithms of today allow automatic detection of teacher gestures. This can be done in a classroom through a non-invasive method that doesn't intervene in the class nor disturb the students. The obtained accuracy rates are similar or better [32] than related works that use much more invasive or expensive [34] methods for the gesticulating subject. We envision thousands of classrooms using accessible (cheap) cameras, processing the recordings in OP locally and then uploading the OP output file to a cloud server to obtain the data about their gestures in the class. This simple procedure has no problem with the size of files, making it very practical, and has no privacy problems. Even if children are recorded, only the OpenPose output files are uploaded, without information from the students.

Through this Automatic Gesture Recognition methodology, uploading and saving classroom files from thousands of teachers, each one with hundreds of hours, could be a completely manageable task. With these amounts of data, Machine Learning algorithms will be able to improve the recognition rates significantly. For future work, we would like to record more real classes and classify more frames manually to enlarge our database and test our automatic gesture recognition method in more K-12 contexts, with the desire of improving our accuracy results. We would also like to explore the sequences of movements of the teacher between two or more frames. This work presents the analysis of static images but studying the sequences between the images could also be a mechanism to improve our accuracy rates significantly.

In conclusion, this paper contributes by introducing a scalable and cost-effective alternative to detect hand gestures using regular video cameras and securing the privacy of students, starting a new area of practical tools to analyze and improve teacher's practices in a nearby future. Combined with other technologies, like speech recognition [54] and text mining of the transcriptions of the teacher discourse [55], a powerful and practical tool can be offered to provide inexpensive, private and timely feedback to teachers about communication features of her teaching practices and its preliminary results prove that recognizing a teacher's gestures in a classroom is possible and is a starting point for many future investigations on gesture detection and gesture-based education.

Acknowledgements. Support from ANID/PIA/Basal Funds for Centers of Excellence FB0003, as well as FONDECYT 3170062 are gratefully acknowledged.

References

1. Sheu, F.R., Chen, N.S.: Taking a signal: A review of gesture-based computing research in education. Comput. Educ. (2014). https://doi.org/10.1016/j.compedu.2014.06.008
2. Patrona, F., Chatzitofis, A., Zarpalas, D., Daras, P.: Motion analysis: action detection, recognition and evaluation based on motion capture data. Pattern Recognit. **76**, 612–622 (2018). https://doi.org/10.1016/j.patcog.2017.12.007
3. Johnson, L., Levine, A., Smith, R., Stone, S.: The Horizon Report 2010. ERIC (2010)
4. Johnson, L., Adams, S., Cummins, M.: Horizon Report 2012 K-12 Edition (2012)
5. Farsani, D.: Making Multi-Modal Mathematical Meaning in Multilingual Classrooms. Thesis (2015)
6. Araya, R., Farsani, D., Hernández, J.: How to attract students' visual attention. In: Verbert, K., Sharples, M., Klobučar, T. (eds.) EC-TEL 2016. LNCS, vol. 9891, pp. 30–41. Springer, Cham (2016). https://doi.org/10.1007/978-3-319-45153-4_3

7. Farsani, D.: Complementary functions of learning mathematics in complementary schools. In: Halai, A., Clarkson, P. (eds.) Teaching and Learning Mathematics in Multilingual Classrooms. LNCS, pp. 227–247. SensePublishers, Rotterdam (2016). https://doi.org/10.1007/978-94-6300-229-5_15
8. Ramseyer, F., Tschacher, W.: Nonverbal synchrony in psychotherapy: coordinated body movement reflects relationship quality and outcome. J. Consult. Clin. Psychol. (2011). https://doi.org/10.1037/a0023419
9. Farsani, D., Barraza, P., Martinic, S.: Non-verbal synchrony as a pedagogical tool for medical education: a clinical case study. Kuwait Med. J. (accepted)
10. Farsani, D.: Mathematics education and the analysis of language working group: making multimodal mathematical meaning. Br. Soc. Res. Learn. Math. **32**, 19–24 (2012)
11. McNeill, D.: Hand and Mind: What Gestures Reveal about Thought. University of Chicago Press, Chicago, IL, US (1992)
12. Kendon, A.: Some relationships between body motion and speech. Stud. Dyadic Commun. **7**(177), 90 (1972). https://doi.org/10.1016/B978-0-08-015867-9.50013-7
13. Sfard, A.: What's all the fuss about gestures? A commentary. Educ. Stud. Math. **70**, 191–200 (2009). https://doi.org/10.1007/s10649-008-9161-1
14. Maschietto, M., Bussi, M.G.B.: Meaning construction through semiotic means: the case of the visual pyramid. Int. Gr. Psychol. Math. Educ. **3**, 313–320 (2005)
15. Streeck, J.: The significance of gesture: how it is established. IPrA Pap. Pragmat. (1988). https://doi.org/10.1075/iprapip.2.1-2.03str
16. Kendon, A.: Gesticulation and speech: two aspects of the process of uterance in M. Group. The relationship of the verbal and nonverbal communication. The Hague: Mouton, pp. 207–228 (1980)
17. Kendon, A.: Do gestures communicate?: a review. Res. Lang. Soc. Interact. (1994). https://doi.org/10.1207/s15327973rlsi2703_2
18. Thompson, L.A., Massaro, D.W.: Children′s integration of speech and pointing gestures in comprehension. J. Exp. Child Psychol. (1994). https://doi.org/10.1006/jecp.1994.1016
19. Graham, J.A., Heywood, S.: The effects of elimination of hand gestures and of verbal codability on speech performance. Eur. J. Soc. Psychol. (1975). https://doi.org/10.1002/ejsp.2420050204
20. Mcneil, N.M., Alibali, M.W., Evans, J.L.: The role of gesture in children's comprehension of spoken language: now they need it, now they don't. J. Nonverbal Behav. **24**, 131–150 (2000). https://doi.org/10.1023/A:1006657929803
21. Riseborough, M.G.: Physiographic gestures as decoding facilitators: three experiments exploring a neglected facet of communication. J. Nonverbal Behav. (1981). https://doi.org/10.1007/BF00986134
22. Alibali, M.W., Nathan, M.J.: Teachers' gestures as a means of scaffolding students' understanding: evidence from an early algebra lesson. Video Res. Learn. Sci. 349–365 (2007)
23. Pea, R., Lindgren, R.: Collaboration design patterns in uses of a video platform for research and education. IEEE Trans. Learn. Technol. **1**, 235–247 (2008). https://doi.org/10.1109/TLT.2009.5
24. Richland, L.E., Zur, O., Holyoak, K.J.: Cognitive supports for analogies in the mathematics classroom. Sci. **316**, 1128–1129 (2007). https://doi.org/10.1126/science.1142103
25. Hiebert, J., et al.: Teaching mathematics in seven countries: Results from the TIMSS 1999 video study. Educ. Stat. Q. **5**, 7–15 (2003)
26. Richland, L.E., Hansen, J.: Reducing cognitive load in learning by analogy. Int. J. Psychol. Stud. **5**, 1–11 (2013). https://doi.org/10.5539/ijps.v5n4p69
27. Valenzeno, L., Alibali, M.W., Klatzky, R.: Teachers' gestures facilitate students' learning: a lesson in symmetry. Contemp. Educ. Psychol. **28**, 187–204 (2003). https://doi.org/10.1016/S0361-476X(02)00007-3

28. Yang, H., Meinel, C.: Content based lecture video retrieval using speech and video text information. IEEE Trans. Learn. Technol. **7**, 142–154 (2014). https://doi.org/10.1109/TLT.2014.2307305
29. Yang, M.T., Liao, W.C.: Computer-assisted culture learning in an online augmented reality environment based on free-hand gesture interaction. IEEE Trans. Learn. Technol. **7**, 107–117 (2014). https://doi.org/10.1109/TLT.2014.2307297
30. AL-Rousan, M., Assaleh, K., Tala'a, A.: Video-based signer-independent Arabic sign language recognition using hidden Markov models. Appl. Soft Comput. J. **9**, 990–999 (2009). https://doi.org/10.1016/j.asoc.2009.01.002
31. Chen, X., Wang, Z.J.: Pattern recognition of number gestures based on a wireless surface EMG system. Biomed. Signal Process. Control **8**, 184–192 (2013). https://doi.org/10.1016/j.bspc.2012.08.005
32. Junker, H., Amft, O., Lukowicz, P., Tröster, G.: Gesture spotting with body-worn inertial sensors to detect user activities. Pattern Recogn. **41**, 2010–2024 (2008). https://doi.org/10.1016/j.patcog.2007.11.016
33. Hachaj, T., Ogiela, M.R.: Full body movements recognition-unsupervised learning approach with heuristic R-GDL method. Digit. Signal Process. A Rev. J. **46**, 239–252 (2015). https://doi.org/10.1016/j.dsp.2015.07.004
34. Farella, E., Pieracci, A., Benini, L., Rocchi, L., Acquaviva, A.: Interfacing human and computer with wireless body area sensor networks: the WiMoCA solution. Multimed. Tools Appl. **38**, 337–363 (2008). https://doi.org/10.1007/s11042-007-0189-5
35. Johnson, L., Smith, R., Willis, H., Levine, A., Haywood, K.: The 2011 Horizon Report (2011)
36. Shih, C.H., Shih, C.T., Chu, C.L.: Assisting people with multiple disabilities actively correct abnormal standing posture with a Nintendo Wii Balance Board through controlling environmental stimulation. Res. Dev. Disabil. (2010). https://doi.org/10.1016/j.ridd.2010.03.004
37. Shih, C.H., Shih, C.T., Chiang, M.S.: A new standing posture detector to enable people with multiple disabilities to control environmental stimulation by changing their standing posture through a commercial Wii Balance Board. Res. Dev. Disabil. (2010). https://doi.org/10.1016/j.ridd.2009.09.013
38. Shih, C.H., Shih, C.J., Shih, C.T.: Assisting people with multiple disabilities by actively keeping the head in an upright position with a Nintendo Wii Remote Controller through the control of an environmental stimulation. Res. Dev. Disabil. (2011). https://doi.org/10.1016/j.ridd.2011.04.008
39. Shih, C.H., Chung, C.C., Shih, C.T., Chen, L.C.: Enabling people with developmental disabilities to actively follow simple instructions and perform designated physical activities according to simple instructions with Nintendo Wii Balance Boards by controlling environmental stimulation. Res. Dev. Disabil. (2011). https://doi.org/10.1016/j.ridd.2011.05.031
40. Shih, C.H., Chen, L.C., Shih, C.T.: Assisting people with disabilities to actively improve their collaborative physical activities with Nintendo Wii Balance Boards by controlling environmental stimulation. Res. Dev. Disabil. (2012). https://doi.org/10.1016/j.ridd.2011.08.006
41. Shih, C.H., Chang, M.L., Mohua, Z.: A three-dimensional object orientation detector assisting people with developmental disabilities to control their environmental stimulation through simple occupational activities with a Nintendo Wii Remote Controller. Res. Dev. Disabil. (2012). https://doi.org/10.1016/j.ridd.2011.10.012
42. Shih, C.H.: A standing location detector enabling people with developmental disabilities to control environmental stimulation through simple physical activities with Nintendo Wii Balance Boards. Res. Dev. Disabil. (2011). https://doi.org/10.1016/j.ridd.2010.11.011

43. Shih, C.H., Chang, M.L.: A wireless object location detector enabling people with developmental disabilities to control environmental stimulation through simple occupational activities with Nintendo Wii Balance Boards. Res. Dev. Disabil. (2012). https://doi.org/10.1016/j.ridd.2011.12.018
44. Nissenson, P.M., Shih, A.C.: MOOC on a budget: development and implementation of a low-cost MOOC at a state university. Comput. Educ. J. **7**(1), 8 (2016)
45. Chan, J.C.P., Leung, H., Tang, J.K.T., Komura, T.: A virtual reality dance training system using motion capture technology. IEEE Trans. Learn. Technol. **4**, 187–195 (2011). https://doi.org/10.1109/TLT.2010.27
46. Hui-Mei, J.H.: The potential of kinect in education. Int. J. Inf. Educ. Technol. **1**, 365–370 (2013). https://doi.org/10.7763/ijiet.2011.v1.59
47. Schneider, J., Borner, D., Van Rosmalen, P., Specht, M.: Can you help me with my pitch? studying a tool for real-time automated feedback. IEEE Trans. Learn. Technol. **9**, 318–327 (2016). https://doi.org/10.1109/TLT.2016.2627043
48. Barakat, R.A.: Arabic gestures. J. Pop. Cult. (1973). https://doi.org/10.1111/j.0022-3840.1973.00749.x
49. Enfield, N.J.: 'Lip-pointing': a discussion of form and function with reference to data from Laos. Gesture. **1**, 185–211 (2002). https://doi.org/10.1075/gest.1.2.06enf
50. Cao, Z., Simon, T., Wei, S.E., Sheikh, Y.: Realtime multi-person 2D pose estimation using part affinity fields. In: Proceedings of the IEEE Conference on Computer Vision and Pattern Recognition, pp. 7291–7299 (2017). https://doi.org/10.1109/CVPR.2017.143
51. Simon, T., Joo, H., Matthews, I., Sheikh, Y.: Hand keypoint detection in single images using multiview bootstrapping. In: Proceedings of the IEEE Conference on Computer Vision and Pattern Recognition, pp. 1145–1153 (2017). https://doi.org/10.1109/CVPR.2017.494
52. Konrad, S.G., Shan, M., Masson, F.R., Worrall, S., Nebot, E.: Pedestrian Dynamic and Kinematic Information Obtained from Vision Sensors. In: 2018 IEEE Intelligent Vehicles Symposium (IV), pp. 1299–1305. IEEE (2018). https://doi.org/10.1109/IVS.2018.8500527
53. Copeland, M., Soh, J., Puca, A., Manning, M., Gollob, D.: Microsof t azure machine learning. Microsoft Azure, pp. 355–380. Apress, Berkeley, CA (2015). https://doi.org/10.1007/978-1-4842-1043-7_14
54. Caballero, D., et al.: ASR in classroom today: automatic visualization of conceptual network in science classrooms. In: Lavoué, É., Drachsler, H., Verbert, K., Broisin, J., Pérez-Sanagustín, M. (eds.) EC-TEL 2017. LNCS, vol. 10474, pp. 541–544. Springer, Cham (2017). https://doi.org/10.1007/978-3-319-66610-5_58
55. Araya, R., Jiménez, A., Aguirre, C.: Context-based personalized predictors of the length of written responses to open-ended questions of elementary school students. In: Sieminski, A., Kozierkiewicz, A., Nunez, M., Ha, Q.T. (eds.) Modern Approaches for Intelligent Information and Database Systems. SCI, vol. 769, pp. 135–146. Springer, Cham (2018). https://doi.org/10.1007/978-3-319-76081-0_12

The Effect of Teacher Unconscious Behaviors on the Collective Unconscious Behavior of the Classroom

Roberto Araya (✉) and Danyal Farsani

Centro de Investigación Avanzada en Educación, Universidad de Chile, Santiago, Chile
roberto.araya.schulz@gmail.com, Danyal.Farsani@gmail.com

Abstract. Normally teachers can consciously control to a great extent the behaviors of their students in the classroom. But additionally, there are unconscious teacher behaviors that also impact the collective behavior of their students. To study this phenomenon, we gather data obtained from mini video cameras mounted on eyeglasses worn by fourth graders. We found that the proportion of scenes where the teacher is pointing his body toward the student is higher than the proportion of scenes when there is mutual gaze, and that this effect is slight pronounced in STEM classes. We also found that this effect is greater among boys than girls, and that is particularly evident at certain distances. More precisely, we found that in STEM classes when a male student is observing the teacher, the teacher is generally pointing their body toward the student (67% of cases). However, with female students, this number is just 46%. However, there is no such difference in non-STEM classes. Moreover, the distance between the student and the teacher also has a significant effect. This is a powerful tool for teachers as it can help them reflect on their strategies, as well as the impact of their unconscious nonverbal behavior in classroom behavior.

Keywords: Collective intelligence · Engagement · Nonverbal communication · Data mining · Visual attention · Body orientation

1 Introduction

Understanding students' behavior, attention, engagement, and collective behavior is very important for improving classroom practices and the quality of teaching. According to [1], over the last few decades "The 'what' of teaching has indeed changed, but when it comes to the how- the pedagogy- few major changes have occurred". He claims that there is one fundamental attribution error which policy makers often make: they focus on the characteristics of the teacher and not on the situation itself. In this paper, we explore the use of video technology to measure the impact on the "how", particularly in terms of students' visual attention. Our goal is to be able to give useful feedback to teachers in order for them to become more effective and generate a more intelligent collective behavior. For every teacher, a critical aspect to achieve effectiveness is student engagement and classroom behavior. However, according to [2] there is little agreement

M. Hernes et al. (Eds.): ICCCI 2020, CCIS 1287, pp. 529–540, 2020.
https://doi.org/10.1007/978-3-030-63119-2_43

on a concrete definition and effective measurement of engagement. [2] underlies that it is critical to identify the grain size: micro level (individual in the moment such as response time and attention location) or macro level (group of learners or course). For student engagement in school [3] identifies two components: a behavioral component, termed participation, and an emotional component, termed identification. At the most basic level participation involves attending school and paying attention to the teacher. In this paper we focus on the latter.

Attention is very broad. It is an adaptation that evolved for filtering critical information for surviving and mating. It is a system functionally specialized [4] that tracks phylogenetically important sensory cues, such as the presence of human and non-human animals. It responds much faster and accurate to animals than to inanimate objects. It also responds particularly well to high-level social cues. Moreover, there is [5] visual and conscious attention, where we can differentiate between shifting spatial attention from maintaining sustained attention. In this paper we only include visual attention, and not the sustained. We have included sustained attention in previous work [6].

For analyzing visual attention we have to consider [7] human eye morphology, and how it is adapted to communicate using gaze signals. Humans possess the largest ratio of exposed sclera in the eye outline. Moreover, [8] concludes that only human beings have highly-visible eye direction and that only humans use this type of information. Furthermore, 12-month-old human infants tend to follow the direction of other people's eyes more than the direction of their head [8].

Does this mean that the teacher's gaze captures students' attention more than any other nonverbal behavior? Eye contact is very important for facilitating communication, understanding, team work and it is a basic nonverbal strategies in teaching [9]. However, studies in other professions, such as the medical profession, have found that the physician pointing their body toward the patient is sometimes more critical than making eye contact. For example, during Doctor-Patient consultations [10], the body orientation signals availability or non-availability for collaboration. The amount of time physicians spend with their bodies pointing toward a patient is positively associated with the patient's level of engagement, nonverbal synchrony and post-visit satisfaction [11]. When a physician points their body away from the patient, it is positively associated with the patient's perception of the physician's dominance, while also being negatively associated with how an external evaluator rates a physician's rapport. On the other hand, [12] reports that the amount of time that doctors maintain eye contact with patients is negatively associated with perceptions of a doctor's rapport with their patients, as well as with the patient's level of satisfaction and understanding. Is there a similar pattern in the classroom in terms of Teacher-Student interactions?

In order to study such phenomena, we use first-person, or egocentric [13] videos (see Fig. 1a). These videos are obtained from mini video cameras mounted on the students' glasses. Eyeglasses are a widely used technological device that have an enormous impact on the everyday life of millions of people. Mounting a mini video camera on the frame of a pair of glasses is not a major change, nor does it inconvenience the students. It is simply a minor add-on to the glasses. This approach would allow any teacher to access and measure students´ visual engagement as these eye-glasses are cheap, affordable and widely accessible, in contrast to eye-tracking devices. In our experience, after a few minutes the students completely forgot that they were wearing the glasses. These devices are also very affordable. Therefore, they can be widely used and proven helpful for teachers and students. In this paper, we use these enhanced glasses to gather information about important collective phenomena, such as teacher–students' visual interaction patterns. Most of these patterns are the product of unconscious behavior. They are hidden to most of us. Students and teachers are often not aware of some of their most critical actions. These are actions that have a major impact on others. Furthermore, not only do students and teachers often fail to realize what they are doing, but also to realize what the other agents in the class are doing and the impact that this has on them. Students and teachers do not receive explicit feedback on these behaviors and, therefore, they do not modify them in order to be more effective and improve the collective intelligence. These unconscious actions are very important in the classroom. For example, teachers try to attract their students' attention. They try different strategies, but many of these are actually unconscious. The question is how can we track such critical variables and how can we estimate their effect?

Fig. 1. a) Student wearing eyeglasses with mini video camera mounted on the frame and with the lenses taken out. 1. b) Classroom where the teacher and a sample of 3 students are wearing the eyeglasses.

Visual technology can provide new opportunities for detecting subtle yet important patterns in classes. Here we analyze two months of videos obtained from cameras mounted on fourth-grade students' glasses. In a previous analysis, [14] found that the gaze from students with a low Grade Point Average (GPA) toward the teacher decreases much more after 40 min than with high GPA students. [6] found that the teacher's gestures have a significant impact on the students' visual attention. In [15], we found that the teacher pointing their body toward the student has a greater impact on the students' visual attention than making eye contact, particularly at certain distances.

In this paper, we tackle a very important questions regarding students' engagement through classroom interaction. We study the impact on students' engagement through their visual attention in STEM classes. These classes are considered to be highly demanding for students. Moreover, new curricula require new, deeper and more cognitively demanding contents and teaching practices [16]. There is now an emphasis on developing and using crosscutting concepts. In addition to this, there are new requirements to increase integration among the different science disciplines, as well as integrating these with mathematics and other subjects [17]. Is the impact of the teacher's nonverbal behavior different in STEM classes than in other classes? Does the impact depend on the student's gender?

2 Literature Review

Proxemics, often referred to as "the science of human space utilization" [18] or how "people regulate themselves in space and how they move in space" [19], have attracted many anthropologists, psychologists and contemporary educators. The term proxemics was coined by an American anthropologist, Edward T. Hall [20], who examined the proxemics of interpersonal communication across different cultures. He categorized people's use of space and the distance they maintain with others into four categories: intimate space (up to 45 cm), personal space (up to 120 cm), social space (up to 370 cm) and public space (more than 370 cm). [15] have renamed these categories as Private, Personal, Professional and Public spaces as they refer to interactions within the context of professional educational. We would like to examine the effects of these four spaces on the students' visual attention.

Particular attention has been given to not only the role of cross-cultural communication, but also how proxemics can be used as a resource in education. Proxemics has often been regarded as a resource that teachers can routinely draw upon, not only in order to regulate a smoother turn-taking process [21], but also to be used as unconscious and non-verbal behavior disciplinary remarks. For example, [22] has shown a proxemics study demonstrating interpersonal distance to be a significant factor in classroom interaction. His analysis of a detailed video recording of a Chinese-American classroom showed that the medium of instruction determined particular patterns of proxemics and interpersonal space. Cantonese not only triggered a closer proxemics space between the interlocutors, but also enabled significantly more turning angles (body orientation) between students and the teacher. This created a more engaging atmosphere and increased student attention. Furthermore, the students were more likely to communicate about topics relating to the lesson. [23] took this idea a step further and analyzed the proxemics behavior between boys and girls from a Persian heritage in the UK. He looked at the multi-modal mathematical messages that British-Iranian learners subconsciously send and receive. Furthermore, he also looked at the ways in which different languages (English and Farsi) affected the students' body orientation and proxemics behavior in classroom interactions. In this sense, English was often employed in order to keep the technical task moving forward, while Farsi was used for making jokes, behavior management and emotional engagement. Therefore, Farsi was a verbal trigger for increasing the turning angle between students. He also showed that girls maintained a closer proximity with a greater turning angle to one another as they were discussing ideas/tasks. In contrast, boys maintained a more distant personal distance, a more acute turning angle and made less eye contact with one another. Although previous studies have shown the different effects that nonverbal teacher behavior can have on students' attention [6], we would like to know if the mutual gaze and teacher body orientation, gender of the student, the curriculum subject, as well as the different space categories in the classroom, affect the students' visual attention. This is a key research gap that we have identified.

3 Method

For two months, a fourth-grade classroom teacher and a sample of three students selected each day were asked to wear for the whole day a mini video camera mounted on eyeglass frames in a primary school in Santiago, Chile (see Fig. 1b). Therefore, the data that emerges in this paper is part of a larger dataset which was set to investigate the unconscious patterns of interactional behavior by the classroom teacher and students [13]. By mounting cameras on the students' glasses, we were able to compute and obtain a better perspective of the class, as seen by the student. Every day, the classroom teacher and three students wore the glasses continuously, with the exception of break times and lunchtime, as shown in Fig. 1. The students were also asked to take their glasses off when they wanted to take a comfort break. The average age of the students was 10.5 years and the class consisted of 21 boys and 15 girls. In total, we obtained nearly 203 h of interactional recordings from the students and the teacher. These video cameras had a recording quality of thirty frames per second (30 fps); for each video, a frame was sampled every second and processed in order to detect the presence of faces. At the end of each day, the recordings were manually downloaded onto a computer. Institutional and written parental consent forms were obtained before starting the study.

In this study, we use the same tools and instruments to analyze the frames that were used in our previous study [6]. Of all the faces that were detected and identified, there was a total of 857 frames (still images) where students were looking at the classroom teacher. Just like before, we were interested in instances where students kept their visual attention on the teacher. There were moments where more than two faces were present in the same frame, e.g. the teacher and someone else who had just arrived late to the lesson. In these cases, we decided to discard the frame as the student's visual attention may have been fixed on the student and not on the teacher. There were other moments where we deliberately discarded frames and did not count them in the analysis. This included cases where the clarity of the frames was low or blurred and we could therefore not discern whether or not the teacher was looking at the student. Putting all of these strict measures in place therefore made our interpretation of the analysis of the frames more effective. In the end, we obtained 646 frames that were used for both qualitative and statistical analysis. Furthermore, given the fact that two consecutive frames from the same video camera where the student is looking at the teacher (i.e. two frames that are only one second apart) do not represent two independent gazes, rather the same gaze that was held for more than one second, we define a 'scene' as two or more consecutive frames coming from the same student's camera. From the total of 646 frames where the students were looking directly at the teacher, we found 311 scenes. The shortest of these contained only one frame, while the longest contained eight frames.

The analysis of proxemics, body orientation and gaze all came from the 646 frames. Analyzing frames or 'reading still images' was an integral part of the analysis, noting what each student and teacher did, moment by moment. When analyzing the gaze, we looked at each frame and noted whether the teacher was looking directly at the observer (the student that was wearing the mini camera). In the analysis of body orientation, we examined each frame and noted whether the teacher's shoulders were pointing toward the observer. Finally, in the analysis of proxemics we measured how far away the teacher was standing or sitting from the observer following a methodology established in [15].

Having obtained an approximate estimation of how far away the observer was from the teacher, we then classified each frame in terms of the proxemics: Private, Personal, Professional and Public space.

4 Results and Discussions

In this section, we first compute the effects on the students' visual attention caused by the teacher making eye contact, as well as pointing their body toward the student. This calculation is made for both STEM and non-STEM classes. Following this, we then analyze the effects according to gender and proxemics.

It is worth noting the difference between a gaze and eye contact. When someone is looking at someone else, then there is a gaze. Eye contact is generated when there is a mutual gaze, i.e. when both parties are simultaneously gazing at each other. When there is direct eye contact, it is traditionally and academically perceived to be a great tool for behavior management, perhaps more so than body orientation. Our results show that this is not the case. [15] examined the effects of body orientation and eye contact on students' visual attention. It was found that body orientation is three times as powerful as mutual gaze in the classroom. Does this effect vary in STEM versus non-STEM classes?

From the total of 311 scenes, 215 were from STEM classes and 96 were non-STEM classes, even though there are more non-STEM classes in fourth grade. This means students pay more visual attention to the teacher in STEM classes. From the 215 scenes that were STEM classes, 41 involved instances where the teacher and learner made eye contact. This is 19.1% of the scenes as shown in Fig. 2. This number increases significantly to 125 scenes in which the teacher's body was pointing toward the observer. This is 58.1% of the scenes as shown in Fig. 3. The difference is statistically significant at 95% confidence level. On the other hand, from the 96 scenes in non-STEM classes, only 22 scenes involved a mutual gaze, which corresponds to 22.9% of the scenes, while in 51 scenes the teacher's body was pointing toward the student, which is 53.1% of the scenes. The difference is statistically significant at 95% confidence level. Moreover, the percentage of scenes with body orientation in STEM classes is higher than in Non-STEM classes, as show in Fig. 3. On the contrary, in STEM classes the percentage of scenes with mutual gaze is lower than in Non-STEM classes, as shown in Fig. 2. However those difference are not statistically significant.

Note that the total number of scenes in which the teacher's body was pointing toward the observer in STEM lessons is 3 times higher than the number of scenes in which eye contact was made (more precisely, 125/41 = 3.05), while in the non-STEM classes it is only slightly more than 2 times higher (more precisely, 51/22 = 2.32). However, this ratio for STEM is not statistically significantly higher than for Non-STEM lessons at 95% confidence level (Fig. 4).

Let us now consider the gender of the observer and mutual gaze effect. On STEM classes, as shown in Fig. 5, the percentage of scenes for observer girls and boys where the teacher is doing eye contact are not much different (15.1% and 22.1%, respectively), and the difference is not statistically different. They are also statistically similar on non-STEM classes (23.1% and 22.9%, respectively). Moreover the percentages seen in STEM and non-STEM classes are also statistically similar.

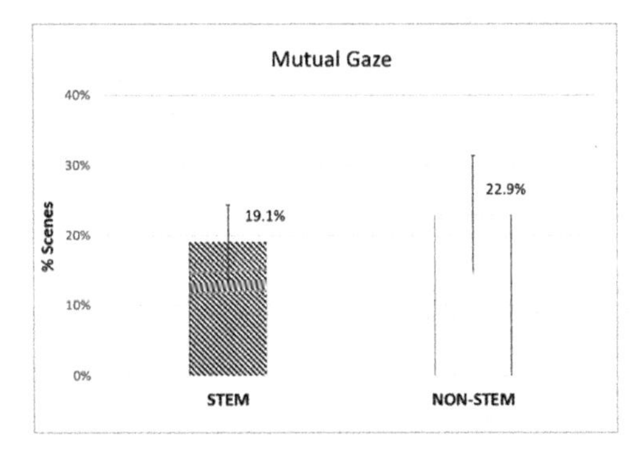

Fig. 2. Percentage of the scenes by subject in which there is a mutual gaze (eye contact).

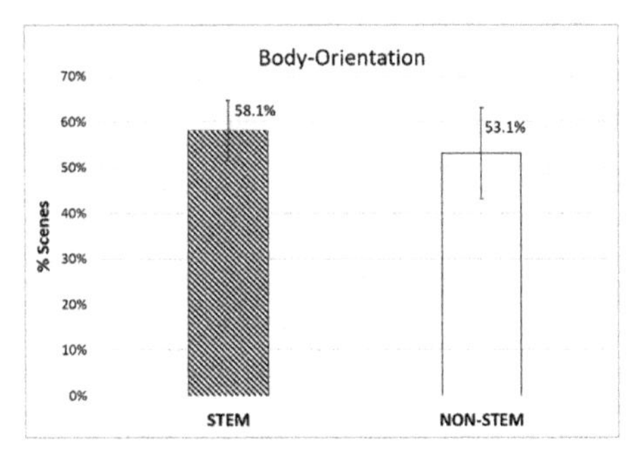

Fig. 3. Percentage of scenes by subject in which the teacher points his body towards the observer.

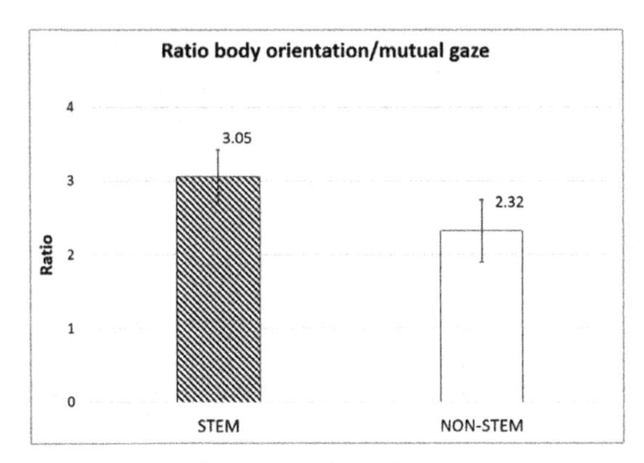

Fig. 4. Ratio of number of scenes with Body-Orientation to number of scenes with mutual gaze.

However, in the case of body orientation it is very different. In STEM classes, the percentage of scenes where the teacher's body is pointing toward the observer is statistically significantly (at 95% confidence level) higher for boys than for girls (67.2% versus 46.2%), as shown in Fig. 6. Surprisingly, for non-STEM classes it is the other way round. In non-STEM classes, the percentage of scenes where the teacher is pointing

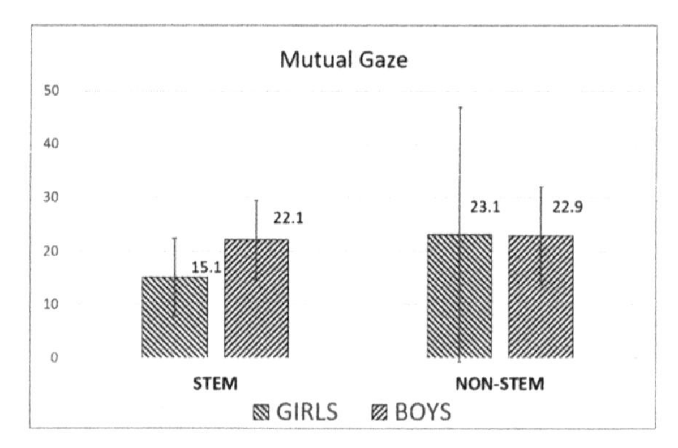

Fig. 5. Percentage of the scenes by curriculum subject and gender in which there is a mutual gaze (eye contact).

his body toward the observer is lower for boys than for girls (51% versus 69%), but the difference is not statistically significant.

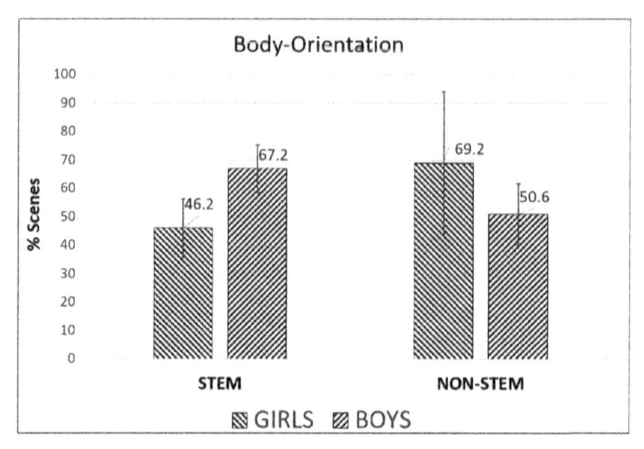

Fig. 6. Percentage of the scenes by subject and gender in which the teacher points his body towards the observer.

We will now examine the combined modulation of gender and different space categories on the effect on students' visual attention. Traditionally, one might have thought that the students' visual attention was focused more on the teacher in the Private space (P1), followed by the Personal space (P2), then the Professional space (P3) and finally the Public space (P4). It is natural to think that the further away a student is sitting from a teacher, for example in P4, the more likely they are to be off task, regardless of their gender. Figure 7 offers us an insight into the black box of classroom practices that has now been made possible by tracking the students' visual attention. There was a total of 311 scenes (coming from 646 frames) where the students capture their classroom teacher in their visual field. This mostly occurred in the professional space, with 167 scenes, followed by the public space with 105 scenes, and then the personal space with 32 scenes. While it appears that the teacher's body orientation did not make much of a difference between boys and girls in P2 and P4, there is a statistically significant difference in P3. Boys in P3 gleaned more information from their teacher's body orientation. Although this relationship does not explain the causation, further multidisciplinary research is needed to explore this interesting phenomenon. However, there is a crystal-clear difference suggesting that the visual attention of male students is more heavily influenced by their teacher's body orientation in the professional space. In P3, when the observer is a boy the teacher's body is pointed toward the observer in 71.7% of the scenes. In the case of girls, this percentage drops to just 50.0%. This difference is statistically significant at 95% confidence level.

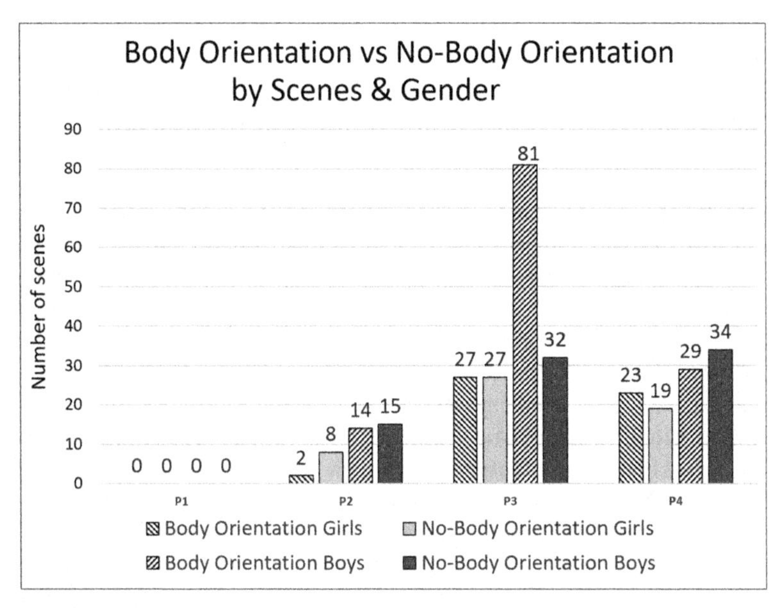

Fig. 7. Number of scenes in the different space categories in which the teacher points his body towards the observer and in which he does not.

5 Conclusions

One of the main challenges teachers face is attracting attention and guiding the collective behavior of their students in the classroom. There are many recommended strategies, but in general they are mainly verbal and consciously used strategies. However, there is a very important universe of unconscious non-verbal teacher behaviors that also have an impact on students. In this paper we have studied two unconscious teacher's behaviors: the mutual gaze between teacher and student, and the body-orientation of the teacher. We have studied their effect on the attention component of students' engagement. This is at the micro level [2] description of engagement and the one that consider student participation [3]. Attention is critical in the definition of boredom [24], when students are not able to successfully engage attention with internal or external information required for participating in satisfying activity. We consider visual attention to the teacher, and particularly visual attention tuned for social interaction [4].

In STEM classes, we found that when a male student is observing the teacher, the teacher is generally pointing their body toward the student (67% of cases). However, with female students this number is just 46%. However, in other curriculum subjects it is the other way round. Moreover, the distance between the student and the teacher also has a significant effect. Furthermore, this effect is different among boys than girls. However, we must consider that these findings come from just two months of video recordings from a single class with 36 students and only one teacher, a teacher who teaches all subjects. Additional data from more classes and teachers is needed in order to confirm these preliminary findings.

Are these body-orientation patterns universal? Contemporary research from evolutionary psychology demonstrates that head and body orientation are critical on animal behavior. Terrestrial predators often send an 'honest signal' [25] suggesting their subsequent course of action [26, 27], i.e., not only what they are going to do, but also in which direction they are going to head. For example, in African elephants, [27] examined whether the frequency of gestures made by an elephant using their head and trunk in order to request food was influenced by indications of an experimenter's visual attention. Elephants signaled significantly less when the experimenter's body was pointing away from the elephant. It seems that there is a strong connection between the importance of visual attention and body orientation for effective communication. Moreover, body orientation may be a more reliable signal of predator intentions as it is more difficult to fake. Gaze is very easy to change and can be used to deceive prey or predators. But why does the main difference come within the professional space? It may be that body orientation signals predator intentions better than gaze, and this is particularly true at this distance. In P2 it may be too late, while in P4 it may be too far away.

Classroom interactions are an important social phenomenon. How students interact with their teacher, as well as among themselves, has an important effect on their achievements. We have already found some interesting patterns regarding social interactions and student engagement. However, what is needed next in order to have a more complete understanding? One central methodological question is what constitutes an explanation of an observed social phenomena in the classroom and how to enhance student engagement? According to [29], perhaps one day people will interpret this type of question by asking: Can you grow these phenomena? In our case, this means describing the micro-specifications of interactions between agents (students and teacher) and studying how emerges a collective behavior in the classroom. What would the plausible rules of interaction be? In order to be able to describe computationally some of the rules we need to measure some of the basic interaction variables. Students' visual attention and mutual gaze are relatively easy to observe, as well as being an important behavior that the teacher tries to control. Furthermore, they can be easily included on a computational model. Similarly body-orientation is an important variable and can also be easily included on a computational model. In this work we have detected that students´ visual attention depends on both, mutual gaze and body orientation of the teacher, and that this relation is modulated by gender, curriculum matter and distance to the teacher. All of them can be included on an agent-based computational model that can simulate emergent collective behavior, and thus help teacher to improve effectiveness in classroom.

Finally, there are some limitations with this study. This is relatively a small scale study, with only one teacher in a specific cultural environment whose personality can significantly influence the whole research. Furthermore, the extent of the study is too small to be generalized. Further research needs to be conducted in a much larger scale, with more teachers and students and, in different cultural contexts.

Acknowledgements. Support from ANID/PIA/Basal Funds for Centers of Excellence FB0003, as well as FONDECYT 3170062 are gratefully acknowledged.

References

1. Cuban, L.: Inside the Black Box of Classroom Practice: Change without Reform in American Education. Harvard Education Press (2013)
2. Azevedo, R.: Defining and measuring engagement and learning in science: conceptual, theoretical, methodological, and analytical issues. Educ. Psychol. **50**(1), 84–94 (2015)
3. Finn, J., Voelkl, K.: School characteristics related to student engagement. J. Negro Educ. **62**(3), 249–268 (1993)
4. Cosmides, L., Tooby, J.: Evolutionary psychology: new perspectives on cognition and motivation. Ann. Rev. Psychol. **64**, 201–229 (2013)
5. Haladjian, H.H., Montemayor, C.: On the evolution of conscious attention. Psychon. Bull. Rev. **22**(3), 595–613 (2014). https://doi.org/10.3758/s13423-014-0718-y
6. Araya, R., Farsani, D., Hernández, J.: How to attract students' visual attention. In: Verbert, K., Sharples, M., Klobučar, T. (eds.) EC-TEL 2016. LNCS, vol. 9891, pp. 30–41. Springer, Cham (2016). https://doi.org/10.1007/978-3-319-45153-4_3
7. Kobayashi, H., Kohshima, S.: Unique morphology of the human eye and its adaptive meaning: comparative studies on external morphology of the primate eye. J. Hum. Evol. **40**, 419–435 (2001)
8. Tomasello, M.: A Natural History of Human Thinking. Harvard University Press. Cambridge, MA (2014)
9. Johnson, D., Johnson, R., Holubec, E.: The Nuts & Bolts of Cooperative Learning. Interaction Book Company (1994)
10. Robinson, J.: Nonverbal Communication and Physician-Patient Interaction: Review and New Directions. The Sage Handbook of Nonverbal Communication, V. Manusov and M.L. Patterson, eds. Thousand Oaks, US: Sage publications, pp. 437–459 (2006)
11. Farsani, D., Barraza, P.S., Martinic, S.: Non-Verbal synchrony as a pedagogical tool for medical education: A Clinical case study. Kuwait Medical Journal, (Accepted)
12. Robinson, J.D.: Getting down to business talk, gaze, and body orientation during openings of doctor-patient consultations. Hum. Commun. Res. **25**(1), 97–123 (1998)
13. Araya, R., Behncke, R., Linker, A., van der Molen, J.: Mining social behavior in the classroom. In: Núñez, M., Nguyen, N.T., Camacho, D., Trawiński, B. (eds.) ICCCI 2015. LNCS (LNAI), vol. 9330, pp. 451–460. Springer, Cham (2015). https://doi.org/10.1007/978-3-319-24306-1_44

14. Araya, R., Hernández, J.: Collective unconscious interaction patterns in classrooms. In: Nguyen, N.-T., Manolopoulos, Y., Iliadis, L., Trawiński, B. (eds.) ICCCI 2016. LNCS (LNAI), vol. 9876, pp. 333–342. Springer, Cham (2016). https://doi.org/10.1007/978-3-319-45246-3_32
15. Farsani, D., Araya, R.: Professional Proxemics: Part I. In: *ECTEL* workshop proceedings (2017)
16. Quinn, H., Schweingruber, H., Keller, T.: A Framework for K-12 Science Education: Practices, Crosscutting Concepts, and Core Ideas. National Academies Press (2012)
17. Honey, M., Pearson, G., Schweingruber, H.: STEM Integration in K-12 Education: Status, Prospects, and an Agenda for Research. National Academies Press (2014)
18. Hockings, P.: Principles of Visual Anthropology. The Hague: Mouton, pp. 507–532 (1995)
19. Collier, J.: Photography and visual anthropology. Principles Visual Anthropol. **2**, 235–254 (1995)
20. Hall, E.T.: The Silent Language. Fawcett, Greenwich (1959)
21. Farsani, D.: Deictic gestures as amplifiers in conveying aspects of mathematics register. In: Proceedings of the 9th Conference of European Society for Research in Mathematics Education, Prague, Czech, pp. 1382–1384 (2015)
22. Collier, M.: Nonverbal Factors in the Education of Chinese American Children: A Film Study. Asian American Studies. San Francisco State University, San Francisco (1983)
23. Farsani, D.: Making multi-modal mathematical meaning in multilingual classrooms. PhD dissertation, Dept. of Education, Univ Birmingham, Birmingham, UK (2015)
24. Eastwood, J., Frischen, A., Fenske, M., Smilek, D.: The unengaged mind: defining boredom in terms of attention. Perspect. Psychol. Sci. **7**(5), 482–495 (2012)
25. Bradbury, J., Vehrencamp, S.: Principles of Animal Communication. Sunderland Massachusetts Sinauer (1998)
26. Book, D.L., Freeberg, T.M.: Titmouse calling and foraging are affected by head and body orientation of cat predator models and possible experience with real cats. Anim. Cogn. **18**(5), 1155–1164 (2015). https://doi.org/10.1007/s10071-015-0888-7
27. Johnson, C., Karin-D'Arc, M.: Social attention in nonhuman primates: a behavioral review. Aquat. Mammal. **32**(4), 423 (2006)
28. Smet, A., Byrne, R.: African elephants (Loxodonta africana) recognize visual attention from face and body orientation. Biol. Lett. **10**(7), 20140428 (2014)
29. Epstein, J.: Generative Social Science. Studies in Agent-Based Computational Modeling. Princeton University Press, New Jersey (2006)

Experience Enhanced Intelligence to IoT

Situational Awareness Model of IoV Based on Fuzzy Evaluation and Markov Chain

Pengfei Zhang, Li Fei(✉), Zuqi Liao, Jiayan Zhang, and Ding Chen

Chengdu University of Information Technology, Chengdu 610225, China
lifei@cuit.edu.cn

Abstract. With the rapid development of the automobile industry, and the continuous advancement of the Internet of Things technology, the emergence of the Internet of Vehicles (IoV) has also led to the security of the IoV. However, scholars have focused on the security research of the components of the IoV, or Research on the safety analysis of the vehicle interior network, or research on vehicle behavior safety. There is almost no research on the overall situation of the entire vehicle at home and abroad. This thesis first analyzes the classic situational awareness model, puts forward the ideal function that the IoV situational awareness should have, and proposes an IoV situational awareness model based on the fuzzy evaluation, AHP hierarchical analysis and Markov chain according to the current situation of the Internet of Vehicles. And design experiments to verify its feasibility.

Keywords: Situational awareness · Car networking · Fuzzy evaluation · AHP analytic hierarchy · Markov chain

1 Introduction

The automobile industry has entered a period of sustained and rapid development. The rapid growth and broad application of information technology have prompted the emergence of intelligent network vehicles and vehicle networking. And because of the advent of 5G technology, the continuous development of mobile Internet technology, and the maturity of artificial intelligence technology, it has promoted the development of digitalization, informationization, networking and intelligence in the automotive industry [1]. While IoV technology is developing rapidly, it also introduces security issues on the Internet into vehicle security. Faced with endless security issues, the security situation of the IoV requires appropriate methods to monitor and analyze. However, today's research on vehicle safety focuses on the car's internal network, or a specific part of the vehicle, for example: Liu Y et al. conducted research on V2V secure communication [2], Liu Yong et al. Research on ECU Safety [3]; or a part of the IoV [4], such as intrusion detection of car CAN (Controller Area Network) [5–7], Kleberger P [8] et al. proposed an in-vehicle network security system that focuses on Vehicle Intranet Network security, which does not consider safety such as in-vehicle OS, IVI, OAT. Some scholars have studied the real-time situational awareness of networked vehicles, but they are not related to information security, for example, for vehicle behavior [9, 10], unmanned

M. Hernes et al. (Eds.): ICCCI 2020, CCIS 1287, pp. 543–557, 2020.
https://doi.org/10.1007/978-3-030-63119-2_44

driving [11]. There is almost no comprehensive research on the overall network security situation of the Internet of Vehicles.

This paper analyzes the classic SA model—Endsley model, the conventional data fusion model—JDL model, and the Tim Bass model based on JDL model. Then we summarizes the SA in other fields, and proposes the ideal functions that the IoV-SA should have. However, due to the immature environment and lack of experimental data, it is difficult to realize ideal functions. Therefore, we propose a model to realize part of the ideal functions. The model determines the weight of essential parts in the vehicle through fuzzy evaluation and AHP (Analytic Hierarchy Process).Then we get single-vehicle situation value, analyze the region-vehicle situation value through the ES(Expert System) analysis and the mathematics method. Finally, we predict the situational value by the Markov chain. The process of the various steps of the model is described in detail later. The final design experiment confirmed its feasibility.

2 Model of IoV Situation Awareness

In 1988, Endsley et al. [12]. Proposed the definition of SA (Situation Awareness) and gave a conceptual model. As shown in Fig. 1., the Endsley model is divided into three parts, collecting data from the environment, situational awareness, and decision making.

The situational awareness part is most famous for the model. It is mainly divided into three steps: *perception*, *comprehension*, and *prediction* [13].

- Perception: Obtain the crucial factors in the environment which is essential to decision-maker. Collect the multi-source heterogeneous data from the environment to identify and assess attributes, states, and related factors that change dynamically over time and space;
- Comprehension: Comprehend the factors from *perception.* Integrate and correlate perceived data and information, analyze their relevance, so that decision-makers can understand and make correct decisions about the current status;
- Projection: Project the future development trend by the perception and comprehension of environmental information.

Although this model is mainly used for anthropogenic factors in the aerospace field, it has a significant influence on the situational awareness technologies in various areas.

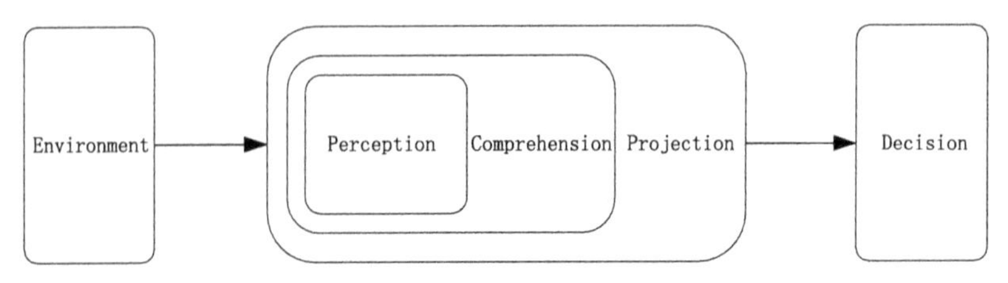

Fig. 1. Endsley model

In 1984, the Joint Directors of Laboratories of the US Department of Defense proposed the JDL model, as shown in Fig. 2. Compared to the Endsley model, the model refines the environment, and the perception, understanding, and prediction become the step of data preprocessing, object refinement, situation refinement, threat refinement, and process refinement. The model was first applied to the battlefield situational awareness in the military field, because of its efficient integration and practical utility, and subsequently gained extensive attention and research.

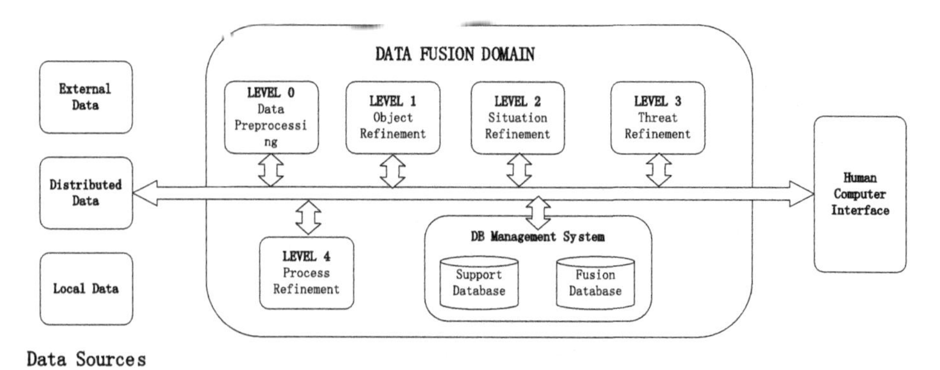

Fig. 2. JDL model

In 1999, Tim Bass [14] and other similar situational awareness in the aerospace field proposed CSA (Cyber Situation Awareness), which used a large amount of heterogeneous data for network situational awareness. For the first time, the JDL model was applied to CSA, as shown in Fig. 3. This model points out the relationship of each part in more detail than the JDL model. Tim pointed out that CSA should focus on solving two problems of congestion control and intrusion detection. Therefore, the Internet of Vehicle Situation Awareness (IoV-SA) should at least address the issue of vehicle traffic problems and IoV security.

Golestan K [9] and others analyzed the situational awareness of the Internet of Vehicles in detail, but the content of the research focused on the detection of vehicle behavior. So far, no one has proposed the vehicle network security situational awareness model related to network security.

According to the study mentioned earlier, the IoV-SA should be based on multi-source heterogeneous data. Analogy Tim Bass's research on network situational awareness, the ideal IoV-SA model, should have three functions: 1. The Traffic Situation of the IoV; 2. The IoV Security Status Situational Awareness; 3. The IoV Situational Awareness.

1. The Traffic Situation of the IoV can correctly monitor and predict the traffic status in detail. This function has been partially implemented by most of the current map software, but it is not perfect.
2. The IoV Security Status Situational Awareness can accurately determine the vehicle's safety status, including the vehicle's internal network security status, the vehicle's wireless network security status (Bluetooth, WiFi), vehicle behavior, and the safety status of driver behavior.

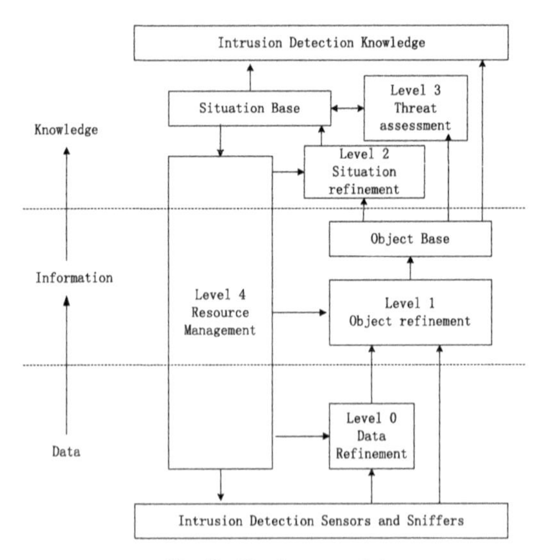

Fig. 3. Tim bass model

3. The IoV Situational Awareness includes the situation of network security status such as V2V (Vehicle-to-Vehicle), V2N (Vehicle-to-Network), V2P (Vehicle-to-Pedestrian), V2I (Vehicle-to-Infrastructure), etc. Perception. As shown in Fig. 4:

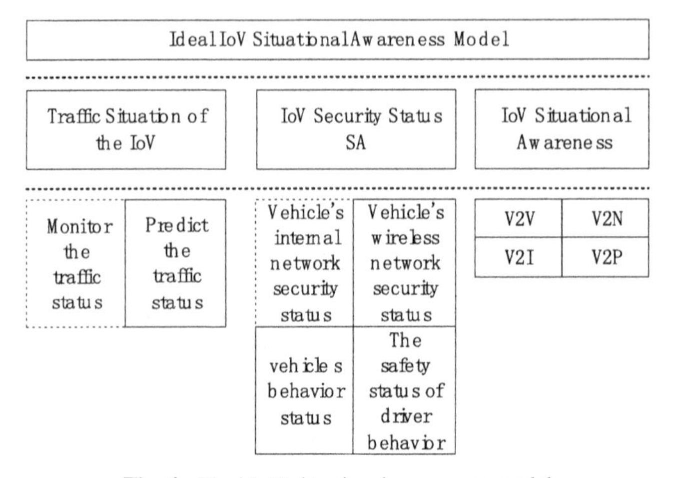

Fig. 4. Ideal IoV situational awareness model

Since the current vehicle networking technology is not perfect, and its implementation details are not standardized, the proposed IoV Situation Awareness model based on this paper only involves the above second and third functions.

Situational awareness systems are generally divided into three steps: perception, comprehension, and projection. Based on these three steps, this paper analyzes the security situation of the entire IoV from point to face. The process is shown in Fig. 5.

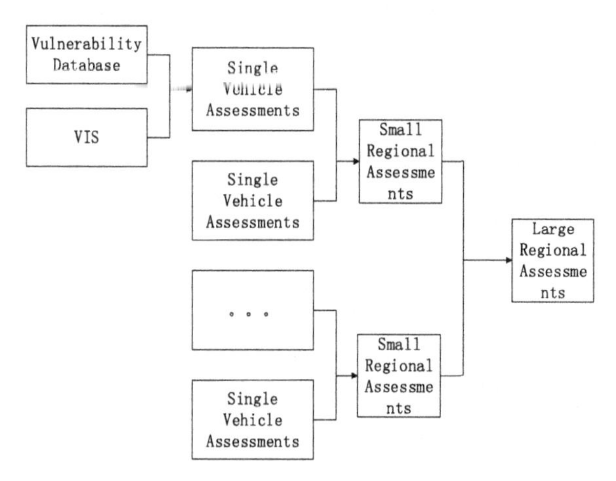

Fig. 5. Entire perception process

Perception is to collect data through the VDB(Vulnerability Database) and the VIS(Vehicle Information System). Comprehension is the safety analysis of a single vehicle, then carry out safety analysis on cars in a small area, and then conduct small-scale security analysis on large areas. And finally projecting the security situation according to the result of comprehension, to achieve the purpose of situational awareness of the IoV.

2.1 Single-Vehicle Assessments

The entire car is very complex. Its security depends on the objects that can be attacked by the network. We call the objects AC(Attacked Carrier).

In the "China Internet of Vehicle Security White Paper" [15], the security of the entire vehicle networking system is analyzed in detail, indicating that the safety of a vehicle and network is affected by the following components: T-Box, CAN (Controller Area Network), OBD (On-Board Diagnostics), ECU (Electronic Control Unit), Vehicle OS (Operating System), IVI (In-Vehicle Infotainment), OAT (Over the Air Technology), Sensors, and Vehicle Key. In this paper, these components that can be attacked by cybers are ACs.

The process of the situation of a single-vehicle is shown in Fig. 6.

The Vehicle Information System (VIS) provides the identification information of the ACs. For example, VIS gives the version number of the T-Box in the vehicle and we can query the VDB through the version number. If there is a vulnerability, directly report the emergency response. If there is no vulnerability, we collect the information through the VIS, and use the information and ES to analyze the security of this AC.

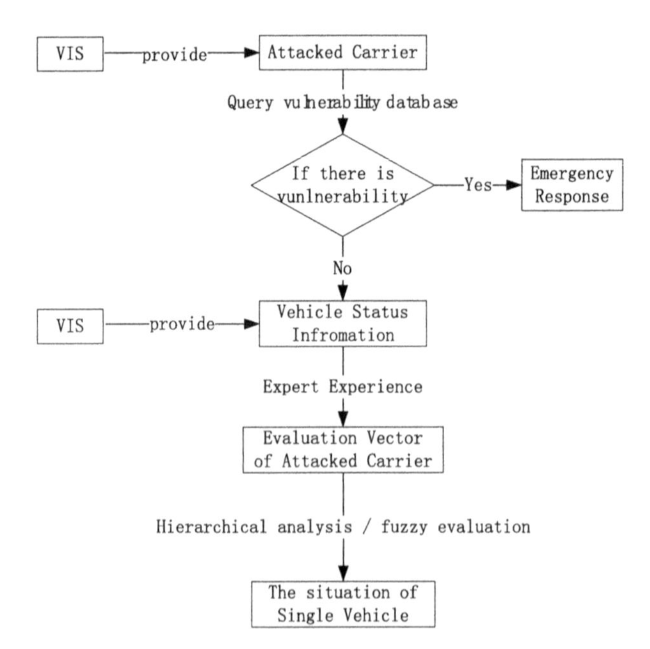

Fig. 6. The process of the situation of a single-vehicle

The method we use to analyze the security of ACs is FE (Fuzzy Evaluation) and AHP. The evaluation matrix of ACs is obtained by fuzzy evaluation, and then the weight of ACs is obtained according to AHP.

Fuzzy Evaluation of Single Vehicle

The fuzzy assessment purpose is to construct a fuzzy evaluation matrix. In the previous section, we mentioned that the primary ACs of the single-car are T-Box, CAN, OBD, ECU, Vehicle OS, IVI, OAT, Sensors, and Vehicle Key. Before considering the security of ACs, you should first confirm the weight of its. Since the IoV has the characteristics of less data, and to avoid the experimental factors, the fuzzy AHP is used to find the weight. Specific steps are as follows:

(1) Select *Factor Set* U and *Comment Set* V.

Factor Set U = {T-Box(u1), CAN(u2), OBD(u3), ECU (u4), Vehicle OS(u5), IVI(u6), OAT (u7), Sensor (u8), Vehicle Key (u9)}

The rating is divided into five levels, that is, the *Comment Set* is V = {normal (v1), comparatively safe (v2), basic safe (v3), unsafe (v4), dangerous (v5)}.

(2) Determine each AC's EV(Evaluation Vector)

The data collected by the VDB and the VIS evaluate each component of the factor set. For example, through the collected information, the probability that T-Box (u1) has 0.05 is normal(v1), the probability of 0.15 is comparatively safe (v2), the probability of 0.15 is basic safety(v3), the probability of 0.6 is unsafe(v4), and the probability of 0.05

is dangerous(v5), then the EV is u1 = (0.05, 0.15, 0.15, 0.6, 0.05). The AC evaluation model here can be expressed as F (VDB, state information collected by the VIS) = EV, where F is based on probability and statistics methods, combined with expert methods, and expert experiences are used in this study.

Similarly, the EVs for the other eight ACs are u2 (0.2, 0.15, 0.15, 0.3, 0.2), u3 (0, 0.1, 0.3, 0.4, 0.2), and u4 (0.5, 0.3, 0.1, 0.05, 0.15), u5 (0.1, 0.2, 0.4, 0.1, 0.2), u6 (0.2, 0.1, 0.3, 0.1, 0.3), u7 (0.2, 0.1, 0.2, 0.3, 0.2), u8 (0.6, 0.1, 0.1, 0.1, 0.1), u9 (0.3, 0.2, 0.25, 0.15, 0.1).

According to the EVs of the above ACs, the single-car fuzzy evaluation matrix R is constructed as follows:

$$R = \begin{bmatrix} 0.05 & 0.15 & 0.15 & 0.6 & 0.05 \\ 0.2 & 0.15 & 0.15 & 0.3 & 0.2 \\ 0 & 0.1 & 0.3 & 0.4 & 0.2 \\ 0.5 & 0.3 & 0.1 & 0.05 & 0.15 \\ 0.1 & 0.2 & 0.4 & 0.1 & 0.2 \\ 0.2 & 0.1 & 0.3 & 0.1 & 0.3 \\ 0.2 & 0.1 & 0.2 & 0.3 & 0.2 \\ 0.6 & 0.1 & 0.1 & 0.1 & 0.1 \\ 0.3 & 0.2 & 0.25 & 0.15 & 0.1 \end{bmatrix}$$

AHP determines the weight of the attacked carriers

For the attacked carriers of the vehicle, the higher the degree of danger, the greater the threat to the whole car. That is to say, the greater the risk, the greater the impact on the car, that is, the greater the weight. Here is AHP is used to determine the weight of each component in a single car.

Determine the judgment matrix, where the numbers 1-9 and their reciprocals are used as the scale, as shown in Table 1 :

Therefore, according to the expert assessment, the judgment matrix of the nine ACs is shown in Table 2:

Now Table 2 is transformed into matrix U_1:

$$U_1 = \begin{bmatrix} 1 & 1/3 & 1/3 & 1/4 & 2 & 3 & 1/4 & 1 & 4 \\ 3 & 1 & 1 & 1/2 & 2 & 3 & 1/4 & 1 & 4 \\ 3 & 1 & 1 & 1/2 & 2 & 2 & 1 & 2 & 3 \\ 4 & 2 & 2 & 1 & 5 & 7 & 2 & 7 & 9 \\ 1/2 & 1/2 & 1/2 & 1/5 & 1 & 1 & 1/3 & 1 & 2 \\ 1/3 & 1/3 & 1/2 & 1/7 & 1 & 1 & 1/2 & 2 & 3 \\ 4 & 4 & 1 & 1/5 & 3 & 2 & 1 & 5 & 3 \\ 1 & 1 & 1/2 & 1/2 & 1 & 1/2 & 1/5 & 1 & 1 \\ 1/4 & 1/4 & 1/3 & 1/9 & 1/2 & 1/3 & 1/3 & 1 & 1 \end{bmatrix}$$

1) Calculate the sum of the columns, the sum of the jth column is $\sum_i^n U_{ij}$;
2) Each element of each column is divided by the sum of each column to obtain a normalized matrix U2;

Table 1 Table.1

Intensity of Importance	Definition	Explanation
1	Equal Importance	Two elements contribute equally to the objective
3	Moderate Importance	Experience and judgment slightly favor one element over another
5	Strong Importance	Experience and judgment strongly favor one element over another
7	Very Strong Importance	One element is favored very strongly over another, its dominance is demonstrated in practice
9	Extreme Importance	The evidence favoring one element over anther is of the highest possible order of affirmation
2, 4, 6, 8		Indicates the median value of the above two adjacent judgments
Reciprocal		If the factor i is compared with j to judge Bij, then the factor j and i are compared and judged as Bji = 1/Bij.

Table 2. Judgment matrix

	u1	u2	u3	u4	u5	u6	u7	u8	u9
u1	1	1/3	1/3	1/4	2	3	1/4	1	4
u2	3	1	1	1/2	2	3	1/4	1	4
u3	3	1	1	1/2	2	2	1	2	3
u4	4	2	2	1	5	7	2	7	9
u5	1/2	1/2	1/2	1/5	1	1	1/3	1	2
u6	1/3	1/3	1/2	1/7	1	1	1/2	2	3
u7	4	4	1	1/5	3	2	1	5	3
u8	1	1	1/2	1/2	1	1/2	1/5	1	1
u9	1/4	1/4	1/3	1/9	1/2	1/3	1/3	1	1

$$
\mathrm{U}_2 = \begin{bmatrix} 0.059 & 0.032 & 0.047 & 0.073 & 0.114 & 0.151 & 0.043 & 0.048 & 0.134 \\ 0.175 & 0.096 & 0.139 & 0.147 & 0.114 & 0.151 & 0.043 & 0.048 & 0.134 \\ 0.175 & 0.096 & 0.139 & 0.147 & 0.114 & 0.101 & 0.17 & 0.095 & 0.1 \\ 0.234 & 0.192 & 0.279 & 0.294 & 0.287 & 0.354 & 0.34 & 0.333 & 0.3 \\ 0.029 & 0.048 & 0.07 & 0.059 & 0.057 & 0.05 & 0.057 & 0.048 & 0.066 \\ 0.02 & 0.032 & 0.07 & 0.042 & 0.057 & 0.05 & 0.086 & 0.095 & 0.1 \\ 0.234 & 0.384 & 0.139 & 0.059 & 0.171 & 0.101 & 0.17 & 0.237 & 0.1 \\ 0.059 & 0.096 & 0.07 & 0.147 & 0.057 & 0.025 & 0.034 & 0.048 & 0.033 \\ 0.015 & 0.024 & 0.047 & 0.032 & 0.029 & 0.017 & 0.057 & 0.048 & 0.033 \end{bmatrix}
$$

3) Each line average gets the weight vector:

$$
W = [0.078, 0.116, 0.126, 0.29, 0.054, 0.061, 0.177, 0.063, 0.035]^T
$$

4) The consistency check determines whether the obtained weight vector $W_{1 \times 9}$ is or not reasonable.

Calculate the vector T:

$$
T = U_2 W_{1 \times 9} = \begin{bmatrix} 0.769 & 1.159 & 1.259 & 2.893 & 0.525 & 0.604 & 1.841 & 0.62 & 0.327 \end{bmatrix}^T
$$

Calculate the intermediate calculation amount, Temp:

$$
\mathrm{Temp} = \frac{1}{9} \sum_{i=1}^{9} \frac{T_i}{W_i} \approx 9.892
$$

CI value:

$$
\mathrm{CI} = \frac{\mathrm{Temp} - 9}{9 - 1} = 0.1115
$$

Look up the table to get RI = 1.45

$$
\frac{CI}{RI} = 0.097 < 0.1
$$

The weight vector $W_{1 \times 9}$ passes the consistency test.

Single vehicle evaluation operation

According to the evaluation matrix of the obtained vehicle, and the weight vector of the safety indicators of each *attack carrier*, the safety evaluation model of the single-vehicle is $M_{1 \times 5} = W_{1 \times 9} R_{9 \times 5}$.

The situational assessment of a vehicle should be a numerical value. Weighting is used here to determine the safety situation value of the final single vehicle. The weighting vector is set to {0.4, 0.3, 0.2, 0.1, 0} according to expert experience. The resulting single-vehicle situation value is [0, 1].

2.2 Zone Evaluation

After assessing the situation of a single vehicle, conduct a regional assessment:

The data obtained at the *agent* has a field for identifying the region, and the region can obtain the region tag. Spo can get the following set:

The set $S = \{V_1, V_2, V_3, V_4 \ldots V_n\}$ represents a set of vehicles situation value in an area. V_i represents the value of one vehicle in the area, but considering that the vehicles has different importance. We divide the vehicles into a categories, and the weight vector of the different categories is $W = [w_1, w_2, \ldots w_a]\left(\sum_{i=1}^{a} w_i = 1\right)$; in one area, if the category i of vehicles has n_i vehicles. In a region, the situation of the jth vehicle in the ith category is expressed as V_{ij}; then the regional situation value Z can be expressed as:

$$Z = \sum_{i=1}^{a} \frac{w_i}{n_i} \sum_{j=1}^{n_i} V_{ij}$$

After the regional situation assessment is completed, the overall situation of the regional situation is summarized and analyzed as follows:

The cities are divided into four levels, the first-tier city, the second-tier city, the third-tier city, and the fourth-tier city. Since the cities have different importance, the weight value of each city is different. Here, the weight of the ith city is as $w_i(w_1 + w_2 + w_3 + w_4 = 1)$, and the area must belong to one of the cities as mentioned earlier. Each city contains a fixed area. Therefore, the number of areas included in the i-class city can be set as N_i, the situation value of the jth region in the i-th city is expressed as Z_{ij}, then the overall situation value is:

$$T = 100 * \sum_{i=1}^{4} \frac{w_i}{N_i} \sum_{j=1}^{N_i} Z_{ij}$$

Finally, the overall situation value of the Internet of Vehicles is $T \in [0, 100]$.

It is mentioned that situational awareness is divided into three steps. The next step in understanding is to analyze the situational values of the obtained regional vehicles and analyze and project based on the obtained value.

2.3 Situation Projection

Assume that a certain regional situation is a value Z, but the value is continuity. Then, discretization the value by equal width method. It is discretized by the following piecewise function S. The security states are: normal, safe, basic, unsafe, and dangerous.

$$S = \begin{cases} \text{normal } 0.8 < Z \leq 1 \\ \text{safe } 0.6 < Z \leq 0.8 \\ \text{basic } 0.4 < Z \leq 0.6 \\ \text{unsafe } 0.2 < Z \leq 0.4 \\ \text{dangerous } 0 \leq Z \leq 0.2 \end{cases}$$

According to the above formula, the situation value of a certain time zone can be divided into the above five levels: normal (S_1), comparative safety (S_2), basic security (S_3), unsafe (S_4), and dangerous (S_5). These five states are represented by 1~5, respectively. p_{ij} represents the probability that the i state to aj state transition.

Then, here can use the *Markov Chain* to predict the regional situation.

First calculate the state transition matrix. The state transition probability is derived from statistical data. For example, if n observations are made within time t, then there are n-1 state transitions, and the transition probability of state i to j can be obtained by the number of transitions of each state: $p_{ij} = \frac{n_{ij}}{n-1}$.

The statistical time period t and the number of observations can be manually adjusted;

The state transition matrix is expressed as follows:

$$\begin{bmatrix} p_{11}\, p_{12}\, p_{13}\, p_{14}\, p_{15} \\ p_{21}\, p_{22}\, p_{23}\, p_{24}\, p_{25} \\ p_{31}\, p_{32}\, p_{33}\, p_{34}\, p_{35} \\ p_{41}\, p_{42}\, p_{43}\, p_{44}\, p_{45} \\ p_{51}\, p_{52}\, p_{53}\, p_{54}\, p_{55} \end{bmatrix}$$

If the state is 5 at time t_1, the probability of a certain state at the next moment is $t_2 = [p_{51}, p_{52}, p_{53}, p_{54}, p_{55}]$.

Then the state probability at time t_3 is

$$t_3 = [p_{11}, p_{12}, p_{13}, p_{14}, p_{15}] \begin{bmatrix} p_{11}\, p_{12}\, p_{13}\, p_{14}\, p_{15} \\ p_{21}\, p_{22}\, p_{23}\, p_{24}\, p_{25} \\ p_{31}\, p_{32}\, p_{33}\, p_{34}\, p_{35} \\ p_{41}\, p_{42}\, p_{43}\, p_{44}\, p_{45} \\ p_{51}\, p_{52}\, p_{53}\, p_{54}\, p_{55} \end{bmatrix}$$

Therefore, it is possible to predict the situation value at the most recent moment, and to analyze the steady state of the Markov chain to derive its probability of developing a final stable situation at that time.

Projection is the last step in the situational awareness of the IoV because the forecast can project the development trend of the future situation in advance, and then can use suitable technical means to deal with the security threat.

3 Experiment

3.1 Experiment Environment

The data comes mainly from a car and car vulnerability library. The related equipment includes tools for collecting relevant data, several PCs for analysis, and associated capture tools. The experimental configuration is shown in Table 3:

Table 3. Experimental configuration

No.	Name	Function/Role	Quantity	Configuration/type
1	car	Vehicle to be tested	1/3	Confidentiality
2	USB-CAN	Collect CAN data	1	USBCAN-II Prop
3	PC-1	Collecting/analyzing data	1	CPU3.20GH、Memory 8G, hard disk 1T, TP-Link wireless network card, 100 M/1000 M adaptive network card, Wireshark capture tool
4	XTOOL	collect OBD2 data	1	5510

3.2 Experimental Result

Firstly, the is collected, and the data of the relevant model car vulnerability database is collected. Then, according to the probability and statistics method, combined with the expert system, the evaluation vectors of nine kinds of components of nine important parts of the car at t_1~t_{10} are obtained, for example. The evaluation matrix of the first period t_1:

$$\begin{bmatrix} 0.133 & 0.303 & 0.245 & 0.167 & 0.152 \\ 0.141 & 0.27 & 0.18 & 0.171 & 0.238 \\ 0.22 & 0.227 & 0.315 & 0.226 & 0.012 \\ 0.119 & 0.206 & 0.393 & 0.107 & 0.175 \\ 0.181 & 0.164 & 0.206 & 0.256 & 0.193 \\ 0.201 & 0.204 & 0.156 & 0.211 & 0.228 \\ 0.139 & 0.306 & 0.217 & 0.166 & 0.172 \\ 0.067 & 0.235 & 0.227 & 0.334 & 0.137 \\ 0.357 & 0.09 & 0.259 & 0.265 & 0.029 \end{bmatrix}$$

According to the experience of experts, the weight of each major component in the car is obtained. For details, see the calculation process in Sect. 3.1.2. The situation value of the i-th period is:

$$S_i = WE_iT$$

The vector W is obtained in the previous step, and E_i is the evaluation matrix of the first step.

The situation values obtained for ten periods are shown in Table 4:

Table 4. Situation values

Time.	T1	T2	T3	T4	T5	T6	T7	T8	T9	T10
Status	0.192	0.214	0.206	0.189	0.208	0.198	0.199	0.202	0.202	0.202

Discretization of data:

Obtaining a security level of ten moments according to the discretization function s; As shown in Table 5:

Table 5. Security level

Time.	T1	T2	T3	T4	T5	T6	T7	T8	T9	T10
Status	Dangerous	Unsafe	Unsafe	Dangerous	Unsafe	Dangerous	Dangerous	Unsafe	Unsafe	Unsafe

From the above, get Fig. 7:

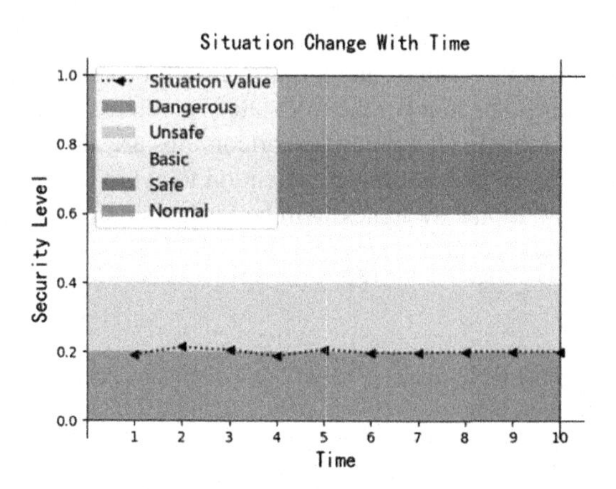

Fig. 7. Situation change with time

It can be seen that the safety level of the car has been fluctuating in both dangerous and unsafe conditions. The reason is that the vehicle's information security is not taken seriously, so the vehicle's vulnerability has not been repaired, so the evaluation is very low, that is, the situation is very low.

According to the statistics, the state transition matrix can be calculated, as shown in Table 6:

It can be seen from the state transition matrix that the probability of a particular state at any time from the security state to the next moment, for example, at time T11 after

Table 6. State transition matrix

	Dangerous	Unsafe
Dangerous	0.25	0.75
Unsafe	0.25	0.75

time T10, the probability of reaching a dangerous state is 0.25, and the probability of remaining unsafe is 0.75. The overall trend is still not safe.

4 Discussion

Internet of Vehicle Security is a direction of information security in the future. The construction of the IoV will bring a series of problems of the Internet to the IoV. The Internet of Vehicles is more complicated than the traditional Internet so that the IoV will face more complicated, changed network attack. This paper puts forward the ideal function that the IoV situational awareness should have for this problem, and proposes a kind according to the current situation of the IoV (the current Internet of Vehicles technology is not perfect, and its implementation details are not standardized) The vehicle network situational awareness model, the vehicle situation data situation sensing method provided by this model is not complicated to calculate, it is convenient to evaluate the network security situation of a certain type of vehicle, it is easy to analyze the safety status of a certain area of the vehicle, and it is also easy for a certain kind of vehicle in a certain area in the future. The trend of the IoV state is predicted, and the data-dependent on the calculation is not high. However, its shortcomings are also obvious. It relies too much on expert experience and insufficient attention to cyber-attack events. Therefore, the vehicle network situational awareness will be further studied for this shortcoming.

References

1. Kombate, D.: The internet of vehicles based on 5G communications. In: 2016 IEEE International Conference on Internet of Things (iThings) and IEEE Green Computing and Communications (GreenCom) and IEEE Cyber, Physical and Social Computing (CPSCom) and IEEE Smart Data (SmartData), pp. 445–448. IEEE (2017)
2. Liu, Y., Wang, Y., Chang, G.: Efficient privacy-preserving dual authentication and key agreement scheme for secure V2V communications in an IoV paradigm. IEEE Trans. Intell. Transp. Syst. **18**(10), 2740–2749 (2017)
3. Yong, L., et al.: Feasibility study of vehicle network vehicle identity authentication based on blockchain technology. Automot. Technol. (6) (2018)
4. Li, F., Zhang, H., Gao, L., Wang, J., Sanin, C., Szczerbicki, E.: A set of experience-based smart synergy security mechanism in internet of vehicles. Cybern. Syst. **50**(2), 230–237 (2019). https://doi.org/10.1080/01969722.2019.1565115
5. Gao, L., Li, F., Xu, X., Liu, Y.: Intrusion detection system using SOEKS and deep learning for in-vehicle security. Cluster Comput. **22**(6), 14721–14729 (2018). https://doi.org/10.1007/s10586-018-2385-7

6. Kang, M.J., Kang, J.W., Tieqiao, T.: Intrusion detection system using deep neural network for in-vehicle network security. PLoS ONE **11**(6), e0155781 (2016)
7. Li, F., et al.: Toward intelligent vehicle intrusion detection using the neural knowledge DNA. Cybern. Syst. **49**(5–6), 412–419 (2018)
8. Kleberger, P., Olovsson, T., Jonsson, E.: Security aspects of the in-vehicle network in the connected car. In: 2011 IEEE Intelligent Vehicles Symposium (IV), pp. 528–533. IEEE (2011)
9. Golestan, K., et al.: Situation awareness within the context of connected cars. Inf. Fusion **29**, 68–83 (2016)
10. Research on vehicle collision hazard situation identification and collision avoidance decision considering driving intention. Wuhan University of Technology (2015)
11. Calhoun, G.L., et al.: Synthetic vision system for improving unmanned aerial vehicle operator situation awareness. In: Enhanced and Synthetic Vision 2005, pp. 219–230 (2005)
12. Endsley, M.: Design and evaluation for situation awareness enhancement. In: Proceedings of the Human Factors Society 32nd Annual Meeting, Human Factors Society, pp. 97–101 (1988)
13. Tianfield, H.: Cyber security situational awareness. In: 2016 IEEE international conference on internet of things (iThings) and IEEE green computing and communications (GreenCom) and IEEE cyber, physical and social computing (CPSCom) and IEEE smart data (SmartData), pp. 782–787. IEEE (2017)
14. Bass, T.: Intrusion detection systems and multisensor data fusion: creating cyberspace situational awareness. Commun. ACM **43**(4), 99–105 (1999)
15. Mobile Internet Security White Paper (2017) released. China Information Security, (10), p. 31 (2017)
16. Nan, Y., Rongbao, K.: Analysis and protection of vehicle network security threats. Commun. Technol. **48**(12), 1421–1426 (2015)
17. Kang, M.J., Kang, J.W., Tieqiao, T.: Intrusion detection system using deep neural network for in-vehicle network security. PLOS one. **11**(6), p. e0155781 (2016)

A Framework for Enhancing Supplier Selection Process by Using SOEKS and Decisional DNA

Muhammad Bilal Ahmed[1(✉)], Cesar Sanin[1], and Edward Szczerbicki[2]

[1] The University of Newcastle, Callaghan, NSW, Australia
muhammadbilal.ahmed@uon.edu.au, cesar.sanin@newcastle.edu.au
[2] Gdansk University of Technology, Gdansk, Poland
edward.szczerbicki@newcastle.edu.au

Abstract. Supplier selection process is one of the significant stages in supply chain management for industrial manufactured products. It plays an integral role in the success of any manufacturing organization and is an important part starting right from selecting raw material to dispatch of finished products. This paper contributes to enhance the supplier selection process by proposing a multi-criteria decision making framework for industrial manufactured products. The proposed framework is based on smart knowledge management technique called Set of experience knowledge structure (SOEKS) and Decisional DNA, which makes the proposed approach dynamic in nature as it updates itself every time a decision is taken.

Keywords: Supplier selection process · Multi-criteria decision making · Set of experience knowledge structure (SOEKS) · Decisional DNA (DDNA)

1 Introduction

Supplier selection process plays an important role for manufactures, as every decision made during this process affects the success of production management [1]. Therefore, companies are focusing more and more on supplier selection process to overcome the challenges of producing high quality products at a lower cost [2]. Single criteria supplier selection approach, in which only low cost suppliers are selected, is almost redundant. In this era, the supplier selection process is a complicated and multi-criteria decision making (MCDM) problem, which requires the consideration of various criteria and sub-criteria [3, 4]. Consequently, manufacturing organizations feel comfortable to work with suppliers who can provide products and services at a required quality level, reasonable cost, and flexible to adopt any changes in design and manufacturing as anticipated [5]. In conjunction with this, companies also pay special attention towards the selection of alternative suppliers [4] to accomplish with uncertain global conditions e.g. recent COVID-19 global supply chain issues. Likewise, the use of multiple suppliers provides more flexibility due to the diversification of the organization's overall requirements and brings up competitiveness among alternative suppliers [6].

M. Hernes et al. (Eds.): ICCCI 2020, CCIS 1287, pp. 558–565, 2020.
https://doi.org/10.1007/978-3-030-63119-2_45

Supply chain management (SCM) has improvised a lot during the third industrial revolution and organizations have been encouraged to implement continuous improvement attitude in their purchasing related matters. As a result of this, the supplier selection process has also evolved significantly in the past forty years. These changes have been beneficial to both the purchasing clients and the suppliers. Decision making process involved in supplier selection depends on multiple qualitative and quantitative criteria [7]. In the past, supplier selection problem has been solved by the researchers by using two different approaches i.e. individual and integrated approaches [8]. The most commonly used individual approaches are: the data envelopment analysis (DEA), mathematical programming, the analytic hierarchy process (AHP), the analytic network process (ANP), neural networks, structural equation modeling, multi-attribute utility theory, dimensional analysis (DA), fuzzy decision-making, genetic algorithms, and, the simple multi-attribute rating technique (SMART), etc. The integrated approaches use more than one approach jointly, e.g. integrated AHP and DEA, integrated AHP and goal programming, etc. [4].

Meanwhile, the world is moving towards Industry 4.0, which is the fourth revolution of the industry. Conventional processes are to be replaced by new concepts, i.e., internet of things (IoT), internet of services (IoS), cyber-physical systems (CPS), mass collaboration, high-speed internet, and affordable 3D printing [9]. These advancements offer enormous opportunities for supply chain intelligence and autonomy establishing stepping stones for Industry 4.0 supply chains (SCs). The supplier selection process is the backbone of the supply chain process, but unfortunately, this has not been realized within Industry 4.0 supply chains [10]. Traditional supplier selection approaches based on MCDM fail to address the supplier selection problem in Industry 4.0, because of the large amount of data produced during product manufacturing in real time [11].

In this paper, we attempt to enhance the supplier selection process by proposing a framework based on smart knowledge management technique called Set of Experience Knowledge Structure (SOEKS or SOE in short) and Decisional DNA [12]. The structure of the paper includes the literature review in Sect. 2, which presents the basic concepts of supplier selection process, supplier selection decision-making criteria, and set of experience knowledge structure and decisional DNA. The proposed framework for the supplier selection process and working algorithm of proposed technique are discussed in Sect. 3. Finally, the conclusions and future work are presented in Sect. 4.

2 Literature Review

2.1 Supplier Selection Process

Supply chain management is the process of managing events related to flow and transformation of goods and services from the source point to usage point [13]. Supplier selection process is one of the important activity in supply chain management. In fact, it is the initial stage of supply chain management and can impact all the consecutive stages [14]. It is the process by which firms identify, evaluate, and contract with their suppliers. The overall objectives of supplier selection process are to reduce purchase risk, maximize overall value to the purchaser, and develop closeness and long-term relationships between buyers and suppliers [15]. To achieve these objectives, supplier selection process needs to consider various quantitative and qualitative criteria [14].

The literature on supplier selection criteria and methods is full of various analytical and heuristic approaches. Some researchers have developed hybrid models by combining more than one type of selection methods. In most of the manufacturing industries, the cost of raw materials and component parts represents the largest percentage of the total product cost. As a result, selecting the right suppliers can lead towards the successful procurement process and can represent a major opportunity for companies to reduce costs across their entire supply chain [6]. Supplier selection methods and criteria are still critical issues for manufacturing industries; this paper proposes a supplier selection method, which uses previous experiential knowledge to solve this issue.

2.2 Supplier Selection Decision-Making Criteria

Identification of decision-making criteria is the backbone of the supplier selection process. From the literature review, it has been found that few of the important decision making criteria for supplier selection process are: cost, quality, delivery, performance history, warranties & claims policies, production facilities and capacity, technical capability, financial position, procedural compliance, repair service, packaging ability, risk factor, reliability, process improvement, and product development process [6].

Recent industrial advancements related to environmental, social, political, and customer satisfaction concerns make these criteria further complex [6]. This all makes the supplier selection process a multi-criteria decision making (MCDM) problem [4]. Whereas, MCDM approaches are formal methods to structure the decision problems with multiple, conflicting criteria or goals [8]. These approaches have been widely used in the fields of transportation, immigration, education, investment, environment, energy, defense, and health care [16].

2.3 Set of Experience Knowledge Structure and Decisional DNA

Set of experience knowledge structure (SOEKS) is a smart knowledge management technique. It collects and analyses formal decision events and uses them to represent experiential knowledge. A formal decision is defined as a choice (decision) made or a commitment to act that was the result (consequence) of a series of repeatable actions performed in a structured manner. A set of experience (SOE, a shortened form of SOEKS) has four components: Variables (V), functions (F), Constraints (C) and Rules (R). Each formal decision is represented and stored uniquely based on these components. Variables are the basis of the other SOEKS components, whereas functions are based upon the relationships and links among the variables. The third SOEKS component is constraints, which, like functions, are connected to variables. They specify limits and boundaries and provide feasible solutions. Rules are the fourth component and are conditional relationships that operate on variables. Rules are relationships between a condition and a consequence connected by the statements 'if/then/else' [17]. The four components of a SOE and its structural body can be defined by comparing it with some important features of human DNA. Just as the combination of its four nucleotides (Adenine, Thymine, Guanine, and Cytosine) makes DNA unique, the combination of its four components (Variables, Function, Constraints, and Rules) makes an SOE unique.

Each formal decision event is deposited in a structure that combines these four SOE components. Several interconnected elements are visible in the structure, resembling part of a long strand of DNA, or a gene. Thus, a SOE can be associated to a gene and, just as a gene produces a phenotype, SOE creates a value for a decision in terms of its objective function. Hence, a group of SOEs in the same category form a kind of chromosome, as DNA does with the genes. Decisional DNA contains experienced decisional knowledge and it can be categorized according to the areas of decisions. Furthermore, just as assembled genes create chromosomes and human DNA, groups of categorized SOEs create decisional chromosomes and DDNA. In short, a SOEKS represents explicit experiential knowledge which is gathered from the previous decisional events [18]. SOE and DDNA have been successfully applied in various fields such as industrial maintenance, semantic enhancement of virtual engineering applications, state-of-the-art digital control system of geothermal and renewable energy, storing information and making periodic decisions in banking activities and supervision, e-decisional community, virtual organizations, interactive TV, and decision-support medical systems, etc. [19].

3 Proposed Framework of Supplier Selection Process

Framework for supplier selection process by using "Set of experience knowledge structure (SOEKS) and Decisional-DDNA" is shown in Fig. 1. It encompasses of user interface, integrator, prognoser, solution layer, and supplier selection decisional DNA (SS-DDNA). The working of each section is explained as follows:

User Interface

Users can interact with the system by inputting data into the user interface. The data can be a simple query for supplier selection problem based on various criteria as discussed above, or it could be an addition of new data, information, or knowledge to enhance the knowledge base of the platform.

Integrator

The integrator receives data from the user interface and acquires information through various applications. It produces sub-solutions according to the objectives of the users.

Furthermore, it transforms the information into a unified language and measurement system. It gathers and organizes the data and transforms it into sets of experience described by the unique set of variables, functions, constraints, and rules. The integrator interacts with the supplier selection decisional DNA (SS-DDNA) for similar sets of experiences and sends the results to the prognoser for further processing.

Prognoser

The prognoser first produces sub-solutions provided by multiple applications according to its established objectives. Then depending on the various scenarios new sets of models can be built by taking into account measurements of uncertainty, incompleteness and imprecision. The prognoser finally produces set of proposed solutions that are sent to the solution layer.

Solution Layer

The solution layer allows the user to select the best solution out of the proposed solutions. Based on the priorities defined by the user it chooses the best solution among the possible solutions provided. The decisional event is then sent to the SS-DDNA.

Supplier Selection Decisional DNA

It is where the sets of experience and formal decisional events are stored and managed. One set of experience represents a decisional DNA gene [18]. The same category of genes are grouped together, which is collectively called as decisional chromosome. Group of such chromosomes; product chromosome, process chromosome, and technology chromosome, constitutes a decisional DNA of the desired process. It also interacts with other components of the framework during the solution process and presents similar experiences which help in finding a reliable solution in less time.

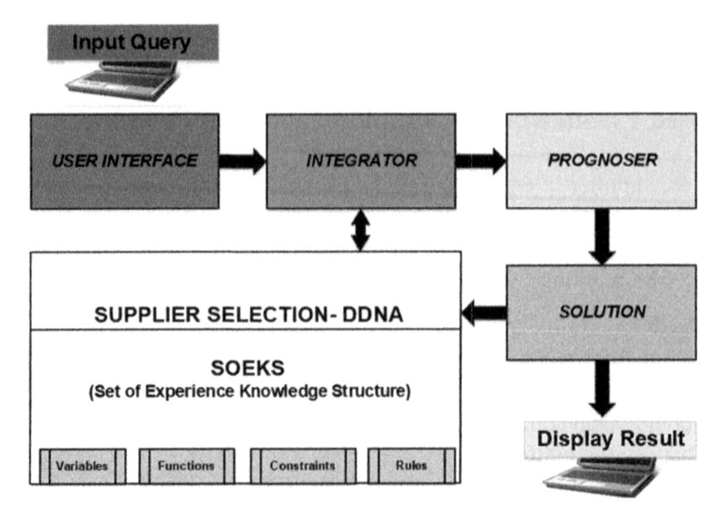

Fig. 1. Framework for the supplier selection process by using SOEKS and Decisional DNA

3.1 Working Algorithm of Proposed Framework

Our proposed supplier selection framework is based on SOEKS and DDNA, which is a smart knowledge management technique and is capable of performing multi-criteria decision making. Various criteria and sub-criteria related to supplier selection process are captured, stored, recalled, and shared from the experiential knowledge in the form of set of experiences (SOEs). Whenever a similar query is presented during the supplier selection process, this stored knowledge is recalled to overcome the problem. It provides a list of proposed optimal solutions according to the priorities set by the user. By the passage of time, the system achieves more expertise in its specific domains as it stores relevant knowledge and experience related to formal decision events. Working algorithm for the proposed supplier selection method is shown in Fig. 2.

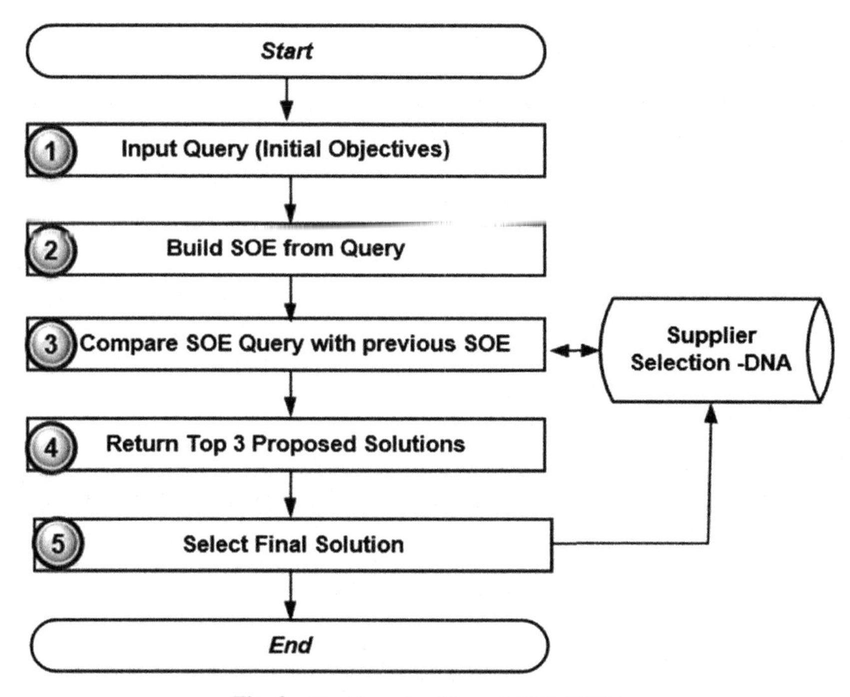

Fig. 2. Working algorithm of SSP-DDNA

The supplier selection process starts by inputting the query into the system based on the initial objectives/criteria. These objectives are entered in the form of variables (qualitative and quantitative criteria), functions, constraints and rules. In Step 2, the entered query is converted to a *SOE.* The built query is compared with available similar SOEs that are ranked according to the common initial objectives and their performance factor (PF). These SOEs are stored in SS-DDNA in a comma-separated values (CSV) file.

The pseudocode for parser reading CSV file for the supplier selection process is shown below:

- *Reads variables, functions, constraints, and rules.*
- *Develops set of variables, set of functions, set of constraints, and set of rules.*
- *Creates a Set of Experience (SOE) = Set of variables + Set of functions +Set of rules.*
- *Form a chromosome of the supplier selection process by collecting SOEs of the same category.*
- *Provide top 3 proposed solutions.*

Finally, a list of proposed top three solutions is returned and the user selects the final solution. This final solution is then stored in the SS-DDNA and can be used for future reference.

4 Conclusion and Future Work

This paper presented a framework to enhance the supplier selection process by using SOEKS and DDNA. The proposed framework supports multi-criteria decision making (MCDMA) approach and uses experiential knowledge of formal decisional events generated during the supplier selection process. The proposed system is dynamic in nature as it updates itself every time a new decision is taken. It will benefit performing the supplier selection process for small and medium enterprises involved in product manufacturing. The next step will be the refinement of the algorithm in more detail and its translation into JAVA platform.

References

1. Ghadimi, P., Azadnia, A.H., Heavey, C., Dolgui, A., Can, B.: A review on the buyer–supplier dyad relationships in sustainable procurement context: past, present and future. Int. J. Prod. Res. **54**, 1443–1462 (2016)
2. González, M.E., Quesada, G., Monge, C.A.M.: Determining the importance of the supplier selection process in manufacturing: a case study. Int. J. Phys. Distrib. Logistics Manage. **34**(6), 492–504 (2004)
3. Agarwal, P., Sahai, M., Mishra, V., Bag, M., Singh, V.: A review of multi-criteria decision making techniques for supplier evaluation and selection. Int. J. Ind. Eng. Comput. **2**, 801–810 (2011)
4. Yadav, V., Sharma, M.K.: Multi-criteria decision making for supplier selection using fuzzy AHP approach. Benchmarking: An Int. J. (2015)
5. Saghafian, S., Hejazi, S.R.: Multi-criteria group decision making using a modified fuzzy TOPSIS procedure. In: International Conference on Computational Intelligence for Modelling, Control and Automation and International Conference on Intelligent Agents, Web Technologies and Internet Commerce (CIMCA-IAWTIC'06), pp. 215–221. IEEE (2005)
6. Pal, O., Gupta, A.K., Garg, R.: Supplier selection criteria and methods in supply chains: a review. Int. J. Soc. Manage. Econ. Bus. Eng. **7**, 1403–1409 (2013)
7. Thiruchelvam, S., Tookey, J.: Evolving trends of supplier selection criteria and methods. Int. J. Automot. Mech. Eng. **4**, 437–454 (2011)
8. Ho, W., Xu, X., Dey, P.K.: Multi-criteria decision making approaches for supplier evaluation and selection: a literature review. Eur. J. Oper. Res. **202**, 16–24 (2010)
9. Ahmed, M.B., Sanin, C., Szczerbicki, E.: Experience-based decisional DNA (DDNA) to support product development. Cybern. Syst. **49**(5–6), 399–411 (2018)
10. Ghadimi, P., Wang, C., Lim, M.K., Heavey, C.: Intelligent sustainable supplier selection using multi-agent technology: theory and application for Industry 4.0 supply chains. Comput. Ind. Eng. **127**, 588–600 (2019)
11. Hasan, M.M., Jiang, D., Ullah, A.S., Noor-E-Alam, M.: Resilient supplier selection in logistics 4.0 with heterogeneous information. Expert Syst. Appl. **139**, 112799 (2020)
12. Sanin, C., Szczerbicki, E.: Towards the construction of decisional DNA: a set of experience knowledge structure java class within an ontology system. Cybern. Syst. Int. J. **38**, 859–878 (2007)
13. Büyüközkan, G., Çifçi, G.: A novel fuzzy multi-criteria decision framework for sustainable supplier selection with incomplete information. Comput. Ind. **62**, 164–174 (2011)
14. Kilic, H.S.: An integrated approach for supplier selection in multi-item/multi-supplier environment. Appl. Math. Model. **37**, 7752–7763 (2013)

15. Taherdoost, H., Brard, A.: Analyzing the process of supplier selection criteria and methods. Procedia Manuf. **32**, 1024–1034 (2019)
16. Wahlster, P., Goetghebeur, M., Kriza, C., Niederländer, C., Kolominsky-Rabas, P.: Balancing costs and benefits at different stages of medical innovation: a systematic review of Multi-criteria decision analysis (MCDA). BMC Health Serv. Res. **15**, 262 (2015)
17. Sanin, C., Szczerbicki, E.: Set of experience: a knowledge structure for formal decision events. Found. Control Manage. Sci. (3), 95–113 (2005)
18. Sanin, C., Szczerbicki, E.: Experience-based knowledge representation: SOEKS. Cybern. Syst. Int. J. **40**, 99–122 (2009)
19. Shafiq, S.I., Sanín, C., Szczerbicki, E.: Set of experience knowledge structure (SOEKS) and decisional DNA (DDNA): past, present and future. Cybern. Syst. **45**, 200–215 (2014)

An Efficient Approach for Improving Recursive Joins Based on Three-Way Joins in Spark

Thanh-Ngoan Trieu[1(✉)], Anh-Cang Phan[2], and Thuong-Cang Phan[1]

[1] Can Tho University, Can Tho, Vietnam
{ttngoan,ptcang}@cit.ctu.edu.vn
[2] Vinh Long University Of Technology Education, Vinh Long City, Vietnam
cangpa@vlute.edu.vn

Abstract. In the evolution of Big Data, efficiently processing large datasets is always a top concern for researchers. A join operation is one of such processing, a common operation appearing in many data queries. This operation generates plenty of intermediate data and data transmission over the network, especially a recursive join operation. Although extremely expensive, a recursive join has a wide variety of domains as database, social network and computer network analyses, compiler, data integration and graph mining. Therefore, this study was carried out to optimize recursive joins based on some solutions in a Spark environment. The solutions leverage the advantages of three-way join operations, Bloom filters, Spark RDD and caching techniques for iterative join computation. These significantly reduce the number of executed iterations and jobs, the amount of redundant data, and remotely accessing persistent data. Our experimental results show that the optimized recursive join is more efficient than a typical one by reducing the number of iterations to half, minimizing data transfer, and thus shorter execution time.

Keywords: Recursive join · Three-way join · Semi-Naive algorithm · Big data analytic · Spark · MapReduce · Bloom filter

1 Introduction

The development of the Internet contributes to create the term Big Data, which has widely been used in the era of Information Technology. Managing, storing, and processing Big Data has been an issue of concern in order to take advantages of valuable existing data sources. One of the most common operations involving data analysis and information retrieval is a join operation that combines two sets of data. A recursive join operation [17,23], a costly and complex type of join operations, is known as a fix-point join because it repeats join operations until there is no new results generated. The implementation of recursive joins becomes more complicated and poses many challenges for processing on large datasets. The well-known processing model to solve this problem is MapReduce.

M. Hernes et al. (Eds.): ICCCI 2020, CCIS 1287, pp. 566–578, 2020.
https://doi.org/10.1007/978-3-030-63119-2_46

The evaluation of a large-scale recursive join in MapReduce is compiled into the repetition of two jobs: a join job and an incremental computation job [31].

A recursive join is a complex operation with a high cost that is used in a wide range of applications such as PageRank [19], Graph mining [21,30], Data mining [11,28], etc. Several studies have been carried out to improve the evaluation of recursive join queries, especially in the context of Big Data. However, those studies have been mainly focused in a MapReduce environment of Hadoop but not in the one of Spark [2,22,31]. Spark has been considered as a next generation large-scale data processing framework thanks to its remarkable in-memory processing capacity. Consequently, this research provides the optimization of large-scale recursive joins using an improved Semi-Naive algorithm in Spark. Semi-Naive is a suitable algorithm for deployment in a distributed parallel processing environment [16].

The structure of this paper is organized as follows. Section 2 presents the background related to the Spark environment and join opperations. Section 3 provides the problem description and our approach for improving recursive joins on large-scale datasets. Section 4 will be the experiments conducted in a Spark cluster. The conclusion of the paper is presented in Sect. 5.

2 Background

2.1 Apache Hadoop and Apache Spark

Apache Hadoop [7] is an open-source framework that supports large dataset processing in a distributed manner over a computer cluster. Hadoop consists of two different parts: storage part with HDFS (Hadoop Distributed File System) and processing part using the MapReduce programming model. HDFS is used to store distributed data and MapReduce part is used for processing those data. HDFS is distributed, scalable, and highly fault-tolerant written in Java. It provides shell commands and Java API for other applications.

Apache Spark [8] is a fast and unified analytic engine for Big Data processing. It is originally developed in the University of California, Berkeley in 2009. The framework provides a lot of features for fast processing of large datasets. Given a task, Spark allows to split it into smaller tasks that can be performed in parallel in a computer cluster. Spark takes advantages of processing power on memory so that it gives high performance for both batch and streaming data, which is up to 100 times faster than Hadoop MapReduce [34].

It is necessary to install additional software for distributed storage since Spark does not provide its own storage systems. That is the reason why users often install Spark on top of Hadoop platform to use HDFS for storage mechanism. The main point makes up differences between Spark and Hadoop MapReduce is that Spark loads data into memory and reuses it instead of having to read/write constantly from/to HDFS as in Hadoop MapReduce.

Resilient Distributed Dataset - RDD - is a core concept in Apache Spark. RDD is a distributed collection of elements that can be stored temporarily on RAM and processed parallel. In Spark, a task is represented by creating new

RDDs, converting existing RDDs, or performing operations on RDDs to get results. Spark is a powerful supporting tool for handling iterative tasks through operations on RDD. The repetition process is to read input data from HDFS, calculate intermediate results with RDD, and save final results to HDFS.

2.2 Bloom Filter

Bloom Filter [12] proposed by Burton Howard Bloom in 1970, is a space-efficient probabilistic data structure that can help to check whether an element is a member of a collection (Fig. 1). There is a probability of false positive elements but never have a false negative element. A false positive element is an element that was found to belong to the collection however it is not. A false negative element is an element said not to belong to the collection but it actually is.

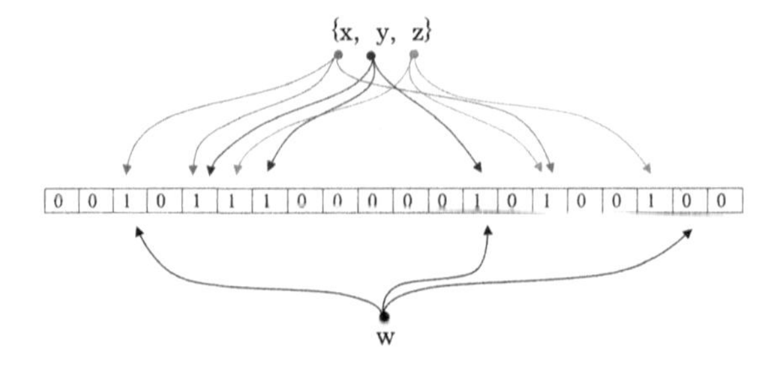

Fig. 1. An example of Bloom filter

An empty Bloom Filter of a collection S - BF_S - is an array of m bits with zero value. There are k hash functions needed to map an element to k positions in the array and k is usually much smaller than m. Adding an element to BF_S can be done by using k hash functions to find k positions in the BF_S array and set the value of those positions to 1. To determine whether an element w is belonged to the collection S, one can use the k hash functions to find k positions in the BF_S array. If all bits of the k positions is 1 then w belongs to S otherwise w is not belong to S.

2.3 Join Operation

Join [24] is an operation for combining tuples between different datasets with a join condition. Processing join operation in a MapReduce environment is a topic of interest since this operation is frequently used in many applications. MapReduce model was first introduced by Google in 2004 [15], has been used in many computational operations on large datasets. There are two main operations in MapReduce model: Map and Reduce. Map function applies on input data producing an intermediate result as a list of key-value pair. Those intermediate data of the same key will be sent to a Reducer for combining values to get a

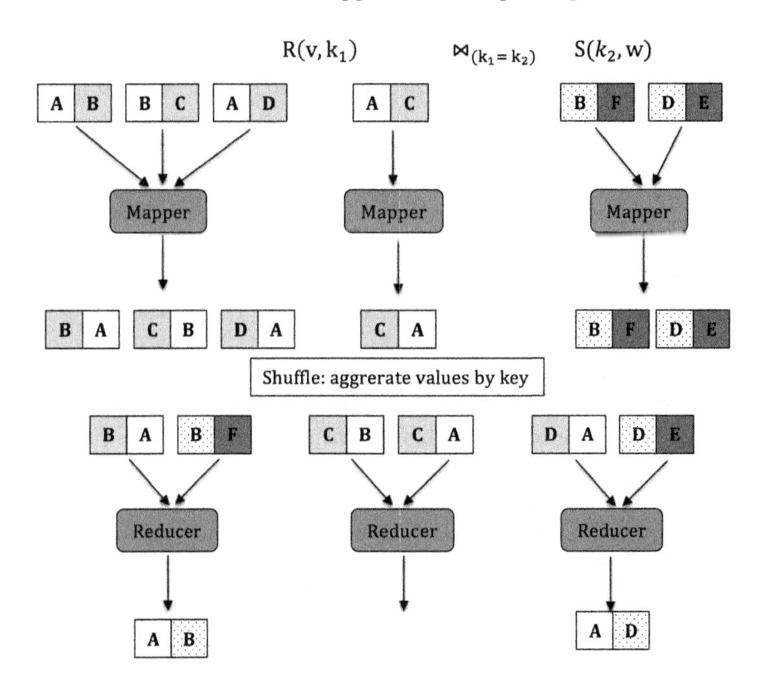

Fig. 2. Join operation in MapReduce environment

result. A join operation in a MapReduce environment can be illustrated by an example in Fig. 2.

Given a join operation between two datasets: $R(v, k_1) \bowtie_{k_1=k_2} S(k_2, w))$, there will be two phases in computing join results. The Map phase is responsible for reading the input datasets R and S. The three Mappers are created as shown in Fig. 2 to process three data blocks (each block will consist of several tuples). Mappers will convert tuples into key-value pairs for join operation. Intermediate datasets with the same key will be sent to the Reducers. A simple reduce function in each reducer takes each value of R and combines with a value of S of the same key to produce the results.

3 Proposed Method for Recursive Joins on Large Datasets

3.1 Recursive Join Operation

Recursive join [17,23] computes the transitive closure of a relation. An example of a recursive join is to discover the relationships of a person in a family database.

$$\text{Ancestor(X, Y)} \leftarrow \text{Parent(X, Y)}$$
$$\text{Ancestor(X, Z)} \leftarrow \text{Ancestor(X, Y)} \bowtie \text{Parent(Y, Z)}$$

A person X is an ancestor of a person Y if X is a parent of Y. A person X is also an ancestor of a person Z if there is a person Y that X is an ancestor of Y and Y is a parent of Z.

There are many algorithms introduced for computing transitive closure of a relation. Those can be divided into two categories: direct algorithms and iterative algorithms. The direct algorithms [5] commonly used are Warshall [33], Warren [32], and Schmitz [29]. The second category was developed in the context of processing recursive queries on a relational database. It can also be categorized into two subsets: linear (Naive [9,10], Semi-Naive [9,10], and Magic-set [14]) and non-linear (Smart [4,16,18]);

Shaw et al. [31] proposed a method for optimization with Semi-Naive algorithm for recursive join operation in Hadoop MapReduce. The implementation of Semi-Naive algorithm for recursive join operation performs two repeated MapReduce jobs: a join job and an incremental computation job (Fig. 3). This method uses Reducer Input Cache mechanism of HaLoop [13] to minimize associated costs. However, a disadvantage of this cache mechanism is that it needs to rewrite the entire cache in each iteration. Moreover, HaLoop is no longer being developed.

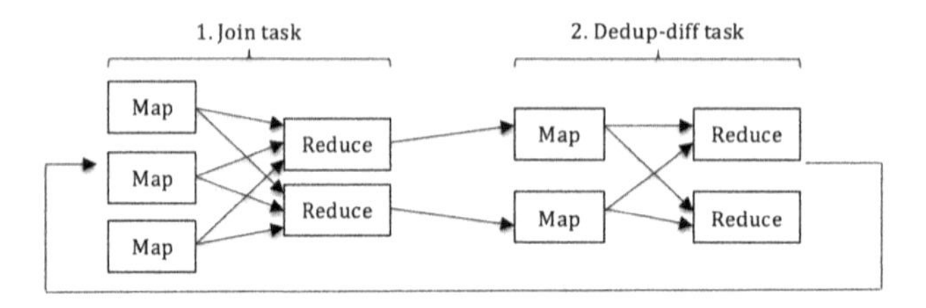

Fig. 3. Two kinds of tasks for Semi-Naive recursive join in Shaw et al. approach.

In a recent study [27], T. C. Phan et al. come up with several ideas for improving recursive joins in the Spark environment. The work uses the iterative processing mechanism provided by Spark RDD (Resilient Distributed Dataset) and the cache mechanism for persisting un-changed dataset on memory that can reduce costs of repeatedly read/write data. It primarily focuses on using Intersection Bloom Filter [24–26] to eliminate redundant data that is not involved in join operations. This significantly reduces communication costs. However, the number of iterations of Semi-Naive is quite a lot, which consumes too much costs in the context of large-scale processing.

Our synthesis approach in this study is an adaptation of the Semi-Naive evaluation commonly used to evaluate a recursive join. We offer a major improvement to a Semi-Naive algorithm in MapReduce by replacing two-way joins with three-way joins to significantly reduce the number of necessary iterations.

3.2 Semi-Naive Algorithm for Recursive Joins

Semi-Naive is a common algorithm for calculating linear transitive closure, which is suitable for evaluating recursive joins in MapReduce and Spark. The details of the algorithm are shown in Table 1. It is based on repetition of two-way joins. Thus, this recursive join algorithm is named as RJ2 for ease of use later.

K denotes an original graph. F and ΔF is initialized with K. F will contain all tuples in the transitive closure of the graph until the end of the algorithm. ΔF includes new tuples being found from the previous iteration. The Semi-Naive algorithm for a recursive join will be performed by two repeated jobs including a join job and an incremental computation job. The join job performs a join operation between two datasets ΔF and K. The second job removes duplicated data found in previous iterations.

Table 1. Semi-Naive algorithm for recursive joins - RJ2

Algorithm 1 - RJ2	
$F_0 = K, \Delta F_0 = K, i = 1;$	(1)
do	(2)
$O_i = \Delta F_{i-1}(c_1, c_2) \bowtie_{c_2=c_{1'}} K(c_{1'}, c_{2'})$;	(3)
$\Delta F_i = O_i - F_{i-1};$	(4)
$F_i = F_{i-1} \cup \Delta F_i;$	(5)
i++;	(6)
while $\Delta F_i \neq \emptyset$	(7)

As shown in Table 1, the most important operation in each iteration is the join computation ($\bowtie$) of the ΔF_{i-1} dataset with the K dataset to produces intermediate result O_i. The intermediate result O_i may contains a lot of tuples which overlaps the previous result. These duplicate tuples should be removed by the second job ($O_i - F_{i-1}$). In the end of each iteration, ΔF_i only contains new tuples used for the next iteration.

Semi-Naive is better than Naive because it avoids repeating join operations between F and K in each iteration. It only allows new data ΔF_{i-1} to participate in the join. The incremental computation job performs removing duplicate tuples of the intermediate result compared with in the previous results.

Consider line number 3 of the algorithm, we can easily see that the K dataset is an un-changed dataset over iterations. The important part of this algorithm is to join the ΔF dataset with the K dataset. It means that the K dataset needs to be read and processing n times if we have n necessary iterations. We proposed a method that can reduce the number of iterations by two using three-way join.

3.3 Recursive Join Optimization Based on Three-Way Join and Filters

Optimization of recursive joins using the three-way join - denoted as ORJ3 - is presented in Table 2. There are some approaches to the optimization. They are as follows:

1. Using three-way joins to improve a recursive join algorithm.
2. Applying Bloom filters to remove non-joining data in the three-way joins.
3. Leveraging the advantages of Spark to compute iterative joins.

Firstly, each iteration of the ORJ3 algorithm in Table 2 is equivalent to two iterations in comparison with the RJ2 algorithm since it does two-way joins in just one iteration.

Table 2. Optimized Semi-Naive algorithm for recursive joins - ORJ3

Algorithm 2 - ORJ3	
$F_0 = K, \Delta F_0 = K, i = 1;$	(1)
do	(2)
$\quad O_i = \Delta F_{i-1}(c_1, c_2) \bowtie_{c_2=c_{1'}} K(c_{1'}, c_{2'}) \bowtie_{c_{2'}=c_{1'}} K(c_{1'}, c_{2'})$;	(3)
$\quad \Delta F_i = O_i - F_{i-1});$	(4)
$\quad F_i = F_{i-1} \cup \Delta F_i;$	(5)
$\quad$ i++;	(6)
while $\Delta F_i \neq \emptyset$	(7)

Here we should consider three-way joins in this proposal. Given an example of join operations on three datasets $R(A, B, X) \bowtie S(B, C, Y) \bowtie T(C, D, Z)$, there are two ways for computing this join operation. For the first one, we can perform a join between R and S then join the result of $R \bowtie S$ with T. In this way, there are two rounds of MapReduce jobs used for the two steps (temporarily called this method as two-round three-way join). For the second one, Afrati and Ullman [1,3] come up with the idea of one-round three-way join meaning that there will be one round of MapReduce jobs for computing three-way join [20].

For one-round three-way join implementation in MapReduce, the number of Reducers is an explicit number $k = k_1 * k_2$. The authors create two hash functions h and g corresponding with the two slot k_1 and k_2. Mappers generate a key-value pair for each tuple in dataset S, and k_1 and k_2 key-value pairs for each tuple in dataset R and T respectively. The key-value pairs of R and T are sent to row and column of the Reducers matrix while the key-value pairs of S are sent to a cell of the Reducers matrix and join operation executes in that cell (Fig. 4).

In study [26] of T. C. Phan et al., the authors show that given a three-way join, one-round three-way join is better than two-round three-way join with a

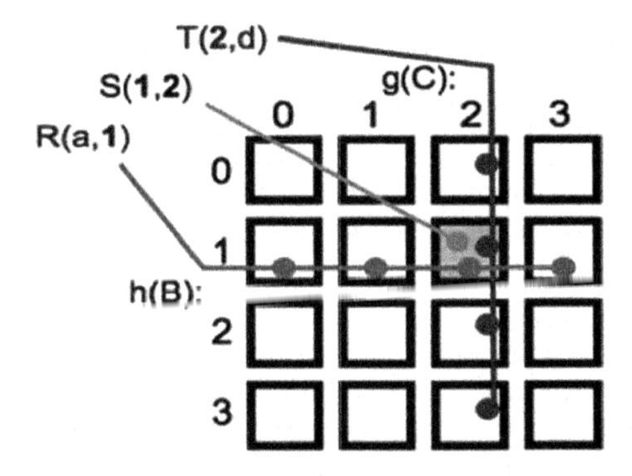

Fig. 4. One-round three-way dimensional join [20]

condition of $r < (|R| * \alpha)^2$, where r is the number of Reducers, $|R| = |S| = |T|$, and α is the probability of two tuples from different datasets that matches join columns. We apply one-round three-way join in our research for optimizing the recursive join. The flowchart for the recursive join in the proposed method is shown in Fig. 5.

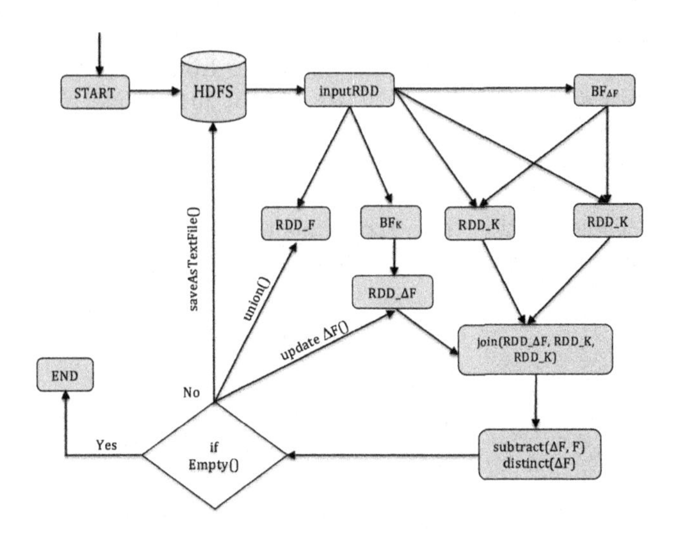

Fig. 5. Flowchart for Optimized Semi-Naive algorithm for recursive joins

Secondly, to reduce the amount of unnecessary data participating in join operation, this study uses Bloom filter (in ORJ3 algorithm) for filtering input data before sending to Reducers. There are two Bloom filters for K dataset and F dataset, BF_K and $BF_{\Delta F}$ respectively. This study creates the two filters with the values $m = 120,000$ and $k = 8$. There is a trade-off between time building up the filters and the false positive probability considering the size of the filters.

This research focuses on the use of ORJ3 algorithm thus we define the small size of the filters.

Creating two filters can be described as follows:

- For each tuple in the input dataset K and F
  - Add the join key of K to the BF_K
  - Add the join key of F to the $BF_{\Delta F}$
- Broadcast BF_K and $BF_{\Delta F}$ to all Reducers

Lastly, the two algorithms ORJ3 and RJ2 are implemented in Spark for leveraging the advantages of iterative operations with Spark RDD.

4 Evaluation

4.1 Spark Cluster

Experiments was conducted in a Spark cluster that contains one master node and 10-worker nodes provided by The Mobile Network and Big Data Laboratory of College of Information and Technology, Can Tho University. Each computing node has the configuration of 5 CPUs, 8 GB RAM, and 100 GB disk. The operating system used in each computer node is Ubuntu 18.04 LTS and the version of the applications needed are Hadoop 3.0.3, Spark 2.4.3, and Java 1.8.

The datasets used for experiments was taken from PUMA benchmarks [6] with several test sets: 1 GB, 5 GB, 10 GB, 15 GB, 20 GB, and 25 GB. The data is in text format with 39 fields separated by comma and 19 characters in each field. There were two approaches being considered: original Semi-Naive algorithm for recursive join (RJ2) and optimized Semi-Naive algorithm for recursive join (ORJ3). The execution time, the number of iterations, and the amount of intermediate data will be recorded for analysis and comparison.

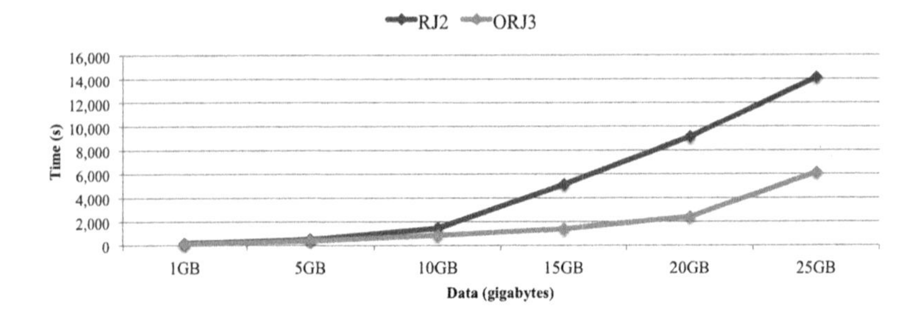

Fig. 6. Processing time between the two approaches

4.2 Results

In Fig. 6, it is clearly shown that there is a big difference between the two approaches in term of execution time (in seconds). Using three-way join instead of two-way join in Semi-Naive algorithm can help to reduce processing time. The use of ORJ3 is more efficient than RJ2 in cases of big datasets. In contrast, when we have small datasets, there was not a clear difference between ORJ3 and RJ2. The reason for this issue is that ORJ3 takes time to build up Bloom filters so that it does not give good performance for small datasets. Table 3 shows the time of execution in more details.

Table 3. Processing time between the two approaches (second)

Approach	Data					
	1 GB	5 GB	10 GB	15 GB	20 GB	25 GB
RJ2	180	531	1,448	5,107	9,151	14,139
ORJ3	82.2	371.7	877	1,405	2,420	6,130

Figure 7 describes the amount of intermediate data that was generated in the join operation and needed to be transferred over the network for the incremental computation job. The number of iterations was reduced by half in ORJ3 thus the amount of intermediate data significantly reduced. In addition, Bloom filter and Spark cache mechanism help to reduce redundant data that does not participate in Join operation for optimization of the recursive join.

The number of iterations in ORJ3 is reduced by two times in comparison with RJ2 (Table 4).

$$N_{ORJ3} = \lceil N_{RJ2}/2 \rceil \tag{1}$$

where, N_{ORJ3} is the number of iterations for ORJ3 and N_{RJ2} is the number of iterations for RJ2.

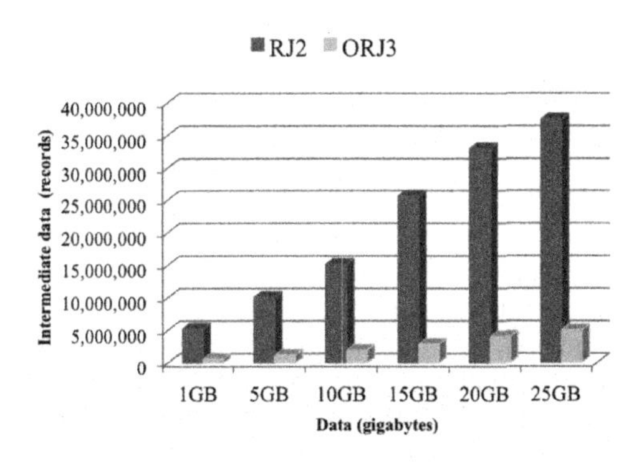

Fig. 7. Intermediate results between the two approaches

Table 4. Number of iterations in each approach

Approach	Data					
	1 GB	5 GB	10 GB	15 GB	20 GB	25 GB
RJ2	4	4	6	9	11	11
ORJ3	2	2	3	5	6	6

5 Conclusion

Recursive joins are very important and popular in a variety of applications such as traversal in social networks, automata-based constructions, data integration and graph mining. However, the joins are costly and complex, even for ordinary datasets because their computation involves the repetition of the join job and the incremental computation job in a distributed environment. For that reason, optimizing recursive joins becomes more meaningful and valuable in the context of Big Data applications. Therefore, this study provides the effective solutions for improving recursive join evaluation in a Spark environment. The solutions include: (1) Providing an improved three-way join algorithm in Spark; (2) Applying Bloom filters for elimination of non-joining data in the joins; (3) Leveraging Spark RDD and Spark cache mechanism on un-change dataset K for iterative join computation; (4) Optimizing recursive joins based on the three-way joins and the filters in Spark.

Our evaluation shows that the proposed method brings better performance than the original one. This is a highly practical contribution because the Semi-Naive algorithm is a popular and widely used algorithm. The work gives a contribution to the development of large data-driven knowledge management systems. However, there is an aspect that is not yet considered in this research. In particular, the PUMA benchmarks gives standard datasets for recursive joins but what happens if we have a skew dataset. This is also a work that our research group would like to study in the near future.

References

1. Afrati, F.N., Ullman, J.D.: Optimizing multiway joins in a map-reduce environment. IEEE Trans. Knowl. Data Eng. **23**(9), 1282–1298 (2011). https://doi.org/10.1109/TKDE.2011.47
2. Afrati, F.N., Borkar, V., Carey, M., Polyzotis, N., Ullman, J.D.: Map-reduce extensions and recursive queries. In: Proceedings of the 14th International Conference on Extending Database Technology, pp. 1–8 (2011)
3. Afrati, F.N., Ullman, J.D.: Optimizing joins in a map-reduce environment. In: Proceedings of the 13th International Conference on Extending Database Technology, EDBT 2010, pp. 99–110. Association for Computing Machinery, New York (2010). https://doi.org/10.1145/1739041.1739056

4. Afrati, F.N., Ullman, J.D.: Transitive closure and recursive datalog implemented on clusters. In: Proceedings of the 15th International Conference on Extending Database Technology, EDBT 2012, pp. 132–143. Association for Computing Machinery, New York (2012). https://doi.org/10.1145/2247596.2247613
5. Agrawal, R., Jagadish, H.V.: Direct algorithms for computing the transitive closure of database relations. In: Proceedings of the 13th International Conference on Very Large Data Bases, VLDB 1987, pp. 255–266. Morgan Kaufmann Publishers Inc., San Francisco (1987)
6. Ahmad, F.: Puma benchmarks and dataset downloads (2011). https://engineering.purdue.edu/~puma/datasets.htm. Accessed 20 Oct 2019
7. Apache: Apache hadoop (2002). https://hadoop.apache.org. Accessed 20 July 2019
8. Apache: Apache spark (2009). https://spark.apache.org. Accessed 20 July 2019
9. Bancilhon, F.: Naive evaluation of recursively defined relations. In: Brodie, M.L., Mylopoulos, J. (eds.) On Knowledge Base Management Systems. TINF, pp. 165–178. Springer, New York (1986). https://doi.org/10.1007/978-1-4612-4980-1_17
10. Bancilhon, F., Ramakrishnan, R.: An amateur's introduction to recursive query processing strategies. SIGMOD Rec. **15**(2), 16–52 (1986). https://doi.org/10.1145/16856.16859
11. Bheemavaram, R., Zhang, J., Li, W.N.: A parallel and distributed approach for finding transitive closures of data records: a proposal, January 2006
12. Bloom, B.H.: Space/time trade-offs in hash coding with allowable errors. Commun. ACM **13**(7), 422–426 (1970)
13. Bu, Y., Howe, B., Balazinska, M., Ernst, M.D.: HaLoop: efficient iterative data processing on large clusters. Proc. VLDB Endow. **3**(1–2), 285–296 (2010)
14. Chen, Y.: On the bottom - up evaluation of recursive queries. Int. J. Intell. Syst. **11**, 807–832 (1996)
15. Dean, J., Ghemawat, S.: MapReduce: simplified data processing on large clusters. Commun. ACM **51**(1), 107–113 (2008)
16. Gribkoff, E.: Distributed algorithms for the transitive closure (2013)
17. Idreos, S., Liarou, E., Koubarakis, M.: Continuous multi-way joins over distributed hash tables. In: Proceedings of the 11th International Conference on Extending Database Technology: Advances in Database Technology, EDBT 2008, pp. 594–605. Association for Computing Machinery, New York (2008). https://doi.org/10.1145/1353343.1353415
18. Ioannidis, Y.E.: On the computation of the transitive closure of relational operators. In: Proceedings of the 12th International Conference on Very Large Data Bases, VLDB 1986, pp. 403–411. Morgan Kaufmann Publishers Inc., San Francisco (1986)
19. Jilani, T.A., Fatima, U., Baig, M.M., Mahmood, S.: A survey and comparative study of different PageRank algorithms. Int. J. Comput. Appl. **120**(24), 24–30 (2015)
20. Kimmett, B., Thomo, A., Venkatesh, S.: Three-way joins on MapReduce: an experimental study. In: IISA 2014, The 5th International Conference on Information, Intelligence, Systems and Applications, pp. 227–232, July 2014. https://doi.org/10.1109/IISA.2014.6878811
21. Leskovec, J., Rajaraman, A., Ullman, J.D.: Mining Social-Network Graphs, 2nd edn., pp. 325–383. Cambridge University Press, Cambridge (2014). https://doi.org/10.1017/CBO9781139924801.011
22. Malewicz, G., et al.: Pregel: a system for large-scale graph processing. In: Proceedings of the 2010 ACM SIGMOD International Conference on Management of Data, pp. 135–146 (2010)

23. Ordonez, C.: Optimizing recursive queries in SQL. In: Proceedings of the 2005 ACM SIGMOD International Conference on Management of Data, SIGMOD 2005, pp. 834–839. Association for Computing Machinery, New York (2005). https://doi.org/10.1145/1066157.1066260
24. Phan, T.C.: Optimization for big joins and recursive query evaluation using intersection and difference filters in MapReduce. Theses, Université Blaise Pascal - Clermont-Ferrand II, July 2014. https://tel.archives-ouvertes.fr/tel-01066612
25. Phan, T.C., d'Orazio, L., Rigaux, P.: Toward intersection filter-based optimization for joins in MapReduce. In: Proceedings of the 2nd International Workshop on Cloud Intelligence, Cloud-I 2013. Association for Computing Machinery, New York (2013). https://doi.org/10.1145/2501928.2501932
26. Phan, T.-C., d'Orazio, L., Rigaux, P.: A theoretical and experimental comparison of filter-based equijoins in MapReduce. In: Hameurlain, A., Küng, J., Wagner, R. (eds.) Transactions on Large-Scale Data- and Knowledge-Centered Systems XXV. LNCS, vol. 9620, pp. 33–70. Springer, Heidelberg (2016). https://doi.org/10.1007/978-3-662-49534-6_2
27. Phan, T.C., Tran, T.T.Q., Phan, A.C.: Optimization of recursive joins on large-scale datasets in spark, pp. 729–742. Vietnam Academy of Science and Technology, Hanoi (2016). https://doi.org/10.15625/vap.2016.00091
28. Save, A.M., Kolkur, S.: Article: hybrid technique for data cleaning. In: IJCA Proceedings on National Conference on Role of Engineers in National Building, NCRENB, pp. 4–8, June 2014
29. Schmitz, L.: An improved transitive closure algorithm. Computing **30**(4), 359–371 (1983). https://doi.org/10.1007/BF02242140
30. Seufert, S., Anand, A., Bedathur, S., Weikum, G.: High-performance reachability query processing under index size restrictions (2012)
31. Shaw, M., Koutris, P., Howe, B., Suciu, D.: Optimizing large-scale semi-Naïve datalog evaluation in hadoop. In: Barceló, P., Pichler, R. (eds.) Datalog 2.0 2012. LNCS, vol. 7494, pp. 165–176. Springer, Heidelberg (2012). https://doi.org/10.1007/978-3-642-32925-8_17
32. Warren, H.S.: A modification of Warshall's algorithm for the transitive closure of binary relations. Commun. ACM **18**(4), 218–220 (1975). https://doi.org/10.1145/360715.360746
33. Warshall, S.: A theorem on Boolean matrices. J. ACM **9**(1), 11–12 (1962). https://doi.org/10.1145/321105.321107
34. Zaharia, M., Chowdhury, M., Franklin, M.J., Shenker, S., Stoica, I., et al.: Spark: cluster computing with working sets. In: HotCloud 2010, no. 10, p. 95 (2010)

Lambda Architecture for Anomaly Detection in Online Process Mining Using Autoencoders

Philippe Krajsic[1(✉)] and Bogdan Franczyk[1,2]

[1] Leipzig University, Grimmaische Str. 12, 04107 Leipzig, Germany
krajsic@wifa.uni-leipzig.de, bogdan.franczyk@ue.wroc.pl
[2] Wrocław University of Economics, ul. Komandorska 118-120, 53-345 Wrocław, Poland

Abstract. The analysis of event data in the context of process mining is becoming increasingly important. In particular, the processing of streaming data in the sense of an real-time analysis is gaining in relevance. More and more fields of application are emerging in which an operational support becomes necessary, i.e. in surgery or manufacturing. For proper analysis a cleanup of the streaming data in a pre-processing step is necessary to ensure accurate process mining activities. This paper presents a blueprint of a lambda architecture in which an autoencoder is embedded that is supposed to allow unsupervised anomaly detection in event streams, like incorrect traces, events and attributes, and thus will help to improve results in online process mining.

Keywords: Anomaly detection · Autoencoder · Lambda architecture · Process mining · Streaming data

1 Introduction

In the spectrum of process mining [1] the problem of the analysis of event data is increasingly getting attention. Existing analysis methods typically assume that the input data is completely free of incorrect data and infrequent behavior, which does not usually correspond to reality [2–4]. Incorrect data in the event streams can lead to incorrect results during further processing. For example, the accuracy of drift detection can be negatively affected by stochastic vibrations due to inaccurate event streams [5, 6]. Approaches that have been conducted in this area using filtering techniques to eliminate erroneous events from event data show an improvement in the quality of process mining techniques which leads to an optimization of the analysis of the processes [7–9].

This work presents an architectural blueprint that not only represents the recognition of anomalies as such in a business setting, but also how such mechanisms can be embedded in an online process mining environment.

In order to take full advantage of the possibilities offered by anomaly detection methods, it is necessary to transfer these methods to an online setting. This enables the use of anomaly detection methods and subsequent process mining techniques in operational support. This allows processes to be influenced in real-time due to deviations in process flow. To enable live anomaly detection, the associated technologies must enable the

M. Hernes et al. (Eds.): ICCCI 2020, CCIS 1287, pp. 579–589, 2020.
https://doi.org/10.1007/978-3-030-63119-2_47

processing of data streams and, in the process mining context, the processing of event streams. The requirements for the processing of streaming data in a real-time setting must be taken into account in the design of the architecture framework.

The contribution of this work to information system research is as follows: An anomaly detection framework in process mining context for embedding in a process mining environment is developed. In addition, the architecture takes into account the processing of data streams on the basis of suitable infrastructures, such as lambda architecture. The anomaly detection itself is based on an unsupervised autoencoder approach. The unsupervised autoencoder approach works without prior knowledge of the respective process and can therefore be used for different use cases independently of the domain. This enables universal use of the anomaly detection framework. In summary, it will answer the following research question:

- **RQ:** How can an architecture with embedded anomaly detection look like in an online process mining environment?

The remainder of the paper is structured as follows: Sect. 2 presents related work in the area of anomaly detection approaches and architecture concepts for streaming data in process mining context. Section 3 describes the lambda architecture approach. Section 4 subsequently describes the general functionality of unsupervised autoencoders for anomaly detection as well as the process of anomaly detection used in this work. In Sect. 5 the architecture with embedded autoencoder in an online process mining setting is presented. Section 6 shows the results of a preliminary conducted experiment. The final Sect. 7 concludes the paper and presents future work.

2 Related Work

Regarding the approaches to filtering methods of event data in the context of process mining, the current state of research can be described as follows. The related work done in the areas of online process mining and anomaly detection is of particular importance for the work presented in this paper. With regard to detection and filtering of anomalies in event logs, there are some approaches described in the literature. In [8] a reference model is used to detect inappropriate behavior and repair the affected log. The approach proposed in [9] is based on an automaton which is modeled on the frequent process behavior recorded in the logs. Events that cannot be reproduced by the automaton are removed. [11] proposes an approach that uses conditional probabilities between activity sequences to eliminate events that are unlikely in a particular sequence. While existing techniques for filtering anomalies from event-logs show that they help improve process quality, they cannot be applied to the filtering of event streams in an online context. In [10] an approach using autoencoders is proposed that reproduces their input and allows anomalies to be identified in event logs without prior knowledge. The autoencoder can also be trained on a noisy dataset that already contains anomalies.

A stand-alone approach for filtering anomalies in event streams is offered in [12]. It proposes an event processor that allows to effectively filter out unwanted events from an online event stream, based on probabilistic automaton. The basic idea of the approach is that dominant behavior achieves a higher probability of occurrence in the automaton than unlikely behavior. Thus, this filtering technique improves process discovery in real-time process mining.

On the other Side of the considered topic are architecture models, which enable the processing of streaming data. This area also represents a focus of research. The current interest in big data approaches has led to a renewed interest in software architectures. Most of the existing work in this context focuses on the development of processing strategies for batch architectures. However, the combination of batch and streaming requirements has so far been given little consideration. The best known approaches in this context are the lambda and kappa architectures.

Lambda architectures have aroused great interest in both industry and science [13] and have also shown that they pose a number of challenges [14, 15]. There are many studies on approaches whose basic structures are essentially based on the concept of lambda architectures. In [16–18] various approaches based on a lambda architecture that enable near-real-time processing of process data are proposed. Offline (batch processing) as well as online processing (stream processing) of the data enables efficient further processing of the data as well as scalability for various big data tasks. The possibility of online processing, thus the processing of streaming data, enables the desired interaction with the running process and in the context of this work the filtering of erroneous event data in real-time.

In contrast to a lambda architecture, in which both batch and stream processing is used, a kappa architecture [19] is a concept that only allows the processing of streaming data. A kappa architecture differs from a lambda architecture in such a way that the batch layer is omitted, which in turn simplifies the implementation of the concept. On the other hand, this type of architecture is not applicable if a batch processing engine is required. However, this is the case when using process mining techniques to process historical data [18]. For this reason, the use of lambda architecture as the primary design pattern was chosen within the scope of this work.

3 Lambda Architecture

The design pattern on which this work is built is based on lambda architecture. The concept of lambda architecture was first introduced by Marz and Warren [13] and is a real-time architecture used in big data settings. The idea behind the lambda architecture is to create typically two streams of input data, which are processed separately and then recombined again. There are three layers: batch layer, speed layer and serving layer, as shown in Fig. 1. One of the streams is processed with a real-time framework that performs various calculations to obtain relevant information. This real-time component is referred to as the speed layer, where the main objectives are low latency and high throughput. The other stream is processed with a batch framework. Since the processing is distributed, it is possible to manage a very large amount of data. The consistency of the data is guaranteed by the constant recalculation of the entire data record in real-time. This batch component is called the batch layer. Finally, both streams are combined in a serving layer to query the results of the batch layer and speed layer. The speed layer stores data as views to perform real-time analysis. The batch layer stores the data in a stable database structure. Real-time data is regularly replaced by batch data. For example, if all data from the previous day has been completely processed in the batch layer, these results replace real-time data that has already been processed.

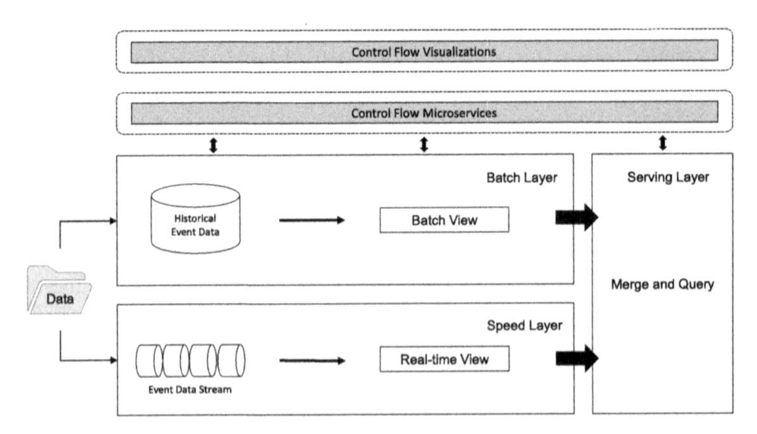

Fig. 1. Lambda architecture with batch layer, speed layer and serving layer

4 Anomaly Detection Using Autoencoders

In the context of this work, an autoencoder in a lambda architecture is used to detect anomalies in a streaming setting. Therefore, the functionality of an autoencoder and the way anomalies are detected in this context are described.

4.1 Autoencoders

An autoencoder basically consists of two components: an *encoder* and a *decoder*. Figure 2 illustrates a conventional autoencoder.

Encoder: The input vectors $\left(x_i \in R^d\right)$ are compressed into m neurons that form the hidden layer. The activation of the neuron i in the hidden layer is given by:

$$h_i = f_\theta(x) = s\left(\sum\nolimits_{j=1}^{n} W_{ij}^{in} x_j + b_i^{in}\right) \tag{1}$$

where x represents the input vector, θ represents the parameters (W^{in}, b^{in}), W represents the encoder weight matrix and b represents the bias vector. To achieve a compression of the input data, the input vector is compressed to a lower dimensional vector.

Decoder: The resulting hidden layer h_i is then decoded back into the original input space R^d. The associated mapping function is as follows:

$$x_i' = g_{\theta'}(h) = s\left(\sum\nolimits_{j=1}^{n} W_{ij}^{hid} h_j + b_i^{hid}\right) \tag{2}$$

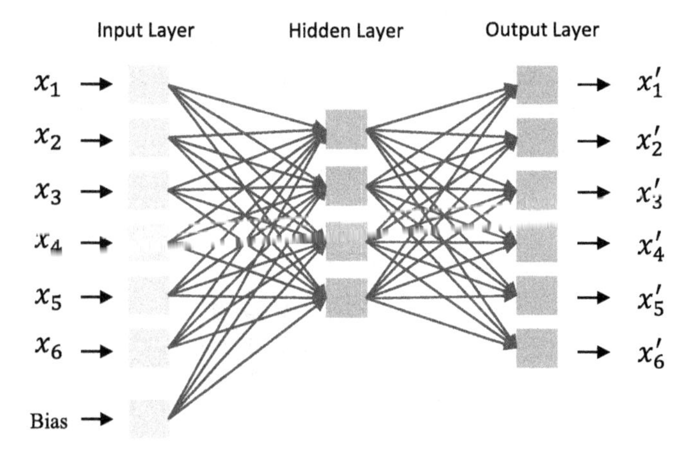

Fig. 2. Architecture of an example autoencoder

The parameter set of the decoder is $\theta' = \{W^{hid}, b^{hid}\}$. The autoencoder is optimized to minimize the average construction error related to θ and θ':

$$\theta^*, \theta'^* = \arg\min_{\theta,\theta'} \frac{1}{n} \sum_{i=1}^{n} \varepsilon\left(x_i, x_i'\right) \tag{3}$$

$$= \arg\min_{\theta,\theta'} \frac{1}{n} \sum_{i=1}^{n} \varepsilon(x_i, g_{\theta'}(f_\theta(x_i))) \tag{4}$$

where ε is the reconstruction error which is defined by:

$$\varepsilon\left(x_i, x_i'\right) = \sum_{j=1}^{d} \left(x_i - x_i'\right)^2 \tag{5}$$

The activation functions f and g should be nonlinear functions to show the nonlinear relationship between the input characteristics. Common activation functions are sigmoid function or rectifier function [20].

The considered autoencoder in this work consists of one input and one output layer with linear units and two hidden layers with rectified linear units.

4.2 General Filter Architecture

Since this work deals with the processing of event streams, a suitable mechanism must be chosen to specify which events are to be passed to the autoencoder for further processing at a certain point in time. Figure 3 illustrates a schematic overview of the anomaly filter architecture. An infinite event stream S is assumed, which contains erroneous data such as incorrect traces, events and other attributes.

Due to the infinity of the stream S, the existence of a finite event window w is assumed, which allows to observe a finite sequence of events at time t. This event window contains new events e that were recently generated. The events contained in the event window w are passed to the autoencoder in the form of batches. These are

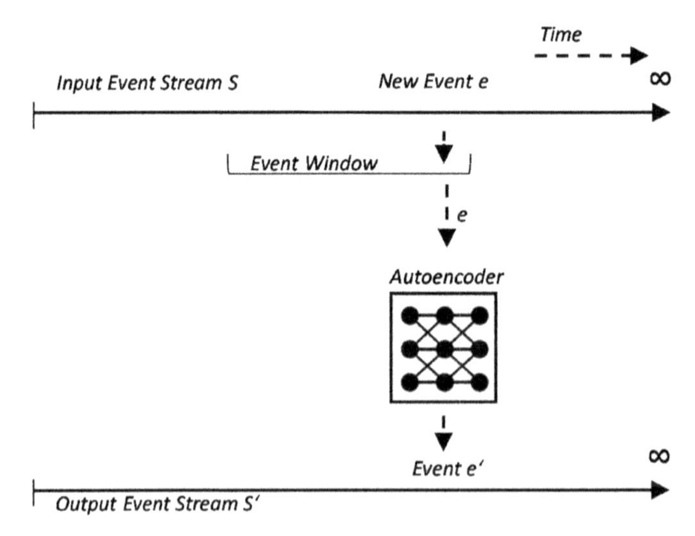

Fig. 3. Schematic architecture of the anomaly filter

then processed and filtered by the autoencoder. In order to decide which events have to leave the event window again, there are a number of stream-based approaches, such as adaptive sliding windows [21] or forward decay methods [22]. For the events received this way, the autoencoder decides which events are correct or incorrect. Incorrect events are discarded and correct events are passed to a filtered event stream S'.

4.3 Method of Anomaly Detection

Since the detection of anomalies is a classification problem, the desired output of a corresponding classification method is a class label. Through the use of neural networks it is possible to get such an output without training the neural network with class labels. A neural network suitable for such a purpose is an autoencoder. Instead of using class labels for training, the original input is used as the target output. To train the autoencoder, the individual activities and users are one-hot encoded in a first step. Each activity is encoded by an n-dimensional vector, where n is the number of individual activities contained in the corresponding training data. This results in a vector for the activities and another vector for the users of each event in a trace. The resulting vectors are then linked to form a single vector. Another way to handle variable size traces is to use n-grams. However, this method complicates the processing of long-term dependencies between events. For such reason one-hot encoding is used for this purpose. To deal with the fixed sized input of a feed-forward network, the vectors have to be pad with zeros to reach the same size as the longest vector in the corresponding training data.

After the mentioned pre-processing steps have been completed the autoencoder can be trained with the backpropagation algorithm [23] using the one-hot encoded event data. Therefore the event data is used both as the input and as the label. In addition, a special noise layer is inserted between the event data and the autoencoder. This noise layer adds Gaussian Noise to the event data. This data is used by the autoencoder as input. The Autoencoder is trained to reproduce its input in order to minimize the mean squared error between the input and its output.

4.4 Classifying Erroneous Data

After the autoencoder has been trained, it can be used to measure the mean squared error between its input and output and thus detect erroneous data. This erroneous data can be either erroneous traces, events or attributes. Therefore it can be assumed that normal traces are reproduced with a lower reproduction error than incorrect traces. To detect an anomaly this way, a certain threshold value τ can be defined. If a certain reproduction error exceeds this threshold, the associated trace can be classified as an anomaly. To define a threshold value, the mean reproduction error over the training data set can be used. The threshold is defined by:

$$\tau = \frac{1}{n}\sum\nolimits_{i=1}^{n} \varepsilon_i \tag{6}$$

where ε_i is the reproduction error for a corresponding trace i and n the number of traces in the data set. This procedure for identifying incorrect traces can be applied analogously to incorrect events and activities by calculating the reproduction error event based.

5 Lambda Architecture-Based Anomaly Filter in an Online Process Mining Setting

After the anomaly detection was described and constructed using an autoencoder, it is now to be adopted into the lambda architecture, as this kind of architecture is well suited for use in an online process mining setting, through its batch and streaming processing. Figure 4 illustrates the embedding of the autoencoder inside the lambda architecture as well as its role in the online process mining context. The figure gives an abstracted overview of the essential architecture components. The batch layer with the provision of historical data and the speed layer with the processing and filtering of streaming data from the real-time process can be recognized. These two layers are merged by the serving layer in which both offline and online process mining take place and lead to the discovery and analysis of process models. The filtered real-time event data used during online process mining is then transferred to the historical data storage which can later be used as historical data for new process mining tasks. The event data used for the process mining activities are generated by human users or digitized and automated processes and extracted and merged from different information systems. After the data has been extracted and collected from the information systems, they can be fed to the corresponding components in a suitable format like the XES format which maintains the compatibility with industry standards. This data serves as raw material for later analyses. The output of the algorithm used in offline process mining (i.e. heuristic miner) is a control flow model. The algorithm used in online process mining produces an update for the process model generated by the offline algorithm. These updates are then passed to the control flow visualization by means of the web sockets technology, that the user is able to see the corresponding updates of the model. In addition, a control flow

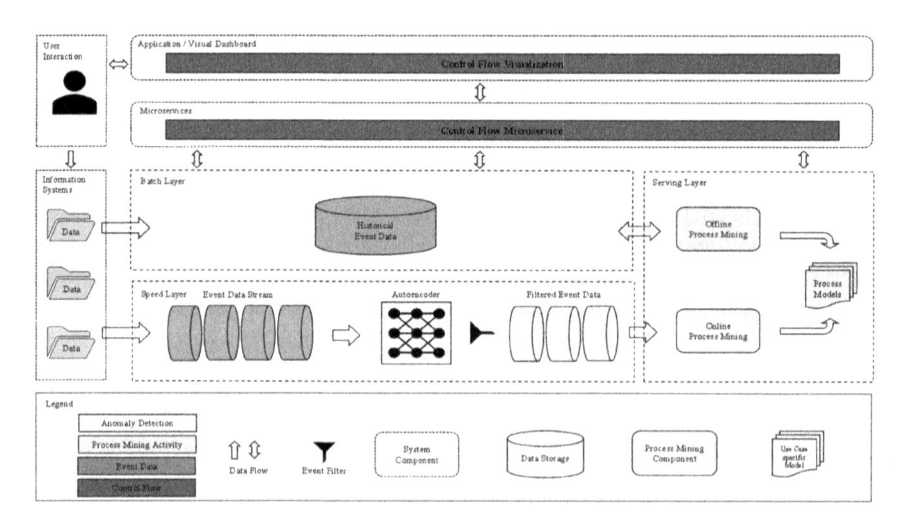

Fig. 4. Integrated autoencoder for filtering anomalies in online process mining

microservice layer is used. This enables a separation of functionalities and simplifies both the implementation and maintenance of the system.

This type of architecture and data processing also poses challenges, in particular by including neural networks, like an autoencoder, in the processing of streaming data. This leads to major problems due to the characteristics of such big data activities [25]. Especially the aspect of velocity brings challenges in a big data streaming environment. Data availability plays a critical role in this environment. Underlying models can change, making the common process of training, testing and prediction more difficult. Furthermore, the use of neural networks in a real-time setting complicates the training of the models. Here it is necessary to reduce the training time drastically due to the short decision times. One more important fact is that an online learning model is not horizontally scalable and cannot be instantiated to any number of instances due to the quickly changing models. This is challenging because this important part, the model, is only vertically scalable, thus a single model instance.

6 Experiment

6.1 Event Data

A preliminary implementation of the autoencoder is evaluated on the public available BPIC19 [26] dataset. BPIC19 consists of over 1.5 million events for purchase orders. The data shows the purchase to pay process.

6.2 Anomalous Event Data

For our experiment erroneous event data was added to the data set. For this purpose, we generated random sequences that correspond to the format of the BPIC19 data set.

6.3 Experimental Setup

Simulating a rapid deployment process, no hyper-parameter tuning was performed. In an first preliminary implementation the autoencoder is trained on batches of size 128 for 100 epochs. This allows an earlier stop if the loss of the validation set did not decrease within the last 10 epochs. Furthermore, the Adam Optimizer [24] was used. The learning rate for the model was initially set and fixed to 0.001. The Autoencoder has two hidden layers with 128 hidden units in the first and 64 hidden units in the second layer.

6.4 Results

Figure 5 shows the confusion matrix of the preliminary conducted experiment. The confusion matrix provides an overview of both the accuracy and the distribution of incorrect predictions.

Predicted	Actual	
	Anomaly	No anomaly
Anomaly	24	11
No anomaly	6	19

Fig. 5. Confusion matrix of the conducted experiment

Table 1. Confusion metrics

Accuracy	Misclassification	Precision	Recall	F1	Specificity
71,6%	28,3%	68,5%	80,0%	73,0%	63,3%

Table 1 lists the corresponding confusion metrics of the experiment. The accuracy of the preliminary model reaches 71,6%. On the other hand, the misclassification amounts to 28.3%. Furthermore, Precision (68.5%), Recall (80%), F1-Measure (73%) and Specificity (63.3%) of the model were measured.

7 Conclusion and Future Work

In this paper a lambda architecture using autoencoder for anomaly detection in an online process mining setting was presented. Therefore the preliminary model of an autoencoder was described that is used to reconstruct its input as output and calculates the mean squared error between input and output to detect incorrect data. The autoencoder as well as the filter architecture are embedded in a lambda architecture to support the processing of streaming data as well as the processing of historical data for process mining purposes. The conducted experiment showed first successful results in detecting anomalies in event data. This construct of technologies is supposed to be used in a real time environment for operational support to reach a cleansed event stream for better process mining results. Therefore future work should include the further development of the autoencoder for the processing of event streams instead of batches. In particular, the focus is on handling streaming data as well as updating and retraining the model with fast changing data. This includes the technical infrastructure of the implementation, like the usage of the right data processing pipelines, as well as the vertical scalability of an online learning model.

Acknowledgements. This work was funded by the German Federal Ministry of Education and Research within the project Competence Center for Scalable Data Services and Solutions (ScaDS) Dresden/Leipzig (BMBF 01IS14014B).

References

1. van der Aalst, W.: Process Mining: Data Science in Action, 2nd edn. Springer, Berlin (2016). https://doi.org/10.1007/978-3-662-49851-4
2. van der Aalst, W., Weijters, T., Maruster, L.: Workflow mining: discovering process models from event logs. IEEE Trans. Knowl. Data Eng. **16**(9), 1128–1142 (2004)
3. Leemans, S.J.J., Fahland, D., van der Aalst, W.M.P.: Discovering block-structured process models from event logs - a constructive approach. In: Colom, J.-M., Desel, J. (eds.) PETRI NETS 2013. LNCS, vol. 7927, pp. 311–329. Springer, Heidelberg (2013). https://doi.org/10.1007/978-3-642-38697-8_17
4. Hassani, M., Sicca, S., Richter, F., Seidl, T.: Efficient process discovery from event streams using sequential pattern matching. In: IEEE Proceedings of SSCI 2015 (2015)
5. Maaradji, A., Dumas, M., La Rosa, M., Ostovar, A.: Detecting sudden and gradual drifts in business processes from execution traces. IEEE Trans. Knowl. Data Eng. **29**(10), 2140–2154 (2017)
6. Ostovar, A., Maaradji, A., La Rosa, M., ter Hofstede, A.H.M., van Dongen, B.F.V.: Detecting drift from event streams of unpredictable business processes. In: Comyn-Wattiau, I., Tanaka, K., Song, I.-Y., Yamamoto, S., Saeki, M. (eds.) ER 2016. LNCS, vol. 9974, pp. 330–346. Springer, Cham (2016). https://doi.org/10.1007/978-3-319-46397-1_26
7. Ghionna, L., Greco, G., Guzzo, A., Pontieri, L.: Outlier detection techniques for process mining applications. In: An, A., Matwin, S., Raś, Z.W., Ślęzak, D. (eds.) ISMIS 2008. LNCS (LNAI), vol. 4994, pp. 150–159. Springer, Heidelberg (2008). https://doi.org/10.1007/978-3-540-68123-6_17
8. Wang, J., Song, S., Lin, X., Zhu, X., Pei, J.: Cleaning structured event logs: a graph repair approach. In: 31st IEEE International Conference on Data Engineering (2015)
9. Conforti, R., La Rosa, M., ter Hofstede, A.H.M.: Filtering out infrequent behavior from business process event logs. IEEE Trans. Knowl. Data Eng. **29**(2), 300–314 (2017)
10. Nolle, T., Luettgen, S., Seeliger, A., Mühlhäuser, M.: Analyzing business process anomalies using autoencoders. Mach. Learn. **107**(11), 1875–1893 (2018). https://doi.org/10.1007/s10994-018-5702-8
11. Sani, M.F., van Zelst, S.J., van der Aalst, W.M.P.: Improving process discovery results by filtering outliers using conditional behavioural probabilities. In: Teniente, E., Weidlich, M. (eds.) BPM 2017. LNBIP, vol. 308, pp. 216–229. Springer, Cham (2018). https://doi.org/10.1007/978-3-319-74030-0_16

12. van Zelst, S.J., Fani Sani, M., Ostovar, A., Conforti, R., La Rosa, M.: Filtering spurious events from event streams of business processes. In: Krogstie, J., Reijers, H.A. (eds.) CAiSE 2018. LNCS, vol. 10816, pp. 35–52. Springer, Cham (2018). https://doi.org/10.1007/978-3-319-91563-0_3
13. Marz, N., Warren, J.: Big Data. Principles and Best Practices of Scalable Realtime Data Systems, Manning (2013)
14. Hoseinyfarahabady, M., Taheri, J., Tari, Z., Zomaya, A.Y.: A dynamic resource controller for a lambda architecture. In: IEEE Proceedings of the International Conference on Parallel Processing (2017)
15. Kiran, M., Murphy, P., Monga, I., Dugan, J., Baveja, S.S.: Lambda architecture for cost-effective batch and speed big data processing. In: IEEE Proceedings International Conference on Big Data (2015)
16. Batyuk, A., Voityshyn, V.: Streaming process discovery for lambda architecture-based process monitoring platform. In: IEEE 8th International Conference on Computer Science and Information Technology (2018)
17. Batyuk, A., Voityshyn, V.: Apache storm based on topology for real-time processing of streaming data from social media. In: IEEE First International Conference on Data Stream Mining & Processing (2016)
18. Batyuk, A., Voityshyn, V., Verhun, V.: Software architecture design of the real-time processes monitoring platform. In: IEEE Second International Conference on Data Stream Mining & Processing (2018)
19. Kreps, J.: Questioning the Lambda Architecture. O'Reilly.com (2014). https://www.oreilly.com/radar/questioning-the-lambda-architecture. Accessed 27 Feb 2020
20. Nair, V., Hinton, G.E.: Rectified linear units improve restricted Boltzmann machines. In: Proceedings of the 27th International Conference on Machine Learning. Association for Computing Machinery (2010)
21. Bifet, A., Gavalda, R.: Learning from time-changing data with adaptive windowing. In: Proceedings of the SDM. SIAM (2007)
22. Cormode, G., Shkapenyuk, V., Srivastava, D., Xu, B.: Forward decay: a practical time decay model for streaming systems. In: IEEE Proceedings of the ICDE (2009)
23. Rumelhart, D., Hinton, G., Williams, R.: Learning representations by back-propagating errors. Nature **323**, 533–536 (1986)
24. Kingma, D.P., Ba, J.: Adam: a method for stochastic optimization. arXiv preprint arXiv:abs/1412.6980 (2014)
25. Augenstein, C., Spangenberg, N., Franczyk, B.: Applying machine learning to big data streams: an overview of challenges. In: 4th IEEE International Conference on Soft Computing and Machine Intelligence (2017)
26. van Dongen, B.F.: Dataset BPI challenge 2019. 4TU. Centre for Research Data (2019). https://doi.org/10.4121/uuid:d06aff4b-79f0-45e6-8ec8-e19730c248f1

Data Driven IoT for Smart Society

Biomedical Text Recognition Using Convolutional Neural Networks: Content Based Deep Learning

Sisir Joshi[1], Abeer Alsadoon[1], S. M. N. Arosha Senanayake[2], P. W. C. Prasad[1(✉)], Abdul Ghani Naim[2], and Amr Elchouemi[3]

[1] Charles Sturt University, 63 Oxford Street, Darlinghurst, NSW 2010, Australia
cwithana@studygroup.com, aalsadoon@studyugroup.com, pwcp1@yahoo.com
[2] Universiti Brunei Darussalam, Gadong BE 1410, Brunei Darussalam
{arosha.senanayake,ghani.naim}@ubd.edu.bn
[3] American Public University System, Charles Town, USA
Amr.elchouemi@mycampus.apus.edu

Abstract. Named Entity Recognition (NER) targets to automatically detect the drug and disease mentions from biomedical texts and is fundamental step in the biomedical text mining. Although deep learning has been successfully implemented, the accuracy and processing time are still major issues preventing it from achieving NMR. This research aims to upgrade the accuracy of classification while decreasing the processing time, by paying more attention to significant areas of NMR. The novel proposed system consists of a Bi-Directional Long Short-Term Memory with Conditional Random Field (BiLSTM-CRF) using dropout strategy to effectively prevent overfitting and enhancing the generalization abilities. The system built includes the attention mechanism and attention fusion for redistributing the weight of samples belonging to each class in order to compensate the problem occurring from data imbalance and to focus only on the critical areas of the observed things and ignoring non-critical areas.

Keywords: Convolutional neural networks · Biomedical text · Text mining · Named entity recognition · Bi-LSTM (Bi-Directional Long Short-Term Memory)

1 Introduction

There has been significant advancements for promoting the performances of chemical NER (Named Entity Recognition). The traditional methods for identifying the biomedical text mentions makes use of various natural language processing (NLP) tools and knowledge resources which are still labour-intensive and skill-dependent task [1]. Several neural network architectures have been proposed for NER task [1, 2]. However, it still requires feature engineering [3]. The latest technology used is the Bi-Directional Long Short-Term Memory with a conditional random field layer (BiLSTM-CRF) model. This model jointly with the other traditional model [1] treats each sentence as a separate

M. Hernes et al. (Eds.): ICCCI 2020, CCIS 1287, pp. 593–604, 2020.
https://doi.org/10.1007/978-3-030-63119-2_48

document and multiple instances of the same taken in different sentences of a document are viewed as independent tagging problems. However, this will lead to tagging inconsistency problem. To solve this issue, an attention-based bidirectional Long Short-Term Memory with conditional random field (Att-BiLSTM-CRF) is used to deal with the tagging inconsistency problems [3].

The deep learning methods have proven its effectiveness in the field of NLP [4, 5]. The BiLSTM-CRF model [6] is a popular and effective deep learning technique for automatically extraction of features including word embedding and character embedding [3]. Compared to other deep learning methods, this model extracts local features in more appropriate way and utilize the contextual information that are futuristic. Forward LSTM processes the input sequence towards the forward direction, whereas Backward LSTM processes the input towards the reverse direction. In addition, CRF aids in capturing the correlations [4]. Bi-LSTM-CRF along with attention mechanism has been adopted successfully in solving NLP problems. It will take document-level global information gained by attention mechanism and resolves the tagging inconsistency problem [3]. It comprises pre-processing stage, feature extraction and classification [5]. However, Still the features like knowledge-based features, character-level features, part-of-speech (POS) features are not considered. The calculation still needs to be improved because of the nesting entities. The output tags have strong restrictions and dependencies.

Current BiLSTM models for identification of named entity use various techniques and algorithms to increase the classification accuracy and lower the processing time. The experiment result of the state-of-art work by [3] shows it achieves F1 score of 73.50 and outperforms the traditional machine learning methods based on the result obtained from embedding dimensions, LSTM units' dimensions, optimization methods and other existing models [2]. The BI-LSTM used capturing of two different feature representations including forward and backward information of sentences [7]. Furthermore, both forward and backward information are merged to gain the hidden layer feature representation and fed to the attention layer. It helps in digging more hidden features. Attention layer helps to focus the model on significant and critical areas rather than other non-critical areas [5]. In addition, the use of the dropout algorithm minimizes overfitting problem and performs better than the traditional Convolution Neural Networks (CNN) and LSTM models [4]. However, the classification accuracy and processing time are affected by the sigmoid activation function which can cause gradient saturation problem [3, 5]. Also, there is ambiguity arose in English words, and misclassifying coordinating junctions. Furthermore, unbalanced datasets are created in relation identification stage. As a result, the improvement is still required.

In the paper [7] addressed the first annotated corpus for supporting of annotation of disabilities and diseases as well as relationship between them. They used the annotated corpus to train the deep learning systems for mentions recognition and relation classification. The evaluation criteria is based on F1-Score that scored 79.13% for disability recognition when word embedding was used. Furthermore, with the addition of case embedding the F1-Score is improved to 81.11%.

In the article [8] developed a model for biomedical Named entity recognition that retrieved the local context information based on n-gram character and word embeddings with the usage of CNN. The model was applied on 3 different datasets making it versatile and not requiring any feature engineering and hand-crafted features. The evaluation was done based on the F1-Score and achieved 87.26%, 87.26% and 72.57% on three datasets BC2, NCBI and JNLPBA respectively [9, 10].

The aim of this research is to improve the classification accuracy by identifying the chemical entity NMR. This work proposes bidirectional long short-term memory with conditional random field (Att-BiLSTM-CRF) algorithm for NMR. The approach is dropout strategy to effectively prevent overfitting and enhance the generalization ability [5]. Moreover, combination of BiLSTM and CRF creates a faster convergence speed and obtain optimal labelling sequence by considering relations between adjacent tags.

2 State of the Art

Good features are highlighted inside the blue broken line in Fig. 1, and the limitations are highlighted inside the red broken line in Fig. 1. [5] proposed a BiLSTM-CRF model that enhanced the dictionary-based and rule based and conventional machine learning method in BNER. He adopted BiLSTM to gain the bidirectional text context information and then developed different attention weight redistribution methods and later fused them to improve the ability of BiLSTM to pay more attention to significant areas only. This solution applies dropout strategy to effectively prevent overfitting and enhance the generalization ability [5]. Therefore, the overall classification accuracy is improved. The use of CRF can better capture the semantic features from text automatically, without many manually labelled data. The BiLSTM model helps to obtain the bidirectional context information and dig more feature effectively [5]. Moreover, the combination of BiLSTM and CRF creates a faster convergence speed and obtain optimal labelling sequence by considering relation between adjacent tags. Compared with the traditional methods, this solution provides more robust and more precise text entity classification on the JNLPBA corpus, with F1-Score of 73.50%. This model consists of three main stages (see Fig. 1): 1) Pre-processing; 2) Feature Extraction; and 3) Classification. But, the **attention-based BiLSTM-CRF model** has 4 layers namely Input (Feature and Embedding), BiLSTM, Attention and CRF [5].

Given an input sequence x = (x1, x2....xt), where t = sequence length.

LSTM hidden state at timestep t is calculated by

$$i_t = \sigma\left(W_i\left[h_{t-1}, x_t\right] + b_i\right) \tag{1}$$

$$f_t = \sigma\left(W_f\left[h_{t-1}, x_t\right] + b_f\right) \tag{2}$$

$$o_t = \sigma\left(W_o\left[h_{t-1}, x_t\right] + b_o\right) \tag{3}$$

$$c_T = \tanh\left(W_c\left[h_{t-1}, x_t\right] + b_c\right) \tag{4}$$

$$c_t = i_t * c_T + f * c_{t-1} \tag{5}$$

$$h_t = o_t * \tanh(c_t) \quad (6)$$

Where,

σ represents the sigmoid activation function.
$\boldsymbol{tanh}$ represents hyperbolic tangent function.
$\boldsymbol{x_t}$ represents the unit input.
$\boldsymbol{i_t}, \boldsymbol{f_t}, \boldsymbol{o_t}$ represent the input gate, forget gate, and output gate at time t
$\boldsymbol{W}$ ***and b*** represent the weights and bias of the input gate, the forget gate, and the output gate respectively.
$\boldsymbol{c_T}$ denotes the current state of the input.
$\boldsymbol{h_t}$ denotes the output at time t.

3 Proposed System

After analysing various classification methods and feature extraction using deep learning for named entity recognition (NER), the advantage and disadvantage of each method are analysed. Based on the analysis, factors affecting the performance of the NER were the word embedding and the character embedding, SoftMax function, loss function, and weight in terms of attention mechanism. From the list of collected classification methods, the solution proposed by [5] is designated as the base for the proposed solution. [5] proposed Dropout regularized Convolutional Neural Network (D-CNN) for preventing the overfitting. Because of overfitting, it is hard for the neural network for generalizing the unknown data and resulting in low performance [3]. Various parameters that are being participated in the training are randomly selected with the implementation of the dropout. Furthermore, the use of LSTM helps to solve the problem of gradient disappearance in RNN by realizing the longer distance information in the sentence using sigmoid activation function and hyperbolic tangent function, and thus, helps to selectively save the context information that will suit to solve the problem for NER i.e. Sequence Labelling. However, there are still some limitations in this work that includes the sigmoid activation function that is prone to the gradient saturation problem [6] which will impact the overall classification accuracy and delay the convergence speed [3]. The input layer has limitation during the embedding stage because it omits other features like knowledge-based features, character-level features, part-of-speech (POS) features before being fed to BiLSTM layer. Attention layer calculation still needs to be improved because of the nesting entities. The output from the CRF layer have strong restrictions and dependencies. To overcome the feature extraction problem, the proposed solution introduces a character level embedding along with other feature embedding function inspired by [3]. Another new feature of this solution is the calculation of the attention mechanism that considers the local attention adapted from the work by [3]. These two new features minimize the drawbacks of the current solution and classification accuracy is improved along with the convergence speed.

The **attention-based BiLSTM-CRF model** has 4 layers namely Input (Feature and Embedding), BiLSTM, Attention and CRF [5].

Input Layer: The input layer is responsible for taking input of each word as the discrete features. Before that, pre-processing of the datasets is performed using tokenization which cut down large chunk of text into sentences and cuts down sentences into list of single word called tokens. Next, pre-trained vectors dictionary is used for searching the original data so that the word would be represented as discrete feature that will helps to express the vocabulary relationship and hence can be input to the embedding layer which is shown in Fig. 1. In the embedding layer, word embedding is done for the pre-processed datasets in which individual word are represented as real-valued vectors that allows to capture syntactic (grammatical structures) and semantic (meaning) representation of the words. For the word embedding, Sandford's publicly available Glove 100-dimensional embedding is used that are trained on 6 billion words from Wikipedia and web texts. In addition, the character level embedding, and feature embedding are also incorporated in the input layer.

BILSTM Layer for Feature Extraction: Basically, it is LSTM (Long Short-Term Memory), but to perform in both forward and backward direction it is named as Bi-directional LSTM (BILSTM). The use of LSTM helps to solve the problem of gradient disappearance in RNN by realizing the longer distance information in the sentence using sigmoid activation function and hyperbolic tangent function, and thus, helps to selectively save the context information that will suit to solve the problem for Named Entity Recognition (NER) i.e. Sequence Labelling. The output from the embedding layer is fed to Bi-LSTM layer where two different feature representations including forward, and backward information of sentences are captured using Forward LSTM and Backward LSTM respectively as shown in the figure. This will help to obtain the bidirectional context information and dig more hidden features.

Attention Layer: It is further divided into two parts: **Attention Mechanism** and **Attention Fusion.**

Attention mechanism is responsible for redistributing the weight for the output of BiLSTM by focusing only on critical areas of the observed things and ignoring the non-critical area. The attention mechanism performs different attention weight redistribution calculation methods for quantifying each word importance within the given sentence. To deal with the attention mechanism problem for state of art solution, local attention mechanism is considered that chooses to focus only on small subset of the source position and removing the drawback from state of art solution which was "to attend all words on the source side for each target word what is expensive and impossible to translate the longer sequence".

Attention Fusion: The output from the attention mechanism will be based on different calculation methods as stated above resulting in different probability matrices P which will be obtained by linear mapping. However, multiple attention matrix will cause information redundancy. To overcome this issue, the similarity of two probability matrix is calculated separately in the attention fusion and then two matrices with most similarity is merged. The calculation is done by adding two most similar matrices and then it is input to the CRF layer.

CRF Layer for Classification: The final output from BiLSTM after undergoing through attention mechanism and fusion will be a probabilistic matrix P representing similar feature. The CRF layer is responsible for the classification and giving output the labelling result L. The size of the P will be the product of number of words and number of class levels. The probability for a prediction sequence L is calculated and later SoftMax function is used to generate all the possible tags sequence L. For training the dataset, maximum conditional likelihood estimates are used. Finally, the output tags sequence with the largest probability is calculated at the decoding phase using Viterbi algorithm. In addition, during the training Dropout is used that prevents the overfitting of the neural network and disable some neurons with certain probability making the model better in terms of generalization.

The attention weight mechanism imitates as the brain that focuses only on significant areas rather than the global areas. The score function used is also known as alignment function for which four alternatives are defined namely Manhattan distance, Euclidean distance, cosine distance and perceptron. The score value of cosine distance and perceptron are larger when the two vectors x_t and x_j are more similar whereas, the score value of Manhattan distance and Euclidean distance are smaller when the two vectors are more similar. The final score is calculated by the maximum scores minus the scores to make their final scores larger when the vectors are more similar. The proposed solution adds the alignment function along with the enhancement in the attention mechanism that given more emphasis on the local areas.

The weight calculation formula for the attention mechanism according to [5]

$$\alpha_i = \frac{\exp(e_i)}{\sum_{k=1}^{n} \exp(e_k)} \tag{7}$$

Where,

α_i = normalized processing results of e_i
e_i = represents the energy function for quantifying word sentence and is calculated by $e_i = \tanh\left(W^T H_i + b\right)$
W represents the weight parameter of the hidden state H_i

The proposed modified solution for the attention weight mechanism consist of the alignment function in the Eq. 1 and is given by:

$$M\alpha_i = \frac{\exp(score(e_i))}{\sum_{k=1}^{n} \exp(\mathrm{score}(e_k))} \tag{8}$$

Thus, the score is a content-based function and has four alternatives Manhattan distance, Euclidean distance, cosine distance and perceptron) according to [3],

$$core\left(x_t, x_j\right) = \left\{ \begin{matrix} W_a\left|x_t - x_j\right| \\ W_a\left(x_t - x_j\right)^T\left(x_t - x_j\right) \\ \frac{W_a\left(x_t . x_j\right)}{\left|x_t\right|\left|x_j\right|} \\ \tanh\left(W_a\left[x_t; x_j\right]\right) \end{matrix} \right\} \tag{9}$$

W_a is the weight matrix and parameter of the model.

The modified attention weight mechanism is now enhanced by adding the Gaussian distribution equation that gives more emphasis on the local attention mechanism rather than the global attention mechanism and is given by:

$$\alpha_{tj} = \frac{\exp(score(x_t, x_j))}{\sum_k \exp(score(x_t, x_k))} \exp\left(-\frac{(s - p_t)^2}{2\sigma^2}\right) \quad (10)$$

Where,

$p_t = \text{S. sigmoid}\left(v_p^T \tanh(W_p x_t)\right)$
S = source sentence length
W_p *and* v_p are model parameters
$\sigma = \frac{D}{2}$, where D is the sentence boundary

The enhanced equation is added with the Gaussian distribution in the Eq. (4) and is given by

$$E\alpha_i = \frac{\exp(score(e_i))}{\sum_{k=1}^{n} \exp(\text{score}(e_k))} \exp\left(-\frac{(s - p_t)^2}{2\sigma^2}\right) \quad (11)$$

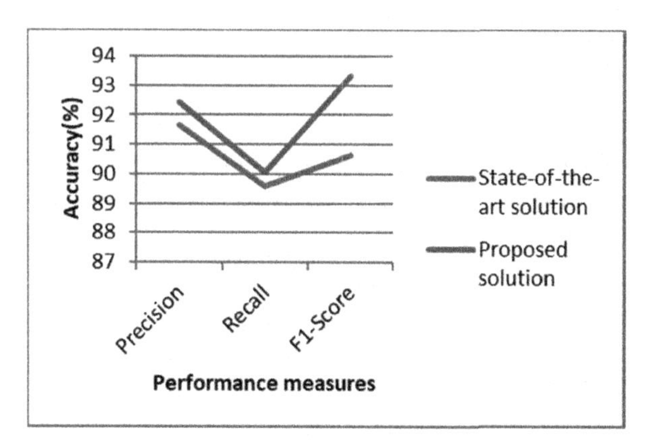

Fig. 1. Accuracy on the dataset of the JNLPBA corpus

4 Results

Python 3.7 is used with the libraries; Keras, Tensorflow, NLTK, Numpy and Anaconda is used for IDE. Tensorflow environment has been activated using CPU integrated graphics. The freely available datasets are used from JNLPBA corpus, CDR corpus and CHEMDNER corpus. Therefore, three datasets in total are used and which vary in the number of samples and data balance degree, which properties are listed in Table 3 and 4. Four-fold cross-validation is used, where the training data is randomly divided into four equal sized sets and one is used for testing and the rest for the training [5]. The entities for the JNLPBA corpus are listed in Table 1. Similarly, the effect of additional feature on the performance of CHEMDNER corpus is listed in Table 2. The model is trained with the following hyperparameter setting of dropout rate of 0.5, initial learning rate of 0.001, gradient clipping of 5, optimizer used is Adam and batch size of 20. The labelling method in NER is BIO method where B(Beginning) indicates the word is a named entity start. I(Inside) indicates that the word is in a named entity, and O(Outside) indicates that the word is outside the named entity. For the CHEMDNER corpus, the four-alignment function are used, and their performances measure are mentioned. In conclusion, three tasks are performed including recognition of named entities in text documents, evaluating pre-trained models with test dataset, training new models with given corpora. For the different three datasets, Metric() method from Python Keras library is used to compute the accuracy and processing time. The average accuracy and processing time are calculated by using the Mean () method from Python Numpy library.

Table 1. Statistics of the three datasets.

Data set name	Number of abstract	Number of sentences	Number of words	Average (words)
JNLPBA	2000	20546	472006	623
CDR corpus	10000	23852	124892	224
CHMED NER corpus	3000	9812	123651	216

JNLP BA	DNA	RNA	Cell type	Cell Line	Pro-tein	Total
Train ing	9534	951	6718	3830	3026 9	5130
Test	1056	118	1921	500	5067	8662

The results obtained from our proposed solution are compared with the state-of-art solution. The results are obtained based on two aspects: 1) based on the three scenarios (i.e. small dataset, large dataset, and unbalanced dataset); 2) based on the different types of embeddings, 3) other results are obtained based on the alignment function for the CHEMDNER corpus. The various stage consists of the following module when executed in python: Backend, Evaluation: To evaluate result of NER, Activations.py: Activations functions, Initilazations.py:, Loader.py: Loads the dataset, Model.py: build the model, Nn.py: the layers of the network architecture, Optimization.py: optimization method, Utils.py:, Train.py: train a basic BiLSTM-CRF model, AttenTrain.py: train a

Att-BiLSTM-CRF model, Tagger.py: tag a document using Bi-LSTM-CRF model and AttenTrain.py: Tag a document using the Att-BiLSTM-CRF model.

Table 2. Feature extraction process including various word dimensions on the JLPBA corpus.

Dimensions	Precision	Recall	F1 score
Dimension-50	69.02	76.00	72.34
Dimension-100	71.57	75.55	73.50
Dimension-200	70.82	75.26	72.97
Dimension-300	70.36	75.87	73.01

Table 3. Feature embedding impact on BC5-CDR corpus which includes the various embedding like (character, word, token-level, POS embedding).

Model	Chemical			Disease		
	Precision	Recall	F1-score	Precision	Recall	F1-score
All embedding	93.73	92.56	93.14	83.98	85.40	84.68
Only one embedding	89.02	86.69	88.03	80.39	79.37	79.87

The results were obtained in the classification stage of the convolutional neural network. The results are analyzed based on the three aspects for all the corpus mentioned above. Based on the number of features considered in the embedding stage, on the use of the alignment function and based on the use of gaussian function there performances measures which are precision, recall and F1 score, are calculated. The state of art solution in the Table 3 by [5] shows that the Bi-LSTM model for the 5 entities of the JNLPBA corpus where the F1-score is 72.62 which is later improved by addition of attention mechanism increasing the overall F1-score of 73.50. Similarly, Table 4 shows the word embedding process of the JNLPBA corpus where the word dimensions are divided into certain values and performance measure are considered.

The result shows that the word dimension of 100 has highest F1-score of 73.50. Next, Table 5 shows the result of the BC5CDR corpus where we have considered two entities a) Chemical and b) Disease. The performance measure are based on the number of feature included in the embedding process which shows that both the chemical and disease entity have higher F1-score of 93.14 and 84.68 by considering all the embedding (Character, Word, Token-level, POS embedding) as compared to the entities where only character embedding is considered having low F1-score of 88.03 and 79.87. Similarly, for the CHEMDNER corpus we have considered the performance measure based on the alignment function and gaussian function. The proposed solution has improved the classification accuracy and processing time by using Eq. (6) to reduce the risk of gradient saturation. The alignment function helps to consider more features and the accuracy is further enhanced with the help of Eq. (9) the gaussian function will help to give local attention increasing the F1-score. And the to reduce the bias in the dataset. This will improve the automatic analysis of the core biomedical entities reducing the cost.

Table 4. Feature embedding impact on CHEMDNER Corpus which includes the various embedding like (character, word, token-level, POS embedding).

Feature consideration in CHEMDNER corpus	Precision	Recall	F-score
One feature embedding	91.31	87.73	89.48
All feature embedding	91.49	89.17	90.31

Table 5. Alignment function impact on the CHEMDNER corpus.

Performance of attention-based model	Precision	Recall	F-score
With alignment function	91.65	90.04	91.23
Without alignment function	91.59	89.67	90.62

5 Discussion

The results show the improvement in classification accuracy and F1 score of the proposed solution compared to the state-of-art solution based on the named entity recognition using alignment function and gaussian function on the attention weight mechanism. The F1 score of 92.44% which is 0.79% more higher than the alignment function [3]. The experimental results show that (i) our attention mechanism that is introduced to capture the document-level correlation information between words has been proved to be effective to alleviate the tagging inconsistency problem; and (ii) our Att-BiLSTM-CRF model is more robust and it is less affected by the removal of manually designed features. Owing to these two advantages, it can still achieve the state-of-the-art performance on the CHEMDNER and CDR corpora with only word and character embeddings (90.84 and 91.96% in F-score, respectively) (Table 6).

Table 6. Gaussian function impact on the CHEMDNER corpus.

Performance of attention-based model CHEMDNER corpus	Precisions	Recall	F-score
With gaussian function	92.44	89.60	93.32
Without gaussian function	91.65	90.04	90.84

The performance measure is calculated using Eq. (12), Eq. (13) and Eq. (14) [5]:

$$\text{Precision(P)} = \frac{True\ Positive(TP)}{True\ Positive(TP) + False\ Positive(FP)} \tag{12}$$

$$\text{Recall(R)} = \frac{True\ Positive(TP)}{True\ Positive(TP) + False\ Negative(FN)} \tag{13}$$

$$\text{F-Score(F1)} = \frac{2 * Precision * Recall}{Precision + Recall} \tag{14}$$

Table 5 shows the result of the BC5CDR corpus where we have considered two entities a) Chemical and b) Disease. The performance measure are based on the number of feature included in the embedding process which shows that both the chemical and disease entity have higher F1-score of 93.14 and 84.68 by considering all the embedding (Character, Word, Token-level, POS embedding) as compared to the entities where only character embedding is considered having low F1-score of 88.03 and 79.87. Similarly, for the CHEMDNER corpus we have considered the performance measure based on the alignment function and gaussian function which shows increase in F1 score from 91.65 to 92.44.

6 Conclusion

Deep learning approaches have been utilized for the named entity recognition however, there still exist limitations in accuracy and processing time due to the external hardware configuration. The classification accuracy and processing time in named entity recognition using the content-based function and the Gaussian function that focuses only on the local critical areas considering all the features in the feature extraction process. The use of Gaussian function in the attention weight alignment function will give priority to the local areas. The use of Gaussian function will focus only on the small subset of the source position per target word. The purposed solution increases the F1 score by considering more than one feature and the Gaussian function for the corpus. The attention mechanism for considering only the local areas makes use of the gaussian function that will concentrate the model only on the local significant areas rather than the global significant areas.

The proposed solution makes use of the content-based function that helps to grasp more feature in the feature embedding process making the higher F1 score. The adoption of BiLSTM along with the attention weight redistribution methods and fusing them improves the ability of BiLSTM to pay more attention in the significant areas. The score function used is also known as alignment function for which four alternatives are defined namely Manhattan distance, Euclidean distance, cosine distance and perceptron.

The attention mechanism along with the content-based function considers the other features like character embedding and provides higher classification accuracy and faster convergence speed.

References

1. Li, L., Jin, L., Jiang, Z., Song, D., Huang, D.: Biomedical named entity recognition based on extended recurrent neural networks. In: 2015 IEEE International Conference on Bioinformatics and Biomedicine (BIBM), pp. 649–652 (2015)

2. Lample, G., Ballesteros, M., Subramanian, S., Kawakami, K., Dyer, C.: Neural architectures for named entity recognition. arXiv preprint arXiv:1603.01360 (2016)
3. Luo, L.: An attention-based BiLSTM-CRF approach to document-level chemical named entity recognition. Bioinformatics **34**, 1381–1388 (2018)
4. Dandala, B., Joopudi, V., Devarakonda, M.: Adverse drug events detection in clinical notes by jointly modeling entities and relations using neural networks. Drug Saf. **42**(1), 135–146 (2019). https://doi.org/10.1007/s40264-018-0764-x
5. Wei, H.: Named entity recognition from biomedical texts using a fusion attention-based BiLSTM-CRF. IEEE Access **7**, 73627–73636 (2019)
6. Dang, T.H., Le, H.Q., Nguyen, T.M., Vu, S.T., Wren, J.: D3NER: biomedical named entity recognition using CRF-biLSTM improved with fine-tuned embeddings of various linguistic information. Bioinformatics **34**(20), 3539–3546 (2018)
7. Fabregat, H., Araujo, L., Martinez-Romo, J.: Deep neural models for extracting entities and relationships in the new RDD corpus relating disabilities and rare diseases. Comput. Methods Prog. Biomed. **164**, 121–129 (2018)
8. Zhu, Q., Li, X., Consea, A., Pereira, C.: GRAM-CNN: a deep learning approach with local context for named entity recognition in biomedical text. Bioinformatics **34**(9), 1547–1554 (2018)
9. Kim, J., Ko, Y., Seo, J.: A bootstrapping approach with CRF and deep learning models for improving the biomedical named entity recognition in multi-domains. IEEE Access **7**(99), 70308–70318 (2019)
10. Cho, H., Lee, H.: Biomedical named entity recognition using deep neural networks with contextual information. BMC Bioinform. **20** (2019). Article number: 735. https://doi.org/10.1186/s12859-019-3321-4

Pattern Mining Predictor System for Road Accidents

Sisir Joshi[1], Abeer Alsadoon[1], S. M. N. Arosha Senanayake[2], P. W. C. Prasad[1](✉), Shiaw Yin Yong[2], Amr Elchouemi[3], and Trung Hung Vo[4]

[1] Study Group Australia, Sydney, Australia
chan999941@yahoo.com, {aalsadoon,cwithana}@studygroup.com
[2] Universiti Brunei Darussalam, Gadong BE 1410, Brunei Darussalam
{arosha.senanayake,shiawyin.yong}@ubd.edu.bn
[3] American Public University System, Charles Town, USA
Amr.elchouemi@ashford.edu
[4] University of Technology and Education, The University of Danang, Da Nang, Vietnam
vthung@ute.udn.vn

Abstract. Road traffic accidents are among the major concerns that are leading for deaths and injuries in the world. Many predictive models use data mining technique to provide semi optimal solutions. Pattern identification and recognition have been used to for road accident predictions based on the critical features extracted depending on the dangerous locations and frequency of occurrences prone for accidents. The aim of this research is to propose a novel predictive model based on the pattern mining predictor which improves the accuracy of accident prediction in frequent accident locations. The proposed system consists of association rules mining technique, which identifies the correlation, frequent pattern and association among the various attributes of the road accident. Clustering technique that discriminates the data based on different patterns and classification technique that classify and predicts the severity of accident. Novel system built leads to an improvement in the accuracy of the accident prediction from 92% to 94%. Furthermore, using selective subset of features decreased the processing time and precision of classification is improved using boosting technique.

Keywords: Road traffic accident prediction · Association rules mining · Road safety · Apriori algorithm · Classification analysis · Data mining · Clustering · Fuzzy logic · Naïve Bayes classifier

1 Introduction

Pattern mining is an emerging area of study of Road Traffic Accident Prediction where prediction is done by identifying the most frequent accident location pattern that are prone to risk. Pattern mining uses different methods and algorithms to find out the relation in very large amount of data set. Mining the data to identify the relation among data is considered as one of the most important technology in previous decades [1]. In the past, various research about road traffic were conducted in many countries and the

M. Hernes et al. (Eds.): ICCCI 2020, CCIS 1287, pp. 605–615, 2020.
https://doi.org/10.1007/978-3-030-63119-2_49

data set from these researches were utilized in this work. The datasets obtained were processed, clustered, classified using data mining algorithms. Association rule mining is one of the popular methodologies which helps to identify the significant relation among the data in large dataset and used for frequent itemset mining [2]. Apriori Algorithm is one of the classical association rule mining technique that we have used to analyse the road traffic accident which yields best rules, but it has major limitation due to lengthy processing time. To overcome this limitation, authors in [3] enhanced the algorithm in order to generate the best rule with the consideration of support count only leading to fast processing time. Classification is used to construct a model from the training dataset to classify the records of unknown class. Naïve Bayes technique is one of the best probability-based method for classification which classifies the data based on the Bayes hypothesis with presumption of independence between pair of variables.

Current studies of data mining in Road Traffic accident uses various technique and algorithm to improve accuracy and processing time to predict the location of accident that are prone to risk, the maximum generate accuracy is 92.45% and processing time of 15 s on average.

The purpose of this research is to develop a novel system to increase the accuracy and precision in the prediction of road traffic accidents and decrease the processing time while predicting to take prevent measures and reduce severity of injuries. The use of single technique is not enough to predict with high accuracy and precision therefore, we use number of data mining techniques that helps in analysis of data and is able to predict road traffic accidents reducing severity of injuries and increase in measures required before, during or after accident.

2 State of the Art

Janani in [4] proposed a system that uses Naïve Bayes classification technique based on Apriori algorithm association mining to identify patterns in order to predict the severity of accident, which further brings out the factor related to the case of accident and a predictive model interfaced with fuzzy based location wise accident frequency. The use of Naïve Bayes Classifier has improved the accuracy of prediction model. The experiment was conducted by clustering dataset generating the rules based on Apriori algorithm, which then decomposed into set of training and test data for predicting the severity of accident as reported in [4]. The outcomes proves an accuracy of 92.45% in predicting accident. This model consists of five stages as illustrated in Fig. 2, i.e. Data Pre-processing, Clustering, Association rules mining, Classification and Prediction.

As per the solution provided in [4], the data was clustered and then rules were generated and further decomposed into training set and test set 70% and 30% respectively. Based on the attribute such as Fatal, Grievous, Damage, and Injury in the data set, class label was created which represent the severity level of accident. Class 0 represents the low severity level and class 1 represents high severity level. Naïve Bayes classifier classified the training data with best accuracy.

Prediction process involves use of fuzzy logic algorithm, which predicts the probability of accident occurrence. The propose model in [4], Fuzzification is done to transform the input to fuzzy values, which is the process in fuzzy domain by inference engine based on knowledge base supplied by domain expert. Then the processed output is transformed back with the defuzzification method that shows the probability.

The Apriori Algorithm is implemented in Association rules mining to extract set of rules, which defines a set of patterns in [4]. However, the processing time can still be reduced in order to extract rules leading to overall processing time minimum.

This model increases the accuracy of prediction by using Naïve Bayes classifier prediction model, which predicts with 92.25% accuracy and 92 precisions, is achieved compared to 90.25% accuracy of decision tree and 88% accuracy of random forest as reported in [4].

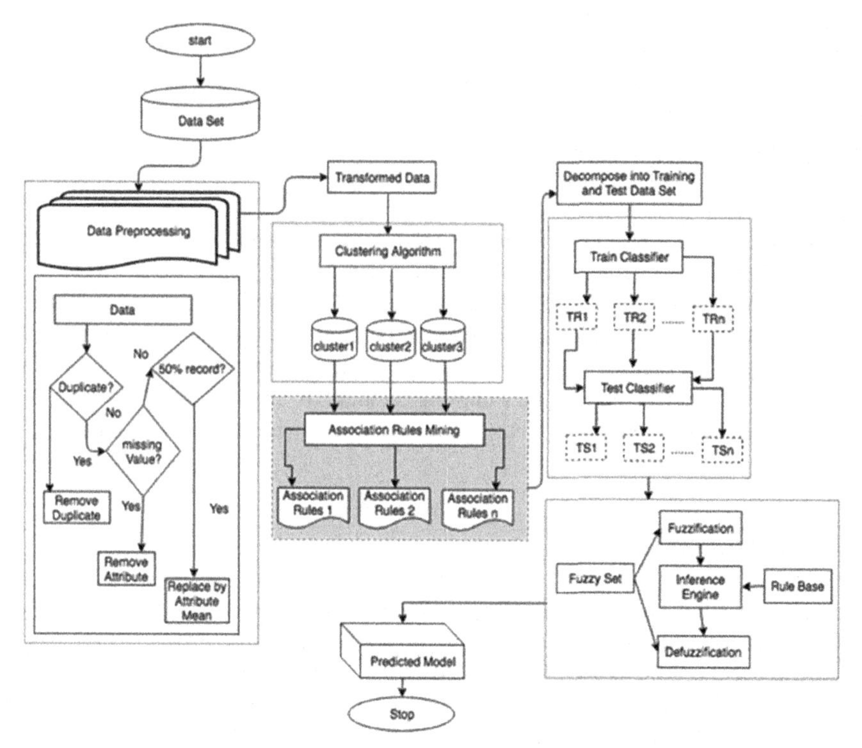

Fig. 1. Naïve Bayes classification technique using Apriori algorithm for prediction the severity of accident proposed as state of the art in [4].

The research work done in [5] investigated classification accuracy on the basic of performance metric to predict the severity of accidents. They proposed a model, which used the Classification and Regression Tree (CART) to analyse road accidents data of Iran and found that not using seat belt, improper overtaking and over speed affect the severity of accidents.

The article [6] discussed the performance of prediction approach in term of predicting model's accuracy. They have discussed that the RTA involved fatal crashes data is directly concerned with nutritional health survey data to analysis of the association of dietary habit of a motor vehicle driver's to road traffic accident by applying Association rule mining algorithms.

A novel rule-based method to predict traffic accident severity according to user's preferences instead of conventional DTs was proposed in [7]. The novel multi-object and rule-based method in [7] outperforms the classification methods such as ANN, SVM, and conventional DTs according to classification metrics like accuracy (88.2%), and performance metrics of rules like support and confidence (0.79 and 0.74, respectively). Proposed method yielded promising result with an increased accuracy of 4.5% from other methods but the obtained rules from this method are not very much effective. Therefore, feature selection method and extraction method must be used to increase the accuracy and improve the effectiveness of the obtained rules.

Authors in [8] collected and analyzed 398 wrong-way driving (WWD) crashes in Illinois and Alabama States. They employed multiple correspondence analysis (MCA) to define the structure of the crash data set and identify the significant contributing factors to crashes. According to the obtained results, driver age, driver condition, roadway surface conditions, and lighting conditions were among the most significant contributors to WWD crashes.

A learning model was built in [9] for predicting accidents on the road using classification analysis and a data mining procedure. The Hadoop framework was proposed to process and analyze big traffic data efficiently and a sampling method to resolve the problem of data imbalance. Based on this, the predicting system first preprocesses the big traffic data and analyzes it to create data for the learning system. The imbalance of created data is corrected using a sampling method. To improve the predicting accuracy, corrected data are classified into several groups, to which classification analysis is applied.

3 Proposed System

After reviewing a range of method for prediction for road traffic accident, we analysed pros and cons of each method. Accuracy, processing time, precision, recall was the main issues to be considered. Proposed solution is the Enhanced Apriori algorithm for extracting best rule with faster iteration compared to the reported algorithm in [4]. Enhanced Apriori algorithm generates the candidate item sets faster, which reduces the processing time. This algorithm reduces the number of candidate item sets that is needing to be scanned and by reducing the number of candidate item sets. Whenever the k of k-itemset and value of minimum support increases, from view of time consumed enhance Apriori algorithm improves the processing time significantly. The use of breadth-first search strategy to count the support of item sets and use of candidate generation function which exploits the downward closure property of support which is one of the best ways to mine the pattern/association rules as reported in [3]. This is a feature adapted from second-best solution. This information reduces the number of transactions to be scanned making it possible to minimize the generation of candidate item sets.

Novel system implemented here consists of five main stages as illustrated in Fig. 1; Data Pre-processing, Clustering, Feature Extraction, Classification and Prediction.

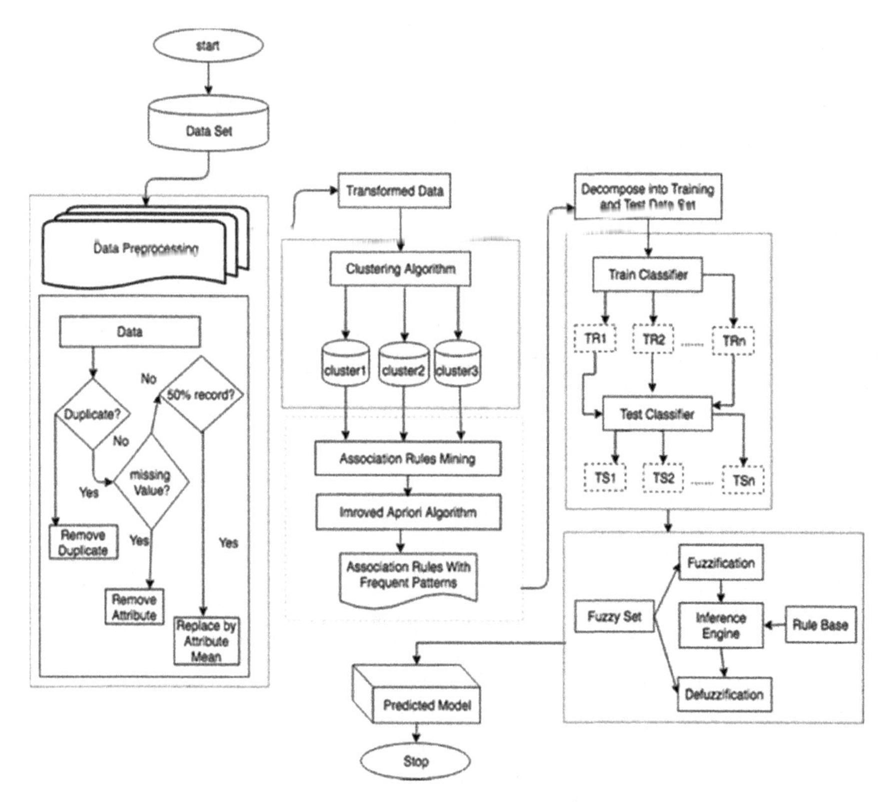

Fig. 2. Flow chart of novel system proposed

Processing time as expressed in [10], Apriori Algorithm is expressed using (1):

$$T = \sum_{i=1}^{n} t_s * m_k + t_c + 1_{k+1} * k + \frac{1}{2} * t_s * n_k / A \tag{1}$$

Where

t_s is the time cost of single scan of database
t_c be the time cost for generating candidate itemset
m_k be the amount of time in candidate itemset
n_k is number of item set in frequent Itemset

A denotes the amount of record in the database k is length of itemset.

From the features of (1), we propose (2) to be used in Improved Apriori Algorithm which is modified Apriori:

$$T = \sum_{i=1}^{n} (2E(|x|)) + k)n\tau \tag{2}$$

Where

n denotes the number of items
E is number of data points
|x| is the average complexity of each data point
m_k is the amount of time in candidate itemset
k is the length of itemset
τ is amount of time per non zero element

As per (2), item set is obtained for each number of iterations prolonging the scan of database which was then addressed in [2] and replaced by (3)

$$T = t_s * m_k + \sum_{i=2}^{k+1} (t_c + 2 * t_x * 1_{k+1}) \tag{3}$$

Where

t_s is the time cost of single scan of database
t_c be the time cost for generating candidate itemset
m_k be the amount of time in candidate itemset
k is the length of itemset

From the features of (3), we propose (4). It is done so because the candidate itemset obtained from overall set rather than obtaining in each iteration reduces processing time. If t_s is the time required for each scan of the database, then total processing time can be calculated as:

$$T = t_s * m_k + \sum_{i=1}^{n} (2E(|x|)) + k)\tau \tag{4}$$

Where

t_s is the time cost of single scan of database
t_c be the time cost for generating candidate itemset
m_k be the amount of time in candidate itemset
k is the length of itemset
E is number of data points
|x| is the average complexity of each data point
τ is amount of time per non-zero element

The main purpose of this change is to reduce the scale of database and reduce itemset generate from candidate set. Reduction of candidate set results in less scanning of database which reduces the processing time significantly thus reducing the time for feature extraction.

Data can be considered as sequence of binary vectors and can be represented by binary matrix which has defined number of rows and columns. The novel equation proposed in

(1) states that data are represented in columns because columns are generally in lesser numbers than rows which allows pointer to go through data faster to cluster data of similar behaviour or data which has occurred frequently. With the help of the proposed equation, association rules are applied much faster to the data. This is possible as data are not visited again.

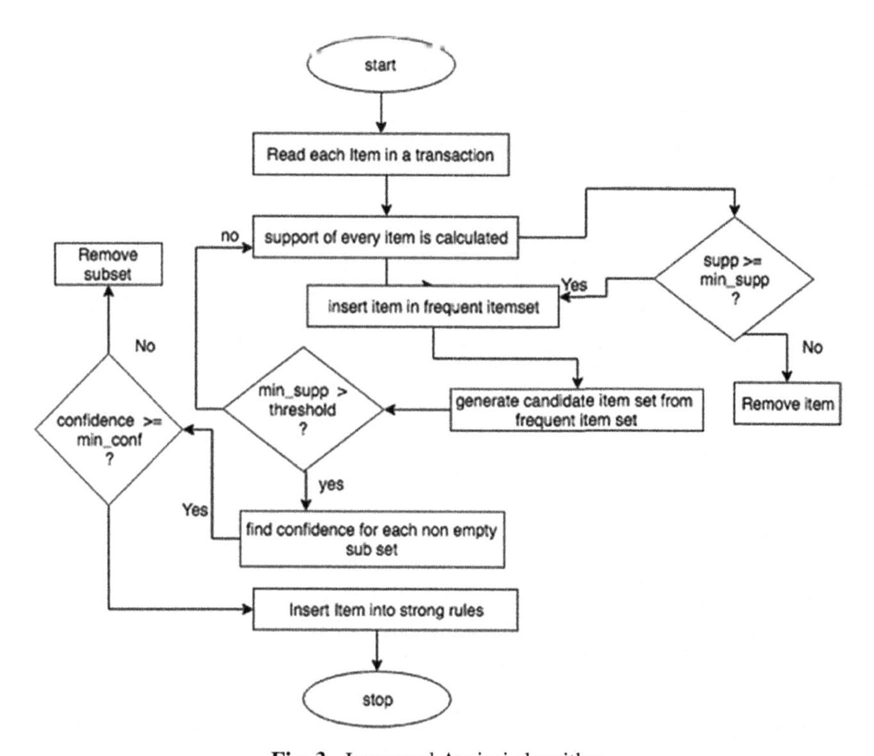

Fig. 3. Improved Apriori algorithm

With general association rules, the data are categorized and grouped with similar behaviour and attributes, which are passed into classifier to derive a prediction model. This proposed system derives a prediction model for the most accident occurring location based on the data supplied. The state of art does not provide the amount of time taken for to apply association rules to generate sets of data, which has similar behaviour, and attribute.

Time is one variant that must be taken under consideration while generating a set of records, which has similar behaviour, and attribute hence, the processing time must be minimized to its very best. With use of Improved Apriori Algorithm, it scans whole database to generate frequently occurring first item set which are then used to find second item set and so on until k number of item sets is reached. Thus, minimizes processing time as previously created cluster is used to create new one.

Current solutions have used association rules without consideration of time to process to form set of records. Our proposed solution has reduced processing time using Apriori algorithm as illustrated in Fig. 3.

Table 1 illustrates the proposed improved Apriori algorithm.

Table 1. Sample implementation of improved Apriori algorithm

Algorithm: Improved Apriori algorithm extracting rules to identify patterns
Begin
Step (1) first item set Lk= find frequent 1 item sets (T) ;
Step (2) For loop k from 2 to while Lk-1!= null;
//Generate the Candidate item set from the Lk-1
Step (3) Ck = candidates generated from L;
//get the item with minimum support in Candidate item set using L1, $(1 \le w \le k)$.
Step (4) x = Get items minimum support(Ck, L1);
// get the target transaction IDs that contain item x.
Step (5) TID = get Transaction ID(x);
Step (6) For each transaction t in TID Do
Step (7) Increment the count of all items in Candidate item that are found in TID;
Step (8) Lk= items in candidate itemset $\geq$ minimum support;
End;

4 Results

Java and IDE Net beans were used for the implementation. A code base program was designed to implement Improved Apriori Algorithm with the sample of 10 dataset. The dataset that was taken as sample to test have different number of data that varies from 500 to 4500 and different number feature, which vary from 45–300. We have specifically taken the sample that have different attributes which covers most of the requirement like accident severity, time and date of the accident, location characteristics etc. The dataset has been taken from different site like data.gov, (Menzies, 2018), which is free online resource and is available on the internet for the student whose focuses on data science. All these datasets are available in CSV and ARFF file format that can be read by many language using data analysis libraries. This library allows for creation of object that can access the method to read the CSV or ARFF file. The data were extracted from java and all of them have been considered. The extracted data were balanced using the XLMiner and then passed through the code, which produces the results as illustrated in Table 2.

The result is compared based on the number of items in the dataset and the number of features extracted along with the amount of time for single scan of dataset. In the dataset, generation of the candidate set is minimized which leads to a smaller number of scanning of dataset therefore decreases the processing time significantly. As shown in Table 2 the number of features extracted is 62, which is significantly less than the number of items. Improved Apriori extracts the candidate item in first scan only and process ahead, which decrease the average processing time extracting the feature as depicted in Table 2. The processing time of proposed solution is 2.119 s, which is 60% less than state of art.

Table 2. Sample implementation of improved Apriori algorithm

Dataset from data.gov	Samples (n)	Items in candidate set (m_k)	State of art		Proposed solution	
			Processing time	Accuracy	Processing time	Accuracy
Road traffic accident data	**915**	**62**	6.001	87.23	2.119	90.33

Samples were compared the state of art and the proposed solutions with the help of graphs and the data reports that are shown in Figs. 4 and 5. The results divided according to number of features extracted and the total amount of time to scan a dataset a single time Here, the results from the sample is presented in terms of average processing time. Processing time is the time in seconds to extract the feature, which is programmed well in java. We have done comprehensive test for 10 datasets. Then the result has calculated by taken the average for all datasets.

These results were compared during different number of transactions. The proposed solution has reduced the processing time by separating the candidate items generating for first scan of dataset as all the transactions are scanned as first it generates the item set with all the ids and support for all the transaction. If another candidate set is to be generated its will not scan all the transaction again but it will look at the item sets that was generated previously in first scan which holds all the id and support, so it can scan for only specific transaction where item exists.

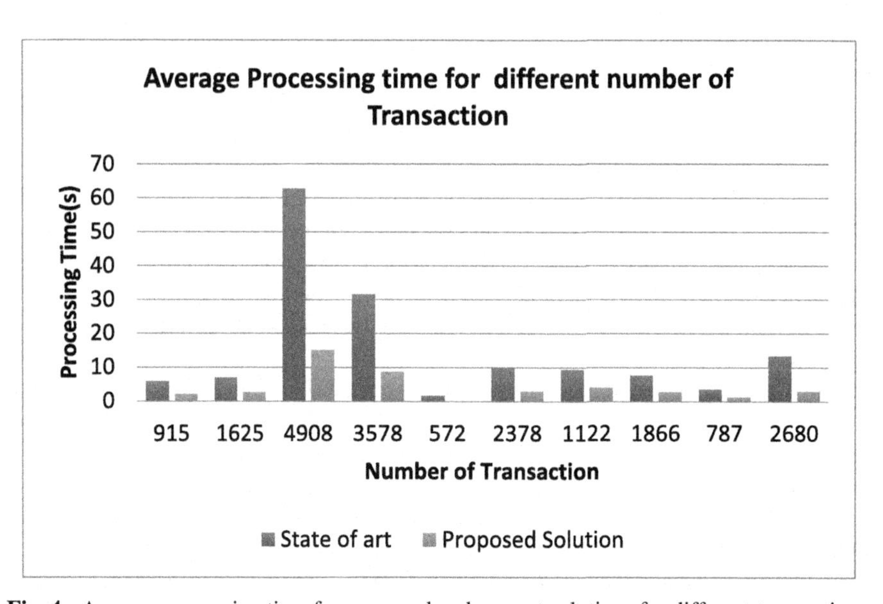

Fig. 4. Average processing time for proposed and current solutions for different transactions.

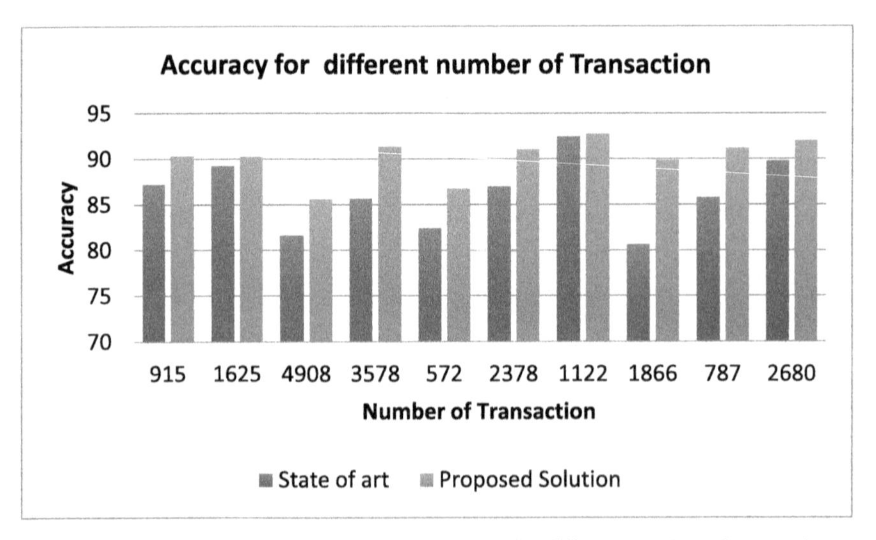

Fig. 5. Accuracy of proposed and current solutions for different number of transactions.

5 Discussion

Results show improved accuracy and decrease in time processing between the current solution and the proposed solution during the generation of prediction model as well as after prediction model are created. An accuracy of 92.75% and improvement in processing time by 67% was achieved, as the cardinal item sets were achieved before iteration rather than on individual iterations. Processing time decreased from 10 s to predict the model for over 50 records was reduced to 3.37 s. Both factors' accuracy and improvement in processing time was calculated and simulated with the help of Java Programming language. Improvement in processing time was calculated with the help of Java for both the state of art for current as well as proposed algorithms.

Use of Apriori Algorithm switched its performance from low to high by the reduction of processing time and helps create prediction model for the traffic road accident. Despite voluminous data, Improved Apriori Algorithm helps in search of most frequent item just once rather than in every repetition. Thus, use of Improved Apriori Algorithm for the prediction improved the processing time and accuracy for the traffic road accident prediction.

For road traffic accident prediction model, Apriori Algorithm has been implemented as a major component, which helped in achieving desirable processing time and accuracy. This research successfully provides a prediction model for road traffic accident with greater accuracy in lesser time. The state of art solution provides prediction model with 92.25% accuracy without ever mentioning the processing time. However, the proposed solution provides accuracy of 92.75% with the significant drop in processing time by 67%. This is verified and validated with the pseudo code for the use of Apriori Algorithm for a faster processing of data.

6 Conclusion

The proposed solution with all the necessary data for traffic road accident prediction model is made easier and faster with the use of Apriori Algorithm. It reduces processing time by 67% even on a large data. the Improved Apriori Algorithm has been tested to produce prediction model with high accuracy by reducing processing time. Information as obtained from the model also helps to plot reasons behind all road traffic accidents and helps in understanding them better.

Improved Apriori algorithm has proven the accuracy 92.75% with decrease in processing time with minimum support. Thus, it performs more efficiently as it finds the frequent occurring item in the database to build up the first item set to generate second and so on in which the item with low minimum support is removed. This process is known to as sampling and doing such a way that the efficiency is increased as processing time is decreased. Furthermore, the algorithm introduces the database mapping which avoids the repetition of database scan.

References

1. Amira, A., Vikas, P., Abdelaziz, A.: Applying association rules mining algorithms for traffic accidents in Dubai. Int. J. Soft Comput. Eng. **5**(4), 2231–2307 (2015)
2. Gu, X., Li, T., Wang, Y., Zhang, L., Wang, Y., Yao, J.: Traffic fatalities prediction using support vector machine with hybrid particle swarm optimization. J. Algorithms Comput. Technol. **12**(1), 20–29 (2017). https://doi.org/10.1177/1748301817729953
3. Arwa, A., Mourad, Y.: Hybrid approach for improving efficiency of apriori algorithm on frequent itemset. Int. J. Comput. Sci. Netw. Secur. **18**(5), 7–10 (2018)
4. Janani, G., Devi, N.: Road traffic accidents analysis using data mining techniques. JITA: J. Inf. Technol. Appl. **14**(2) (2018). https://doi.org/10.7251/jit1702084j
5. Tavakoli Kashani, A., Shariat-Mohaymany, A., Ranjbari, A.: A data mining approach to identify key factors of traffic injury severity. PROMET Traffic Transp. **23**(1) (2011). https://doi.org/10.7307/ptt.v23i1.144
6. Mulay, P., Mulatu, S.: What you eat matters road safety: a data mining approach. Indian J. Sci. Technol. **9**(15) (2016). https://doi.org/10.17485/ijst/2016/v9i15/92119
7. Hashmienejad, S., Hasheminejad, S.: Traffic accident severity prediction using a novel multi-objective genetic algorithm. Int. J. Crashworthiness **22**(4), 425–440 (2017). https://doi.org/10.1080/13588265.2016.1275431
8. Jalayer, M., Pour-Rouholamin, M., Zhou, H.: Wrong-way driving crashes: a multiple correspondence approach to identify contributing factors. Traffic Inj. Prev. **19**(1), 35–41 (2017). https://doi.org/10.1080/15389588.2017.1347260
9. Park, S., Kim, S., Ha, Y.: Erratum to: highway traffic accident prediction using VDS big data analysis. J. Supercomput. **72**(7), 2832 (2016). https://doi.org/10.1007/s11227-016-1655-5
10. Krishnaveni, S., Hemalatha, M.: A perspective analysis of traffic accident using data mining techniques. Int. J. Comput. Appl. **23**(7), 40–48 (2011). https://doi.org/10.5120/2896-3788

Artificial Neural Network Approach to Flood Forecasting in the Vu Gia–Thu Bon Catchment, Vietnam

Duy Vu Luu[1], Thi Ngoc Canh Doan[2], and Ngoc Duong Vo[3](✉)

[1] The University of Danang – University of Technology and Education, Da Nang, Vietnam
ldvu@ute.udn.vn
[2] The University of Danang – University of Economics, Da Nang, Vietnam
canhdth@due.edu.com
[3] The University of Danang, University of Science and Technology, Da Nang, Vietnam
vnduong@dut.udn.vn

Abstract. Flooding in the Vu Gia-Thu Bon catchment has destroyed critical facilities, such as infrastructure and housing. This study develops an application of an Artificial Neural Network (ANN) to forecast the flow at the Nong Son gauging station in the catchment. The ANN model uses rainfall data at upstream locations to estimate flows at downstream point. Architectures of the ANN model are adjusted to calculate flooding. Daily rainfall at Tra My, Tien Phuoc, Hiep Duc and Nong Son between 1991 and 2010 is used to predict flooding at Nong Son. The analysis shows that the ANN is a reliable method to forecast the flood in the Vu Gia-Thu Bon catchment, where there is a lack of a wide range of accurate data, particularly hydrological, meteorological and geological data.

Keywords: Artificial Neural Network (ANN) · Flood forecasting · Vu Gia-Thu Bon catchment

1 Introduction

Flooding's occurrence around the world has increased vulnerabilities of individuals and households. One study of Navrud et al. [1] mentioned the 2007 floods directly cost approximately US$ 200 per household in Quang Nam province in Vietnam, and cost Quang Nam province US$ 53 million. Therefore, many solutions have been presented to prevent the flood disasters, such as flood mitigation and prevention, and flood forecasting based on hydrological models. To reduce negative impact of water scarcity and flooding, Sanz et al. [2] developed a coupled two-dimensional distributed hydrologic and hydraulic model.

Real world systems can be presented by a hydrological model in a simple way [3]. The hydrological models have three main kinds, including conceptual models, physical models, and empirical models. For the conceptual model, it uses perceived systems, such as reservoirs, to represent the vital elements of environment, such as rainfall and evaporation. For example, to simulate rainfall and evaporation in a catchment, the model uses

M. Hernes et al. (Eds.): ICCCI 2020, CCIS 1287, pp. 616–625, 2020.
https://doi.org/10.1007/978-3-030-63119-2_50

a reservoir that is filled by rainfall and is dried by evaporation. Semi-empirical equations are used in this model [4]. Field data and calibration are used to fix model parameters. In terms of calibration, it uses the observed data. The physical model is an ideal method [5] in which physical principles are used to present the hydrological processes in time and space. Regarding calibration process, it does not require meteorological and extensive hydrological data. However, more parameters, such as physical characteristics, are needed in the evaluation process [6]. The model requires a wide range of hydrological and meteorological data for the calibration process because it has many hydrological variables, such as soil moisture content and dimensions of flow [7]. An empirical model considers collected data, and adopt mathematical equations to simulate relationships between inputs and outputs, without analyzing hydrological processes. One example of empirical models is Artificial Neural Networks (ANNs). ANNs consider the hydrological process as a 'black box'. Inputs are usually rainfall or flow, and outputs are usually flow. ANNs are efficient for dynamic problems [8]. In addition, ANNs tend to be simpler and cheaper than physically based models [9].

Physical models and conceptual models have many advantages, but they are built for a wide range of climatic areas, not a specific region. Therefore, when employing the models, many input data must be accurate, such as geological data. However, some of the data are not usually existing in many developing countries, such as Vietnam. Therefore, the ANNs seem to be an alternate solution because they do not require the need for collecting many kind data [10].

As a result, because the ANN model is a potential approach for quick and flexible data integration and model development, the project adopts an ANN model to forecast flooding along the Vu Gia-Thu Bonriver system in Vietnam. It is predicted to develop a flood warning system for certain sections of the Vu Gia-Thu Bon catchment.

In artificial neural networks, there are many different architectures and algorithms adopted in flooding forecasting systems. Thirumalaiah and Deo [11] mention all training algorithms should be tested in a specific application. Moreover, there is a lack of ANN application for the Vu Gia-Thu Bon catchment. They are motivation for this research to take advantage of an ANN-watershed runoff prediction (ANN-WRP) model for flooding forecasting at the Vu Gia-Thu Bon river system.

2 Artificial Neural Network

2.1 Background

An ANN is a means of computation inspired by the structure and operation of the human brain and nervous system, which allows it to learn, master and reveal the relationship hidden in the data (Fig. 1). For example, the model is able to present complex nonlinear input/output time-series relationships in a river, without analyzing the physical characteristics of that process [12]. In general, an ANN system is a network of processing neurons called nodes that are embedded in layers and connected via links. Each neuron connects with several inputs and a number of outputs.

ANNs provide a new and interesting solution to many complex problems. Therefore, there are a wide range of applications of ANN models in many fields, including pattern classification, prediction and financial analysis, and control and optimization.

For pattern classification, ANNs have a strong ability to detect infer hidden in pictures. Its applications include handwritten numeral recognition through fusion of features and machine learning techniques [13] and detection, identification, and tracking of objects hidden from view [14]. In prediction problems, ANN models can detect geotechnical properties [15] or floods [16]. Optimization problems involve reservoir operation [17], and international diversified portfolio [18].

There are some different ANNs' architectures, such as single-layer and multilayer networks, recurrent and self-organizing networks. However, the Mulit-Layer Perceptron (MLP) is the most popular and is used in this paper. This architecture includes an input layer, an output layer and one or more hidden layers. The nodes in different layers are either fully or partially interconnected. The relations between nodes in the neighboring layers are weighted. Strength of their relations is represented by the weights. Moreover, the strength can be changed. Positive weight values mean excitatory connections while negative values present inhibition connections. On the other hand, there is non-existent when the zero weights occur. In the input layers, input data is infused into neurons. The neurons in the hidden layers adopt data from neurons in the previous layers, and process data. Finally, the outputs are produced in the output layer. Input from every node is multiplied by a weight. Then in each node in the hidden layer, the weighted signals are summed, and a bias data is added. After that, a nonlinear transfer function is adopted to transfer the combine input into the output. This means that ANNs are not limited to linear function.

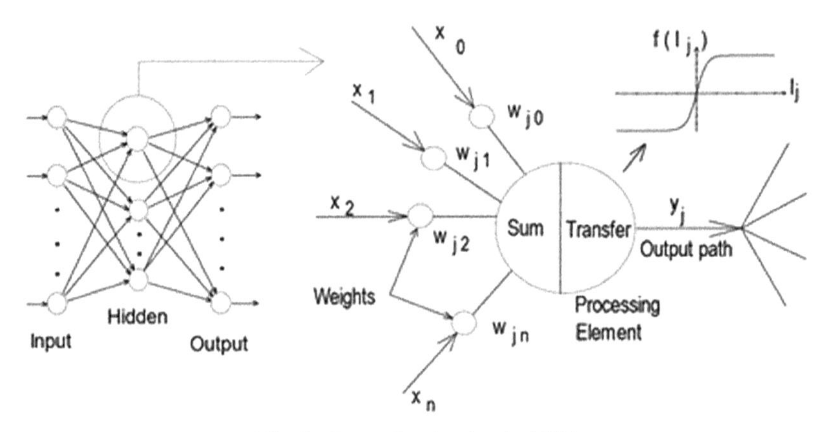

Fig. 1. Example of a simple ANN

The mathematical operation of a neuron is given as the below equations:

$$I_j = \sum w_{ji} x_i + \theta_j \quad (1)$$

$$yj = (Ij) \quad (2)$$

In which y = the output of a neuron j; f = an activation function; x_i = an input of the vector of inputs ($i = 1, 2, \ldots, n$); w_i = the weight.

In non-inear regression ANN, hidden layers can use Log-sigmoid transfer function (LOGSIG) or Hyperbolic tangent transfer function (TANSIG) while Purelin Transfer function is assigned to the output layer. However, there is a similarity between LOGSIG and TANSIG in flood forecasting [19]. Therefore, this project focus on LOGSIG.

2.2 Training

The training process aims to give targets and input data to the model and tailor internal parameters, including weight and biases, based on model assessment, such as error measures in order to improve performance of ANNs. There are many methods to train a neutral network architecture, for example, the error backpropagation algorithm, conjugate gradients and cascade correlation. The error backpropagation algorithm is used in the majority of researches based on the MLP [20], and this study uses it.

There are four main steps in the training process. First, the weights randomly receive small values. Secondly, the output is generated based on the previous weight in produced. Afterwards, the global error function, which is the error surface in n-dimensional space, is determined by following equation:

$$E = \frac{1}{2}(\mathrm{od} \ - \mathrm{op})^2 \tag{3}$$

Where E = the global error function; Od is the desired output; Op is the predicted output.

Finally, this method uses the error between the target output and the network output to fed backward via the network, which allows it to adjust weights to minimize the error. The equation is used for improving weights.

The training process is initiated and stopped when an acceptable minimum error is reached. The method starts randomly determining an area in the error surface. Then, the first-order derivative of the error function is used to direct weight changes down error gradients. However, the weight change rate should be controlled. If the rate is too high, training can move directly from one "nonoptimal set of weight" to another one. If the rate is too low, training can be stuck in a local error.

2.3 Nodes

The numbers of neurons in the input layers and output layers are based on the number of input and output variables in a given system. The best way to determine the number of layers and neurons in hidden layers is via trial and error [21].

The number of nodes of hidden layers should be less than the below limit [22], which allows ANNs to estimate continuous functions:

$$N_h = 2N_i + 1 \tag{4}$$

Where Nh is the number of nodes of hidden layers; Ni is the number of inputs.

However, Rogers and Dowla [23] suggest the following upper limit should be satisfied to avoid ANNs not overfitting the training data.

$$N^h \leq \frac{N^{tr}}{N^i + 1} \tag{5}$$

Where N^{tr} is the number of training samples.

If there are inadequate neurons in the hidden layers, the neural network has less changes to identify the intricate relationships between the indicator inputs and the target outputs. In contrast, if there are too many hidden layer neurons, the network may overfit the training data, which can require a large computational time, and cause a poor generalization performance [20].

3 Study Area

The Vu Gia-Thu Bon river system is one of the largest river systems in the central region of Vietnam, and one of the nine major river systems of the country. It originates from the eastern side of the Truong Son mountain range and end at the Vietnamese East Sea near the cities of Da Nang and Hoi An.

Fig. 2. Vu Gia Thu Bon catchment.

The catchment is nearly 10,350 km^2, the length of the main river is 205 km, providing the main source of water for living and production for residents in Quang Nam province and Da Nang city. This catchment has very high rainfall intensity with an annual average of 2000 mm to 4000 mm. From over 60% to 80% of annual rainfall occur between September to December. Therefore, the Vu Gia-Thu Bon river system is also an annual flood threat to the areas (Fig. 2).

Every year, floods cause many losses of people and properties of the areas, for example 1964, 1998 and 1999 historical floods. Flood from 17th to 25th November 1998 in Quang Nam caused 38 deaths, damage of VND 145 billion. 1999 floods in Quang Nam killed 118 people, injured 399 people and caused about VND 758 billion in damage. In addition, the IPCC shows that global warming can lead to extremer weather events, such as global sea level rise, flood and drought disasters in the basin in the future.

4 Model Development

The ANN-WRP model is built to recognize and generate the relationship between rainfall and runoff data the Nong Son gauging station in the Vu Gia–Thu Bon catchment. The project uses daily rainfall and runoff data because there is a lack of quality rainfall and flow data for less than 24 h in this area. All data are supplied by the Mid-Central Region Centre for Hydro-meteorological Forecasting. The input data are rainfall in Tra My, Tien Phuoc, Hiep Duc and Nong Son stations, and runoff data in Nong Son are output from 1991 to 2010. 70%, 15% and 15% of the data set are divided into training, validation and testing data set, respectively.

The time-lags between input and output data are ignored because the one interval of daily data adopted is 24 h, too big. The project did test one-day time-lag but the result was not good.

The ANN-WRP model is unable to extrapolate over the range of training data set [24]. Under climate change view, the risk of a record-breaking flood has potential to occur. Standardizing all data is ideal solution to extrapolation. All input data are normalized based on the following equations.

$$\bar{X}_i = \frac{\sum X_i}{N} \tag{6}$$

$$Std = \sqrt{\frac{\sum (X_i - \bar{X}_i)^2}{N - 1}} \tag{7}$$

$$S_i = \frac{X_i - \bar{X}_i}{Std} \tag{8}$$

Where X_i is input data, $\bar{X}_i$ presents average of X_i; Std is standard deviation of X_i; S_i is normalized X_i; N is the total number of the target.

For selecting the best ANN model, the study evaluates four scenarios using different architectures. The architectures and the results of scenarios are shown in Table 1.

Selection of the best ANN model is a complicated task. Trial and error procedure is the most popular solution [25]. Therefore, this study uses it to select the suitable ANN architecture with optimum performance. Mean squared error (MSE) and the coefficient of determination (R2) are used to estimate performance different structures. MSE is a good tool to assess the goodness of fit at high flows [26]. MSE is always non-negative, and the MSE value getting closer to zero is better. The coefficient of determination determine how well target can be explained and predicted by the model. The R2 value ranges from 0 to 1. If the R2 value equals to 1, it is a perfect fit and the model is reliable for forecasting. On the other hand, a value of 0 shows that the model fails in explaining and predicting the observed variation.

The structure is selected if MSE is the smallest and R2 is the largest and its structure has minimum layers and neurons.

Table 1. Four scenarios and model efficiency results

Scenario	Structure of ANN-WRP	Training		Validation		Testing	
		MSE	R^2	MSE	R^2	MSE	R^2
1	4-4-1	0.35	0.817	0.3	0.823	0.45	0.833
2	4-8-1	0.35	0.839	0.25	0.833	0.4	0.800
3	4-4-4-1	0.35	0.815	0.48	0756	0.27	0.829
4	4-8-8-1	0.30	0.832	0.30	0.793	0.3	0.883

5 ANN Model Results

The project tries to find the most optimum transfer function and architecture for the ANN-WRP model in the catchment. The model performs well and are selected if R2 is the largest, MSE is the smallest and there are minimum layers and neurons. Based on the Table 1, all three layers models perform well when they have high correlation (R2 $\geq$ 0.8) and MSE values of all the training and testing are close to 0. However, if there are four layers and middle layers have more than 4 nodes, overfitting occurs. The scenario 3 is the optimum option because it only requires minimum layers and neurons but also performs well (R2 > 0.8 and MSE = 0.4).

Figure 3 and Fig. 4 compare the relationship between the forecasted runoff and the observed runoff in flooding in 2003, 2007 and 2009. The ANN-WRP model show the ability to generate the same pattern with the observed data.

However, the developed model usually underestimates the flow value. The observed maximum flow is recorded at 8,410 m^3/s on 12th Dec 2007 while the modelled maximum flow is 7,707.77 m^3/s on the same day. On 3rd Dec 2009, the observed gauge and the predicted value are 3,630 m^3/s and 2,620.92 m^3/s respectively. The reason for the

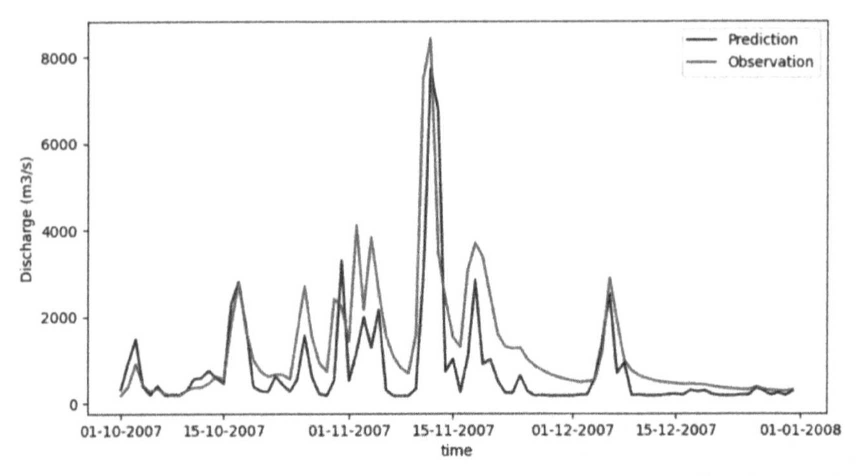

Fig. 3. Observed and ANN predicted flows at the Nong Son gauging station from 01/10/2007 to 31/12/2007

discrepancies can be soil becomes saturated in the periods when rain season in this catchment lasts from September to December. Saturated soil allows saturation excess overland flow occur immediately when it rains. This results in observed foods are usually larger than estimated counterparts.

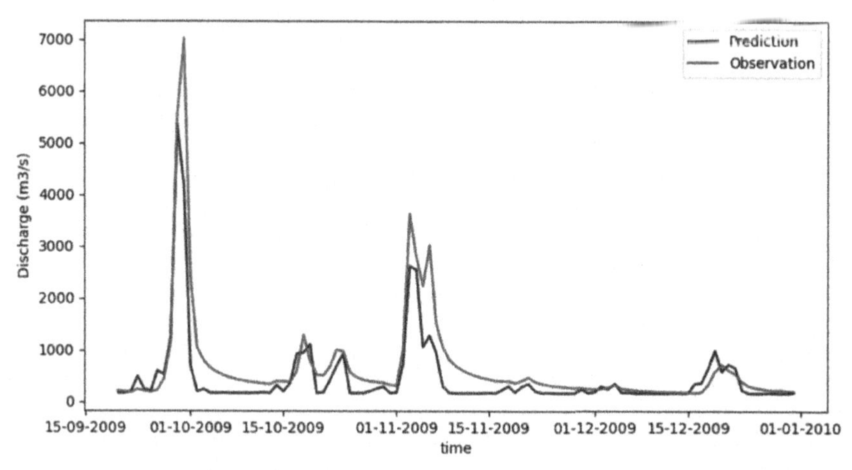

Fig. 4. Observed and ANN predicted water flows at the Nong Son gauging station from 01/10/2009 to 31/12/2009

6 Conclusion

This study aims to predict flows in the Vu Gia-Thu Bon river system based on the ANN-WRP model. This method has advantages because it does not require several variables to generate accurate results while there is a lack of data in this catchment.

This study proves the ability of the ANN-WRP model can generate daily flows based on daily rainfall. Although the are differences between modeled data and observed data, the model predict trends in flood magnitude, based on limited daily data. Therefore, in developing countries, where it is difficult or impossible to collect data, the ANN method is one of the most viable solutions for flood forecasting. Moreover, because this method does not require topographical and hydrological information, it decreases the amount of time spent on collecting and analyzing the data.

Acknowledgement. This work was supported by The University of Danang, University of Science and Technology, code number of Project: T2020-02-47.

References

1. Navrud, S., Huu Tuan, T., Duc Tinh, B.: Estimating the welfare loss to households from natural disasters in developing countries: a contingent valuation study of flooding in Vietnam. Glob. Health Action **5**(1), 17609 (2012)

2. Sanz Ramos, M., Amengual, A., Bladé i Castellet, E., Romero, R., Roux, H.: Flood forecasting using a coupled hydrological and hydraulic model (based on FVM) and highresolution meteorological model. In: E3S Web of Conferences, Volume 40 (2018): River Flow 2018-Ninth International Conference on Fluvial Hydraulics, pp. 1–8 (2018)
3. Wheater, H., Sorooshian, S., Sharma, K.D.: Hydrological Modelling in Arid and Semi-Arid Areas. Cambridge University Press, Cambridge (2007)
4. Devia, G.K., Ganasri, B., Dwarakish, G.: A review on hydrological models. Aquat. Procedia **4**(1), 1001–1007 (2015)
5. Vo, N.D., Gourbesville, P.: Application of deterministic distributed hydrological model for large catchment: a case study at Vu Gia Thu Bon catchment, Vietnam. J. Hydroinformatics **18**(5), 885–904 (2016)
6. Abbott, M.B., Bathurst, J.C., Cunge, J.A., O'Connell, P.E., Rasmussen, J.: An introduction to the European Hydrological System—Systeme Hydrologique Europeen, "SHE", 1: history and philosophy of a physically-based, distributed modelling system. J. Hydrol. **87**(1–2), 45–59 (1986)
7. Abbott, M., Bathurst, J., Cunge, J., O'connell, P., Rasmussen, J.: An introduction to the European Hydrological System—Systeme Hydrologique Europeen,"SHE", 2: structure of a physically-based, distributed modelling system. J. Hydrol. **87**(1–2), 61–77 (1986)
8. Thirumalaiah, K., Deo, M.: Real-time flood forecasting using neural networks. Comput.-Aided Civ. Infrastruct. Eng. **13**(2), 101–111 (1998)
9. Campolo, M., Andreussi, P., Soldati, A.: River flood forecasting with a neural network model. Water Resour. Res. **35**(4), 1191–1197 (1999)
10. Wurbs, R.A.: Computer models for water resources planning and management. Army Engineer Inst for Water Resources Fort Belvoir VA (1994)
11. Thirumalaiah, K., Deo, M.: River stage forecasting using artificial neural networks. J. Hydrol. Eng. **3**(1), 26–32 (1998)
12. Filipova, V.: Urban flooding in Gothenburg-A MIKE 21 study. TVVR12/5010 (2012)
13. Sharma, A.K., Thakkar, P., Adhyaru, D.M., Zaveri, T.H.: Gujarati handwritten numeral recognition through fusion of features and machine learning techniques. Int. J. Comput. Syst. Eng. **3**(1–2), 35–47 (2017)
14. Musarra, G., et al.: Detection, identification, and tracking of objects hidden from view with neural networks. In: Advanced Photon Counting Techniques XIII, vol. 10978, p. 1097803. International Society for Optics and Photonics (2019)
15. Taleb Bahmed, I., Harichane, K., Ghrici, M., Boukhatem, B., Rebouh, R., Gadouri, H.: Prediction of geotechnical properties of clayey soils stabilised with lime using artificial neural networks (ANNs). Int. J. Geotech. Eng. **13**(2), 191–203 (2019)
16. Falah, F., Rahmati, O., Rostami, M., Ahmadisharaf, E., Daliakopoulos, I.N., Pourghasemi, H.R.: Artificial neural networks for flood susceptibility mapping in data-scarce urban areas. In: Spatial Modeling in GIS and R for Earth and Environmental Sciences, pp. 323–336. Elsevier (2019)
17. Silva Santos, K.M., Celeste, A.B., El-Shafie, A.: ANNs and inflow forecast to aid stochastic optimization of reservoir operation. J. Appl. Water Eng. Res. **7**(4), 314–323 (2019)
18. Bayramoglu, M.F., Basarir, C.: International diversified portfolio optimization with artificial neural networks: an application with foreign companies listed on NYSE. In: Machine Learning Techniques for Improved Business Analytics, pp. 201–223. IGI Global (2019)
19. Dorofki, M., Elshafie, A.H., Jaafar, O., Karim, O.A., Mastura, S.: Comparison of artificial neural network transfer functions abilities to simulate extreme runoff data. Int. Proc. Chem. Biol. Environ. Eng. **33**, 39–44 (2012)
20. Dawson, C., Wilby, R.: Hydrological modelling using artificial neural networks. Prog. Phys. Geogr. **25**(1), 80–108 (2001)

21. Shamseldin, A.Y.: Application of a neural network technique to rainfall-runoff modelling. J. Hydrol. **199**(3–4), 272–294 (1997)
22. Abarghouei, H.B., Hosseini, S.Z.: Using exogenous variables to improve precipitation predictions of ANNs in arid and hyper-arid climates. Arab. J. Geosci. **9**(15) (2016). Article number: 663. https://doi.org/10.1007/s12517-016-2679-0
23. Rogers, L.L., Dowla, F.U.: Optimization of groundwater remediation using artificial neural networks with parallel solute transport modeling. Water Resour. Res. **30**(2), 457–481 (1994)
24. Rajaee, T., Mirbagheri, S.A., Zounemat-Kermani, M., Nourani, V.: Daily suspended sediment concentration simulation using ANN and neuro-fuzzy models. Sci. Total Environ. **407**(17), 4916–4927 (2009)
25. Heidari, E., Sobati, M.A., Movahedirad, S.: Accurate prediction of nanofluid viscosity using a multilayer perceptron artificial neural network (MLP-ANN). Chemometr. Intell. Lab. Syst. **155**, 73–85 (2016)
26. Karunanithi, N., Grenney, W.J., Whitley, D., Bovee, K.: Neural networks for river flow prediction. J. Comput. Civ. Eng. **8**(2), 201–220 (1994)

Ensuring Comfort Microclimate for Sportsmen in Sport Halls: Comfort Temperature Case Study

Bakhytzhan Omarov[1](✉), Bauyrzhan Omarov[2], Abdinabi Issayev[3], Almas Anarbayev[1], Bakhytzhan Akhmetov[3], Zhandos Yessirkepov[1], and Yerlan Sabdenbekov[3]

[1] International University of Tourism and Hospitality, Turkistan, Kazakhstan
bakhytzhanomarov@gmail.com
[2] Al-Farabi Kazakh National University, Almaty, Kazakhstan
[3] Khoja Akhmet Yassawi International Kazakh-Turkish University, Turkistan, Kazakhstan

Abstract. The opinion about the difficulties of maintaining appropriate climatic conditions at sports facilities is quite popular among athletes of various disciplines. This problem can lead to health and economic problems. First of all, users of the facilities are not provided with proper conditions for sports, which can lead to bruises. Secondly, the cost of using the object increases. With the increasing number of sports facilities, maintaining internal thermal comfort in them, while ensuring low operating costs, is becoming increasingly important. The article presents the work on modeling the microclimate in a multifunctional sports hall with the most maximum mode of its use and a detailed analysis of the maintenance of thermal comfort in the hall and cognitive-utilitarian conclusions.

Keywords: Comfort microclimate · Optimization · Sport halls · Comfort temperature

1 Introduction

The air environment in buildings, regulated by many parameters of exposure to exogenous and endogenous factors, determines the working and living conditions, health, and comfort of a person. Providing a "healthy" and comfortable air environment is very expensive, because expensive, technically complex, and multi-functional engineering systems are used [1]. In order to remove only 1 kW of excess heat in buildings or premises (to maintain air temperature), could cost in the region of 300–600 USD [2].

For the above mentioned reasons, today indoor environment quality and comfort have become a topic of relevance, and for this reason heating, ventilation, and air conditioning (HVAC) systems have become popular in many buildings. Reducing the power consumption of these systems, while maintaining a suitable comfort level, is of great interest, and has not yet been completely resolved. Traditionally, HVAC systems have not completely ensured the comfort, as they just attempted to maintain the conditions

M. Hernes et al. (Eds.): ICCCI 2020, CCIS 1287, pp. 626–637, 2020.
https://doi.org/10.1007/978-3-030-63119-2_51

within certain limits. Therefore, comfort optimization depends entirely on the way the system is tailored to the needs of the user.

However, according to some research [3–5], HVAC systems can be regarded as multiple-input multiple-output problems, as they work with interrelated variables to extract sets of output values. Moreover, they are affected by a wide variety of uncertainty parameters as user preferences, occupants' activities, and outdoor environmental parameters that can change their usual operations. Consequently, HVAC control problems can be seen as multi-criteria tasks that are characterized with the help of complex analytical expressions.

Despite conventional PID controllers providing rational solutions, they are not able to completely control the indeterminacy of the dynamics of HVACs that can be readily described by linguistic variables and rules [6]. Fuzzy logic controllers (FLC) act as viable alternatives to traditional controllers, as they do not require mathematical modeling [4], and they are ready to handle different criteria, as they represent the dynamics of the HVAC system in accordance with knowledge. FLCs appear to be a viable solution for conventional controllers, as they are able to handle different criteria that represent the dynamics of the HVAC system. Moreover, they do not require math modeling, and FLC's higher efficiency, and lower power consumption, than PIDs were demonstrated in [7].

Analysis of the structure of energy consumption of residential, public and administrative buildings [8–10] shows that most of the energy consumed by thermal energy is accounted for by thermal energy (Fig. 1). For this reason, and also taking into account the high cost of this resource, energy-saving measures aimed at reducing the consumption of heat energy are most often implemented in buildings.

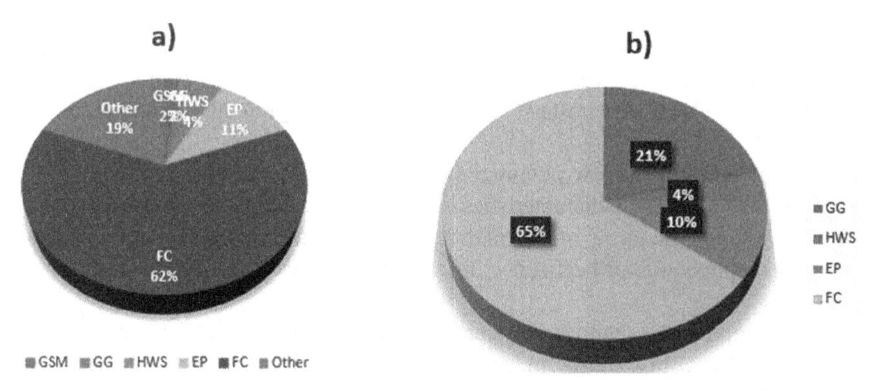

Fig. 1. The distribution of heat energy by types of heat losses through the elements of the building structure is shown in Fig. 1.3 [11].

According to research [33], buildings built in Russia before 1990 according to standard projects have significant potential in the field of energy conservation, because in the Soviet period, the policy of "cheap energy carriers" was carried out and scientific and technical documentation on thermal protection of buildings in construction was insufficiently developed (Fig. 2).

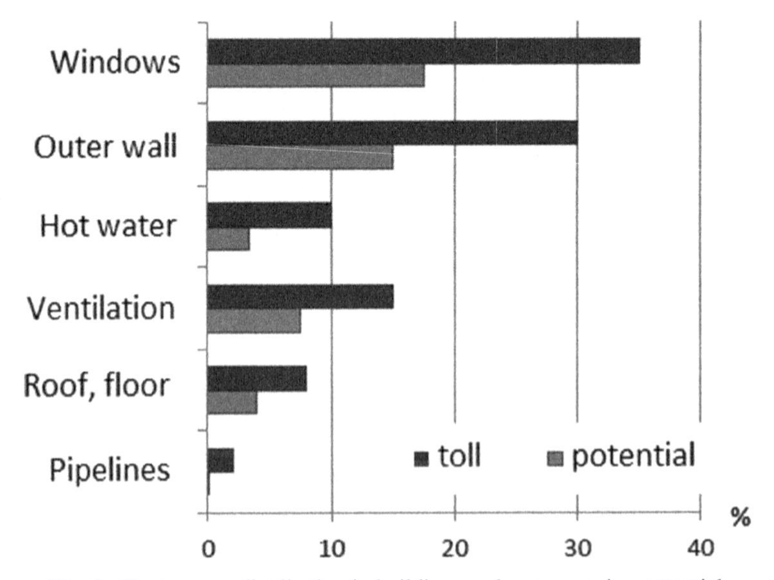

Fig. 2. Heat energy distribution in buildings and energy saving potential.

The potential for improving energy efficiency in systems that form the microclimate of buildings (heating, ventilation and air conditioning), determines the variety of energy-saving measures in this area.

In work [109] three approaches to energy saving in systems of power supply of buildings are allocated:

- increase of accuracy of regulation of the consumed heat energy (units of automatic control of heat load, thermostats, regulation of the flow rate and temperature of the supply air);
- use of heat energy from low-potential sources (heat utilizers, heat pumps);
- improvement of thermal characteristics of enclosing structures and elements of power supply systems (increase of thermal resistance of thermal transfer of enclosing structures, application of heat-reflecting screens, increase of sealing of buildings).

Since a significant potential for energy saving is in the modernization of building envelope structures (Fig. 3), the most appropriate is the introduction of energy-saving measures that increase the heat-protective characteristics of walls, Windows and floors of the building. Insulation and hermetization of buildings due to the imposition of thermal insulation or replacement of individual elements of enclosing structures naturally leads to a reduction in heat losses and, as a result, a decrease in the required amount of heat for heating. However, studies show [11–13], in buildings with natural ventilation worsen hygienic conditions of stay of people due to changes of microclimate and air exchange conditions of: decreasing the flow of fresh air, increased relative humidity, increased the likelihood of mould growth on the inner surface of the enclosing structures, which also adversely affects the quality of indoor air [14]. The problem of air quality deterioration in the use of window units with high tightness in European countries is solved by additional measures that contribute to an increase in air flow (installation of air valves, mechanical ventilation systems) [15].

2 A Case Study of Thermal Comfort in the Indoor Sport Hal

2.1 The Rules Governing the Parameters of the Microclimate in the Premises (the European Union)

EN 13779 standard defines indoor climate quality categories (I – high level; II – normal level; III – satisfactory level; IV-other). When determining the parameters of the microclimate, Stan-dart refers to EN 15251 "Indoor Environmental Input Parameters for Design and Assessment of Energy Performance of Buildings Addressing Air Quality, Thermal Environment, Lighting and Acoustics" ("Initial parameters of the microclimate for the design and evaluation of energy efficiency of buildings in relation to air quality, thermal comfort, lighting and acoustics") [16].

This standard considers the perception of human microclimate parameters in terms of the thermal balance of his body. This approach is used to calculate the Predicted Mean Vote (PMV) - the predicted average estimate of the level of thermal comfort [17]:

$$\mathrm{PMV} = \left(0.303\mathrm{e}^{-0.036}\mathrm{q}_{\mathrm{MT}}+0.028\right)\left(\left(\mathrm{q}_{\mathrm{MT}} = \mathrm{q}_{\mathrm{p}}\right) - 3.05 * 10^{-3}(5733 - 6.99) - \mathrm{p}_{\mathrm{B}}\right)$$
$$- 0.42\left(\left(\mathrm{q}_{\mathrm{MT}} - \mathrm{q}_{\mathrm{p}}\right) - 58.15\right) - 1.7 * 10^{-5}\mathrm{q}_{\mathrm{MT}}\left(5867 - \mathrm{p}_{\mathrm{B}}\right) - 0.0014\mathrm{q}_{\mathrm{MT}}(34 - \mathrm{T}_{\mathrm{B}})$$
$$- 3.96 * 10^{-8} fo\left((T_0 + 273)^4 - (Tw, p + 273)^4 - fo\alpha(\mathrm{T}_0 - \mathrm{T}_{\mathrm{B}})\right) \quad (1)$$

$$\mathrm{T}_0 = 35.7 - 0.028\left(\mathrm{q}_{\mathrm{MT}} - \mathrm{q}_{\mathrm{p}}\right) - 0.155\mathrm{R}_0$$
$$\left(3.96 * 10^{-8} fo\left((T_0 + 273)^4 - (Tw, p + 273)4 - fo\alpha(\mathrm{T}_0 - \mathrm{T}_{\mathrm{B}})\right)\right); \quad (2)$$

$$\alpha = 2.38(\mathrm{T}_0 - \mathrm{T}_{\mathrm{B}})^{0.25}\mathrm{by}\ 2.38(\mathrm{T}_0 - \mathrm{T}_{\mathrm{B}})^{0.25} > 12.1\mathrm{w}_{\mathrm{e}}^{0.5.\,and}$$
$$\alpha = 2.1\mathrm{w}_{\mathrm{e}}^{0.5}\mathrm{by}\ 2.38(\mathrm{T0} - \mathrm{T}_{\mathrm{B}})^{0.25} < 12.1\mathrm{w}_{\mathrm{e}}^{0.5}$$

$$fo = 1 + 0.29\mathrm{R}_0\mathrm{by}\mathrm{R}_0 \leq \left(0.078\mathrm{m}^2 * {}^{\circ}\mathrm{C}/\mathrm{W}\right) \text{ and}$$
$$fo = 1.05 + 0.1\mathrm{R}_0\mathrm{by}\mathrm{R}_0 > \left(0.078\mathrm{m}^2 * {}^{\circ}\mathrm{C}/\mathrm{W}\right); \quad (3)$$

Where *qMT* – human metabolism, W/m2;

qp – energy costs for performing work to employees, W/m2;
pв – partial pressure of water vapor, Pa;
Tв – air temperature, °C;
fo – coefficient of covering a part of the body with clothes in relation to naked skin;
Tw,p – the average radiation temperature, °C;
α – convective heat transfer coefficient, W/(m2·⁰C); *Ro* – thermal insulation coefficient of clothing; *wo,в* – relative air mobility in the room, m/s:

$$W_{O,\mathrm{B}} = W_{\mathrm{B}} + 0.005(q_{MT} - 58) \quad (4)$$

here w_{θ} – the average mobility of the air in the room, m/s.

On the basis of PV, the Predicted Percentage of Dissatisfied (PPD) is calculated – the predicted percentage of people dissatisfied with the quality of the microclimate:

$$\mathrm{PPD} = 100 - 95\mathrm{e}^{-(0.03353\mathrm{PMV}^4 + 0.2179\mathrm{PMV}^2)} \quad (5)$$

Since the feeling of heat in people comes with a different combination of parameters of the microclimate, the value of this indicator will always be different from zero.

Depending on the values of PMV and PPD, rooms are divided into 4 categories according to the quality of the microclimate (Table 1).

Table 1. Categories of premises depending on the quality of the microclimate

Category	PPD, %	PMV, %	Climate comfort level
I	< 6	$-0.2 < \mathrm{PMV} < 0.2$	High: recommended for areas where there are very sensitive people with special requirements: the elderly, the disabled, sick people, small children
II	< 10	$-0.5 < \mathrm{PMV} < 0.5$	Normal: should be used fornewly constructed and reconstructed buildings
III	< 15	$-0.7 < \mathrm{PMV} < 0.7$	Satisfactory: can be used for existing buildings
IV	> 15	$\mathrm{PMV} < -0.7$ or $0.7 < \mathrm{PMV}$	Other: the category can only be used for a limited time of year

In addition to recommendations for PMV and PPD values, EN 15251 contains recommended values for temperature, humidity, energy consumption, thermal insulation of clothing, fresh air intake, which consists of the consumption of air per person (depending on the quality category of the microclimate and PPD) and the consumption per 1 m2 of floor (depending on the allocation of hazards in the environment), etc.

To avoid air quality reduction during the implementation of energy-saving measures, it is necessary to determine the actual air exchange of the premises of the object and to forecast changes in air exchange after the implementation of energy-saving measures. Air quality is an important indicator when analyzing the level of comfort of the environment in the room.

2.2 The Principle of Air-Tightness

Leaking air leads to heat leakage, which in turn leads to excessive energy consumption. [18] Uncontrolled movement of air and its loss through the enclosing structures (walls) of buildings and structures can not be called ventilation.

To provide buildings and structures with low energy consumption, such walls and enclosing elements are necessary that would have an airtight property, so that the ventilation of the building and structure was controlled [19].

Only with the provision of all the above mentioned thermal insulation of buildings and structures can be effective. The Congress on environmental protection has developed requirements for air tightness of buildings and structures. These recommendations are an indication only for new construction [20].

Energy consumption of new buildings and structures that meet these guidelines should be reduced by 2/3, which corresponds to the following indicators: 50 kWh/m2/year (this value may vary depending on the area and the size of the living area) [21].

According to these requirements, it is necessary to conduct a mandatory test to ensure the airtightness of the premises before putting objects into operation [22].

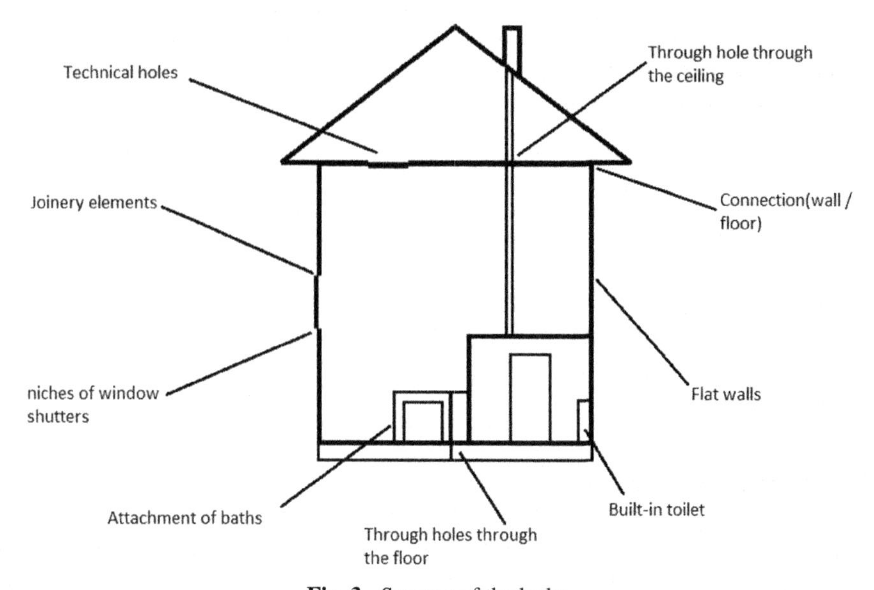

Fig. 3. Sources of the leaks.

Sources of air leaks:

1. Flat walls. Facing facades provides air tightness of vertical walls, but non-plastered areas (garage walls, roof, joints, etc.) allow air to pass through.
2. Connection (wall/floor). When installing floor slabs, voids may form along its perimeter. It is necessary to seal this circuit to prevent the movement of air flow through the joints.
3. Niches of window shutters. Leaks may form in the places where mobile shutters are installed.
4. Joinery elements. Joiner's elements should be sealed in places of fastenings from outside and from inside by means of mastic.

5. Through holes through the floor/ceiling. These holes must be carefully treated.
6. Electricity. Air leaks in places of laying of electric wires and switches are insignificant, but places of laying of electric networks through walls and protecting designs can serve as a source of air filtration.
7. Fixing baths. Failure to isolate the mounting holes may result in significant air loss.
8. Built-in toilets.
9. Technical holes.

3 Math Model

To obtain information about the state parameters of the HVAC system, a variety of sensors are used, including temperature sensors, humidity sensors, air flow pressure sensors, subject occupancy sensors, air quality sensors humidity, air quality sensors. All received data will be saved in the database (DB). This database will allow you to create specific solutions for climate management. In addition, each solution is unique in the data warehouse and has a unique set of identifiers.

The temperature at any time is determined by the formula

$$T = T_0 + \frac{Q}{c \cdot m} dt, \tag{6}$$

Where T 0 – temperature at the previous time, (°C);

Q – thermal balance of the object, W;
c – the heat capacity of the object, W/(kg*°C);
m – the mass of an object, kg;
dt – the account interval, c.

4 Experiment Results

In this study, the mixing ventilation system was modeled. It is organized with the help of air distributors that supply air to the room with air jets that have high speed and turbulence, causing intense air circulation. As a result, the fresh air of the supply jet is mixed with the air of the room. If complete mixing occurs, the air parameters (temperature, relative humidity, speed of movement), as well as the content of pollutants will be the same at any point of the serviced room at a certain distance from the inflow site.

The initial data and theoretical calculations of the expected parameters of the microclimate adopted during the study are shown in Table 1.

In the present article, the study simulated the following situation: airflow is determined by the rate of supply of sanitary standards (80 m^3/h per person), the supply air is supplied with ambient air temperature at parameter A.

Ideally, the results of the theoretical calculations given in the table above and the results of the numerical simulation should converge, i.e. the moisture content of the air, the air temperature and the carbon dioxide content in the exhaust should be equal.

Table 2. The source data and the theoretical calculations of the expected parameters

Source data	
Hall type	Universal (Group strength training)
Type of load	Moderate severity (II a)
Number of persons	25
Air consumption, m3/h	2000
Air temperature at the tributary, 0 C	22
Moisture content of the air at the inflow g/kg	10
CO_2 content at the inflow kg/kg	0.00061
Room temperature, 0 C	25
Air density, kg/m3	1.19
Heat gain	
From 1 person explicit/full, W	73/194
From the people in the room a clear/complete, W	1825/4850
From equipment, W	0
From lighting, W	1680
From solar radiation, W	2210
Total explicit/full, W	5788/8934
Moisture access	
From 1 person, kg/h	0.185
From people, kg/h	4.625
From other sources, kg/h	0
Moisture loss, kg/h	0
Total, kg/h	4.625
CO2 gain	
From 1 person, kg/h	0.07
From people, kg/h	1.75
From other sources, kg/h	0
Total, kg/h	1.75
Calculation of the expected parameters	
Temperature, 0 C	30.7
CO2 on the hood, kg/kg; ppm	0.0013/876.41
Moisture content on the hood, g/kg	11.88
Process beam, kJ/kg	6954

a) Plan of the testing sport hall b) Sport hall

Fig. 4. Testing facility

Description of the Facility. We provided experiments in the Sport Hall of A. Yasawi International Kazakh-Turkish University during the sportsmen did different type of sport games. The experiments were carried out on the previously described test facility (Fig. 4).

Figure 5a shows the result of the experiment with the proposed fuzzy-PID controller. To bring the monitored zone to a given point from different values of the initial temperature, it takes between 30–60 min. It becomes shorter if the initial temperature is higher, close to target temperature, and the air conditioning zone is well sealed (windows are not open). If the initial temperature is lower and the door and windows are often opened, the controller takes longer to reach the set temperature. Eventually, the controller did not take more than sixty minutes to achieve a room temperature of up to 21°C, which corresponds to 80% bandwidth satisfaction in accordance with ASHRAE 55-2010 [8] under all conditions. Everywhere the statement that the PID controller is fuzzy starts to working and operate the heater at 8:00, in our experiments it is one hour before the starting working time.

Figure 5b shows the results of temperature changes for a single day on the premises, January 17, 2017. The initial room temperature is about 15 °C, and rises rapidly to the set value when heated. During working hours from 9:00 to 18:00, the room temperature is usually maintained between 20 °C and 21 °C. From this it follows that the temperature curve inside the room is relatively stable and there are no abrupt changes during working hours. This means that the controller with a fuzzy-PID controller controls the room temperature well.

Figure 5c demonstrates the indoor air temperature that observed in May 2017, when average outdoor temperature was 25 °C in daytime, and 11.5 °C at night. The day when the experiment conducted is May 18, 2017, that time the highest outdoor temperature was 29 °C. In this temperature the controller keeps the desired temperature due to working power of conditioning system. When the working day starts initial temperature was about 20 °C, from 8.00 the PID controller starts to regulating. There is the highest temperature can be observed between 12.00 and 14.00 because of opening the window that time. Despite, does not exceed 26 °C, although it needs more working power of conditioner and requires more power consumption (Table 2).

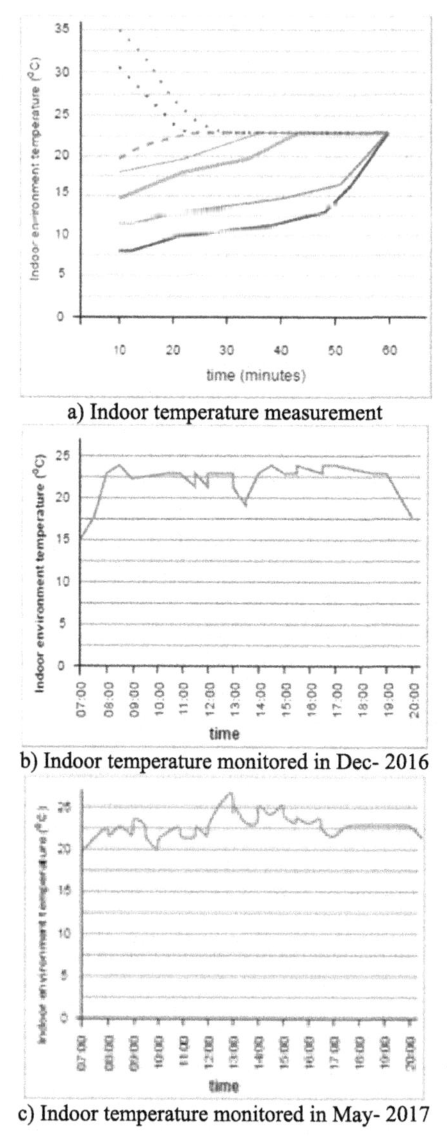

a) Indoor temperature measurement

b) Indoor temperature monitored in Dec- 2016

c) Indoor temperature monitored in May- 2017

Fig. 5. PID parameters regulating process

5 Conclusion

Ensure a comfortable climate in a building, considering energy-efficient operation, attracts a great deal of attention worldwide. In this paper, we proposed a model to ensure a comfortable indoor environment for sportsmen. To obtain high accuracy in controlling the comfort parameters, mathematical models of the parameters are investigated, after which decoupling strategies for comfort parameters are considered. By applying the developed mathematical model, the proposed fuzzy based self-adjusted PID controller was designed.

Experimental results demonstrated the effectiveness of the proposed controller. Comfort values of the indoor environment parameters were according to international standards for indoor environment comfort as specified by ASHRAE and ISO. In future, in order to get higher accuracy, we will improve the system by using a presence detector, and training the system in GPU.

References

1. Escandón, R., Ascione, F., Bianco, N., Mauro, G.M., Suárez, R., Sendra, J.J.: Thermal comfort prediction in a building category: artificial neural network generation from calibrated models for a social housing stock in southern Europe. Appl. Thermal Eng. **150**, 492–505 (2019)
2. Altayeva, A., Omarov, B., Suleimenov, Z., Im Cho, Y.: Application of multi-agent control systems in energy-efficient intelligent building. In: 2017 Joint 17th World Congress of International Fuzzy Systems Association and 9th International Conference on Soft Computing and Intelligent Systems (IFSA-SCIS), pp. 1–5. IEEE, June 2017
3. Omarov, B., Altayeva, A., Im Cho, Y.: Smart building climate control considering indoor and outdoor parameters. In: Saeed, K., Homenda, W., Chaki, R. (eds.) CISIM 2017. LNCS, vol. 10244, pp. 412–422. Springer, Cham (2017). https://doi.org/10.1007/978-3-319-59105-6_35
4. Moshkalov, A.K., Baimuhkanbetov, B.M., Baikulova, A.M., Anarbayev, A.K., Ibrayev, A.Z., Mynbayeva, A.P.: The content-structure model of students' artistic selfdevelopment through the use of information and communications technology. Astra Salvensis—Rev. Hist. Cult. VI (12), 363–383 (2018)
5. Hilliard, T., Swan, L., Qin, Z.: Experimental implementation of whole building MPC with zone based thermal comfort adjustments. Build. Environ. **125**, 326–338 (2017)
6. Kabrein, H., Yusof, M.Z.M., Hariri, A., Leman, A.M., Afandi, A.: Improving indoor air quality and thermal comfort in office building by using combination filters. In: IOP Conference Series: Materials Science and Engineering, vol. 243, No 1, p. 012052. IOP Publishing, September 2017
7. Omarov, B., et al.: Agent based modeling of smart grids in smart cities. In: Chugunov, A., Misnikov, Y., Roshchin, E., Trutnev, D. (eds.) EGOSE 2018. CCIS, vol. 947, pp. 3–13. Springer, Cham (2019). https://doi.org/10.1007/978-3-030-13283-5_1
8. Altayeva, A., Omarov, B., Im Cho, Y.: Towards smart city platform intelligence: PI decoupling math model for temperature and humidity control. In: 2018 IEEE International Conference on Big Data and Smart Computing (BigComp), pp. 693–696. IEEE, January 2018
9. Buyak, N.A., Deshko, V.I., Sukhodub, I.O.: Buildings energy use and human thermal comfort according to energy and exergy approach. Energy Build. **146**, 172–181 (2017)
10. Gaonkar, P., Aadhithan, N.A., Bapat, J., Das, D.: Energy budget constrained comfort optimization for smart buildings. In: 2017 IEEE Region 10 Symposium (TENSYMP), pp. 1–5. IEEE, July 2017
11. Lin, C.M., Liu, H.Y., Tseng, K.Y., Lin, S.F.: Heating, ventilation, and air conditioning system optimization control strategy involving fan coil unit temperature control. Appl. Sci. **9**(11), 2391 (2019)
12. Hao, J., Dai, X., Zhang, Y., Zhang, J., Gao, W.: Distribution locational real-time pricing based smart building control and management. In: 2016 North American Power Symposium (NAPS), pp. 1–6. IEEE, September 2016

13. Brissette, A., Carr, J., Juneau, P.: The occupant comfort challenge of building energy savings through HVAC control. In: 2017 IEEE Conference on Technologies for Sustainability (SusTech), pp. 1–7. IEEE, November 2017
14. Ding, Y., Wang, Q., Kong, X., Yang, K.: Multi-objective optimisation approach for campus energy plant operation based on building heating load scenarios. Appl. Energy **250**, 1600–1617 (2019)
15. Shektibayev, N.A., et al.: A model of the future teachers' professional competence formation in the process of physics teaching. Man India **97**(11), 517–529 (2017)
16. Omarov, B., Orazbaev, E., Baimukhanbetov, B., Abusseitov, B., Khudiyarov, G., Anarbayev, A.: Test battery for comprehensive control in the training system of highly Skilled Wrestlers of Kazakhstan on National wrestling "Kazaksha Kuresi". Man India **97**(11), 453–462 (2017)
17. Narynov, S., Mukhtarkhanuly, D., Omarov, B.: Dataset of depressive posts in Russian language collected from social media. Data Brief **29**, 105195 (2020)
18. Zhang, C., Kuppannagari, S.R., Kannan, R., Prasanna, V.K.: Building HVAC scheduling using reinforcement learning via neural network based model approximation. In: Proceedings of the 6th ACM International Conference on Systems for Energy-Efficient Buildings, Cities, and Transportation, pp. 287–296, November 2019
19. Omarov, B., et al.: Fuzzy-PID based self-adjusted indoor temperature control for ensuring thermal comfort in sport complexes. J. Theor. Appl. Inf. Technol. **98**(11) (2020)
20. Haniff, M.F., Selamat, H., Khamis, N., Alimin, A.J.: Optimized scheduling for an air-conditioning system based on indoor thermal comfort using the multi-objective improved global particle swarm optimization. Energy Effi. **12**(5), 1183–1201 (2018). https://doi.org/10.1007/s12053-018-9734-5

Applications of Collective Intelligence

Smart Solution to Detect Images in Limited Visibility Conditions Based Convolutional Neural Networks

Ha Huy Cuong Nguyen[1,2(✉)], Duc Hien Nguyen[1,2], Van Loi Nguyen[1,2], and Thanh Thuy Nguyen[1,2]

[1] Vietnam-Korea University of Infomation and Communication Technology, Danang, Vietnam
nhhcuong@vku.edu.vn, {ndhien,nvloi}@vku.udn.vn

[2] Faculty of Computer Science, VNU University of Engineering and Technology, Hanoi, Vietnam
nguyenthanhthuy@vnu.edu.vn

Abstract. Decrease in visibility causes many difficulties in vision, tracking. Current classic object detection techniques do not give satisfying results in less visibility. It is essential to detect and recognize the objects under such conditions and devise a better object detection mechanism. The paper proposes a solution to this problem by using a multi step approach that uses Saliency techniques and modern object detection algorithms to obtain the desired results. The distorted image is enhanced via a deep neural network for visibility enhancement. The image frame of a better quality undergoes saliency techniques so that less visible objects are visible. Faster Region-based Convolutional Neural Network (R-CNN) then runs on the saliency output to yield bounding boxes for all the objects. The coordinates of the bounding boxes are then applied on the original image thus detecting all the objects in a distorted image with less visibility.

Keywords: Saliency · Deep neural network · R-CNN · Visibility enhancement

1 Introduction

Currently, in countries around the world, the weather conditions are not guaranteed for people when circulating on the road. For example the following case: people are moving on the road with different means of transport such as motorcycles, cars, motorcycles or other rudimentary vehicles. Sudden heavy rains occur making it very difficult for people to travel, or due to work demands people move in the late night, or early morning. All of the problems that stand out now are the many accidents that often happen. According to statistics, the number of accidents occurring is exponentially higher than in normal weather conditions.

M. Hernes et al. (Eds.): ICCCI 2020, CCIS 1287, pp. 641–650, 2020.
https://doi.org/10.1007/978-3-030-63119-2_52

Therefore, the solution to detect obstacles in stormy and difficult weather conditions is practical and humane in the research process and this is considered a very positive solution for all society.

Less visibility in images can be due to various reasons. These could be environmental conditions like mist or other reasons like pollution, smoke and dust which could lead to poor image quality. Pixelated and blur images are also types of distorted images. It is very difficult to do object detection in such images and many state of the art object detection algorithms like Faster R-CNN algorithm do not give the desired output when working with such images. These conditions make some or all objects in an image less visible and hence bounding boxes are not obtained around these objects in an image. As shown in Figs. 1 (a) and (b), Faster R-CNN Algorithm is unable to detect all the objects. The objects in the images are not clearly visible due to smoke and mist. These poor results point to the need of devising a solution such that existing detection algorithms can be extended to such images wherein all the objects are detected accurately.

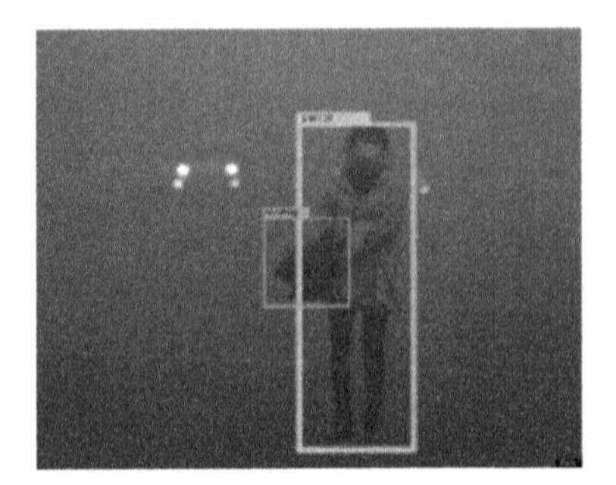

Fig. 1. (a) and (b) : Failure of faster R-CNN detection algorithm to detect all objects

Saliency map is a tool for image segmentation which expresses the image in a more meaningful manner. It extracts the contours from the image and highlights the objects that are not clear in the image frame, in a better manner. It can be clearly shown that the objects that were not detected and recognized by Faster R-CNN have clear and distinct boundaries in their respective Saliency maps. The Saliency Map of the images are shown in Figs. 2 (a) and (b). The intensities of white shaded regions for each pixel the saliency map points to the fact that an object exists. These intensities are incorporated in the Saliency Matrix.

The results in Figs. 2 (a) and (b) thus provide an additional proof to our notion of using Saliency Map for detecting objects that were not detected by Faster R-CNN Algorithm. In this paper, we propose various techniques to enhance the visibility in such poor quality images, using Saliency techniques and then finally implementing the Faster R-CNN algorithm on our saliency output. The object detection done on the saliency output can yield accurate dimensions of bounding boxes that can be drawn upon the original image for accurate object detection.

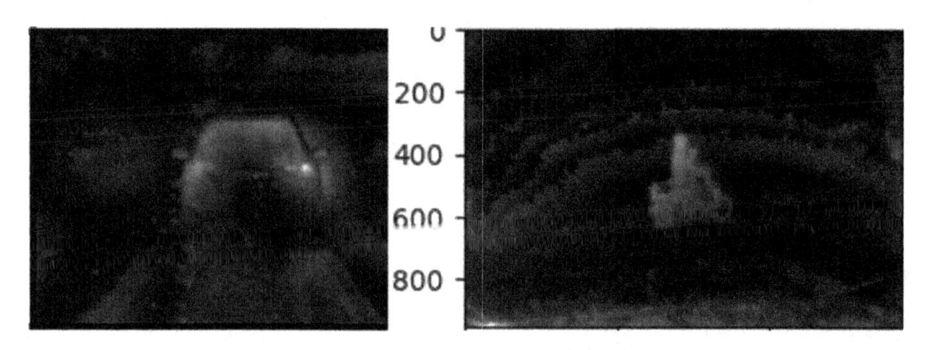

Fig. 2. (a) and (b) : Saliency Map of the images from the dataset used

2 Related Works

Kopf *et al.* [1] propose a system for enhancement of images by combining them with geo-referenced terrain and urban models. It allows a variety of tools like dehiring and relighting the image and has been incorporated in the visibility enhancement section. Numerous works have also proposed similar techniques under foggy and hazy circumstances [3,7,10]. Lu *et al.* [2] devised a new method of Object Saliency Detection in foggy images. Fog can be one of the reasons causing less visibility in images and therefore covariance feature description matrix method has been used as one of the saliency technique for obtaining a good saliency output. Islam et al. [4] propose a ranking based saliency object detection technique and is able to detect multiple objects depending on various parameters. This technique lays the foundation of creation of saliency map in our study. A detailed insight of the Fast R-CNN and Faster R-CNN object detection mechanism have been proposed by [5] and [6]. The implementation fails to give desired results in case of images in less visibility and therefore our solution proves to overcome these drawbacks.

R-CNN family includes R-CNN, Fast R-CNN, Faster R-CNN, and Mask R-CNN that are popular object detection.

Two research groups of Zhao [20] and Girshick [36] proposed model (R-CNN). This is a basic technique that uses Neural networks to detect objects, but this technical solution requires a lot of time. processing time. The input to the technique using the R-CNN model is an image that is extracted into small dimensions called area suggestions for ease of handling. The R-CNN model uses a selective search method to extract reference ranges, areas that can be divided into groups of three objects or a set of objects. Selective search provides a range of candidate suggestions. Next, all of these regional proposals are packaged and sent to the cumulative Neural Network (CNN) [33]. Lastly, they use a Support Vector Machine (SVM) to classify the presence of the object [29].

In recent years, researchers have improved the CNN model that processes each image of one input one after another, announcing the R-CNN improvement model. The R-CNN model has a faster computation time, the research method uses a region of aggregated group of the same size to create a fixed-size vector of regional proposals. Quick R-CNN provides all images based on CNN training techniques to create cumulative feature maps. The softmax function is used in various multiclass classification methods [20,23,34].

3 Proposed Work

3.1 Overview

This research proposes a solution for object detection in distorted images with less visibility. Initially, the distorted image is first preprocessed via visibility enhancement mechanism using a deep neural network and is discussed in Sect. 3.2. After the image has been enhanced, saliency region detection techniques are used to obtain a saliency map wherein the objects are detected. The possibility of having an object can be seen with white spots in contrast to the black background of the obtained Saliency map, which is discussed in Sect. 3.3. The Saliency map is then processed via an object detection algorithm i.e. Faster R-CNN that has been trained on saliency maps of the dataset mentioned. This technique is discussed in Sect. 3.4. The model for implementation of Faster R-CNN was trained using a dataset comprising of saliency maps of the dataset, PASCAL VOC 2020 available on the Internet. Thus, the R-CNN in our research created bounding boxes over the saliency obtained after preprocessing of the distorted image. Finally, the coordinates of all the bounding boxes generated by the Faster R-CNN Algorithm on the saliency map are saved and these boxes are then marked on the originally distorted image. The final output consists of the original distorted image with bounding boxes around all the distinct objects including the less visible ones.

3.2 Deep Neural Network for Visibility Enhancement

The deep neural network accepts a distorted image as input, models the corresponding distorted component and yields an enhanced version of the image as output. The network has an input layer, output layer and some hidden layers between the two. Work has been proposed [1] that a deep photo system can be established, by using the large amounts of data from the geographic information system (GIS). This information can greatly help in achieving operations like visibility enhancement. Hussain et al. [3] focus on finding the optimal combination of weights for approximating the network function to the given function *f*. We have used a deep neural network to solve the enhancement problem so that the images can be processed further using salient detection techniques. The network has been trained on several images with less visibility like those in foggy conditions, or excessive air pollution. For learning the generalized visibility function, the distorted images are split into non overlapping blocks of size $N \times N$ pixels and is normalized in range for optimizing the results of DNN. Rasterization is performed for transforming 2D Block into 1D vector as our DNN takes 1D vector as input. The weights are randomized, and input is forward propagated for generating propagation output activation. Previous work [3] used the hyperbolic tangent transfer function as in Eq. (1).

$$F = \frac{e^{x} - e^{-x}}{e^{x} + e^{-x}} \tag{1}$$

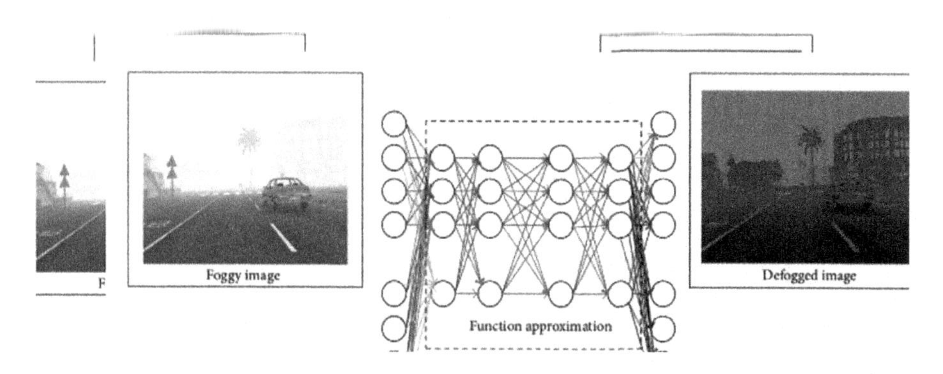

Fig. 3. Deep neural network for image enhancement

This transfer function is similar to sigmoid and produces output in a closed range from −1 to +1. It is efficient because of broader output space. An error function is calculated since the output produced by the input differs from the target output. This function is minimized in order to find the solution of the learning process. Hussain *et al.* [3] obtain the desired set of weights via back-propagation algorithm. The back-propagation is an iterative gradient descent algorithm in which the output error signals are transmitted backward from the output layer to each node in the hidden layer that immediately contributed to the output layer. This backward propagation is continued, layer by layer, until each node in the network has received an error signal that describes its relative contribution to the overall error. This back-propagation of errors and the updating of weights continue until the value of error function becomes sufficiently small. The network training process is completed and the weights are stored such that the network successfully produces the desired output for new input patterns (Fig. 3).

3.3 Saliency Detection Techniques

Islam *et al.* [4] uses a deep neural network to detect saliency objects by considering the relative rank of salience of these objects and is able to detect all saliency regions in an image. The detection network is an extension to the convolution-deconvolution pipelines, which include a feed-forward network for initial coarse-level prediction. A stage-wise refinement mechanism is used in which predictions of finer structures are gradually reinstated. The feature extractor converts the image to a rich feature representation, while lost contextual information is recovered by the refinement stages. The network thus becomes capable of yielding accurate predictions along with the ranking.

Training the Network. The repository [8] was cloned and dataset was prepared with correct annotations and bounding boxes. The same architecture as the PASCAL VOC 2020 dataset transformed into the format shown in Fig. 4.

```
|-- data/
    |-- Annotations/
        |-- *.txt (Annotation files)
    |-- Images/
        |-- *.png (Image files)
    |-- ImageSets/
        |-- train.txt
```

Fig. 4. R-CNN architecture

3.4 Object Detection on Distorted Image

Technical solution includes the following steps:

- Step 1. Saliency Map with Bounding Boxes generated by Faster R-CNN is taken as input (As per the previous section).
- Step 2. Coordinates of all Bounding Boxes are saved.
- Step 3. Bounding Boxes having same coordinates are formed over the distorted image.

4 Results and Discussions

The experimental results are shown in Fig. 5(a), (b). Figure 5(a), (b) explains the causes of the losses, the losses are considered valid and the error rate in 5 training cycles with the learning rate of 0.0005. The accuracy we achieved was 97%. The proposed classification process was based on PyTorch's deep learning library. We have trained the Resnet100 model with all the middle-aged people (about 40–60 years old) in include all kinds of 2D CNN, 3D CNN, and CRNN [38].

The model was evaluated by calculating precision and recall, using True Positives (TP), True Negatives (TN), False Positives (FP), False Negatives (FN) in the confusion grid matrix.

Fig. 5. The image post implementation of DNN for visibility enhancement

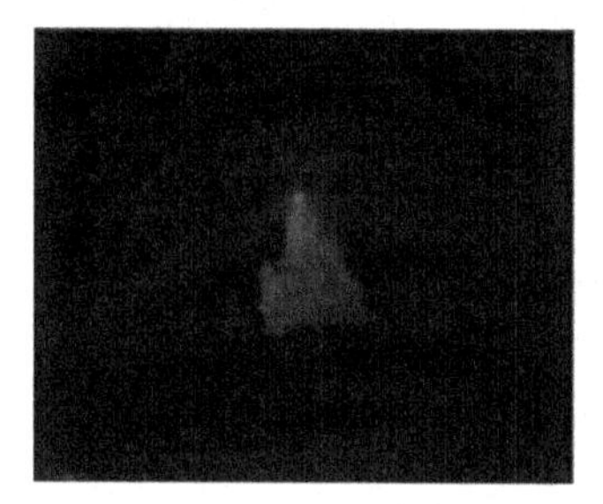

Fig. 6. Shows saliency map generated post implementation of saliency techniques

Figure 5(a) shows the image taken as input for carrying the research. Figure 5(b) shows the image post implementation of DNN for visibility enhancement. Figure 6(a) shows saliency map generated post implementation of saliency techniques. Figure 6(b) shows bounding boxes over distinct objects in the saliency map after implementation of faster R-CNN. Figure 7 shows bounding boxes over distinct objects in Fig. 6(a).

In the paper, the refining and release techniques were applied to all layers of the input image data dataset trained from the improved R-CNN model. In order to maintain the calculation results through the trained steps, the set of typical classes of the trained base model. Experiments conducted to extract some of the selected data a subset compared to the initially divided classes. In the experiment script in all the proposed method retains deep learning rate is 0.00005 for the initial class and 0.000055 for the final layer. After 20 min of training, the accuracy achieved is 98.33 %.

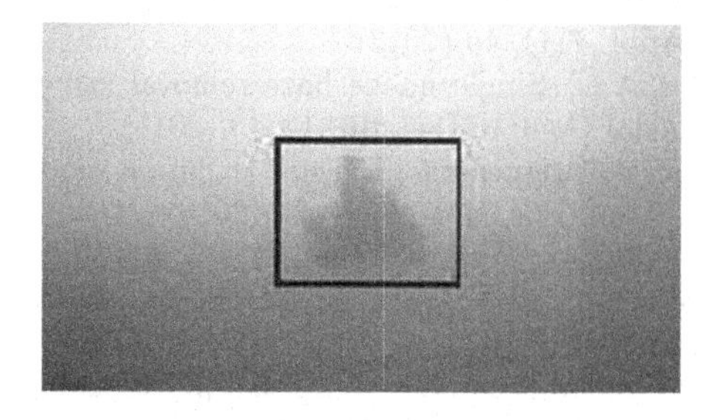

Fig. 7. Shows bounding boxes over distinct objects in Fig. 5(a)

5 Conclusion

In this paper, we proposed one research comprising of a three-step process has given good results for distorted images. A lot of the method research has been tested on images with fog, smoke, and dust. The output obtained in all such cases is accurate and as per the expectation. Thus, implementation of visibility enhancement DNN followed by the implementation of faster R-CNN on obtained saliency image can detect all the objects in poor quality images. With the published technical solution, we hope to apply and deploy it in developing countries with many difficulties. In particular, support for nearsighted people, eyes do not regulate well when participating in traffic in poor weather conditions. The study proposes future scope into video frames allowing real-time object detection in less visibility.

References

1. Kopf, J., et al.: Deep photo: model-based photograph enhancement and viewing. In: ACM Transactions on Graphics, Proceedings of SIGGRAPH Asia, vol. 27, pp. 116:1–116:10 (2008)
2. Lu, W., Sun, X., Li, C.: A new method of object saliency detection in foggy images. In: Zhang, Y.-J. (ed.) ICIG 2015. LNCS, vol. 9217, pp. 206–217. Springer, Cham (2015). https://doi.org/10.1007/978-3-319-21978-3_19
3. Hussain, F., Jeong, J.: Visibility enhancement of scene images degraded by foggy weather conditions with deep neural networks. J. Sens. **2016**, Article ID 3894832, 9 (2016)
4. Amirul Islam, M.D., Kalash, M., Bruce, N.D.B.: Revisiting salient object detection: simultaneous detection, ranking, and subitizing of multiple salient objects. In: Computer Vision and Pattern Recognition (2018). arXiv:1803.05082
5. Girshick, R.: Fast R-CNN. arXiv:1504.08083 (2015)
6. Ren, S., He, K., Girshick, R., Sun, J.: Faster r-cnn: towards real-time object detection with region proposal net- works. arXiv preprint arXiv:1506.01497 (2015)
7. Guo, J., Ren, T., Bei, J., Zhu, Y.: Salient object detection in RGB-D image based on saliency fusion and propagation. In: Proceedings of the 7th International Conference on Internet Multi-media Computing and Service, p. 59. ACM (2015)
8. Song, H., Liu, Z., Du, H., Sun, G., Le Meur, O., Ren, T.: Depth-aware salient object detec-tion and segmentation via multiscale discriminative saliency fusion and bootstrap learning. IEEE Trans. Image Process. **26**(9), 4204–4216 (2017)
9. Guo, F., Tang, J., Cai, Z.: Fusion strategy for single image dehazing. Int. J. Digital Content Technol. Appl. **7**(1), 19 (2013)
10. Ansia, S., Aswathy, A.L.: Single image haze removal using white balancing and saliency map. Procedia Comput. Sci. **46**, 12–19 (2015)
11. Erdem, E.: A region covariance-based visual attention model for RGB-D images. Int. J. Intell. Syst. Appl. Eng. **4**(4), 128–134 (2016)
12. Qu, L., He, S., Zhang, J., Tian, J., Tang, Y., Yang, Q.: RGBD salient object detection via deep fusion. IEEE Trans. Image Process. **26**(5), 2274–2285 (2017)
13. Du, J.: Understanding of object detection based on CNN family and YOLO. In: Journal of Physics: Conference Series, vol. 1004, p. 012029. IOP Publishing (2018)

14. Erdem, E., Erdem, A.: Visual saliency estimation by nonlinearly integrating features using region covariances. J. Vis. **13**(4), 11–11 (2013)
15. Chen, L.-C., Papandreou, G., Kokkinos, I., Murphy, K., Yuille, A.L.: Deeplab: semantic image segmentation with deep convolutional nets, atrous convolution, and fully connected crfs. TPAMI **40**(4), 834–848 (2018)
16. Lakshmanaprabu, S.K., Mohanty, S.N., Shankar, K., Arunkumar, N.: Optimal deep learning model for classification of lung cancer on CT images. Future Gener. Comput. Syst. **19**(1), 374–382 (2019)
17. Rajeshwari, P., Abhishek, P., Srikanth, P., Vinod, T.: Object detection: an overview. Int. J. Trend Sci. Res. Dev. (IJTSRD) **3**(1), 1663–1665 (2019)
18. Tian, Y., Yang, G., Wang, Z., et al.: Apple detection during different growth stages in orchards using the improved YOLO-V3 model. Comput. Electron. Agric. **157**, 417–426 (2019)
19. Wu, X., Sahoo, D., Hoi, S.C.H.: Recent advances in deep learning for object detection. arXiv:1908.03673v1 [cs.CV], August 2019
20. Zhao, Z.-Q., Zheng, P., Xu, S., Wu, X.: Object detection with deep learning. arXiv:1807.05511 [cs.CV], April 2019
21. Alexey, A.B.: Apple detection during different growth stages in orchards using the improved YOLO-V3 model (2018). https://github.com/AlexeyAB/Yolo
22. Kim, S., Ji, Y., Lee, K.: An effective sign language learning with object detection based ROI segmentation. In: Second IEEE International Conference on Robotic Computing (IRC), Laguna Hills, CA, pp. 330–333 (2018)
23. Nwankpa, C., Ijomah, W., Gachagan, A., Marshall, S.: Activation functions: comparison of trends in practice and research for deep learning. arXiv:1811.03378 [cs.LG], November 2018
24. He, K., Gkioxari, G., Dollár, P., Girshick, R.: Mask R-CNN. arXiv:1703.06870 [cs.CV], January 2018
25. Munera, S., Amigo, J.M., Blasco, J., Cubero, S., Talens, P., Alexios, N.: Ripeness monitoring of two cultivars of nectarine using VIS-NIR hyperspectral reflectance imaging. J. Food Eng. **214**(3), 29–39 (2017)
26. Bargoti, S., Underwood, J.: Deep fruit detection in orchards. In: IEEE International Conference on Robotics and Automation (ICRA), Singapore, pp. 3626–3633 (2017). https://doi.org/10.1109/ICRA.2017.7989417
27. Zhang, Y., Sohn, K., Villegas, R., Pan, G., Lee, H.: Improving object detection with deep convolutional networks via bayesian optimization and structured prediction. arXiv:1504.03293 [cs.CV], January 2016
28. Lu, Y., Javidi, T., Lazebnik, S.: Adaptive object detection using adjacency and zoom prediction. arXiv:1512.07711 [cs.CV], April 2016
29. Santagapita, P.R., Tylewicz, U., Panarese, V., Rocculi, P., Dalla Rosa, M.: Non-destructive assessment of kiwifruit physic-chemical parameters to optimize the osmotic dehydration process: a study on FT-NIR spectroscopy. J. Biosyst. Eng. **142**(2), 101–129 (2016)
30. Ren, S., He, K., Girshick, R., Sun, J.: Faster R-CNN: towards real-time object detection with region proposal networks. arXiv:1506.01497v3 [cs.CV], January 2016
31. Redmon, J., Divvala, S., Girshick, R., Farhadi, A.: You only look once: unified, real-time object detection. arXiv:1506.02640 [cs.CV], May 2016
32. Simonyan, K., Zisserman, A.: Very deep convolutional networks for large-scale image recognition. In: ICLR (2015)
33. O'Shea, K., Nash, R.: An introduction to convolutional neural networks. arXiv:1511.08458 [cs.NE], December (2015)

34. Girshick, R.: Fast R-CNN. arXiv:1504.08083 [cs.CV], September 2015
35. Erhan, D., Szegedy, C., Toshev, A., Anguelov, D.: Scalable object detection using deep neural networks. In: The IEEE Conference on Computer Vision and Pattern Recognition (CVPR), pp. 2147–2154 (2014)
36. Girshick, R., Donahue, J., Darrell, T., Malik, J.: Rich feature hierarchies for accurate object detection and semantic segmentation. arXiv:1311.2524 [cs.CV], October 2014
37. Jia, K., Wang, X., Tang, X.: Image transformation based on learning dictionaries across image spaces. IEEE Trans. Pattern Anal. Mach. Intell. **35**(2), 367–380 (2013)
38. Common Objects in Context. http://cocodataset.org/
39. Open Images Dataset V5. https://storage.googleapis.com/openimages/web/index.html

Experience Report on Developing a Crowdsourcing Test Platform for Mobile Applications

Nguyen Thanh Binh[1](✉), Mariem Allagui[2], Oum El-Kheir Aktouf[2], Ioannis Parissis[2], and Le Thi Thanh Binh[3]

[1] The University of Danang - Vietnam Korea, University of Information and Communication Technology (VKU), Da Nang, Viet Nam
ntbinh@vku.udn.vn

[2] Grenoble Alpes Univ. Grenoble INP, LCIS, Valence, France
mariem.allagui@grenoble-inp.org,
oum-el-kheir.aktouf@lcis.grenoble-inp.fr,
ioannis.parissis@grenoble-inp.fr

[3] The University of Danang, University of Science and Education, Da Nang, Viet Nam
lttbinh@ued.udn.vn

Abstract. Crowdsourcing-based testing is a recent approach where testing is operated by volunteer users through the cloud. This approach is particularly suited for mobile applications since various users operating in various contexts can be involved. In the field of software engineering, crowd-testing has acquired a reputation for supporting the testing tasks, not only by professional testers, but also by end users. In this paper, we present TMACSTest (Testing of Mobile Applications using Crowdsourcing). This platform provides the important features for crowdsourcing testing of mobile apps by means of the following functionalities: It allows mobile app providers to register and upload mobile apps for testing, and it allows volunteering Internet users to register and test uploaded mobile apps. Expected behavior is that uploaded mobile apps are tested by many different Internet users in order to cover different runtime platforms and meaningful geographical locations.

Keywords: Mobile application testing · Crowdsourced testing · Crowd-based testing

1 Introduction

Nowadays, mobile applications are becoming increasingly popular. Some studies estimate that by 2020 there will be 378 billion downloads of mobile applications. However, some statistics consider that about 70% of users uninstall mobile apps due to a bug [1]. Therefore, mobile apps need to be tested across a full range of operating contexts before their release on the market to ensure a high-quality user experience. Mobile apps are usually expected to work on many devices with various screen settings and operating

M. Hernes et al. (Eds.): ICCCI 2020, CCIS 1287, pp. 651–661, 2020.
https://doi.org/10.1007/978-3-030-63119-2_53

systems, in different locations, and across a host of carriers and networking technologies. As a result, mobile app testing has become difficult and pricey in terms of resources.

There are a multitude of challenges and issues in mobile app testing [2, 14, 15]. Some of them are:

- Higher testing cost from using real mobile devices and testers.
- Mobile apps are commonly supported by various wireless connectivity infrastructures and service plans.
- This requires mobile networking service costs and infrastructure support in a lab-based testing environment.
- Mobility provides location-based services, which need to be tested in many real user environments on different geographical locations with different languages.

Thus, we need new cost-effective test methods and tools to ensure quality of mobile apps.

Recently, inspired by the crowdsourcing concept-crowdsourcing is an emerging distributed problem solving model by online workers [3]. A new testing approach named crowdsourced testing has gained much attention in the software engineering community [4, 5].

Crowdsourced testing uses an online platform to assign software test tasks to a group of online testers. And it has become an imperative approach and has gained a reputation for the software engineering community [6]. However, with regards to test efficiency, two main questions have been arisen: (1) how to ensure an efficient test execution by crowdsourced testers, who are not necessarily mobile applications or even software tester experts, and (2) how to allow the analysis of test reports and how to ensure that specific test coverage measures are reached.

This paper presents an ongoing development of a web application, called TMACSTest, that uses crowdsourcing to test mobile applications. As a web application testing platform, TMACSTest connects mobile apps developers with thousands of crowdsourced mobile testers to test those applications. To answer question (1) above, TMACSTest provides users with specific mobile application models to allow efficient application of model-based testing to mobile applications. Question (2) is taken into account by providing an aggregation procedure of test result reports with an in-depth analysis to deduce relevant test results and test coverage measures. By doing so, TMACSTest makes the usage of crowdsourced test for mobile applications easier and more efficient.

The remaining parts of this paper are organized as follows. In Sect. 2, we present the background on crowdsourced testing. In Sect. 3, we describe the developed solution for crowdsourced testing provided by TMACSTest. Section 4 provides the details of TMACSTest development. Finally, Sect. 5 concludes the paper.

2 Background

2.1 Basics on Crowdsourced Testing

Crowdsourcing is an distributed problem-solving and production model. It was used to describe how businesses were using the Internet to outsource work to the crowd. The term "crowdsourcing" was coined in 2006 by Howe and Robinson. Crowdsourcing used in software testing is called "Crowdsourced Testing" or "Crowd Testing" [7].

Crowdsourced testing activities including performance testing, usability testing, GUI testing… This method differs from the traditional testing approach in that the testing is carried out by testers from different areas, testers who can be professional testers, novice testers, real application users, etc.

We use mobile crowdsourced testing to refer to test activities for mobile applications that use freelance testing engineers or end-users communities on diverse wireless connectivity infrastructures to ensure the quality of mobile applications in terms of functions, behavior, performance and service quality, as well as special features of mobile applications such as mobility, usability, security and compatibility.

Mobile crowdsourced testing has three following main components [7] (Fig. 1):

- The crowd mobile testers: individuals who perform the test. They work as freelance mobile testers for selected mobile application testing projects. They use their own mobile devices which were pre-configured real-world wireless network infrastructures to carry out the tasks.
- The crowd mobile seekers: people submit projects for testing. They hire free mobile testers to complete their published mobile app testing projects and tasks, and provide them with incentives and payments based on their services.
- An intermediation platform: building a link between the crowd mobile testers and the crowd mobile searchers. This serves as a community support tool that allows customers to express their needs and the individuals and companies that make up the crowd to meet these needs.

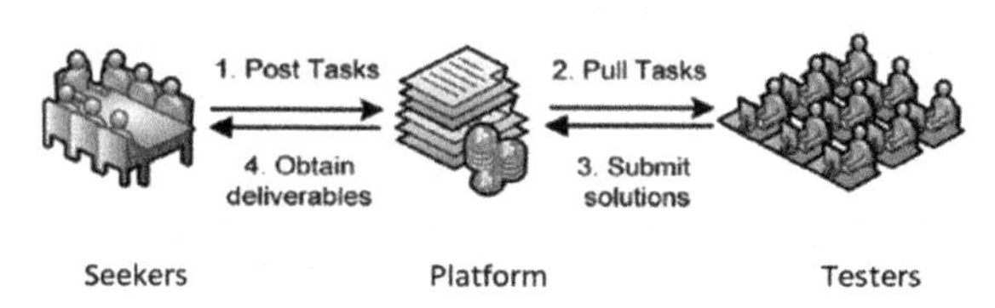

Fig. 1. Components in mobile crowdsourced testing

Today, a number of mobile crowdsourced testing platforms have been developed. These crowdsourced testing platforms provide a cloud-based infrastructure to connect mobile app developers with thousands of crowdsourced mobile testers. Some of these platforms areUtest, MyCroud QA, Pay4Bug, Crowdsourcedtesting, Testin, QA infotech and 99Test.

The particularity of Utest is that it offers to testers articles and tools (Forum, courses) to improve their technical skills in testing applications. The Utest platform emerged in 2007 and is based on the principle of crowd-based testing. Since its foundation, the platform has undergone strong growth given the large number of registered testers and developers and the quality of service delivered for testing web, mobile and desktop applications [8].

This platform offers developers three types of tests, the Beta test, the compatibility test and the functional test. The Testin platform is used by leading companies in their field (IBM, DELL, Intel, Samsung…) [9].

This platform gives developers the ability to test several types of applications: websites, mobile apps, video games and desktop applications. This platform provides developers three types of tests: functional test, user test and location test [10].

Crowdsourcing is the collection from a crowd of people working at diverse workplaces. Particularly, the process is online, leading to good results. Testers involved in crowdsourcing are paid only when their work is finished or when the bug is found. Crowdsourcing testing is the testing approach is called "taas: test as a service".

2.2 Technical Aspects of Crowdsourced Testing

Testing is necessary to effectively implement the software application or product. It is extremely important to ensure that the application will not lead to any failures as it can be very costly in the future or in later stages of development. The easiest and fastest way to test software applications is to use cloud-based testing tools available online [13].

There are some differences when compared to traditional methods such as testers who are from different locations, testers who do not belong to any specific organization, testers may not be professional or experienced, etc. Crowd testing provides benefits, cost effectiveness and makes the software reliable and mostly error free.

There are various reasons for which testers should go with crowdsourced testing [11]. The reasons are listed below:

- Crowdsourced testers can test an application in many different environments, simultaneously with different internet bandwidths, with different devices, with different test streams, etc.
- Crowdsourced testing performed by testers are unbiased because they do not belong to any organization.
- The crowdsourced model can complete the test in a short amount of time because a large number of testers participate in the testing process around the globe.
- Crowdsourced testers are being paid based on the valid errors they find, so this is a very cost-effective test.
- There is no time limit for crowdsourced testers because they belong to different time zones and they can work even after office hours or weekends or work late at night, which leads to quick order for a quality product.

Figure 2 indicates that an application the app seeker can issue a test task that including the mobile applications under test and expected test requirements. These requirements often contain the detailed instructions to test for mobile application testing and the test result. Each task is distributed online as a web page within a time window which setting by the publisher. All crowd employees can choose published tasks to work according to their preferences and other motivations. During testing, testers use their own resources, such as their mobile devices to test mobile applications as required by the task and send test reports containing errors with details [16].

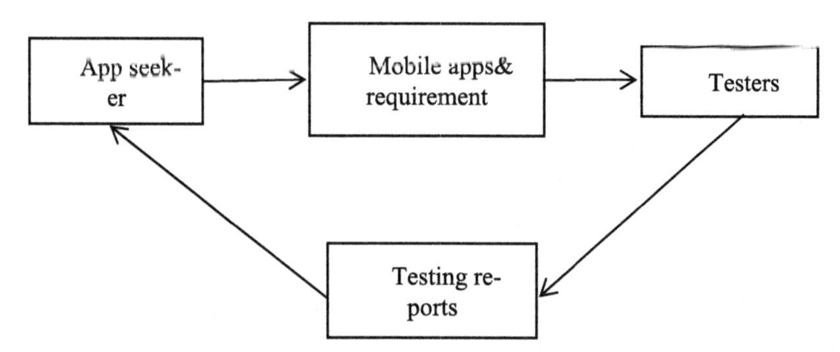

Fig. 2. Crowdsourced testing model for mobile apps

3 Development of a Test Platform: TMACSTEST

This section reports the development of mobile application testing with crowdsourcing (TMACSTest). Particularly, we illustrate two main reasons for developing this new platform: Providing mobile app models for performing model-based testing and providing an automated analysis of bug reports. Although work on these aspects is still ongoing, we report below the main directions that our research is currently investigating. In subsect. 3.1 below we make the link with mobile app development and the component-based software paradigm. Then we provide some conclusion regarding the testing models that can be used for mobile apps. In subsect. 3.2, the need for test results analysis is introduced.

3.1 Model-Based Testing for Mobile Applications

Component-based software development has proved for many years to be an effective and efficient way for software development, allowing component reuse and system evolution. Software components and libraries have historically offered tested and optimized solutions to various issues in software development. The concept of component-based software engineering already exists in mobile app development today – as UI widgets for native Android/iOS or as commercial Marketplace modules for Appcelerator. These components could provide developers with the building blocks needed to bridge the gap between web and native apps today. As a result, this approach has been investigated in the development of mobile applications [18]. For example, Android application development identifies specific main types of components within a mobile application:

- Activities: these are components that handle the user interaction;

- Services: these components handle background processing associated with a mobile application;
- Broadcast receivers: these components deal with the communication between the OS (Android) and the mobile applications;
- Content providers: are components that handle database management issues.

In addition to these basic component types, Android mobile application development distinguishes other components like views (UI elements), fragments (portions of user interfaces), etc.

Component-based models have function that follow us to understand the individual behavior of application components as well as interactions among them. They also provide a simple and the ultimate way to develop specific model-based testing approaches.

Software component testing refers to testing activity that examines component along with its design, generates component tests, identifies component faults and evaluates component reliability. Therefore, component testing plays a significant role for the development of a quality component based software product [4].

Software testing techniques developed for software components are mainly [2]:

- The Component Meta-data way;
- UML based test model for Component;
- Component Interaction Graph (CIG);
- Built-in-tests in components (BIT);
- Component Interaction Testing (CIT).

Table 1. below provides the main features of these testing techniques.

Table 1. Summary of component-based software testing techniques

Technique	Testing criteria
The Component Meta-data way	Attach additional information with the components
UML-based test model	Use Sequence and Collaboration diagrams to extract faults
Component Interaction Graph	Detect faults in the interfaces and interaction among components
Built-in-tests in components	Built-in-test as member function in source code
Component Interaction Testing	Capturing assumptions as formal test requirements

Using such models supports application analysis and helps to focus on the toughest and most important features while generating test cases or deriving test strategies. One specific concern of mobile applications is the location features and corresponding services [19].

Our issue concerns the testing of mobile app component services given a user/app location. Consequently, as a first step in our research, we intend to work on UML-based test model for component endowed with location features. Indeed, the proposed model aims to cover the location aspects of a given mobile app.

3.2 Test Results Analysis

The test results generated by all the testers for a test task would be submitted to the crowdtesting platform.

How to aggregate the test results together without conflict and provide useful analysis for the application or service developer is also a challenge to a mobile crowdtesting platform. Work on this topic should be based like preceding fault analysis approaches developed in [20].

4 Describing the Details of TMACSTest Development

This section reports the requirements specification phase and the design phase of TMACSTest.

4.1 The Requirements Specification Phase

In TMACSTest, there are two actors: tester and provider. The relationship between them is shown in the use case diagram (Fig. 3).

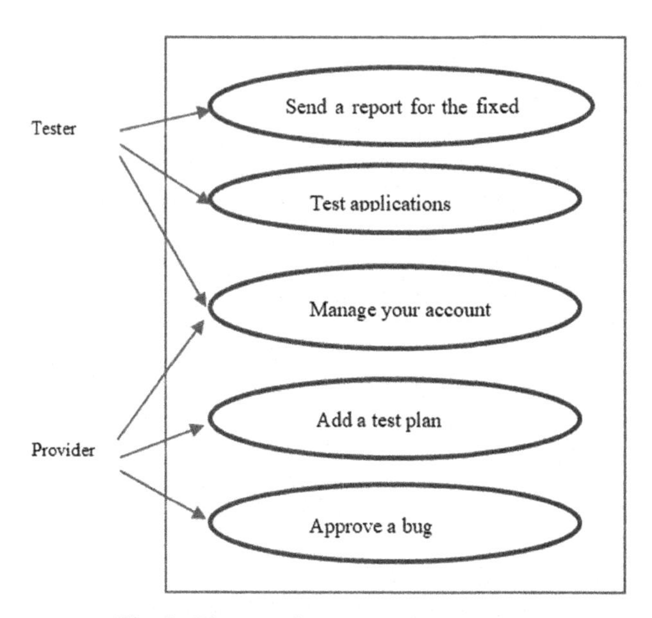

Fig. 3. Diagram of use cases of TMACSTest

Providers can release crowdsourced tasks for mobile app testing. Testers registered on the TMACSTest platform can then choose test tasks of interest to perform in their own hardware and software environments, and submit test reports for their completed tasks. The large number of participants naturally covers various hardware and software environments, which makes it possible to achieve a high operating contexts coverage.

4.2 The Design Phase

Component technologies are perceived as an important means to keep software architectures flexible. Though the ways for implementation of the concept vary greatly, we applied in TMACSTest the 3-tier architecture concept. Figure 4 below shows the detailed architecture of the application.

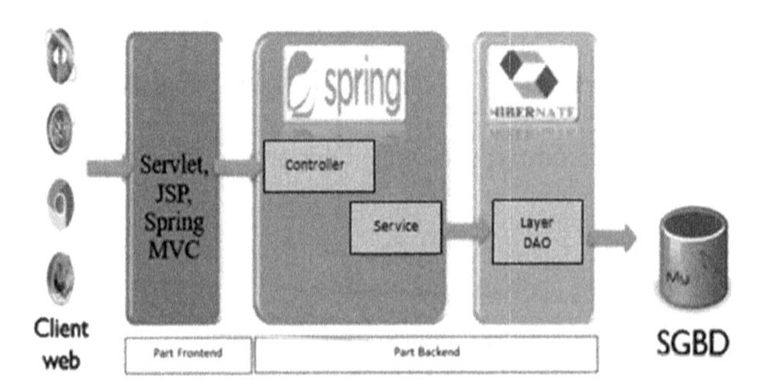

Fig. 4. Architecture of TMACSTest

TMACSTest is composed of two main parts:

- Backend part concerns the business layer and persistence layer of data previously represented in the 3-tier application architecture [12]. It is in this part that lies the core business of the application and the various treatments representing the application's functionalities.
- Frontend part is the web interface of the application. Web pages are linked to the application's controllers to satisfy the user's needs.

We used several technologies such as JAVA/JEE, Spring, Hibernate, Maven, Junit, MySQL as database and Cloud Foundry as a cloud host (Fig. 5).

TMACSTest has the following features:

- Configuration of the database associated with the application;
- Access to add, modify and even delete parts of the application;
- Functional web interface:
- Homepage (static page - Fig. 6, register page -Fig. 7).
- Provider page: the provider can add an application and display the list of applications (Fig. 8).
- Tester page contains the tester profile, the bug list, the test cycle list, and the addition of a bug for a test cycle (Fig. 9).

Fig. 5. Used technologies

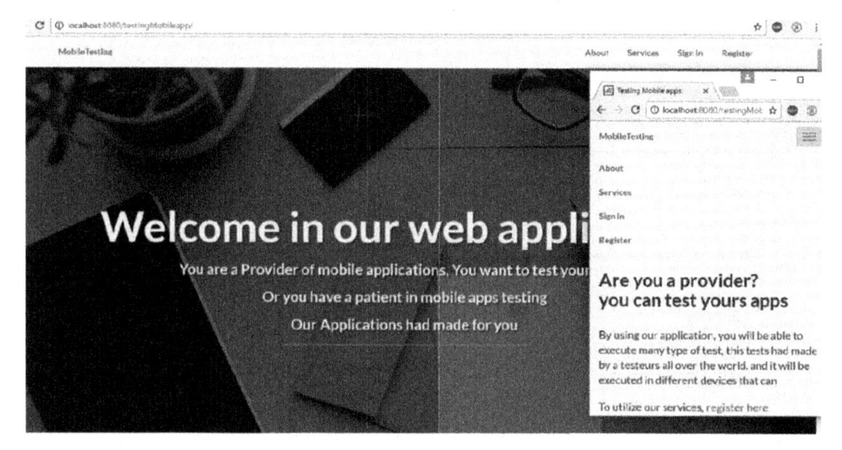

Fig. 6. Application homepage

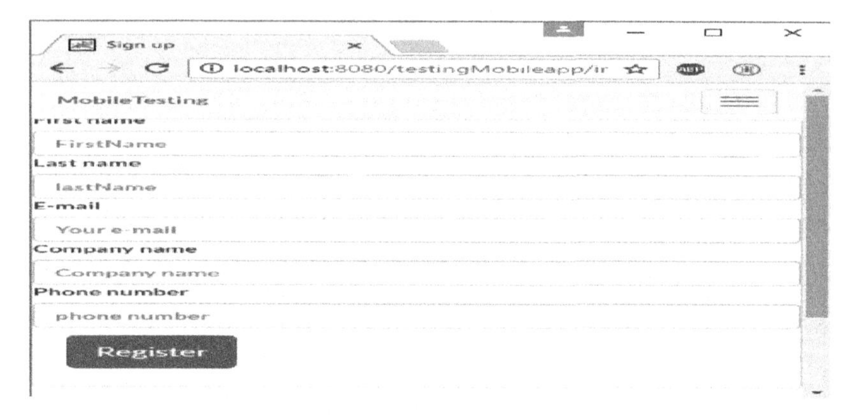

Fig. 7. Supplier registration form

Fig. 8. Adding an application

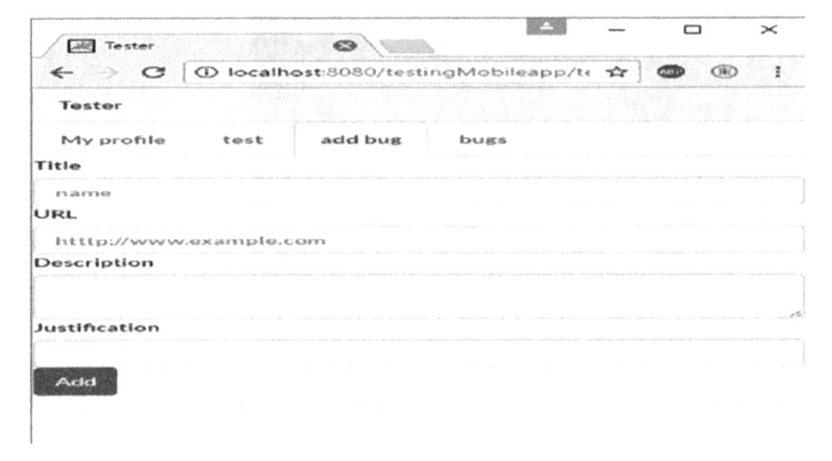

Fig. 9. Adding a bug

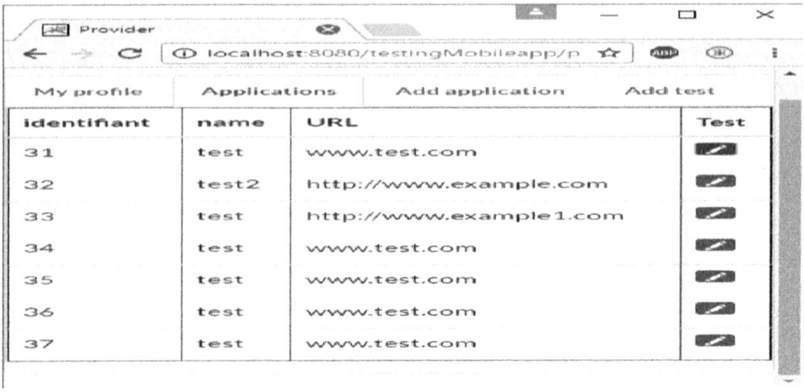

Provider

localhost:8080/testingMobileapp/p

My profile | Applications | Add application | Add test

identifiant	name	URL	Test
31	test	www.test.com	
32	test2	http://www.example.com	
33	test	http://www.example1.com	
34	test	www.test.com	
35	test	www.test.com	
36	test	www.test.com	
37	test	www.test.com	

Fig. 10. Vendor application list

5 Conclusion and Future Work

TMACSTest is an intermediary linking the mobile application providers who need to test these applications before releasing them on the market, and the testers who test these applications (Fig. 10).

There are currently several perspectives for this work to come. First, we need to complete ongoing experimental work and provide corresponding results. Then, we aim to introduce the use of non-relational databases. These databases offer significant power for heterogeneous data processing against relational databases and use a large distribution of data and associated processing across multiple servers [17]. They are used primarily for distributed applications in the cloud.

References

1. Liang, C.-J.M., et al.: Caiipa: automated large-scale mobile app testing through contextual fuzzing. In: 20th International Conference on Mobile Computing and Networking, pp. 519–530 (2014)
2. Gao, J., Bai, X., Tsai, W.T.: Mobile application testing: a tutorial. Computer **47**, 46–55 (2014)
3. Mao, K., Capra, L., Harman, M., Jia, Y.: A survey of the use of crowdsourcing in software engineering. J. Syst. Softw. **126**, 57–84 (2016)
4. Chen, Z., Luo, B.: Quasi-crowdsourcing testing for educational projects. In: 36th International Conference on Software Engineering, pp. 272–275 (2014)
5. Liu, D., Bias, R.G., Lease, M., Kuipers, R.: Crowdsourcing for usability testing. Am. Soc. Inf. Sci. Technol. **49**, 1–10 (2012)
6. LaToza, T.D., van der Hoek, A.: Crowdsourcing in software engineering: models, motivations, and challenges. IEEE Softw. **33**, 74–80 (2016)
7. Speidel, D., Sridharan, M.: A framework and research agenda for crowdsourced testing (2013)
8. uTest. https://www.utest.com
9. TestCloud. https://www.xamarin.com/test-cloud
10. Crowdsourced Testing. https://crowdsourcedtesting.com
11. Guide to crowdsourced testing. https://www.softwaretestinghelp.com/guide-to-crowdsourced-testing
12. Gao, J., Tsai, W.T., Paul, R., Bai, X.Y.: Mobile testing-as-a-service (mobile taas) - infrastructure, issues, solutions and needs. In: IEEE International Symposium of High Assurance Systems Engineering, pp. 158–167 (2014)
13. Tsai, W.T., et al.: A cloud-based Taas infrastructure with tools for SaaS validation, performance and scalability evaluation, pp. 464–471 (2012)
14. Blokland, K., Mengerink, J., Pol, M.: Testing cloud service (2013)
15. Vashistha, A., Ahmed, P.: SaaS multi-tenancy isolation testing-challenges and issues. Int. J. Soft Comput. Eng. (IJSCE), **2**(5), 49–50 (2012). ISSN: 2231–2307
16. Naganathan, V., Sankarayya, S.: Overcoming challenges associated with SaaS testing. Infosys (2011)
17. Crowdsourced Usability Testing. http://alexcrockett.com/wp-content/uploads/downloads/Books/Crowdsourced_Usability_Testing.pdf
18. Orso, A., Rothermel, G.: Software testing: a research travelogue (2014)
19. Xie, M., Wang, Q., Yang, G., Li, M.: COCOON: crowdsourced testing quality maximization under context coverage constraint. In: IEEE 28th International Symposium on Software Reliability Engineering (2017)
20. Khedam, R., Aktouf, O., Parissis, I., Boughazi, S.: Monitoring of RFID failures resulting from LLRP misconfigurations. In: SoftCOM, pp. 1–6 (2013)

Vision Based Facial Expression Recognition Using Eigenfaces and Multi-SVM Classifier

Hla Myat Maw(✉), Soe Myat Thu, and Myat Thida Mon

University of Information Technology, Yangon, Myanmar
{hmyatmaw,soemyatthu,myattmon}@uit.edu.mm

Abstract. Facial Expression Recognition (FER) has become one of the most popular areas of research in computer vision and biometrics authentication and it has achieved a lot of enthusiasm from researchers. The Vision based Facial Expression Recognition system intends to classify the facial expression of a given image. In this paper, the proposed system automatically classifies the facial expression. The system is composed of feature extraction and expression classification. In preprocessing, Hybrid filter (Median and Gabor) and Histogram Equalizations, is used to reduce noise and enhance images. Feature extraction is to extract feature vectors from face images using the Eigenfaces approach, based on Principal Component Analysis (PCA). To classify facial expression, extracted feature vectors are fed into a Multiclass Support Vector Machine (Multi-SVM) classifier. Experiments are performed on the standard dataset of the Japanese Female Facial Expression (JAFFE) and achieved 80% accuracy. The proposed system showed satisfying performance comparing with other methods and effects state-of-the-art performance on the JAFFE dataset.

Keywords: Facial Expression Recognition (FER) · Principal Component Analysis (PCA) · Multiclass Support Vector Machine (Multi-SVM)

1 Introduction

Facial expression is a sort of expression of the emotional state and the human face has a high power of expression. In communication and emotional state, it gives a powerful and flexible nature expression. Utilizing facial expressions can express internal emotion and provide important clues to social communication. Mehrabian [1] delivered out that 55% by outward appearance, 38% by paralanguage (vocal part), and 7% by phonetic language (verbal part) is communicated by human communication information. Thus facial expressions are the most significant data for face-to-face conversation perception of the feelings. In academia and industry, facial expression recognition (FER) has recently attracted because of its wide range of applications such as security, healthcare, human-computer interface, and robotics. In recent years, numerous progresses have been made. But a challenging problem remains due to high head position, illumination variation, occlusion, motion blur, etc. in facial expression recognition.

There are six basic facial expressions such that angry, disgusted, happy, neutral, sad and surprised. Extracting relevant facial features that allow for the identification of

M. Hernes et al. (Eds.): ICCCI 2020, CCIS 1287, pp. 662–673, 2020.
https://doi.org/10.1007/978-3-030-63119-2_54

proper and precise expressions is important. Human facial expression recognition and classification by computer is an essential point in the vision community for developing an automatic facial expression recognition (AFER) system. A lot of research has been performed in recent years about the computer perception of human facial expressions. Any AFER system's performance relies upon its method of feature extraction is typically focused on appearance or geometric features.

Features of the appearance define the complexion of the face and its changes, whereas features of the geometric define the structure of the face by applying specific feature points from various facial parts.

In recent years, the common type of deep learning method is Convolutional Neural Networks (CNNs) used for face and facial expression recognition. Deep learning methods require large datasets and take time to train networks as are the disadvantage of deep learning. And also deep learning requires a high-performance environment. The proposed system does not need deep learning models because of not using Graphical Processing Units (GPU). And the system used a small standard dataset such as JAFFE.

Recognition of facial expression performance is based mainly on the accurate extraction of components of the facial feature. It depends on challenging problems. One of the most problematic is the illumination variation. The proposed method used the preprocessing techniques [12] including a median filter for noise reducing and Gabor filter for edge enhancing. Gabor filter parameters (wavelength, orientation, etc.) changing gets to the enhanced edges. Histogram equalization is used to enhance filtered face images and to eliminate the effect caused by the illumination condition. After preprocessing, Principal Component Analysis (PCA) is used for feature extraction and Multi-Support Vector Machines (multi-SVMs) is used for facial expression classification. Multi-SVM's effectiveness depends on kernel selection, kernel's parameters, and soft margin parameter C. The Radial Basis Function (RBF) kernel Multi-SVM is used to recognize facial expressions and predict the class label. Experiments are carried out using faces from the JAFFE image dataset. Classification accuracy is more reliable than other methods.

The remainder of the proposed paper is set out as follows: The related works are represented in Sect. 2. Section 3 outlines the PCA approach using eigenfaces and Multi-Class Support Vector Machine. Section 4 describes the proposed approach. The experimental results are contrasted to other state-of-the-art methods, and discussions are discussed in Sect. 5. Conclusion and future work are presented in Sect. 6.

2 Related Work

Dea et al. [7] (2015): The authors presented the technique, using the Euclidean distance classifier, which is a model of identification of facial expression based on eigenfaces to identify the different emotions. The self-made expression dataset consists of different people with five emotions like anger, fear, happiness, sorrow, and surprise is used. Face detection in the images by using the HSV (Hue-Saturation-Value). The results showed that the highest recognition rate of 93.1% is for joy, likewise, surprise and anger have a good recognition rate of 91% and 86.2% respectively and the recognition rate of 78.9% and 77.7% is for sorrow and fear. But a similarity exists between sorrow and fear.

Jameel et al. [8] (2016): The authors identified the survey paper reviews several papers using some of the techniques from 2001 to 2012 for recognition of facial expression. And the authors concluded that creating an authentic database is very difficult while making it comparatively easy to create a semi-authentic database. Semi-supervised learning approaches are very useful for data labeling. It should be capable of detecting micro-expressions with pose for system effectiveness.

Shan et al. [9] (2017): The authors presented the Automatic Facial Expression Recognition Method, using a deeper convolutional neural network (CNN) to find deeper feature representations of facial expression. The system included four modules (input, pre-processing, recognition, and output). The efficiency of the recognition is simulated and analyzed using the Japanese Female Facial Expression Database (JAFFE) and the Cohn-Kanade Extended Database (CK +). The accuracy of performance leads 76.74% and 80.30% for the JAFFE and CK + respectively.

Islam et al. [10] (2018): The authors proposed a human emotion recognition system to recognize the facial expression of humans. The following parts: face detection included in preprocessing, facial parts segmentation, feature extraction, dimensional reduction, and classification are included in the system. In feature extraction, uses two dimensional Gabor filters to extract features. The Principal Component Analysis (PCA) is used for the reduction of the dimensions and Multiclass Support Vector Machine (SVM) classifier to classify complex problems in high dimensional spaces. The performance of the proposed method is checked across three datasets: JAFFE, CK + RaFD.

Meng et al. [11] (2019): The authors presented Frame Attention Networks for Facial Expression Recognition in Videos, in an end-to-end system, would automatically highlight certain discriminative frames. The system consists of two modules, an embedding feature module, and the frame attention module. A Deep Convolutional Neural Network (CNN), embedding module is embedded facial imaging feature vectors. Multiple weights of attention are learned to use for adaptive aggregation of the feature vectors. A single discriminative video representation is formed with these feature vectors. Datasets CK + and AFEW8.0 are used as tests. It achieves state-of-the-art performance on CK +.

3 Background Theory

3.1 Principal Component Analysis (PCA)

Principal Component Analysis is a statistical method applied to eliminate dimensions in fields of pattern recognition and computer vision, and used to extract features and represent data. Using this, vectors of the facial features are obtained.

Kirby and Sirovich (1990) have developed the eigenfaces method for face recognition using PCA and M.A Turk and Alex Pentland (1991) have implemented it. [3]. Eigenfaces based on PCA represent each image as a matrix, a set of eigenvectors that represents the principal components of the matrix, called eigenfaces. Eigenvectors are the basis for calculating variance between multiple faces. A matrix of covariance gives us details on these feature vectors of data. In a face space with lower dimensionality, similar faces can be represented.

The basic steps for computation of eigenfaces are performed by the following algorithm.

Algorithm:

Step 1: An $N^2 *1$ size vector can be represented by an N*N size face image. In Training images, m images are represented as $I_1, I_2,.., I_m$.

Step 2: An average of the training face images is obtained with

$$\varphi = \frac{1}{m}\sum_{i=1}^{m} I_i \tag{1}$$

Where $i = (1, \ldots, m)$

Step 3: A vector displays a normalizing face.

$$\emptyset_i = I_i - \varphi \tag{2}$$

Step 4: Calculating Training set's covariance matrix (C) can be described by

$$C = \sum_{i=1}^{N} \emptyset_i \emptyset_i^T = AA^T \quad (N^2 x N^2) \tag{3}$$

Where $A = [\emptyset_1 \emptyset_2 \ldots \emptyset_M] \ \left(N^2 x M\right)$

Step 5: The matrix is huge for computing u_i eigenvectors of the matrix AA^T. Hence, Computing the eigenvectors v_i of $A^T A$.

$$A^T A x_i = \lambda_i x_i \tag{4}$$

$$A A^T A x_i = A \lambda_i x_i \tag{5}$$

$$CA x_i = \lambda_i A x_i \tag{6}$$

Where $u_i = Ax_i$ (Eigen vectors)

λ_i Eigenvalues

Step 6: Keep only K Eigenvectors corresponding to the K largest Eigenvalues. These Eigenvalues are called principal components.

3.2 Multi-Support Vector Machines (Multi-SVMs)

Support Vector Machine is a supervised learning technique in machine learning. In a high dimensional space, it constructs a hyper-plane or set of hyper-planes. It does a good separation which has the maximum distance from any class to the nearest training data point. It was originally designed for a binary classifier but the real classification issues require more than two classes. Multi-SVMs are utilized to determine problems in class n. Multi-SVMs use two kinds of methods which are one to one, and one to all. For each pair of classes, one to one method constructs one classifier. One to all method constructs one classifier for each class. SVM is mostly used in applications such as classification and regression. To increase its performance it uses four kernel types. These are a linear, polynomial, function of Radial Basis (RBF), and function of the sigmoid. The linear kernel maps high-dimensional data and divides linearly. The polynomial kernel learns from the nonlinear models and solves their similarity as well. The function of the RBF kernel maps the unique feature to data of high dimensions.

A multi-class SVM classifier is used for the classification of different expressions. One to all SVM classifiers with the RBF kernel, the function is used to obtain a proper rate in facial expression recognition and classification experiments. In the RBF kernel, the classification rate is determined by the standard deviation parameters.

SVM algorithm divides the training data by a hyper-plane specified by the type of kernel function used in the feature space. It finds the maximum margin hyper-plane, defined as the sum of the hyper-plane distances from the closest data point in each of the two classes. Multi-class classification is carried out in combination with a voting system by a cascade of binary classifiers.

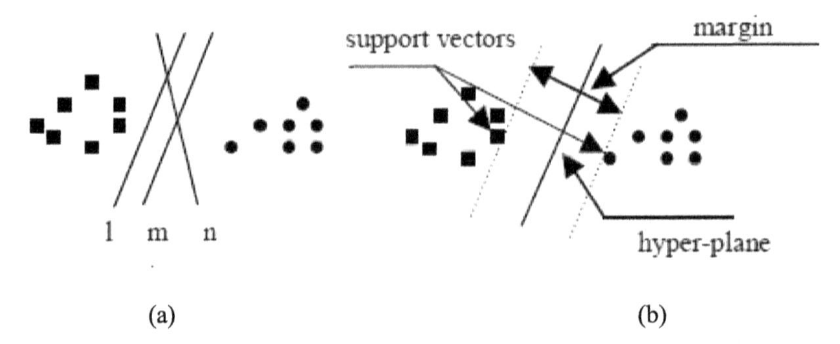

Fig. 1. (a) Arbitrary hyper-planes l, m and n (b) Optimal separating hyper-plane

4 Proposed System

Face analysis (Face recognition and facial expression recognition) has attained much interest from the science and industrial communities in recent decades according to its potential utility through applications in security and access control, human-computer interaction, law enforcement, and surveillance. The efficiency of the recognition system for facial expression relies on challenges. One of the most problems is illumination variation. The research motivation is intended to eliminate the illumination variation situation and improve the performance of facial expression recognition (Table 1).

Framework for the system proposed is structured in 4 parts, Data Acquisition, Pre-processing, Feature Extraction, and Facial expression classification. Firstly, images are taken from standard data set (JAFFE) and separate Training and Testing data set. Then use preprocessing technique [12]. It uses a Median filter to reduce noise and 2D Gabor filter to enhance edge, and then uses histogram equalization for image enhancement. Feature vectors are extracted using the eigenfaces approach and Multi-class support

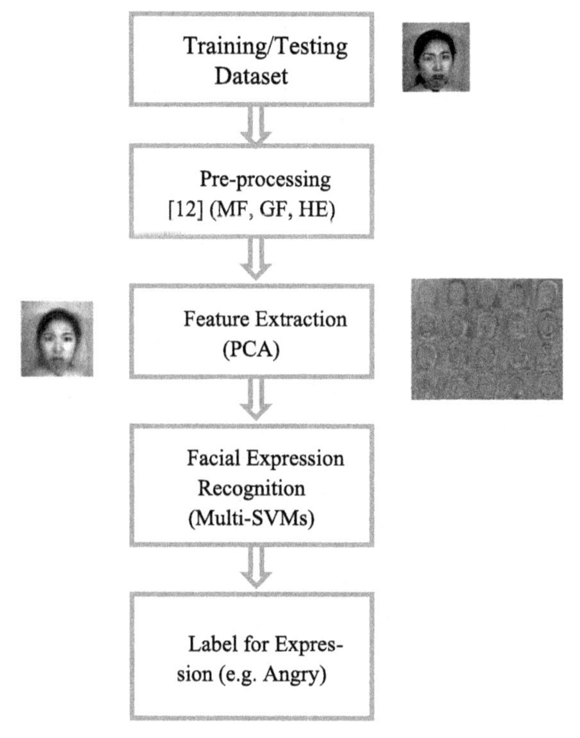

Fig. 2. The framework of the proposed system

vector machines using the Radial Basis Function (RBF) function are used to classify facial expression and predict class labeling (Fig. 1).

The system used the standard dataset of the Japanese Female Facial Expression (JAFFE).

Step1. Create Training and Testing datasets

This uses 143 images for training and uses 70 images for testing.

Step2.Image pre-processing

a. Loading images from the training/testing datasets.
b. Converting color image to gray image, if it is in color.
c. Reducing noise by using hybrid filters and histogram equalization is used for enhancing the image.

Step3. Extraction feature

a. Transform each two dimensional image into a one-dimensional image vector.
b. Obtaining the mean image.
c. Normalizing face by finding the difference between each image and the mean image.
d. Computing covariance matrix C.
e. Finding eigenvector of C.

Step4. Classification of facial expressions

a. Projecting the whole image-focused into the face space.
b. Loading images of test face from the test set.
c. Pre-processing images of test face.
d. Extracting feature vectors of the test face images.
e. Using the extracted features to construct Multiclass SVM model with RBF kernel to predict the class of expression of the test faces.

5 Experimental Results

The experiments are carrying by using JAFFE (Japanese Female Facial Expression), which is a standard face dataset. The accuracy of recognition of facial expression is defined with the average rates of recognition and the confusion matrices. Within its columns, the confusion matrix represents the predicted expressions within its rows against the percentages of the actual expressions. The percentages of correctly categorized facial expressions are diagonal entries, while the off-diagonal entries relate to misclassifications. The performance of the proposed approach is compared to the state of the art methods using the JAFFE dataset (Fig. 2).

5.1 Dataset

Japanese Female Facial Expression Database (JAFFE) [16]: It includes 213 images of 7 facial expressions (6 specific expressions + 1 neutral) represented by 10 female models

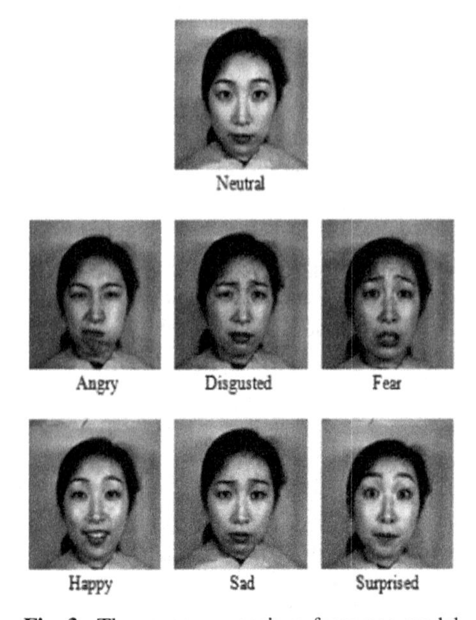

Fig. 3. The seven expressions from one model

in Japan. Each model has about twenty images and each expression includes two to three images. The seven expressions are angry, disgusted, happy, neutral, sad, and surprised respectively. Figure 3 shows the seven expressions from one model and sample face images from the dataset are given in Fig. 4.

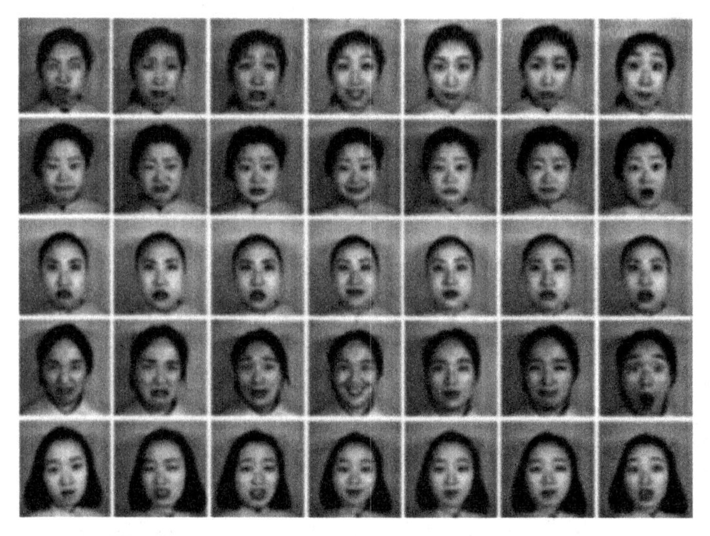

Fig. 4. Face sample images from the dataset JAFFE

Table 1. Recognition accuracy tested on the JAFFE dataset

Expression	Train images	Test images	Predicted expression	Accuracy
Angry	20	10	8	80%
Disgusted	19	10	8	80%
Fear	22	10	7	70%
Happy	21	10	7	70%
Neutral	20	10	10	100%
Sad	21	10	7	70%
Surprised	20	10	9	90%
Total	143	70	56	80%

The confusion matrix of the expected dataset results is shown in Fig. 6. The proposed system's average precision is nearly 80%. Table 2 shows the comparison with methods of the state of arts on the dataset of JAFFE (Fig. 5).

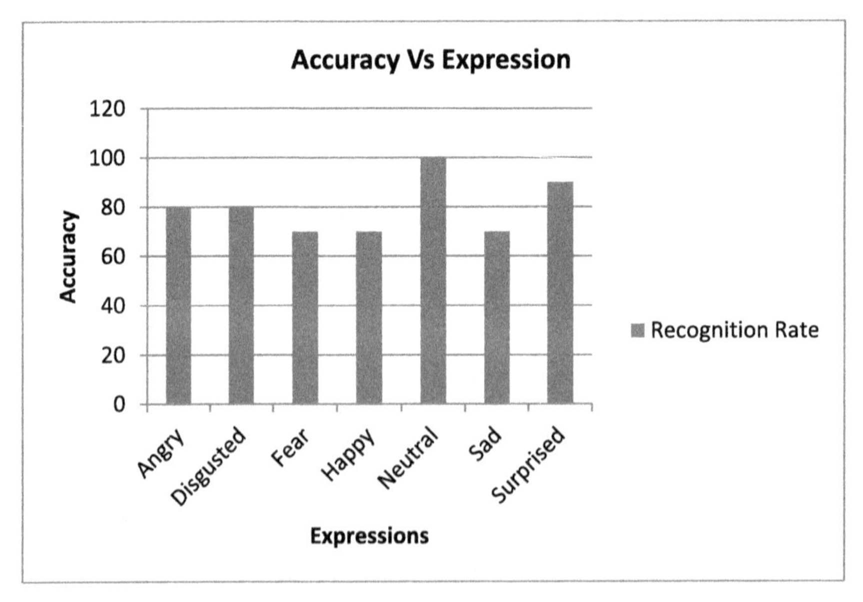

Fig. 5. Recognition rate of the proposed system tested on the JAFFE dataset

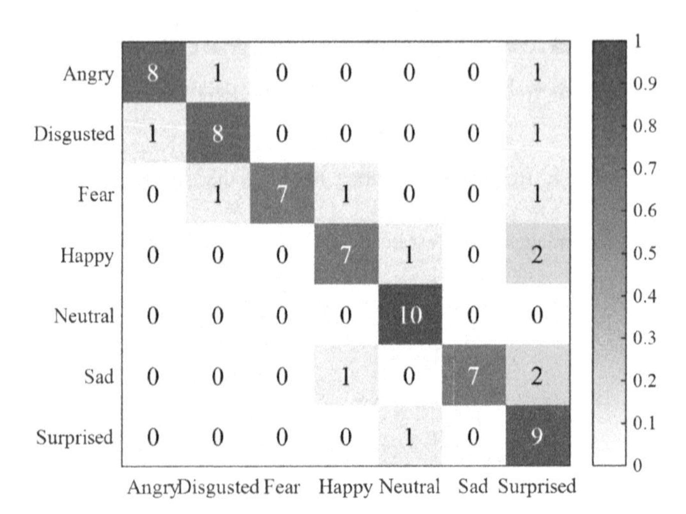

Fig. 6. Confusion matrix of the predicted outcomes tested on the dataset (JAFFE)

Here Neutral is seen to have the highest recognition rate of 100 percent. Surprised also has a high 90% recognition rate. Angry and disgusted have a recognition rate of 80%. Happy, Fear and Sad have a recognition rate of 70%.

A review of the Confusion matrix for the JAFFE dataset advises that the best-recognized class is Neutral followed by Surprised. Angry and disgusted which are confused with each other. The difference between those expressions is hard to discriminate against. The maximum confused between Happy, Fear and Sad.

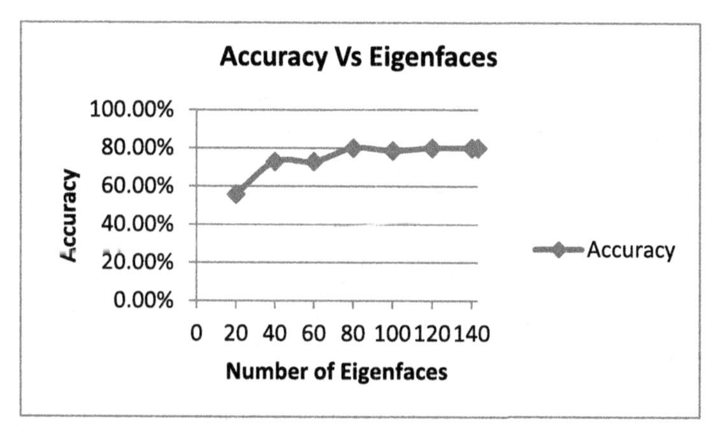

Fig. 7. Accuracy analysis is performed by a varying number of Eigenfaces

Figure 7 shows that classification accuracy depends on the number of eigenfaces. Number of eigenfaces, is less than half of the training set, is chosen the accuracy rate is not good. To increase the accuracy rate, the number of eigenfaces is set to at least half of the training set.

Table 2. Evaluation of the proposed method compared to the state of the art methods on the JAFFE dataset

	Method	Accuracy rate
Murthy, G.R.S. et al. [13]	Modified PCA	75.00%
Ghadekar, P.P. et al. [14]	Local Directional Number Pattern (LDNP) SVM	82.14%
Bhat A. et al. [15]	PCA,SVM	78.00%
Shan K. et al. [9]	Deep convolutional neural network (CNN)	76.74%
Proposed system	PCA, Multi-SVM	80.00%
	Preprocessing, PCA, Multi-SVM	84.29%

5.2 Experimental Environment

The performance evaluations of this system are implemented on Matlab R2015a and an Intel® Celeron® Processor N230 (2.41 GHz) and 2 GB DDR3 of memory. JAFFE dataset is used to analyze and evaluate the system. The datasets include a small amount of data (213 images). Multi-SVM is suitable for use as a classifier instance of a deep learning model that needs a vast amount of data and a high-end system to increase the performance of a deep learning algorithm.

6 Conclusion

The proposed system for facial expression recognition is PCA based which uses eigenfaces to extract the feature vector and to classify facial expression using multiclass SVM. The experiment is tested on the standard dataset (JAFFE) by varying the number of eigenfaces. The results show that promising results are better success rates. This work is purposed not only for known face images but also the unknown face images.

The future work will be considered for increasing the recognition rate, where the system will be modified and tested using other datasets.

References

1. Mehrabian, A.: Communication without words. Psychol. Today **2**(4), 53–56 (1968)
2. Ekman, P., Friesen, W.V., Hager, J.C.: Facial Action Coding System. Consulting Psychologists Press, Palo Alto, CA (1978)
3. Turk, M.A., Pentland, AP.: Eigenfaces for recognition computer vision and pattern recognition. In: Proceedings of CVPR, pp. 586–591 (1991). https://doi.org/10.1109/cvpr.1991.139758
4. Bian, Z., Zhang, X.: Pattern Recognition, 2nd edn. Tsinghua University Press, Beijing (2000)
5. Jonsson, K., Kittler, J.: Support vector machine for face authentication. Image Vis. Comput. **20**(5–6), 369–375 (2002). https://doi.org/10.1016/s0262-8856(02)00009-4
6. Jonsson, K., Matas, J.: Learning support vectors for face verification and recognition. In: 4th IEEE International Conference on Automatic Face and Gesture Recognition (CAFGR), pp. 208–213. France (2000). https://doi.org/10.1109/afgr.2000.840636
7. Dea, A., Sahaa, A., Dr. Palb, M.C.: A humanfacial expression recognition model based on eigen face approach. In: International Conference on Advanced Computing Technologies and Applications (ICACTA), pp. 282–289 (2015). https://doi.org/10.1016/j.procs.2015.03.142
8. Jameel, R., Singhal, A., Bansal, A.: A comprehensive study on facial expressions recognition techniques. In: 6th International Conference on Cloud System and Big Data Engineering, pp. 478–483. IEEE, India (2016). https://doi.org/10.1109/confluence.2016.7508167
9. Shan, K., Guo, J., You, W., Lu, D., Bie, R.: Automatic facial expression recognition based on a deep convolutional-neural-network structure. In: 15th International Conference on Software Engineering Research, Management and Applications (SERA), pp. 123–128. IEEE, UK (2017). https://doi.org/10.1109/sera.2017.7965717
10. Islam, B., Mahmud, F., Hossain, A.: Facial expression region segmentation based approach to emotion recognition using 2D gabor filter and multiclass support vector machine. In: 21st International Conference of Computer and Information Technology (ICCIT). IEEE, Bangladesh (2018). https://doi.org/10.1109/iccitechn.2018.8631922
11. Meng, D., Peng, X., Wang, K., Qiao, Y.: Frame attention networks for facial expression recognition in videos. In: IEEE International Conference on Image Processing (ICIP), pp. 3866–3870. IEEE, Taiwan (2019). https://doi.org/10.1109/icip.2019.8803603
12. Maw, H.M., Lin, K.Z., Mon, M.T.: Preprocessing techniques for face and facial expression recognition. In: 33rd International Technical Conference on Circuits/Systems, Computers and Communications (ITC-CSCC), pp. 377–380. Thailand (2018)
13. Murthy, G.R.S., Jadon, R.S.: Effectiveness of eigenspaces for facial expressions recognition. Int. J. Comput. Theory Eng. **1**(5), 638–642 (2009)
14. Dr. Ghadekar, P.P., Alrikabi, H.A., Dr. Chopade, N.B.: Efficient face and facial expression recognition model. In: International Conference on Computing Communication Control and automation (ICCUBEA), pp. 1–8. IEEE, India (2016). https://doi.org/10.1109/iccubea.2016.7860053

15. Bhat, A., Veigas, J.P.: Efficient implementation on human face recognition under various expressions using LoG, LBP and SVM. Int. J. Eng. Sci. Comput. (IJESC) **7**(7), 14052–14055 (2017)
16. The Japanese Female Facial Expression (JAFFE) Database. http://www.kasrl.org/jaffe.html

An Effective Vector Representation of Facebook Fan Pages and Its Applications

Viet Hoang Phan[1], Duy Khanh Ninh[1](✉), and Chi Khanh Ninh[2]

[1] The University of Danang, University of Science and Technology, Danang, Vietnam
hoangvietit15@gmail.com, nkduy@dut.udn.vn
[2] The University of Danang – Vietnam - Korea, University of Information and Communication Technology, Danang, Vietnam
nkchi@vku.udn.vn

Abstract. Social networks have become an important part of human life. There have been recently several studies on using Latent Dirichlet Allocation (LDA) to analyze text corpora extracted from social platforms to discover underlying patterns of user data. However, when we wish to discover the major contents of a social network (e.g., Facebook) on a large scale, the available approaches need to collect and process published data of every person on the social network. This is against privacy rights as well as time and resource consuming. This paper tackles this problem by focusing on fan pages, a class of special accounts on Facebook that have much more impact than those of regular individuals. We proposed a vector representation for Facebook fan pages by using a combination of LDA-based topic distributions and interaction indices of their posts. The interaction index of each post is computed based on the number of reactions and comments, and works as the weight of that post in making of the topic distribution of a fan page. The proposed representation shows its effectiveness in fan page topic mining and clustering tasks when experimented on a collection of Vietnamese Facebook fan pages. The inclusion of interaction indices of the posts increases the fan page clustering performance by 9.0% on Silhouette score in the case of optimal number of clusters when using K-means clustering algorithm. These results will help us to build a system that can track trending contents on Facebook without acquiring the individual user's data.

Keywords: Topic modeling · Latent Dirichlet Allocation · Interaction index · Facebook fan pages · Social network analysis

1 Introduction

Nowadays, social networks have become an essential part of human life. Based on a recent research of Statista, more than 3.5 billion people on earth have at least one account on a social platform in 2019 [1]. With rapid growth of users, comes giant amount of data. This can bring a lot of opportunities for those who can discover patterns inside of the user data and find out meaningful usages of them.

M. Hernes et al. (Eds.): ICCCI 2020, CCIS 1287, pp. 674–685, 2020.
https://doi.org/10.1007/978-3-030-63119-2_55

In Vietnam, Facebook is the social network having the largest number of users [1]. Posts on Facebook can come from individuals (particularly famous figures) or from organizations in a form of what is called a fan page. Because of the great ability of sharing posts to the fans (i.e., people who follow pages), fan pages are playing an important role in spreading information, news, and facts on Facebook. If we can model the topics of posts on popular pages, we will have a good chance to find out trending contents on Facebook.

In recent years, there have been a lot of researches on using Latent Dirichlet Allocation (LDA) to cluster the scientific documents [2, 3] and news articles [4, 5]. For social networks document analysis, there were some studies about modeling the topic on Twitter [6, 7] or favorite topics on Facebook posts [8]. This research focuses on modeling Facebook fan pages by using the method of topic modeling from documents (i.e., the fan page's posts).

In this paper, we propose a solution of modeling the topic of documents with LDA combined with calculating the interaction index of the Facebook posts to find an effective vector representation of Facebook fan pages. Then we apply this representation to analyze topic distribution of each fan page and to find out groups of similar fan pages. The proposed solution shows the effectiveness on clustering the fan pages into subsets by increasing the clustering performance than modeling using just LDA. The fan page representation also helps point out the similarities between fan pages and give us an idea about what is happening on Facebook in a particular period of time.

This paper is organized as follows. Section 2 reviews past studies leading to the motivation of our work. Section 3 describes our proposed solution. Experiments and results are given in Sect. 4. Section 5 presents the conclusion and future work of the current research.

2 Related Work and Motivation

Topic modeling using LDA is not a new technique in Natural Language Processing. LDA uses an unsupervised learning model, therefore it is a good technique for document classification, especially on unlabeled datasets such as social network's textual data. There were several researches taking this advantage of LDA to model and analyze Twitter conversations [6] or favorite topics of young Thai Facebook users [8]. The main focus of these studies is the modeling and mining the topics from the text corpus of social network users. Their proposed methods can help us to obtain the topics in which a part of users interested in, for example educational workers and students at National University of Colombia [6] or students at Assumption University in Thailand [8]. However, when we wish to discover the major contents of a social network on a large scale such as finding trending topics among users of a nation, the available approaches exhibit their limitations, that is it is almost impossible to collect published data (i.e., the posts) of every person on the social network because it goes against privacy rights as well as takes a lot of time and resource to collect and process the data.

This paper tackles this problem by focusing on a class of special users that have much more impact on social networks than other individuals, which are key opinion leaders (KOLs) and popular organizations. A post of a KOL or a well-known organization, usually on their fan pages, may lead the opinions, represent for the thoughts, and attract the interests of many people which follow them on social networks. Therefore, instead of collecting data from each regular account on a social network, we only need to get and analyze data from a number of influential accounts of KOLs and organizations, thereby achieving the equivalent effectiveness in capturing the trends of the social network on a large scale. In this paper, we selected the most reputable Facebook fan pages in Vietnam for topic mining and other data analyses.

3 A Novel Solution for Facebook Fan Page Modeling

3.1 Observations

To know what a Facebook fan page is talking about, we have to analyze the contents of its posts. In this research, we are only interested in the textual part (called the document hereinafter) and the interactional part (i.e., reactions and comments) of a post. If we can extract the topics of every document in the text corpus of a fan page, we probably can figure out the most popular topics of a fan page.

Assuming that each document in a fan page's corpus has its own topic probability distribution or, in other words, each document can be represented by a fixed-dimensional vector depending on what its content is about. For example, if a document has its topic proportions of 30% about sport, 50% about technology, 20% about politics, the topic distribution vector of this document will be [0.3;0.5;0.2]. In practice, the results of vectorizing documents are not clearly visible like the above example, but usually hidden in the textual data. We need a solution to combine the topic distribution vectors of the documents in a fan page's corpus to find the topic distribution vector representing the fan page.

After studying about Facebook data properties, we realized that different posts (thus their corresponding documents) have different degrees of importance to the topic proportions of a fan page. The posts that receive more interactions from users are likely to contribute more to the composition of the topics of a fan page and to the distinction among fan pages.

3.2 Proposed Process Diagram

Figure 1 presents the proposed process flow. Firstly, the raw data of fan pages are collected from crawlers, from which textual data and interactional data are extracted. After that, the textual data is pre-processed by removing page signatures, special characters, icons, and stop words. Pre-processed documents are then applied LDA-based topic modeling process, returning topic distribution vectors of all documents of each fan page's corpus. Meanwhile, the interactional data is used to calculate the interaction indices of all documents of each fan page. Finally, the vector representation for every fan page is obtained by combining the topic distribution vectors and interaction indices of all documents of the page. How the combination is done is described in details in Sect. 3.4.

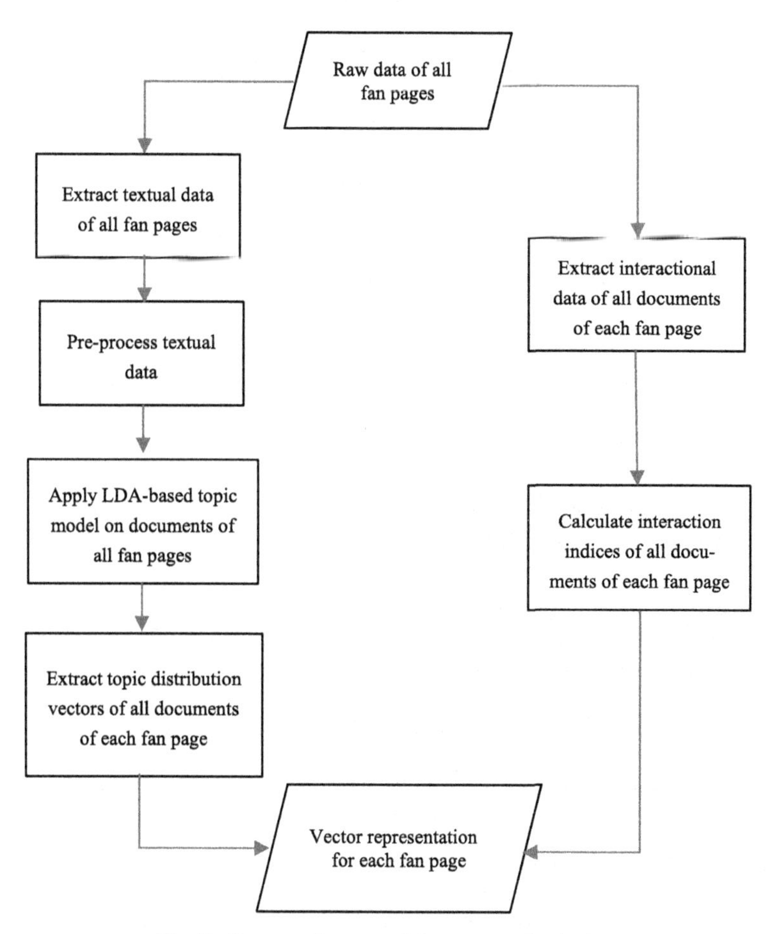

Fig. 1. Process diagram of the proposed solution.

3.3 Topic Modeling Using LDA

LDA is a method widely employed for modeling the topics of documents in a corpus, which was proposed by Blei et al. in 2003 [9]. This method assumes that each document in the corpus is a probability distribution of topics and each topic is a probability distribution of words in the vocabulary of the corpus. Given a corpus *D*, LDA assumes that the corpus can be generated by the following process [10] (Fig. 2):

Step 1. For each topic *k* in *K* topics, draw a distribution over words in the vocabulary:

$$\varphi(k) \sim Dirichlet(\beta)$$

Step 2. For each document $d \in D$:
a) Draw a distribution over topics of the document:

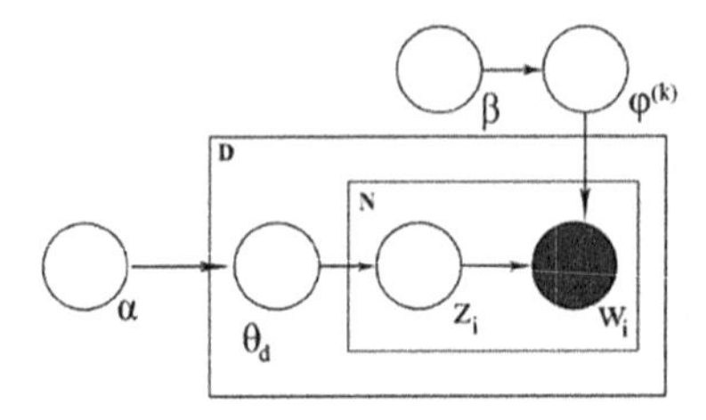

Fig. 2. Graphical representation of LDA model (modified from [9]).

$$\theta_d \sim Dirichlet(\alpha)$$

b) For each word $w_i \in d$:
i. Draw a topic assignment: $z_i \sim Discrete(\theta_d)$
ii. Choose the word: $w_i \sim Disctete\left(\varphi^{(z_i)}\right)$

where K is the number of latent topics in the corpus and $\alpha,\ \beta$ are the parameters of the corresponding Dirichlet distributions.

The above process results in the following joint distribution [10]:

$$p(\mathbf{w}, \mathbf{z}, \theta, \varphi|\alpha, \beta) = p(\varphi|\beta)p(\theta|\alpha)p(\mathbf{z}|\theta)p(\mathbf{w}|\varphi_z) \tag{1}$$

where **w** is the vocabulary and **z** is the topic assignment for each word in w.

3.4 Fan Page Representation Using Vector

Figure 3 illustrates the process to obtain the topic distribution vector for a particular document in a fan page's text corpus by using LDA. LDA model gives us two outputs, the cluster of words for each topic and the topic assignment for each word in the corpus. Therefore, we can know exactly how many times a topic appears in the document or, in other words, how many times a topic has been assigned to any word in the document by counting. We then get the topic distribution vector of the document by calculating the probability of each topic being assigned to a word in that document. Consequently, we can generate the topic distribution vector for each fan page in some way.

We propose a simple way to calculate the topic distribution vector for each fan page by summing over the topic distribution vectors of all documents in its corpus. However, each document in the sum should be associated with a weight reflecting how interactive its corresponding post is, as presented in Sect. 3.1. Therefore, we additionally propose to use the number of reactions (e.g., like, haha, angry, etc.) and the number of comments on each post as the parameters to compute the weight of that post, thus its document, in making of the topic distribution of a fan page.

Let $V = \{v_1; v_2; \ldots; v_n\}$ the set of topic distribution vectors of the documents of a fan page's corpus; n is the number of documents of the corpus; t_i, r_i, c_i are respectively

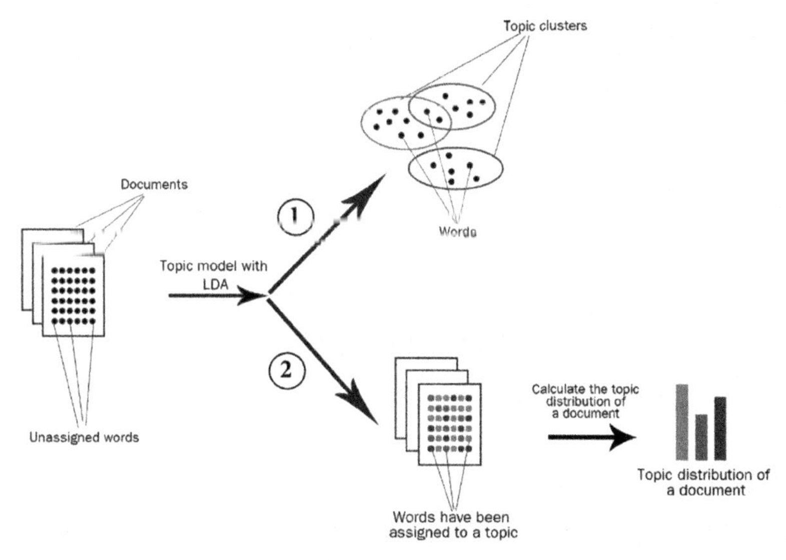

Fig. 3. Process to obtain the topic distribution of a document by using LDA.

the interaction index, number of reactions and number of comments of the ith document ($1 \leq i \leq n$). The interaction index of the ith document can be calculated as

$$t_i = \eta r_i + \mu c_i, \tag{2}$$

where η, μ respectively represents the relative importance between reactions and comments in the interaction index. Since comments are considered more valuable than reactions in terms of the degree of interaction, we experimentally set $\eta = 1$ and $\mu = 3$.

Let P the topic distribution vector represented a fan page. P can be calculated as the weighted sum of topic distribution vectors of all documents of the page, i.e.

$$P = w_1 v_1 + w_2 v_2 + \ldots + w_i v_i + \ldots w_n v_n, \tag{3}$$

where the weight of each document is its interaction index normalized among all documents of the fan page, i.e.

$$w_i = \frac{t_i}{\sum_{i=1}^{n} t_i}. \tag{4}$$

Finally, we can rewrite the vector representation of the page as:

$$P = \frac{\sum_{i=1}^{n} (t_i \times v_i)}{\sum_{i=1}^{n} t_i}. \tag{5}$$

4 Experiments and Results

4.1 Data

The data for this project was crawled from the top fan pages that have the biggest fanbase in the Media category of Vietnamese Facebook (according to the ranking of socialbakers.com in October, 2019 [11]). Details of the dataset are described as below:

- Number of fan pages: 100
- Number of posts (documents): 27,226
- Number of unique words (segmented by the pyvi toolkit [12]): 56,135
- Total number of words: 743,725
- Timespan: during October, 2019

4.2 Experimental Settings

All experiments were conducted using the scikit-learn toolkit [13]. We used LDA model for the document topic modeling process with the following parameters:

- Number of topics: $K = 20$
- Parameters of Dirichlet distributions: $\alpha = \beta = \frac{1}{K} = 0.05$

If the number of topics is too small, there will be little diversity among topic distributions of the corpus. On the contrary, if the number of topics is too big, it will be difficult to interpret what the topics are about since the topics are not obvious anymore. Therefore, we set the number of topics K to 20 in the experiments.

4.3 Topic Modeling Results

Table 1 presents the topic modeling results based on LDA method by showing the top 10 keywords of 20 topics. We can observe that several topics represent quite well about hot events or issues happening in October, 2019. For example, Topic 8 is clearly about the football match between national teams of Vietnam and Malaysia inside the World Cup 2022 qualification round with keywords such as "việt_nam" (*Vietnam*), "malaysia", "trận" (*match*); Topic 5 can be associated with the protest escalation in Hong Kong with the keywords such as "hồng_kông" (*Hong Kong*), "biểu_tình" (*demonstrate*), "dân_chủ" (*democracy*); or Topic 3 can be identified as the air pollution spike in Hanoi due to the keywords such as "không_khí" (*air*), "ô_nhiễm" (*pollution*), "hà_nội" (*Hanoi*), etc. Other topics about daily issues also can be easily identified from their keywords such as Topic 0 (about fashion and music), Topic 6 (about technology and cellphone), Topic 14 (about food and restaurant), to name just a few.

4.4 Fan Page Modeling Results

Based on the outputs of LDA, we can infer the topic distribution vector of a document. Since a fan page can publish multiple documents with different topics, we can represent

Table 1. Top 10 keywords of 20 topics found by LDA (English translation in parentheses).

Topic 0	Topic 1	Topic 2	Topic 3	Topic 4
đẹp *(beauty)* thời_trang *(fashion)* nghệ_sĩ *(artists)* nữ *(female)* diễn_viên *(actor/actress)* ca_sĩ *(singer)* nàng *(she/her)* mv *(stands for Music Video)* nhạc *(music)* khán_giả *(audience)*	đồng *(currency)* công_ty *(company)* tiền *(money)* đầu_tư *(invest)* hàng *(product)* giá *(price)* trung_quốc *(China)* thương_mại *(commerce)* kinh_tế *(economy)* triệu *(million)*	công_an *(police)* vụ *(case)* án *(crime)* điều_tra *(investigate)* cảnh_sát *(police)* bắt *(arrest)* tp *(stands for city)* tỉnh *(province)* đối_tượng *(criminal)* thông_tin *(information)*	không_khí *(air)* ô_nhiễm *(pollution)* bệnh *(disease)* hà_nội *(Hanoi)* môi_trường *(environment)* bác_sĩ *(doctor)* sức_khỏe *(health)* bệnh_viện *(hostpital)* ung_thư *(cancer)* thuốc *(pharmacy)*	tập *(episode)* phim_truyện *(drama)* khám_phá *(discovery)* thế_giới *(world)* tập_h *(episode)* vtv *(stands for Vietnam Television)* tiếng *(language)* việt_nam *(Vietnam)* phim *(movie)* đi *(go)*
Topic 5	**Topic 6**	**Topic 7**	**Topic 8**	**Topic 9**
trump tổng_thống *(president)* trung_quốc *(China)* đảng *(party)* biểu_tình *(demonstrate)* hồng_kông *(Hong Kong)* chính_trị *(politics)* dân_chủ *(democracy)* điều_tra *(investigate)* hoa_kỳ *(the US)*	việt_nam *(Vietnam)* game ứng_dụng *(application)* công_nghệ *(technology)* sự_kiện *(event)* thông_tin *(information)* diễn *(perform)* iphone trải_nghiệm *(experience)* điện_thoại *(cellphone)*	trung_quốc *(China)* việt_nam *(Vietnam)* tàu *(ship)* biển *(sea)* quốc_tế *(international)* khu_vực *(region)* biển_đông *(East Sea)* nga *(Russia)* mỹ *(the US)* máy_bay *(aircraft)*	việt_nam *(Vietnam)* trận *(match)* malaysia đội *(team)* trận_đấu *(match)* hlv *(coach)* cầu_thủ *(footballer)* đội_tuyển *(team)* bóng *(ball)* sân *(stadium)*	thvl (*stands for Vinh Long Television)* kênh *(channel)* đôi *(team)* full tình_yêu *(love)* fanpage youtube lỡ *(miss)* cực *(very)* phát_sóng *(broadcast)*
Topic 10	**Topic 11**	**Topic 12**	**Topic 13**	**Topic 14**
mẹ *(mother)* đi *(go)* vợ *(wife)* tao *(me)* tiền *(money)* chồng *(husband)* bé *(baby)* đứa *(kid)* thuốc *(medicine)* đời *(life)*	yêu *(love)* đi *(go)* chẳng *(not)* đừng *(don't)* sống *(live)* phụ_nữ *(woman)* đàn_ông *(man)* hai *(two)* hạnh_phúc *(happy)* cuộc_sống *(life)*	trường *(school)* học *(study)* trẻ *(kid)* học_sinh *(student)* lớp *(class)* đại_học *(university)* tiếng *(language)* giáo_dục *(education)* phụ_huynh *(parents)* bé *(baby)*	xe *(vehicle)* đường *(street)* đi *(go)* cháy *(burn)* hà_nội *(Hanoi)* dân *(resident)* chạy *(run)* giao_thông *(traffic)* tàu *(ship)* xảy *(happen)*	đi *(go)* món *(food)* rau *(vegetable)* bánh *(cake)* lắm *(very much)* ngon *(delicious)* đồ *(food)* thịt *(meat)* gà *(chicken)* quán *(food stall)*
Topic 15	**Topic 16**	**Topic 17**	**Topic 18**	**Topic 19**
thi *(compete)* giải *(prize)* tham_gia *(participate)* quà *(gift)* chương_trình *(program)* câu *(question)* việt_nam *(Vietnam)* trao *(give)* giải_thưởng *(award)* may_mắn *(lucky)*	đón *(wait)* tập *(episode)* htv *(stands for Ho Chi Minh City Television)* full hấp_dẫn *(hot)* đừng *(do not)* chủ_nhật *(sunday)* chồng *(husband)* link gameshow	tỉnh *(province)* dự_án *(project)* xây_dựng *(construction)* tp *(city)* ubnd *(stands for People's Committee)* quy_định *(rule)* thông_tin *(information)* dân *(resident)* đầu_tư *(investment)* tổ_chức *(organization)*	phim *(movie)* đi *(go)* dám *(dare)* tốt_đẹp *(nice)* tặng *(give)* xe *(car)* cướp *(robbery)* đầu_bếp *(chief)* kết_quả *(result)* chuyên_nghiệp *(professional)*	ảnh *(picture)* vàng *(yellow)* mùa *(season)* đi *(go)* hoa *(flower)* thời *(time)* lịch_sử *(history)* hàng *(store)* phố *(street)* thu *(autumn)*

the page based on the topic distribution vectors of its documents. The page's topic distribution vector is defined as the weighted sum of the document vectors as described in Sect. 3.4. Thus it has the same dimension of 20 with the document vectors (due to $K = 20$).

As an example, the resulting topic distribution vector of the fan page for "Báo Đời Sống Pháp Luật" (*Law and Life Journal*) is displayed in Fig. 4. As can be observed, the topic probability distribution attains notable peaks at three topics, which are: Topic 2 – a justice-related topic with the keywords such as "cảnh sát" (*police*), "vụ" (*case*), and "điều tra" (*investigate*); Topic 10 – a family-related topic with the keywords such as "mẹ" (*mother*), "vợ" (*wife*), and "tiền" (*money*); Topic 13 – a transportation-related topic with the keywords such as "xe" (*vehicle*), "đường" (*street*), and "giao thông" (*traffic*).

This result is quite reasonable because justice, family, and transportation are the most concerns of this journal.

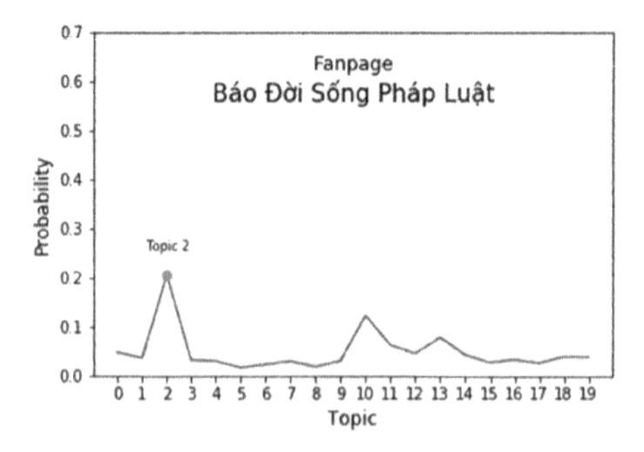

Fig. 4. Topic distribution of fan page "Báo Đời Sống Pháp Luật" (*Law and Life Journal*).

4.5 Fan Page Clustering Results

With the resulting vector representations of fan pages, we can group them into different clusters so that the pages in each cluster have similar topic distributions and the resulting clusters are well separated each other. We have tried to cluster the topic distribution vectors of all fan pages in the dataset with the K-mean Clustering algorithm. With the optimal number of clusters of 12 (see the results in Table 2), we got several example results as follows.

Table 2. Silhouette scores comparison between two methods of fan pages modeling.

Method	# of clusters						
	2	4	6	8	10	12	14
LDA with interaction indices	0.1398	0.1606	0.1896	0.2328	**0.2669**	**0.3008**	**0.2748**
LDA only	**0.1404**	**0.1821**	**0.1965**	**0.2345**	0.2370	0.2759	0.2523
Method	# of clusters						
	16	18	20	22	24	26	28
LDA with interaction indices	**0.2665**	0.2462	**0.2503**	**0.2756**	**0.2580**	**0.2556**	**0.2324**
LDA only	0.2623	**0.2525**	0.2403	0.2279	0.2273	0.2029	0.2322

Cluster 1 includes several fan pages such as "Giải trí TV" *(Entertainment TV)*, "HTV3 - DreamsTV", "Kênh Nhạc Việt" *(Vietnamese Music Channel)*, "VTV Giải trí VTV6" *(VTV Entertainment VTV6)*. All of these are the pages of entertainment channels (Fig. 5).

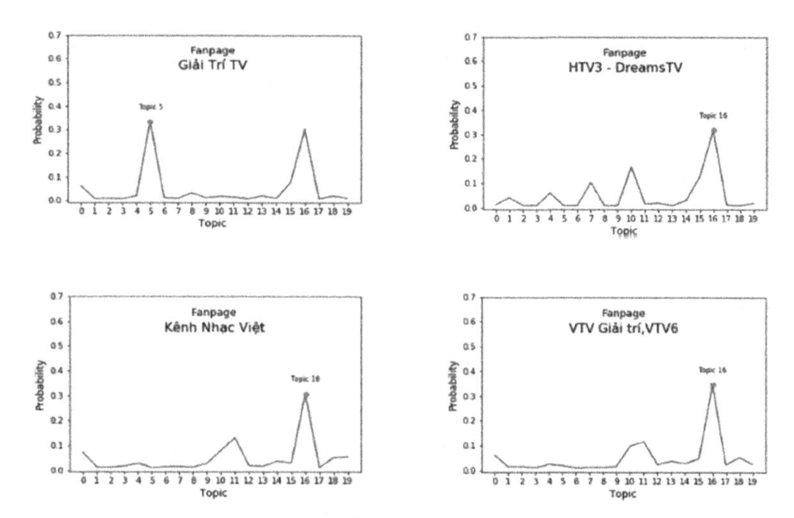

Fig. 5. Topics distribution vectors of the fan pages belonging to Cluster 1.

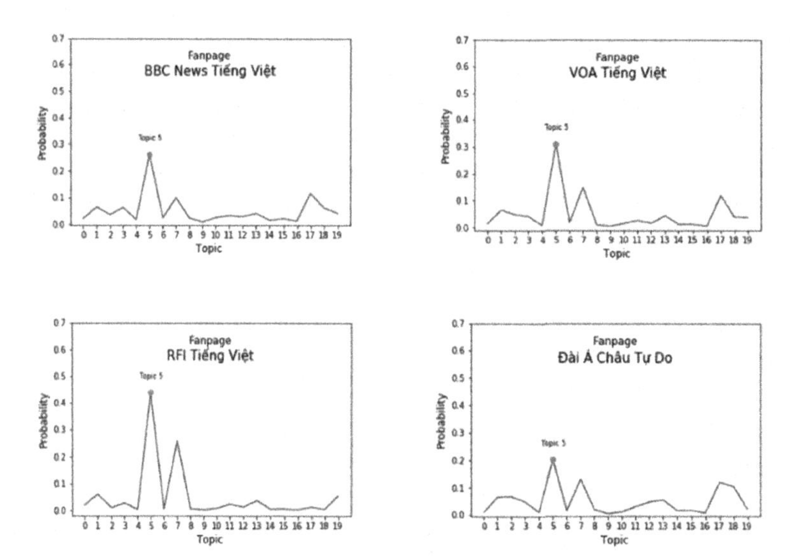

Fig. 6. Topics distribution vectors of the fan pages belonging to Cluster 2.

Cluster 2 includes fan pages of broadcasters about news and politics such as "BBC Tiếng Việt" *(BBC Vietnamese)*, "Đài Châu Á Tự Do" *(Radio Free Asia)*, "RFI Tiếng Việt" *(RFI Vietnamese)*, "VOA Tiếng Việt" *(VOA Vietnamese)* (Fig. 6).

As can be seen on Fig. 5 and Fig. 6, those fan pages having similar topic distributions were grouped quite well thanks to their vector representations.

To quantitatively evaluate the clustering performance, we used Silhouette score [14] to measure how well the clusters are separated to each other. The higher the score, the better clustering process. We compared the clustering performance between the two vector representations of fan pages: our proposed method (LDA-based topic distributions combined with interaction indices, i.e., each document has a different weight in Eq. (3)) and conventional one (LDA-based topic distributions only, i.e., all documents have the same weight in Eq. (3)). The results in Table 2 show that when the number of clusters is high enough (more than 8), our proposed method outperforms the conventional one on Silhouette score. In particular, both of the two fan page representation methods achieve optimal clustering performance when the number of clusters is set to 12. In that case, our proposed method improves 9.0% on Silhouette score compared to the conventional one (0.3008 vs. 0.2759).

5 Conclusion

In this paper, we have proposed a method to represent a fan page by a vector using LDA-based topic modeling on all fan pages in the corpus combined with interaction index analysis of their posts. Experiment results showed that this representation can be used to cluster a set of fan pages effectively and obtained better clustering performance than the conventional one just based on LDA. The proposed vector representation of fan pages also showed its effectiveness in figuring out hot topics as well as regular issues posted on Facebook in a fixed period of time. The main benefit of our approach to fan page modeling and mining is that it helps us to follow trending contents on this social platform on a large scale without collecting the data of regular individual users. In the future, we will apply other models that focus more on the segmentation of documents such as lda2vec [15] to find out how positive or negative different fan pages talk about the same topic. We also want to extend the proposed method so that the time factor is included to reflect how the relationship between fan pages changes over time.

Acknowledgments. This research is funded by the University of Danang – University of Science and Technology under grant number T2017-02-93.

References

1. Datareportal, "Social Media Users by Platform," 2019. Available: https://datareportal.com/social-media-users. Accessed 19 Nov 2019
2. Yau, C.-K., Porter, A., Newman, N., Suominen, A.: Clustering scientific documents with topic modeling. Scientometrics **100**, 767–786 (2014)
3. Kim, S.-W., Gil, J.-M.: Research paper classification systems based on TF-IDF and LDA schemes. Hum. centric Comput. Inf. Sci. **9**(1), 1–21 (2019). https://doi.org/10.1186/s13673-019-0192-7
4. Pengtao, X., Eric, P.X.: Integrating document clustering and topic modeling. In: Proceedings of the Twenty-Ninth Conference on Uncertainty in Artificial Intelligence, AUAI Press, Virginia, United States, pp. 694–703 (2013)
5. Gialampoukidis, I., Vrochidis, S., Kompatsiaris, I.: A hybrid framework for news clustering based on the DBSCAN-martingale and LDA. MLDM 2016. LNCS (LNAI), vol. 9729, pp. 170–184. Springer, Cham (2016). https://doi.org/10.1007/978-3-319-41920-6_13
6. Eliana, S., Camilo, M., Raimundo, A.: Topic modeling of Twitter conversations (2018)

7. Zhou, T., Haiy, Z.: A text mining research based on LDA topic modelling. In: International Conference on Computer Science, Engineering and Information Technology, pp. 201–210 (2016)
8. Jiamthapthaksin, R.: Thai text topic modeling system for discovering group interests of Facebook young adult users. In: 2016 2nd International Conference on Science in Information Technology (ICSITech), pp. 91–96. IEEE 2016
9. Blei, D.M., Ng, A.Y., Jordan, M.I.: Latent dirichlet allocation. J. Mach. Learn. Res. **3**, 993–1022 (2003)
10. Darling, W.M.: A theoretical and practical implementation tutorial on topic modeling and gibbs sampling. In: Proceedings of the 49th annual meeting of the association for computational linguistics: Human language technologies, pp. 642–647 (2011)
11. Social Bakers Vietnamese Statistics. Available: https://www.socialbakers.com/statistics/facebook/pages/total/vietnam. Accessed 19 Nov 2019
12. pyvi toolkit. Available: https://pypi.org/project/pyvi/. Accessed 19 Nov 2019
13. Pedregosa, F., et al.: Scikit-learn: machine learning in python. J. Mach. Learn. Res. **12**, 2825–2830 (2011)
14. Rousseeuw, P.J.: A graphical aid to the interpretation and validation of cluster analysis. J. Comput. Appl. Math. **20**, 53–65 (1987)
15. Moody, C.E.: Mixing dirichlet topic models and word embeddings to make lda2vec. arXiv arXiv:1605.02019 (2016)

Natural Language Processing

Wordnet – a Basic Resource for Natural Language Processing: The Case of plWordNet

Agnieszka Dziob(✉) and Tomasz Naskręt

Wroclaw University of Science and Technology, Faculty of Computer Science and Management, Wroclaw, Poland
{agieszka.dziob,tomasz.naskret}@pwr.edu.pl

Abstract. This paper presents a wide scope of wordnet applications on the example of applications of plWordNet – a wordnet of Polish. Wordnets are large lexical-semantic databases functioning as primary resources for language technology. They are machine-readable dictionaries. Thus, they are indispensible for tasks such as basic flow of text processing, text mining, word sense disambiguation, information extraction and retrieval. On a larger scale, wordnets are used in research, education and business. In this paper a few examples of specific plWordNet applications are described in detail.

Keywords: Wordnet · plWordNet · Language technology. · NLP

1 Introduction

Wordnet is a hierarchically organised lexical-semantic database in which nouns, verbs, adjectives and adverbs are grouped into sets of cognitive synonyms (synsets), each expressing a distinct concept. Synsets are interlinked by means of conceptual-semantic and lexical relations. The first wordnet in the world was Princeton WordNet (henceforth, PWN) created for the English language at Princeton University and developed since the mid 1980 s [31]. The main element of PWN structure is synset, defined as a set of synonyms. Single words as well as synsets are linked with each other by lexical (at word level) and conceptual-semantic (at synset level) relations. Fundamental synset-level relations are hyponymy (hierarchy building), meronymy, and troponymy (the latter for verbs only). Basic word-level relations are derivational relations.

Unfortunately, since 2006, when version 3.1 was released, PWN has no longer been actively developed. Currently, works are continued within the English WordNet project [30]. Version 3.1 of PWN is available, containing 146,533 nouns, 25,061 verbs, 30,072 adjectives and 5,592 adverbs [8]. PWN is the core on which many wordnets for other languages have been based. These have been created by various methods such as automatic or semi-automatic translation of PWN synsets and lexical units or complete manual projection.

M. Hernes et al. (Eds.): ICCCI 2020, CCIS 1287, pp. 689–700, 2020.
https://doi.org/10.1007/978-3-030-63119-2_56

PWN is available under an open license. Although WordNet originates from psycholinguistic research on language acquisition by children, it has demonstrated its full potential in the NLP [31]. In 2004, J. Morato and others [32] carried out a statistical analysis of the use of PWN based on publications available in databases such as: LISA [6], INSPEC [3], IEEE [5], ResearchIndex (now CiteSeerX) [1] and The Library of the University of Carlos III of Madrid [7] for the period 1994–2003. The bibliographic databases used by them include publications mainly from the area of engineering, especially computer and information science. They focus less on the use of PWN in Humanities and Social Sciences and in non-academic applications.

Since the analysis of Morato et al., we can observe the increase of interest in the use of wordnets in the field of natural language processing. The aim of this paper is to present the applications of a wordnet for the Polish language – plWordNet (henceforth, plWN), developed at the Wrocław University of Science and Technology since 2005 [39], with its latest official 4.1 version published in 2019 [14]. plWN is equally oriented towards applications in computer and information science, as well as in humanities and social sciences and in business. This is largely motivated by the fact that since 2013 plWN has become an element of the Polish part of CLARIN (Common Language Resources and Technology Infrastructure) [2], providing tools and resources for natural language processing outside the field of engineering sciences. It focuses on their popularization in the field of Humanities and Social Sciences.

The aim of this paper is to present plWN from the perspective of its versatile applications. To this end, we use the broadest possible data of plWN citations, collected from Google Scholar [4], for the period between 2005–2019, which allowed for manual verification of its usage in the last 15 years. Of 407 works that cite plWN we have chosen those that described its usage in research or existing applications. The second source of data are questionnaires filled in by users while downloading the plWN database. They have been collected since 2015, yet they do not give the full picture of applications, but rather indicate the potential of usage of national wordnets. We will strive to show how the range of applications of wordnets has changed in relation to the ones described in [32], and to present them in the perspective of Polish language processing.

2 WordNet Implementation

The following areas of application of PWN emerge from [32]: wordnet-internal and wordnet-external ones. The former refer to building wordnet editing systems and multilingual resources. The latter cover information extraction and retrieval, document structuring and categorisation, audio and video retrieval. The authors also point out developmental trends which include: development of systems that allow to determine the degree of inter-lingual equivalence, the use of wordnets in information extraction systems (e.g. semantic search engines) to improve their efficiency, creating and categorising semantic ontologies[1], and other knowledge resources in the Internet as well as audiovisual and multi-media systems.

Morato et al. see WordNet's success as mainly due to its open licence and big potential of use in NLP. While in 2004 there was potential, in 2019 the multitude of applications in NLP became a reality. The CiteSeerX database alone shows almost 29,000 results for 'wordnet'. Although plWN did not receive such a large number of citations, we will try to present on its example a cross-section of applications of wordnet type databases – actual, confirmed by citations, and potential, indicating trends in NLP development using wordnets.

3 plWordNet in Brief

Similarly to Princeton WordNet, plWN defines meanings using lexical-semantic relations [39], but the central element is lexical unit [28] (henceforth, LU). This stems from the specificity of the Polish language, whose very rich morphology results in the fact that derivational relations (of formal and semantic character) are as significant as purely semantic relations between synsets. Compared to PWN , plWN is also characterized by a greater number of different relations at the level of lexical units and synsets.

In addition to LUs, synsets and relations, plWN has, like PWN , additional areas of description: glosses (abbreviated definitions) and examples of use (currently not for all lexical units, successively completed). Moreover, plWN has a very rich network of connections with external resources (see 4.1.). Since 2012 the mapping of Polish synsets on the English ones [45] has been carried out, and since 2018 it has been experimentally extended to the level of LUs [46]. These are pioneering works on a global scale.

Table 1 contains abbreviated statistics of the version 4.1 of plWN [14] published in 2019, covering Polish lemmas (words), LUs, synsets, as well as their mapping on PWN , which includes the number of inter-lingual synonyms for synsets and strong equivalence for LUs, relations particularly useful in applications related to translation. The table also covers the statistics on connections to external sources.

Unlike many wordnets created by the translation of PWN, plWN has been built manually, with the support of NLP tools and based on large corpora of text [39], so it gives an excellent picture of the Polish language vocabulary. Moreover, it is currently the largest electronic dictionary for Polish language. The initial set of lexical-semantic relations, taken mostly from PWN , was extended in terms of typical elements of dictionary description in Polish and potential applications in NLP. The authors were also guided in the choice of relations by other principles, according to which they were supposed to be established in the Polish lexicographic tradition and be frequent in the Polish lexical system [28].

[1] As early as 2004, as Morato et al. mentioned, ontologies and semantic web were one of the most dynamically developing areas of wordnet applications.

Table 1. Basic statistics of plWN 4.1 (http://plwordnet.pwr.edu.pl)

Elements	Verbs	Nouns	Adv.	Adj.	All
plWN 4.1 Lemmas	20 430	134 674	8 042	29 349	**192 495**
plWN 4.1 Lexical Units	43 701	178 167	14 088	54 410	**290 366**
plWN 4.1 Synsets	32 102	133 747	11 295	47 035	**224 179**
Links to PWN (Synsets):	**280 502**				
Inter-lingual synonymy:	46 883				
Links to PWN (LUs):	**10 244**				
Strong equivalence	**9 860**				
Links to SUMO	**206 426**				
Links to Wikipedia	**44 432**				

In the current version of plWN there are about 200 types and subtypes of relations between synsets, LUs and external lexicons. Among them one can mention universal relations for this type of relational lexicons, e.g. hyponymy (*jamnik* 'dachshund' → *pies* 'dog'), antonymy (*zmniejszać* 'to decrease' ↔ *zwiększać* 'to increase'), meronymy (*rodzynka* 'a raisin' → *bakalie* 'dried fruit and nuts'), cause (for verbs) (*powodować* 'to cause' → *stawać się* 'to become'). However, plWN also contains many Polish-specific relationships, especially between LUs: feminity (*lekarka* 'a woman-doctor' ← *lekarz* 'a doctor'), markedness, which says that the unit has some emotional load (*guziczek* 'a little button' ← *guzik* 'a button') or role and role inclusion typical for nouns derived from verbs and verbs derived from nouns (e.g. role: *drukować* 'to print' ← *drukarka* 'a printer'; role inclusion: *nocować* 'spend a night' ← *noc* 'a night').

4 Applications of plWordNet

4.1 plWordNet Development

Wordnet Editing System. The plWN graph-based editing system (WordnetLoom) [35] has three editing perspectives – units, synsets, and synset relations, which allow linguists to work on both levels in an integrated way. It has been recently enriched with a graphical presentation module (Wordnet Viewer) and an integrated linguist support module, which allows to preview changes made to the database, and presents language material from text corpora. In addition, the WordNet Weaver module has been implemented, allowing for partial automation of the plWN expansion process [39]. At the same time, a separate diagnostic module was developed, which allows to correct typical errors in plWN structure [40].

Since 2012, due to the linking of plWN with PWN by inter-lingual relations [45], it has become necessary to adapt the existing system to the requirements of linking with other language resources and the requirements of increasingly large teams working in an integrated way. In 2018, WordnetLoom 2.0 was presented with an extensive graphical module and limited remaining editing perspectives, linking to a modified [35] design database. WordnetLoom is currently used, in addition to the Polish team, also by teams building wordnets for languages: Portuguese [35], Danish [38] and African languages [17].

This system has been developed with a module for tracking changes in the plWN database integrated with the developed diagnostic system, based on formal cohesion determinants of plWN and, partly also, content-related ones [34]. In 2018, a method of content-related evaluation of relationship cohesion for plWN was presented on the example of verbs [13]. Works on content-related and formal diagnostics are currently underway.

Extension of Resources. The mapping of plWN onto PWN has been carried out since 2012 [45]. Throughout the mapping, there has been observed a stable tendency for inter-lingual hyponymy to outgrow inter-lingual synonymy in terms of its frequency for all parts of speech. Lexicographers noted that in many cases inter-lingual hyponymy was introduced due to coverage gaps in Princeton WordNet. This observation motivated an experimental extension of PWN. Eventually, the size of this extension, called enWordNet, reached 11 294 LUs [47]. plWN is currently the part of the Open Multilingual Wordnet [9], compiled using inter-lingual relations (mainly inter-lingual synonymy) between individual wordnets and PWN .

Multi-word Expression (henceforth, MWE) units have been included in plWN, whose definition and selection from language material are presented in [29]. Semantically described in plWN , MWE are syntactically described using WCCL syntax adapted to the description of the common language in MWELexicon, currently containing over 50k units belonging to all parts of speech described in plWN in different structural types. The work on MWE was the motivation to develop the description of terminology in plWN [29]. However, from 2016 to 2019 works were carried out to combine plWN with external terminological resources and ontologies, including Wikipedia, DBPedia, SUMO and YAGO ontologies and thesauri, conducted semi-automatically, i.e. on automatic propagation of manual connections, based on plWN structure and gloss. They were based, apart from the mentioned description of MWE, also on previous studies on automatic linking of plWN with the SUMO ontology and many works on linking plWN and Wikipedia [41].

Since 2015, plWN has been supplemented with the emotive annotation for LUs [53], conducted manually by two annotators (a psychologist and a linguist), supervised by a superannotator, decisive of discrepancies in the annotation. The set of 100,000 manual annotations was extended by means of automatic propagation using the Classifier-based Polarity Propagation method [20]. The plWN emotive annotation became the basis for a sense-tagged sentiment resource comparison [10], aimed at developing methods of assessing the quality of emotive annotation in wordnets.

Walenty, a comprehensive valency dictionary of Polish developed at the Institute of Computer Science, Polish Academy of Sciences, containing currently over 18,000 units, is built from two levels: syntactic, describing syntactic schemes of units for the Polish language, and semantic, assigning meaning to units. These meanings correspond to the meanings of LUs from plWN [18]. In 2019, a semi-automatic projection of Walenty semantic layer on the plWN structure was carried out in such a way that relations corresponding to Walenty selection preferences were introduced into plWN , connecting synsets containing units that are described in Walenty with the use of these preferences [18].

The multitude of different sources by which plWN was extended makes it a valuable component of coTagging, parsing, stemming, normalizationmbined resources, such as the above mentioned Open Multilingual Wordnet, but also created for the Polish language: Lexical Platform [42] and Multisłownik [37].

4.2 plWordNet for Systems of Natural Language Processing

Text Mining. At the text analysis level, plWN is used to recognize and classify elements of text, such as MWEs [43], named entities [16], events [21], temporal expressions [22], spatial expressions [25], and terminology [33]. Many applications are related to word sense disambiguation. Among them we can mention the research [19], in which at first the method for Polish language based on Page Rank was used, and then plWN combined with SUMO ontology was used to improve its effectiveness. In another solution, cf. [48], to improve the results was used a combination of models trained on text corpora and information extracted from plWN, including relations and glosses.

Development trends in NLP indicate growing interest in using wordnets to build intelligent e-commerce and opinion mining tools. The article [49] presents a sample analysis of sentiment in the discourse of Polish politicians conducted using plWN. The system of opinion analysis applicable to texts on different fields was presented in the article [24]. The article [36] presents an approach combining the extraction of key words from the Polish text with the analysis of sentiment. The article [50] presents a comprehensive system for features extraction and opinion mining, allowing for the analysis of Internet comments, to which the synonyms and antonyms of plWN were used. plWN was also used in many cases as a basis or one of the main sources to build other text mining tools. These include tools for comprehensive text analysis such as Cluo [23] and LEM [26], for stylometric analysis [15] and for the analysis of text complexity [12].

Knowledge Extraction and Building Semantic Web. The analysis of structure at the sentence and text level is applicable to the construction of semantic search engines. In 2016 plWN became one of the main sources for building the first such search engine for Polish language NEKST [11]. The real challenge are Question Answering systems, which are used in semantic search engines showing results for questions asked in natural language, as well as for building intelligent assistants (including voice). The use of plWN in QA systems is shown in [44]. The RAFAEL system described by the author, based on the Deep Entity Recognition (DeepER) procedure, alternative to Named Entity Recognition (NER). The operation of DeepER is based on combining knowledge resources the information available for Named Entities in plWN and Wikipedia.

Intelligent systems of the semantic search and answer to questions in natural language are, besides e-commerce, a branch of NLP, for which we can find great interest in the use of plWN (see Table 1) and predict that it will grow. New possibilities are created by combining plWN with Linked Open Data resources, saved in SKOS format, which is described in [27]. Building such a system in the SKOS/LMF standard, using data contained in plWordNet, is described in [52].

Multi-modal Data. Another area, where growing interest in wordnets can be indicated, is multi-modal data analysis. So far plWN has been used to describe and classify visual data [51] and to the semantic annotation of indicating gestures. Table 2 shows the development of this range of uses, especially in the area related to speech analysis.

Other Applications. Since 2015, users downloading the plWN database are asked to fill in a short form, in which they indicate what they plan to use it for. They are specific, practic applications, not described in the scientific publications. The data collected so far indicate the following applications, declared by representatives of science, commercial companies and individuals interested in technological development. They are presented in Table 1.

At the text mining level, plWordNet's application is used by its users to analyse groups of texts such as customer service reports, software requirements, design requirements, CVs, e-mails, inquiries, economic documents, debates, job offers, transcriptions of telephone conversations, technical texts translation, analysis of intentions in dialogues, analysis of press releases, research on communication in social networks (very often devoted to sentiment research) and analysis of changes taking place in the labour market on the basis of e.g. job advertisements.

Another category are works of users declaring the application of plWordNet, devoted to varieties of language or discourse, among which one can mention analyses of parliamentary data, public debate and opinion, or political or religious discourse, research on changes in business language or media information.

As already mentioned, there is a growing interest in using NLP tools and resources in the e-commerce market in Poland. The results of text mining research with the use of plWN and the plWN database itself are, according to the declarations, used for analyzing opinions about products, automatic categorization of purchase and sale announcements, for describing products and their grouping (classification) in an online store, creating advanced product search engine (the quality of which is improved by the ability to recognize synonyms in user queries), as well as creating SEO texts, indexing and searching, creating material indexes. One of the more advanced applications are recommendation systems, based on opinion mining research.

Table 2. plWordNet implementations declared in user's forms.

Field	Example
Text analysis	Tagging, parsing, stemming, normalization of texts, control of correctness and error detection (also automatic, in real time)
Text mining	Sentiment analysis, opinion mining, semantic features analysis, paraphrases of texts, stylometry, automatic classification and categorisation of texts and theirs parts, similarity of texts, analysis of social nets, anty-plagiarism system, automatic summaries, personalizing texts, keywords and phrases extraction, word sense disambiguation, content analysis, named entity recognition
Knowledge and information extraction	Semantic nets, mapping onto Linked Open Data, ontology extracting, generating and interlinking, semantic search, intelligent search for products in e-commerce (e.g. synonyms recognition), automatic generation and classification of questions and answers, Question Aswering systems, anonymization, full-text search, digital shadow analysis, automatic learning of knowledge bases, query expansion, analysis of links between persons (so-called PEP: politically exposed person) Knowledge Graph constructing, fake news detection
Multilingual data and language learning	translation and building systems of machine translation, building multi-lingual resources, linking of wordnets, tools for native and second language learning, multi-lingual research, automatic generation of tasks and language games for language learners, translation of artificial languages
Speech recognition and intelligent systems	Chat boots, intelligent assistants, intelligent systems of remote Control, machine learning, language modelling, communication in natural language with robots, systems for blocking undesirable content
Data visualisation	Building clouds of words and graphs
Linguistic research	regular polysemy, semantic nests (semantic nodes), semantic classes, Lexicology and lexicography, metaphores in language, psycholinguistics, corpus linguistics, valency, predicate-argument structures, semantic roles, multiword lexical units, terminology, proper names, semantic relations in language, derivational morphology, verbal aspect in Polish
Clinical research	Neurolinguistics, research on mental disabilities (including autism and dementia) and aging, building medical databases, analysis of medical patients autodescriptions, psychological research on trauma
Other research in NLP field	Building and training language models and semantic similarity measures, SPAM and pishing detection
others	on field of experimental philosophy, crosswords generator, creating UUID abbreviations, mnemonic password generator

5 Conclusions and Further Works

As predicted by Morato et al., our data indicate the greatest interest in plWN in areas such as combining data for many languages and creating a database of combined data with non-dictionary resources, including the expansive development of links with ontologies, especially domain-specific ones, the use of text mining documents for semantic search, as well as the development of multi-modal systems (especially in the area related to speech processing). Other areas that we can point out are the interest in describing and processing terminology (single and multi-word) and the analysis of sentiment, both in applications related to commercial activities and for the analysis of various areas of social life (e.g. by associations working for the benefit of societies). The above summary shows that plWN has numerous applications covering both research and development of tools and systems used in non-academic applications. Aware of the potential and limitations of the resource, as well as trends in the application of NLP tools, we anticipate developments in the following areas:

1. further diagnostics with the use of WordnetLoom diagnostic model and numerical data extracted from text corpora (for Polish and English), which will significantly improve the quality of data contained in plWN;
2. extending plWN with other relations, typical for the Polish language, which are well-established in lexicography and at the same time useful in commercial applications (e.g. for building intelligent e-commerce search engines);
3. expanding the coverage of terminological units and their relations, especially in the area of law and medicine
4. linking plWN to low-level ontologies will increase the use of grouping and classification of domain-specific texts.

At the same time, it should also be noted that the importance of national languages is growing, and that the future expansion of both plWN and wordnets for many other languages other than English can also be seen.

References

1. CiteSeerX. citeseerx.ist.psu.edu. Accessed 14 Jan 2020
2. CLARIN homepage. www.clarin.eu. Accessed 14 Jan 2020
3. Engineering Research Database. www.elsevier.com/solutions/engineering-village/content/inspec. Accessed 14 Jan 2020
4. Google Scholar. https://scholar.google.com/. Accessed 14 Jan 2020
5. Institute of Electrical and Electronics Engineers. www.ieee.org. Accessed 14 Jan 2020
6. Library and Information Science Abstracts. www.proquest.com/products-services/lisa-set-c.html. Accessed 14 Jan 2020
7. Library of the university of carlos III of madrid. www.uc3m.es/Home. Accessed 14 Jan 2020
8. WordNet. https://wordnet.princeton.edu/. Accessed 14 Jan 2020

9. Bond, F., Foster, R.: Linking and extending an open multilingual wordnet. In: Proceedings of the 51st Annual Meeting of the Association for Computational Linguistics. **1**, pp. 1352–1362 (2013)
10. Bond, F., Janz, A., Piasecki, M.: A comparison of sense-level sentiment scores. In: Proceedings of the 10th Global Wordnet Conference, pp. 363–372 (2019)
11. Czerski, D., Ciesielski, K., Dramiński, M., Kłopotek, M., Łoziński, P., Wierzchoń, S.: What NEKST?—semantic search engine for polish internet. In: De Tré, G., Grzegorzewski, P., Kacprzyk, J., Owsiński, J.W., Penczek, W., Zadrożny, S. (eds.) Challenging Problems and Solutions in Intelligent Systems. SCI, vol. 634, pp. 335–347. Springer, Cham (2016). https://doi.org/10.1007/978-3-319-30165-5_16
12. Dębowski, Ł., Broda, B., Nitoń, B., Charzyńska, E.: Jasnopis-a program to compute readability of texts in polish based on psycholinguistic research. Nat. Lang. Process. Cogn. Sci. pp. 51–61 (2015)
13. Dziob, A., Piasecki, M.: Dynamic verbs in the wordnet of polish. Cogn. Stud. (18) (2018)
14. Dziob, A., Piasecki, M., Rudnicka, E.: plWordNet 4.1-a linguistically motivated, corpus-based bilingual resource. In: Proceedings of the 10th Global Wordnet Conference, pp. 353–362 (2019)
15. Eder, M., Piasecki, M., Walkowiak, T.: An open stylometric system based on multilevel text analysis. Cognitive Studies| Études cognitives (17) (2017)
16. Graliński, F., Jassem, K., Marcińczuk, M., Wawrzyniak, P.: Named entity recognition in machine anonymization. Recent Advances in Intelligent Information Systems, pp. 247–260 (2009)
17. Griesel, M., Bosch, S., Mojapelo, M.L.: Thinking globally, acting locally-progress in the african wordnet project. In: Proceedings of the 10th Global Wordnet Conference, pp. 191–196 (2019)
18. Hajnicz, E., Bartosiak, T.: Connections between the semantic layer of Walenty valency dictionary and plWordNet. In: Proceedings of the 10th Global Wordnet Conference, pp. 99–107 (2019)
19. Kędzia, P., Piasecki, M., Orlińska, M.: Word sense disambiguation based on large scale Polish CLARIN heterogeneous lexical resources. Cognitive Studies (15) (2015)
20. Kocoń, J., Janz, A., Piasecki, M.: Context-sensitive sentiment propagation in WordNet. In: Proceedings of the 9th Global Wordnet Conference, pp. 333–338 (2018)
21. Kocoń, J., Marcińczuk, M.: Generating of events dictionaries from polish wordNet for the recognition of events in polish documents. In: Sojka, P., Horák, A., Kopeček, I., Pala, K. (eds.) TSD 2016. LNCS (LNAI), vol. 9924, pp. 12–19. Springer, Cham (2016). https://doi.org/10.1007/978-3-319-45510-5_2
22. Kocoń, J., Marcińczuk, M.: Supervised approach to recognise Polish temporal expressions and rule-based interpretation of timexes. Nat. Lang. Eng. **23**(3), 385–418 (2017)
23. Maciołek, P., Dobrowolski, G.: Cluo: web-scale text mining system for open source intelligence purposes. Comput. Sci. **14**(1), 45–62 (2013)
24. Marchewka, A., et al.: Recognition of emotions, valence and arousal in large-scale multi-domain text reviews pp. 274–280 (2019)
25. Marcińczuk, M., Oleksy, M., Wieczorek, J.: Preliminary study on automatic recognition of spatial expressions in Polish texts. In: Sojka, P., Horák, A., Kopeček, I., Pala, K. (eds.) TSD 2016. LNCS (LNAI), vol. 9924, pp. 154–162. Springer, Cham (2016). https://doi.org/10.1007/978-3-319-45510-5_18

26. Maryl, M., Piasecki, M., Walkowiak, T.: Literary exploration machine: a Web-based application for textual scholars. In: Selected papers from the CLARIN Annual Conference, (147) pp. 128–144 (2018)
27. Maziarz, M., Piasecki, M.: Towards mapping thesauri onto plWordNet. In: Proceedings of the 9th Global WordNet Conference (GWC 2018), pp. 45–53 (2018)
28. Maziarz, M., Piasecki, M., Rudnicka, E.: Słowosieć-polski wordnet. Proces tworzenia tezaurusa. Polonica **34**, 79–98 (2014)
29. Maziarz, M., Szpakowicz, S., Piasecki, M.: A procedural definition of multi-word lexical units. In: Proceedings of the International Conference Recent Advances in Natural Language Processing, pp. 427–435 (2015)
30. McCrae, J.P., Rademaker, A., Bond, F., Rudnicka, E., Fellbaum, C.: English wordnet 2019–an open-source wordnet for english. In: Proceedings of the 10th Global Wordnet Conference, pp. 245–252 (2019)
31. Miller, G.: WordNet: An Electronic Lexical Database. MIT Press (1998)
32. Morato, J., Marzal, M.A., Lloréns, J., Moreiro, J.: WordNet applications. In: Proceedings of 2nd Global Wordnet Conference, pp. 270–278 (2004)
33. Mykowiecka, A., Marciniak, M.: Combining wordnet and morphosyntactic information in terminology clustering. In: Proceedings of COLING 2012, pp. 1951–1962 (2012)
34. Naskręt, T.: A collaborative system for building and maintaining wordnets. In: Proceedings of the 10th Global Wordnet Conference, pp. 323–328 (2019)
35. Naskręt, T., Dziob, A., Piasecki, M., Saedi, C., Branco, A.: WordnetLoom-a multilingual wordnet editing system focused on graph-based presentation. In: Proceedings of the 9th Global Wordnet Conference, pp. 191–200 (2018)
36. Nowaczyk, A., Jackowska-Strumiłło, L.: Rozpoznawanie emocji w tekstach polskojęzycznych z wykorzystaniem metody słów kluczowych. Informatyka, Automatyka, Pomiary w Gospodarce i Ochronie Środowiska **7**(2), 102–105 (2017)
37. Ogrodniczuk, M., Bronk, Z., Kieras, W.: Multisłownik: linking plWordNet-based lexical data for lexicography and educational purposes. In: Proceedings of the 9th Global Wordnet Conference, pp. 368–375 (2018)
38. Pedersen, B.S., Nimb, S., Olsen, I.R., Olsen, S.: Merging danNet with princeton wordnet. In: Proceedings of the 10th Global Wordnet Conference, pp. 125–134 (2019)
39. Piasecki, M., Broda, B., Szpakowicz, S.: A Wordnet from the ground up. Oficyna Wydawnicza Politechniki Wrocławskiej Wrocław (2009)
40. Piasecki, M., Burdka, Ł., Maziarz, M., Kaliński, M.: Diagnostic tools in plWordNet development process. In: Vetulani, Z., Uszkoreit, H., Kubis, M. (eds.) LTC 2013. LNCS (LNAI), vol. 9561, pp. 255–273. Springer, Cham (2016). https://doi.org/10.1007/978-3-319-43808-5_20
41. Piasecki, M., Kaliński, M., Indyka-Piasecka, A.: Disambiguating wikipedia articles on the basis of plWordNet lexico-semantic relations. In: Castro, F., Gelbukh, A., González, M. (eds.) MICAI 2013. LNCS (LNAI), vol. 8265, pp. 228–239. Springer, Heidelberg (2013). https://doi.org/10.1007/978-3-642-45114-0_18
42. Piasecki, M., Walkowiak, T., Rudnicka, E., Bond, F.: Lexical Platform-the first step towards user-centred integration of lexical resources. Cognitive Studies| Études cognitives (18) (2018)
43. Piasecki, M., Wendelberger, M., Maziarz, M.: Extraction of the multi-word lexical units in the perspective of the wordnet expansion. In: Proceedings of the International Conference RANLP, pp. 512–520 (2015)
44. Przybyła, P.: Boosting question answering by deep entity recognition. arXiv preprint arXiv:1605.08675 (2016)

45. Rudnicka, E., Maziarz, M., Piasecki, M., Szpakowicz, S.: A strategy of mapping Polish Wordnet onto Princeton WordNet. In: Proceedings of COLING 2012: Posters, pp. 1039–1048 (2012)
46. Rudnicka, E., Piasecki, M., Bond, F., Grabowski, Ł., Piotrowski, T.: Sense equivalence in plWordNet to Princeton WordNet mapping. Int. J. Lexicography **1**, 1–30 (2019)
47. Rudnicka, E.K., Witkowski, W., Kaliński, M.: Towards the methodology for extending princeton wordnet. Cogn. Stud.| Études cognitives (15), pp. 335–351 (2015)
48. Rutkowski, S., Rychlik, P., Mykowiecka, A.: Estimating senses with sets of lexically related words for Polish word sense disambiguation. In: Proceedings of the 10th Global Wordnet Conference, pp. 118–124 (2019)
49. Rybiński, K.: Political sentiment analysis of Polish politicians. e-politicon, **24**, 162–195 (2017)
50. Twardowski, B., Gawrysiak, P.: Domain dependent product feature and opinion extraction based on e-commerce websites. In: Zgrzywa, A., Choroś, K., Siemiński, A. (eds.) Multimedia and Internet Systems: Theory and Practice, pp. 261–270. Springer, Berlin Heidelberg, Berlin (2013). https://doi.org/10.1007/978-3-642-32335-5_25
51. Wróblewska, A.: Polish corpus of annotated descriptions of images. In: Proceedings of the 11th Int. Conference on Language Resources and Evaluation (2018)
52. Wróblewska, A., Protaziuk, G., Bembenik, R., Podsiadły-Marczykowska, T.: Associations between texts and ontology. In: Bembenik R., Skonieczny L., Rybinski H., Kryszkiewicz M., Niezgodka M. (eds.) Intelligent Tools for Building a Scientific Information Platform. Studies in Computational Intelligence, vol 467. Springer, Berlin, Heidelberg (2013). https://doi.org/10.1007/978-3-642-35647-6_20
53. Zaśko-Zielińska, M., Piasecki, M., Szpakowicz, S.: A large wordnet-based sentiment lexicon for Polish. In: Proceedings of the International Conference RANLP, pp. 721–730 (2015)

KEFT: Knowledge Extraction and Graph Building from Statistical Data Tables

Rabia Azzi[1(✉)], Sylvie Despres[2], and Gayo Diallo[1(✉)]

[1] BPH Center INSERM U1219, Team ERIAS, Univ. Bordeaux, 33000 Bordeaux, France
{rabia.azzi,gayo.diallo}@u-bordeaux.fr
[2] LIMICS - INSERM UMRS 1142, Univ. Paris 13, Sorbonne Université, 93017 Bobigny Cedex, France
sylvie.despres@univ-paris13.fr

Abstract. Data provided by statistical models are commonly represented by textual, tabular or graphical form in documents (reports, articles, posters and presentations). These documents are often available in PDF format. Even though it makes accessing a particular information more difficult, it is interesting to process the PDF documents directly. We present KEFT, a solution in the statistical domain and we describe the fully functional pipeline to constructing a knowledge graph by extracting entities and relations from statistical Data Tables. We showcase how this approach can be used to construct a knowledge graph from different statistical studies.

Keywords: Information extraction · Knowledge graph · Table recognition · PDF document

1 Introduction

Statistics are used in almost all areas of human activities with the purpose of describing and explaining hidden phenomenons with data. This is the case in life sciences for instance, engineering, management and economy [1]. In statistics, data are first collected, processed, analyzed and finally results are reported in various ways, including raw texts, tables, or graphics. Tabular forms are suitable for detailing statistical results by the mean of grouped information of the same nature. Statistics from scientific studies are often available in PDF format as for scientific publications that report them. Handling them from PDF documents is often cumbersome.

A statistical table is constituted of a set of cells organized into rows and columns, containing words or numbers. Such a table can be organized either (i) with a simple entry, allowing to report about a variable of interest for a given population or (ii) with a double entry, making it possible to simultaneously study two variables about a given population. Whatever the structure of the table is, understanding its content relies on a similar process and is rather human

M. Hernes et al. (Eds.): ICCCI 2020, CCIS 1287, pp. 701–713, 2020.
https://doi.org/10.1007/978-3-030-63119-2_57

oriented and not machine easy interpretable. Indeed, elements within cells could have implicit meaning [2]. That is why identifying and interpreting the relation between different cells in a table are generally considered as subjective, according to the user capability and background regarding the domain being described. This may lead to interpretation biases.

Currently there is a large body of literature on (PDF-based) information extraction approaches and systems available (see Sect. 2). We consider the main challenge in Knowledge Extraction from statistical studies is that there are two issues that need to be simultaneously dealt with. The first is how to reduce such biases which may occurs for statistical tables (structured text) interpretation. The second is how to use unstructured text that contains useful knowledge. This is an important issue because the aim is to combine all the extracted knowledge to enrich existing data sources.

In the work described here, we aim at contributing to resolve the first issue. To do so, we propose the Knowledge Extraction From statistical Data Tables (KEFT) approach. Its main idea is to extract and structure implicit knowledge contained into statistical tables as the mean of a Knowledge Graph (KG). This KG will allow later on (i) to provide a structured description which makes explicit the given domain knowledge being studied; (ii) to enable a flexible extension of the interpretation scope thanks to the formal description and the subsequent reasoning that could be possible; and (iii) to help building a generic KG by aggregating different graphs obtained from different related statistical studies.

Although various approaches have been proposed to extract the relevant statistical information from PDF documents, they require a pre-processing phase which involves converting the input PDF document into a tag format structure such as HTML and XML, or a plain text format for extracting information [3–6]. In contrast, KEFT relies directly on the PDF document to extract relevant statistical information.

2 Related Work

Statistical knowledge is generally implicit in the tables and usually the goal is to extract this knowledge in order to exploit it. One of the characteristics of statistical tables is a synthetic and visual representation of statistical results containing related information. The first issue is the extraction of relevant information and the link between them. A second issue is that statistical studies are often published as free text documents. This led us to consider an approach to convert the PDF to an exploitable format while preserving the structure of the tables. The shape of the PDF files varies, which results to the implementation of treatment methods adapted to each of them [7]. Implemented approaches so far are customized and are based on structural elements such as title, sections, figures, tables, etc. In any case, to automate the extraction of information, it is necessary to convert these files into a format that can be machine-processable.

In this context, [8] propose two methods of table recognition based on unsupervised learning techniques and heuristics that automatically detect the location and structure of tables in a PDF article. The approach is accomplished in two steps: (i) algorithmically identifying the bounding boxes of the individual tables from a set of labeled text blocks; (ii) extracting the tabular structure in the form of a rectangular grid of table cells from the set of terms contained in these table regions. TEXUS, proposed in [9] makes use of XML and CSV format to keep extracted cells, rows and columns of PDF tables. A more scalable approach is described in [10]. It consists in extracting information from research papers by exploiting both XML and text formats intelligently. To do so, various patterns and rules are used. Finally, [11] propose a strategy to extract a personalized KG from all versions of documents (PDF, XML, etc.). We can observe from the surveyed literature that there is a limitation related to the processing format and to the identification of knowledge.

3 The KEFT Knowledge Extraction System

The overview of the KEFT workflow is depicted in Fig. 1. There are three major phases which consist in respectively, locating, extracting and building the KG. A pre-processing phase is involved and is dedicated to processing statistical results published in PDF format. More specifically, it converts such a document into an exploitable structure format.

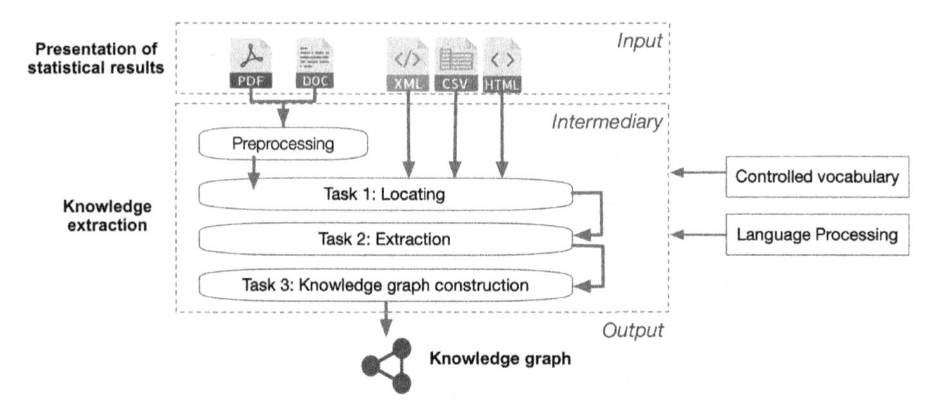

Fig. 1. KEFT: knowledge extraction form statistical data tables. The result of data pre-processing is a DataFrame. The task 1, 2 and 3 are used to locate and extract relevant knowledge using a controlled vocabulary and natural language processing (NLP) techniques. As of result, a KG is produced.

The first task is to convert the documents to process in order to be able to exploit the contents of the tables. As they are various tools that are provided for this pre-processing task, we did selected *Tabula-py* based on the comparison in Table 1 and according to the following criteria: (1) the output format preserves the tabular structure during the conversion process; (2) information regarding the position of the table in the document is available. Thus, it could be possible to perform the reverse engineering starting from the elements extracted; (3) the conversion process is done in an automated way.

Table 1. Comparative table of PDF conversion tools.

	Use as	Preserved layout	Output format	Table extraction	Processing
Tabula-py	Python library	Yes	CSV-TSV-JSON	Yes	Automatic
camelot-py	Python library	Yes	CSV-HTML-JSON	Yes	Automatic
PDFMiner	Python library	No	Text-HTML	No	Automatic
pdftohtml	Linux command	No	HTML-XML	No	Automatic
Pdf2htmlEX	Linux command	Yes	HTML	No	Automatic
PDFOnline	Web-API	Yes	HTML	No	Automatic
GROBID	Web-API	Yes	XML	No	Automatic

3.1 Locating Information

One of the challenges is related to the recognition of tables containing information relating to a study context. During the visual analysis of a chart, it is relatively easy for a human to easily identify the subject being studied. This process of deduction is executed by identifying some terms in the title, legend, etc. The automation of this process is often not trivial, even when the system can locate tables by recognizing TABLE tags in the HTML document. The issue is how to automatically determine whether the information described in a table relates to the context of the study. Rather than locating information in a similar way to [12] or finding heuristics as suggested [13] and [14], we opted for supervised approach using a controlled vocabulary of the domain. We identify text strings containing vocabulary terms. We can, for example, determine that a table is relevant if it contains the terms of the vocabulary in the title or body of the table. Contrary to the two approaches that involve important work of heuristics and rules development, our approach is generic and requires only a slight pretreatment. For example, it is enough to convert the document without studying the disposition of each type of table present in the document to construct heuristics adapted to each one.

3.2 Extraction of Knowledge

After identifying the relevant tables (containing information about the context of the study) in the document, the next step is to extract this information and structure it to explain the links between that information. Extracting information from a table requires two steps: (i) determining the relevant columns that is obtained using a controlled vocabulary. Thus, instead of extracting the set of columns and ending up with a large number of triplets to be processed, we target the columns to be extracted. The advantage of this approach is that it allows accessing all the information contained in the tables, for that it is enough to add the term to the controlled vocabulary and redo the extraction; (ii) the construction of paraphrases of interpretation of the data which is obtained by taking into account the title of each line, the headings of the relevant columns and the cells being at the intersection between the line and the header. The KEFT extraction approach consists in three steps:

- **Extracting the relevant tables.** This step consists in two phases: (a) recognition of the tables; (b) verification of the relevance of each table. Tables recognition is used to identify and retrieve all tables in the document using the format of tags. The purpose of verifying the relevance of the tables is to determine whether the extracted table deals with the subject of the study being considered. The principle is to tag each term of the vocabulary recognized in the table. Thus, if it contains terms from the vocabulary, it is declared as relevant and stored as an associative array (*page_number* $\Longrightarrow$ *title* $\Longrightarrow$ *content*). Otherwise, it is discarded. Relevance is determined by the number of tags made on the board, so the user can set a threshold value from which a table is considered relevant.
- **Extracting the relevant columns.** This step consists in extracting the columns of the table whose headings contain the terms of the controlled vocabulary. However, table headers can appear on multiple lines with empty headers, making the task more difficult. To solve this issue, we start by identifying the header templates used in the tables. Then, an extraction rules for these models are built. For example, if the header of the column has two rows, then the processing rule is to duplicate the title of the first line in the second line. Extracted tables often have empty cells and irrelevant lines. Cleaning and formatting of the tables must therefore be applied.
- **Cleaning and formatting tables.** It refers to the set of techniques used to transform row data from the original data source. For example, substitution of the empty cells which are due to the structuring of the tables on several levels. The processing makes it possible to copy the contents of the preceding cell into the next empty cells of the same column. To perform this treatment, one must consider the type of cell data in order not to introduce bias. Therefore, this processing is only applied on cells containing alphanumerical characters.

3.3 Knowledge Graph Construction

At the end of the two previous steps, we obtain a collection of clean and usable tables. The purpose of this third step is to describe all relevant tables using the RDF (Resource Description Framework) language. The chosen approach is described in Fig. 2. Each row i of the table is considered as a blank node marked $\_{:}xi$. A row i is described by a set of triples $(S_i, P_k, O_(i,k))$, where: $\_{:}xi$ is a blank node associated with the row i; P_k denotes a literal, from the contents of the header of column k and $O_(i,k)$ denote a typed literal, from the contents of the cell at the intersection between the row i and the column k. Each title F of a table T is segmented to identify the elements described and the term designating the link to connect the elements of the title F. This treatment is performed using the Stanford CoreNLP pos-tagging tool[1]. From the element and

[1] https://stanfordnlp.github.io/CoreNLP/index.html.

the link identified in the title F, a triple is constructed according to the model (S_t, P_t, O_t) where: S_t denotes the element identified in the title; P_t denotes the link between the title and the content of the table and O_t denotes the blank node for each row of table T. To illustrate the approach described in Fig. 2 with a real case, consider the example in Fig. 4

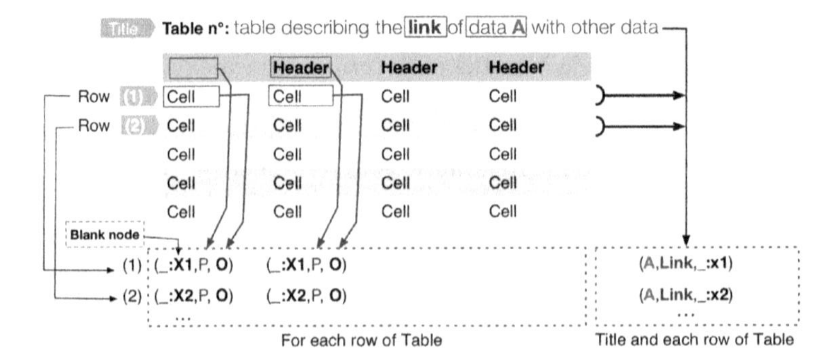

Fig. 2. Knowledge extraction approach.

4 Experiment

To conduct our experiment, we rely on a study conducted by [15] in the field of epidemiology as use case. The objective of this study was to highlight the interactions between cardiovascular risk factors. The results of the study, obtained by applying regression tests, are summarized in different tables, figures and texts. The whole study is published as a PDF document, in which different tables describe the associations between cardiovascular risk factors.

In order to construct the controlled vocabulary, we scrutinised the literature about cardiovascular risk factors (CRF) knowledge resources. We found that the interactions between cardiovascular risk factors have not been addressed per se [16–18]. To overcome this limitation, we developed a controlled vocabulary using relevant terms related to the cardiovascular disease used by experts in the field (see Fig. 3). The aim is at using it to identify, in a precise way, data appearing in a given statistical table. Developing the controlled vocabulary followed these steps [19]: (i) select the related work in related sources (for example, BioPortal or Pubmed); (ii) extract the concepts from the studies according to whether or not their semantic types are associated to the cardiovascular risk factors; (iii) remove the terms which belong to non-interaction factors to meet the current needs of the knowledge base (in our case realised manually by experts in the field); (iv) remove the source whose numbers of terms are very few or subjects are not related with our scope. After that, identify the structure that makes the terms usable. A well-designed controlled vocabulary requires to structure relevant terms into a hierarchy.

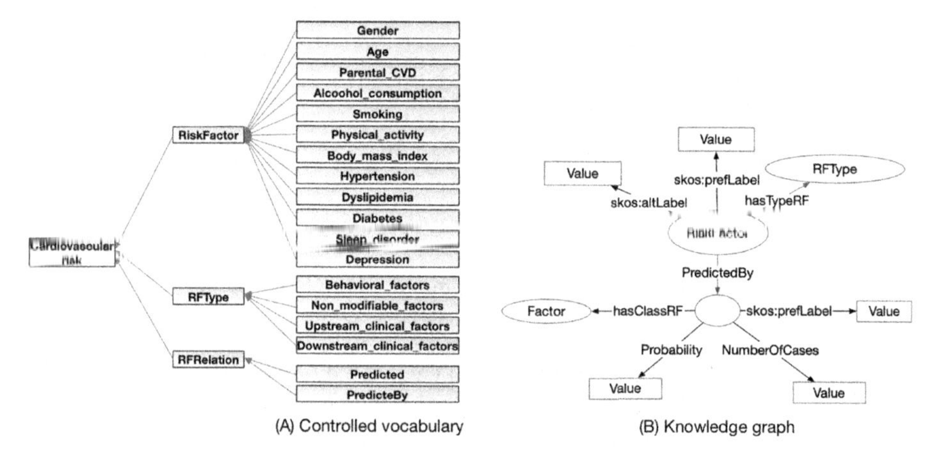

Fig. 3. Controlled vocabulary for cardiovascular risk factors. In figure (A): `RiskFactor` lists the cardiovascular risk factors most often studied in the literature. However, this list is not exhaustive; `RFType` lists the different categories associated with the cardiovascular risk factor; `RFRelation` lists the different relationships between risk factors. For the moment, we have identified only two relationships. This category will be enriched later. In figure (B) the knowledge graph that we can build from this vocabulary.

To illustrate our approach, we chose one table in the PDF document (see Fig. 4). All the tables describing associations between cardiovascular risk factors have the same structure.

Risk factor — link

Title: Table 4. Risk of smoking according to predictive factors at baseline.

		No. of cases	Unadjusted HR (95% CI)	Unadjusted p	Multi-adjusted HR (95% CI)	Multi-adjusted p
Body mass index	Optimal	508	1.00		1.00	
	Overweight	466	1.29 (1.13–1.46)	<0.0001	1.29 (1.12–1.47)	0.0002
	Obesity	67	1.25 (0.96–1.60)	0.09	1.27 (0.97–1.64)	0.08
Depression	No	786	1.00		1.00	
	Yes	255	1.19 (1.03–1.37)	0.02	1.18 (1.02–1.37)	0.02

Row (1), Row (2), Blank node

(1) (_:x1,Class, body_mass_index) (_:x1,Label, Optimal) (_:x1,NumberOfCases,508) ... (_:x1,P,null) (smoking,Predicted_by,_:x1)
(2) (_:x2,Class, body_mass_index) (_:x2, Label, Overweight) (_:x2, NumberOfCases,466)...(_:x2,P,0.0002) (smoking,Predicted_by,_:x2)
...

Fig. 4. Knowledge extraction approach applied to table.

The obtained result is shown in Fig. 5. We got three blank nodes (_:x1, _:x2, _:x3). Taking the example of the blank node _:x1 associated with the first row of the table, five triples are produced. Four triplets generated from row "1" and the two columns "Probability" and "NumberOfCases" that are: (_:x1, 'Class', 'Body_mass_index'), (_:x1, 'Label', 'Overweight'), (_:x1, 'NumberOfCases', '466'), (_:x1, 'Probability', '0.0002'). The fifth triple constructed from the title and the row "1" is ('Smoking', 'PredictedBy', _:x1).

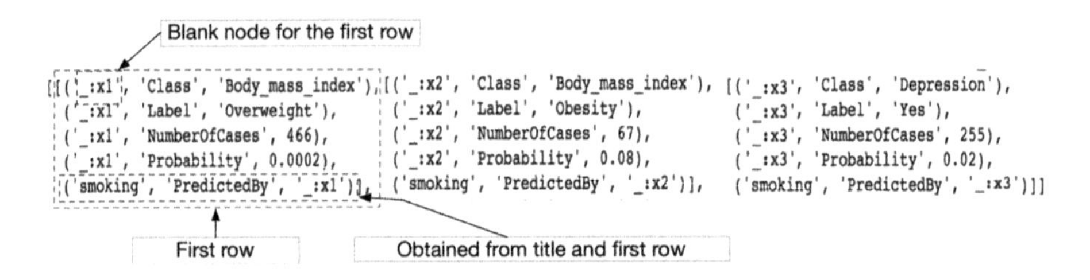

Fig. 5. Extracted triples from our example in Fig. 4

Once we extracted the knowledge from each table (see Fig. 2), we get al.l the triplets for each table (for example, see Fig. 5). The next step for our framework is to build the knowledge graph using the model described in Fig. 3(B). For our example in Fig. 5, the result is described in Fig. 6(A).

The previous section details the knowledge extraction process from statistical tables available in PDF format. It is carried out in two stages: (i) locating relevant the information in the tables; (ii) extracting information in the form of RDF triples.

5 Evaluation

To the best of our knowledge, this work is the first to address KG building from statistical data tables. Existing KG building approach such as MTab [20], ITEM [21], MantisTable [22], Mentor [23] and T2KG [24] are used to match or enrich and improve KG by employing tabular data using Knowledge base as DBpedia[2]. Comparing them requires significant additional data preprocessing work. As a result, this will be addressed in a future work.

Instead, to allow a preliminary evaluation of KEFT, two settings are used: the first is related to the extraction of tables while the second addresses the relevance of extracted triples.

Evaluation of the Extraction Process. We conducted an evaluation based on an expert's interpretation of the selection of statistical tables which corresponds to the topic of the addressed study in a given PDF document. To do this, two datasets available in PDF format are used: the first dataset (D1) is about the statistical model of risk factors in the cardiovascular diseases domain [15]. It contains 32 tables in PDF documents and among them 13 tables should be extracted. ; the second dataset (D2) is about the international survey of foreign exchange and derivative transactions[3]. It contains 13 tables in PDF documents and among them 3 tables should be extracted.

Three traditional metrics, Precision, Recall and F-measure, are used to evaluate the extraction process. Let T_fac be the number of tables identified as dealing

[2] https://wiki.dbpedia.org/.

[3] https://www.banque-france.fr/sites/default/files/media/2016/11/24/enquete-triennale-principaux-resultats.pdf.

with the interactions between cardiovascular risk factors (D1) and T_vat the number of tables dealing with transaction activity volume (D2). For each dataset, the extraction process is performed with and without a controlled vocabulary. Precision, Recall and the F-measure are calculated thereafter. Table 2 reports the results of the evaluation. We observe that on the two datasets Precision is low when the extraction process is realized without a controlled vocabulary. This low value is due to the large number of extracted arrays. In contrast, Precision and F-measure on D1 increase when a controlled vocabulary is used. For the dataset D2, Precision and Recall increase as the number of terms increases in the vocabulary. These results indicate that relying on an adapted vocabulary is crucial for a better extraction process.

Table 2. Experimental results for D1 and D2.

Dataset	With vocabulary	Precision	Recall	F-measure
D1	No	0.40	0.90	0.55
	Yes (20 terms)	**1.0**	**0.98**	**0.99**
D2	No	0.50	0.45	0.47
	Yes (10 terms)	**0.76**	**0.69**	**0.72**
	Yes (20 terms)	**0.89**	**0.80**	**0.84**

The evaluation on D1 provides an almost perfect performance. Naturally, the good precision derives from the fact that: (1) the approach developed on the dataset (D1); (2) all the tables describing the interactions between cardiovascular risk factors in the document have the same structure; (3) the vocabulary used for the extraction process is adapted to the domain. It is quite surprising to observe the good performance on the dataset (D2). The observation that could be made is that the restrictive and precise selection of the searched patterns using the vocabulary impacts positively the results.

Evaluation of the Relevance of Extracted Triples. To evaluate the relevance of extracted triples, we applied the extraction process to a third dataset (D3) in the field of cardiovascular disease. D3 is a set of statistical tables describing interactions between cardiovascular risk factors. These results were obtained from the SUVIMAX study[4] conducted by the French EREN research group[5] in the field of nutritional epidemiology. D3 is the result of the examination of all the relationships among the major risk factors for cardiovascular disease in this cohort. Cox regression tests consistently verified prospective associations between behavioral and clinical risk factors (gender, age, parental history of cardiovascular disease, non-moderate alcohol consumption, smoking, physical inactivity, obesity, hypertension, dyslipidemia, diabetes, sleep disorder, depression,

[4] https://eren.univ-paris13.fr/index.php/en/epidemiological-studies.html.
[5] https://eren.univ-paris13.fr/index.php/en/.

unhealthy diet). It contains 30 tables in PDF documents and among them 13 tables should be extracted.

The process performed on D3 shows variations and similarities with respect to the model of risk factors obtained from D1. For example, in Fig. 6 the risk factor *Smoking* is predicted by *Depression* and *Obesity* in D1 while it is predicted by *Age*, *Gender* and *Alcohol Consumption* in D3. A second example is shown in Fig. 7. It shows that the factor *Obesity* is predicted by the factors *Dyslipidemia*, *PhysicalInactivity* and *Smoking* in D1 and D3. On the other hand, a variation is observed for the factors *Depression* and *Hypertension* in D1, while it is predicted by *Age*, *Gender* and *Alcohol Consumption* in D3.

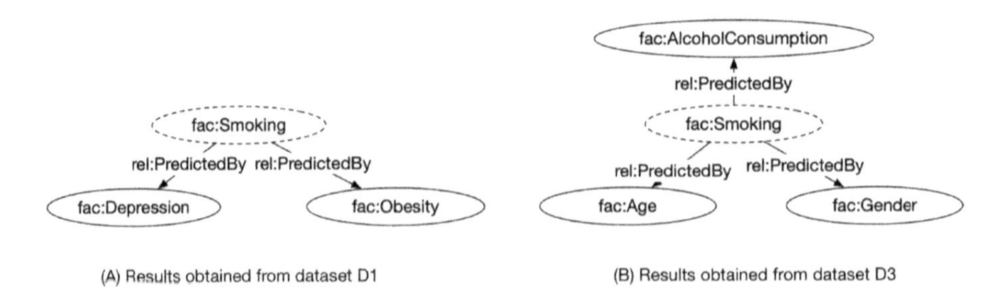

Fig. 6. Results obtained for the risk factor "Smoker" for datasets D1 and D3.

Three explanations make it possible to understand these variations:

- similarity: is related to factors and relationships that are already validated by scientific evidence;
- difference: is related to factors and relationships for which the use of experts or literature is needed to validate them;
- lack: correspond to knowledge gaps, they are about the new factors and links that are not yet reported in the literature.

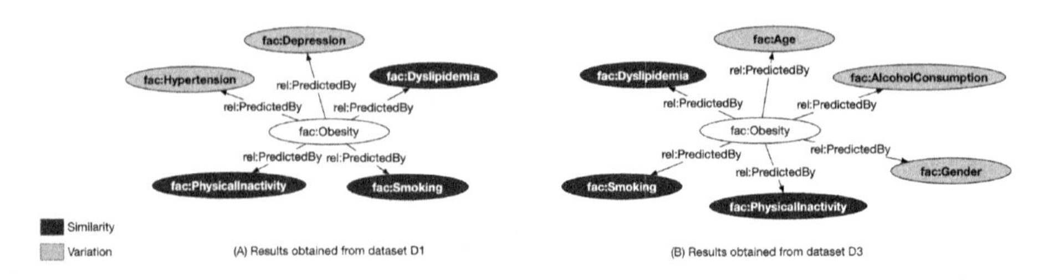

Fig. 7. Results obtained for the "Obesity" risk factor for datasets D1 and D3.

The aim is that this description of open problems will help to inspire and provide foundations for constructing a KG. For example, in Fig. 7 which describes the risk factor "Obesity", the common factors between these two models corresponds to *"Dyslipidemia, Smoking and Physical Inactivity"*. They also show the missing concepts in the different models. A researcher can have various research questions while reviewing this graph. Some of these research questions include risk factors addressed in the literature or literature validation of newly included interactions.

6 Conclusion

In this paper, we have described the KEFT approach designed and implemented to extract relevant knowledge from statistical data rendered in tabular form and published in PDF format. KEFT provides a knowledge graph which models relevant information from the underlying statistical model. Two contributions are made in particular: (i) how to determine the relevance information contained in statistical tables and in which format to extract them; (ii) how to move from an unstructured PDF format to a suitable format for semantic processing of information. The main advantage of the approach is the ability to extract implicit knowledge represented in statistical tables for different domains. The originality is to associate a conceptual model with the tables in a PDF document. In addition, we showcase how this approach can be used to construct a knowledge graph from different statistical studies.

Preliminary experiments on different datasets are encouraging and still need to be improved. We currently aim to extend and further improve the KEFT system in several dimensions. Therefore we are considering to: (i) evaluate the relevance of extracted triples, with a large dataset to show the performance of the algorithm; (ii) compare the performance of KEFT with other approaches for demonstrating the usefulness of our approach and improving our system; (iii) use unstructured text that contains useful knowledge. Rather to focus on the statistical tables, it could be relevant to exploit the complete content of the document that being processed (text, figure, etc.). In addition, the approach is currently used on statistical studies in the field of cardiovascular disease and leads to different conceptual models. The idea is to merge these models by using the knowledge acquisition process to develop a generic model of the interactions between risk factors for cardiovascular disease.

References

1. Saporta, G.: Probabilités, analyse des données et Statistique, 3rd edn. Technip, Paris (2011)
2. Heiberger, R.-M., Holland, B.: Statistical Analysis and Data Display, 2nd edn. Springer, New York (2015). https://doi.org/10.1007/978-1-4939-2122-5
3. Klahold, A., Fathi, M.: Knowledge discovery from text (KDT). Computer Aided Writing, pp. 83–115. Springer, Cham (2020). https://doi.org/10.1007/978-3-030-27439-9_8

4. Lu, W., Zhang, Z., Lou, R., Dai, H., Yang, S., Wei, B.: Mining RDF from tables in Chinese encyclopedias. In: Li, J., Ji, H., Zhao, D., Feng, Y. (eds.) NLPCC -2015. LNCS (LNAI), vol. 9362, pp. 285–298. Springer, Cham (2015). https://doi.org/10.1007/978-3-319-25207-0_24
5. Tekli, J., Charbel, N., Chbeir, R.: Building semantic trees from XML documents. J. Web Semant. **37**(38), 1–24 (2016)
6. Souili, A., Cavallucci, D., Rousselot, F.: Natural language processing (NLP) – a solution for knowledge extraction from patent unstructured data. Procedia Eng. **131**, 635–643 (2015)
7. Ronzano, F., Saggion, H.: Knowledge extraction and modeling from scientific publications. In: González-Beltrán, A., Osborne, F., Peroni, S. (eds.) SAVE-SD 2016. LNCS, vol. 9792, pp. 11–25. Springer, Cham (2016). https://doi.org/10.1007/978-3-319-53637-8_2
8. Klampfl, S., Jack, K., Kern, R.: A comparison of two unsupervised table recognition methods from digital scientific articles. D-Lib Mag. **20** (2014). CNRI Acct
9. Rastan, R., Paik, H.-Y., Shepherd, J., Ryu, S.-H., Beheshti, A.: TEXUS: table extraction system for PDF documents. In: Wang, J., Cong, G., Chen, J., Qi, J. (eds.) ADC 2018. LNCS, vol. 10837, pp. 345–349. Springer, Cham (2018). https://doi.org/10.1007/978-3-319-92013-9_30
10. Ahmad, R., Afzal, M.-T., Qadir, M.-A.: Information extraction from PDF sources based on rule-based system using integrated formats. In: Sack, H., Dietze, S., Tordai, A., Lange, C. (eds.) SemWebEval 2016. CCIS, vol. 641, pp. 293–308. Springer, Cham (2016). https://doi.org/10.1007/978-3-319-46565-4_23
11. Gentile, A.L., Gruhl, D., Ristoski, P., Welch, S.: Personalized knowledge graphs for the pharmaceutical domain. In: Ghidini, C., et al. (eds.) ISWC 2019. LNCS, vol. 11779, pp. 400–417. Springer, Cham (2019). https://doi.org/10.1007/978-3-030-30796-7_25
12. Ermilov, I., Auer, S., Stadler, C.: User-driven semantic mapping of tabular data. In: Proceedings of the 9th International Conference on Semantic Systems, pp. 105–112. ACM. New York (2013). https://doi.org/10.1145/2506182.2506196
13. Shigarov, A.-O.: Table understanding using a rule engine. Procedia Eng. Expert Syst. Appl. **42**, 929–937 (2015)
14. Shigarov, A., Altaev, A., Mikhailov, A., Paramonov, V., Cherkashin, E.: TabbyPDF: web-based system for PDF table extraction. In: Damaševičius, R., Vasiljevienė, G. (eds.) ICIST 2018. CCIS, vol. 920, pp. 257–269. Springer, Cham (2018). https://doi.org/10.1007/978-3-319-99972-2_20
15. Meneton, P., et al.: A global view of the relationships between the main behavioural and clinical cardiovascular risk factors in the GAZEL prospective cohort. PLoS ONE **911**, 01–20 (2016). Public Library of Science (PLoS). Andrea Icks
16. Barton, A., Rosier, A., Burgun, A., Ethier, J.-F.: The cardiovascular disease ontology. In: Frontiers in Artificial Intelligence and Applications, vol. 267, pp. 409–414. IOS Press (2014)
17. Divakar, H.-R., Prakash, B.-R., Mamatha, M.: An ontology based system for healthcare people to prevent cardiovascular diseases. Int. J. Recent Technol. Eng. **8**, 983–988 (2019). Blue Eyes Intelligence Engineering and Sciences Engineering and Sciences Publication - BEIESP
18. Gedzelman, S., Simonet, M., Bernhard, D., Diallo, D., Palmer, P.: Building an ontology of cardio-vascular diseases for concept-based information retrieval. In: 2005 Computers in Cardiology, Lyon, France, pp. 255–258. IEEE (2005). https://doi.org/10.1109/CIC.2005.1588085

19. Wu, M., Liu, Y., Kang, H., Zheng, S., Li, J., Hou, L.: Building a controlled vocabulary for standardizing precision medicine terms. CoRR, abs/1807.01000, pp. 1–4 (2018)
20. Nguyen, P., Kertkeidkachorn, N., Ichise, R., Takeda, H.: MTab: matching tabular data to knowledge graph with probability models. In: CEUR Workshop Proceedings (eds.) Proceedings of the 14th International Workshop on Ontology Matching Co-Located with the 18th International Semantic Web Conference (ISWC 2019), Auckland, New Zealand, 26 October 2019, vol. 2536, pp. 191–192. CEUR-WS.org (2019)
21. Guo, X., Chen, Y., Chen, J., Du, X.: ITEM: extract and integrate entities from tabular data to RDF knowledge base. In: Du, X., Fan, W., Wang, J., Peng, Z., Sharaf, M.A. (eds.) APWeb 2011. LNCS, vol. 6612, pp. 400–411. Springer, Heidelberg (2011). https://doi.org/10.1007/978-3-642-20291-9_45
22. Cremaschi, M., Rula, A., Siano, A., De Paoli, F.: MantisTable: a tool for creating semantic annotations on tabular data. In: Hitzler, P., et al. (eds.) ESWC 2019. LNCS, vol. 11762, pp. 18–23. Springer, Cham (2019). https://doi.org/10.1007/978-3-030-32327-1_4
23. Cannaviccio, M., Ariemma, L., Barbosa, D., Merialdo. P.: Leveraging Wikipedia table schemas for knowledge graph augmentation. In: Proceedings of the 21st International Workshop on the Web and Databases-WebDB 2018, pp. 1–6. ACM Press (2018). https://doi.org/10.1145/3201463.3201468
24. Kertkeidkachorn, N., Ichise, R.: T2KG: an end-to-end system for creating knowledge graph from unstructured text. In: AAAI Workshops (eds.) The Workshops of the The Thirty-First AAAI Conference on Artificial Intelligence, San Francisco, California, USA, 4–9 February 2017, vol. WS-17, pp. 743–749. AAAI Press (2017)

Devising a Cross-Domain Model to Detect Fake Review Comments

Chen-Shan Wei[1], Ping-Yu Hsu[1], Chen-Wan Huang[1], Ming-Shien Cheng[2(✉)], and Grandys Frieska Prassida[1]

[1] Department of Business Administration, National Central University, No. 300, Jhongda Road, Jhongli City, Taoyuan County 32001, Taiwan (R.O.C.)
984401019@cc.ncu.edu.tw
[2] Department of Industrial Engineering and Management, Ming Chi University of Technology, No. 84, Gongzhuan Road, Taishan District, New Taipei City 24301, Taiwan (R.O.C.)
mscheng@mail.mcut.edu.tw

Abstract. The online reviews not only have huge impact on consumer shopping behavior but also online stores' marketing strategy. Positive reviews will have positive influence for consumer's buying decision. Therefore, some sellers want to boost their sales volume. They will hire spammers to write undeserving positive reviews to promote their products. Currently, some of the researches related to detection of fake reviews based on the text feature, the model will reach to high accuracy. However, the same model test on the other dataset the accuracy decrease sharply. Relevant researches had gradually explored the identification of fake reviews across different domains, whether the model built using comprehensive methods such as text features or neural networks, encountering the decreasing of accuracy. On the other hand, the method didn't explain why the model can be applied to cross-domain predictions. In our research, we using the fake reviews and truthful reviews from Ott et al. (2011) and Li, Ott, Cardie, and Hovy (2014) in the three domain (hotel, restaurant, doctor). The cross-domain detect model based on Stimuli Organism Response (S-O-R) combine LIWC (Linguistic Inquiry and Word Count), add word2vec quantization feature, overcoming the decreasing accuracy situation. According to the research result, in the method one SOR calculation of feature weight of reviews, the DNN classification algorithm accuracy is 63.6%. In the method two, calculation of frequent features of word vectors, the random forest classification algorithm accuracy is 73.75%.

Keywords: Fake reviews · Stimuli-Organism-Response (S-O-R) framework · word2vec

1 Introduction

A study by Klaus and Changchit (2017) proposed that when the Internet review comment system is reliable, consumers' purchase decisions reflect this system. Thus, electronic commerce companies focus heavily on comment quality management to

M. Hernes et al. (Eds.): ICCCI 2020, CCIS 1287, pp. 714–725, 2020.
https://doi.org/10.1007/978-3-030-63119-2_58

prevent malicious or fake comments from reducing the quality of the comment system and influencing consumers' purchase decisions.

Studies on identifying true and fake comments have used language types to extract their characteristics, which results in over-reliance on terminology from the same batch of data. Thus, once the data are changed to those of another batch, the accuracy of original identification methods performing well decreases substantially. Universal identification rules cannot be formed, and future applications are limited.

For scholars currently conducting interdisciplinary research, Li, Ott, Cardie, and Hovy (2014) developed general rules for data collected by Ott et al. (2011). Word calculation was implemented using language characteristic models from unigram, linguistic inquiry and word count (LIWC), and part of speech (POS), and evaluation rules were established combining support vector machines (SVMs) and the synthetic, augmentative, generative, experiential (SAGE) model. Ren and Ji (2017) used an artificial neural network (ANN) to identify fake comments. Finally, increases in interdisciplinary accuracy were higher than those in the experiment results of Li et al. (2014). However, no clear explanation has emerged for why words can be applied to interdisciplinary predictions.

Aslam et al. (2019) proposed that although fake comments in the filters of Yelp have similar language with that of true comments on the platform, the fake comments in the Yelp filters still leave psychological tracks when written. Therefore, this study applied psychological theories and the stimuli-organism-response (S-O-R) framework and established a classification model applicable across disciplines. Thus, the problem of substantially reduced accuracy during interdisciplinary identification from former studies can be overcome.

This paper organized as follow: (1) Introduction: research background, motivation and purpose. (2) Related work: review of scholars researches on the interdisciplinary identification of fake comments and stimulus-organism-response (S-O-R) framework. (3) Research methodology: content of research process in this study. (4) Research result: two method results and discussion. (5) Conclusion and future research: contribution of the study, and possible future research direction is discussed.

2 Related Work

2.1 Studies on Interdisciplinary Identification of Fake Comments

Previously, during interdisciplinary identification of true and fake comments, Li et al. (2014) attempted to capture general differences between fake and true comments from language usage. These differences were used to help clients make purchase decisions and maintain the quality of platform comment systems. The three types of language characteristic models comprised LIWC, POS, and unigram, and the two analysis models comprised SVM and SAGE. The study applied psychological theories but yielded unfavorable results. The maximum accuracy in the interdisciplinary section reached 78.5%, which was much lower than predictions within the same domain. In addition, Ren and Ji (2017) used an ANN, which presented relatively favorable identification of subtle language characteristics for identifying fake comments. The general regression neural network model had the most favorable performance and reached 83.5% accuracy in the restaurant domain and 57% in the doctor domain, which were both higher than the 78.5% and 55% by Li et al. (2014). However, models were established through language characteristic methods without explained reasons for interdisciplinary identification. Therefore, this study used psychological S-O-R theory to establish an interdisciplinary model.

2.2 Stimulus-Organism-Response (S-O-R) Framework

S-O-R theory was initially proposed by psychologist Woodworth in 1929. It posits that environmental stimuli change people's minds or individual statuses, and behavioral intentions or avoidance responses are generated. With the rise of online electronic commerce, studies have increasingly focused on stimulation from the Internet environment on consumers' purchases. Eroglu, Machleit, and Davis (2001) studied high and low correlation clues influencing online consumers' behavior. In addition, atmospheric stimulation was defined as cumulative hints seen and heard on the website, such as words, links, colors, animations, and sounds. Studies have indicated that Internet interfaces influence the emotions and cognitive status of Internet consumers and thus influence purchase behavior.

Apart from discussing stimulation's influence on consumers, some scholars have studied stimulation's internal influence on individuals. Adelaar, Chang, Lancendorfer, Lee, and Morimoto (2003) explored the relationship between the media's influence on personal emotions and impulsive purchases. The PAD (pleasure, arousal, and dominance) emotional state model and ANCOVA tests were used to reveal that personal emotions are positively correlated with impulse purchases. Menon and Kahn (2002) discussed the Internet atmosphere's influence on consumers' emotions and subsequent behavior. Experiencing pleasurable feelings when online shopping positively influences consumers' behavior.

For Internet environments, S-O-R theory is primarily the basic structure for discussing how stimulation factors on websites influence consumers' purchase behavior. With changes in consumers' purchase behavior, Internet comments have become a factor changing consumers' behavior and an indispensable link of the consumer purchase process. With Internet comments confirmed as a factor stimulating consumers' behavior, this paper discusses an identification model for fake comments from the S-O-R theory framework.

3 Research Methodology

According to the study structure, to establish an interdisciplinary identification model for comments, the following steps were conducted. Figure 1 presents the study structure and process. First, the comment databases established by Ott et al. (2011) and Li et al. (2014) were collected as original data of true and fake comments. The S-O-R psychological theory was used as the background, and writers' psychological states when writing comments were simulated. Related weighting values were calculated, and two methods were derived. Method 1 combines the 2015 LIWC dictionary with S-O-R theory and selects 14 categories of conforming dictionaries for the first comment study. Each comment's weighting values corresponding to the 14 categories were calculated. To study the relationship between selected words in the comments and the 14 categories from dictionaries, this study extended to Method 2. A second interdisciplinary identification method was conducted through frequency characteristic extraction of word2vec word vectors to achieve the expected method results.

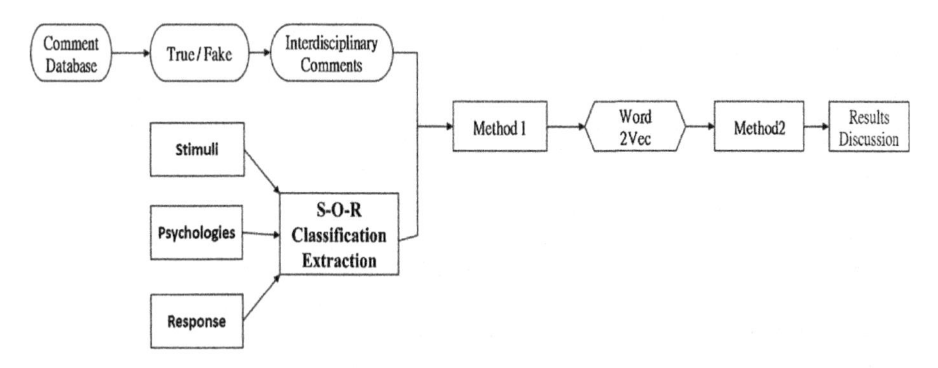

Fig. 1. Research process

3.1 S-O-R Category Filter Method

The LIWC dictionary was chosen to select categories conforming to definitions from the S-O-R framework. Gatautis and Vaiciukynaite (2013) stated that Internet elements in a virtual environment comprise website design, Internet communication elements, website content, and website instructions. Consumers' purchase results are easily influenced by website content. Comments on the Internet are also included in this. The five sensory processes are the receiving media for users to absorb information. Consequently, the perceptual process dictionary and its subcategories see, hear, and feel in LIWC were used.

In the S-O-R framework, organism is an internal process between external stimulation and final personal choice. we noted that activities include cognition, physiology, feeling, and thought. Emotions created after an individual internalizes information influence consumption behavior. Wolfe's psychological theory divides an individual's mind into two sections: desire and cognition. From this, the creation of online shopping experience is related to the external stimulation of individual cognition. Thus, this study used the cognitive process dictionary and the subcategories insight, causation, discrepancy, tentative, certainty, and differentiation.

For desire, and Kavanagh, Andrade, and May (2005) proposed that the generation of desire often results in pleasure or unhappy emotions. Boujbel and d'Astous (2015) defined desire as when consumers begin wanting to possess an object or behavior and begin imagining the sense of pleasure it brings after this desire is fulfilled and the sense of loss from when this desire is not satisfied. Because LIWC currently does not provide emotional dictionaries related to desire, this study implemented the WordNet expansion of desire-related dictionaries to identify hyponyms with desire as a vocabulary concept.

According to the word meaning of desire by Boujbel and d'Astous (2015), which is to strongly feel or have a desire for something or feel unsatisfied, the first and second desire dictionaries were constructed separately.

The response in the S-O-R framework represents the final outcomes and decisions of consumers, who can be approached or avoided. Chang, Eckman, and Yan (2011) defined approach behavior as positive actions or performance of an action such as staying, browsing, and making purchases. LIWC did not have dictionaries for reactions; therefore, the action dictionaries provided by Jeff Gardner were used to represent dictionaries of the reaction category.

3.2 Method 1: S-O-R and the Characteristic Weights of Comments

Before calculation, the comments underwent world pretreatment by the Natural Language Toolkit in Python. To pretreat English word exploration data, mostly stemming, lemmatization, normalization, and noise cancellation were implemented. In this study, only noise cancellation was adopted. Symbols or numbers that may have interfered with the analysis were deleted, all capital letters were converted into small letters, and words were segmented. The former three treatments were not adopted because the authenticity, writing styles, and words used in manual writing were hoped to be reserved in this study.

The characteristic weight values of every article were those calculated for the 14 categories selected from LIWC by using S-O-R theory. Consequently, 14 weight values resulted for each comment. This study used the R language to calculate values as fractions. The numerator was the number of words from a single category dictionary appearing in one comment, and the denominator was the total words used in that comment.

3.3 Method 2: Frequency Characteristics of Word Vectors

In Method 2, the relationship between words used in each comment and the 14 category dictionaries were discussed. Accordingly, the frequency characteristics of word vectors were developed. Word2vec was used to generate word vectors. After neural network training to learn a large number of articles, word2vec can simplify words into vectors in space. Thus, distances between vectors can be combined with the quality "words with close semantic meanings are also close in space" to represent similarity levels between word semantics.

The word2vec training database used in this study is an open English database downloaded from Wikipedia. Wikipedia data was downloaded in the xml format and converted into the text format. In addition, Python was adopted to train 300-dimension word vectors, and 1,500,000 words were ultimately trained. After the training was complete, every word could be expressed as a vector composed of 300 dimensions.

The average word vectors of a single comment and category were calculated using the word2vec word vector method. First, trained word vectors were used to calculate Euclidean distances between all words in one single comment and all words in a category. Then, the average word vector of one category could be calculated by measuring the average of all values. In addition, an extra cell, characteristic word extraction, was added. It can be a fake weight keyword or a truth weight keyword, and thus one weight word can be given according to the comment content. When conducting the classification algorithm, the S-O-R and comment characteristic weights from the original Method 1 were added with the word vector frequency characteristic values from Method 2.

4 Research Results

4.1 Data Collection

In the literatures of Ott et al. (2011) and Li et al. (2014), the review data of the three domains of hotel, restaurant and doctor were collected. Real reviews come from online review sites, fake reviews come from Amazon's mass intelligence platform (AMT) and experts and scholars.

Among the fake reviews, hotel domain: Ott et al. (2011) collected a total of 400 reviews all from AMT and Li et al. (2014) collected a total of 540 reviews of which 140 were written by experts and 400 articles were collected from the research of Ott et al. (2011); restaurant domain, a total of 320 reviews were collected, among them, 200 articles came from Amazon's AMT, 120 articles came from restaurant employees; doctors domain, a total of 232 reviews were collected, of which 200 articles came from Amazon's AMT and 32 articles came from hospital doctors.

In real reviews, Ott et al. (2011) collected a total of 400 reviews in the hotel sector, from TripAdvisor. Li et al. (2014) scholars collected 200 reviews each in the restaurant and doctor domains, all from TripAdvisor review sites. This study uses reviews from these two researches, the data collection can be seen in Table 1.

Table 1. Data collections

	True reviews	Fake review (AMT/expert)	Source
Hotel	400	400/140	Ott et al. (2011) Li et al. (2014)
Restaurant	200	200/120	Li et al. (2014)
Doctor	200	200/32	Li et al. (2014)

4.2 S-O-R Category Data

Some examples of the 14 categories and the number of words in each category are listed in Table 2.

Table 2. Volume of words of 14 categories

Categories	Example	No. of words
S (Stimulate)		
Perceptual	ache, aching, acid, acrid, appear…	433
see	appear, appearance, appeared, appearing…	126
hear	audibl, audio, boom, cellphone, choir…	92
feel	ache, aching, brush, burn, burned…	128
O (Organism)		
cogproc	abnormal, absolute, absolutely, accept…	797
insight	accept, accepts, accepted, accepting…	256
causation	activate, affect, affected, affecting, affects	133
Discrepancy	abnormal, besides, could, couldn't…	82
tentant	allot, almost, alot, ambigu, any…	179
certainty	absolute, absolutely, accura, all…	113
differentation	actually, adjust, against, aint, ain't…	77
desire1	abjure, abnegate, absence, ache…	351
desire2	ache, actuation, aim, allurement…	136
R (response)		
action	accomplish, achieve, attain, benefit…	32

4.3 Comments and S-O-R Word Characteristic Weights

The characteristic weights were calculated for every article in three domains. The weight value was between 0 and 1, and this showed the word usage weight ratio of comment in the category. Tables 3, 4 and 5 show the result.

4.4 Method 1: S-O-R and Characteristic Weights of Comments

The identification accuracy of Method 1 results are presented in Table 6. Eight classification algorithms were used: k-nn, naïve bayes (NB), decision tree, random forest (RF), gradient boosting machine, SVM, XGBoost (XGB), and deep neural networks (DNNs). DNNs had the highest performance in the hotel and restaurant domains with an accuracy of 62.08% and 63.6%, respectively.

4.5 Method 2: Word Vector Frequency Characteristics

Tables 7, 8 and 9 present an average word vector summary of the three domains, v1–v14 were the following 14 categories in order: action, differ, discrepancy, hear, certainty, see, feel, causation, desire2, tentant, insight, desire1, perceptual, and cogproc, D1–D400 were 400 fake comments, and T1–T400 were 400 true comments, D and T Average were the average distances of each category between fake and true comments.

Table 3. Average characteristic weights in hotel field

S (Stimulate)		
Category	Fake	True
insight	**0.0070**	0.0059
tentant	**0.0277**	0.0257
causation	**0.0061**	0.0049
Discrepancy	**0.0142**	0.0117
certainty	**0.0201**	0.0158
differ	0.0159	**0.0228**
desire1	**0.0228**	0.0218
desire2	**0.0110**	0.0073
cogproc	**0.0850**	0.0798
O (Organism)		
Category	Fake	True
Perceptual	**0.0177**	0.0146
see	**0.0097**	0.0076
hear	0.0015	**0.0031**
faal	**0.0054**	0.0036
R (response)		
Category	Fake	True
action	**0.0088**	0.0070

Table 4. Average characteristic weights in restaurant field

S (Stimulate)		
Category	Fake	True
insight	**0.0077**	0.0071
tentant	**0.0328**	0.0324
causation	**0.0066**	0.0056
Discrepancy	0.0109	**0.0115**
certainty	**0.0230**	0.0150
differ	0.0221	**0.0298**
desire1	**0.0228**	0.0182
desire2	**0.0099**	0.0067
cogproc	**0.0965**	0.0930
O (Organism)		
Category	Fake	True
Perceptual	0.0175	**0.0188**
see	0.0060	**0.0076**
hear	0.0023	**0.0038**
faal	**0.0035**	0.0030
R (response)		
Category	Fake	True
action	0.0088	**0.0089**

Table 5. Average characteristic weights in doctor field

S (Stimulate)		
Category	Fake	True
insight	0.0146	**0.0187**
tentant	0.022	**0.0330**
causation	0.0108	**0.0110**
Discrepancy	0.0144	**0.0154**
certainty	0.0178	**0.0214**
differ	0.0180	**0.0235**
desire1	0.0145	**0.0209**
desire2	**0.0162**	0.0121
cogproc	**0.0408**	0.1138
O (Organism)		
Category	Fake	True
Perceptual	0.0180	**0.0215**
see	**0.0099**	0.0086
hear	0.0066	**0.0049**
faal	**0.011**	0.0076
R (response)		
Category	Fake	True
action	**0.0115**	0.0158

Table 6. The Method 1 result

Classification algorithm	Hotel	Restaurant	Doctor	Average
KNN	54.16%	57.02%	55.35%	55.51%
NB	59.15%	61.98%	60.11%	60.41%
DT	57.08%	57.02%	60.71%	58.27%
RF	59.16%	59.50%	61.90%	60.19%
GBM	57.08%	57.85%	61.30%	58.74%
SVM	47.91%	47.93%	63.09%	52.98%
XGB	58.33%	60.33%	63.09%	60.58%
DNN	62.08%	63.60%	61.90%	62.53%

The average word vector tables of the hotel and restaurant domains revealed that the values of D Average were smaller than those of T Average, which signified that words used for writing fake comments were closer to the categories selected with S-O-R theory. In the original assumptions, fake comment writers may have used the S-O-R

Table 7. Average word vector summary in hotel domain

Hotel	v1	v2	v3	v4	v5	v6	v7	v8	v9	v10	v11	v12	v13	v14
D1	3.692	3.686	3.612	3.856	3.572	3.802	4.017	3.804	3.697	3.748	3.697	3.854	3.940	3.703
D2	3.692	3.693	3.635	3.857	3.587	3.799	4.014	3.810	3.704	3.752	3.700	3.855	3.942	3.710
D3	3.604	3.599	3.504	3.733	3.472	3.686	3.908	3.721	3.580	3.638	3.592	3.743	3.819	3.602
:	:	:	:	:	:	:	:	:	:	:	:	:	:	:
D398	3.840	3.827	3.749	3.970	3.719	3.930	4.098	3.936	3.842	3.885	3.862	3.980	4.031	3.849
D399	3.644	3.616	3.572	3.805	3.508	3.719	3.946	3.735	3.638	3.689	3.644	3.790	3.869	3.644
D400	3.755	3.742	3.683	3.892	3.610	3.789	4.002	3.853	3.727	3.808	3.771	3.875	3.936	3.762
T1	3.703	3.669	3.641	3.836	3.567	3.755	3.969	3.800	3.687	3.744	3.713	3.831	3.900	3.705
T2	3.712	3.679	3.642	3.835	3.576	3.756	3.979	3.797	3.689	3.752	3.714	3.841	3.901	3.710
T3	3.765	3.738	3.718	3.893	3.654	3.819	4.026	3.856	3.751	3.813	3.782	3.895	3.958	3.776
:	:	:	:	:	:	:	:	:	:	:	:	:	:	:
T398	3.779	3.752	3.710	3.888	3.654	3.830	4.044	3.864	3.760	3.814	3.779	3.905	3.966	3.777
T399	3.675	3.654	3.600	3.801	3.547	3.747	3.955	3.768	3.662	3.715	3.675	3.812	3.880	3.675
T400	3.762	3.736	3.705	3.885	3.653	3.821	4.039	3.856	3.748	3.809	3.773	3.893	3.961	3.771
D Average	**3.692**	**3.678**	**3.619**	**3.828**	**3.568**	**3.765**	**3.982**	**3.792**	**3.678**	**3.740**	**3.695**	**3.832**	**3.905**	**3.697**
T Average	3.729	3.704	3.662	3.860	3.605	3.793	4.008	3.822	3.717	3.773	3.735	3.865	3.933	3.733
|DA - DT|	0.037	0.026	0.043	0.032	0.037	0.028	0.026	0.03	0.039	0.033	0.04	0.033	0.028	0.036

Table 8. Average word vector summary in restaurant domain

Restaurant	v1	v2	v3	v4	v5	v6	v7	v8	v9	v10	v11	v12	v13	v14
D1	3.654	3.640	3.585	3.792	3.532	3.738	3.937	3.751	3.635	3.704	3.655	3.790	3.858	3.659
D2	3.757	3.724	3.690	3.887	3.621	3.820	4.026	3.846	3.730	3.791	3.756	3.883	3.945	3.753
D3	3.699	3.680	3.614	3.816	3.547	3.768	3.970	3.800	3.662	3.725	3.690	3.823	3.887	3.691
:	:	:	:	:	:	:	:	:	:	:	:	:	:	:
D200	3.584	3.594	3.483	3.700	3.451	3.679	3.881	3.707	3.561	3.622	3.576	3.738	3.795	3.586
D201	3.596	3.603	3.501	3.710	3.479	3.693	3.885	3.732	3.582	3.634	3.599	3.756	3.800	3.606
D202	3.673	3.634	3.588	3.797	3.541	3.736	3.932	3.762	3.646	3.700	3.666	3.802	3.852	3.665
T1	3.726	3.713	3.641	3.832	3.604	3.800	4.020	3.822	3.694	3.762	3.714	3.853	3.920	3.722
T2	3.731	3.715	3.673	3.866	3.634	3.814	4.001	3.848	3.740	3.775	3.747	3.883	3.927	3.747
T3	3.830	3.841	3.782	3.948	3.746	3.900	4.053	3.947	3.815	3.889	3.867	3.966	3.974	3.860
:	:	:	:	:	:	:	:	:	:	:	:	:	:	:
T198	3.982	3.973	3.927	4.066	3.881	4.036	4.219	4.083	3.950	4.011	3.993	4.102	4.134	3.990
T199	3.930	3.890	3.866	4.044	3.809	3.977	4.130	4.013	3.902	3.956	3.945	4.044	4.057	3.930
T200	3.875	3.858	3.810	3.990	3.778	3.940	4.126	3.967	3.863	3.918	3.889	4.009	4.044	3.886
D Average	**3.736**	**3.719**	**3.655**	**3.858**	**3.609**	**3.809**	**4.002**	**3.835**	**3.712**	**3.770**	**3.736**	**3.869**	**3.919**	**3.736**
T Average	3.776	3.748	3.703	3.897	3.655	3.842	4.032	3.869	3.758	3.811	3.781	3.907	3.953	3.777
|DA - DT|	0.04	0.029	0.048	0.039	0.046	0.033	0.03	0.034	0.046	0.041	0.045	0.038	0.034	0.041

Table 9. Average word vector summary in doctor domain

Doctor	v1	v2	v3	v4	v5	v6	v7	v8	v9	v10	v11	v12	v13	v14
D1	3.526	3.587	3.295	3.433	3.345	3.526	3.717	3.717	3.371	3.476	3.509	3.617	3.605	3.500
D2	3.857	3.816	3.764	3.965	3.694	3.923	4.133	3.912	3.818	3.879	3.805	3.962	4.055	3.823
D3	3.596	3.591	3.483	3.719	3.458	3.699	3.921	3.695	3.580	3.627	3.549	3.744	3.875	3.579
.	:	:	:	:	:	:	:	:	:	:	:	:	:	:
D355	3.597	3.608	3.494	3.747	3.470	3.722	3.948	3.701	3.591	3.655	3.561	3.756	3.861	3.593
D356	3.620	3.602	3.541	3.808	3.490	3.751	3.970	3.712	3.637	3.685	3.597	3.784	3.898	3.620
D357	3.727	3.705	3.628	3.831	3.586	3.780	3.984	3.823	3.693	3.747	3.698	3.851	3.913	3.711
T1	3.675	3.682	3.610	3.882	3.554	3.819	4.014	3.773	3.696	3.762	3.669	3.847	3.954	3.690
T2	3.536	3.551	3.451	3.676	3.405	3.648	3.866	3.658	3.529	3.581	3.510	3.705	3.785	3.536
T3	3.646	3.649	3.552	3.777	3.520	3.752	3.971	3.755	3.640	3.692	3.619	3.807	3.889	3.644
:	:	:	:	:	:	:	:	:	:	:	:	:	:	:
T199	3.640	3.628	3.535	3.766	3.503	3.733	3.947	3.746	3.609	3.673	3.602	3.768	3.868	3.626
T200	3.541	3.583	3.425	3.644	3.392	3.654	3.857	3.678	3.509	3.568	3.495	3.702	3.770	3.531
T201	3.523	3.561	3.422	3.662	3.403	3.672	3.870	3.672	3.526	3.572	3.501	3.708	3.790	3.532
D Average	3.625	3.608	3.535	3.782	3.484	3.736	3.945	3.714	3.616	3.674	3.597	3.772	3.873	3.616
T Average	**3.613**	**3.600**	**3.520**	**3.760**	**3.470**	**3.721**	**3.931**	**3.710**	**3.600**	**3.656**	**3.584**	**3.760**	**3.856**	**3.604**
|DA-DT|	0.012	0.008	0.015	0.022	0.014	0.015	0.014	0.004	0.016	0.018	0.013	0.012	0.017	0.012

Table 10. The Method 2 result

	Hotel	*Restaurant*	*Doctor*
KNN	66.66%	65.28%	69.64%
NB	68.75%	65.29%	66.66%
DT	63.75%	64.46%	65.47%
RF	73.75%	73.55%	71.42%

structure to promote products in the store. Thus, consumers accepted related stimulation and purchased products after reading the comments.

However, different outcomes were observed in the medical domain, and the D Average values were all larger than the T Average values in every category. Therefore, the words used in the true comments were related to selected dictionaries with S-O-R theory. The differences between D Average and T Average in the hotel, restaurant, and medical domains were 0.033, 0.039, and 0.014, respectively. This finding can be used as reference when discussing word usage differences when writing comments on the three domains.

Table 11. Result comparison of Method 1 and 2

Doctor	Method 1 (SOR characteristic weight)	Method 2 (SOR characteristic weight + Word vector)
KNN	55.35%	64.28%
NB	60.11%	60.07%
DT	60.71%	63.09%
RF	61.90%	66.07%
GBM	61.30%	67.26%
SVM	63.09%	67.26%
XGB	63.09%	69.06%
DNN	61.90%	63.25%
Average	60.93%	65.04%

Method 2 indicated the classification results of adding 15 words vector frequency characteristics with S-O-R characteristic values from Method 1. The method results are listed in Table 10. The classification algorithm RF are the greatest performance.

Table 11 presents the accuracy of true and fake comment identification in the medical domain after training the data from two other domains and excluding data from the medical domain. In the method results of Method 2, the XGB classification algorithm performed the most satisfactorily (69.06%), followed by GMB and SVM (67.26%). These results all were higher than the 55.7% produced by Ren and Ji (2017) by using ANNs and the 64.7% by Li et al. (2014) by using SAGE.

All classification algorithms were compared with Method 1 and listed. All algorithms of Method 2 except for NB exhibited higher performance than did those of Method 1. The average value of all classification algorithms was 60.93%, and this improved by 4.11% when Method 2 was used. From this experiment, the addition of word vector frequency characteristics could enhance the accuracy of the extraction of the word characteristic frequency used in true and fake comments. Thus, establishing an interdisciplinary identification model for fake comments can be accomplished.

5 Conclusion and Future Research

Online shopping has a new consumption model. Online reviews are of great significance to consumers, e-commerce platforms, and stores. Due to the habit of consumers to refer to reviews before buying, shop owners will write fake reviews to promote products as a boost. In the face of real and fake comments on the Internet, e-commerce platforms have an obligation to protect consumers and maintain the quality on the platform. Therefore, the ability to develop a system that can quickly identify the authenticity of online reviews is an important issue for online platforms.

This research is aimed at the reviews in the three domains of hotel, restaurant, and doctor, and attempts to develop a model that can discriminate true and fake reviews across domains. After using SOR theory to select the corresponding category fonts from the LIWC font to calculate the feature values, word2vec is used to increase the frequent features of the word vector, and it becomes a cross-domain model that can effectively discriminate between true and fake reviews.

Many deficiencies were discovered after the end of the research. We have provided 2 points below in terms of limitations and recommendations:

1. In this research, the labeled reviews are used to focus on the three areas of hotel, restaurant, and doctor. In the future, if data from other domains can be used as training materials for modeling, we can explore whether data in different domains can also be applied rule.
2. In future research, because the idioms used in each generation are different. Therefore, it is suggested to expand or update the related category fonts based on the SOR theory to maintain or improve the discrimination accuracy, and at the same time explore whether the category fonts will change over time.

References

Adelaar, T., Chang, S., Lancendorfer, K.M., Lee, B., Morimoto, M.: Effects of media formats on emotions and impulse buying intent. J. Inf. Technol. **18**(4), 247–266 (2003). https://doi.org/10.1080/0268396032000150799

Aslam, U., Jayabalan, M., Ilyas, H., Suhail, A.: A survey on opinion spam detection methods. Int. J. Sci. Technol. Res. **8**(9) (2019)

Boujbel, L., d'Astous, A.: Exploring the feelings and thoughts that accompany the experience of consumption desires. Psychol. Mark. **32**(2), 219–231 (2015). https://doi.org/10.1002/mar.20774

Chang, H.-J., Eckman, M., Yan, R.-N.: Application of the Stimulus-Organism-Response model to the retail environment: the role of hedonic motivation in impulse buying behavior. Int. Rev. Retail Distrib. Consum. Res. **21**(3), 233–249 (2011)

Eroglu, S.A., Machleit, K.A., Davis, L.M.: Atmospheric qualities of online retailing: a conceptual model and implications. J. Bus. Res. **54**(2), 177–184 (2001)

Gatautis, R., Vaiciukynaite, E.: Website atmosphere: towards revisited taxonomy of website elements. Econ. Manag. **18**(3) (2013)

Kavanagh, D.J., Andrade, J., May, J.: Imaginary relish and exquisite torture: the elaborated intrusion theory of desire. Psychol. Rev. **112**(2), 446 (2005)

Klaus, T., Changchit, C.: Toward an understanding of consumer attitudes on online review usage. J. Comput. Inf. Syst. **59**(3), 277–286 (2017). https://doi.org/10.1080/08874417.2017.1348916

Li, J., Ott, M., Cardie, C., Hovy, E.: Towards a general rule for identifying deceptive opinion spam. Paper Presented at the Proceedings of the 52nd Annual Meeting of the Association for Computational Linguistics (Volume 1: Long Papers) (2014)

Liu, W., Jing, W., Li, Y.: Incorporating feature representation into BiLSTM for deceptive review detection. Computing **102**(3), 701–715 (2019). https://doi.org/10.1007/s00607-019-00763-y

Menon, S., Kahn, B.: Cross-category effects of induced arousal and pleasure on the Internet shopping experience. J. Retail. **78**(1), 31–40 (2002)

Ott, M., Choi, Y., Cardie, C., Hancock, J.T.: Finding deceptive opinion spam by any stretch of the imagination. Paper Presented at the Proceedings of the 49th Annual Meeting of the Association for Computational Linguistics: Human Language Technologies-Volume 1 (2011)

Ren, Y., Ji, D.: Neural networks for deceptive opinion spam detection: an empirical study. Inf. Sci. **385–386**, 213–224 (2017). https://doi.org/10.1016/j.ins.2017.01.015

Low Resource Languages Processing

Towards the Uzbek Language Endings as a Language Resource

Sanatbek Matlatipov[1(✉)], Ualsher Tukeyev[2], and Mersaid Aripov[1]

[1] National University of Uzbekistan, University Street, 14, 100174 Tashkent, Uzbekistan
mr.sanatbek@gmail.com, mirsaidaripov@mail.ru

[2] Al-Farabi Kazakh National University, Al-Farabi Ave., 71, 050040 Almaty, Kazakhstan
ualsher.tukeyev@gmail.com

Abstract. The Uzbek language belongs to low-resource languages. It is very important to increase the number of language resources such as dictionaries, corpora (monolingual and bilingual) for the Uzbek language. Dictionaries may be different kinds: monolingual, orthographical, bilingual, grammar special dictionaries: stems dictionaries, affixes dictionaries, etc. For different NLP tasks of agglutinative languages, such as morphological analysis, information retrieval, machine translation (segmentation preprocessing) in some cases needs a dictionary of words' endings. In this paper, we proposed the first electronic dictionary of Uzbek words' endings in variants for morphological segmentation preprocessing useful for neural machine translation. The resource analysed by the initial version of the Lexicon free stemming tool [3] created by authors. For creation of Uzbek words' endings' electronic dictionary, it was used a combinatorial approach inferring apply for part of speech of the Uzbek language: nouns, adjectives, numerals, verbs, participles, moods, voices.

Keywords: Uzbek language · Electronic dictionary · Words' endings · Language resource · Lexicon free stemming tool

1 Introduction

The Uzbek language is used by more than 27 million people according to Wikipedia, who live in Uzbekistan, Afghanistan, Tajikistan, Kyrgyzstan, Kazakhstan, Turkmenistan, Russia, Turkey, Iran, China. The language is agglutinative where words are generated by adding affixes(mainly suffixes) to the base. It can be derived as a new word by adding affixes. Furthermore, in many cases a single Uzbek word may correspond to a *many-word sentence* or phrase in a non-agglutinative language [1]. Let's consider the following complex word/sentence as an example: the word *"uchrashaolmaganlardanmisiz?"*, corresponds to *"Are you one of those who couldn't meet?"* in English. However, there are simple words which are ambiguous. For example, *the noun* meaning of the *"olma"* corresponds to *"apple"*, but, *verb* meaning of *"ol+ma(negation ending)"* corresponds to *"don't take"*. As the Uzbek language belongs to low-resource languages, because there are lack of language resources, such as monolingual/bilingual corpora, electronic

M. Hernes et al. (Eds.): ICCCI 2020, CCIS 1287, pp. 729–740, 2020.
https://doi.org/10.1007/978-3-030-63119-2_59

dictionaries etc. For different NLP tasks, such as morphological analysis, information retrieval, machine translation (segmentation preprocessing) in some cases needs a dictionary of word endings [2, 3]. In this paper, we propose the current version of *Uzbek endings* dictionary and its usage in the initial version of the *Lexicon free stemming tool*[1]. Combinatorial inferring of language word endings method [4] is used in the proposed work to create Uzbek endings language resource.

2 Related Works

One of the important tasks in neural machine translation is source text segmentation. That task is connected with the decreasing vocabulary of NMT, also influential on solving a problem of rare and unknown words. The state-of-the-art of source text segmentation for NMT is the BPE (byte pair encoding) method proposed in work [5]. BPE method based on the combinations of part of word and choose more frequency combinations. One of the well known methods of segmentation is Morfessor open-source software method. Morfessor segments words according to their morphological structure [6]. There are works focused on the morphological analysis of natural language words developed by Koskenniemi K. (Koskenniemi, 1983; Oflazer, 1994; Kessikbayeva and Cicekli, 2014 [7–9], based on finite-state transducers (FST). Some of the works have been done for Uzbek language as well [10]. These methods for morphological segmentation are needed, transforming a little.

3 Motivation

The creation and use of a complete system of Uzbek word endings has the following grounds:

1. The Uzbek language belongs to low-resource languages, so the creation of language resources is important.
2. Although neural machine translation has shown impressive results for many world languages, it does not solve the problem of low-resource languages. Therefore, the development of tools perfecting the use of NMT for low-resource languages is relevant.
3. There are other areas of natural language processing, such as information retrieval, sentiment analysis, summarization, etc., which require knowledge of the subject area, in this case, deep semantic knowledge of the Uzbek language. Therefore, the creation of a language resource that allows improving the processing of the Uzbek language is necessary.
4. Development of a complete system of endings of the Uzbek language will allow in one step (referring to the table of the system of endings of the language) to perform: segmentation of the word into suffixes; perform morphological analysis of the word. Since the ending system will have a decision for each ending of the language to segment into suffixes or suffixes with their morphological characteristics (for morphological analysis).

[1] https://uz-kaz-nlp-tools.herokuapp.com/ - API based web service is created by author1.

5. The development of a complete system of word endings of the Uzbek language allows to guarantee the analysis of almost any word of the Uzbek language, as this is determined by the inferring of the complete system of word endings of the language.

4 The Inferring of Uzbek Word Endings' Complete System

The Uzbek language as other Turkic languages have two part word endings: nominal endings (nouns, adjectives, numerals) and verbal endings (verbs, participles, gerunds, mood and voice).

4.1 Inferring Nominal Endings

The nominal endings of the Uzbek language have five types base affixes: The stem of word is denoted by **S**. Case suffixes - **C;** Plural suffixes - **K;** Possessive suffixes - **T;** Diminutive-endearment suffixes - **L;** Noun formation suffixes - **P**.

Consider all types of base affixes placements (affixes sequence) variants: of the one type, of two types, of three types, and of four types. Number of placements determined by the formula:

$$A_n^k = n!/(n-k)!.$$

Then, the number of placements will be determined as follows:

$A_5^1 = 5!/(5-1)! = 5, A_5^2 = 5!/(5-2)! = 20, A_5^3 = 5!/(5-3)! = 60, A_5^4 = 5!/(5-4)! = 120, A_5^5 = 5!/(5-5)! = 120$. All possible placements number is 325.

The word endings placements for one type are in the Table 1.

Table 1. The word endings placements for one type placements

Suffix type	Suffixes		Count
C type	1. Genitive 2. Dative 3. Accusative, 4. Locative, 5. ablative	1. -ning, 2. -ga, -ka, -qa, 3. -ni, 4. -da, 5. -dan	7
K type	-lar		1
T type	-im/m, -ing/ng, -si/i, -imiz/miz, -ingiz/ngiz		10
L type	-cha		1
P type	1. people; 2. places; 3. Abstract things	1. -chi, -dosh, 2. -zor, -iston, -xona; 3. -lik, -chilik	7

The word endings placements for two types are the following:

CK, CL, CP, CT, **KC**, KL, KP, **KT**, **LC**, **LK**, LP, **LT**, **PC**, **PK**, PL, **PT**, **TC**, TK, TL, TP.

The analysis of the semantics of the two types of endings placements shows that bold placements are valid, and the remaining placements belong to unacceptable endings. For example, TK – will be unacceptable: after the possessive endings plural endings are not used, CK – unacceptable: after case endings are not accepted to put the plural, CT – unacceptable: after case endings are not put possessive endings. Thus, the number of valid (correct) placements of two types of endings is 9. In the following, we decided to show some of the tables with few examples for combinations. Full list of the description tables as below and segmentation table as resource is made available on GitHub[2].

Inferring of endings for placements type KC (Plural-Case) present in Table 2. Combinations in placements KC: (one suffixe K)*(5 suffixes C) = 5 endings KC. Examples for placement KC are presented in Table 2.

Table 2. Inferring of endings for placements type KC (Plural-Case).

Suffix type K	Suffixes type C		Examples	Number of endings
-lar	1. genitive 2. dative 3. accusative 4. locative 5. ablative	-ning -ga -ni -da -dan	Kitob+**lar+ning** - books' Kitob+**lar+ga** - to the books Kitob+**lar+ni** - the books Kitob+**lar+da** - on the books Kitob+**lar+dan** - from the books	1*5 = 5

Combinations in placements TC: (10 suffixe T)*(5 suffixe C) = 50 endings. Combinations in placements LC: (1 suffixe L)*(5 suffixe C) = 5 endings; Combinations in LK: (1 suffixe L)*(1 suffixe C) = 1 endings; Combinations in LT: (1 suffixe L)*(5 suffixe T) = 5 endings; For example, in Table 3 present Inferring of endings for placement type TC (Possessive - case).

Table 3. Inferring of endings for placements type TC (Possessive - case).

Suffix type T	Suffixes type C		Example	Number of endings
1. -im, -m 2. -ing, -ng 3. -si, -i 4. -imiz, -miz 5. -ingiz, -ngiz	1. genitive 2. dative 3. accusative 4. locative 5. ablative	1. -ning 2. -ga 3. -ni 4. -da 5. -dan:	1. Kitob+**im+ning** - my book's 2. Igna+**m+ga** - to my needle 3. Bola+**si+ning** - his/her son's 4. his/her son 5. Kitob+**i+da...**	50

Combinations in placements PC: (7 suffixe P)*(5 suffixe C) = 35 endings; Moreover, Combinations in placements PK: (4 suffixe P)*(1 suffixe K) = 4 endings; Combinations in placements PT: (7 suffixe P)*(5 suffixe C) = 35 endings. For example, in Table 4 present inferring of endings for placements type PC (Noun formation suffixes - case).

[2] https://github.com/SanatbekMatlatipov/uzbek-word-affixes.

Table 4. Inferring of endings for placement type PC (noun formation suffixes - case).

Suffixes type P		Suffixes type C		Examples	Count
1. people 2. places 3. abstract things	1. -chi, -dosh 2. -zor, -iston, -xona 3. -lik, -chilik	1. genitive 2. dative 3. accusative 4. locative 5. ablative	-ning -ga/-ka/-qa -ni -da -dan	Kitob+**chi+ning** –books' Sinf+**dosh+ga** O'zbek+**iston+da** Ota+**lik+ka**	35

The word endings placements of the three types are as follows:

KCT, KCL, KCP, **KTC**, KTL, KTP, LCK, LCT, LCP, **LKC**, **LKT**, LKP, **LTC**, LTK, TCK, TCL, TCP, PCK, PCT, PCL, **PKC**, **PKT**, PKL, **PTC**, PTK, PTL

Determination of permissible placements of three types of endings do the following rule: *if the placement of the three types have invalid placement of two types, this placement - unacceptable.*

Therefore, the permissible endings placements of three types is 7 (in bold) (Tables 5, 6 and 7).

Table 5. Inferring of endings for placement type KTC (Plural-Possessive-Case).

Suffix type K	Suffixes type T	Suffixes type C		Examples	Number of endings
-lar	-im -ing -imiz -ingiz -i	1. genitive 2. dative 3. accusative 4. locative 5. ablative	-ning -ga -ni -da -dan	1. Kitob+**lar+im+ning** 2. Kitob+**lar+ing+ga** 3. Kitob+**lar+imiz+ni** 4. Kitob+**lar+ingiz+da** 5. Kitob+**lar+i+dan**	25

Combinations in placements KTC: (1 suffixe K)*(5 suffixe T)*(5 suffixe C) = 25 endings.

The endings placements of the four types are as follows:

KTCL, KTCP, LKCP, LKCT, **LKTC**, LKTP, LTCK LTCP, PKCL, PKCT, **PKTC**, PKTL, PTCK, PTCL

Determination of permissible placements of four types of endings do the following rule: *if the placement of the four types have invalid placement of three types, this placement - unacceptable.*

Therefore, the permissible endings placements of four types is **2** (in bold) (Table 8).

The endings placements of the five types are as follows:

LKTCP PKTCL

Table 6. Inferring of endings for placements type LTC (Diminutive-Possessive -Case).

Suffix type L	Suffixes type T	Suffixes type C		Example	Number of endings
-cha	-m -ng -miz -ngiz -si	1. genitive 2. dative 3. accusativ 4. locative 5. ablative	-ning -ga -ni -da -dan	Mushuk+**cha+m+ga** - to my kitty Mushuk+**cha+ng+ning** Mushuk+**cha+miz+ni** Mushuk+**cha+ngiz+da** Mushuk+**cha+si+dan**	1*5*5 = 25

Combinations in placement LTC: (1 suffixe L)*(5 suffixe T)*(5 suffixe C) = 25 endings; Combinations in placement LKC: (1 suffixe L)*(1 suffixe K)*(1 suffixe C) = 5 endings; Combinations in placement LKT: (1 suffixe L)*(1 suffixe K)*(5 suffixe T) = 5 endings.

Table 7. Inferring of endings for placement type PKC (Noun formation-Plural -Case)

Suffixes type P		Suffix type K	Suffixes type C		Examples	Number of endings
people places	-chi, -dosh -zor, -xona	-lar	1. genitive 2. dative 3. accusativ 4. locative 5. ablative	-ning -ga -ni -da -dan	ish+**chi+lar+ning** Sinf+**dosh+lar+ga** Bog'+**zor+lar+ni** Issiq+**xona+lar+dan** Issiq+**xona+lar+da**	20

Combinations in placement PKC: (4 suffixe P)*(1 suffixe K)*(5 suffixe C) = 20 endings; Combinations in placement PKT: (4 suffixe P)*(1 suffixe K)*(5 suffixe T) = 20 endings; Combinations in placement PTC: (7 suffixe P)*(6 suffixe T)*(5 suffixe C) = 210 endings.

Table 8. Inferring of endings for placements type PKTC (noun formation-plural-case)

Suffixes type P		Type K	Suffix type T	Suffixes type C		Example	Number of endings
people places abstract things	-chi, -dosh -zor, -iston, -xona -lik, -chilik	-lar	-im -ing, -imiz, -ingiz, -i	1. genitive 2. dative 3. accusativ 4. locative 5. ablative	-ning -ga -ni -da -dan	Sinf+**dosh+lar+im+ning**- my classmates' Gul+**chi+lar+imiz+dan**- from our flower makers	7*5*5 = 175

Combinations in placement PKTC: (7 suffixe P)*(1 suffixe K)*(5 suffixe T)*(5 suffixe C) = 175 endings; Combinations in placement LKTC: (1 suffixe P)*(1 suffixe K)*(5 suffixe T)*(5 suffixe T) = 25 endings.

Determination of permissible placements of five types of endings do the following rule: *if the placement of the five types have invalid placement of four types, this placement - unacceptable.*

Therefore, the permissible endings placements of four types is **0**.

Total permissible ending's placements of one type of - 26, of two types - 145, of three types - 310 of four types - 200, of five types - 0. So, the total number of valid types of ending's placements in the nominal words is **345**.

4.2 Verbal Endings

The set of endings of Uzbek language to the verbal stems includes the following types: - system of verb endings; - system of participles endings; - system of verbal adverbs endings; - system of moods endings; - system of voices endings.

The system of verb endings includes the following types:

- Tense suffixes (8 types)

1. Present-future tense: *Kel+a+man;*
2. Present continuous tense: Kel+ayap+man, kel+moqda+man
3. Indefinite/simple past tense: kel+di+m
4. Recent past/present perfect tense: kel+gan+man
5. Distant past/past perfect tense: kelgan edi+m;
6. Past continuous tense: kelar edi+m;
7. Future tense: kel+ar+man;
8. Future intentional tense: kel+moqchi+man;

- Person (4 types): 1 person, 2 person, 2 person(respect), 3 person
- Question (so'roq) suffix (-mi) =>(1 type);
- Negation (bo'lishsizlik) suffix (-ma, -mas) =>(2 type);

The total number of possible types of verb endings is - 319. In the following, only one tense combination description tables is shown as an example[3] (Tables 9 and 10):

Table 9. Verb after consonant letter - undoshdan keyin

Example	1 person	2 person	2 person (respect)	3 person	Number of endings
O'tir- Yur- Ket-	-aman	-asan	-asiz	-adi	8
	-amiz	-asizlar	-asizlar	-adilar	
Total					8

3 https://github.com/SanatbekMatlatipov/uzbek-word-affixes.

Table 10. Verb after vowel letter - unlidan keyin

Examples	1 person	2 person	2 person (respect)	3 person	Number of endings
O'qi- kuyla- qara- kuyla-	-yman plural: -ymiz	-ysan plural: -ysizlar	-ysiz plural: -ysizlar	-ydi plural: -ydilar	4 4
Total					8

Present/Future Simple Tense (Hozirgi zamon) Suffixes
Furthermore, there are 32 endings of verbs with question forms.[4]

4.3 Inferring Participle Endings

The system of participle endings includes the following types:

1. participle's base affixes (denoted R)

   1.1. -gan, -qan, kan, -(a)yotgan, -(y)yotgan, -(a)digan, -(y)digan, -(a)r, -(ma =>**negative**)s, -ajak, -mish, -gusi, -g'usi, -asi; **(number of endings are 14)**

2. case affixes (denoted **C**);
3. plural affixes (denoted **K**);
4. possessive affixes (**T**);

Then, possible variants of affixes types sequences (participle's base affixes for all variants is the same) will be:

- one type of affixes sequences: **RK, RT, RC;**
- two types of affixes sequences: **RKT, RTC, RKC,** RCT, RTK, RCK;
- three types of affixes sequences: RCKT, RCTK, RKCT, **RKTC**, RTCK, RTKC

Let's consider a semantic permissibility of variants of the endings.

All variants of participle's endings on one type of the endings are semantically permissible.

The analysis of semantics of placements of two types of the participle's endings shows that the placements allocated by bold font are permissible, and other placements are carried to unacceptable. Thus, the quantity of types of the endings of participles is 7.

In the following, only case of participle combination description table is shown as an example (See footnote 3) (Table 11):

[4] https://github.com/SanatbekMatlatipov/uzbek-word-affixes/blob/master/.data_resources/THE%20UZBEK%20LANGUAGE%20ENDINGS%20AS%20A%20LANGUAGE%20RESOURCE%20(2).docx.

Table 11. Inferring of endings for participle placements type RC

Suffixes type R	Suffixes type C		Total number of endings
-gan, -qan, kan	1. genitive	-ning	58
-(a)yotgan, -(y)yotgan	2. dative	-ga	
-(a)digan, -(y)digan	3. accusative	-ni	
-(a)r, -(ma =>negative)s,	4. locative	-da	
-ajak, -mish, -gusi, -g'usi, -asi	5. ablative	-dan	

Let's consider types of the endings of verbal adverbs. They are represented by the endings of transitive future time for which follows personal endings: VJ, where V - the base ending of a verbal adverb, J - personal endings. For the given class we shall allocate only the following base endings: -b, -gani, -qani. Thus, we count that quantity of types of the endings of a verbal adverb is 1.

Let's consider the endings of moods, namely, conditional, imperative, desirable. The endings of an indicative mood coincide with the endings of verbs in the present, the past and the future.

The type of the endings of declinations is similar adverbs, i.e. the base endings of moods which personal endings follow. Thus, we consider that there are three types of the endings of moods: conditional, imperative, desirable.

There are 4 types of voices endings, namely, **Passive voice(-(i)l; -(i)n), Reflexive voice suffix(-(i)n; -i(l), -lan), Reciprocal voice suffix** (**-(i)sh,- lash), Causative voice suffix** (**-(i)t, -tir, -ir, -ar, -iz**), also are determined under the previous scheme: the base endings of voices for which follow personal endings.

So, the total of types of the endings of words with verbal bases is **402**.

The total of the endings of words with nominal bases plus total of types of the endings of words with verbal bases equal **747**.

5 The Complete System of Uzbek Language Endings as a Language Resource

For different NLP tasks of agglutinative languages, such as morphological analysis, information retrieval, machine translation (segmentation preprocessing) in some cases needs a dictionary of words' endings. Dictionaries of word endings for morphological analysis require breaking the ending into affixes, which are accompanied by their grammatical characteristics. It should be noted that for agglutinative languages, each affix has its exact grammatical purpose. Therefore, the formation of such accompanying morphological characteristics does not present a particular problem.

The paper proposed the first electronic dictionary of Uzbek words' endings in variants for morphological segmentation preprocessing useful for neural machine translation. Example of the part of the dictionary of Uzbek words' endings for morphological segmentation present in Table 12.

Table 12. The segment of dictionary of Uzbek words' endings for morphological segmentation

#	Uzbek endings	Suffixes segmentation	Types count
	-larning	lar@@ning	2
	-larga	lar@@ga	2
	-larda	lar@@da	2

Table 13. Segmentation analysis with percentage by using our stemming tool

Resource	Words count	Correct count	Incorrect count
News-1[a]	369	351(95.2%)	18(4.8%)
News-2[b]	937	882(94%)	55(6%)
News-3[c]	90	87(96.6%)	2(3.6%)
News-4[d]	201	193(96%)	8(4%)

[a]https://kun.uz/uz/news/2020/04/30/buyuk-britaniya-koronavirusdan-olimlar-soni-boyicha-uchinchi-oringa-chiqdi.
[b]https://kun.uz/uz/news/2020/04/29/odamlar-charchadi-prezident-karantindan-bosqichma-bosqich-chiqish-boyicha-navbatdagi-qadamlarni-qoyish-vaqti-kelgani-haqida-gapirdi.
[c]https://kun.uz/uz/news/2020/04/30/andijonda-koronavirusga-chalingan-18-kishi-sogaydi.
[d]https://kun.uz/uz/news/2020/04/30/tiv-charter-reyslar-va-chet-eldagi-ozbekistonliklarga-yordam-masalasi-boyicha-izoh-berdi.

5.1 Lexicon Free Stemming Tool for Segmentation Analysis

We created initial version of API based[5] web service for lexicon free segmentation by using the lexicon free stemming algorithm [3] and created above the complete set of Uzbek words' endings. As a resource, we segmented the most popular news website data in Uzbekistan by using our tool. Then, we manually checked the result by reviewing one-by-one[6]. The stemming service tool has been mainly experimented using lexicon free stemming tools for short text consisting of overall almost 1600 words (Table 13). The incorrect segmentation happened due to word ambiguities as some endings can be attached to words as root. For example, the word *"kuni (that day)"* should be segmented

[5] https://uz-kaz-nlp-tools.herokuapp.com/.
[6] https://github.com/SanatbekMatlatipov/uzbek-word-affixes/blob/master/.data_resources/Manual_analysis.xlsx.

as *"kun+i"*. However, there is an ending *"-ni"* which makes doubts in the algorithm and segments it as *"ku+ni"*, wrongly.

6 Conclusion and Further Work

The paper presents a new language resource: the Uzbek language endings. The set of Uzbek words' endings is inferred by combinatorial approach methods including also harmony rules of Uzbek language. This new Uzbek language resource was created for the main Part-of-Speech of the Uzbek language: nouns, adjectives, numerals, verbs, participles, moods, voices. The testing of the Uzbek language endings for morphological segmentation show errors level approximately 5%. This Uzbek words' endings language resource plan to register in ELRA Catalogue of Language Resources and will be used for a segmentation preprocessing of Uzbek source data in neural machine translation.

Acknowledgments. This work was carried out under the project "Development of the interdisciplinary master program on computational Linguistics at Central Asian universities. Number 585845-EPP-1-2017-1-ES-EPPKA2- CBHE-JP" (15/10/2017 to 14/10/2020) funded by the Education, Audiovisual and Culture Executive Agency of European Union.

References

1. Matlatipov, G., Vetulani, Z.: Representation of uzbek morphology in prolog. In: Marciniak, M., Mykowiecka, A. (eds.) Aspects of Natural Language Processing. LNCS, vol. 5070, pp. 83–110. Springer, Heidelberg (2009). https://doi.org/10.1007/978-3-642-04735-0_4
2. Tukeyev, U., Sundetova, A., Abduali, B., Akhmadiyeva, Z., Zhanbussunov, N.: Inferring of the morphological chunk transfer rules on the base of complete set of Kazakh endings. In: Nguyen, N.-T., Manolopoulos, Y., Iliadis, L., Trawiński, B. (eds.) ICCCI 2016. LNCS (LNAI), vol. 9876, pp. 563–574. Springer, Cham (2016). https://doi.org/10.1007/978-3-319-45246-3_54
3. Tukeyev, U., Turganbayeva, A., Abduali, B., Rakhimova, D., Amirova, D., Karibayeva, A.: Lexicon-free stemming for Kazakh language information retrieval. In: IEEE 12th International Conference on Application of Information and Communication Technologies, AICT 2018, Kazakhstan, Almaty, 17–19 October 2018, pp. 95–98 (2018)
4. Tukeyev, U.: Automaton models of the morphology analysis and the completeness of the endings of the Kazakh language. In: Proceedings of the International Conference on "Turkic Languages Processing" TURKLANG-2015, Kazan, Tatarstan, Russia, 17–19 September 2015, pp. 91–100 (2015)
5. Sennrich, R., Haddow, B., Birch, A.: Neural machine translation of rare words with subword units. In: Proceedings of the 54th Annual Meeting of the Association for Computational Linguistics, vol. 1, pp. 1715–1725 (2016)
6. Creutz, M., Lagus, K.: Unsupervised discovery of morphemes. In: Proceedings of the ACL-2002 Workshop on Morphological and Phonological Learning, vol. 6, pp. 21–30 (2002)
7. Koskenniemi, K.: Two-level morphology: a general computational model for word-form recognition and production. Ph.D. thesis, University of Helsinki (1983)
8. Oflazer, K.: Two-level description of Turkish morphology. Literary Linguist. Comput. **9**(2), 137–148 (1994)

9. Kessikbayeva, G., Cicekli, I.: Rule based morphological analyzer of Kazakh language. In: Proceedings of the 2014 Joint Meeting of SIGMORPHON and SIGFSM, Baltimore, Maryland, USA, pp. 46–54 (2014)
10. Madatov, Kh.: A prolog format of uzbek WordNet's entries. In: Human Language Technology as a Challenge for Computer Science and Linguistics, pp. 316–320 (2019)

Inferring the Complete Set of Kazakh Endings as a Language Resource

Ualsher Tukeyev(✉) and Aidana Karibayeva

Al-Farabi Kazakh National University, al-Farabi Avenue, 71, 050040 Almaty, Kazakhstan
ualsher.tukeyev@gmail.com, a.s.karibayeva@gmail.com

Abstract. The Kazakh language belongs to low-resource languages. For application of actual modern branches as artificial intelligence, machine translation, summarization, sentiment analysis, etc. to the Kazakh language needs increasing the number of electronic language resources. Although neural machine translation (NMT) has shown impressive results for many world languages, it does not solve the problem of low-resource languages. Therefore, the development of resources and tools perfecting the use of NMT for low-resource languages is relevant. For perfect use of NMT for the Kazakh language needs bilingual parallel corpora, but also needs a perfect method of the segmentation source text. By the opinion of authors, one of the effective ways for source text segmentation is morphological segmentation. The authors propose to use for morphological segmentation of Kazakh text a table of a complete set of Kazakh words' endings. In this paper is described the inferring of the complete set of Kazakh words' endings. Development of the table of the complete set of word' endings of the Kazakh language will allow in one-step (by reference to the table of endings of the language) to perform the segmentation of the word's ending into suffixes. The complete set of endings of the Kazakh language allows guaranteeing the analysis of any word of the Kazakh language, as this is determined by the inferring of the complete system of words' endings of the language.

Keywords: The Kazakh language · Morphological segmentation · Words' ending · Language resource

1 Introduction

The field of natural language processing (NLP) is rapidly developing in connection with the development of the Internet.

The NLP area includes machine translation, information retrieval, summarization, sentiment analysis, etc. Currently, an actively developing area is machine-learning methods for solving the above problems of NLP. Machine learning methods require a significant volume of source data. However, many languages do not yet have sufficient text data on the Internet. Such languages belong to the class of low-resource languages.

The Kazakh language belongs to the Turkic group of languages. About 13 million people use the Kazakh language according to Wikipedia: Kazakh speakers live in Kazakhstan, Russia, China, Uzbekistan, Mongolia, and Turkmenistan. As regards language resources, the Kazakh language may safely be classified as a low-resource language.

M. Hernes et al. (Eds.): ICCCI 2020, CCIS 1287, pp. 741–751, 2020.
https://doi.org/10.1007/978-3-030-63119-2_60

In machine learning methods, in particular, in neural machine translation (NMT), a vocabulary of words of the processed source text for learning is formed. NMT automatically compiles any distinguishing word of the source text into a vocabulary. As a result, almost all words from the text fall into the vocabulary, which overflows the memory of the NMT. To solve this problem, word segmentation of the source text is used. The word segmentation allows solving the problems of rare and unknown words too. Currently, the main segmentation method is the BPE (Byte Pair Encoding) method [1]. For agglutinative languages, morphological segmentation based on the suffix system of these languages can be used. This can be done in two ways: to perform morphological analysis of each word of the source text, then breaking the word into the stem and suffixes; or segmentation of words based on a complete language ending system. In this paper, we propose the inference of a complete system of endings of the Kazakh language based on the combinatorial approach proposed in [2].

2 Related Works

Sennrich et al. developed method BPE (byte pair encoding) segmentation based on a compression method [1]. The benefit of BPE is that rare words segmented into frequent subsequences, which allows building the translation of unknown words, which is an actual task in NMT.

There are several modifications of BPE, in part, Tacorda et al. proposed the use of the controlled byte pair encoding (CBPE) method [3]. Wu and Zhao presented generalized BPE segmentation and experimented on the German–English and Chinese–English language pairs. They compare different types of substring goodness measures and came to the conclusion that a segmentation strategy is sensitive to language pairs [4].

Ataman et al. used BPE segmentation in two experiments. In the first, BPE used on the source language side, in the second - it used on the target language side. In the Turkish translation with BPE they received unreliable translation and segments were ambiguous [5].

There is Morfessor open-source software for unsupervised morphological analysis. Morfessor segments words according to their morphological structure [6]. Evolutionarily, there are three main versions of Morfessor: Morfessor Baseline, Morfessor Categories-ML, and Morfessor Categories-MAP.

There is a large number of works devoted to the study of two-level morphology model developed by Koskenniemi K. (Koskenniemi, 1983; Oflazer, 1994; Beesley and Karttunen, 2003; Kairakbay, 2013; Kessikbayeva and Cicekli, 2014.) [7–11]. These methods are based on finite-state transducers (FST) and they are mainly focused on the morphological analysis of natural language words. However, they did not solve the problem of morphological segmentation in the form required for NMT.

3 Motivation

The creation and use of a complete system of Kazakh endings has the following grounds:

1. The Kazakh language belongs to low-resource languages, so the creation of language resources for it is important.
2. Although neural machine translation has shown impressive results for many world languages, it does not solve the problem of low-resource languages. Therefore, the development of tools perfecting the use of NMT for low-resource languages is relevant.
3. There are other areas of natural language processing, such as information retrieval, sentiment analysis, summarization, etc., which require knowledge of the subject area, in this case, deep semantic knowledge of the Kazakh language. Therefore, the creation of a language resource that allows improving the processing of the Kazakh language is necessary.
4. Development of a complete system of endings of the Kazakh language will allow in one-step (by reference to the table of endings of the language) to perform: segmentation of the word into suffixes; perform morphological analysis of the word.
5. The development of a complete system of endings of the Kazakh language allows guaranteeing the analysis of almost any word of the Kazakh language, as this is determined by the inferring of the complete system of endings of the language.

4 Methods of Inferring the Set of Kazakh Endings

The words of the Kazakh language can be divided into two large groups: words with a nominal base and words with a verb base. Words with a nominal base include nouns, adjectives, numerals, and adverbs. Words with a verb base include verbs, participles, voices, moods. In this paper, to derive endings of the language, it is proposed to use a combinatorial approach, taking into account the semantic admissibility of placements of basic suffixes of the language and the peculiarities of harmonizing the sounds of the language.

The nominal endings of the Kazakh language have four types of base affixes: Plural affixes (denoted by K); Possessive affixes (denoted by T); Case affixes (denoted by C); Personal affixes (denoted by J). The number of different placements of base types suffixes is determined by the formula:

$$A_n^k = \frac{n!}{(n-k)!} \tag{1}$$

Then, the number of placements will be determined as follows:

$$A_4^1 = \frac{4!}{(4-1)!} = 4, A_4^2 = \frac{4!}{(4-2)!} = 12, A_4^3 = \frac{4!}{(4-3)!} = 24, A_4^4 = \frac{4!}{(4-4)!} = 24 \tag{2}$$

The first step is checking the semantical validity of each placement of base types of Kazakh suffixes. The endings placements of one type (K, T, C, J) are semantically valid.

The semantically valid placements for the two types are 6 from 12 possible suffixes type placements: KT, TC, CJ, KC, TJ, KJ.

The semantically valid placements of the three types are 4 from 24 possible suffixes type placements: KTC, KTJ, TCJ, KCJ.

The semantically valid placements of the four types are 1 from possible 24 possible suffixes type placements: KTCJ.

The set of endings of the Kazakh language to the verbal stems includes the following types: - a system of verb endings; - a system of participles endings; - a system of verbal adverbs endings; - a system of moods endings; - a system of voices endings.

The system of endings of verbs include the following types: 8 types of tense; 4 types of person; number: single and plural. Each of these types have some set of suffixes. Different variants of combinations (placements) of the suffix's types are defined using a combinatorial approach taking into account the sequences of types, the harmonizing rules of the sequences sounds of the types. Last sound (vowels: solid or soft; consonants: deaf or voiced) in placement define the choice of the type of following suffix type (solid or soft).

For the participle endings are used a similar combinatorial approach as for nominal bases words. The system of participle endings includes the following types: - participle base affixes (denoted R); - plural affixes (denoted K); - possessive affixes (T); - case affixes (denoted C); - personal affixes (denoted J). Then, having considered possible variants of affixes types placements (participle base affixes for all variants is the same) on the semantic permissibility, we define set of placements: - for placements from one type affixes permissible sequences; RK, RT, RC, RJ - for placements from two type affixes permissible sequences; RKT, RTC, RCJ, RKC, RKJ - for placements from three type affixes permissible sequences; RKTC, RKCJ - for placements from four type affixes permissible sequences: 0. Thus, the quantity of permissible placements of the endings of participles is 11.

For the system of endings of verbs, verbal adverbs, moods, and voices are used the same combinatorial approach to determine various options for combining (placing) affixes taking into account the sequences of types, person suffixes, the harmonizing rules of the sequences sounds of the types. More detailed information on the combination of affixes types for a verb-based word is presented in the next section.

5 Inferring the Set of Kazakh Endings

5.1 Inferring of Endings for Nominal Base Words

For the semantical valid placements of nominal base words the combination method is taking into account:

- the type of stem (last character: vowel/consonant; vowel: solid/soft; consonant: deaf/voiced-sonorous);
- the kind of the last syllable (solid/soft) of each suffix for corresponded type;
- for the placement is used links (sequences) between the kind of features: solid-solid, soft-soft, deaf-solid, voice-soft.

Inferring of endings for placement KT (Plural – Possessive) is presented in Table 1. Combination of endings in placements KT equal to: K*T = (6 suffixes of K) *(5 suffixes of T) = 30 endings.

Table 1. Inferring of endings for placement KT (Plural – Possessive).

	Suffixes of K	Suffixes of T		Number of endings
		Singular	Plural	
Examples	dar- der- lar- ler- tar- ter-	ym, im	ymyz, imiz	6*5 = 30
		yń, iń	yńyz, ińiz	
		yńyz, ińiz	yńyz, ińiz	
		y, i	y, i	
ana-	-lar-	ym,yń,yńyz,y	ymyz	5
ini-	-ler-	im,iń,ińiz,i	imiz	5
at-	-tar-	ym,yń,yńyz,y	ymyz	5
it-	-ter-	im,iń,ińiz,i	imiz	5
ań-	-dar-	ym,yń,yńyz,y	ymyz	5
pán-	-der-	im,iń,ińiz,i	imiz	5

The example part is showing that the harmonic rules are working: for each type of stem is correspondent the same type of K suffix and for that kind of suffix is correspondent the same type of T suffix.

Due to the limited volume of the article, we will not be able to show the inference of endings for each type of placement of types of endings in a table.

Combinations of endings in placements KC equal to: K*C = (6 suffixes K) *(one suffix from each 6 cases) = 36 endings.

Combinations of endings in placements KJ equal to: K*J = (6 suffixes K) *(one suffix from each of 3 persons J) = 18 endings.

Combinations of endings in placements TC equal to: T*C = (24 suffixes T) *(one suffix from each 6 cases C) = 144 endings.

Examples: ana-m-nyń; ana-m-a; ana-m-dy; ana-m-da; ana-m-nan; ana-m-men.

Inferring of endings for placement TJ (Possessive-Personal) equal to: there are 7 available links T1-J2:8; T1-J3:8; T2-J1:8; T3-J1:8; T4-J1:8; T4-J2:8; T4-J3:8. Each TJ link consist from (4 suffixes T)*(2 suffixes J: singular, plural) = 8 endings. Therefore placement TJ consists T*J = (7 TJ links)*(8 endings) = 56 endings.

Inferring of endings for placement CJ (Case-Personal). There are 4 available cases: dative, locative, ablative, instrumental. Each available case consists from (possible suffixes C)*(3 suffixes of persons J)*(2 numbers: singular, plural). Possible suffixes for dative - 8, for locative - 6, for ablative - 6, for instrumental - 3. Therefore available endings of placement CJ is equal: 8*3*2 + 6*3*2 + 6*3*2 + 3*3*2 = 138 endings.

Inferring of endings for placement KTC (Plural – Possessive- Case) is presented in Table 2. Combinations of endings in placements KTC is following: (there are 6 suffixes K)*(4 persons of suffixes T) * (6 cases of C) = 6*4*6 = 144.

Table 2. Inferring of endings for placement type KTC (Plural – Possessive- Case).

Suffixes type K	Suffixes type T	Suffixes type C		Number of endings
- dar	- ym, im	1. nom.	-	6*4*6 = 144
- der	- yń, iń	2. gen.	-dyń, diń, nyń, niń, tyń, tiń;	
- lar	- yńyz, ińiz	3. dat.	-a, e, ǵa, ge, qa,qe, na, ne;	
- ler	- y, i	4. acc.	-n, na, ne, dy, di, ty,ti;	
- tar		5. loc.	-da, de, nda, nde, ta, te;	
- ter		6. abl.	-dan, den, nan, nen, tan, ten;	
		7.instr.	-ben, men, ten	

Combinations of endings in placement KTJ. There are 7 possible links T1-J2:6; T1-J3:6; T2-J1:6; T3-J1:6; T4-J1:6; T4-J2:6; T4-J3:8. Each TJ links consist from (one suffix T)*(6 suffixes J: singular, plural) = 6. T*J = (7 TJ links)*(6 endings) = 42 endings. Therefore numbers of endings of placement KTJ is (7 endings K)*(42 placements TJ) = 252 endings.

Inferring of endings for placement KCJ (Plural – Case- Personal). Number of endings in placement KCJ equal to: K*C = (6 suffixes K) *(4 cases C: dat, loc, abl, inst)*(3 singular + 3 plural of J) = 144 endings. Example: ana-lar-ǵa-myn.

Inferring of endings for placement TCJ (Possessive-Case-Personal). Number of endings in placements TCJ are: T*C*J = (16 suffixes of T) *(one suffix from each of 4 cases C: C3, C5, C6, C7)*(3 singular + 3 plural of J) = 384 endings. Example: ana-m-a-myn.

Inferring of endings for placement KTCJ (Plural – Possessive- Case-Personal). Number of endings in placements KTCJ equal to: (6 suffixes K)*(16 of suffixes T) * (6 cases of C) = 6*16*6 = 576 endings. The harmony rules of the Kazakh language are working in all cases depending on solidity/softness.

The full description of the inference of endings for nominal base words are uploaded to github repository (https://github.com/walsher46?tab=repositories).

5.2 Inferring of Endings for a Verbs Base Words

Words with a verb base include verbs, participles, voices, moods.

Inferring for verbs by tenses and forms are presented in tables below (Tables 3, 4, 5, 6, 7 and 8).

Transitive present tense (Auspaly osy śaq). Number of endings in Transitive present tense equal to: (3 suffixes of verb)*(4 personal suffixes)*2 (singular, plural) = 24 suffixes.

Past simple tense (Jedel ótken śaq). Number of endings in Past simple tense: (4 suffixes of verb: -ty, dy; -ti, di)*(4 personal suffixes)*2 (number: singular, plural) = 32 suffixes. Personal suffixes are the same as 'Personal' in 'Simple Present tense'.

Table 3. Simple Present tense

	1 person	2 person	2 person (polit)	3 person	Number of endings
otyr- jatyr- týr- júr-	-myn, min	-syń, siń	-syz, siz	–	12 (different)
	plural: myz, miz	plural: -syńdar, sińder	plural: -syzdar, sizder	plural: -	
				Total	12

Table 4. Complex Type Present tense

Examples	Suffixes	1 person	2 person	2 person (polit)	3 person	Number of endings
jaz- kúl-	-yp,ip,p + otyr,týr,júr-	-myn,min plural: -myz, miz	-syń, siń plural: -syńdar, sińder	-syz, siz plural: -syzdar, sizder	- plural: -	6 + 6 6
bar- kel-	-a,e + jatyr-					
qara-	-p + otyr,týr,júr-					
					Total	18

Table 5. Transitive present tense (Auspaly osy śaq)

Examples	Suffixes	1 person	2 person	2 person (polit)	3 person	Number of endings
jaz- kúl- bar- kel- qara-	-a,e,ı-	-myn,min plural: -myz, miz	-syń, siń plural: -syńdar, sińder	-syz, siz plural: -syzdar, sizder	-dy,di plural: -dy,di	3*4 = 12 3*4 = 12
					Total	24

Negative type (Bolymsyz túri). Number of endings in Past simple tense-Negative equal to: (6 suffixes of verb)*(4 personal suffixes)*2 (number: singular, plural) = 48 suffixes.

Previous past tense (býryngy ótken śaq). Number of endings in 'Previous past tense': (4 suffixes of verb: -ǵan, qan;-gen, ken)*(4 personal suffixes)*2 (number: singular, plural) = 32 endings. Personal suffixes are the same as Personal in Simple Present tense.

Ancient past tense (Ejelgi ótken śaq). Number of endings in 'Ancient past tense': (4 suffixes of verb)*(4 personal suffixes)*2 (number: singular, plural) = 32 endings.

Table 6. Negative type (Bolymsyz túri)

Examples	Suffixes	1 person	2 person	2 person (polit)	3 person	Number of endings
bar- jaz- aıt- kúl- kel- júz- ket-	-ma,ba,pa - dy - -me,be,pe - di -	-m; -m;	-ń; -ń;	-ńyz; -ńiz;	- -	12 12
	-ma,ba,pa - dy - -me,be,pe - di -	plural: -q; -k;	plural: -ńdar; -ńder;	plural: -ńyzdar; -ńizder;	plural: - -	12 12
					Total	48

Table 7. Ancient past tense (Ejelgi ótken śaq)

Examples	Suffixes	1 person	2 person	2 person (polit)	3 person	Number of endings
aıt- jaz- kúl- bar- kel- qara-	-yp, p - -ip, p -	-pyn; -pin;	-syń; -siń;	-syz; -siz;	-ty -ti	8 8
	-yp, p - -yp, p -	plural: -pyz; -piz;	plural: -syńdar; sińder;	plural: -syzdar; -sizder;	plural: -ty; -ti;	8 8
					Total	32

Table 8. Intent future tense (Maqsatty keler śaq)

Examples	Suffixes	1 person	2 person	2 person (polite)	3 person	Number of endings
aıt- jaz- kúl- bar- kel- qara- tóle-	-maq,baq,paq -mek,bek,pek	-pyn; -pin;	-syń; -siń;	-syz; -siz;	-śy -śi	12 12
	-maq,baq,paq -mek,bek,pek	plural: -pyz; -piz;	plural: -syńdar; -sińder;	plural: -syzdar; -sizder;	plural: -śy -śi	12 12
					Total	48

Negative type (Bolymsyz túri). Number of endings in Ancient past tense -Negative: (6 suffixes of verb: -ma, ba, pa - p; -me,be,pe - p)*(4 same personal suffixes)*2 (number: singular, plural) = 48 endings. For choosing suffixes for word stem and choosing suffixes of persons in Ancient past tense are working harmony rules of the Kazakh language.

Transitive past tense (Auyspaly ótken śaq). Number of endings in 'Transitive past tense' equal to: (4 suffixes of verb: *-atyn, ıtyn; -etin, ıtin*)*(4 personal suffixes same as Simple present)*2 (number: singular, plural) = 32 endings.

Negative type (Bolymsyz túri). Number of endings in 'Transitive past tense' in negative form are following: (6 suffixes of verb: *-ma, ba, pa – ıtyn; -me,be,pe - ıtin*)*(4 same personal suffixes)*2 (number: singular, plural) = 48 endings. Example: ait-pa-ıtyn-myn.

Assumed future tense (Boljaldy keler śaq). Number of endings in 'Assumed future tense' equal to: (4 suffixes of verb: *-ar, r;-er, r*)*(4 personal suffixes same as Simple present)*2 (number: singular, plural) = 32 endings. Example: ait-ar-myn.

Negative type (Bolymsyz túri). Number of endings in 'Assumed future tense' in negative form are following: (4 suffixes of verb: -ar, r + emes; -er, r + emes) + (6 same personal suffixes)*2 (number: singular, plural) = 16 endings. Example: ait-ar emes-pyn.

Intent future tense (Maqsatty keler śaq). Number of endings in 'Intent future tense' equal to: (6 suffixes of verb)*(4 personal suffixes)*2 (number: singular, plural) = 48 endings. For choosing suffixes for word stem and choosing suffixes of persons in 'Intent future tense' are working harmony rules of the Kazakh language. Example: ait-paq-pyn.

Number of endings in 'Transitive past tense' in negative form have: (6 suffixes of verb) + (6 personal suffixes)*2 (number: singular, plural) = 18 endings. Example: ait-paq emes-pyn.

Transitive future tense (Auyspaly keler śaq). Number of endings in Transitive future tense equal to: (4 suffixes of verb: -a, ı;-e, ı-)*(4 personal suffixes same as Transitive Present tense)*2 (number: singular, plural) = 32 endings. Example: ait-a-myn.

Negative type (Bolymsyz túri). Number of endings in Transitive future tense - Negative: (6 suffixes of verb: -ba, ma, pa- ı-; -be, me, pe- ı-)*(4 same personal suffixes)*2 (number: singular, plural) = 48 suffixes. Example: ait-pa-ı-myn.

Inferring for participles, verbal adverbs, derivative adverbs, moods, voices tables are uploaded to github repository (https://github.com/walsher46?tab=repositories).

The total number of complete system Kazakh endings are 4727.

6 The Complete System of Kazakh Endings as a Language Resource

The complete system of endings of the Kazakh language are used as a language resource in neural machine translation. The complete system of endings helps to solve the problem of morphological segmentation in the neural machine translation of the Kazakh language. According to the complete system of endings, the segmentation algorithm will divide the source word in the Kazakh language into the word's stem and the ending.

A complete list of the Kazakh endings were created, sorted by the length of the endings. The maximum length of the word endings is 13 characters. The algorithm based on the complete system of endings are divided into two parts. The first part divides a word to stem and ending. In the next part, the end is divided into the suffixes by separating with @@. The stem of the word must consist of at least two letters.

Table 9 illustrates the examples of endings with their internal segments. The pattern of endings can be from a single type of ending (like only K, only T, only C, only J) to a combination of four types of endings (KCTJ).

Table 9. Example of ending types in the table of Kazakh words' endings with segmentation on suffixes

Type of ending	Complete ending	Internal segmentation of ending
K-T-C-J	дарымызбенмін	дар@@ ы@@ мыз@@ бен@@ мін
K-T-C	ғандарымыздан	ған@@ дар@@ ы@@ мыз@@ дан
J-K	ңыздар	ңыз@@ дар
K-T	тарың	тар@@ ың
T-C	мнен	м@@ нен
C	ның	ның

The Kazakh language ending system is useful in problems of morphological analysis, for the task of segmentation, to determine the basis of the word. In neural machine translation is one of the actual tasks, such as reducing the volume of the dictionary. A large dictionary can lead to a memory error since adding all word forms to the dictionary increases a required volume of memory. Initially, the dictionary was created according to the most frequent words of the corpus, which included frequency word forms. But not all words of the corpus can be covered with a dictionary of frequent word forms, therefore the segmentation of words is sufficient for this problem.

By the use of the complete system of Kazakh endings, the volume of the dictionary in the neural machine translation system was reduced twice.

In the experimental part for testing of created language resource of the Kazakh endings, 100 sentences of the Kazakh language (number of words - 1693) were taken and a check was made for the incorrect segmentation of the words. The result was 7% segmentation errors from the text. These errors are connected with the ambiguity of words and with named-entity words wrong recognizing.

7 Conclusion and Future Work

The paper presents the language resource of the Kazakh language endings divided on suffixes, important for solving the problem of segmentation of source Kazakh text for neural machine translation learning. For the creation of the set of Kazakh endings was used a combinatorial approach for inferring all endings for words with nominal bases and with verb bases. This language resource of Kazakh endings plans to register in ELRA Catalogue of Language Resources in order that, in the future, other researchers could use it for the segmentation of the Kazakh language in a machine translation and other NLP applications. In the future, we plan to use the described in this paper approach for other Turkic languages and create endings' language resources for these languages.

Acknowledgements. This work was carried out under grant No. AP05131415 "Development and research of the neural machine translation system of Kazakh language", funded by the Ministry of Education and Science of the Republic of Kazakhstan for 2018-2020.

References

1. Sennrich, R., Haddow, B., Birch, A.: Neural machine translation of rare words with subword units. In: Proceedings of the 54th Annual Meeting of the Association for Computational Linguistics, vol. 1, pp. 1715–1725 (2016)
2. Tukeyev, U.: Automaton models of the morphology analysis and the completeness of the endings of the Kazakh language. In: Proceedings of the International Conference "Turkic Languages Processing" TURKLANG 2015, Kazan, Tatarstan, Russia, 17–19 September, pp. 91–100 (2015)
3. Tacorda, A.J., Ignacio, M.J., Oco, N., Roxas, R.E.: Controlling byte pair encoding for neural machine translation. In: 2017 International Conference on Asian Language Processing, pp. 168–171 (2017)
4. Wu, Y., Zhao, H.: Finding better subword segmentation for neural machine translation. In: Sun, M., Liu, T., Wang, X., Liu, Z., Liu, Y. (eds.) CCL/NLP-NABD -2018. LNCS (LNAI), vol. 11221, pp. 53–64. Springer, Cham (2018). https://doi.org/10.1007/978-3-030-01716-3_5
5. Ataman, D., Negri, M., Turchi, M., Federico, M.: Linguistically motivated vocabulary reduction for neural machine translation from Turkish to English. Prague Bull. Math. Linguist. **108**(1), 331–342 (2017)
6. Creutz, M., Lagus, K.: Unsupervised discovery of morphemes. In: Proceedings of the ACL 2002 Workshop on Morphological and Phonological Learning, vol. 6, pp. 21–30 (2002)
7. Koskenniemi, K.: Two-level morphology: a general computational model for word-form recognition and production. Ph.D. thesis, University of Helsinki (1983)
8. Oflazer, K.: two-level description of Turkish morphology. Literary Linguist. Comput. **9**(2), 137–148 (1994)
9. Beesley, K.R., Karttunen, L.: Finite-State Morphology. CSLI Publications, Stanford University (2003)
10. Kairakbay, B.: A nominal paradigm of the Kazakh language. In: 11th International Conference on Finite State Methods and Natural Language Processing, pp. 108–112 (2013)
11. Kessikbayeva, G., Cicekli, I.: Rule based morphological analyzer of Kazakh language. In: Proceedings of the 2014 Joint Meeting of SIGMORPHON and SIGFSM, Baltimore, Maryland USA, pp. 46–54 (2014)

A Multi-filter BiLSTM-CNN Architecture for Vietnamese Sentiment Analysis

Lac Si Le[1,2], Dang Van Thin[1,2](✉), Ngan Luu-Thuy Nguyen[1,2], and Son Quoc Trinh[1,2]

[1] University of Information Technology, Ho Chi Minh City, Vietnam
17520669@gm.uit.edu.vn, {thindv,ngannlt,sonqt}@uit.edu.vn
[2] Vietnam National University, Ho Chi Minh City, Vietnam

Abstract. Feedback is information about reactions to a product or a person's performance of a task. It is powerful as it serves as a guide to assist people to know how others perceive their performance and helps them meet standards. This paper concentrates on the use of natural language processing and deep learning. It combines the advantages of these approaches to perform sentiment analysis on student and customer feedback. Furthermore, word embedding is also applied to the model to add complementary effectiveness. The preliminary findings show that the use of BiLSTM-CNN–the first to catch the temporary information of the data and the second to extract the local structure thereof–outperformed other algorithms in terms of the F1-score measurement, with 93.55% for the **Vietnamese Student's Feedback Corpus (VSFC)** and 84.14% for the **Vietnamese Sentiment (VS)**. The results demonstrate that our method is an improvement compared to the best previously proposed methods on the two datasets.

Keywords: Sentiment Analysis · Deep learning · Bidirectional long short-term memory · Convolution neural network · Vietnamese language

1 Introduction

Sentiment Analysis (SA) is the process of identifying and classifying text into different sentiments–for example, positive, negative, or neutral–or emotion sentiments–such as happy, sad, angry, or disgusted–to determine the human's attitude toward a particular subject or entity [18].

Feedback from the learner is important for educators and other stakeholders to understand whether or not the educational program is effective and the school is living up to its mission and vision. Feedback assists with: (1) self-assessment and (2) the professionalization of teaching in higher education. The collection of learner feedback identifies the aspects that cause the student's opinions (positive, negative, neutral). In order to understand the exact sentiment of the students, a textual feedback technique [1] is used. In this textual form, the learner is given

M. Hernes et al. (Eds.): ICCCI 2020, CCIS 1287, pp. 752–763, 2020.
https://doi.org/10.1007/978-3-030-63119-2_61

a set of questions, and they need to answer it in sentences. This helps both the academic administration and the instructors to overcome the issues related to their organization. In this paper, the student feedback comprising varied opinions is collected using forms.

Feedback from customers is information provided by customers on how happy or unhappy they are with a specific product or service and on their general experience with a brand. It gives businesses an opportunity to change their actions and improve customer experience accordingly. This data can be gathered through different types of surveys (e.g. prompted feedback) but also with the aid of web tracking tools, view counts, and customer comments posted online (unprompted feedback). These sources are critical to gaining a full picture of how clients perceive brands.

Through experimentation and reviews of the results, we attempt to improve the accuracy of sentiment analysis, which typically uses a combination of Convolutional neural network (CNN) and Long short-term memory (LSTM) approaches. We adopt an improvement approach that uses Bidirectional long short-term memory (BiLSTM) rather than LSTM. This is because unidirectional LSTM information flows backward to forward. Conversely, BiLSTM information not only flows backward to forward but also forward to backward using two hidden states. Hence, BiLSTM understands context better.

To make our comparison with the traditional approaches, we used the same domain data, and selected the most effective and productive model. The results show that our system is a considerable improvement in comparison to the best previous systems, according to the F1-Measure (the weighted average of precision and recall), recall, and precision of the five sets of experiments. The main contributions of this paper are:

- For Vietnamese sentiment analysis, we proposed a BiLSTM-CNN-Multi (using multiple filters) model on two different large-volume domains. The strengths of BiLSTM and CNN were combined into one using sequence embedding.
- We re-implemented the basic deep learning models and the best of the previous work for each dataset to assess our proposed model's performance more visually and practically. Our model outperforms the individual models–CNN and LSTM–and also the combined models–Multi-channel LSTM-CNN, CNN-LSTM, and LSTM-CNN-Multi.
- We analyzed and evaluated the impact of the two datasets on our model's performance and figured out the error analysis of our model's outcomes.

The rest of this paper is structured as follows. Section 2 summarizes the literature review on sentiment analysis. In Sect. 3, we provide a detailed description of the ensemble model, the architecture for the LSTM, BiLSTM, and CNN models used in our work and the word-embedding combination. The evaluation procedure and results are presented in Sects. 4 and 5. In Sect. 6, we discuss error analysis. Finally, Sect. 7 concludes the research work and describes the potential development route to improve our proposed model.

2 Related Work

Frequently, the previous approaches used feature extraction techniques combined with supervised algorithms [5,9,12,22] to perform sentiment tasks. Supervised methods are based on classification of the training (such as Support Vector Machine, Random Forest, and Naive Bayes) utilizing various combinations of features including word N-grams, Part-Of-Speech (POS) tags, and text context information features, such as capital words, hashtags, emoticons, etc. Such approaches are focused on the presence of lexical or syntactical characteristics that convey the details about sentiment directly. In certain instances, however, the sentiment of a text is directly related to the meaning of its context.

Deep learning approaches are now well developed, and they have become especially effective in image and voice processing tasks. Such methods provide automatic feature extraction, more abundant representation capabilities, and better performance than traditional feature-based techniques [2].

Commonly, for high-resources language, sentiment classification is benchmarked on the movie review and Twitter domain. Various methods have been developed, [17] performed sentiment analysis of movie reviews. Their results indicated that machine learning techniques are better than simple methods using human-generated baselines (i.e. there are certain words that people tend to use to express strong emotions, so it can be enough simply to list these terms introspectively and rely solely on them to identify the texts). They received approximately 83% in term of the F1-score. [10] applied a dynamic CNN for Twitter and movie reviews sentiment classification. They proposed a CNN architecture, unigram dense, low-dimensional word vectors (initialized to random values), and positing latent as inputs. Their experiments demonstrated that dynamic CNNs perform better than those depended on bigram and unigram models. [6] discovered a new deep CNN that takes advantage of character-level to sentence-level information to implement sentiment analysis in short texts; the method achieved a prediction accuracy of 86.4% on the Stanford Twitter Sentiment corpus. [8] presented a system that assigns scores indicating positive or negative opinion to a distinct entity in the text corpus (news and blogs).

There have been numerous works that apply supervised methods to Vietnamese sentiment analysis in recent years. [25] introduced a lexicon-based method for sentiment analysis using Facebook data for the Vietnamese language by focusing on two core components in a sentiment system. They built a dictionary of Vietnamese emotions (VED) including five sub-dictionaries–noun, verb, adjective, and adverb–and proposed features based on the English emotional analysis method adapted to traditional Vietnamese language. It was then used to classify the user's emotional message using the support vector machine classification method. [7] proposed the Maximum Entropy classifier in-depth learning approach for the education field. In contrast, [19] proposed a classification result of 30,000 Vietnamese sentences (including 15,000 positive Vietnamese documents and 15,000 negative Vietnamese documents extracted from millions of Vietnamese websites, Vietnamese Facebook, and social networks) by using the Valence-Totaling Model for Vietnamese (VTMfV) and the Vietnamese Sentiment

Dictionary (VSD_JM). They achieved an accuracy of 63.9% for their Vietnamese testing data set. Although their model's accuracy was not high, their model was still a new contribution to Vietnamese sentiment classification and sentiment classification of other languages.

On the other hand, there are few works that use the deep learning approach. [28] employed a LSTM and CNN to produce information channels for Vietnamese sentiment analysis, from reviews or comments of commercial web pages in Vietnam. This strategy catches global as well as local dependencies in a sentence. Their model obtained better performance than a Support vector machine (SVM), CNN, and LSTM on Vietnamese datasets. Very recently, a study of student feedback with supervised learning algorithms was performed by [16]. Particularly, four algorithms are employed in their analysis, including Maximum Entropy, Naive Bayes, BiLSTM, and LSTM. According to comparison of results, they concluded that Bi-LSTM surpasses the others in both tasks (sentiment analysis and topic classification). These findings are particularly important for resource-limited languages, where the amounts of data typically necessary for deep learning are non-existent.

As previously explained, there's not a lot of research focusing on using deep learning technique for low-resource domains yet. This is because such research involves extensive training data, capable architecture, and the need for the domain-specific data used for deep learning models, which is lacking. After examining different models of deep learning, we propose the BiLSTM-CNN-Multi model in the analysis of sentiment, which shows an improved results than the previous state-of-art approaches.

3 Methodology

3.1 Data Pre-processing

Reviews are usually composed of incomplete expressions, a lot of noise, and poorly-structured phrases because of the frequent presence of acronym, irregular grammar, ill-formed words, and non-dictionary terms. Unstructured data and noise affects the performance of sentiment classification [24]. Thus, a typical text classification framework with pre-processing is one of the critical components. This is a step towards cleaning up and preparing classification information. It includes four follow-up steps:

- **Step 1:** Delete special characters and punctuation. This is appropriate because Word2vec was designed to model primarily word-to-word similarity in word analogy tests. In this case, these are not useful.
- **Step 2:** Remove numbers. Generally, the numbers carries no sentiment information, so they are useless when measuring sentiment and are removed from the text to optimize the material.
- **Step 3:** Remove most frequent words (TF-High) and remove words that occur once, i.e., singleton words (TF1) based on [21] combined with Vietnamese stop-words[1].

[1] Vietnamese-stopwords: https://github.com/stopwords/vietnamese-stopwords.

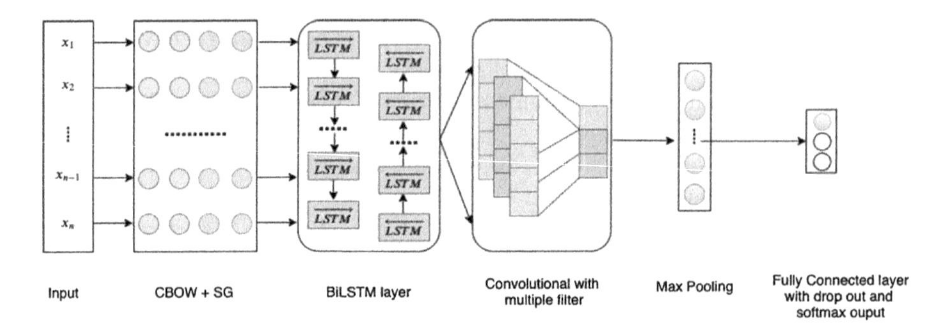

Fig. 1. The BiLSTM-CNN architecture for sentiment analysis classification

- **Step 4:** The ad-hoc tokenizer (pre-trained tokenizer) then partitions each string of characters into tokens[2] made up of (lower-cased) letters separated by spaces.

3.2 BiLSTM-CNN Architecture

Every model has its own approach of catching the information target into feature vectors as an input. The CNN applies convolutional filters in particular to catch local dependencies between neighboring words. However, the filter size constraint makes it difficult for the CNN system to capture all of the phrase/document dependences. We could take into account a CNN-constructed representation vector as concatenating local value. At the same time, a memory cell for storing data for a long time is implemented in the BiLSTM model. This is because information not only flows backward to forward but also forward to backward using two hidden states. Consequently, a BiLSTM designed function vector carries the overall relationships of an entire sentence or document. For these two vectors, we suppose, one with a local dependencies and one with a full dependencies, to provide mutual support to boost classification efficiency.

We propose an architectonic model closely based on [3]. Our model architecture is shown in Fig. 1. Given an input, we use pre-trained embeddings from Word2Vec [20]. This embedding representation is then fed to the BiLSTM to generate feature vectors. To generate feature maps, the feature vectors from BiLSTM are fed into the CNN layer. 1-max pooling operator is next used to obtain the highest value on each feature map. Afterward, such values are combined in a CNN function to synthesize a high-level feature in the neural layer of the network. Finally, the output layer of the model includes a fully connected layer with sigmoid activation. The dimensionality of the fully-connected layer is determined by the number of categories in the classification task. The first-order gradient optimization algorithm Rectified Adam (RAdam) [13][3] is used during the training phase to determine the template parameters. RAdam is a new variation of

[2] PyVi: https://pypi.org/project/pyvi/.
[3] RAdam: https://github.com/LiyuanLucasLiu/RAdam.

the classic Adam optimizer that provides an automated, dynamic adjustment to the adaptive learning rate, based on the researchers' detailed study into the effects of variance and momentum during training.

3.3 Pre-trained Word Embedding

Recent analyses have demonstrated the use of embedded words can dramatically boost model efficiency in numerous natural language processing (NLP) tasks and speech processing [4,14,23,29]. Diverse approaches have been introduced to estimate word embeddings through neural networks.

There are two types of word embeddings that are easy to implement and that effectively use Word2Vec[4], including skip-gram (SG) and Continuous Bag-of-Words models (CBOW) [15].

In this study, all the word embeddings are calculated using the same training data. In addition to these word embeddings evaluations, combination through concatenation is applied to find an effective embedding that can achieve better performance for our proposed models.

4 Experiments and Results

4.1 Dataset

In this section, we provide a quick overview of the two datasets used in our work, VSFC [26] and VS [28].

Table 1. Statistic summary of two datasets

Dataset	Quantity of sentences			Distribution of sentiment label (%)		
	Training	Validating	Testing	Positive	Neutral	Negative
VS	12,000	1,780	3,720	34.21	33.54	32.25
VSFC	11,426	1,538	3,166	49.62	4.4	45.98

VSFC: Student feedback includes a description of actions, words, gestures, and attitudes towards teaching activities, which is either negative or positive or neutral. They present their perspective on subjects, lessons, homework, assignments, syllabuses, etc. The summary of dataset distribution is provided in Table 1.

VS: This contains 17,500 comments and reviews from e-commercial pages in Vietnam (e.g. TinhTe.vn, Tiki.vn, Lazada.vn, etc.) on different items, such as computers, books, mobile, food, manually labelled by three annotator(s) on the basis of positive/ negative/ neutral. The statistical summary of the dataset is presented in Table 1.

[4] Gensim: https://radimrehurek.com/gensim/models/word2ve.

4.2 Hyper-parameters

We keep 1510 learning examples from the VSFC dataset as validation data to determine the hyper-parameters. Generally, we used domain embedding from word2vec for our models with a size of vocabulary and dimension of 10000 and 200, respectively, trained with 20 epochs and a batch-size of 40 for each model. The model-specific details are as follows.

- **CNN:** The number of filters using in the CNN layer was 128 with kernel sizes k = 2. The dropout rate was 0.55, and the learning rate of RAdam optimizer [11] was 0.0001 because CNN training tends to be unstable.
- **CNN-LSTM:** Setting of CNN layer was the same as CNN implementation above; in addition, the LSTM layer was implemented with 300 hidden units.
- **LSTM-CNN-Sing:** We did a reverse of the CNN and LSTM layer on this model with implementation details the same as CNN-LSTM.
- **LSTM-CNN-Multi:** Instead of using a single filter in CNN layer, we added multiple filters to use on the CNN layer, including 128, 256, and 512 filters with kernel sizes, sequentially, of 7, 5, and 2.
- **BiLSTM-CNN-Sing:** After our experimentation, we proposed another approach using BiLSTM rather than LSTM. BiLSTM was implemented with 300 hidden units and a returning sequence. The dropout and recurrent dropout rate are 0.25 and 0.1, respectively.
- **Multi-channel LSTM-CNN:** We re-implemented based on [28]. By comparing its performance with our proposed models, we can gain a deep understanding, which is significant when we compared with other models on a different domain.
- **BiLSTM-CNN-Multi:** Following the idea of using multi-filters in the CNN layer, we did the same implementation based on the LSTM-CNN-Multi.
- **LSTM and BiLSTM:** We used the DeepLearning4j[5] framework for the implementation of the LSTM and BiLSTM algorithms. The framework is a library written in the language of Java Programming.

For the VS dataset, to fine-tune our model's hyper-parameters, we scanned a grid for 30%. Next, we created our corpus-specific word embeddings through the efficient tool Word2Vec[6] rather than Quan-Hoang Vo's–the author of the Multi-channel LSTM-CNN–approach of using the embedding layer. On this main point, Word2Vec is currently among the most accurate and usable word embedding methods that can convert words into meaningful vectors. Notably, pre-trained word embedding has a significant impact on research on sentimental analyses [27].

4.3 Model Performance and Comparison

Table 2 reports the results of sentiment prediction on the VSFC dataset. Our model's performances on the VS dataset are described in Table 3. Generally, we

[5] DeepLearning4j: https://deeplearning4j.org.
[6] Gensim: https://radimrehurek.com/gensim/.

Table 2. Results of our architecture compared to other methods on the VSFC dataset. The symbol * indicates the result is statistically significant at the level of 0.05 and the scores are the average of 5 runs. The best performance is marked in bold.

	Precision	Recall	F1-Score
LSTM*	88.40%	87.10%	87.60%
MaxEnt [26]	–	–	87.00%
BiLSTM*	90.80%	93.40%	92.00%
CNN*	91.88%	92.07%	91.97%
CNN-LSTM*	91.60%	91.62%	91.61%
LSTM-CNN-Sing*	93.21%	93.05%	93.13%
LSTM-CNN-Multi*	93.25%	93.10%	93.17%
BiLSTM-CNN-Sing*	93.21%	93.40%	93.30%
BiLSTM-CNN-Multi*	**93.51%**	**93.61%**	**93.55%**
Multi-channel LSTM-CNN*	89.30%	90.00%	89.64%

can observe that the BiLSTM-CNN-Multi model is in all cases superior to other approaches, with 93.55% and 84.14% of the F1-score, respectively, on the VSFC and VS datasets. From the average, our proposed model achieves a maximum improvement in F1-Score of 8.55% and a minimum improvement of 2.36% over the other methods.

It is notable that the VS dataset includes many grammatical errors and is pretty informal. These directly impact the performance of word segmentation and the learning of sentiment in the VS dataset. The VSFC dataset, however, is much more structured because of it comprises shorter texts and fewer grammatical errors.

Our model provides better performance in terms of empirical results. It demonstrates the efficiency of our proposed BiLSTM and CNN with multiple filter approach.

5 Result

Here we present the Precision, Recall, and F1-score of our model on two Vietnamese datasets: VSFC and VS. Table 2 shows the results of our architecture compared with relevant methods on the VSFC dataset, while Table 3 shows the results on the VS dataset. In [26], the Maximum Entropy algorithm is proposed using the Standford Classifier[7] on the VSFC dataset, while [28] proposed a novel model that combine the effectiveness of the LSTM and CNN on the VS dataset.

From Table 2, it is clear that our variant approaches, BiLSTM-CNNN-Sing and BiLSTM-CNN-Multi, achieve results of 93.30% and 93.55% of the F1-score, respectively, when using BiLSTM rather than LSTM; this indicates a significant

[7] The Stanford Classifier: https://nlp.stanford.edu/software/classifier.shtml.

improvement in performance. Our results demonstrated that BiLSTM-CNN-Multi achieves better performance on average in terms of F1-Measure, recall, and precision than the other models. Notably, the BiLSTM-CNN-Multi took significantly longer in the training process (total params: 6,658,275) compared to the BiLSTM-CNN-Sing (total params: 2,687,927). Based on our experiments with the VSFC dataset, we now firmly contend our proposed models will perform similarly well on different domains. Our experiment results on the VS dataset will be coming after in Table 3.

The main conclusion that can be drawn is that the ensemble technique performs better than traditional approaches. This is because ensembling combines the strengths of each approach, and additionally, the concatenating in word-embedding influences the performance of the sentiment task.

Table 3. Results of our architecture compared to other methods on the VS dataset. The symbol * indicates the result is statistically significant at the level of 0.05 and the scores are the average of 5 runs. The best performance is marked in bold.

	Precision	Recall	F1-Score
CNN*	79.30%	78.00%	73.00%
CNN-LSTM*	66.30%	65.20%	65.30%
Multi-channel LSTM-CNN [28]	80.30%	79.67%	79.67%
LSTM-CNN-Sing*	77.18%	76.11%	76.50%
LSTM-CNN-Multi*	80.17%	76.15%	78.10%
BiLSTM-CNN-Sing*	77.10%	76.00%	76.45%
BiLSTM-CNN-Multi*	**84.19%**	**84.09%**	**84.14%**

6 Discussion

Sentiment analysis, generally, has problems recognizing phenomena like sarcasm and irony, negations, jokes, and exaggerations–the kind of aspects that would also be difficult for a person to identify from written text. This can skew the results. Notably, like all opinions, the sentiment is inherently subjective from person to person, and can even be outright irrational. This is critical to remember when mining a large and relevant sample of data to attempt to measure sentiment.

We manually inspected some of the cases where incorrect predictions were shown to determine the weakness of our proposed model. The following analyses are typical examples of this.

The first example was in the VFSC dataset: "*thuc su mjh lo dung roi chu vs lai minh o xa ko bjet gjo bao hanh hay sua o dau. gj ma mua va dug dc hon 2tuan dun bep co hjen tuong nứt những đầu đốt và càng ngày càng to ra. và đun cũng lâu nữa. cho hỏi giờ mang đi bảo hành ở đâu đây????*" (Honestly, it happened that I have used it already, and I live pretty far from the city center so I don't know where I can have it repaired. Even worse, it'd only been 2 weeks when some cracks appeared on the stove and grew in number. And it also took longer to boil water. So, please tell me where I can get this fixed). Our model predicted this as neutral; in fact, it was labeled as negative. Firstly, without diacritical marks, the word "dung" could be understood as "đúng" (right) or "dùng" (using). According to the reviewer, they intended to use this word as "dùng". Secondly, nearly half of the sentence is written in Vietnamese teen code[8]. This provides our model with a challenge. To improve this aspect, we suggest some techniques for text pre-processing (like adding accents) and syntax data should be used to avoid such ambiguity and misinterpretation.

A second example was in the VS dataset: "*Bộ 2 nồi thủy tinh VS-346-PVG rất chắc chắn nấu nhanh tuy nhiên do 2 quai nhỏ nên thao tác rữa nồi không thoải mái nhìn chung rất hài lòng về sản phẩm này*" (Set of 2 VS-346-PVG glass pot is good for fast cooking, but the washing of the pot is uncomfortable due to the two small straps. It is generally very satisfied with this product). Our model's predict label was neutral; in fact, the correct label was positive. The review contains both negative sentiment "không thoải mái" (uncomfortable) and positive sentiment "rất chắc chắn nấu nhanh" (sturdy glass pots and cooks fast). As we have observed, the model tends to attribute these reviews with neutral labels. It is difficult for our model to determine the polarity of such sentiments.

7 Conclusions and Future Work

We conducted a sentiment analysis of two datasets using ensemble learning algorithms in this research. After comparing the results, the main conclusion that can be drawn is that the BiLSTM-CNN using multiple filter model–the first to catch the temporary information of the data and the second to extract the local structure–outperforms other models for effectiveness in sentiment analysis tasks. It achieved 93.55% on **VSFC** and 84.14% on **VS** in terms of the F1-scores.

In future, we plan to improve the results of ambiguous instances containing both positive sentiment and negative sentiment as well as the algorithms by applying other ones or embedding feature extraction and the deep learning model. Further, we intend to implement imbalanced techniques of data-processing to boost performance to enhance the outcomes. Additionally, word-embedding in large volumes is expected to give us a significant improvement in the transformation of our model into another domain. Generally, we contend that this sequence ensemble model can be applied to advance prediction accuracy in several deep learning-based text processing tasks.

Acknowledgments. Our thanks especially go to the authors of **VSFC** and **VS** for providing Vietnamese datasets; these provide us with invaluable data for experiments

[8] A term for Vietnamese abbreviations of young people in Vietnam. Even each young group has its own creative way to differentiate itself from the rest.

with our models. This research is funded by the University of Information Technology - Vietnam National University Ho Chi Minh City under grant number B2019-26-01.

References

1. Agrawal, P.K., Alvi, A.S.: Textual feedback analysis. In: 2015 International Conference on Computing Communication Control and Automation, pp. 457–460. IEEE (2015)
2. Araque, O., Corcuera-Platas, I., Sanchez-Rada, J.F., Iglesias, C.A.: Enhancing deep learning sentiment analysis with ensemble techniques in social applications. Expert Syst. Appl. **77**, 236–246 (2017)
3. Chen, N., Wang, P.: Advanced combined LSTM-CNN model for twitter sentiment analysis. In: 2018 5th IEEE International Conference on Cloud Computing and Intelligence Systems (CCIS), pp. 684–687. IEEE (2018)
4. Collobert, R., Weston, J., Bottou, L., Karlen, M., Kavukcuoglu, K., Kuksa, P.: Natural language processing (almost) from scratch. J. Mach. Learn. Res. **12**(Aug), 2493–2537 (2011)
5. Da Silva, N.F., Hruschka, E.R., Hruschka Jr., E.R.: Tweet sentiment analysis with classifier ensembles. Decis. Support Syst. **66**, 170–179 (2014)
6. Dos Santos, C., Gatti, M.: Deep convolutional neural networks for sentiment analysis of short texts. In: Proceedings of COLING 2014, the 25th International Conference on Computational Linguistics: Technical Papers, pp. 69–78 (2014)
7. Duyen, N.T., Bach, N.X., Phuong, T.M.: An empirical study on sentiment analysis for Vietnamese. In: 2014 International Conference on Advanced Technologies for Communications (ATC 2014), pp. 309–314. IEEE (2014)
8. Godbole, N., Srinivasaiah, M., Skiena, S.: Large-scale sentiment analysis for news and blogs. ICWSM **7**(21), 219–222 (2007)
9. Hagen, M., Potthast, M., Büchner, M., Stein, B.: Twitter sentiment detection via ensemble classification using averaged confidence scores. In: Hanbury, A., Kazai, G., Rauber, A., Fuhr, N. (eds.) ECIR 2015. LNCS, vol. 9022, pp. 741–754. Springer, Cham (2015). https://doi.org/10.1007/978-3-319-16354-3_81
10. Kalchbrenner, N., Grefenstette, E., Blunsom, P.: A convolutional neural network for modelling sentences. arXiv preprint arXiv:1404.2188 (2014)
11. Kingma, D.P., Ba, J.: Adam: a method for stochastic optimization. arXiv preprint arXiv:1412.6980 (2014)
12. Kiritchenko, S., Zhu, X., Mohammad, S.M.: Sentiment analysis of short informal texts. J. Artif. Intell. Res. **50**, 723–762 (2014)
13. Liu, L., et al.: On the variance of the adaptive learning rate and beyond. arXiv preprint arXiv:1908.03265 (2019)
14. Luong, M.T., Socher, R., Manning, C.D.: Better word representations with recursive neural networks for morphology. In: Proceedings of the Seventeenth Conference on Computational Natural Language Learning, pp. 104–113 (2013)
15. Mikolov, T., Chen, K., Corrado, G., Dean, J.: Efficient estimation of word representations in vector space. arXiv preprint arXiv:1301.3781 (2013)
16. Nguyen, P.X., Hong, T.T., Van Nguyen, K., Nguyen, N.L.T.: Deep learning versus traditional classifiers on Vietnamese students' feedback corpus. In: 2018 5th NAFOSTED Conference on Information and Computer Science (NICS), pp. 75–80. IEEE (2018)

17. Pang, B., Lee, L., Vaithyanathan, S.: Thumbs up? Sentiment classification using machine learning techniques. In: Proceedings of the ACL-02 Conference on Empirical Methods in Natural Language Processing, vol. 10, pp. 79–86. Association for Computational Linguistics (2002)
18. Pang, B., Lee, L., et al.: Opinion mining and sentiment analysis. Found. Trends® Inf. Retrieval **2**(1–2), 1–135 (2008)
19. Phu, V.N., Chau, V.T.N., Tran, V.T.N., Duy, D.N., Duy, K.L.D.: A valence-totaling model for Vietnamese sentiment classification. Evolving Syst. **10**(3), 453–499 (2019)
20. Rong, X.: word2vec parameter learning explained. arXiv preprint arXiv:1411.2738 (2014)
21. Saif, H., Fernandez, M., He, Y., Alani, H.: On stopwords, filtering and data sparsity for sentiment analysis of twitter. In: Proceedings of the Ninth International Conference on Language Resources and Evaluation (LREC 2014), pp. 810–817 (2014)
22. Saif, H., He, Y., Alani, H.: Semantic sentiment analysis of Twitter. In: Cudré-Mauroux, P., et al. (eds.) ISWC 2012. LNCS, vol. 7649, pp. 508–524. Springer, Heidelberg (2012). https://doi.org/10.1007/978-3-642-35176-1_32
23. Socher, R., et al.: Recursive deep models for semantic compositionality over a sentiment treebank. In: Proceedings of the 2013 Conference on Empirical Methods in Natural Language Processing, pp. 1631–1642 (2013)
24. Sulea, O.M., Zampieri, M., Malmasi, S., Vela, M., Dinu, L.P., Van Genabith, J.: Exploring the use of text classification in the legal domain. arXiv preprint arXiv:1710.09306 (2017)
25. Trinh, S., Nguyen, L., Vo, M., Do, P.: Lexicon-based sentiment analysis of Facebook comments in Vietnamese language. In: Król, D., Madeyski, L., Nguyen, N.T. (eds.) Recent Developments in Intelligent Information and Database Systems. SCI, vol. 642, pp. 263–276. Springer, Cham (2016). https://doi.org/10.1007/978-3-319-31277-4_23
26. Van Nguyen, K., Nguyen, V.D., Nguyen, P.X., Truong, T.T., Nguyen, N.L.T.: UIT-VSFC: Vietnamese students' feedback corpus for sentiment analysis. In: 2018 10th International Conference on Knowledge and Systems Engineering (KSE), pp. 19–24. IEEE (2018)
27. Vo, N., Hays, J.: Generalization in metric learning: should the embedding layer be embedding layer? In: 2019 IEEE Winter Conference on Applications of Computer Vision (WACV), pp. 589–598. IEEE (2019)
28. Vo, Q.H., Nguyen, H.T., Le, B., Nguyen, M.L.: Multi-channel LSTM-CNN model for Vietnamese sentiment analysis. In: 2017 9th International Conference on Knowledge and Systems Engineering (KSE), pp. 24–29. IEEE (2017)
29. Zheng, X., Chen, H., Xu, T.: Deep learning for Chinese word segmentation and POS tagging. In: Proceedings of the 2013 Conference on Empirical Methods in Natural Language Processing, pp. 647–657 (2013)

Computational Collective Intelligence and Natural Language Processing

Causality in Probabilistic Fuzzy Logic and Alternative Causes as Fuzzy Duals

Serge Robert[1(✉)], Usef Faghihi[2(✉)], Youssef Barkaoui[2(✉)], and Nadia Ghazzali[2(✉)]

[1] University of Québec at Montréal, Case postale 8888, succursale Centre-ville, Montréal, Québec, Canada
robert.serge@uqam.ca

[2] University of Québec at Trois-Rivières, 3351, boul. des Forges, Trois-Rivières, Québec, Canada
{usef.faghihi,youssef.barkaoui,nadia.ghazzali}@uqtr.ca

Abstract. Causal inference is an essential part of human reasoning. Computer scientists often use inferential logic and probability theories to find the causes of the events. In causal inference, researchers are interested in finding the relationship between two observable events. However, most of the time, we can only perceive the co-occurrences of events or usual successions of events. So, we must be careful in trying to establish causality between events. Also, many alternative causes or conjoint causes can be at work. So, we need good tools to deal with the complex problem of causality. In this paper, we will explore probabilistic fuzzy logic (PFL) as a sophisticated alternative to establish causal inference and fuzzy duals as tools for the expression of alternative possible causes.

Keywords: Probabilistic fuzzy logic · Reasoning · Artificial intelligence · Causality and causal inference

1 Introduction

Faghihi [2–4] and Pearl [1] suggest that machines will only be intelligent once they can reason. Causal reasoning is the use of knowledge to explain what is already observed in order to predict the future. Pearl [1], for example, suggests the use of inferential logic, with 3 levels of causal hierarchy: 1) Association, to identify interrelated phenomena; example: what if we do X?, 2) Intervention, to predict the consequence(s) of an action; example: How does the duration of my planned life change if I became a vegetarian? 3) Counterfactual [5–8], to reason about hypothetical situations and possible outcomes; example: Would my grandfather still be alive if he had not smoked? To introduce counterfactual capability in computers, Pearl [1] uses causal inference which represents causal rules using causal diagrams. We will explain this using the Smoke-Tar-Cancer (Fig. 1) example from [1]. Suppose we would like to study the effects of Smoking on Cancer. In Fig. 1, the direct causes of Cancer are shown as Tar and Smoking Gene.

At this point, we have no way to observe whether a Smoking Gene does exist. Thus, Smoking is influencing Cancer through Tar and there is no arrow between Gene and Tar.

M. Hernes et al. (Eds.): ICCCI 2020, CCIS 1287, pp. 767–776, 2020.
https://doi.org/10.1007/978-3-030-63119-2_62

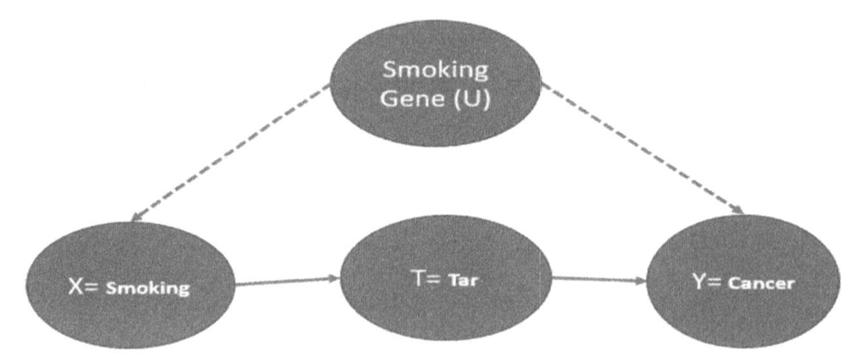

Fig. 1. A causal diagram of the Smoking and Cancer example from [1], page 225

Given the Smoking Gene is not observable, we cannot use the mere probability methods to solve this problem.

To do so, Pearl suggests cutting the arrow between Smoking Gene and Smoking. That is, making the Smoking (X) constant, which gives us P(y|do(x)) in Fig. 2. P(y|x) $\neq$ P(y|do(x)), as do(x) means that we intervene and set the number of cigarettes a person smokes per-day to a fix number.

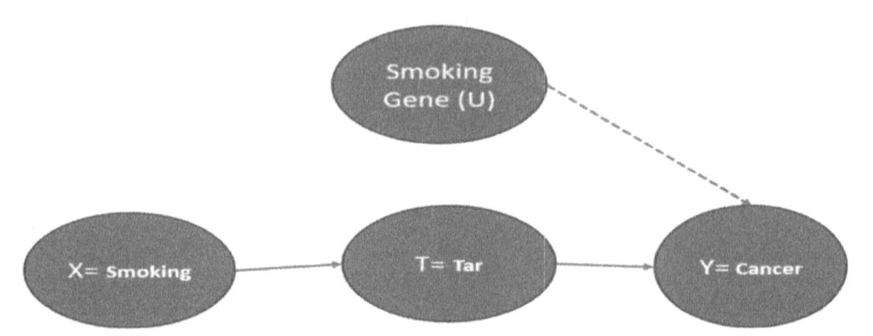

Fig. 2. P(y|x) $\neq$ P(y|do(x), "The probability of a patient having lung cancer Y = y, given that we intervene and ask the person to smoke a pack of cigarette per day (set the value of X to x) and subsequently observe what happens." [1]

Furthermore, in order to solve this problem Pearl suggests the following rules:

Rule1: P(Y|do(X), T, U) = P(Y|do(X), T)

The probability distribution of a variable Y will not change after eliminating the variable U, if we observe that the variable U is unrelated to the variable Y.

Before explaining Rule 2, we need to explain very briefly what is a back-door criterion. In a directed acyclic graph (DAG) (i.e., Figure 1), a set of variables T fulfills the back-door criterion if:

We cannot find a node in T that is a descendent of X_i, T blocks all paths between X_i and X_j that has an arrow into X_i. Accordingly, conditional on T, do(X) is equal to X:

Rule 2: $P(Y|do(X), T) = P(Y|X, T)$

Furthermore, suppose that X and Y are two disjoint subsets of nodes in Fig. 1. T can fulfill back-door criterion for (X, Y), if it fulfills the criterion for any pair of (X_i, X_j) where $X_i \in X$ and $X_j \in Y$. In Fig. 2, to prevent back-door effect on Smoking, the back-door arrow between Smoking Gene and Smoking is deleted.

Rule 3: $P(Y|do(X)) = P(Y)$

When there is no causal path between X and Y, we can eliminate do(X) from the above formula.

Using the three rules explained briefly above, Pearl claim to solve the Smoke-Cancer problem in Fig. 1. For more details, readers are encouraged to consult [1].

Pearl's inferential logic uses a probabilistic approach. Probability theories are associated with events, and these events can occur or not. A bottle of wine with 50% probability of containing arsenic means that there is a 50% chance that the wine is completely toxic, and 50% chance that the wine is completely non-toxic.

Furthermore, the probability deals with the uncertainty inherent to our knowledge of facts and whether they are true or false. There is no gradient possible about the reality of the facts. For instance, to find out whether smoking will cause cancer, using probability theories, we must compare people who smoke all the time or who never smoke. That is, one needs to calculate the conditional probability of having cancer given the presence or absence of smoking.

However, without additional knowledge of causal structure, probability theories cannot generalize hypothetical scenarios and interventions. Furthermore, there might be hidden confounders. Perhaps a gene corresponding to cancer causes a person to smoke (Fig. 1). Having the gene cancer, the person is more likely to Smoke and have Cancer. The mere conditional probability cannot say whether two quantities are causally related.

Pearl's approach to causation cannot answer the following problem: given that you smoke a little, what is the probability that you have cancer to a certain degree? Another problem with Pearl's inferential logic approach is we cannot simulate the real-world problems using DAG.

An alternative to what Pearl is suggesting is probabilistic fuzzy logic [9]. In fuzzy logic, objects can belong to a set with a degree of membership which fits into the closed interval [0, 1], so that its membership can be partial and not only total or null.

With fuzzy logic, wine with a fuzzy 0.5 of toxicity means that the wine will contain 50% of toxic elements such as arsenic and 50% non-toxic. So, in probabilistic fuzzy logic (PFL) processes sources of uncertainty which are: 1) randomness, or 2) probabilistic uncertainty, and fuzziness. PFL can both manage the uncertainty of our knowledge (by the use of probabilities) and the vagueness inherent in the world's complexity (by the fuzzification of the data) [9]. In other words, the use of PFL allows to represent various degrees of dependency between events, before applying probabilities to these degrees of dependency.

PFL has been used to solve many engineering problems. Among others, it is used for security intrusion detection [10, 11]. In the following, we will explain how PFL can be used to create a causal model similar to [1].

2 Probabilistic-Fuzzy Logic (PFL)

What we propose is to keep Pearl's association, intervention and counterfactual, but with fuzzification of each of these steps.

1) Fuzzy association: for instance, events A and B can each have a qualitative value such as "a lot of A", "a little bit of B", "a lot of A and B at the same time", and so on. Then, we will make quantitative fuzzifications of these qualitative values, relying on the judgment of the experts in the domain, assigning values to the variables, so that event A might have a fuzzy value of 0.5, or 0.8 and so on. Then, the fuzzy dependency between fuzzy event A and fuzzy event B will be calculated using fuzzy implication rules such as, for example, the Lukasiewicz implication rule. This way, the associations between events will be fuzzified [9].
2) Fuzzy intervention rule can be built as min $(1, 1 - A + B)$, which is Lukasiewicz implication rule and which corresponds to a fuzzy version of Pearl's do(x): in other words, what will happen to fuzzy B if we apply fuzzy C to the relation between A and B? For example, given that a lot of A produces a little bit of B, what happens to B when we add an important amount of C to A?
3) **Fuzzy counterfactual:** In classical logic, a relation of causality is represented by a conditional $(A \supset B)$ (if A, then B), which corresponds to $(\sim A \vee B)$. The dual connective of the disjunction $(\vee)$ is the conjunction (&). So, the dual of $(A \supset B)$ is $(\sim A \ \& \ B)$. This is exactly the way by which an alternative cause can be expressed, which happens when the usual cause A does not occur, but B nonetheless occurs, given an alternative cause [12]. Different possible ways for representing t-conorms (fuzzy disjunctions) in fuzzy logic are the max function, the sum function or the upper-bounded sum. With the max function, $(\sim A \vee B)$ becomes in fuzzy calculation (max $(1 - A, B)$) which corresponds to a Kleene's disjunction. With the sum function, $(\sim A \vee B)$ becomes $(1 - A+B - ((1 - A) \times B))$, or $(1 - A+AB)$, which is a Reichenbach's disjunction. Finally, the upper-bounded sum is min $(1, 1 - A + B)$, that is a Lukasiewicz's disjunction. Generating the dual of each of these three functions, we produce T-norms (or fuzzy conjunctions) and we obtain three fuzzy versions of $(\sim A \ \& \ B)$, that is three functions for the calculation of the truth-value of an alternative cause. First, a min function $\min(1 - A, B)$ as the dual of the max; then, a product function $((1 - A) \times B)$, or $(B - AB)$ as the dual of the sum function; and, finally, the lower-bounded difference $\max(0, 1 - A + B - 1)$ or $\max(0, B - A)$ as the dual of the upper-bounded sum. We have tried these three functions for the calculation of alternative causes: 1) $\min(1 - A, B)$, 2) $(B - AB)$ and 3) $\max(0, B - A)$ [9]. After the establishment of this fuzzy system, it can be made probabilistic using Bayes rule, as Pearl does in [1]. Applying Bayes rule to each of our fuzzy conditional including their alternative causes, will make a system more flexible than Pearl's. It will result in a Probability Fuzzy Logic (PFL) system in which any new event can be taken into account and will provide learning. Learning occurs by modifications to any of the constituents of the system. When the system converges to satisfying conclusions on a given case, it will defuzzify these conclusions in order to provide crisp decisions for actions.

Furthermore, by representing "degrees of being" (e.g., a little toxic or much toxic) as well as degrees of certainty (e g. probabilities of being a little or very toxic), PFL can both manage the probabilistic uncertainty of our knowledge and the fuzzy dimension inherent to the world complexity.

Being symbolic, explainability comes by default with PFL. What is more, we get more flexible reasoning by taking PFL chains, making probability relations between fuzzy nodes in fuzzy causal chains. As another example, in PFL we can answer questions such as given that you smoke a little bit, what is the probability that you have cancer and to what degree? PFL turns our qualitative reasoning into a quantitative form; it formalizes it, makes it more rigorous and precise [9].

At this point, we would like to mention that we could not find any real experiments with data in [1] to replicate them and compare them with our model. The only experiment we found so far is Microsoft DoWhy[1] project where authors generated some data and applied Pearl's rules. Thus, we used the same data as Microsoft DoWhy and applied PFL to them.

3 Results

In this section, we applied our model to the DoWhy(see footnote 1) project data. The DoWhy project is sponsored by Microsoft and implements Pearl's inferential logic rules to find whether a treatment is the cause of some specific Outcome in Table 1.

Table 1. Treatment, outcome and w_0 columns from Microsoft DoWhy project

Treatment	Outcome	W_0
9.907534	20.117281	3.819063
4.486652	9.514061	−1.180991
8.327747	16.448445	1.814381
1.789252	3.726993	−3.978817
6.573413	12.925626	0.550170

The DoWhy database has three columns. Treatment, Outcome, and w0 where w0 can be a confounder to both Treatment and outcome. The treatment values are between [0–12) which means 12 is not appeared in the data.

As we mentioned above, in Fuzzy logic, using an expert's knowledge, one needs to define intervals. However, the challenge in this experiment is that there is no expert to help us with the interval's definition. Thus, before applying our model to the data in Table 1, the first step was to find interval values in the Treatment column that correspond to the Outcome.

[1] https://microsoft.github.io/dowhy/example_notebooks/dowhy_confounder_example.html.

To do so, we used the fuzzy c-means (Fig. 2) clustering algorithm[2] [12] to cluster Treatment values. Fuzzy c-means is an unsupervised algorithm. In Fig. 2, every partition has two values. The word Center represents the number of clusters (i.e., centers = 2, two clusters) and fuzzy partition coefficient (FPC). FPC is a value between [0, 1] with 1 being the best indicating how cleanly our data is described by a model.

It is worth mentioning that in Fig. 3 as the number of clusters increases, the FPC values decrease. The best cluster is with center = 2 which has an FCP value of 0.88. In order to translate human language, fuzzy logic uses the following words a little, very, very very, extremely, etc.

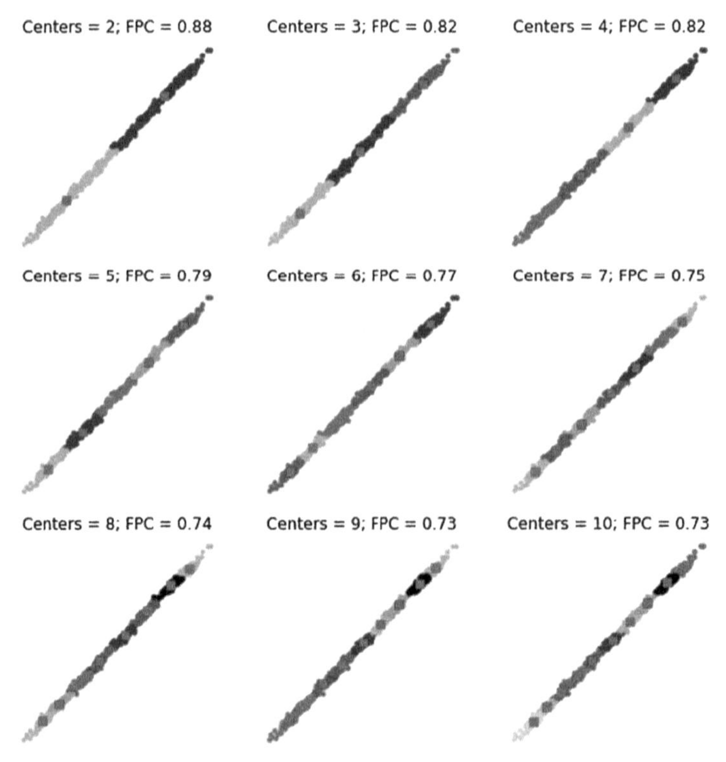

Fig. 3. Represents clustered Treatment data using fuzzy c-means clustering (fcc) algorithm. Fcc uses different cluster numbers to partition data. The values under each partition represent fuzzy partition coefficient (fpc), which is a value between [0, 1]

In our case, in order to have three following categories such as a little, very, extremely (explained above), we fixed the numbers of clusters for the algorithm into three. One can interpret three clusters as little, very and extremely.

---

[2] https://pythonhosted.org/scikit-fuzzy/auto_examples/plot_cmeans.html#example-plot-cmeans-py.

The algorithm output for three categories are the following intervals [0–4.3], (4.3–7.7], (7.7–12). We also changed the number of clusters to more than three categories which caused the computation more intensive and did not really changed the results.

Before applying our PFL algorithm, we had to divide our data into training and validation. Out of 10000 records, we randomly selected 600 records for test and the remaining 9400 records were used for training. It is worth mentioning that we tried our data as 80%, 90% training and 20%, 10% test respectively and the results were almost the same.

Using the clustered data, we then applied the fuzzification and defuzzification steps explained in the previous sections for the Treatment and w0. To calculate the defuzzified values for the Outcome, we used a *cause_function* (source code on GitHub[3]). To generate the value of the Outcome, the *cause_function* takes a value for Treatment and w0 as input and applies among, others, one of the following fuzzy implication rules such as min(1 – A, B).

In the *cause_function()*(see footnote 3), we implemented eight Fuzzy rules to Treatment and w0, out of which three are explained in the Probabilistic-fuzzy logic section (above). For the rest of the rules readers are encouraged to see [10]. For instance, for a Treatment value of 11, and a w0 of 4, the *cause_function* will outputs, the following vector [19, 20, 20, 20.0555, 20, 20, 19.8888, 20.5]. It is worth mentioning that the values in the vector have small variations which mean w0 is not the cause of the Outcome. In the next step and in order to create PFL, our model convert these defuzzified values into probabilities.

Since DoWhy project uses inferential logic and probability methods, the estimated value for the correlation between Treatment and Outcome is computed using the following formula:

$$E(outcome = high|T = high) - E(outcome = high|T = low) = 13.39$$

Furthermore, the left part of Fig. 4 shows the DoWhy project results where there is a clear correlation between the treatment and the outcome. The authors interpreted the slope of the line as a causal effect of the treatment on the outcome.

Our model (Fig. 4, right) used the same observed data as showed in Fig. 4, left. It predicts the causal variation which corresponds to the slope of the linear regression and it's 95% confidence interval, of the DoWhy data. Our model's causal variation value is 1.2876 which is very similar to the DoWhy project result (Fig. 4, left).

To obtain this result, we chose 27 equidistributed points of coordinates Treatment and w0 respectively from each sample (interval of Treatment). We then, applied the min dual fuzzy function min (1 – w0, Treatment).

In the DoWhy project, the authors used a linear model to calculate the expected values for the effects of the treatment on the outcome. However, their current model does not support the expected value for continuous variables[4].

[3] https://github.com/joseffaghihi/Fuzzy-Causal-DoWhy-project/.

[4] https://github.com/microsoft/DoWhy/issues/86.

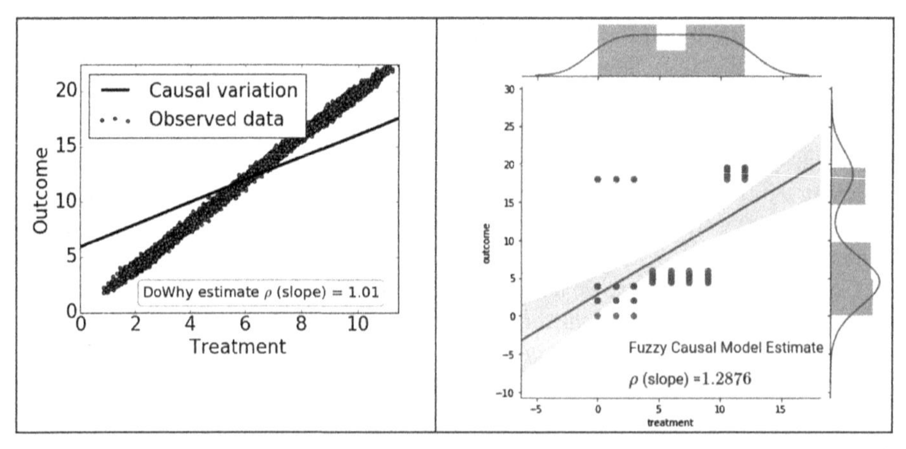

Fig. 4. Left, the DoWhy project(see footnote 2). Right, our model: a regression formula used in order to predict correlation/cause slope between treatment and outcome

Given the continuous nature of the DoWhy project data and intervals used in probabilistic fuzzy logic, to calculate expected values, we used Riemann double sum midpoint method. The conditional expected values for the Outcome given the Treatment is calculated as below:

$$
\begin{aligned}
&E[c < 0 < d|a < T < b] \\
&\qquad = \int_c^b 0 * P(O|T \in [a, b]) \\
&\qquad = \int_c^d \int_a^b 0 * f_{o|t}(O, T) \\
&\qquad \approx \sum\nolimits_{O \in Outcome} \sum\nolimits_{T \in Treatment} O * f_{o|t}(O, T)\Delta A
\end{aligned}
$$

Where ΔA is the area of the parallelepiped with the length and width ΔT and ΔO respectively. ΔO is calculated by subtracting d and c values (above formula) of the interval divided by the number of bins which is 10000 in our case.

The height corresponds to O * $f_{o|t}$(O, T) in the above formula which is calculated using the joint distribution function multiplied by the Outcome.

Figure 5, shows the results after applying the Riemann double sum midpoint method to the data.

For instance, for the treatment values [4–12], the outcome values are almost [5–7.5], [18–20]. However, for the small treatment values (between 0–4), the confounding variable affects the outcome which is between 3–17.5.

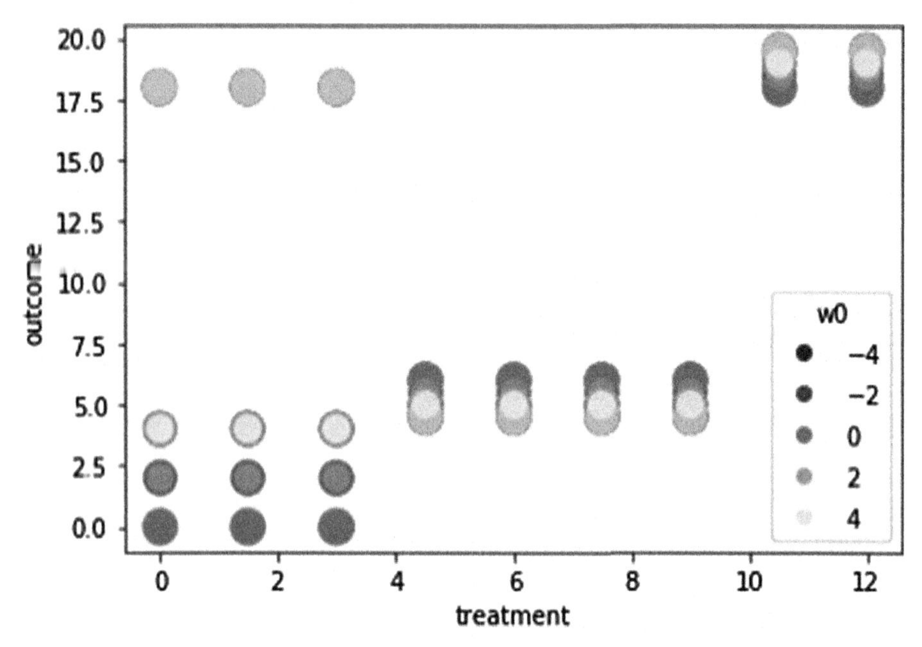

Fig. 5. Riemann double sum midpoint method applied to continues values.

4 Conclusion

In this paper, we showed a brief comparison between probabilistic fuzzy logic (PFL) and inferential logic. With probabilistic fuzzy logic, we can represent various degrees of causal dependency between events and probabilistic measures of our belief in these degrees of dependencies, which is more flexible and subtler than Pearl's account. Also, using causal PFL models, we could improve causal chains with dual rules, that can represent alternative causes, the ones that happen when the effect occurs as the result of a different cause. This way, we can find the hidden patterns in data better and more concise than what Pearl suggests.

We would like to mention that we are aware of the membership functions drawback in PFL which can be different should a problem given to different experts. However, using scaling techniques for the intervals and membership functions, PFL gives the same results.

One problem with the Pearl's current model is that not all real-world problems can be simulated using directed acyclic graphs. For instance, in a network having many nodes, the nodes can send and receive data mutually. In our next work, we will adapt our model so that it can work for nodes that affect each other mutually.

References

1. Pearl, J., Mackenzie, D.: The Book of Why: The New Science of Cause and Effect. Basic Books (2018)
2. Faghihi, U., Robert, S., Poirier, P., Barkaoui, Y.: From association to reasoning, an alternative to pearl's causal reasoning. In: Proceedings of AAAI-FLAIRS 2020. North-Miami-Beach, Florida (2020)
3. Faghihi, U., Franklin, S.: The LIDA model as a foundational architecture for AGI. In: Wang, P., Goertzel, B. (eds.) Theoretical Foundations of Artificial General Intelligence, Atlantis Thinking Machines, vol 4, pp. 103–121. Atlantis Press, Paris. Springer (2012). https://doi.org/10.2991/978-94-91216-62-6_7
4. Faghihi, U., Estey, C., McCall, R., Franklin, S.: A cognitive model fleshes out Kahneman's fast and slow systems. Biol. Inspired Cogn. Archit. **11**, 38–52 (2015)
5. Rubin, D.B.: Estimating causal effects of treatments in randomized and nonrandomized studies. J. Educ. Psychol. **66**, 688 (1974)
6. Rubin, D.B.: Assignment to treatment group on the basis of a covariate. J. Educ. Stat. **2**, 1–26 (1977)
7. Rubin, D.B.: Bayesian inference for causal effects: the role of randomization. Ann. Stat. 34–58 (1978)
8. Rubin, D.B.: Randomization analysis of experimental data: the fisher randomization test comment. J. Am. Stat. Assoc. **75**, 591–593 (1980)
9. Yager, R.R., Zadeh, L.A.: An Introduction to Fuzzy Logic Applications in Intelligent Systems. Springer, Berlin (2012)
10. Zhao, D.-M., Wang, J.-H., Wu, J., Ma, J.-F.: Using fuzzy logic and entropy theory to risk assessment of the information security. In: 2005 International Conference on Machine Learning and Cybernetics, pp. 2448–2453. IEEE (2005)
11. Cheng, P.-C., Rohatgi, P., Keser, C., Karger, P.A., Wagner, G.M., Reninger, A.S.: Fuzzy multi-level security: an experiment on quantified risk-adaptive access control. In: 2007 IEEE Symposium on Security and Privacy (SP 2007), pp. 222–230. IEEE (2007)
12. Robert, S., Brisson, J.: The Klein group, squares of opposition and the explanation of fallacies in reasoning. Log. Univers. **10**, 377–392 (2016)

Enhance Trend Extraction Results by Refining with Additional Criteria

Ei Thwe Khaing(✉), Myint Myint Thein(✉), and Myint Myint Lwin(✉)

University of Information Technology, Yangon, Myanmar
{eithwekhaing,myintmyintthein,myintmyintlwin}@uit.edu.mm

Abstract. Traders make their investment based on stock trends or price directions to get more profit. Many researchers have tried to retrieve the interesting features for trends on large financial data such as news. Data obtained from stock market is highly volatile and correlated. It is characterized with high dimensionality to make prediction of stock trends a challenging. Feature extraction is an important part in developing a fully automated stock market prediction system. Features in stock news have Multiple Interdependent Nature (MIN) which are the relationship between two or more features. Rule and Syntactic Feature based relation extraction have already proposed to get trend related features. Our previous work is still required to solve MIN for stock trend prediction. This paper proposes the additional criteria to improve the trend extraction on MIN. These criteria are date conversion, main verb identification, stock reference and advanced trend extraction. According to the experiment, the trend extraction with additional criteria on stock news gets more extracted trends than the extraction without criteria. The stock trend extraction results with additional criteria become more consistent with the real stock price movement.

Keywords: Stock market prediction · Text mining · Named Entity Recognition · Relation extraction · Trend extraction

1 Introduction

Prediction of stock trends provides benefits for traders and investors in the economy and stock market prediction [1, 2]. Many researches have used news contents or stock price data as features to predict stock trends. This system can help the prediction of stock trends automatically extracting the useful knowledge out of textual resources. Feature Extraction is essential to extracting knowledge from huge volumes of data. Feature extraction for stock news is a difficult task because news are unstructured and contain irrelevant words which bad impact the trend prediction result. Feature extraction methods on text, which are Bag of Words, Noun Phrases, N-gram, word combination and Named Entities are used to extract trend related features in many researches. The named entities provide relevant and semantic features for stock trend prediction. In our previous work, the relevant features or named entities are extracted by Name Entity Recognition (NER) for stock trend extraction. NER is used to identify and classify unique

M. Hernes et al. (Eds.): ICCCI 2020, CCIS 1287, pp. 777–788, 2020.
https://doi.org/10.1007/978-3-030-63119-2_63

features from unstructured and irrelevant data. The five unique features in our published paper [3] are extracted using word and pattern based methods to get relevant features for stock trends on the news contents. Five types of named entities are extracted from news using corpus and patterns. These entities are Stock, Location, Date, Money and Quantity. These named entities and other words (verbs and price keyword) are used as important inputs for trend extraction. Relationships between named entities are essential to extract stock trends. The features related to stock trends are interdependent features on unstructured, time dependency, irrelevant and complex data. The interdependent features means features which are dependent on each other. Each feature are dependent on other features to get stock trends. For example, the stock feature depends on its price feature and the price depends on the describing date. Sometimes, features in a sentence relates to other features in another sentence or adjacent sentence. This nature is called MIN. It curses negative effects of the trend extraction on the real news contents because trend is not correctly extracted. The published paper [4] tried to extract the trends on the news contents which are unstructured, time dependency, irrelevant and complex but all trends cannot be extracted from the news contents with MIN. The correctness of trend extraction in the whole news are dependent on the extracted trends from the sentences.

In this paper, the trend extraction is extended by adding criteria to improve the accuracy of extraction and solve MIN problem. There are four types of the additional criteria which are mostly related to the extracted trend in common data nature of news contents. They are proposed and explained as follow:

Criteria I: *Date conversion.* The specific date expressions is normalized to extract trends from date comparison of the news contents.

Criteria II: *Main verb identification.* Main verbs identify to get the relevant features for trend extraction.

Criteria III: *Stock reference.* It resolves the reference used within the stock with a same sense.

Criteria IV: *Advanced trend extraction.* Trends are extracted from the two adjacent sentences by combining the previous three criteria

The remainder of the paper is structured as follows: Sect. 2 describes the related works feature extraction of the stock market prediction in textual data. Section 3 shows proposed system. Section 4 explains the additional criteria. Section 5 describes the experimental results of stock trend extraction. Section 6 summarizes the paper.

2 Related Works

According to the data structure in data mining applications, the most algorithms for text categorization are used to extract relevant features. These algorithms in data mining require to transform from the textual contents of the documents to numeric representation. Feature extraction and feature selection have used in preprocessing process of text data. Bag-of-words is the most common and the simplest techniques for extracting features from text data. It forms a word presence feature set from all the words of an instance or sentence. N-gram, Term Frequency-Inverse Document Frequency (TF-IDF),

Noun Phrases and Named Entities are mostly applied in the text representation for the feature extraction. TF-IDF method uses the frequencies of single discrete words to represent text and is not capturing any interrelation or semantics between words. Therefore, the relationships between words with stock trends are not extracted using TF-IDF. Feature selection has been applied to reduce the large number of features using dimension reduction methods. The main objective of feature selection is to eliminate irrelevant features that provide few or less important items of information.

R. Schumaker et al. [5] applied the three different types of feature extraction in text representation. There are Bag of Words, Noun Phrases and Named Entities to estimate a stock price on financial news articles. Proper Nouns had the better performance than Named Entities but other words such as verbs, time and numbers don't consider in the text representation. M. Hagenau et al. [6] proposed stock prediction approach to guess stock price on textual news. Bag of words (BoW), N-gram, Noun Phrases and 2 word combination are used to extract feature in text representation. The BoW model is not capturing semantic features and relationships between words. Noun Phrases and N-gram are used to get features in the word combinations of words. The combination between unwanted words causes low performance and words with irrelevant features.

The entity recognition approaches are rule-based and machine learning-based in NER. Rules can represent knowledge visually and naturally, closest to the human's way of thinking and get high accuracy in specific domain. Machine learning can automatically extract patterns by generalizing from a given set of examples. It needs to create training dataset. Many researches [7–9] show that Conditional Random Field (CRF) is better for machine learning-based NER task than other methods such as Maximum Entropy Markov Model and Hidden Markov Model. CRF is a probabilistic framework for labeling and segmenting structured data, such as sequences, trees and lattices. It is a conditional probability distribution over label sequences given a particular observation sequence. CRF is an undirected graph-based model that considered words that not only occur before the entity but also after it. The position of a word or entity in our news contents is indefinite or disorder and the word ranges in each entity are various. CRF is not compatible for NER on our data nature.

Stanford CoreNLP and Apache OpenNLP are mostly used as open source tools for features extraction on textual document or news articles. These tools provide tokenization, Part-of-Speech (POS) tagging, Parser, lemmarization, Named Entity Recognition, co-reference and so on. These tools cannot support automatic feature extraction and training resources in our testing data. Although, Apache OpenNLP uses to tokenize and identify POS tags in data preprocessing of our research because OpenNLP is fast and adds own creation model. This paper explores the effect of trend extraction on the stock news. A huge number of published news for stock trend prediction are used on text mining and time series analysis. The trend extraction is still an interested topic on text mining for the stock market prediction. For feature extraction in text representation of stock news, Named Entities are used to extract features because they are suitable to get relevant features on our testing data. Therefore, the named entities used to identify the relevant or semantic features and reduce irrelevant features for stock trend prediction. Trends were extracted from the relations of the named entities in stock news.

3 Proposed System

This system demonstrates three processes to extract trends for stock price movements on the news articles. These processes are data preprocessing, NER and relation extraction. They are applied to extract stock trends from news articles. The below subsections are explained about the whole system of the trend extraction.

3.1 Data Preprocessing

Data preprocessing is important part of the information extraction. Its processes are sentence detection, tokenization, Part-Of-Speech (POS) tagging, lemmatization and stop word removal. Apache OpenNLP tool is used to detect sentence, tokenizes words, do POS tags and select root words in the data preprocessing. The results of this process are root words with POS tags in each sentence. Before data preprocessing, numeric expressions as three, which describe date expressions on the news are manually transformed to the number as digit 3. The prepositions between date expressions is also omitted. Stop words are identifiers, propositions and conjunctions.

3.2 Name Entity Recognition (NER)

NER extracts relevant information from unstructured text data and categorizes it into groups such as people, organization, etc. The financial news are a huge amount of words, heterogeneous data types and unstructured text data. NER is used to get useful information by identifying and classifying named entities using words with POS tags in the data preprocessing. Rule-based approach in NER is used to get unique features for stock trends on disorder and various range of words. NER recognizes the five types of unique features or named entities that concerned with stock trends with tri-gram that means one to three words. These entities are stock, location, date, money and quantity. Two types of nature in NER are used to extract named entities. The first one is word based features using corpus to support stock data in Myanmar. The second one is pattern based features using regular expressions concerned with POS tags, e.g. noun [NN], verb [VB], and context structure, e.g. start with a capital letter of a word in a sentence. Stock and location entities are extracted from corpus and date, money and quantity are extracted using regular expression rules.

3.3 Relation Extraction

Stock trends have extracted from the news using rule and syntactic feature based methods. Rules on the relationships between named entities were described in our previous work. Some sentences can extract trend by matching rules. For example 1, a sentence is "*The price of mung beans has been gradually increasing since the end September.*", trend for this sentence can be extracted to match rules. Although some sentences are matched rules, trends cannot be extracted. For example 2, other sentence is "*Mung bean was priced below K600,000 per ton on 22 October 2018 and the rate bounced back to K988,000 per ton on 6 November, 2018 on the back of increasing demand from India.*", its sentence matches a rule as {S + VB + M1 + Q1 + D1 + VB + M2 + Q2 + D2} but stock trend is not defined. The context structure or syntactic features in the rules are considered to extract trends in example 2.

The nature of the trend related features has MIN in stock market prediction. The limitation of our previous work cannot be extracted all consistent trends on the whole news because of MIN. This paper tries to solve the limitation and add the additional criteria in the trend extraction. The aim of this system is to get more accurate trend prediction results with criteria. The process flow of the trend extraction with the additional criteria is shown in Fig. 1.

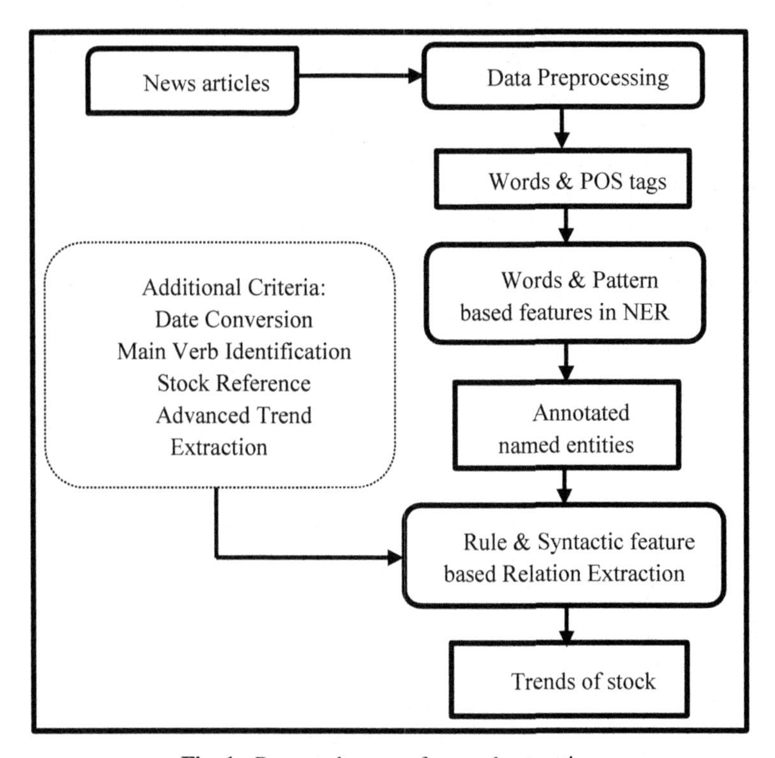

Fig. 1. Proposed system for trend extraction

4 Additional Criteria

There are four types of the additional criteria to support trend extraction and solve MIN structure in our previous work. The detail descriptions of these criteria are explained in the following subsections.

4.1 Criteria I: Date Conversion

The stock trends on news articles are occasionally based on time series nature. The task of mapping from a textual phrase describing a complex date or duration to a temporal representation is called "Date Conversion". It needs to identify stock trends on time expressions. Date conversions are directly or indirectly to learn the interpretation of temporal expressions. The textual phrases (e.g. 2 August) are automatically converted to temporal expressions. Specific patterns for the automatic date conversion are identified using Date Format in java. In Java Date Format, 'dd' stands for day describing 2 digits, 'MM' stands for month describing 2 digits and 'yyyy' stands for year describing 4 digits. Other time expressions (e.g. 3 months, last week, this week and earlier week) are not specified date using Date Format in java. SUTime [10] in Stanford is a tool to automatically convert from date expression to specific date type. In SUTime, the date expression for '3 months' is expressed as the period 'P3M' and "current" describes as "PRESENT_REF". The definite temporal expressions require to extract trends in date comparison of news contents. Therefore, date conversion is proposed to get the definite date expressions in the trend extraction.

This paper proposes indirect date conversion, which are described token and string pattern with explanations. Token patterns are specified over tokens using Java regular expression library. The example of the token patterns is a simple rule specifying a mapping from a token that matches the regular expression {[0–9]$^+$ $TimeUnit} to the predefined duration type depending on $TimeUnit. This pattern uses [0–9] as the numerical expressions (e.g. 0, 1, 2, 3, …, 9), '+' means one or more digits and $TimeUnit as the words relating to date expressions (e.g. date, week, month and year). String patterns are matched over text using regular expressions. The example of string patterns is "this week" that refer to the current week and "last week" that refer to the previous week. Table 1 shows the results of the indirect date conversion with different date format depend on 4 February 2019.

Table 1. Results of date conversion on 4 February 2019

Date contents	Pattern types	Date format	Our results
3 months	Token	MM/yyyy–MM/yyyy	12/2018–02/2019
2 years	Token	yyyy–yyyy	2019–2020
Current	String	MM/dd/yyyy	02/04/2019
This week	String	MM/dd/yyyy–MM/dd/yyyy	02/03/2019–02/09/2019
End January	String	MM/dd/yyyy–MM/dd/yyyy	01/23/2019–01/31/2019
Early May	String	MM/dd/yyyy–MM/dd/yyyy	05/01/2019–05/07/2019
Last 2 weeks	Token+String	MM/dd/yyyy–MM/dd/yyyy	01/20/2019–01/26/2019

4.2 Criteria II: Main Verb Identification

Verbs are the most commonly used to describe action roles and to link the part of the sentence. The verb are different forms in the grammar rules of English language. It appears in different forms in a sentence depending upon the nature and grammatical perspective of the subject (subject-verb agreement), tenses, moods, voices, different structures, modals, etc. Stanford CoreNLP and Apache OpenNLP tools don't support to identify main verb on news contents but they classify POS tags for verb. The POS tags for helping and auxiliary verbs of the tenses are various and are irrelevant features to extract trend. This system proposes to extract the main verb by removing helping or auxiliary verb based on words with POS tags. Table 2 describes the result of the main verbs on the words with POS tags.

Table 2. Results of main verb identification

Words	POS tags	Main verbs after lemma
Was fetching	VBD VBG	Fetch
Was priced	VBD VBN	Price
Has been increasing	VBP VBN VBG	Increase
Are primarily produced	VBP RB VBN	Produce
Is lower compared	VBZ JJR VBD	Compare
Has been gradually increasing	VBZ VBN RB VBG	Increase

4.3 Criteria III: Stock Reference

In computational linguistics, co-reference resolution [11] refers to resolving references used within the text with a same sense. The resolution methods depend on the different semantic and syntactic forms in which references can occur. The scope of resolution is word references on the single or different sentences. Stock expressions in our testing data are necessary to provide specific stock' prices in the trend extraction. The open source tools don't support stock references for our data because unstructured text data depend on the writers' opinions or views. Stock is identified using rules on below three types of descriptions.

1. Use 'it' pronoun
2. Use possessive and demonstrative pronoun
3. Don't clearly present pronoun

Rule based stock references are easy to create and maintain in our testing dataset because the existing resources or corpus don't support our stock data. Table 3 describes results of the stock reference on three types of nature.

4.4 Criteria IV: Advanced Trend Extraction

In our previous work, stock trends are extracted from each sentence using the relation extraction methods. The relation extraction methods used rules and syntactic features on the relationships between named entities. The entity relationships on stock news

Table 3. Results of Stock Reference

Type	Sample sentences	Descriptions
1	**Black sesame** was cultivated in the summer, and **it** was priced at K220,000 per bag last year	A pronoun 'it' use to point to stock entity 'black sesame'
2	Such **sesame** has been found to contain acid residue, and **its price** is, thus, only over K40,000 per viss," he added	The possessive expression 'its price' refers to a stock entity 'sesame'
3	"Prior to the Thingyan Festival, **mung beans** were priced at K760,000 per ton. The price has now increased to around K845,000	In 2nd sentence, pronoun does not clearly present. According to the Type 3, the price in 2nd sentence assume the mung bean's price according to previous sentence

are unstructured, time dependency and complex. Rule based approach is appropriate to extract relevant features on stock news because training resources for our data is not existed. Each rule has context words, POS tags and named entity types in each sentence. The previous work cannot extract trends in two adjacent sentences that are described price movements. Therefore, the advanced trend extraction is proposed to extract trends in the two adjacent sentences.

This criteria is considered by combining the result of criteria I, II and III on the two adjacent sentences. Stock trend extraction is applied the previous relation extraction methods with these criteria. The rules and syntactic features are used to extract trends based on the extracted named entities. The sentences in news have the same stock information and they are checked whether or not the contents of other entities such as price, quantity and date. If two contents of date, money and quantity entities are contained, the trend extraction on the date comparison influences on price difference with the same quantity. Trend is identified from the result of the different price. The result of advanced trend extraction is shown in Fig. 2.

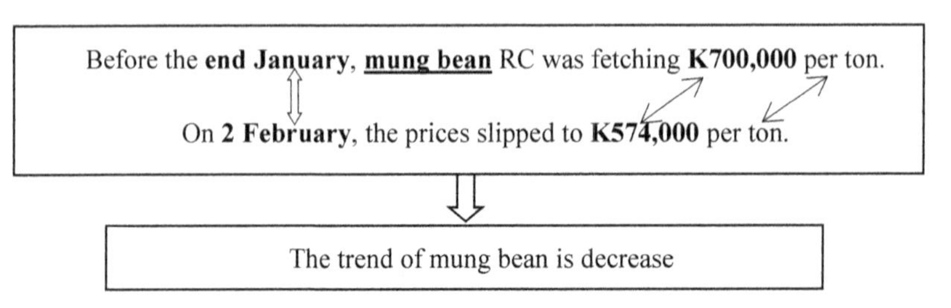

Fig. 2. Result of advanced trend extraction

5 Experiments

Test data are collected the news articles which are published from the Global New Light of Myanmar (GNLM) [12] and stock price in Agriculture & Market Information Agency (AMIA) [13] and Ba Yint Naung [14] websites in Myanmar. There are 26 news articles that provide trend information concerning with the Myanmar Beans, Pulses and Sesame in 2019. Other 26 news from 1 January 2020 to 30 June 2020 are added in test data. The stock price are collected from AMIA and Ba Yint Naung on all posted date except holiday. Table 4 describes the trend extraction results on the sample sentences of news with the four types of the additional criteria.

In above table, the sample sentences of a news article on 4 February 2019 are used in the example of trend extraction. The title of the sample news is "India halts pulses imports; mung bean prices hit three-month low", can download from the link [15]. In Table 4, 1+2 means the combination of the 1st sentence and 2nd sentence as two adjacent sentences. The final results in the example of trend extraction are decrease in 1+2 and 2+3 according to the advanced trend extraction and stock reference. Trend in sentence 4 is increase. In the experiments, the predicted trends for the whole news are identified probability on the number of the extracted trends. If the extracted trends depend on time

Table 4. Example of Trend extraction on news

	Sentence No.		Trend
Input sentences	1	Before the end January, mung bean RC was fetching K700,000 per ton	
	2	On 2 February, the prices slipped to K574,000 per ton	
	3	Mung bean was priced K576,000 per ton on 18 October	
	4	Mung bean was priced below K600,000 per ton on 22 October 2018 and the rate bounced back to K988,000 per ton on 6 November, 2018 on the back of increasing demand from India	
Criteria I	1	end January → 01/23/2019-01/31/2019	
	2	2 February 02/02/2019	
	3	18 October → 10/18/2018	
	4	22 October 2018 → 10/22/2018, 6 November 2018 → 11/06/2018	
Criteria II	1	was fetching → fetch	
	2	slipped → slip	
	3	was priced → price	
	4	was priced → price	
Criteria III	1	mung bean → stock entity	
	2	prex (plural of price) → mung bean's price (using stock reference)	
Criteria IV	1 + 2 2 + 3	If Stocks are same but stock information doesn't exist, stock reference is used If number of Date is 2, these dates are converted and compared If number of Quantity and Money is 2 and same Quantity, trends are extracted depending on the difference of Money	
Final trend results	1	Before the end January, mung bean RC was fetching K700,000 per ton	D
	2	On 2 February, the prices slipped to K574,000 per ton	D
	3	Mung bean was priced K576,000 per ton on 18 October	I
	4	Mung bean was priced below K600,000 per ton on 22 October 2018 and the rate bounced back to K988,000 per ton on 6 November, 2018 on the back of increasing demand from India	

series, the trend on the nearest current date is considered higher priority than the trend on past. According to Table 4, the trend of mung bean is decrease on 2 February.

The trends shown in Table 5 depend on the stock price difference on predicted date. There are five types of the predicted date on the number of days after news released.

Table 5. Trends on different predicted date of Mung Bean's price in AMIA

Date dd/MM/yyyy	Price	Predicted date after news released				
		Next 1 day	Next 2 day	Next 3 day	Next 4 day	Next 5 day
01/28/2019	22670					
01/29/2019	21900	D				
01/31/2019	19400	D	D			
02/01/2019	18300	D	D	D		
02/04/2019	18600	I	D	D	D	
02/06/2019	19350	I	I	D	D	D

The experimental results for stock price prediction are shown in Figs. 3 4, 5, 6, 7. In this results, 52 news between January 2019 and June 2020 are tested. Each chart is divided by the five types of the predicted date. These charts describe the stock price prediction on the real stock price movements. For the stock price prediction, trend extraction are considered with or without additional criteria in our previous work. Two types of results in price prediction are correct or incorrect depending on matching with the extracted trends and the real price movements on the predicted date. The two types of prediction results is described in horizontal line. The vertical line is shown the number of the price prediction. When amount of test data are large, the correctness of trend extraction results is almost the same the previous results on small data in 2019. In Fig. 3, the stock price prediction with criteria is greater correct results than the price prediction without criteria and is smaller incorrect results than without criteria. In Figs. 4, 5, 6, 7, the stock price prediction with criteria is more correct results than the price prediction without criteria and is less incorrect results than without criteria.

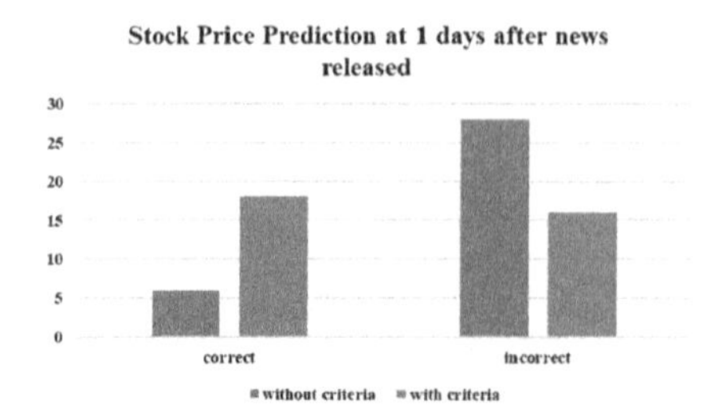

Fig. 3. Price prediction after next 1 day

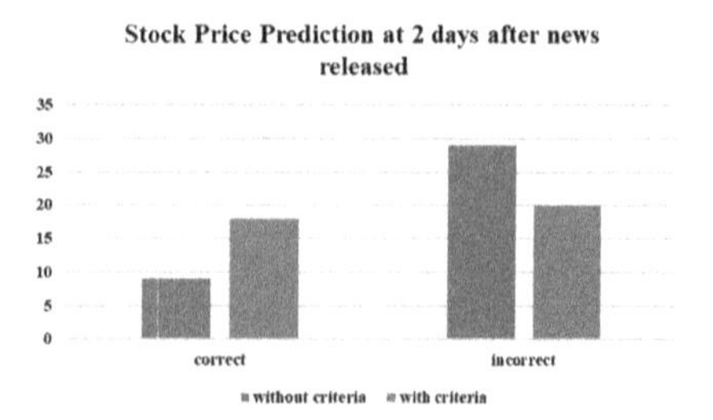

Fig. 4. Price prediction after next 2 days

According to the stock price prediction on the different predicted date, more consistent results are getting when date that news released are near the predicted date. As

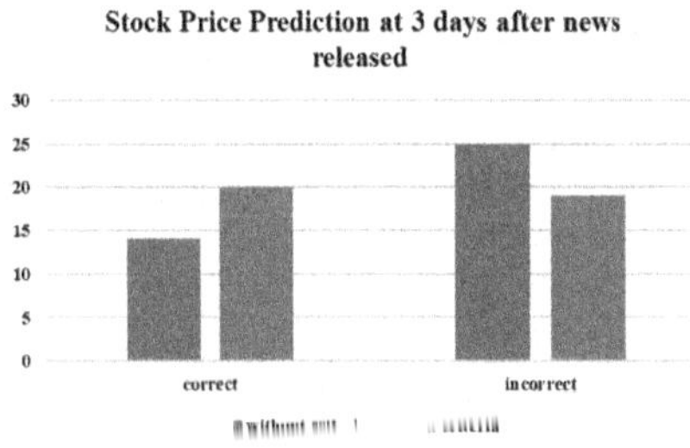

Fig. 5. Price prediction after next 3 days

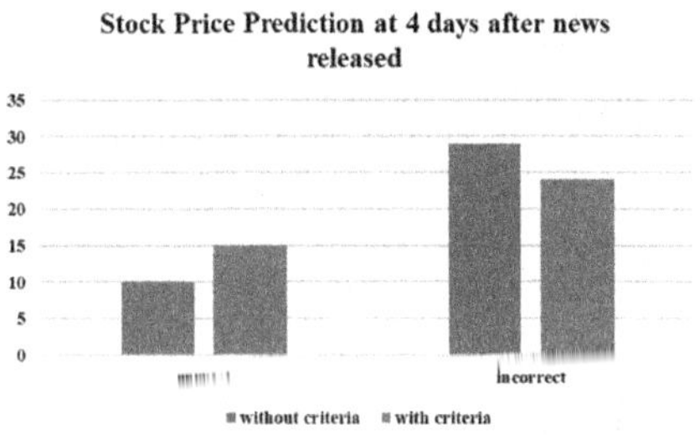

Fig. 6. Price prediction after next 4 days

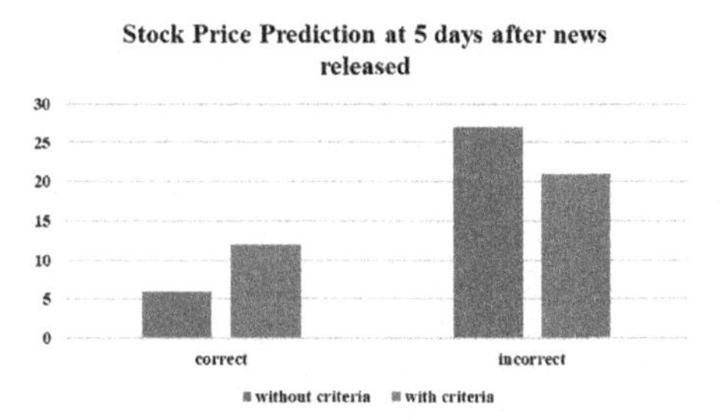

Fig. 7. Price prediction after next 5 days

the experimental results, the trend extraction with the additional criteria provides better results than the previous research work.

6 Conclusion

The stock trend extraction is an important part of the stock market prediction. The aim of this paper is to automatically extract the stock trends from the news articles in the stock market. This paper presents additional criteria to extract trends and to improve our research. Therefore, the extracted stock trends will be applied in the prediction of stock prices. An automatic trend prediction system for Myanmar stock market will be developed in future work.

Acknowledgement. I would like to thank supervisors, professors and rectors at University of Information Technology. I would like to regard with participating as a member in Data Analysis and Management (DAM) Lab. Our research supports in part by the DAM lab at the University of Information Technology.

References

1. Gunduz, H., Cataltepe, Z., Yaslan, Y.: Stock daily return prediction using expanded features and feature selection. Turk. J. Electr. Eng. Comput. Sci. **25**(6), 4829–4840 (2017)
2. Xiao Zhong, X., David Enke, D.: Forecasting daily stock market return using dimensionality reduction. Expert Syst. Appl. **67**, 126–139 (2017)

3. Khaing, E.T., Thein, M.M., Lwin, M.M.: Myanmar stock named entity recognition for high dimensional data in local business news articles. In: The 13th Conference on Project Management (PROMAC 2019), vol. E13, pp. 789–796, November 2019
4. Khaing, E.T., Thein M.M., Lwin, M.M.: Stock trend extraction using rule-based and syntactic feature-based relationships between named entities. In: 2019 International Conference on Advanced Information Technologies (ICAIT), pp. 78–83. IEEE, November 2019
5. Schumaker, R.P., Hsinchun Chen, H.: Textual analysis of stock market prediction using breaking financial news: the AZFin text system. ACM Trans. Inf. Syst. (TOIS) **27**(2), 1–19 (2009)
6. Hagenau, M., Liebmann, M., Neumann, D.: Automated news reading: stock price prediction based on financial news using context-capturing features. Decis. Support Syst. **55**(3), 685–697 (2013)
7. Singh, S.: Natural language processing for information extraction. arXiv preprint arXiv:1807.02383 (2018)
8. Song, S., Zhang, N., Huang, H.: Named entity recognition based on conditional random fields. Cluster Comput. **22**(3), 5195–5206 (2017). https://doi.org/10.1007/s10586-017-1146-3
9. Lafferty, J., McCallum, A., Pereira, F.C.: Conditional random fields: probabilistic models for segmenting and labeling sequence data (2001)
10. Chang, A.X., Manning, C.D.: SUTIME: a library for recognizing and normalizing time expressions. In: LREC, pp. 3735–3740, May 2012
11. Sukthanker, R., Poria, S., Cambria, E., Thirunavukarasu, R.: Anaphora and coreference resolution: a review. Inf. Fusion **59**, 139–162 (2020)
12. GNLM. http://www.globalnewlightofmyanmar.com. Accessed 29 May 2020
13. AMIA Webpage in facebook. Accessed 29 May 2020
14. Naung, B.Y.: Webpage in Facebook. Accessed 18 July 2020
15. News1 link. https://www.globalnewlightofmyanmar.com/india-halts-pulses-imports-mung-bean-prices-hit-three-month-low/. Accessed 29 May 2020

Image Captioning in Vietnamese Language Based on Deep Learning Network

Ha Nguyen Tien, Thanh-Ha Do(✉), and Van Anh Nguyen

VNU University of Science, Hanoi, Vietnam
tienhapt@gmail.com, hadt_tct@vnu.edu.vn, vananhnt57@gmail.com

Abstract. Image captioning is an underlying and crucial problem in artificial intelligence. This is challenging since it requires advanced research in computer vision to detect objects and the correlation of these objects in the image, and text-mining to convert these relationships into words. Although some based deep-learning and machine translation approaches have been achieved state-of-art results in English recently, it is missing an approach to generate the caption in Vietnamese, which is a local language in Vietnam with complex grammar and variable meaning in simple words. Moreover, machine translation is affected negatively by a significant issue called unknown words, which is caused by both large vocabulary size and unbalanced dataset. In this paper, we propose a new approach to generate the Vietnamese caption of the image and also a simple and effective solution to tackle unknown words problem of machine translation. In general, the results of these methods achieved in the self-build testing database are promising.

Keywords: Image captioning · Vietnamese caption · Unknown words · Machine translation

1 Introduction

In the last few years, social networks have been more popular and used by billions of people to share daily life moments. Within this big digital database, there is a large number of images do not go along with a clear description, but humans do not doubt to understand this information easily and correctly.

However, the computer with incredible computation capability is considered as the best supportive and useful tool for humans, but it cannot understand the meaning of the context. Therefore, our motivation is to study and develop a model to take computational advantage of the computer, combine with artificial intelligence in automatically generating a Vietnamese caption for scene image. The obtained results can be integrated into the application to disabled people aware of their surroundings.

In general, to make the computer understand these contents, it needs to recognize the objects over images and then represent the objects' relations.

M. Hernes et al. (Eds.): ICCCI 2020, CCIS 1287, pp. 789–800, 2020.
https://doi.org/10.1007/978-3-030-63119-2_64

Firstly, objects and their positions are important information used in object recognition. Nevertheless, the computer sees a raw image by the color measurement of each pixel, which contains both interesting information about the object in the image and noise. To reduce the noise and still keep the main characteristics of objects, images are represented as other forms as a feature vector or feature matrix. Essentially, there are two different directions used to represent image, which are image processing techniques and deep-learning networks. In image processing techniques, the feature extraction methods such as Radon Transform [1], Local Binary Patterns [2], the Histogram of Oriented Gradients [3] or combination of such features are widely used to present the objects. In general, the limitation of these traditional techniques is that they are designed to deal with a specified family of images, therefore they cannot adapt well to new kinds of images. In deep learning network-based techniques, the parameter of models is trained based on a training set, even when it is a complex and diverse database. Therefore, these models could adapt well with any images, as long as they have the same distribution with the training set. The state-of-the-art deep neural networks, which are CNN [4], Mask CNN [5], R-CNN [6], etc, achieved good performances in object detection and object classification.

Secondly, after encoding the image as a feature vector, it is taken to another model to capture and describe the connection of objects over an image in a certain language. The dataset trained on this model contains both images and their corresponding sentences (or caption) in the specific language. In recent years, the studies of this task in the English language [7–10] have good performance and prospects. Essentially, the model in this topic needs to be large and diverse to cover multiple objects and their action. However, due to the lack of available datasets in the Vietnamese language, we face many problems and limitations to develop an end-to-end model to describe an image by a Vietnamese caption. Moreover, building a qualifying dataset will take a lot of effort and time of many workers. Consequently, we propose a new multi-model that combines the English caption generation model and English-Vietnamese translation model to overcome the problem and inherit the available dataset for each model.

The English caption generation model used in this paper is NeuralTalk2 model[1] that built on the Flickr8K/Flickr30K/COCO database and the sentence descriptions collected with Amazon Mechanical Turk. The NeuralTalk2's architecture includes: The deep neural network model that infers the latent alignment between segments of sentences and the region of the image that they describe and the multi-modal Recurrent Neural Network architecture that takes an input image and generates its description in text format.

For machine translation, two approaches get researchers' attention being Phrase-based Statistic Machine Translation (PSMT) [11] and Neural Machine Translation (NMT) [12]. In which, NMT has a lot of advantages: does not need to select candidate translations; parallel processing easily; more fluid and more meaningful. Therefore, we decide to develop NMT model, following by an unknown words processing to improve the quality of obtained translation.

---

[1] https://github.com/karpathy/neuraltalk2.

Overall, the main contributions of this paper are:

1. Developing the Vietnamese automatically caption model for scene image by combining the English caption model and the English-Vietnamese translation model.
2. Proposing an unknown word processing model as the post-processing stage of the neural machine translation model to improve the quality in English-Vietnamese translation
3. Conducting a testing database to evaluate the performance of the Vietnamese caption system for scene images[2]

The paper is organized as follows: Sect. 2 summarize the related works; Sect. 3 describes detail of the proposed Vietnamese caption model; Experiment results is presented in Sect. 4, and we conclude and present future work with Sect. 5.

2 Related Works

Image captioning is one of the most essential topics in scene understanding. It has been widely studied and applied to real-life applications such as helping disabled people to aware of the dangerous situation in front, searching relative images based on provided sentence fragments.

The studies in this topic have given outstanding results, which can be mentioned as the following: in [13], Farhadi *et al.* proposed an approach to link image space and sentence space into a meaning space. The relative between an image and sentence is measured by a score and this meaning space.

However, Fariha *et al.* submitted a different idea, which uses multi-task learning in [14]. There are two main tasks in this system: caption construction and activity identification within the image. The second task not only aims to improve the shared layer representation, in fact, it also helps to improve the performance of all the system.

Mathews *et al.* [8] proposed another model using natural language processing techniques and frame semantics, which is a novel encoder-decoder model. After map the image into a semantic term representation using the term generator, the language generator takes these terms as the input to generate a caption in the target style.

In [10], Gourisaria *et al.* developed a model using different variations of CNNs, LSTMs, and GRUs. Their work focuses on the generation of annotations for Underwater images. They also train the model on a new self-built dataset named PESEmphasis5k. In detail, this model incorporates seven simple stages being (i) dataset construction; (ii) feature extraction; (iii) data pre-processing; (iv) model definition; (v) model evaluation; (vi) generating captions; and (vii) developing new manual evaluation method.

In general, the state-of-the-art techniques for image captioning use a large database including both images and descriptions, which is most of them in

[2] http://www.mediafire.com/file/bmj0w3wdmyfwzf4/Captiontestset.rar/file.

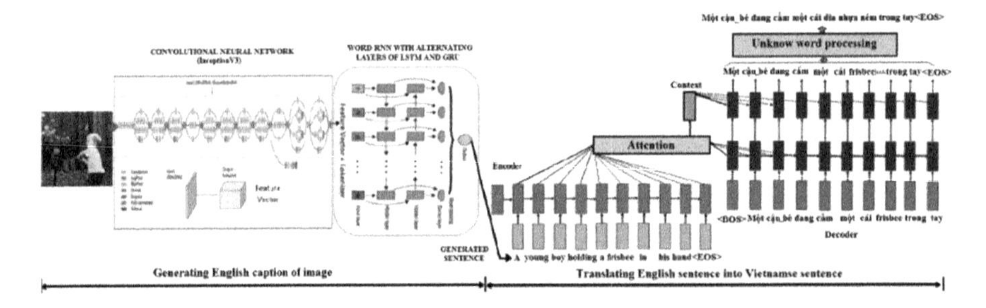

Fig. 1. The image captioning model by Vietnamese

English. For the Vietnamese language, the corpus of images and Vietnamese sentences describing the relationship between objects in these images is unavailable, even when we try to create the self-built dataset, it is not big enough. To overcome the problem, we propose a multi-model system, which combines the automatic English caption model for the image with the English-Vietnamese neural machine translation model. By this approach, we could not only overcome the lack of data in Vietnamese but also inherit the English caption data corpus for image and the English-Vietnamese sentence-aligned bilingual corpus. In our method, the unknown words are also processed to improve the performance of system.

3 Generating Vietnamese Language Caption for Scene Images

The Fig. 1 is the diagram of the general Vietnamese caption model that including three models: the image captioning model, the English-Vietnamese neural machine translation model, and the unknown word processing model.

The Image Captioning Model

Some image captioning models give remarkable results such as LRCN [15], Mao *et al.* [16], MS Research [17], Google NIC - Vinyals *et al.* [18]. However, the BLUE n-grams score of the work of the research group at Stanford gives the best result compared to others when evaluating in 1000 test images. Therefore, we decided to develop our model based on the Stanford group's work. That work developed an inter-modal to maximum the correspondences between language and visual data. This model is a combination of Convolution Neural Networks (CNN) and bi-directional recurrent neural networks (bi-RNN) - a structured objective that aligns the two modalities through a multi-modal embedding [18]. In which, CNN is considered as a feature extractor to extract over regions of images, and the output is a vector feature. Bi-RNN takes this vector as an input of the first step. Each output at each step is considered as a probability of word and input of the next step.

The Neural Machine Translation Model

The translation problem is defined as following: for any sentence in source language S given, find a sentence T in target language such that $argmax_T p(T|S)$. In fact, the conditional probability $p(T|S)$ can be directly modeled using a neural machine translation system [19].

The architecture of Neural Machine Translation (NMT) consists of two recurrent neural networks (RNNs), the first one (*or Encoder*), is used to encode the input text S to a latent vector θ and the second one, named Decoder, is used to generate translated text in output from that latent vector to product T. NMT does not translate well in long sentences [20] because this model tends to forget the older input text after a certain step. The attention mechanisms [19] can handle this disadvantage of NMT. However, this is not an issue for image captioning since almost all captions of the image are short sentences. Consequently, we decided to use NMT in this paper.

The neural machine translation system takes a conditional language modeling view of translation by modeling the probability of a target sentence T $(t_{1:n})$ given a source sentence S $(s_{1:m})$ as

$$p(T|S) = p(t_{1:n}|S) = \prod_{i=1}^{n} p(t_i|\{t_{1:i-1}\}, S, \theta) \tag{1}$$

The Encoder maps each source word to a word vector and processes these into a sequence of hidden vectors $h_1, \cdots, h_m$. The Decoder aims to combine an RNN hidden representation of previously generated words $\{t_1, \cdots, t_{i-1}\}$ with source hidden vectors to predict scores for each possible next word.

The Unknown Word Processing Model

Machine translation technologies have been developed and given the remarkable results. Besides the achievements, understand context accuracy could be challenging since it relates to cultural nuances, contextual content clues, and local slang. Furthermore, our work aims to generate the caption in the Vietnamese language, which associate with Vietnamese culture. Multiple abstract words can not translate word-by-word but must to translate base on context. In other words, the accuracy of our method is affected by the quality of the machine translation model.

Aware of this issue, we propose a model to process the unknown word problem in Vietnamese translation, which helps to improve the final result of our system. The main idea of the unknown word processing model includes 3 stages: the first stage is based on the source English sentence to replace the unknown word with the corresponding English word. The second stage is segmentation and annotation of the Vietnamese sentence. The last stage is searching for the English word in the Vietnamese translation in the post-tagged Vietnamese-English bilingual dictionary to replace by the corresponding Vietnamese word. The Algorithm 1 presents the detail of the proposed unknown word processing model.

Algorithm 1: Handle the unknown words

Input: The Vietnamese sentence that has unknown words (S)
Output: The Vietnamese sentence that some unknown words is replaced (called by S_o)
Begin
Segment(S); Tag(S);
while *(not EOF(S))* **do**
  Get W in S;
  if $(W = W_e$ *) and* (W_e *in DicEV)* **then**
    Replace(W,W_v);
S_o=S;
End

In Algorithm 1, the elements W, $DicEV$, W_e, W_v are defined as following:

- W: a word of S
- $DicEV$: the post-taged Vietnamese-English bilingual dictionary
- W_e: the post-taged English word that occurs in Vietnamese sentence
- W_v: the post-taged Vietnamese word that is W_e's translation in $DicEV$.

4 Experiment Results

The Training Databases

In this paper, there are three databases are used to train 1) captioning model 2) the English-Vietnamese translation model, and 3) the unknown word processing model. Details of 3 datasets as following:

- Captioning model uses 113,000 instances taken from the MSCOCO dataset[3]. Each instance has an image, its labeled 45 regions and at least 5 captions (some images have more). The annotation process is done by 9 different workers.
- Machine Translation model is trained on Vietnamese-English bilingual corpus [21] consists of over 800,000 sentence pairs, and our self-collected of more than 270,000 sentence pairs[4].
- The unknown word processing model use VncoreNLP that was build by Vu et al. in 2018 [22].[5] The post-taged Vietnamese-English bilingual dictionary $DicEV$ has more than 400,000 entries

Measure: BLEU Metric

BLEU (*BiLingual Evaluation Understudy*) proposed by Papineni *et al*, 2002 [23] is the standard metric to measure the quality of the machine translation model

[3] http://cocodataset.org/#download.
[4] https://github.com/Tienhavn/generalcorpus.
[5] https://github.com/vncorenlp/VnCoreNLP.

when scores the overlap between a translation output and a reference target on a scale of 0 to 1. This method is used frequently since its result is considered to be the correspondence between a machine's output and that of a human.

$$BLEU = BP.e^{\left(\sum_{n=1}^{n} w_n log pn\right)} \tag{2}$$

in which

- pn: the geometric average of the modified n-gram precisions.
- w_n: the positive weight.
- BP: the brevity penalty calculated by

$$BP = \begin{cases} 1 & \text{if} \quad c > r \\ e^{1-\frac{r}{c}} & \text{if} \quad c < r \end{cases} \tag{3}$$

with c is the length of the candidate translation and r is the effective reference corpus length.

Evaluation

Because this is the first work to generate an automatically Vietnamese caption for scene image and there is no any public testing database. To overcome this problem, we came up with two solutions: 1) separately evaluating each model 2) conducting a new testing database for the image-Vietnamese caption. Following the first one, we could inherit the public dataset for both image-English caption and English-Vietnamese translation. However, the distribution of the dataset could have an impact on the quality of the model either negatively or positively, and two kinds of public datasets could be very distinct. For these reasons, we decided to conduct a new testing database including 500 images and their corresponding described sentences in Vietnamese languages. This database is generated and evaluated by a language specialist.

The testing phrase includes two main steps as following:

- Step 1: Generate 500 English captions of 500 scene images using the image caption model
- Step 2: Use the NMT model to translate 500 English captions into Vietnamese captions.

To verify the performance of the proposed approach, we compare obtained translation with translation done by English experts using BLEU measurement. After evaluating, the proposed method gets 88.96 BLEU score.

To find out the suitable machine translation system for automatically generating caption of images, we trained two MT systems: Moses and OpenNMT toolkit as representatives for phrase-based and neural-based machine translation approaches, respectively. The set-up is to use the same as Vietnamese-English sentence-aligned bilingual corpus but different sentence pairs: 300, 000, 600, 000, 894, 665. The quality of the proposed approach when using difference machine translation model is evaluated over 500 captions from English to Vietnamese. The result is shown in Table 1 in which:

- **SMT:** The phrase-based statistical machine translation system
- **SMT+UnP:** The model that combines SMT and the unknown word processing module
- **NMT:** The English-Vietnamese neural machine translation system
- **NMT+UnP:** The model that combines NMT and the unknown word processing module

Table 1. BLEU of translating 500 captions from English to Vietnamese.

Corpus size *(sentence pairs)*	SMT	SMT+UnP	NMT	NMT+UnP
300,000	45.17	47.07	55.13	56.73
600,000	43.74	45.04	56.58	57.78
894,665	43.70	44.80	88.96	**89.76**

Table 2. Example of the effectiveness of using unknown word processing

Result without UnP	**Result with UnP**
Một chiếc xe buýt **double decker** đỏ trong một bãi đỗ xe	Một chiếc xe buýt **hai boong** đỏ trong một bãi đỗ xe

The experimental results show that the quality of the NMT system gets better when training with a bigger dataset. In particular, by using the proposed unknown word processing method, the quality of translations in all experimental systems is significantly improved: the average increase value of the BLEU score is 1.32 BLEU score. As indicated in the Fig. 2: with the scene image, the English captioning system's output or the input of the translation model is the sentence "*A red* ***double decker*** *bus parked in a parking lot*". Without using UnP, the word **double decker** cannot be translated while by using UnP they are translated to "hai boong" in Vietnamese (means "*two deckers*" in English).

We also use the Google translation system to translate these 500 captions from English to Vietnamese to compare to our result, the plot is presented

Table 3. Some illustrations for generating caption in Vietnamese language for scene images. The English sentences are translation of Vietnamese sentences done by language experts

Một người phụ nữ đứng trên vỉa hè
(*A woman is standing on the sidewalk*)

Một thành phố đông đúc với đèn giao thông trên nó
(*A crowded city with traffic lights on it*)

Một vài con bò đang đứng trong một cánh đồng
(*A few cows are standing in a field*)

Con chim nhỏ đứng trên bãi cỏ
(*A small bird is standing on the grass*)

Một người đàn ông và một người phụ nữ ngồi trên ghế bành
(*A man and a woman are sitting on an armchair*)

Nhà bếp có bếp và tủ lạnh
(*The kitchen has a stove and a refrigerator*)

Một chiếc bánh và một đĩa
(*A sandwich and a plate of salad*)

Một con diều đang bay trên bầu trời một ngày đầy nắng
(*A kite is flying in the sky on a sunny day*)

Một người phụ nữ nuôi con hươu cao cổ trong vườn thú
(*A woman is feeding a giraffe in the zoo*)

Một người đàn ông cưỡi ván trượt xuống sườn núi
(*A man is riding skateboard down the mountainside*)

in Fig. 2. The unknown word processing module has not promoted its ability because the number of unknown words in here is not much. Table 3 shows some results of the proposed model.

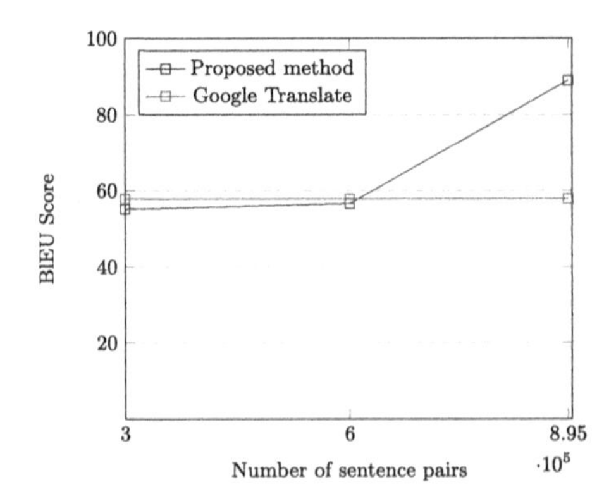

Fig. 2. The performances of generating captions of images obtained by our model and by using google translation

5 Conclusions and Future Work

In this paper, we present a new approach to generate the Vietnamese language caption for scene images based on multi-models built from the deep-learning networks and machine translation. We also propose the unknown word processing method to improve the quality of Vietnamese captions for scene images. Besides, we also conduct the dataset to evaluate the performance of the proposed model. Generally, the experimental results indicated that the proposed method is promising.

In the future, we will research to find out how we can integrate the neural machine translation model with the image captioning model directly on Vietnamese languages and developed the unknown word processing module.

References

1. Radon, J.: On the determination of functions from their integral values along certain manifolds. IEEE Trans. Med. Imaging **5**(4), 170–176 (1986)
2. Ojala, T., Pietikainen, M., Maenpaa, T.: Multiresolution gray-scale and rotation invariant texture classification with local binary patterns. IEEE Trans. Pattern Anal. Mach. Intell. **24**, 971–987 (2002)
3. Dalal, N., Triggs, B.: Histograms of oriented gradients for human detection. In: Proceedings of Conference on Computer Vision and Pattern Recognition (CVPR 2005), San Diego, CA, USA, pp. 28–39, June 2005

4. Jaswal, D., Vishvanathan, S., Kp, S.: Image classification using convolutional neural networks. Int. J. Sci. Eng. Res. (2014)
5. Wei, X.-S., Xie, C.-W., Wu, J.: Mask-CNN: localizing parts and selecting descriptors for fine-grained image recognition. Pattern Recogn. **76**, 704–714 (2018)
6. He, K., Gkioxari, G., Dollár, P., Girshick, R.: Mask R-CNN. CoRR, vol. abs/1703.06870 (2017)
7. Al-muzaini, H., Al-Yahya, T.N., Benhidour, H.: Automatic Arabic image captioning using RNN-LSTM-based language model and CNN. Int. J. Adv. Comput. Sci. Appl. **9** (2018)
8. Mathews, A., Xie, L., He, X.: SemStyle: learning to generate stylised image captions using unaligned text. In: Proceedings of Conference on Computer Vision and Pattern Recognition (CVPR 2018), pp. 8591–8600 (2018)
9. Rahman, M., Mohammed, N., Mansoor, N., Momen, S.: Chittron: an automatic bangla image captioning system. Proc. Comput. Sci. **154**, 636–642 (2018)
10. Gourisaria, H., Rama, S., Jayanthi, A., Gangey, T., Penikalapati, R.: Generating captions for underwater images using deep learning models. In: Conference on Artificial Intelligence: Research, Innovations and its Applications (2019)
11. Setiawan, H., Li, H., Zhang, M., Ooi, B.C.: Phrase-based statistical machine translation: a level of detail approach. In: Dale, R., Wong, K.-F., Su, J., Kwong, O.Y. (eds.) IJCNLP 2005. LNCS (LNAI), vol. 3651, pp. 576–587. Springer, Heidelberg (2005). https://doi.org/10.1007/11562214_51
12. Bahdanauand, D., Cho, K., Bengio, Y.: Neural machine translation by jointly learning to align and translate. In: Proceedings of the 3rd International Conference on Learning Representations (2015)
13. Farhadi, A., et al.: Every picture tells a story: generating sentences from images. In: Daniilidis, K., Maragos, P., Paragios, N. (eds.) ECCV 2010. LNCS, vol. 6314, pp. 15–29. Springer, Heidelberg (2010). https://doi.org/10.1007/978-3-642-15561-1_2
14. Fariha, A.: Automatic image captioning using multitask learning. In: Proceedings of the 29th Conference on Neural Information Processing Systems, Barcelona, Spain (2016)
15. Donahue, J., et al.: Long-term recurrent convolutional networks for visual recognition and description. IEEE Trans. Pattern Anal. Mach. Intell. **39**(4), 677–691 (2017)
16. Mao, J., Xu, W., Yang, Y., Wang, J., Yuille, A.L.: Explain images with multimodal recurrent neural networks. CoRR, vol. abs/1410.1090 (2014)
17. Fang, H., et al.: From captions to visual concepts and back. In: Proceedings of Conference on Computer Vision and Pattern Recognition (CVPR 2015), Washington, USA, pp. 1473–1482 (2015)
18. Vinyals, O., Toshev, A., Bengio, S., Erhan, D.: Show and tell: a neural image caption generator. CoRR, vol. abs/1411.4555 (2014). http://arxiv.org/abs/1411.4555
19. Luong, T., Pham, H., Manning, C.D.: Effective approaches to attention-based neural machine translation. In: Proceedings of the Conference on Empirical Methods in Natural Language Processing, Lisbon, Portugal, pp. 1412–1421 (2015)
20. Wu, Y., et al.: Google's neural machine translation system: bridging the gap between human and machine translation. CoRR, vol. abs/1609.08144 (2016)
21. Ngo, H.: Building an English-Vietnamese bilingual corpus for machine translation, pp. 157–160, November 2012

22. Vu, T., Nguyen, D.Q., Nguyen, D.Q., Dras, M., Johnson, M.: VnCoreNLP: a Vietnamese natural language processing toolkit. In: Proceedings of the 2018 Conference of the North American Chapter of the Association for Computational Linguistics: Demonstrations, pp. 56–60. Association for Computational Linguistics, New Orleans, June 2018. https://www.aclweb.org/anthology/N18-5012
23. Ward, T., Kishore, P., Roukos, S., Zhu, W.-J.: BLEU: a method for automatic evaluation of machine translation. In: Proceedings of the 40th Annual Meeting of the Association for Computational Linguistics (ACL), pp. 311–318, July 2002

Textual Clustering: Towards a More Efficient Descriptors of Texts

Ayoub Bokhabrine, Ismaïl Biskri(✉), and Nadia Ghazzali

Université Du Québec à Trois-Rivières, Trois Rivières CP 500, Canada
{ayoub.bokhabrine,ismail.biskri,nadia.ghazzali}@uqtr.com

Abstract. In this paper, we present how itemsets can be used as a discriminating descriptor in a textual clustering process. We implemented a platform named "IDETEX" capable of extracting itemsets from textual data and using them for the experimentation in different types of clustering methods, such as K-Medoids, Hierarchical clustering and Self-Organizing Map (Kohenon). To some extent the experimentations performed reveal promising results with different classifiers either "Hierarchical", "Non-hierarchical" or "Neural network".

Keywords: Textual clustering · Frequent itemsets · IDETEX

1 Introduction

Before applying a clustering method, the identification and the evaluation of descriptors that distinguishes one cluster from another must be the first step. This step will affect the quality of the obtained results. The choice of the descriptors influences the behavior of the clustering method, the presence of one type of descriptor rather than another being decisive in identifying the cluster to which a text belongs.

In several works in textual clustering, the texts are divided into segments (portions of text), themselves represented by vectors whose dimensions are words. Two segments will belong to the same cluster if the words they share appear with comparable frequencies. However, it is common for two segments, dealing with two different topics, to share the same words with different meanings, given the context of their use. Words only can not bring a lot of knowledge. Taking into account only their presence or absence is sparsely informative and its usefulness in textual clustering is not always satisfactory.

In fact, an association between words in the same segment can bring out some of its "semantic peculiarities". Our assumption is that these peculiarities can help improve the accuracy of the similarity between segments underlying the construction of clusters. The concept of itemset can express the association between words.

In this paper, we are going to show that is possible to generate better results while combining itemsets extractions and clustering methods. Itemsets will be considered as textual descriptors. Based on several experimentations, we observe that the use of itemsets as descriptors of textual data generate better results than using words.

In the following sections we will introduce the concept of association rules, present different clustering methods used in this paper, then explain our methodology followed by an interpretation of results, and finally sum up with the conclusion and perspectives.

M. Hernes et al. (Eds.): ICCCI 2020, CCIS 1287, pp. 801–810, 2020.
https://doi.org/10.1007/978-3-030-63119-2_65

2 Association Rules

Mining data using association rules makes it possible to identify relationships between two or more data from a very large database. It is also an approach which ultimately makes it possible to identify frequent patterns if-then called association rules. For example, if a housewife buys butter every time she buys bread and cheese, the noted association rule (bread, cheese) → butter can be established. This approach is applied to various domains and various types of data, for example, in music mining (Rompré et al. 2017), in image mining (Alghamdi et al. 2014) and in text mining (Zaïane and Antonie 2002).

Let us have a look, now, on some important definitions that will allow us to explain our approach.

Let $\boldsymbol{I} = \{i_1, i_2, \ldots, i_j, \ldots, i_d\}$ be a set of distinct d items. One item may correspond to a word when it comes to extracting lexical associations from texts. It may correspond to a sequence of pixels extracted from images or music descriptors, in case of respectively image or music mining process. From $\boldsymbol{I}$, 2^d possible itemsets (which are subsets of $\boldsymbol{I}$) can be generated (Tan et al. 2002).

Let $\boldsymbol{T} = \{t_1, t_2, \ldots, t_k, \ldots, t_n\}$ be a set of transactions. A transaction t_k is a set of items that occur in the same "event". In case the items are words, sentences, for example, could be considered as transactions. Simply, if the items are descriptors of a document, then a transaction is a subset of the set $\boldsymbol{I}$.

The main objective of extracting association rules is to identify, as much as possible, hidden relationships between items, under the supervision of a user. As shown previously, with d items, the search space is theoretically equivalent to all of the possible 2^d itemsets. This could appear to be too large. However, if the items are words, many combinations of items will simply not appear. We can, also, use some measures, like the support.

The support of an itemset X, noted S(X) is the ratio of the number σ(X) (1.2) σ(X) that contain X by the number n of all transactions of T (1.1). The itemset X is considered frequent when its support is greater than or equal to a threshold set by a user according to his/her goal.

$$S(X) = \sigma(X) / n \tag{1.1}$$

$$\sigma(X) = |\{t_k \mid X \subseteq t_k,\ t_k \in T| \tag{1.2}$$

$$S(X \to Y) = \sigma(X \to Y)/S(X) \tag{1.3}$$

$$\sigma(X \to Y) = |\{t_k | X \subseteq t_k \wedge Y \subseteq t_k, t_k \in T\}| \tag{1.4}$$

An association rule noted X → Y expresses a co-occurrence association between the two frequent itemsets X and Y with X ∩ Y = Ø. In this rule X is the antecedent while Y is the consequent. We can use some measures, like the support, to ensure that an association rule is of quality. The support of X → Y noted S(X → Y) is the ratio of the number σ(X → Y) (1.4) of transactions that contain X and Y by the number σ(X) (1.2) of transactions that contain X. An association rule is considered of quality when its support is greater than or equal to a threshold set by a user. The reader might have a look on (Le Bras et al. 2010; Geng and Hamilton 2006; Tan et al. 2002) to get an overview of several studies on evaluation measures of the quality of the association rules. For three decades, several studies have focused on the application of association rules for classification purposes (Liu et al. 1998; Zaïane and Antonie 2002; Bahri and Lallich 2010). The different classifiers that result from this work produce results that are able to compete with those obtained using other approaches such as decision trees (Mittal et al. 2017).

The extraction of association rules is usually done using the Apriori algorithm (Agrawal and Srikant 1994) or FP-Growth (Han et al. 2000). Other algorithms are presented in (Fournier-Viger et al. 2017). The two main challenges when extracting association rules are computational cost management of memory and time required to seek for frequent itemsets. The most interesting way to address this problem is to control the reduce the set of items without affecting the quality of the results.

In our work, we need to define the notion of a superset and redefine the notion of a transaction in order to target and extract the most descriptive frequent itemsets.

A superset is an itemset defined in relation to another itemset. For example, if we have the itemsets {a, b, c}, {a, b}, {a, c} or {b, c}, we consider {a, b, c} as a superset to avoid the redundancies.

A transaction becomes a set of itemsets. Let a set of Itemsets = {itemset_1, ..., itemset_i, ..., itemset_l}, then transactions can be represented as follows:

$$Transaction_1 = \{itemset_{10}, itemset_{16}, itemset_{20}\}$$

$$Transaction_2 = \{itemset_{20}, itemset_{19}\}$$

$$Transaction_3 = \{itemset_8, itemset_2, itemset_7\}$$

These transactions allow us to construct a binary matrix where the transactions are considered as vectors and the frequent itemsets (supersets) as a descriptor, their intersection represents either the existence of the itemsets as described by 1 or in the opposite case it is considered 0.

The binary matrix is generated based on the results of the completion of the process of itemsets extraction, following that, the binary matrix is utilized as the input of the classifier. Thus, these classifiers attempt to encounter the similarity between transactions and finally cluster them in separated clusters.

3 Clustering Methods

Literature is well supplied in automatic classifiers. Each classifier is distinguished by its conceptual and algorithmic structure. Each has both advantages and limitations. Several factors influence the performance of classifiers, including the nature of the data under study and the research objectives pursued. Nevertheless, the quality of a classification is evaluated according to the homogeneity of the resulting clusters and its ability to identify the relevant information within the targeted segments. The goal is to highlight the similarity between these segments.

Our study proposes to use unsupervised classification methods or clustering on textual data. Thus, we considered three different types of clustering methods. The first one is based on Ward's Criterion which is the ascending hierarchical (Ward 1963). The second one is non-hierarchical, based on K-Means, called K-Medoids. The third one is neuronal based on Kohonen's Self-Organizing Map (SOM).

The Hierarchical clustering (Hclust) aims to generate a set of partitions of the objects to be classified. Each partition is made up of separate clusters, where each of them contains the most similar objects. Thus, it starts with the partition by treating each object as a cluster. Then, it generates a new partition by performing repeatedly the two steps: (1) identify the two most similar clusters, and (2) merge them together to generate a new cluster. It continues until all objects are in the same cluster. And the result is represented by a graph called a dendrogram.

The K-Medoids (Jin and Han 2011) considers a partitioning algorithm, which is a variant of the K-Means algorithm. However, the K-Medoids is more robust to noise. Regarding the K-Means, the representatives of the clusters are the centroids while in the K-Medoids, data are chosen to be the medoids. A medoid is the object of a cluster, whose average dissimilarity to all the objects in the cluster is minimal. The difference between K-Means and K-Medoids can be compared to the distinction between mean and median: where mean represents the average value of data, while median minimizes dissimilarity of all data items in order to guarantee equal distribution. The K-Medoids algorithm initially processes with the K representative objects as medoids. After determining the set of medoids, each object of the dataset is assigned to the nearest medoid.

The SOM (Self-Organizing Map) algorithm (Kohonen 1995) is based on unsupervised competitive learning. It provides a topology preserving the mapping of high dimensional space in map units called neurons. The neurons generally form a two-dimensional network allowing mapping of a high dimensional space on two dimensions while preserving the relative distance between the points. SOM can therefore analyze large data. This network also makes it possible to recognize and characterize data that has never encountered before. New data is assigned to the nearest neuron. In this paper we used the R package Sombrero.

4 Methodology

What we want to prove, in our research, is that using itemsets/supersets to describe a text that gives better results than using unique words. To do this, we consider transactions are portions of text called segments. Segments can be set of consecutive sentences or paragraphs. The segments are expressed by means of vectors whose dimensions are the frequent itemsets/supersets they contain. The more frequent itemsets/supersets two segments share, the greater is the degree of their similarity, our assumption being co-occurrent frequent itemsets/supersets within segments are representative of texts.

In a concrete way, we present our approach in 5 steps:

- The first step is segmenting the document to prepare it for the extraction of frequent itemsets/supersets. The document is represented by a set of segments. These segments are the transactions, and the words contained in these segments are the items of the transactions. The vocabulary of French is roughly estimated to 500 000 words (in our case items), which can generate $2^{500\,000}$ itemsets. It would be computationally expensive to process $2^{500\,000}$ itemsets, this is why, as detailed below, we must reduce the lexicon and limit the vocabulary extracted from a text to a few thousands of words.

- The second step is dedicated to the reduction of the vocabulary by removing some not informative words like stop words or by replacing the inflected forms of words by their lemma. To do this, a list of more than 500 stop words of French and a lexical database for French are used. In addition, we can delete numbers and punctuation characters and convert uppercase letters in lowercase letters.
- The third step allows extracting frequent itemsets with using the Apriori algorithm. We want to identify a small set of frequent itemsets by an iterative process. To do this, the user must initialize, first, the minimum support. This minimum support is decreased by 0.1, each time the number of frequent itemsets is less than 10. The iteration stop condition is when the number of items in the obtained itemsets is greater than a threshold set by the user (e.g. number of items = 3) or the minimum support is less than 0.1. At this point, the algorithm only keeps supersets and eliminates their subsets (as explained in Association rules section).
- The fourth step is dedicated to the comparison between the segments, according the number of frequent supersets shared, in order to identify which segments are similar.
- The fifth step consists of clustering the similarity matrix using all three clustering methods such as Hierarchical method, K-Medoids clustering and SOM clustering (Self-Organizing Map). The outputs of those methods are plotted to visualize the quality of the clustering and assisting the user to make interpretation of the results.

5 Experimentation and Discussion

The purpose of this section is to present and discuss the results obtained from two different experimentations using textual documents written in French. The first experimentation concerns a well-adapted document with few paragraphs. The second one analyzes a real large document.

In order to evaluate our proposed approach, we have developed a platform called IDETEX in C# to import documents, pre-process them and extract itemsets/supersets using our above-mentioned methodology. RStudio was used for clustering and visualization.

5.1 First Experimentation

This experimentation concerns a small document of 22 paragraphs related to 3 various thematics, segmented in the form of each segment contains one paragraph. Segments 1 to 4 converse on sport particularly “the biography of Michael Jordan”. Segments 5 to 8 cover Information Technology (IT) thematic on “Microsoft company”. The last segments 9 to 22 contain only music thematic such as “W.A. Mozart”. Using this well-adapted data where we know the nature of the data and the number of different thematics allows us to evaluate our approach as to its ability to detect three distinct clusters corresponding to the three different thematics.

To do this, we first pre-processed the document and then extracted the itemsets/supersets according to the methodology described earlier. Thus, we measured the discriminating power of frequent supersets. We compared the clustering results when the descriptors 1) are the frequent supersets and 2) are the words.

Our previous knowledge of the nature and structure of our data allows us to analyze the accuracy of the clustering results using three different types of classifiers.

Table 1. Comparative table using supersets as descriptors of textual data

	HClust	K-Medoids	Sombrero
Cluster 1	1, 2, 3, 4	1, 2, 3, 4	1, 2, 3, 4
Cluster 2	5, 6, 7, 8	5, 6, 7, 8	5, 6, 7, 8
Cluster 3	9, 10, 11, 12, 13, 14, 15, 16, 17, 18, 19, 20, 21, 22	9, 10, 11, 12, 13, 14, 15, 16, 17, 18, 19, 20, 21, 22	9, 10, 11, 12, 13, 14, 15, 16, 17, 18, 19, 20, 21, 22

Table 1 shows the results of three different types of clusters by considering frequent supersets as descriptors of our textual data, herewith cluster 1 combines only the segments 1 to 4 which cover sport, while cluster 2 contains the segments 5 to 8 which is IT thematic and the last cluster 3 is formed by the segments 9 to 22 which is music.

The three different clusters contain the same segments as resulted previously. Thus, the three different clustering methods "hierarchical", "nonhierarchical" and "neuronal" give the same results while using itemsets/supersets as descriptors of our textual data.

On the flip side we are utilizing the same data meanwhile considering unique words as descriptors of our textual data (see Table 2 below).

Table 2. Comparative table using words as descriptors of textual data

	HClust	K-Medoids	Sombrero
Cluster 1	2, 3	2	2, 3
Cluster 2	5, 6, 7, 8	3	1, 4, 5, 6, 7, 8
Cluster 3	1, 4, 9, 10, 11, 12, 13, 14, 15, 16, 17, 18, 19, 20, 21, 22	1, 4, 5, 6, 7, 8, 9, 10, 11, 12, 13, 14, 15, 16, 17, 18, 19, 20, 21, 22	9, 10, 11, 12, 13, 14, 15, 16, 17, 18, 19, 20, 21, 22

Table 2 shows a heterogeneous clustering while using only unique words as descriptors.

Hclust subdivides the thematic 1 (sport) in two. Thus segments 2 and 3 are combined in one cluster (cluster 1) while segments 1 and 4 are merged with thematic 3 (music) in another cluster (cluster 3). Thematic 2 (IT) is preserved in cluster 2.

K-Medoids generates a large cluster by grouping the IT and music thematics as well as two segments (1 and 4) relating to sport. The other two clusters consist of a single segment each concerning sport. Thus, sport thematic is separated in three different clusters.

Sombrero clustering shows different results from the previous ones. However, like Hclust, Sombrero divides thematic 1 (sport) in two and associates it with thematic 2 (IT) whereas Hclust linked thematic 1 to thematic 3 (music). More precisely, segments 2 and 3 form one cluster (cluster 1) and segments 1 and 4 are merged with thematic 2 (IT) to form cluster 2. Thematic 3 (music) is preserved in the same cluster (cluster 3).

Based on the first experimentation when itemsets/supersets are used as the descriptors of text, we observed the homogeneity and persistency of the results regardless of these three types of clustering "hierarchical", "nonhierarchical" and "neuronal". Nevertheless, whilst using words as descriptors of text the three classifiers are not able to identify exactly the three thematics.

5.2 Second Experimentation

In order to evaluate our proposed approach, several experiments were carried out while using our platform IDETEX. The experimentation was performed on the book « La civilisation des Arabes, Gustave Le Bon, 1884 » written in French, presents a larger and more complex document whose purpose is to demonstrate the relevance of the use of frequent itemsets/supersets as descriptors.

The document has 6 chapters. The segmentation of these chapters is carried out by considering each segment formed by 6 paragraphs:

- Chapter 1 and 2 contain the segments from 1 to 6 that respectively cover topics "Arabia" et "The Arabs".
- Chapter 3 contains segments 7 to 16 that deal with the subject "the Arabs before Muhammad".
- Chapter 4 contains segments 17 to 27 that narrate the story of "Muhammed and birth of the Arab Empire".
- Chapter 5 contains segments 28 to 32 that talk about "the Koran".
- Chapter 6 contains segments 33 to 42 that talk about "the conquests of the Arabs".

Consequently, our cluster analysis will focus on 42 segments.

To determine the optimal number K of clusters, we used the R package NbClust (Charrad et al. 2014). The NbClust suggests $K = 4$. Table 3 shows the results generated by using the three different clustering methods.

The segmentation of the document was based on 6 paragraphs per each segment, the results are not easy to interpret. In this matter, we seek to identify the segments that always merge together regardless of the clustering methods. Such groups of segments, called Strong forms, will indicate a certain consistency or regularity in the clusters thus obtained. Table 3 shows a better regularity of common segments within all three clustering methods. In cluster 1, we notice that the segments 1, 2, 4 are detected. Segments 30, 39, 40, 41 show up on cluster 3. Cluster 4 has the common segments 5, 6, 7, 8 10, 13, 28, 42. The last group of segments 11, 14, 15, 16, 17, 18, 19, 20, 22, 24, 25, 26, 27, 29, 31, 32, 36 appear in cluster 4 and cluster 2. This regularity of group of segments that we extract shows a consistency of results that requires more advanced tools to explain it.

Table 3. Comparative table using Itemset as descriptor of textual data

	HClust	K-Medoids	Sombrero
Cluster 1	1, 2, 4, 33	1, 2, 3, 4, 9	1, 2, 3, 4, 33, 34, 35
Cluster 2	3, 9, 34, 35	33, 34, 35	9, 11, 14, 15, 16, 17, 18, 19, 20, 22, 24, 25, 26, 27, 29, 31, 32, 36
Cluster 3	30, 39, 40, 41	30, 37, 39, 40, 41	12, 21, 23, 30, 37, 38, 39, 40, 41
Cluster 4	5, 6, 7, 8, 10, 11, 12, 13, 14, 15, 16, 17, 18, 19, 20, 21, 22, 23, 24, 25, 26, 27, 28, 29, 31, 32, 36, 37, 38, 42	5, 6, 7, 8, 9, 10, 11, 12, 13, 14, 15, 16, 17, 18, 19, 20, 21, 22, 23, 24, 25, 26, 27, 28, 29, 31, 32, 36, 38, 42	5, 6, 7, 8, 10, 13, 28, 42

On the other hand, considering only words as descriptors of the document, the results, in Table 4, especially with Hclust and K-Medoids generate three clusters with only one segment each (segments 9, 22, 30). Cluster 4 gathers all remained segments, such a cluster cannot represent a strong form since it includes almost all segments together. Sombrero shows a different character of clustering. We noticed that sombrero gather the segments 9, 22, 30 on the same cluster in contrast to Hclust and K-Medoids.

Table 4. Comparative table for words as descriptor of textual data

	HClust	K-Medoids	Sombrero
Cluster 1	9	9	1, 2, 3, 4, 7, 11, 31, 34
Cluster 2	22	22	9, 22, 23, 30
Cluster 3	30	30	12, 13, 38, 39, 40, 41, 42
Cluster 4	1, 2, 3, 4, 5, 6, 7, 8, 10, 11, 12, 13, 14, 15, 16, 17, 18, 19, 20, 21, 23, 24, 25, 26, 27, 28, 29, 31, 32, 33, 34, 35, 36, 37, 38, 39, 40, 41, 42	1, 2, 3, 4, 5, 6, 7, 8, 10, 11, 12, 13, 14, 15, 16, 17, 18, 19, 20, 21, 23, 24, 25, 26, 27, 28, 29, 31, 32, 33, 34, 35, 36, 37, 38, 39, 40, 41, 42	5, 6, 8, 10, 14, 15, 16, 17, 18, 19, 20, 21, 24, 25, 26, 27, 28, 29, 32, 33, 35, 36, 37

Preliminary analysis of the results shows the regularity and the consistency of the generated clusters while using itemsets/supersets as descriptor of textual data.

During our experimentation, when varying the number of clusters K (e.g. K = 2 or K = 3), we observed that using itemsets/supersets as descriptors of textual data generate homogenies and well separated clusters. On the other hand using unique words generate similar results, the large cluster keeps including the remaining segments.

We notice that Sombrero shows a higher capacity to generate balanced clusters even while using only unique words.

6 Conclusion and Perspectives

The concept of frequent itemsets and its variant, frequent supersets, is very useful as discriminating descriptors of texts. Indeed, using itemsets/supersets instead of words generates homogenies and well separated clusters.

The relationship between words in itemsets/supersets represents the semantic context which allows to distinguish the uses of the same word in two different segments with different topics, what a unique word as descriptor of text does not allow.

The experiments carried out on two corpora with using three clustering methods confirmed our assumptions.

Despite the interesting results obtained, some challenging limits need to be addressed in our future work. For example: what is the most optimal value for the different thresholds of minimum supports? what is the most optimal number of clusters?

Moreover, we propose as perspectives to explore itemsets/supersets with other classification and/or clustering methods (among others deep clustering), to experiment our approach on other languages like Arabic and to Develop advanced tools to better explain the nature of data and result for further use eventually for a training process.

References

Agrawal, R., Imielinski T., Swami, A.: Mining association rules between sets of items in large databases. In: Proceedings of the 1993 ACM SIGMOD International Conference on Management of data, Washington, D.C, pp. 207–216 (1993)

Agrawal, R., Srikant, R.: Fast algorithms for mining association rules. In: 20th International Conference on Very Large Database, San Francisco, CA, pp. 487–499 (1994)

Alghamdi, R.A., Taileb, M., Ameen, M.: A new multimodal fusion method based on association rules mining for image retrieval. 17th IEEE Mediterranean Electrotechnical Conference "MELECON", pp. 493–499. Beirut, Lebanon (2014)

Bahri, E., Lallich, S.: Proposition d'une méthode de classification associative adaptative. 10eme journées Francophones d'Extraction et Gestion des Connaissances, EGC 2010, pp. 501–512 (2010)

Charrad, M., Ghazzali, N., Boiteau, V., Niknafs, A.: NbClust: an R package for determining the relevant number of clusters in a data set. J. Stat. Softw. Foundation for Open Access Statistics Press, 61(6) (2014)

Fournier-Viger, P., Lin, J.C.W., Vo, B., Chi, T. T., Zhang, J., Le, H.B.: A survey of itemset mining. Wiley Interdisc. Rev. Data Min. Knowl. Discov. **7**(4), e1207 (2017)

Geng, L., Hamilton, H.J.: Interestingness measures for data mining a survey. ACM Comput. Surv. (CSUR). **38**(3), 9–11 (2006)

Han, J., Pei, J., Yin, Y.: Mining frequent patterns without candidate generation. ACM Sigmod Record **29**(2), 1–12 (2000)

Huy, T.N., Shao, H., Tong, B., Suzuki, E.: A feature-free and parameter-light multi-task clustering framework. Knowl. Inf. Syst. **36**, 16–20 (2013). https://doi.org/10.1007/s10115-012-0550-5

Jin, X., Han, J.: K-medoids clustering. In: Sammut, C., Webb, G.I. (eds.) Encyclopedia of Machine Learning. Springer, Boston, MA (2011). https://doi.org/10.1007/978-0-387-30164-8_426

Ward Jr, J.H.: Hierarchical grouping to optimize an objective function. J. Am. Stat. Assoc. **58**(301), 236–244 (1963)

Le Bras, Y., Meyer, P., Lenca, P., et Lallich, S.: Mesure de la robustesse de règles d'association.: QDC (2010)

Liu, B., Hsu, W., Ma, Y.: Integrating classification and association rule mining. In Knowledge Discovery and Data Mining, New York City, NY.: American Association for Artificial Intelligence Press, pp. 80–86 (1998)

McCallum, A., Nigam, K.: A comparison of event models for naive bayes text classification. In: AAAI workshop on learning for text categorization, American Association for Artificial Intelligence Press, pp. 41–48 (1998)

Mittal, K., Aggarwal, G., Mahajan, P.: A comparative study of association rule mining techniques and predictive mining approaches for association classification. Int. J. Adv. Res. Comput. Sci. 8(9) (2017)

Rompré, L, Biskri, I., Meunier, J-G.: Using association rules mining for retrieving genre-specific music files. In: Proceedings of FLAIRS 2017, AAAI Press, pp. 706–711 (2017)

Tan, P.N., Kumar, V., Srivastava, J.: Selecting the right interestingness measure for association patterns. In: Proceedings of the eighth ACM SIGKDD international conference on Knowledge discovery and data mining. New York.: ACM Press, pp. 32–41 (2002)

Kohonen, T.: Self-Organizing Maps. Springer, Heidelberg, Berlin (1995)

Zaïane, O.R., et Antonie, M.L.: Classifying text documents by associating terms with text categories. In: Proceedings of the 13th Australasian database conference-Volume 5, pp. 215–222 (2002)

Artificial Intelligence in Detecting Suicidal Content on Russian-Language Social Networks

Sergazy Narynov[1], Kanat Kozhakhmet[1,2], Daniyar Mukhtarkhanuly[1], Aizhan Sambetbayeva[3], and Batyrkhan Omarov[3,4,5(✉)]

[1] Alem Research, Almaty, Kazakhstan
narynov@alem.kz
[2] Open University of Kazakhstan, Almaty, Kazakhstan
[3] Al-Farabi Kazakh National University, Almaty, Kazakhstan
batyahan@gmail.com
[4] International Information Technology University, Almaty, Kazakhstan
[5] Khoja Akhmet Yassawi International Kazakh-Turkish University, Turkistan, Kazakhstan

Abstract. Due to the anonymity of online media and social networks, people tend to Express their feelings and suffering in online communities. To prevent suicides, it is necessary to detect messages about suicides and user perceptions of suicides in cyberspace using natural language processing methods. We focus on the social network Vkontakte and classify users' messages with potential suicide and without suicidal risk using text processing and machine learning methods.

In this paper, we tell about suicidal and depressive ideation detection in Russian Language. For this purpose, we create a dataset that consists of 64,000 posts that collected from Russian language social network Vkontakte. The dataset was applied to eight machine learning algorithms.

Keywords: Suicidal ideation detection · Suicide · Machine learning · Social networks · Suicidal content

1 Introduction

Currently, one of the most pressing problems facing mental health professionals is the problem of suicidal behavior, especially in the adolescent population. At the same time, suicide is a serious public health problem and mental health in particular. According to who data, the overall level of suicidal activity has increased by 60% over the past 45 years [1], and the number of suicide cases in the world in recent decades has reached about 1 million people a year. Suicide is the 14th leading cause of death and accounts for 1.5% of all deaths in the world, which corresponds to a global death rate of 16.7 people per 100,000 population and is equivalent to the death of one person every 40 s [2, 3].

According to world statistics, the number of suicides exceeds the number of victims of murder, terrorist acts and wars combined [4]. Every 20 s, one person commits suicide, and every 2 s, someone unsuccessfully tries to take their own life. At the same time, men are three times more likely than women to resort to suicide, but women are 2–3

M. Hernes et al. (Eds.): ICCCI 2020, CCIS 1287, pp. 811–820, 2020.
https://doi.org/10.1007/978-3-030-63119-2_66

times more likely to attempt suicide. This is based on the increased demonstrativeness of women [5].

Content of this paper as following: Next section make a review about suicide problem. Section 3 describe the dataset development process that contains suicidal post collection from Vkontakte [6] social network, description of the developed dataset. Section 4 shows the experiment results and evaluation process. In addition, we give online link to the dataset in open source and make a conclusion of our research.

2 Background

Suicide is an extremely complex and multifaceted human phenomenon, caused by a variety of causes and circumstances that sometimes contradict each other. If we try to generalize, we can say that suicide is associated with both individual psychological and social, cultural and existential factors [7].

Traditionally, several forms of suicidal behavior are distinguished: suicidal thoughts, suicidal attempts, and completed suicides [8]. Suicidal thoughts are relatively widespread, including among teenagers. About a third of teenagers have experienced them [9]. Suicidal attempts are one of the most accurate predictors of future suicide. The tragedy of completed suicides is not only the death of a teenager, but also the severe consequences for his micro - and macro-social environment.

To prevent suicide, Facebook announced in early March 2017 that it would use artificial intelligence to identify users on the verge of suicide and offer them support. The social network contains detailed information about more than two billion of its users. It will be processed by algorithms, and then pattern recognition technology will highlight posts "with a high probability of containing suicidal thoughts" and transmit them to a team of analysts for human intervention [10].

A couple of weeks later, the Russian network Vkontakte announced similar plans. However, there is an emphasis on image analysis: "dangerous" content, according to the developers, will simply stop getting into users' feeds. First of all, we are talking about the symbols of "death groups". Page authors of "dangerous" images are blocked: the next time you sign in, the social network user asked what prompted him to post a particular picture, and depending on the response, send for help, or psychologists, or support [11].

Attempts to teach technology to fight suicide have been going on for a long time: back in 2007, a group of researchers from the Queen Victoria University of new Zealand tried (quite successfully) to analyze user records MySpace.com to identify those who are on the verge of suicide. Among other things, it is known that machine learning algorithms are much more effective than professional doctors can distinguish a real suicide note from a fake one (78% accuracy against 63%), and suicidal tendencies can be indicated, for example, by the duration of vowel sounds during conversation. These searches do not arise from scratch: a recent study found that in fifty years of studying suicide, scientists have not made any visible progress in predicting it. Traditional risk factors identified over the past half century—depression, stress, and substance use-provide no more accuracy than random guessing. According to the study's author, Joseph Franklin of the University of Florida, "information about previous suicide attempts helps improve the accuracy of the forecast in much the same way that buying a second lottery ticket helps win the jackpot" [12].

The main methodological features of the study were described in detail earlier [13]. In this work, we used the answers of teenagers to questions concerning 28 types of NHS that occurred in the last 6 months. The proposed event options included (in the specified order): Illness of a family member; Appearance of a new family member; Mother/father starts or stops working; problems with parents; Bullying or harassment; Death of a pet; unemployment of parents; Increased workload at school; Change in the financial situation of parents; Theft of personal property; Grades worse than expected; Serious conflict with a teacher, Transfer to another school; Sexual problems; marriage of a sister/brother with whom you are emotionally close; a family Member abusing alcohol or drugs; being taken To the police; Serious quarrel with a close friend; Minor legal violation; Important interview, exam; Failed exam; Serious injury or illness; Separation from a boyfriend/girlfriend; divorce of parents; Death of a close friend; Death of a close family member; Pregnancy; Other. We also analyzed the answers to the question "How often do you feel stress? Stress means a situation in which you feel tense, anxious, irritated, anxious or confused" with the suggested answer options "Never or several times a year, monthly, weekly, several times a week, most days a week". This question is the key one in the P88 questionnaire. The data analysis was performed using the methods of variational and nonparametric statistics. The total number of respondents included in this analysis, taking into account the quality of filling out questionnaires, was 589 [14].

3 Materials and Methods

Suicide can be considered as one of the most serious social health problems in modern society. Many factors can lead to suicide, such as personal problems such as hopelessness, severe anxiety, schizophrenia, alcoholism, or impulsivity; social factors such as social isolation and excessive exposure to death; or negative life events, including traumatic events, physical illnesses, affective disorders, and previous suicide attempts. Thousands of people around the world are suicidal every year, making suicide prevention a critical global public health mission.

According to world statistics, Kazakhstan ranks 3rd–4th in the world in the number of suicides [15], as well as 1st in the number of suicides among girls aged 15 to 19 years worldwide [16–18].

Suicidal thoughts or suicidal thoughts are people's thoughts about suicide. It can be considered as an indicator of the risk of suicide. Suicidal thoughts include fleeting thoughts, extensive thoughts, detailed planning, role-playing, and incomplete attempts. According to a who report [19], an estimated 788,000 people worldwide committed suicide in 2015. Moreover, a large number of people, especially teenagers, have been reported with suicidal ideas. Thus, one possible approach to effective suicide prevention is early detection of suicidal ideas.

3.1 Keywords

What do you mean "keywords, confirming the possibility of determining the post like a suicidal"? There is a certain set of words that are often used by people who have decided to commit suicide. Basically, these words are directly related to the idea of life and death, but sometimes it happens that in posts that are written by people who are in a drooping, suicidal mood, try to avoid using words that directly mean their attempt at suicide. Nevertheless, they try to use synonyms of these same words, thereby giving us the opportunity to find their posts, using more and more new sets of keywords.

From the previous topic, we identified key words that are associated with suicide. For example, coffin, life, end, etc. These keywords will help you search for suicidal posts in social networks.

As you find suicidal posts, the database of keywords will be updated, thereby providing a more accurate definition of suicidal posts.

In order to more accurately find suicidal posts based on keywords, a meeting with experts in this field will be held to expand the database of keywords and understand the characteristic signature of suicides.

Set of available keywords: kill, die, goodbye, life is shit, coffin, don't love yourself, hate yourself, want to die, hang yourself, suicide, suicide, unrequited love, hate school, die, die, die, don't want to live, to the next world, die, heaven, hell, I'm guilty, depression, die, etc.

3.2 Data Collection

Before classifying information as suicidal or depressive, it is necessary to determine the criteria for "danger". One solution is to define a set of keywords. This method of determining the types of information was used in the developed software package. For this purpose, a set of keywords was compiled, which was used for analyzing information in the social network Vkontakte. The software package, based on the presence or absence of the specified keywords in the text, concludes that the text is suitable for further research. Figure 1 shows the entire data collection, analysis, and job classification scheme.

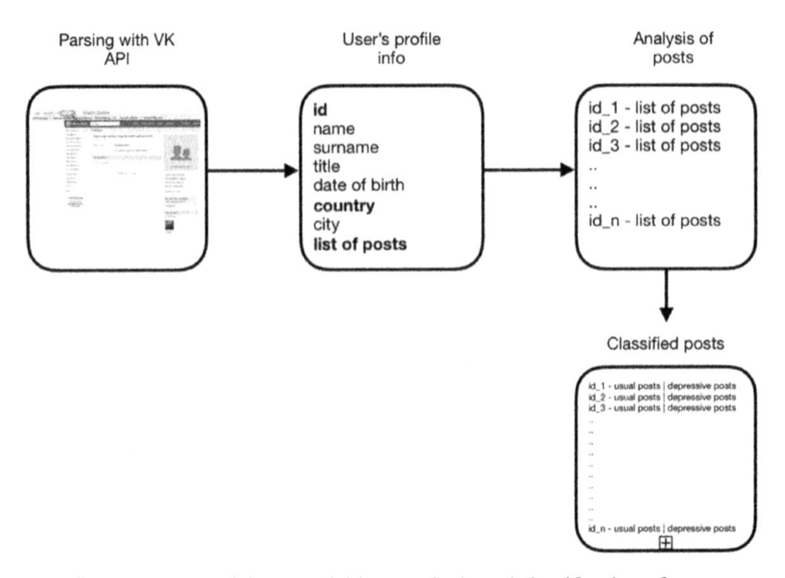

Fig. 1. Scheme of data acquisition, analysis and classification of posts

3.3 Dataset

We collected about 64,000 posts in Russian language that labeled to depressed and not depressed. 32,021 posts are non-depressed posts, when 32,018 are depressed posts. During the applying machine learning models, we divided the dataset into 2 parts as 80% for training and the remaining 20% for testing. Next four figures, from Fig. 2 to Fig. 5 describe characteristics of the developed dataset for depressive post detection.

Figure 2 illustrates characteristics of collected data.

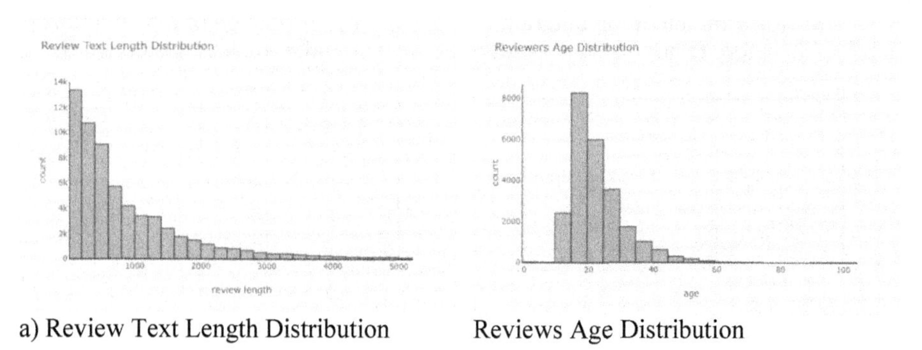

a) Review Text Length Distribution Reviews Age Distribution

Fig. 2. Illustrates characteristics of collected data.

Figure 3 illustrates distribution of top unigrams and bigrams of the collected dataset.

Figure 4 illustrates distribution of unigrams and bigrams taking into account only depressive posts.

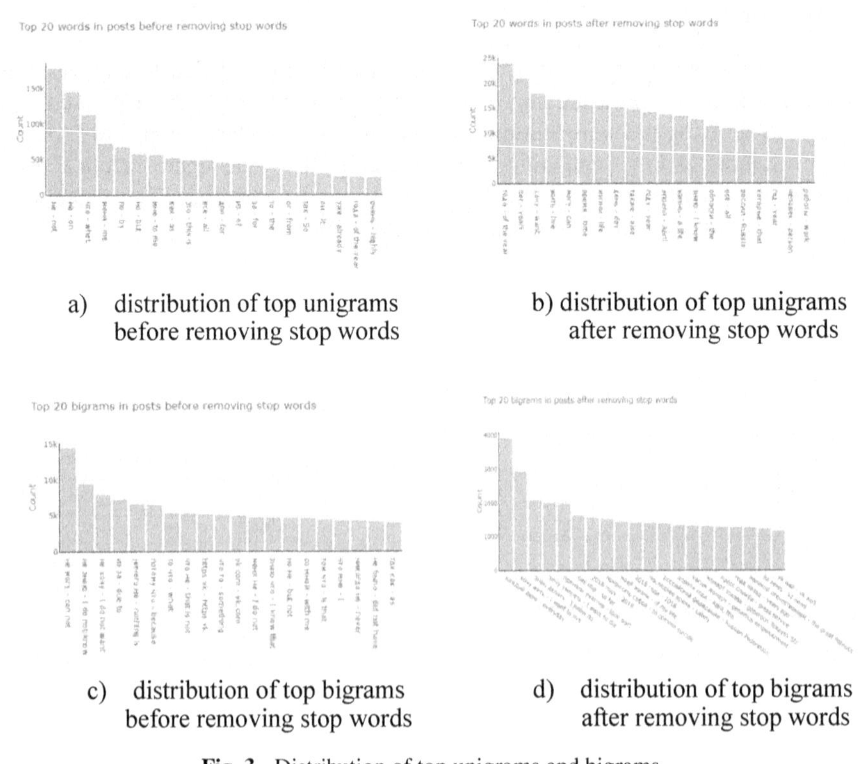

a) distribution of top unigrams before removing stop words

b) distribution of top unigrams after removing stop words

c) distribution of top bigrams before removing stop words

d) distribution of top bigrams after removing stop words

Fig. 3. Distribution of top unigrams and bigrams

Figure 5 illustrates the distribution of post length by each label.

4 Experiment Results

In this section, we tell about applying the developed dataset and results of machine learning based classification.

First, all texts were lemmatized-the process of removing only the endings and returning the basic or dictionary form of a word, which is known as a Lemma. The Yandex “Mistem” lemmatizer was used to lemmatize words in the context of the Russian language, as it demonstrated excellent results. Subsequently, the nltk library for stop words was used to delete the stop word, which reduced potential noise in the data. Numbers, special characters, and not Cyrillic letters were also removed.

Second, preprocessed texts were vectorized—the process of representing texts in a vector space for arithmetic operations on the entire data structure. The vector view saves time. For text vectorization, the IDF TF and Word2Vec models were used.

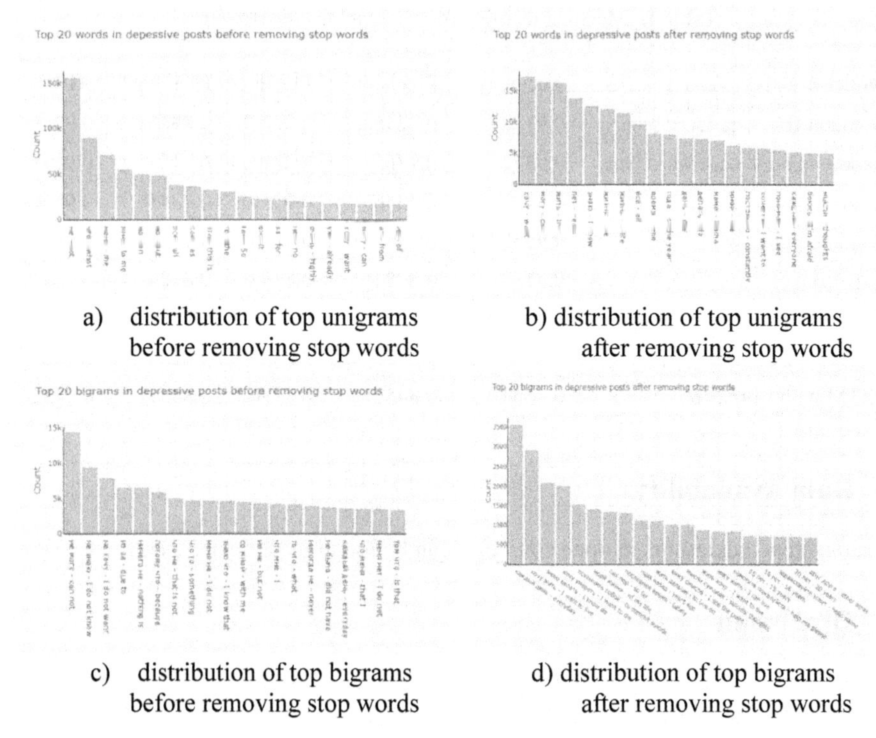

a) distribution of top unigrams before removing stop words

b) distribution of top unigrams after removing stop words

c) distribution of top bigrams before removing stop words

d) distribution of top bigrams after removing stop words

Fig. 4. Distribution of top unigrams and bigrams taking into account only depressive posts

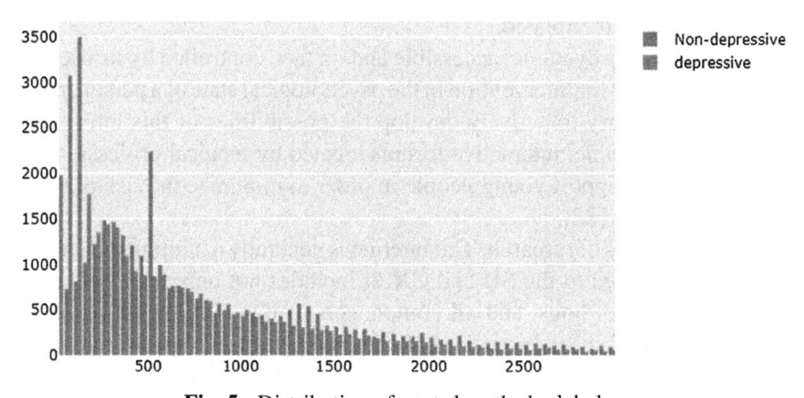

Fig. 5. Distribution of posts lengths by labels

The developed dataset applied to test eight machine learning algorithms. Table 1 demonstrates the results of each method. As evaluation parameters of methods we take accuracy, precision, recall, f1 score, and Area Under The Curve (AUC).

Table 1. Machine learning for suicidal ideation detection

Model	Accuracy	Precision	Recall	F1 score	AUC
Gradient Boosting word2vec	0.9178	0.9245	0.9427	0.8945	0.9748
Random Forest with word2vec	0.9372	**0.9765**	0.9215	0.9154	0.9578
Gradient Boosting with tf-idf	0.9435	0.9346	0.9215	0.9345	0.9576
Random Forest with tf-idf	**0.9537**	0.9462	**0.9876**	**0.9642**	**0.9623**
Naïve Bayes	0.8132	0.8254	0.8352	0.8124	0.9348
SVM	0.8635	0.8524	0.8864	0.8754	0.9467
XGBoost with tf-idf	0.8456	0.8325	0.8571	0.8254	0.9425
LSTM with word2vec	0.8654	0.8452	0.8576	0.8457	0.9472

5 Data Availability

The developed dataset is in open source. It can be found in [20] and [21].

6 Discussion

People now reflect their psychological state in social networks in the form of images, posts, and groups that they subscribe to. This is sufficient and even excessive to determine the psychological state of the child. With the help of one post, the psychologist can determine about 40 parameters of the child's psychological state. It is very important to note that only information that is publicly available is used. We see only what man has allowed everyone to see. We do not violate the Constitution or the boundaries of personal territory. Technologically, it is not possible to take information from closed accounts. Ethical standards are not violated.

Since the Internet is dynamic, accessible and, in fact, controlled by its users, and can also be an effective tool for intervention in the psychological state of a person, researchers agree that it is necessary to actively develop the possibilities of this intervention in a positive way. For example, interactive forums created by medical professionals can be a way to inform and support young people in order to minimize the risk of suicide and self-harm among them [22].

The main directions of research. The Internet is generally a multimedia environment, and so is content related to the SU and CX. It includes not only text, but also photos, videos, and music recordings, and all components of this multimedia content are the subject of attention of specialists [23, 24]. The subject of our special interest are the posts in the social networks and methods of its automated (computerized) research and monitoring.

7 Conclusion

The volume of text continues to grow along with the popularization of social networks. And suicide prevention remains an important task in our modern society. Therefore, it is crucial to develop new methods for detecting online texts containing suicidal ideas, in the hope that suicide can be prevented.

In this paper, we investigated the problem of detecting suicidality in online user content. We claim that most of the work in this area has been done by psychological experts with statistical analysis that is limited by costs and the issue of confidentiality in obtaining data. By collecting and analyzing anonymous online data from the active social network Vkontakte, We provide a huge amount of knowledge that can complement the understanding of suicidal thoughts and behavior. Although we applied feature processing and classification methods to our carefully constructed data sets, Vkontakte. We evaluated, analyzed, and demonstrated that our structure can achieve high performance (accuracy) in contrast to suicidal thoughts in normal messages in online user content.

Acknowledgements. This research was supported by grant of the program of Ministry of Education of the Republic of Kazakhstan BR05236699 "Development of a digital adaptive educational environment using Big Data analytics". We thank our colleagues from Suleyman Demirel University (Kazakhstan) who provided insight and expertise that greatly assisted the research. We express our hopes that they will agree with the conclusions and findings of this paper.

References

1. World Health Organization. (2014). Preventing suicide: a global imperative. Geneva, Switzerland (2014). http://apps.who.int/iris/bitstream/10665/131056/1/9789241564779_eng.pdf?ua=1 &ua = 1
2. https://vk.com/topic-78863260_30603285 - About VK Social Network
3. Brown, J.: Is social media bad for you? The evidence and the unknowns. BBC Future, January 2018
4. Naslund, J.A., Aschbrenner, K.A., Bartels, S.J.: How people with serious mental illness use smartphones, mobile apps, and social media. Psychiatr. Rehabil. J. **39**(4), 364 (2016)
5. Escobar-Viera, C.G., et al.: Passive and active social media use and depressive symptoms among United States adults. Cyberpsychol. Behav. Soc. Netw. **21**(7), 437–443 (2018)
6. VK.com – Vkontakte Social Network
7. Waytz, A., Gray, K.: Does online technology make us more or less sociable? A preliminary review and call for research. Perspect. Psychol. Sci. **13**(4), 473–491 (2018)
8. Villalonga-Olives, E., Kawachi, I.: The dark side of social capital: a systematic review of the negative health effects of social capital. Soc. Sci. Med. **194**, 105–127 (2017)
9. Wang, J.L., Gaskin, J., Rost, D.H., Gentile, D.A.: The reciprocal relationship between passive social networking site (SNS) usage and users' subjective well-being. Soc. Sci. Comput. Rev. **36**(5), 511–522 (2018)
10. Elhai, J.D., Tiamiyu, M., Weeks, J.: Depression and social anxiety in relation to problematic smartphone use: the prominent role of rumination. Internet Res. **28**(2), 315–332 (2018)
11. Verduyn, P., Ybarra, O., Résibois, M., Jonides, J., Kross, E.: Do social network sites enhance or undermine subjective well-being? A critical review. Soc. Issues Policy Rev. **11**(1), 274–302 (2017)
12. Woods, H.C., Scott, H.: # Sleepyteens: social media use in adolescence is associated with poor sleep quality, anxiety, depression and low self-esteem. J. Adolesc. **51**, 41–49 (2016)
13. Primack, B.A., et al.: Use of multiple social media platforms and symptoms of depression and anxiety: a nationally-representative study among US young adults. Comput. Hum. Behav. **69**, 1–9 (2017)

14. Frison, E., Eggermont, S.: Exploring the relationships between different types of Facebook use, perceived online social support, and adolescents' depressed mood. Soc. Sci. Comput. Rev. **34**(2), 153–171 (2016)
15. Twenge, J.M., Joiner, T.E., Rogers, M.L., Martin, G.N.: Increases in depressive symptoms, suicide-related outcomes, and suicide rates among US adolescents after 2010 and links to increased new media screen time. Clin. Psychol. Sci. **6**(1), 3–17 (2018)
16. Andreassen, C.S., et al.: The relationship between addictive use of social media and video games and symptoms of psychiatric disorders: a large-scale cross- sectional study. Psychol. Addict. Behav. **30**(2), 252 (2016)
17. Tan, Y., Chen, Y., Lu, Y., Li, L.: Exploring associations between problematic internet use, depressive symptoms and sleep disturbance among southern Chinese adolescents. Int. J. Environ. Res. Public Health **13**(3), 313 (2016)
18. Primack, B.A., et al.: Use of multiple social media platforms and symptoms of depression and anxiety: a nationally-representative study among US young adults. Comput. Hum. Behav. **69**, 1–9 (2017)
19. Lin, L.Y., et al.: Association between social media use and depression among US young adults. Depression Anxiety **33**(4), 323–331 (2016)
20. Narynov, S., Mukhtarkhanuly, D., Omarov, B.: Dataset of depressive posts in Russian language collected from social media. Data Brief **29**, 105195 (2020)
21. https://data.mendeley.com/datasets/838dbcjpxb/1
22. Ersoy, M.: Social media and children. In: Handbook of Research on Children's Consumption of Digital Media, pp. 11–23. IGI Global (2019)
23. Olenik-Shemesh, D., Heiman, T.: Cyberbullying victimization in adolescents as related to body esteem, social support, and social self-efficacy. J. Genet. Psychol. **178**(1), 28–43 (2017)
24. Guntuku, S.C., Yaden, D.B., Kern, M.L., Ungar, L.H., Eichstaedt, J.C.: Detecting depression and mental illness on social media: an integrative review. Curr. Opin. Behav. Sci. **18**, 43–49 (2017)

Correction to: A Simple Method of the Haulage Cycles Detection for LHD Machine

Wioletta Koperska, Artur Skoczylas, and Paweł Stefaniak

Correction to:
Chapter "A Simple Method of the Haulage Cycles Detection for LHD Machine" in: M. Hernes et al. (Eds.): *Advances in Computational Collective Intelligence*, CCIS 1287, https://doi.org/10.1007/978-3-030-63119-2_27

In the originally published version of the chapter 27, the first names and surnames of the authors were used in an incorrect order. This has been corrected.

The updated version of this chapter can be found at
https://doi.org/10.1007/978-3-030-63119-2_27

M. Hernes et al. (Eds.): ICCCI 2020, CCIS 1287, p. C1, 2021.
https://doi.org/10.1007/978-3-030-63119-2_67

Author Index

Printed in the United States
· Bookmasters